Trap Magmatism and Ore Formation in the Siberian Noril'sk Region

Modern Approaches in Solid Earth Sciences

VOLUME 3

Series Editors

Yildirim Dilek, *Department of Geology and Environmental Earth Science, Miami University, Oxford, OH, U.S.A.*
Franco Pirajno, *Geological Survey of Western Australia, The University of Western Australia, Perth, Australia*
M.J.R. Wortel, *Faculty of Geosciences, Utrecht University, The Netherlands*

For further volumes:
http://www.springer.com/series/7377

V.V. Ryabov • A.Ya. Shevko • M.P. Gora

Trap Magmatism and Ore Formation in the Siberian Noril'sk Region

Volume 1. Trap Petrology

Springer

V.V. Ryabov
Russian Academy of Sciences
Institute of Geology and
Mineralogy SB RAS
Novosibirsk, Russia

A.Ya. Shevko
Russian Academy of Sciences
Institute of Geology and
Mineralogy SB RAS
Novosibirsk, Russia

M.P. Gora
Russian Academy of Sciences
Institute of Geology and
Mineralogy SB RAS
Novosibirsk, Russia

This is a revised and updated version of the book in Russian "Магматические образования Норильского района. Т. 1. Петрология траппов. Новосибирск, издательство Нонпарель, Т. 2. Атлас магматических пород. Новосибирск, издательство Нонпарель", published by the Directorate of the Institute of Geology and Mineralogy, Siberian Branch, Russian Academy of Sciences, and by Nonparel Publishers, 2000, vol 1, 407 pages, ISBN 5-93089-008-0 and in 2001 vol 2, 600 pages, ISBN 5-93089-009-09.

This monograph was translated as follows: The main part of Volume 1 and the entire Volume 2 were translated by Tatiana Korneva; the second and fourth chapters of Volume 1 were translated by Elena Godovikova and Tamara Votentsova, respectively. Special gratitude goes to Nathan Tetlaw for editing the English version of the monograph.

ISSN 1876-1682
ISSN 1876-1690 (electronic)
Volume 1: ISBN 978-94-007-5021-0
ISBN 978-94-007-5022-7 (eBook)
Volume 2: ISBN 978-94-007-6880-2
ISBN 978-94-007-6881-9 (eBook)
Set ISBN: 978-94-017-8699-7
DOI 10.1007/978-94-007-5022-7
DOI 10.1007/978-94-007-6881-9
Springer Dordrecht Heidelberg New York London

Library of Congress Control Number: 2014936419

Credit cover figure: bubaone/istockphoto.com

Printed on acid-free paper

Springer is part of Springer Science+Business Media (www.springer.com)

To all the geologists who have contributed to the study of trap magmatism and ore deposits of the Noril'sk region in connection with the 300th anniversary of the Mining and Geological Service of Russia.

Preface to the English Edition

Volume 1: Trap Petrology

The Noril'sk Region is one of the most extensively studied parts of the Siberian Platform, attracting geological researchers from various Russian scientific organizations over the decades and, more recently, foreign researchers. This interest stems from the unique giant sulfide Ni–Cu–(PGE) deposits hosted in thick volcanic sequences and intrusions of varying composition. The Noril'sk Intrusions feature almost every Rock type that has been observed in other parts of the Siberian Traps. Because of this, the Noril'sk Region has been used to formulate genetic models for Trap magmatism (Russian geologists refer to continental flood-volcanic sequences as "Traps"), differentiation of basaltic melts, petrogenesis, and mineralization.

Early work by Kotul'sky (1948), Godlevsky (1959), Korovyakov et al. (1963), Urvantsev (1959), as well as their colleagues and followers, laid the foundation for ore–magmatic concepts, nomenclature and classification of Traps of the Siberian Platform, and theories for their origin. Following these early investigations, other geologists have continued study of the Traps and have contributed their own understanding of the problems of Trap magmatism and formation. Many of the early concepts proposed did not stand up to scrutiny and were rejected as new data were accumulated. Other concepts, which were initially accepted, have been updated and improved by geologists, and now these new genetic concepts may supersede the traditional ones. However, problems still remain.

Following the discovery of the Talnakh and Oktyabr'sky deposits in 1960s–1970s, interest in the Noril'sk Region increased; the deposits were studied by almost all the leading institutes in Russia. The volume of geological and analytical information sharply increased, and numerous papers and monographs on Noril'sk targets were published. Despite this, however, the principal problems of Trap magmatism and mineralization remain and are still discussed in the literature. There is no consensus among geologists over the principal genetic models that form the basis of their hypotheses. The lack of consensus and the large number of competing hypotheses are not solely due to the complexity of the problem but also due to the lack of knowledge about the investigation targets, primarily caused by an underestimation of their complexity, a disdainful attitude to well-known geological facts, and the lack of convincing arguments in favor of the proposed concepts that would prevent differing interpretations. The large volume of precise analytical data on the Siberian Traps obtained from laboratories in the USA and Canada (and since recently available in literature), and interpretations of these data, unfortunately, have not shed light on genetic models and have actually increased the number of competing hypotheses. These high-precision analyses need to be integrated with ground-truthed information collected by Noril'sk geologists over the years, rather than used in genetic models that were constructed for other regions (Likhachev 1997; Zolotukhin 1997; Ryabov 1999a, b). These integrated genetic models, based on high-quality and professionally collected data, should prove robust. This high-quality data will never become obsolete, unlike some fashionable hypotheses, and should lay the foundation for experiments and serve as a basis for ore–magmatic and physicochemical modeling of natural systems. To support this, it was necessary to create a database of detailed petrographic and geochemical data of standard Rock types and reference sequences for igneous complexes of the Noril'sk Region. These data serve as a reference to compare with new observations and allow users to exclude unjustified genetic models, as well as forming the basis for both experimental physicochemical simulation of natural processes and evaluation of criteria for exploration in order to reveal ore deposits.

The scheme for the formal subdivision of traps of the Noril'sk Region created by Noril'sk geologists forms the basis of this book. This scheme is now used for geological investigations, and, as new data are collected, it undoubtedly will be upgraded. Being aware that a number of ore–magmatic problems include the subdivision of Traps, comagmatism, and differentiation of magma, it appears to be reasonable to assess the data in accordance with the applicable subdivisions of magmatic styles of the region. In essence, we attempt to sum up the extent that magmatic ores have been studied petrologically and the state of genetic models for traps of the northwestern Siberian Platform at the start of the twenty-first century.

The basis of this investigation is the petrographic and geochemical understanding of principal igneous Rock types of the Noril'sk Region in order to demonstrate that these data provide unique and self-evident solutions to the problems of petrogenesis and mineralization. The results of the investigations are presented in two volumes: the first includes text and the second contains illustrations.

In the first volume, the state of the main problems of the genesis of igneous Rocks with reference to Traps and related ore deposits is discussed, as well as short petrological descriptions of igneous complexes in the region, the mineral and geochemical diversity of the Rocks, and aspects of the differentiation of basaltic melts and mineralization are described. Taking into account the vast number of publications on the petrology of Traps of the Noril'sk Region, primary attention in the monograph is given to earlier unknown phenomena, as well as other aspects that are of great importance for solving genetic problems. Some exotic geologic targets such as the Mikchandinsky differentiated cover, the magnetite lava flow of the Putorana Plateau, the magmatogenic breccia of Kharaelakh, and others are described in detail.

The second volume contains an atlas of Rock indications of igneous Rock types, formally identified reference Rocks from all igneous complexes of the region, as well as photographs of thin sections of typical Rocks and analytical tables of Rocks and minerals from the key sections of sedimentary units and intrusions. Each Rock type has been geochemically and petrographically analyzed, thereby providing a formal identity, complete with a photograph of the thin section.

Photomicrographs of the Rocks in this book will be a useful aid in visualizing the diversity of Rock types in the Traps, each photograph reflecting a unique combination of minerals. The diversity in Rock types is so large that it is impossible to find two Rock samples that are completely identical in nature even from within the same region or within the same Rock type. The chemical composition of Rock types and the major and trace elements content reveal the changes in magma composition and the distribution of chemical elements between magma phases. Such data can be used by geologists for solving problems in the early stages of investigations, followed by more detailed study of Rock-forming components. Other data, such as the size and idiomorphism of individual crystals, the pattern of phenocrysts in the groundmass, the degree of Rock crystallinity, crystal morphology, crystallization order, crystal zoning, coexisting crystal phases, intergrowth patterns, crystal internal structure, the nature of alteration, and the composition of minerals and their major- and trace-element abundances, bear important genetic information. Many of these attributes are discussed in genetic mineralogy, with specific reference to the petrogenesis of igneous Rocks and their use in regard to physicochemical problems, and these aspects should be observed when considering genetic petrography. All of the above mentioned characteristics of the Rocks reveal details about the parent magma and crystallization conditions. The task facing geologists is to find the set of most informative features for solving the problems of initial melt composition, mechanisms and conditions of differentiation of the melt, crystallization peculiarities, fluid regime of the magmatic system, and ore formation. There is good reason to believe that progress in the development of microprobe analyses will simultaneously improve analyses in genetic petrography, which will assist in problems associated with ore and magmatic geology. The authors have tried to perform investigations like these on the genetic petrography of magmatic complexes in the Noril'sk Region.

The authors hope that the information presented in this book will enable geologists to augment or correct some petrogenetic concepts and to reject others completely.

The monograph incorporates data from Rock samples collected during expeditions and the resulting geologic investigations carried out by Ryabov in the Noril'sk Region from 1963. Investigations of the Avamsky complex, Dzhaltul Intrusion, and microprobe analyses of volcanic Rocks were performed by the authors in collaboration.

Preparation of the monograph involved the authors assuming responsibility for different aspects: the text was written by V.V. Ryabov; electron microprobe analyses were performed jointly; A.Ya. Shevko prepared samples used in analytical studies, created a reference library of igneous Rocks, created the programs for calculating mineral formulae, and performed thin section photography; M.P. Gora created the database of analytical results and the digital version of the monograph.

The authors would like to take this opportunity to express their sincere gratitude to all Noril'sk geologists for their generosity and assistance in performing fieldwork. The authors also like to gratefully acknowledge V.S. Shatsky for assistance in the production of the photomicrographs, G.N. Anoshin for his help in geochemical investigations, and V.N. Sharapov for critical discussion of the investigation results. The special thank goes to O.N. Simonov, president of "Taimyrkomprirodresursy," for his long-term interest in the investigations of the Noril'sk Region and his assistance toward performing them.

This work was written by request of the Committee of Natural Resources of Taymyr National District. The English version was supported by RFBR Grant 09-05-12015-Ofi_m "Thermochemical model of the Tarim and Siberian super-plumes forms a basis for the prediction and estimation of the origin of the Cu–Ni–PGE, precious metal and rare element deposits."

We would like to acknowledge the leader of the project, N.L. Dobretsov, academician RAS and N.P. Pokhilenko, academician RAS, director of the Institute of Geology and Mineralogy SB RAS.

Our special gratitude to Vladimir Igorevich Strzhalkovsky, director general of Norilsk Nickel, who permanently shows an active interest in scientific ore-geological investigations in the Noril'sk Region and who had made provision for Financial Support-for translation of the present monograph into English.

The monograph by V.V. Ryabov, A.Ya. Shevko, and M.P. Gora entitled *Magmatic Formations of Noril' sk Region* was published in Russian in 2000. This book was republished in 2001 with insignificant supplements. The authors deeply appreciate Professor F. Pirajno who initiated publication of the English version of the book, as well as proposed its new title *Trap Magmatism and Ore Formation in the Siberian Noril' sk Region*. Upon the recommendation of sponsors of publication of this book in English, some new data on the Noril'sk Region obtained by the authors in recent years, as well as additional references, are included into the English version as Sects. 2.14, 2.15 and 2.16 and 5.8 and are reflected in the extended references.

References

Godlevsky MN (1959) Traps and ore-bearing intrusions of the Noril'sk region. Gosgeoltekhizdat, Moscow, 68 p (in Russian)

Korovyakov IA, Nelyubin AE, Raikova ZA, Khortova LK (1963) Origin of the Noril'sk trap intrusions hosting sulfide copper-nickel ores. In: Trudy VIMS (ed Gon'shakova VI), New Series, issue 9, Gosgeoltekhizdat, Moscow, 102 p (in Russian)

Kotul'sky VK (1948) The state-of-art problem of genesis of copper-nickel sulfide deposits. Sov Geol 29:11–24 (in Russian)

Likhachev AP (1997) Trap magmatism and Pt–Cu–Ni ore formation in the Noril'sk region. Otechestv Geol 10:8–19 (in Russian)

Ryabov VV (1999a) Fluid regime of trap magmatism and ore formation (petrological aspect). Russ Geol Geophys 40(10):1437–1452

Ryabov VV (1999b) Ore geology of the Noril'sk region on the threshold of millennium. In: Simonov ON (ed) Abstracts of the regional symposium: trends in geological-exploration works on nickel, copper and platinoids in the Taimyr Autonomous Area in 2000–2005. Taimyrkomprirodresursy, Noril'sk, pp 25–26 (in Russian)

Urvantsev NN (1959) The Yenisei ore field. Collected papers on Arctic region geology. Trudy NIIGA (ed Markov FG), vol 102, issue 10. Nedra, Leningrad, pp 28–48 (in Russian)

Zolotukhin VV (1997) Mafic pegmatoids of the Noril'sk ore-bearing intrusions and the problem of genesis of the Platinum-copper-nickel mineralization of the Noril'sk type (ed Polyakov GV). SB RAN NITS OIGGM, Novosibirsk, 88 p (in Russian)

Contents

List of Abbreviations

Rock Abbreviations

A	aphanite
AB	aphyric basalt
AK	ankaramite
AN	it is an essential plagioclase Rock (anorthosite, labradorite)
AUG	augitite
B	basalt
BO	basalt with olivine dendrites
C.f.	$100 \times (FeO + Fe_2O_3)/(FeO + Fe_2O_3 + MgO)$
CG	contact gabbrodolerite
cr	chromespinellides in the form of an impregnation
CT	contact trachydolerite
D	diorite, gabbrodiorite, prismatic granular gabbrodolerite
DB	basalt with pyroxene dendrites
FG	ferrogabbro
G	gabbrodolerite, non-olivine gabbrodolerite (olivine = 0–3%), dolerite
Gl	glass
	Glc – partially crystallized globules glass
	Glg – globules with emulsion liquating glass
	Glh – homogeneous glass
	Gli – interstitial glass
Go	olivine-bearing gabbrodolerite (olivine = 3–10%)
GP	gabbropegmatite
Gq	quartz-bearing gabbrodolerite
Gqo	quartz- olivine-bearing gabbrodolerite
GR	granophyre
GS	gabbrosyenite
L	melt
LM	limburgite
LG	leucogabbro
M	metasomatite
MA	mandelstein
MB	magmatogenic breccia
MD	metadiorite
O	olivine gabbrodolerite (olivine = 10–20%)
OA	olivine-anhydrite Rock
OB	olivine porphyritic basalt
OG	olivine gabbro ore-bearing identification is put for the Rocks containing not less of 7–10% of ore minerals, and for the intrusions which do not have commercial ores

OTD	olivine trachydolerite (olivine = 10–20%)
P	picritic gabbrodolerite
PB	picritic basalt (olivine = 20–60%)
PG	pegmatoid gabbrodolerite
PTD	picritic trachydolerite
S	syenite
SB	basalt with clinopyroxene spherocrystals
SV	sulfide vein
T	taxitic gabbrodolerite
TB	trachybasalt
TD	trachydolerite
TDo	olivine-bearing trachydolerite (olivine = 3–10%)
TDP	pegmatoid trachydolerite
Tf	tuff
Tl	taxite-like gabbrodolerite
TR	troctolite gabbrodolerite (olivine = 10–60%)

Minerals Abbreviations

Amf	amphibole
Ap	apatite
Arm	armalcolite
Bi	biotite
Bw	bowlingite
Chl	chlorite
Cpx	clinopyroxene
	Ca-tcher – Ca-tschermakite component
	En – enstatite component
	Fs – ferrosilite component
	Wo – wollastonite component
Cr	spinel group mineral (chromite series)
Cub	cubanite
Di	diopside
f	FeO × 100/(FeO + MgO) at%
Fsp	potash feldspar
Gr	garnet
H	halite
Hem	hematite
Ilm	ilmenite
Leu	leucite
Mh	mooihoekite
Mt	spinel group mineral (magnetite series)
MW	magnesiowüstite
Ol	olivine
	Fa – fayalite component
	Fo – forsterite component
Opx	orthopyroxene
	En – enstatite component
	Fs – ferrosilite component
	Wo – wollastonite component

Per	periclase
Phl	phlogopite
Pl	plagioclase
	Ab – albite component
	An – anorthite component
	Or – orthoclase component
Plg	palagonite
Po_h	hexagonal pyrrhotite
Po_m	monoclinic pyrrhotite
Pn	pentlandite
Pr	perovskite
Px	pyroxene
Qu	quartz
Rut	rutile
Sp	spinel group mineral (spinel series)
Tn	talnakhite
Tr	troilite
Usp	ulvospinel component
Wü	wüstite

Concentrations of oxides and S in the chemical analyses of Rocks are given in wt%, chemical elements in ppm. In microprobe analyses of minerals, all concentrations are given in wt%.

LOI	loss on ignition

Internal Intrusion Structure Abbreviations

CZ	central zone
LZ	lower contact zone
UZ	upper contact zone

Intrusive Abbreviations

CH	Mountain Chernaya
DDI	Daldykansky
ENI	East Noril'sk
KHI	Khungtukun
LFI	Lower Fokinsky
LNI	Lower Noril'sk
LTI	Lower Talnakh
MMI	Maymechinsky
MRI	Mountain Morongo
N-I	Noril'sk-I
N-II	Noril'sk-II
PCI	Mountain Picritovaya
PGI	Mountain Pegmatitovaya
PTI	Mountain Putanaya
PVI	Pyasinsko-Vologochansky
RNI	Mountain Ruinnaya
Sk	Skaergaard
TTI	Tulaek-Taas River
UTI	Upper Talnakh
ZI	Zubovsky

Figure 1.6

Buffers:

HM	hematite–magnetite
MW	magnetite–wustite
QFM	quartz–fayalite–magnetite
QIF	quartz–iron–fayalite
WI	wustite–iron

Figure 2.29

TS	Tungussky sineclise
N	Noril'sky Province
K	Kamensky Province
WB	western board
NB	northern board
EB	east board
MK	Maymecha–Kotuysky Province

1 Principal Issues Surrounding Trap Magmatism of the Siberian Platform

Abstract

The Noril'sk region has a special place in the basalt field of the Siberian Platform. It is characterized by a complex geodynamic environment, the most entire sequence of the volcanogenic succession, great variety of hypabyssal intrusions of varying composition, and with different degree of differentiation. In this region, the Noril'sky-type layered intrusions and related largest sulfide Pt–Cu–Ni and Pt low-sulfide deposits such as the Oktyabrsky, Talnakh, and Noril'sk-I are located. Geological investigations in the region are carried out over many decades, but in spite of this, the main problems related to the origin of the volcanogenic succession, layered intrusions, and ore deposits are still relevant. In Sect. 1.1, features of geological setting and magmatism in the northwest of the Siberian Platform are described. A great deal of attention is drawn to the problems of Trap magmatism and ore formation. In Sect. 1.2, the present notions of geologists about the composition of parental magma of Siberian Traps, their chamber differentiation, and the origin of high-magnesian Traps, pegmatoids, and sulfide ores of the Noril'sk deposits are considered. In connection with these matters, a short review of the published experimental data in the field of differentiation and crystallization of silicate and ore liquids, as well as petrogenesis and ore formation that form the basis for genetic models, is given.

The large-scale magmatism (about 1.5 million km^3) that formed the Siberian Traps (Russian geologists refer to continental flood-volcanic sequences as "Traps") took place on the Siberian Platform and created a wide variety of volcanic and sub-volcanic (hypabyssal) rocks, and related ore deposits. The northern margin and northwestern corner of the Siberian Platform is of particular interest to geologists as it features almost the complete diversity of Trap-formed rocks and concetrations of the various mineral resources including Cu–Ni ores, native iron and copper, magnetite, and Iceland spar.

1.1 Geological Features of the Northwestern Part of the Siberian Platform

The Noril'sk region is located in the northwest of the Siberian Platform (Fig. 1.1). It is bounded by major tectonic features, the deep Yeniseysky Fault to the west, the Yenisey–Khatangsky Rift to the north, and by the junction of Tungussky sineclise with Anabarskaya Craton to the east. Being located at the junction of such large-scale structures, the region has a vast diversity of geologic formations and complex tectonic environment (Malich et al. 1987).

The cover sequence of the Siberian Platform comprises Upper Proterozoic terrigenous carbonate deposits; Upper and Middle Proterozoic sulfate-bearing, argillaceous, carbonate deposits; Upper Paleozoic and Mesozoic volcanogenic rocks; and clastic formations of Mesozoic–Cenozoic age. The northwestern and northern margins of the Siberian Platform are confined by bounding structures: the Turukhansky uplift; the Dudinsky and Khantaysko–Rybninsky uplifts composed of sediments of Upper Proterozoic and Lower Paleozoic age; the Noril'sky, Vologochansky, Kharaelakhsky, Ikensky, Bol'sheavamsky, and Boyarsko–Del'kansky troughs filled with volcanogenic unit of Permian–Triassic age. Toward the center of the Platform, the zone of troughs is separated from the largest structural unit of the Platform, the Tungussky sineclise, by a

V.V. Ryabov et al., *Trap Magmatism and Ore Formation in the Siberian Noril'sk Region: Volume 1. Trap Petrology; Volume 2. Atlas of Magmatic Rocks*, Modern Approaches in Solid Earth Sciences 3, DOI 10.1007/978-94-007-5022-7_1,

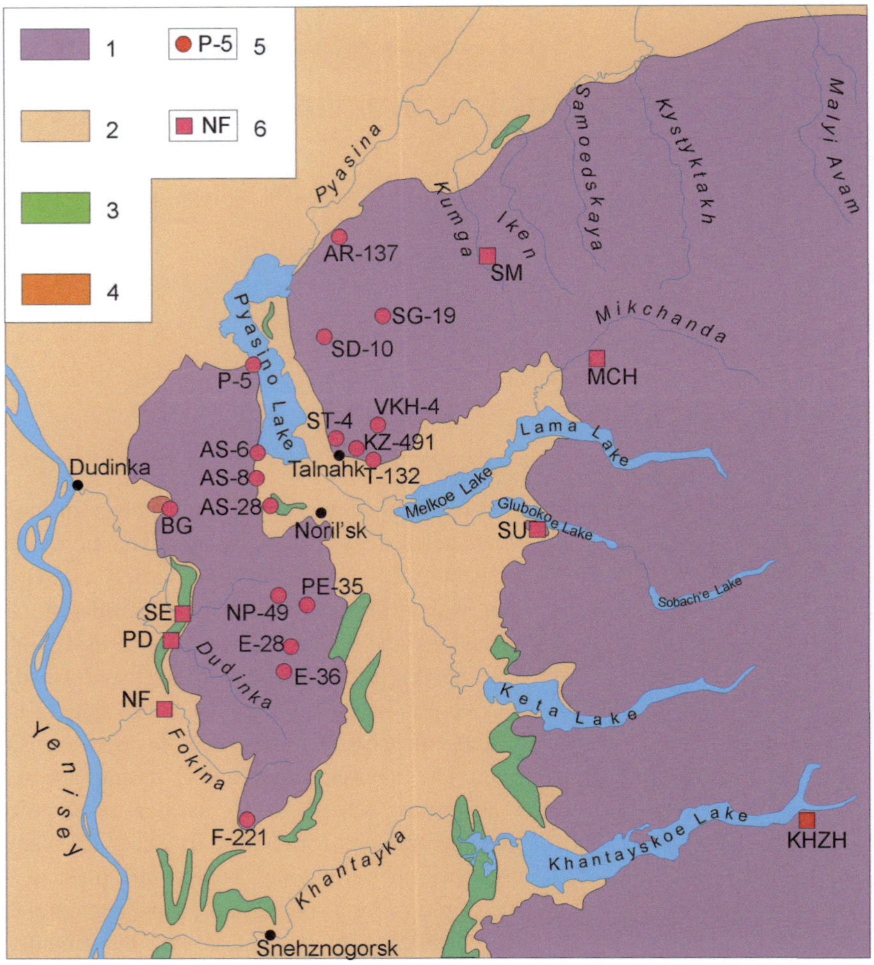

Fig. 1.1 Schematic geological map of the northwestern part of the Siberian Platform. *1* Volcanogenic rocks, *2* sedimentary deposits, *3* mafic intrusions, *4* granitoid intrusions, *5–6* Locations of reference cross sections: along drill holes (*5*) and in exposures (*6*). Additional information on the arrangement of drill holes in the Talnakh region is given in Fig. 4.36

system of faults (Ayano-Ambardakhsky to the north and Imangdino-Letninsky to the west). The major structures in the region are the Yeniseysky north-trending fault in the west, the Yenisey–Kotuysky fault in the north, and Taymyro–Baykalsky fault in the east.

Geological investigations of the immense Siberian Platform, carried out over many years, have revealed vast igneous intrusions of Proterozoic, Paleozoic, and Mesozoic ages that feature unique compositional peculiarities. These investigations have enabled geologists to outline the nature of magmatic evolution in the different petrographic provinces, igneous complexes, and orthomagmatic ore fields. The problems associated with the classification and nomenclature of Traps, their structural–tectonic zonation, and lithofacies features have been described by many geologists (Lurje et al. 1962, 1976; Malich et al. 1965, 1987; Vilensky 1967; Dodin 1967, 1982; Vilensly and Oleinikov 1970; Masaitis 1973, 1974; Oleinikov 1979; Zolotukhin 1984a, b; Zolotukhin et al. 1986).

Within each cycle, a sequence of evolving differentiation stages can be observed. More specifically, a trend from minor trachybasalt intrusion to large-scale Traps (basalt and trachybasalt) is followed by localized emplacement of alkaline ultramafic rocks, kimberlites, and nepheline syenite intrusions. These rocks were most widely developed in the Middle Paleozoic–Early Mesozoic.

The petrogenetic evolution of igneous rocks is accompanied by a regular change in their composition, especially toward increasing concentration of titanium, calcium, magnesium, and alkali elements. In individual basalt flows, the concentration of calcium and magnesium increases with decreasing concentration of silica and alkali elements (Masaitis 1973). This petrochemical trend is most obvious in the northwestern region of the Siberian Platform as successive lava flows form subalkaline trachybasalt formations to picrite or sub-ultramafic formations. In the case of intrusive rocks, the recognition of an evolutionary petrogenetic series is related to the abundances of alkalies, silica, and potassium in the melts (Vilensky 1967).

Petrochemical features of rocks from single magmatic formations, subformations, and particular geologic bodies form petrographic provinces that are characterized by a certain diversity of rocks and differentiation extent, as well as compositional variations. Investigations have shown that within a single province, oxide concentration remains stable but will vary between igneous provinces (Vilensky 1967).

Table 1.1 Composition of main rock-forming minerals in differentiated intrusions from some Trap provinces (Feoktistov 1978)

Site intrusions	Plagioclase (% An)	Clinopyroxene (% Fs)	Orthopyroxene (% Fs)	Olivine (% Fo)
Trap sills of Siberian platform	10–95	9–37	13–32	8–86
Dolerite sills of Karoo, RSA	25–80	11–37	15–80	0–85
Sill Palisade, New Jersey, USA	37–66	14–36	15–70	7–82
Dolerite sills of Tasmania	26–75	5–25	15–30	–
Dolerite sills of eastern Antarctic	10–83	9–50	17–31	–
Skaergaard Intrusion, eastern Greenland	30–66	21–57	19–56	0–67

Hereinafter: a crossed-out section indicates absence of a mineral

Such behavior is observed in other Trap provinces around the world (Sobolev 1986). As a whole, the composition of parental magmas that form the Siberian Traps varies from quartz tholeiite to silica-undersaturated olivine tholeiite (Feoktistov 1978). This compositional variation is responsible for the petrochemical and mineralogical features of rocks in different Trap provinces around the world (Table 1.1), as well as the mode of differentiation and metallogeny.

Most geologists are inclined to believe that the trend of differentiation in present-day magmas is toward olivine-rich rocks during the initial stages of intrusions and to enrichment of the magma in iron or silica and alkali elements during the final stages of crystallization. If features of the present-day chamber differentiation can be established directly by studying geologic targets, then the composition of a parental melt and reasons of its diversity, as well as the mechanisms of generation and deep differentiation of magma, could be resolved.

1.2 Magmatic and Ore Geology: State of the Main Problems

1.2.1 Parental Magma

The diversity of rocks of the Siberian Traps is believed to result from variations in the initial composition of melts and the degree of differentiation in deep-seated chambers and in feeder zones. The parental magma is assumed to have had a pyrolite, pyroxenite, or eclogite (quartz and garnet) composition, and variations in magma composition are likely related to inhomogeneities in mantle material or the depth of magma chambers and varying degrees of partial melting.

Sobolev (1986) wrote that the diversity of Trap rocks results from abyssal and shallow chamber differentiation of magmatic melts of basaltic composition. Trap magmas, in his opinion, lie within the tholeiitic petrochemical series. The evolution of these melts in magmatic chambers leads to the formation of a wide spectrum of rock types, from picrite to olivine-free gabbroic and pegmatitic rocks. Lebedev (1957, 1958) challenged the concept that the entire diversity of intrusive rocks can be explained by one (basalt) or two (tholeiite and olivine basalt) parental magmas. Godlevsky (1959, 1961, 1967) pointed out that the magma that formed the Noril'sk intrusions, though being similar to ultramafic magma, is derived from deeper levels than traditional Trap magmas and that melts that formed trachybasalt sills have a different feeder unrelated to tholeiitic Traps. Studies of different igneous complexes on the Siberian Platform (Masaitis 1973, 1974) indicate that there were multiple parental melts. Other geoscientists dismiss the abyssal differentiation of melts as they think that Trap rocks have a composition that is close to abyssal melts (Kutolin 1972), and the variety of intrusive rocks is a result of differentiation in the magma chamber (Feoktistov 1984).

A number of geoscientists believe that abyssal differentiation was one of the most important factors in creating the diversity of rocks found in Traps and for the composition of source magma. Zolotukhin (Zolotukhin 1964, 1981; Ryabov and Zolotukhin 1978; Zolotukhin and Laguta 1985; Zolotukhin et al. 1986) suggested that the parental magma compositions for the Siberian Traps were richer in magnesium than the average Trap composition elsewhere in the world and were similar to the weighted average composition of the Noril'sk intrusions. Fractional crystallization of such magnesium-rich liquids in deep-seated chambers, in his opinion, predetermined the successive fractionation creating different compositions and isolating daughter melts that were parental for particular igneous rocks including, for example, intrusive complexes (Lurje et al. 1962). Most likely, the diversity of igneous rocks in the Siberian Platform is due to magma chamber depth, differing degrees of partial melting of pyrolite, and further abyssal differentiation of the magma during ascent to the present-day high-level chamber and differentiation within it (Zolotukhin and Vasil'ev 1975, 1976). These authors proposed that the magmatic composition of picritic basalts and the Noril'sk intrusions is "the result of rather appreciable melting of mantle rock," and "its considerable abyssal differentiation is unlikely." Evidence of present-day high-level chamber differentiation was identified with those of abyssal and intermediate chamber differentiation.

According to a model by Likhachev (1978, 1988) and Godlevsky and Likhachev (1979), the fractionation of melts begins when crystallization differentiation and gravitational separation force the liquid fraction to move vertically in a magma column. This resulted in an igneous layered column extending for about 100 km in which magnesian ore-bearing and unmineralized portions of melt alternate.

The significant role of the pre-chamber stage of melt differentiation is described in the genetic reconstructions by Yakitian geologists (Oleinikov and Tomshin 1976; Oleinikov 1979, 1981). The process of melt evolution is divided into stages responsible for the state of magma in deep, intermediate, and present-day chambers. Fractional crystallization of melt at each stage influenced the composition of the rocks and was reflected in the occurrence of specific mineral parageneses. In particular, crystallization of chrysolite resulted in high-aluminous magmas with an anorthosite crystallization trend typical of potentially ore-bearing intrusions, for example, Noril'sk intrusions. The mechanism of ore formation is believed to be the result of the interaction of silicate melt with reducing fluids in an intermediate-level chamber and is assumed to have resulted in the metal enrichment of magmas and the formation of droplets of ore-liquid segregations (Oleinikov 1981). High sulfur activity in the fluid phase contributes to the sulfurization of metal-rich liquid and the preservation of typomorphic features of magmatic deposits rich in sulfide (e.g., Noril'sk deposits).

1.2.2 Chamber Differentiation of Basaltic Melts

1.2.2.1 Origin of Layered Intrusions

Layered intrusions have always been of interest to geologists. These intrusions have been studied for many decades in different parts of the world, but the problem of their formation is far from being solved. The formation of layered intrusions and related mineralization is typically treated in the context of crystallization–gravity differentiation or multiphase intrusion of melts with different compositions, liquid immiscibility, and fluid–magmatic differentiation. The problems of the genesis of layered intrusions are discussed in Gurulev (1965), Gorbunov (1968), Chernyshev (1971), Kozlov (1973), Zharikov (1976), Marakushev (1979), Sharkov (1980), Cawthorn (1996), Likhachev (2006), and Naldrett (2004).

A classical example of fractional crystallization differentiation of tholeiite magma is the Skaergaard Intrusion, which is well described in some papers (Wager and Deer 1939; McBirney and Nakamura 1974; Morse et al. 1980; Cawthorn 1996). Other classic examples are the dolerites of the Karoo (Insizwa) Region of South Africa (described by Walker and Poldervaart (1949)), and the dolerites of Tasmania (Grey Lake; studied by Edwards 1942). Excellent examples of layered intrusions are the Bushveld and Stillwater intrusions, which have some traits in common with the Noril'sk intrusions such as the nature of concealed layering and ore mineralization (Cameron 1970; Wager and Brown 1968; Likhachev 1975; Naldrett and Cabri 1976).

A great number of mafic intrusions on the Siberian Platform have been studied, such as the Khungtukun, Maymechinsky, Khanarsky, and Arylakhsky intrusions in the north (Lebedev 1955; Staritsyna et al. 1972; Duzhikov et al. 1988; Ryabov et al. 1985c); the Anakitsky, Nizhny-I and II, Kureysky, Upper Gorbiachinsky, Nekongdokon, Dzhaltul, Khuringda River, and Fat'yanikha River intrusions in the west (Reverdatto 1963; Vilensky et al. 1964; Zolotukhin and Vasil'ev 1967; Kavardin et al. 1968; Oleinikov 1979; Zolotukhin 1984a, b); the Alamdzhakh, Velingna, Kholomolakh, Vavukan, and Ust'-Khann'ya intrusions in the east (Masaitis 1958; Vilensky 1967; Oleinikov 1979; Frenkel et al. 1988); and the Tolstomysovsky, Usol'sky, and Padunsky intrusions, and other sills in the south (Feoktistov 1978). It is practically impossible to list all the available publications; we can only note that the degree of differentiation described for these intrusions differs. Most authors share Sobolev's opinion (Sobolev 1986) about the leading mechanism of fractional crystallization in Trap formation. This mechanism was repeatedly illustrated by different geologists examining the layered intrusions of the Siberian Platform. The main indicators of the completed fractionation process are common changes in rock composition in the Trap, namely, an increase in concentration of alkaline elements and silica content from the base to the top and a corresponding decrease in Mg content. This is also reflected in the mineralogy; concealed layering is formed due to the increased alkalinity and higher Fe content of ferromagnesian minerals (Fig. 1.2), as observed in the type sections of some layered intrusions (Masaitis 1958; Nesterenko and Almukhamedov 1973; Zolotukhin et al. 1975; Feoktistov et al. 1975; Zolotukhin and Shchedrin 1977; Distler et al. 1979).

Geochemical studies helped to establish the elements (and their ratios) that are sensitive to melt differentiation, and that are important for defining rock types, intrusion types, and for determining their ore potential (Dodin 1967; Nesterenko and Almukhamedov 1973; Arkhipova 1975; Natorkhin et al. 1977; Vilensky 1978a; Feoktistov 1978; Oleinikov 1979).

It is reported in a number of publications that the melt evolution in the majority of Traps follows the Skaergaard scheme or Fenner's trend (Masaitis 1958; Reverdatto 1963; Vilensky et al. 1964; Zolotukhin 1964; Staritsyna et al. 1972; Nesterenko and Almukhamedov 1973). However, the Noril'sk intrusions are an exception; their differentiation

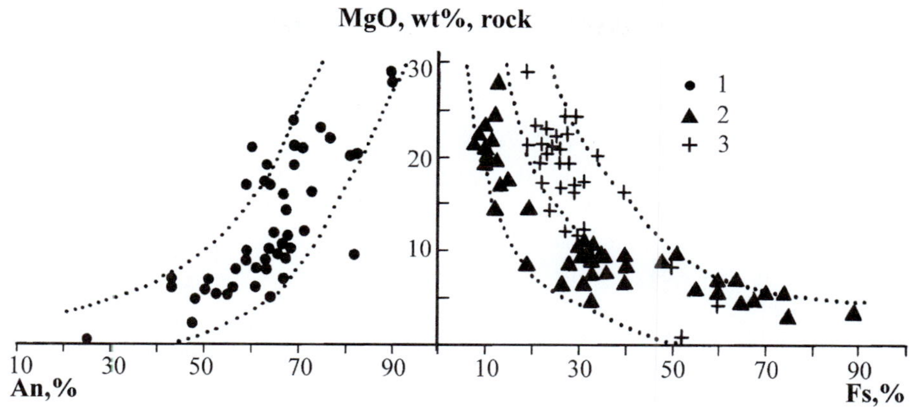

Fig. 1.2 Plot of plagioclase and pyroxene compositions versus the MgO content in rocks. The compositions of minerals are taken from Ryabov and Zolotukhin (1977). *1* Plagioclase, *2* clinopyroxene, *3* orthopyroxene

trend is similar to Bowen's reaction series but is complicated by the participation of volatile elements and dense accumulation of olivine at the base of the intrusions (Zolotukhin 1964; Godlevsky 1967; Vilensky 1967). For these bodies, Stepanov (1975) suggests independent differentiation differing from generally accepted theories.

The magmatism of the northwestern Siberian Platform has been extensively studied. Abundant empirical material has been accumulated through numerous studies of Traps and sulfide ores in the Noril'sk Region, but despite this fact, the region remains in dispute among geologists. It is fair to say that any genetic model related to Trap magmatism and ore formation can be first tested on samples from Noril'sk. A dozen monographs and hundreds of papers have been devoted to the Noril'sk ore-bearing intrusions, but an explicit integration of all the collected data is still lacking. Thus, the most relevant problems of the initial melt composition in the Noril'sk intrusions, mechanisms of differentiation of this melt, nature of mafic pegmatoids, ores and others are still discussed. The solution to these problems is vital for understanding the prospectivity of the region and the development of any mineral resources.

Many geologists believe that the layered structure of the Noril'sk intrusions is a product of the fractional crystallization of the original melt, whose initial composition corresponded to the weighted average composition of the intrusions and was close to that of olivine gabbrodolerite and troctolitic gabbrodolerite (Korovyakov et al. 1963; Zolotukhin 1964, 1978a, b; Vilensky 1967; Kavardin et al. 1968; Urvantsev 1970; Dodin and Batuev 1971; Staritsyna et al. 1972; Zolotukhin et al. 1975; Nekrasov and Gorbachev 1978; Oleinikov 1979; Genkin et al. 1981; Dneprovskaya et al. 1985). The parental magma is assumed to contain about 12 wt% MgO (Zolotukhin and Laguta 1985). In genetic models, some authors place a strong emphasis on fractional crystallization differentiation in deep-seated chambers, intermediate chambers, and feeders or magmatic columns (Zolotukhin and Vasil'ev 1975, 1976; Likhachev 1977; Oleinikov 1979; Zolotukhin 1984b). This gives an insight into the diversity of compositions of various layered intrusions, antidromic eruption sequence, the varying ore saturation in intrusions, and the scale and composition of ores.

Along with the generally accepted notion of melt immiscibility, geologists pointed to a number of geologic facts that could not be explained within the framework of fractional crystallization theory. These include sharp contacts between different types of rocks, alternation of early crystal differentiates and flow structures in picrite gabbrodolerite, inverted position of the heavy fraction of magma in intrusion section, presence of picrite gabbrodolerite near the top of intrusions, lack of correlation between the thicknesses of olivine-bearing layers and more fractionated layers, and the discrepancy between the weighted average composition for intrusions and that for contact facies. The requirements for deducing a petrogenetic model of the Noril'sk intrusions are available in Smirnov (1966), Zolotukhin and Ryabov (1977), and Koptev-Dvornikov et al. (1976). An alternative to the fractional crystallization theory is the two- or multiphase mechanism of formation of layered intrusions (Rogover 1959; Smirnov 1966; Ivanov et al. 1971a, b). The characteristic structure of intrusions and their specific composition were explained through the mixing of two magmas in a single conduit (deep ultramafic and shallow mafic melts) that underwent further fractional crystallization in the present-day high-level chambers (Malich et al. 1987; Tuganova 1988). However, in contrast to this, some authors believe that an abyssal melt will separate into immiscible mafic and ultramafic liquids that will rise together and crystallize in a chamber in this immiscible form (Marakushev et al. 1982). One of the main arguments in favor of this hypothesis is inconsistency of the composition of contact rocks of intrusions with the weighted average composition of the intrusions (Table 1.2). This is not the prerogative of the Noril'sk intrusions but is typical of other layered intrusions as well (Marakushev 1979).

Table 1.2 Average chemical composition of the stratified series of mafic rocks and contact facies of stratified intrusions (It is led 100%, on Marakushev 1979)

Oxides	1	2	3	4	5	6
SiO_2	4.21	49.83	50.58	51.53	50.97	48.52
TiO_2	1.09	1.35	0.66	0.34	0.45	1.18
Al_2O_3	15.64	15.98	15.24	18.70	17.74	17.38
Fe_2O_3	3.57	3.70	1.04	0.28	0.26	1.33
FeO	7.55	8.03	10.08	9.05	9.94	8.52
MnO	0.17	0.17	0.23	0.47	0.15	0.16
MgO	8.14	7.57	8.31	6.85	7.75	8.70
CaO	12.31	9.89	11.31	10.97	10.53	11.49
Na_2O	2.07	1.98	2.24	1.58	1.88	2.39
K_2O	1.09	1.27	0.19	0.14	0.24	0.25
P_2O_5	0.16	0.23	0.12	0.09	0.09	0.08

Note: 1, 2—Non-olivine gabbrodolerites and contact gabbrodolerites of Noril'sky-type intrusions; 3, 4—the same for Bushveld Intrusion; 5—microgabbro of the Stillwater Intrusion; 6—the same for Skaergaard Intrusion

The genetic models of Zotov (1979, 1982) show that transmagmatic fluids played the principal role in the formation of Noril'sk intrusions and associated mineralization. Zotov believes that 50–75 vol% of the Upper Talnakh Intrusion was formed due to replacement of host rocks by magma. Earlier, when studying intrusions in the Yeniseysky ore province, including the Noril'sk intrusions, Goryainov (1969, 1971) and Gulin and Sukhov (1973) came to the conclusion that the stratified structure of layered intrusions is a reflection of metasomatic zoning, which develops under the influence of fluids on undifferentiated intrusions.

1.2.3 Nature of High-Magnesian Traps

Knowledge of the nature of high-magnesian Traps can help resolve the problems of abyssal magma formation, initial composition of Trap magma, origin of picrite–troctolite gabbrodolerites in present-day chambers, and the relationship between gabbrodolerites of the Noril'sk intrusions and high-Mg rocks within other Traps. The problem becomes more urgent owing to the empirically established relationships of sulfide mineralization with layered magnesian intrusions of the Noril'sky-type and associated picrite gabbrodolerites. In addition, a great number of layered magnesian intrusions studied on the Siberian Platform (Karbyshev et al. 1978; Ryabov 1984b; Zolotukhin 1984a) have horizons of picrite and troctolite gabbrodolerites, but only few of them are ore-bearing. Therefore, the task of geologists is to reveal the nature not only of initial melts but also the mechanism for the formation of layered intrusions, their differentiation, and the petrogenesis of olivine-rich rocks—picrite and troctolite gabbrodolerites.

Most geologists assign the formation of picrite and troctolite gabbrodolerites in layered intrusions to fractional crystallization with gravitational accumulation of olivine. This mechanism traditionally holds true for most layered intrusions. In the Noril'sk-I and Talnakh intrusions, the presence of olivine-rich rocks is considered to be one of the most important features of potentially ore-bearing intrusions (Likhachev 1977; Karbyshev and Zolotukhin 1978; Ryabov 1984b). It is assumed that picrite horizons can form in melts with a g content of 10–12 wt%, that is, those similar in composition to magnesian basalts (Naldrett and Cabri 1976). The appearance of ultramafic mineral assemblages is also probable in melts with MgO content not lower that 10 wt% (Stepanov 1975, 1977).

Comparing the magnesian intrusions from the Siberian Platform, Zolotukhin (1982) found that the initial composition of the Noril'sky-type and Morongovsky Complex intrusions formed two end members of differentiation of magnesian basalt magma. The Morongovsky Complex intrusions crystallize following the Skaergaard scheme, which is typical (Zolotukhin 1982) of the deeper parts of the Noril'sky-type intrusions, and some intrusions (Noril'sk-I, Zub Mountain, and Morongo Mountain intrusions) form links of a single chain of submergin magmatic column (Zolotukhin 1978a). It is therefore assumed that the occurrence of Morongovsky Complex intrusions with olivine-rich rocks may indicate the existence of the Noril'sky-type ore-bearing intrusions. Discovery of a number of high-Mg intrusions in the middle course of the Yenisey River and their similarity with Morongovsky Complex intrusions make this region prospective for ore-bearing intrusions (Zolotukhin 1978a). Vilensky (1967, 1978b) suggested that the high-Mg content in the melt is the result of, rather than the reason for, formation of Ni-bearing intrusions, and that this Mg content increases owing to mineralization and involvement of fluids.

In spite of the comprehensive analysis of all the necessary conditions for the occurrence of picrite gabbrodolerites in layered intrusions, some geological facts do not fit the assumed mechanism of fractional crystallization (Rogover 1959; Smirnov 1966; Ivanov et al. 1971a, b; Stepanov 1975, 1977). Godlevsky (1967) pointed to the dual nature of the Noril'sk intrusions, namely, that their middle parts are similar in some geochemical features to other Traps around the world, whereas the lower parts are compositionally similar to ultramafic rocks. Other authors point out the wide compositional variations of picrite–troctolite gabbrodolerites in different intrusions, compared with variations from within a single intrusion and even along a single sequence (Ryabov and Yakobi 1981; Ryabov 1984b). The presence of picrite and chromite-rich mineralization at the top of the Noril'sk intrusions is inexplicable in terms of fractional crystallization differentiation (Ivanov et al. 1971b; Ryabov 1984b).

Some scientists regard them as intratelluric segregations; others see them as relics of the initial magnesian melt "frozen" to the upper contact, and still others suggest interaction of melts with fluids (Ryabov 1983, 1985).

1.2.4 Pegmatoids of the Noril'sk Intrusions

The problem of pegmatoid genesis has been discussed by geologists for many decades, and a variety of theories exist as to their formation; Fersman (1940) regarded pegmatites as formed from residual melts, Loewinson-Lessing (1933; Loewinson-Lessing and Struve 1963) saw them as the products of pneumatolytic processes, and Zavaritsky (1947), Korzhinsky (1973), and Sobolev (1986) thought they were formed as igneous rocks metasomatized by the fluids remaining after crystallization. Zavaritsky (1947) points out the ore-generating nature of pegmatites, placing them between igneous rocks and ore veins, whereas Uspensky (1968) expresses confidence in metasomatic origin of gabbro-pegmatites.

Most geologists believe the gabbro-pegmatites in the Traps are generated from a magma fraction enriched with volatile elements. This theory is supported by Tomkeieff (1939), who studied the Great Whin Sill in northern England; Shannon (1924), who examined Goose Creek in Virginia; Edwards (1942), the explorer of Tasmanian dolerites; and Walker and Poldervaart (1949), who studied the Poliseyd Sill and Karoo dolerites in South Africa. This origin is also suggested for pegmatites of the Traps of the Siberian Platform (Sobolev 1986; Lebedev 1955, 1957), although Sobolev (1986) explains the pegmatite as the result of fractional crystallization and Lebedev (1955, 1957) emphasizes the immiscibility of the pegmatite and dry Trap magmas.

The genesis of pegmatoid bodies as a result of liquid immiscibility of dry and water-saturated melts was proposed by Tomkeieff in 1928 for the Whin Sill Intrusion (Sobolev 1986). Sobolev (1986) assumed liquid immiscibility was higher in melts enriched with volatile elements, which occurs at a very late stage in magma crystallization. In his later works, however, he denied the possibility of silicate–silicate liquid immiscibility in nature (Zavaritsky and Sobolev 1961; Sobolev and Kostyuk 1975).

The immiscibility of dry and water-saturated melts was proposed as an explanation for gabbro-pegmatites schlieren observed in Traps of the Siberian Platform (Lebedev 1957). The alkaline mafic (analcime-bearing gabbro) schlieren are regarded by Masaitis (1958) as a result of liquid immiscibility. Metasomatic diorite and picrite in differentiated picritic lava flows are also regarded to be an example of silicate–silicate immiscibility (Marakushev 1979; Marakushev et al. 1982; Ryabov 1988a, 1989a, b, c). Natorkhin et al. (1977) and Dodin et al. (1987), while studying the Talnakh Intrusion, also describe the signs of silicate–silicate and ore–silicate liquid immiscibility.

Though the Noril'sk intrusions have been well studied, and fierce disputes arose over the genesis of observed pegmatoids, the wide variety of pegmatoidal forms provoked numerous (and often opposing) theories over the conditions during their genesis and their age. It is difficult to name a researcher of the Noril'sk intrusions who would not have published on genesis of taxitic gabbrodolerites and leucogabbro, which are considered as pegmatoids (Godlevsky 1959; Rogover 1959; Korovyakov et al. 1963; Vilensky et al. 1964; Zolotukhin 1964; Likhachev 1965; Smirnov 1966; Arkhipova and Natorkhin 1968; Dodin and Batuev 1971; Ivanov et al. 1971b; Borisov 1972; Lyul'ko et al. 1972; Zolotukhin et al. 1975; Stepanov 1975; Tarasov 1976; Mitenkov et al. 1977; Zotov 1979; Oleinikov 1979).

Some geologists consider taxitic gabbrodolerites as the result of interaction between magma and the surrounding country rock (Godlevsky 1959; Korovyakov et al. 1963; Vilensky et al. 1964; Mitenkov et al. 1977; Stepanov 1975). Other geologists, focusing on intersecting contacts of taxitic gabbrodolerite and other intrusive rocks, concluded that the pegmatoids formed later than other rocks in the layered series. Smirnov (1966) and some others consider the taxitic gabbrodolerites to be a product of a secondary melt intrusion that had earlier segregated in a deep magmatic chamber. Zolotukhin and Vasil'ev (1967), Goryainov (1969), Ivanov et al. (1971b), Tarasov (1976), and Natorkhin et al. (1977) regard the genesis of taxitic gabbrodolerites as the interaction of the melt with volatile elements in a high-level chamber. The characteristic feature of this process, according to some geologists, is decreasing silica abundance, whereas others see increasing abundance of silica, the third are speaking about degeneration of melt with involvement of volatiles (Korzhinsky 1973; Natorkhin et al. 1977; Zolotukhin 1984a, b).

There is an alternative hypothesis that leucogabbro and ophito-taxitic gabbrodolerites are the products of crystallization at depth (Korovyakov et al. 1963; Likhachev 1965, 1977; Tarasov 1976; Oleinikov 1979). The suggestion is based on the existence of loaf-like segregations that range from leucogabbro bodies within gabbro to veins of picrite gabbrodolerites that cut leucogabbro. Contrary to this, it has been observed that gradual transitions from melanocratic to leucocratic taxitic gabbro and leucogabbro are widespread in the top and bottom horizons of pegmatoids (Zolotukhin and Vasil'ev 1967; Ryabov 1969). Tarasov (1976) regards the leucogabbro segregations in picrite gabbrodolerites as xenoliths, whereas Ryabov (1989a, b, c) indicates they are homeogenous inclusions. The accumulations and trails of high-Mg olivine grains that are widespread in taxitic gabbrodolerites belong to intratelluric minerals according to some geologists (Vorontsov et al. 1986; Korovyakov

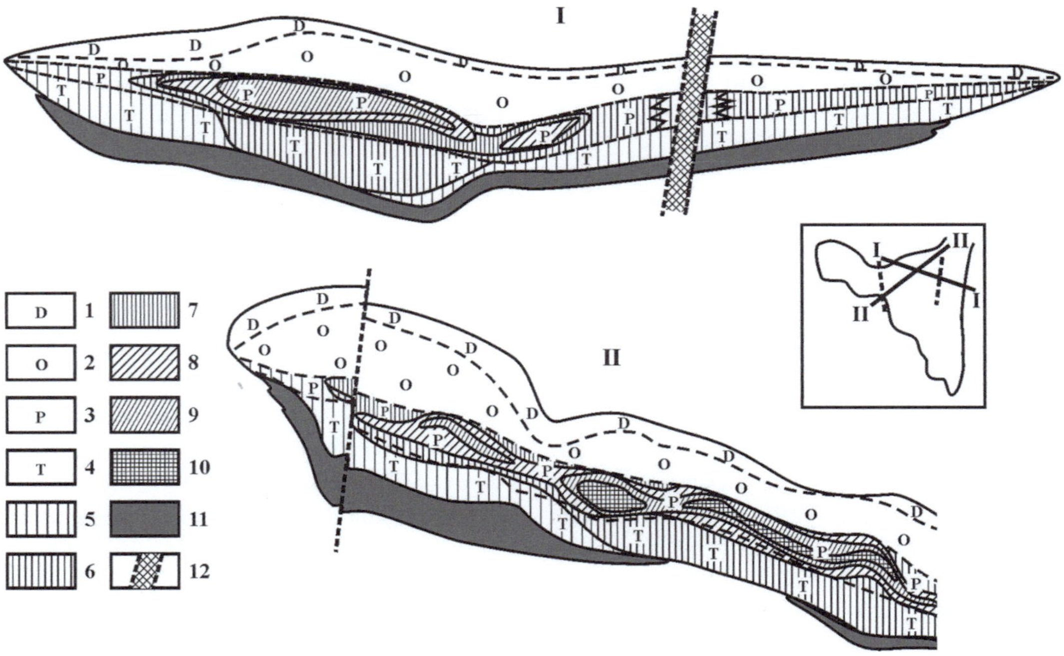

Fig. 1.3 Distribution of sulfide mineral associations in vertical sections of the Talnakh differentiated intrusion (From Genkin et al. 1981). Locations of Sections I and II are shown in the *inset* (schematic plan of intrusion). *1* diorite, gabbro–diorite, gabbrodolerite, *2* olivine-free, olivine-bearing, and olivine gabbrodolerite, *3* picritic gabbrodolerite, *4* taxitic gabbrodolerite, *5–10*: sulfide mineral associations: *5* Pn + Cp + Po_m, *6* Pn + Cp + Po_h, *7* Pn + Cub + Cp + Po_h, *8* Pn + Cub + Cp + Tr + Po_h, *9* Pn + Cub + Cp + Tr, *10* Pn + Cub + Tn(Mh, Cp_c) + Tr, *11* bodies of massive sulfide ores, *12* disjunctive dislocations

et al. 1963), whereas others (Ryabov and Pavlov 1984) regard them as picritic gabbrodolerites relics, or as metasomatized xenoliths of sedimentary and magmatic rocks (Lyul'ko et al. 1972; Zotov 1979).

Zolotukhin 1964, 1997; Zolotukhin and Vasil'ev 1967; Zolotukhin et al. 1975 assume the formation of taxitic gabbrodolerites to be a result of secondary near-contact remelting (magmatic replacement) of intrusive rocks accompanied by metasomatism. A number of geologists consider the metasomatic processes to be of more importance in formation of leucogabbro and taxitic gabbrodolerites (Goryainov 1969, 1971; Borisov 1972; Gulin and Sukhov 1974; Zotov 1979).

Considering the important connection between ore mineralization and mafic pegmatoids various layered intrusions, such as the Bushveld Igneous Complex in South Africa (Scoon and Mitchell 1994) and the high mineralization of taxitic gabbrodolerite horizons in the Noril'sk intrusions, the problem of the genesis of mafic pegmatoids has direct impact on the genesis of ore deposits (Zolotukhin 1997).

1.2.5 Origin of Sulfide Ores of the Noril'sk Deposits

The copper–nickel sulfide deposits, as a rule, are confined to mafic–ultramafic complexes located in platform structures and orogenic belts. They are characterized by a connection with intrusive bodies and their metamorphosed contact zones, as well as by a chalcopyrite–pentlandite–pyrrhotite composition. These characteristics serve as a basis for the comparison of deposits from different regions and, when combined with additional information from these regions, can be used in the development of a generalized model of ore genesis (Godlevsky 1968; Gorbunov 1968; Chernyshev 1971; Kozlov 1973; Likhachev 1977, 1988; Zolotukhin 1979, 1988; Distler and Duzhikov 1988; Naldrett and Cabri 1976).

The geology of the Noril'sk deposits, structure of the ore bodies, mineral composition of ores, variations in ore components, and mineralization parameters have been considered in numerous publications by many researchers (Godlevsky 1959; Rogover 1959; Smirnov 1966; Goryainov 1971; Dodin et al. 1971; Kavardin and Mitenkov 1971; Genkin et al. 1981, and many others). All ore types can be represented by the associations of a few main mineral species with wide variations in their compositions. For example, the paragenetic mineral assemblages, represented by changing ratios in pyrrhotite and chalcopyrite species, pentlandite, and cubanite, have allowed researchers to trace zoning in a disseminated ore horizon (Fig. 1.3). The latent layering of ore-bearing horizons of intrusion manifests itself in the variation of sulfur content in minerals, the change of pentlandite structure, as well as in distribution of ore minerals in sulfides (Genkin et al. 1981).

Though the Noril'sk deposits have been well studied, discussions regarding their genesis have continued for many years. There are a number of hypotheses on the origin of ores: liquid immiscibility, infiltration–metasomatism, or metamagmatic sulfurization. Most popular among geologists is the liquid immiscibility concept by analogy with the Sudbury Intrusion (Vilensky et al. 1964).

Sobolev, in his work published in 1936 and republished in 1986, observed the similarity of the Noril'sk deposit to Sudbury mineralization and assumes that sulfide formation took place in the pegmatite stage. In the same publication (Sobolev 1986, p. 27), he writes: "...as sulfides, separated from silicate magma as an immiscible liquid, remain in a liquid state for a long time due to their mobility they will easily pass to MgO-poor dolerite, as well as to lateral rocks within the intrusion and the local area."

The liquid immiscibility hypothesis was developed by Kotul'sky (1948), while its physicochemical basis was formulated by Godlevsky (1959, 1961, 1967, 1968, 1971). As for the timing and emplacement of these immiscible liquids into ore minerals with a silicate melt, opinions of geologists differ. Some suppose it took place during stratification in present-day chambers (Korovyakov 1960), in present-day and intermediate chambers (Godlevsky 1959; Oleinikov 1979), in deep chambers (Likhachev 1977; Godlevsky and Likhachev 1979; Genkin et al. 1981; Stepanov 1981; Dodin et al. 1987; Distler and Duzhikov 1988), or in deep and present-day chambers (Kotul'sky 1948).

Ores connected with komatiite are usually used as proof of a high mantle source for the sulfide melt. The possibility of a sulfide-rich melt forming in abyssal conditions can be explained by the melting of sulfides that are disseminated in mantle rocks (Godlevsky and Likhachev 1977, 1979) and their carrying over in the form of fine emulsion of sulfide liquid, as it is offered for world class copper–nickel sulfide deposits (Naldrett and Cabri 1976).

Korzhinsky (1973) made the assumption that sulfide ore formation is a result of metamagmatic processes, and these ideas are shared by other geologists (Natorkhin et al. 1977). Following on from this work, Zotov (1979, 1982) considered the process of transmagmatic fluid interaction with magmatic melt in deep hypabyssal chambers. Changes in the physicochemical parameters of the intruding melt and fluids promote the formation of ore liquid which later accumulates under the influence of gravity at the base of intrusions. Thus, the mantle serves as the main source of ore material, while abyssal fluids act as the transporting agents for metals.

Almukhamedov and Medvedev (1982) presented the concept of magmatogenic sulfurization. It provides for the formation of sulfide-rich liquids through the interaction of magma with sulfur in the course of melt migration through sulfur-bearing rocks in the Earth's crust.

The infiltration autometasomatic hypothesis of Noril'sk mineralization was substantiated by Zolotukhin (1964, 1965, 1971, 1979, 1988). The basis of his concept is the theory of infiltration metasomatism and the acid–basic evolution of postmagmatic solutions, developed by Korzhinsky (1973). Korzhinsky (1973) connects pegmatite formation to mineralization, assigning the dominant role to volatile elements in genetic reconstructions. He believes that the presence of sulfur in fluids leads to sulfurization that peaks at the high-temperature postmagmatic stage, when ore formation of all types takes place. In his late works, he continues to develop the concept but considers it more correct to name it a sulfurization–metasomatic process (Zolotukhin 1988, 1997).

Ideas essentially close to Zolotukhin's infiltration–metasomatism concept are expressed by supporters of the hydrothermal, hydrothermal–metasomatism (postmagmatic) hypothesis of mineralization of Eliseev (1959), Kavardin et al. (1968), Goryainov (1971), and Tarasov (1976). It is necessary to note that the authors consider rich sulfide ores to be epigenetic, while poor interstitial mineral ores are supposed to be magmatic and syngenetic.

1.3 Brief Review of Experimental Research in Mafic Rock Petrogenesis

One of the main causes for the variety of observed magmatic rocks is the degree of differentiation of the melt. Numerous opinions on melt conditions, melt differentiation, mechanisms for melt differentiation, and other processes of petrogenetic significance exist (Loewinson-Lessing 1935; Grigor'ev 1946; Zavaritsky and Sobolev 1961; Zharikov 1976; Marakushev 1978, 1976, 1981). The most popular among geologists is crystallization–gravity (fractional) differentiation along with the more recently popular mechanism of magmatic (fluid–magmatic, liquid immiscibility) differentiation. Let us consider the validity of these two types of differentiation.

1.3.1 Experimental Data on Fractional Crystallization

The theoretical basis for basaltic magma fractional crystallization was established by Bowen (1928). Despite the fact that all the concepts discussed in this publication have been updated and revised, the laws of crystallization established by Bowen have kept their value till now. Nowadays, new data for oxide and silicate systems, and systems with volatile elements have been obtained, the role of iron oxides and oxygen fugacity has been thoroughly studied, and the ways of crystallization for systems with ternary solid solutions and a quantitative estimation of fractional crystallization have been studied. After the long-term rejection of the possibility of liquid immiscibility processes in nature, its experimental and geological proof is now established, and geologists have realized its petrogenetic importance as a mechanism of pre-crystallization differentiation (Roedder 1979), using data from ultramafic melts (peridotitic komatiites).

As far as the principles for crystallization of multicomponent melts are the same as those of binary, ternary, and quaternary experimental systems, their study allows us to more accurately understand the processes of magma crystallization.

The fractional crystallization of basaltic magmas was studied by Bowen on diopside–forsterite–silica, anorthite–forsterite–silica, and anorthite–albite–diopside ternary systems, an analysis that conveniently follows the generalized sequence of Trap crystallization (Zavaritsky and Sobolev 1961). Considering gravity separation during the early phase and using two-component systems, one can easily explain the crystallization of olivine-enriched rocks, silica-rich rocks containing free quartz from a basaltic magma, the wide occurrence of pyroxene–plagioclase associations in undifferentiated Traps, zoning in minerals of alternate composition, and the trend to increasing alkalinity in layered intrusions (Sobolev 1986). Bowen's reaction series presented a way in which the origin of a variety of igneous rocks could be traced to common basaltic magma, which in due course evolves through differentiation to a residual melt of granite composition (Bowen trend). Fenner, and later Sobolev (1986), while studying Traps, observed that basaltic magma was enriched in iron during crystallization (Fenner trend). In the geological literature, these two trends of crystallization were opposed until Kennedy (1948, 1955) proved experimentally that the value and character of changes in oxygen fugacity was one of the main factors that determined the direction of melt differentiation. This meant that low partial pressure of oxygen during crystallization or a high initial $FeO:Fe_2O_3$ ratio in basaltic magma leads to an increase in the iron content of the final products of differentiation and to the generation of ferrogabbro (Fenner trend). On the contrary, high partial pressure of oxygen or low initial iron oxides ratio promotes formation of granite residues (Bowen trend). Kennedy's conclusions have been confirmed and developed later in numerous works of Muan (1955, 1983) and Osborn (1959, 1979). The investigations conducted by them have shown that the direction of melt evolution depends on the system either being open for oxygen, in which case oxygen pressure remains stable or increases (Bowen trend), or closed for oxygen, in which case oxygen pressure decreases in the course of melt crystallization (Fenner trend).

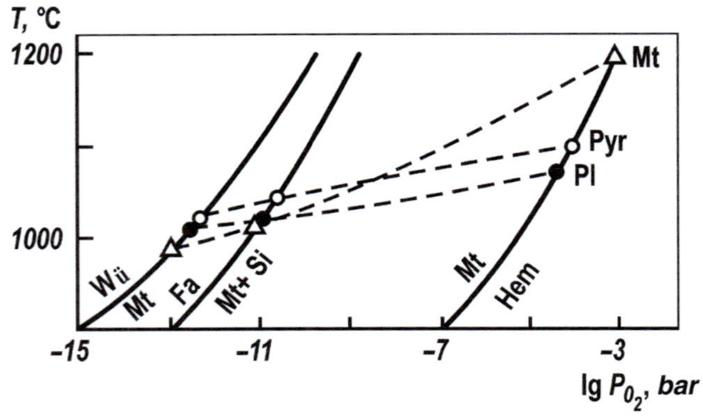

Fig. 1.4 Crystallization sequence of minerals from tholeiitic basalt depending on pressure of oxygen (After Osborn 1964). *Mt* magnetite, *Pyr* pyroxene, *Pl* plagioclase. *Solid lines* show oxygen pressure values (depending on temperatures) regulated by various buffers. Buffer components include *Wu* wustite, *Mt* magnetite, *Fa* fayalite, *Si* silica, *Hem* hematite

The investigation of a $MgO–FeO–Fe_2O_3–CaAl_2O_8–SiO_2$ system (Roeder and Osborn 1966) allowed them to show quantitatively that in case of fO_2 exceeding 10^{-9}, the end product crystallized from basaltic magma will be an andesite melt, whereas with lower oxygen partial pressure, ferrogabbro will be formed. The oxide ore minerals form as early crystal phases in oxidizing conditions and as late phases in reducing conditions (Fig. 1.4). In the first case, high fO_2 (10^{-7} atm) is required. In case of low fO_2, magnetite is absent, and instead ferrous olivine develops. Various modes of fO_2 evolution allowed Osborn (1959) to set out, by analogy to the Bowen reaction series, two series for melt crystallization at low and high fO_2.

The methods of melt crystallization, known for binary and ternary systems, can be illustrated by a quaternary diagram (Fig. 1.5), on which data for incongruently melting compounds ($MgSiO_3–FeSiO_3$) of dry melts are combined with the mafic mineral solid solutions (Muan 1983).

Experimental studies of the dependence of the structure of olivine on fO_2 and temperature are presented by Hill and Roeder (1967), Roeder and Emslie (1970), Nitsan (1974), and Nikol'sky (1978). The degree of iron oxidation is connected with oxygen fugacity by the following equation: Log $X_{FeO}/ X_{FeO} = -0/20$ Log fO_2 – 1.04 (Roeder and Emslie 1970). Using diagrams made by Roeder and Emslie, in particular Fig. 1.6, one can see the dependence of olivine composition of fO_2 and temperature changes.

The Skaergaard Intrusion is regarded as a classical example of fractional crystallization, and geologists often use data available from the intrusion to compare with other Trap intrusions (Zolotukhin 1964; Feoktistov 1978). A comparison of crystallization features from the Skaergaard and Noril'sk intrusions was undertaken by Nekrasov and Gorbachev (1978). The distinction in initial

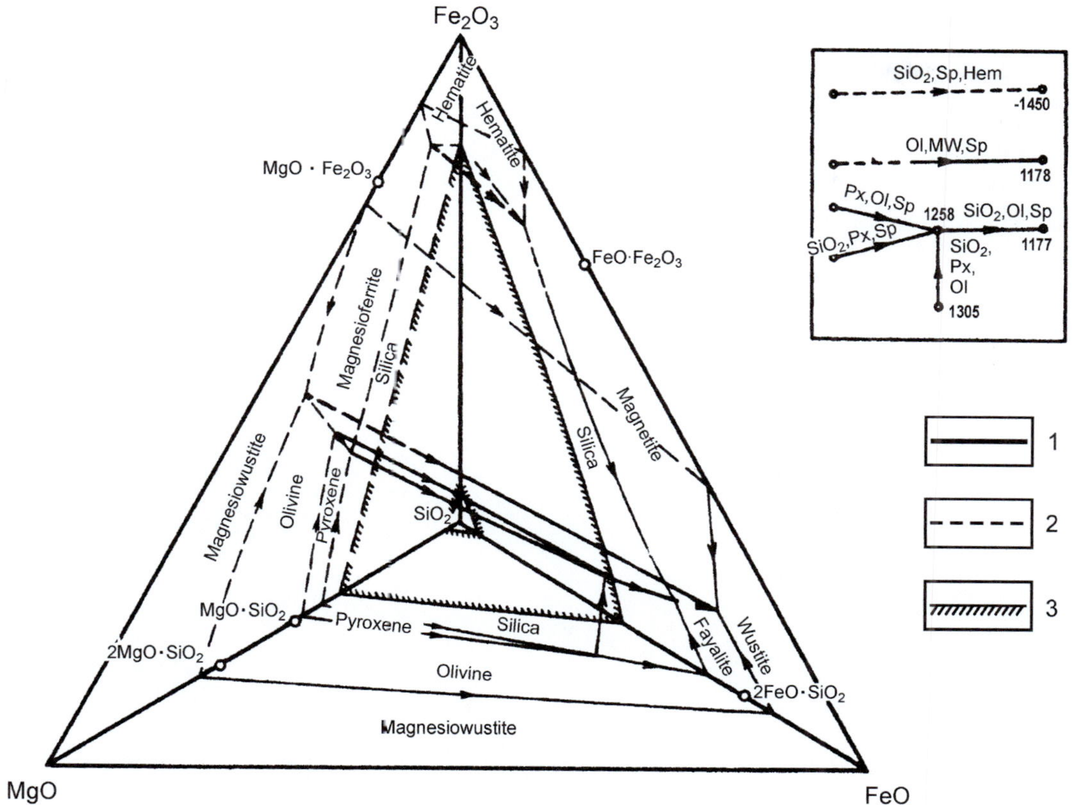

Fig. 1.5 Diagram showing liquidus phase relationships in the system MgO–FeO–Fe_2O_3–SiO_2 according to Muan (1983) data. Boundary curves: *1* on tetrahedron faces, *2* inside tetrahedron, *3* area of liquid immiscibility. *Inset*: simplified representation of evolutes of univariant liquidus curves: *Sp* spinel, *Hem* hematite, *Ol* olivine, *MW* magnesiowustite, *Pyr* pyroxene

melt compositions, according to these authors, predetermined their quite unlike differentiation, just as it is shown in diagrams (Fig. 1.7) plotted with the use of An–Ab–Fo–Di and An–Ab–Fo–Q experimental systems.

One of the methods of studying layered intrusions is physicochemical modeling, by mathematical methods and computer software, of magma crystallization with the use of field and experimental data. Using the scheme of Natan and Van Kirck (1978), a program to model silicate melt crystallization was created for dry systems, and later for water-saturated systems (Feoktistov 1984). Presently, it seems possible to estimate the speed and dynamics of fractional crystallization given an initial melt composition, the presence or absence of volatile elements, and consideration of the cooling regime and viscosity changes, and also to predict mineralogical, petrochemical, and geochemical laws of intrusion structure, and the conditions of its occurrence (Sharapov and Cherepanov 1986; Frenkel et al. 1988; Ariskin 1999). Modeling of fractional crystallization dynamics for various layered intrusions is presented by Dneprovskaya et al. (1985), Koptev-Dvornikov et al. (1976), Frenkel et al. (1988). The model allows for a number of possible solutions to the thermodynamic problem, depending on a set and composition of phases, temperature, and heat and mass transfer equations (diffusion, gravitational separation of crystals and liquid, convection of heterogeneous liquids). With the proposed model, the authors explain the peculiarities in internal structure and composition of layered intrusions, degree of differentiation of tholeiite magmas, and pegmatite formation (Ariskin 1999).

The rich empirical material collected over a long period of time has formed the basis for local and regional forecast of potentially ore-bearing intrusions, prospective ore targets, by mathematical and logically derived mathematical methods (Dodin and Sadikov 1967; Dmitriev et al. 1968; Natorkhin et al. 1977; Zolotukhin 1978a; Sukhov 1985; Karbyshev 1986).

1.3.2 Crystallization Temperatures of Traps

Data on recent volcanism, as well as experimental and model data, indicate that the temperature of a basaltic magma is close to 1,100–1,200°C (Godlevsky 1959; Yoder and Tilley 1962; Natorkhin et al 1977). Experiments on the melting and

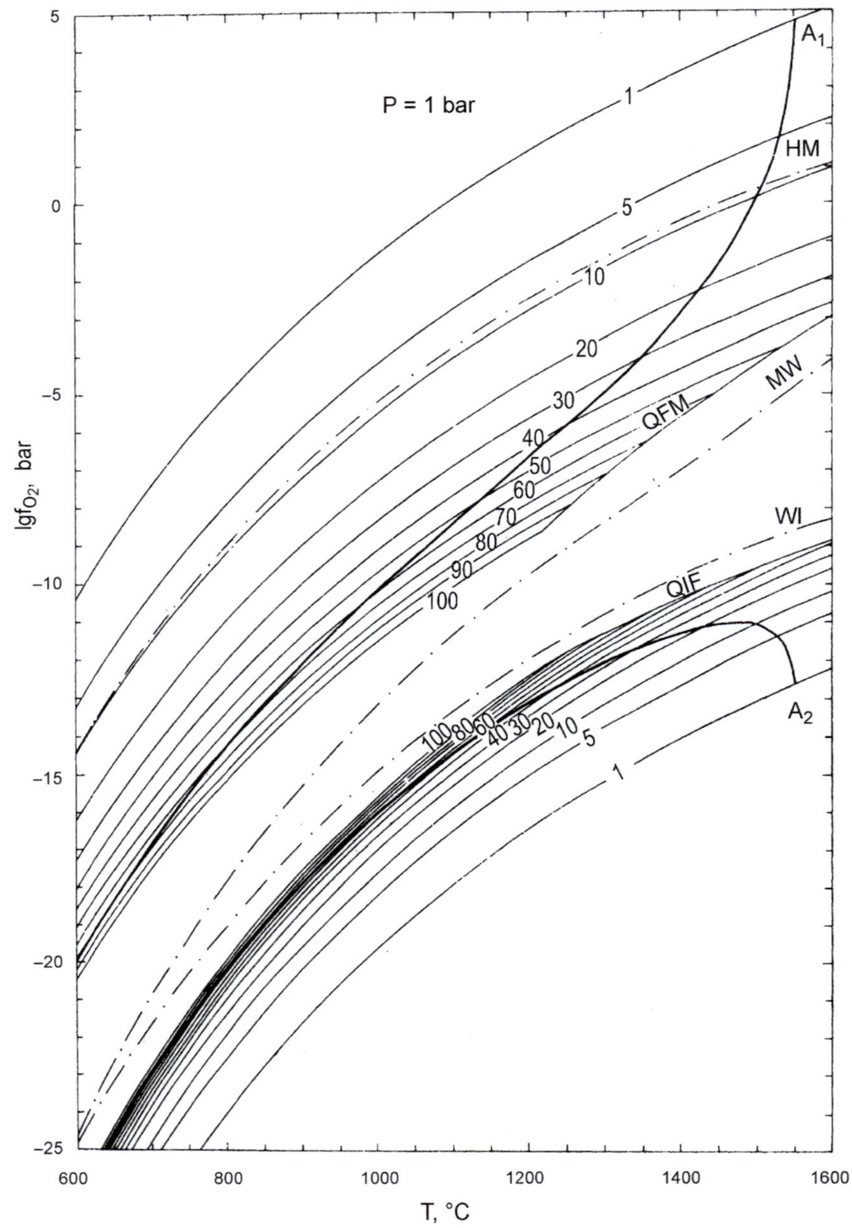

Fig. 1.6 Equilibrium isolines versus temperature and logarithm of oxygen volatility at overall pressure of 1 bar. Figures on the curves mark the fayalite content in total olivine (Nikol'sky 1978). A_1 and A_2 are plots showing changes in free energy

crystallization of rocks revealed that the rocks ranging from picrite to granophyre and gabbroic pegmatite were crystallized at 1,360–800°C (Zolotukhin 1964; Nesterenko and Almukhamedov 1973; Feoktistov 1978; Almukhamedov and Medvedev 1982).

Studies of fluid inclusions in minerals give important information about the physicochemical conditions during crystallization of magmatic melts. For example, the study of inclusions in pyroxene of the Noril'sk intrusions allowed Bulgakova (1971) to evaluate the temperatures of crystallization of a layered series and gabbroic pegmatoids to be 1,130–1,280°C and 1,070–1,300°C, respectively. The analyses of inclusions in minerals of the Talnakh Intrusion conducted by Vortsepnev (1978) showed that their temperatures of homogenization from various magmatic rocks were at 1,120–1,140°C, 1,120–1,170°C, and 1,060–1,070°C in clinopyroxene, plagioclase, and quartz from diorite, respectively.

Fluid inclusion studies of minerals by Kovyazin and Shcherbakova (Zolotukhin et al. 1988), which embraced different types of olivine-rich rocks of the Siberian Platform, attracted much interest. It was established that olivine in picrite from the Noril'sk intrusions may have a high crystallization temperature (1,440–1,470°C), whereas the homogenization of inclusions in plagioclase of these rocks was at 1,250°C. In picrite–troctolite intrusion of Sukhoy Bakhty

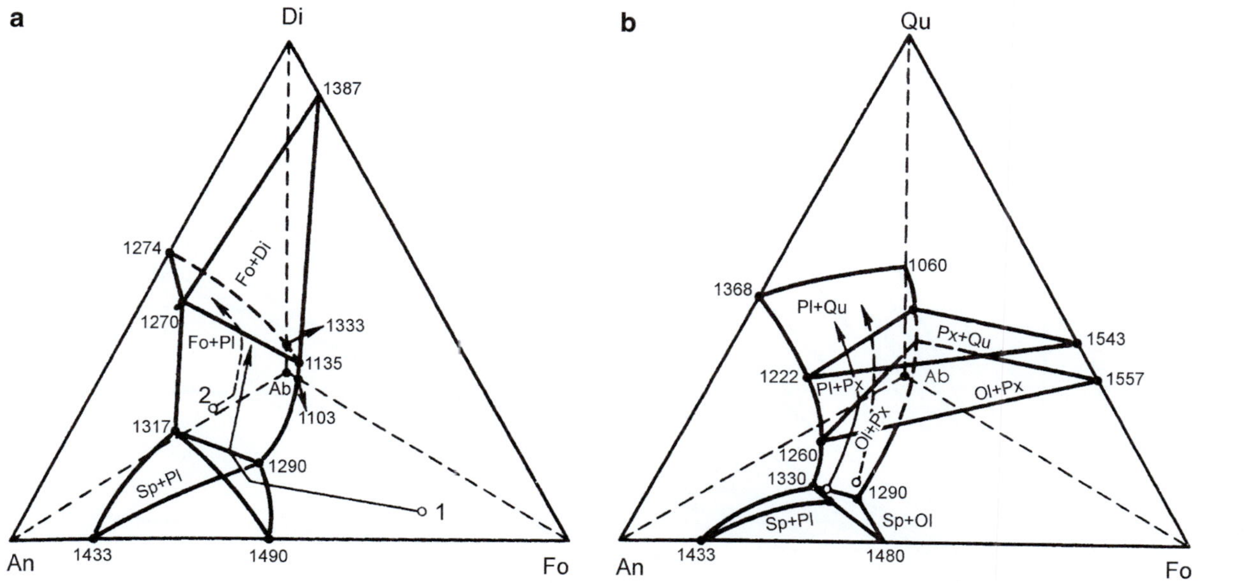

Fig. 1.7 Fusion diagrams for () An–Ab–Fo–Di and (b) An–Ab–Fo–Q systems. *Arrows* indicate ways of melt crystallization in intrusions of Noril'sk (1) and Skaergaard (2) types. Figures show temperature in degrees Celsius (Nekrasov and Gorbachev 1978)

and Intrusion of the Fat'yanikha River, crystallization temperatures were 1,350–1,400°C and 1,360°C for olivine and 1,280–1,300°C and 1,180°C for plagioclase, respectively. From the Pankov (1986) data, the homogenization temperatures of inclusions in plagioclase from the Ust-Khan'insky, Toulbinsky, and Amovsky intrusions indicate gradual crystallization between 1,410–1,450°C and 1,160–1,220°C.

Based on a study of inclusions in rock-forming minerals from "normal Traps" in the valleys of the Taimur, Tutonchan, and Nizhnyaya Tunguska Rivers, Zolotukhin and Podgornykh (1998) indicated that the basaltic magma was superheated to as much as 1,600°C. The authors state that this may be explained by the melt having a low viscosity, which also explains the formation of extensive thin basalt flows and dolerite sills, the coarse-grained poikilophitic texture of the basalt, and the common lack of porphyritic texture in contact facies of Traps.

The temperature conditions for the onset of differentiation of immiscible liquids in Traps are known from three localities. Data on inclusions in olivine from the Ust'-Khaininsky Intrusion indicates the silicate–silicate immiscibility began at temperatures lower than 1,110–1,170°C (Pankov 1983). Guided by data on the homogenization of inclusions in plagioclase from the mesostasis of the Putoran flood basalts, the onset of silicate–silicate immiscibility may be considered at temperatures lower than 1,190–1,200°C (Tikhonenkov et al. 1985). Based on inclusions in pyroxene, Krasov (Ryabov et al. 1985b) established that the ore–silicate immiscibility took place at 1,100–1,180°C. It should be noted that temperatures for the onset of differentiation obtained by different authors agree with experimental data of other researchers (Roedder 1979; Philpotts 1982).

1.3.3 Experimental Data on Differentiation in Silicate and Ore Liquids

Liquid immiscibility as a possible mechanism of formation of various rocks and ores from the same parental was proposed by Durocher (1857), and it was used later by a number of researches (e.g., Rosenbush 1887) for explanation of the magmatic differentiation. As Loewinson-Lessing and Struve (1963) reported, Vogt (1923) accepted liquid immiscibility for explanation of formation of magmatic ore deposits. Daly (1933) and Niggli (1920) stated that the diversity of rocks could be due to so-called assimilation of rocks by magma, that is, a change in the magma composition promoted its differentiation.

However, experimental investigations performed at Carnegie Institute allowed Greig (1928) to only obtain segregation of melts in binary systems at temperatures greater than 1,700°C, and the results of experiments plotted on the SiO_2–RO–(R_2O + Al_2O_3) diagram showed significant discrepancies between the immiscibility field obtained and the composition of native magmatic rocks. This evidence suggests the immiscibility to be an unrealistic process for natural environments.

The first experimental verification of liquid immiscibility in silicate liquids at a temperature around 1,200°C was achieved by Grigor'ev (1935) who smelted schicht material containing volatile elements. The liquids obtained from

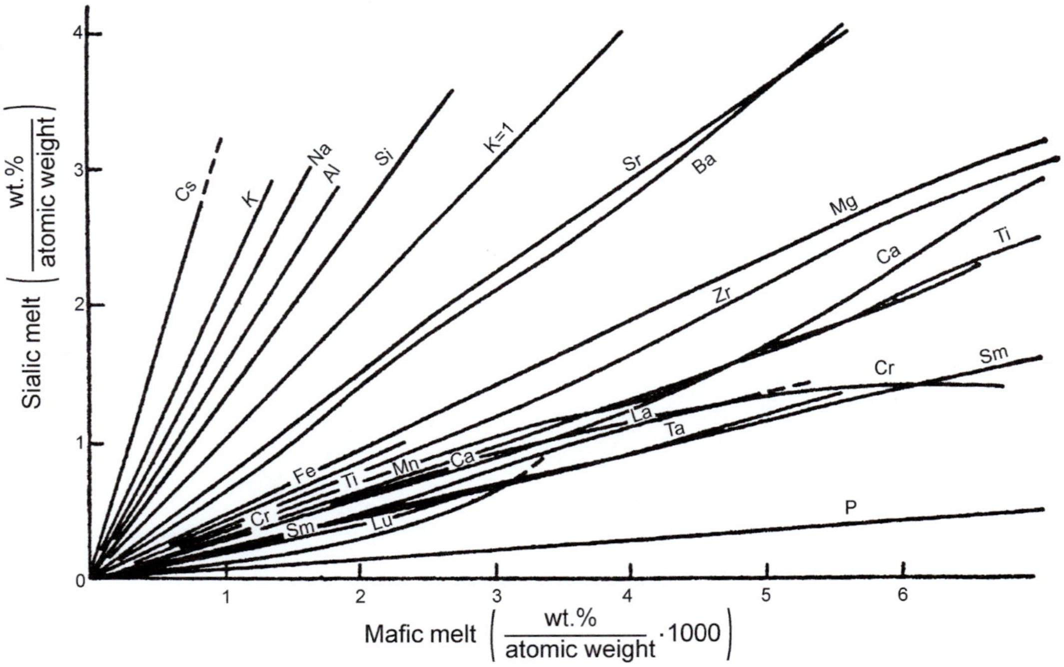

Fig. 1.8 Diagram of chemical element distribution between silicic melt and mafic melt based on experimental data (From Watson 1976). $K = 1$ (even distribution)

the schicht material were close in composition to alkali-magnesian basalt, syenite, missourite-bearing rock, and peridotite, that is, those forming typical Inagli-type mafic intrusions.

Roedder (1951) determined the region of stable low-temperature (1,100°C) liquid immiscibility in the leucite–fayalite–SiO_2 system. Later, Roedder and Weiblen (1971) observed a glass emulsion with immiscible liquid features common for lunar basalts and plotted the composition data on a Greig diagram. This demonstrates the close similarity between the experimental and natural data (Roedder 1979). At present, the number glass finds with immiscible liquid features has been significantly increased. These glasses have also been found in Siberian Traps (Pankov 1983; Ryabov et al. 1985b; Tikhonenkov et al. 1985). The most comprehensive report on this matter is in a monograph by Ryabov (1989c).

In the last 10–15 years, the number of theoretical and experimental publications on immiscible liquids has grown (Warren and Pinkus 1940; Fisher 1950; Esin and Gel'd 1966; Filippovich and Dmitriev 1971; Andreev et al. 1974; Anfilogov 1975; Carstens 1979; Varshall 1981; Babushkin and Zhunina 1983; Pavlov 1983; Roedder 1979). As Bondar' and Toropov (1963) have noted, "previously the existence of immiscible liquid areas was considered to be an exceptional phenomenon, now it has been shown that the formation of micro-inhomogeneities is a specific feature of both artificial and natural melts."

Liquid immiscibility (otherwise known as fluid phase differentiation) is a process of segregation of a liquid into two or more liquids with a distinguishable interface (Andreev et al. 1974). In the opinion of Esin and Gel'd (1966), the principal factors for the origin of immiscibility in ore–magmatic melts are related to the structure of anionic groups and the metal oxide concentration, both of which are responsible for initiation of liquid segregation, extension of the immiscibility area on phase diagrams, and will increase with increasing cation charge to cation radius ratio. The cause of fluid immiscibility in silicate melts is that it is energetically favorable for the positively charged metal ion to come close together with any negatively charged oxygen ions surrounding it (Filippovich and Dmitriev 1971). Experiments (Watson 1976; Krivenko et al. 1984) allowed the authors to understand the common changes in chemical element distribution throughout the immiscibility process (Fig. 1.8). Works by Vogt (1923), Grigor'ev and Iskyul (1937), and Olshansky (1951) are fundamental in the field of ore–silicate immiscibility. These researchers determined the solubility of sulfides in a silicate melt and showed trends in solubility changes. The experiments of Olshansky (1948, 1951) were directly related to the formation of Noril'sk ores. Studying the FeS–FeO–SiO_2 system, Ol'shansky established areas where the existence of immiscible silicate–sulfide and sulfide–metal liquids was possible, and in the earliest work, showed the viscosity of a sulfide liquid and its tendency toward segregation. This system has now been worked out

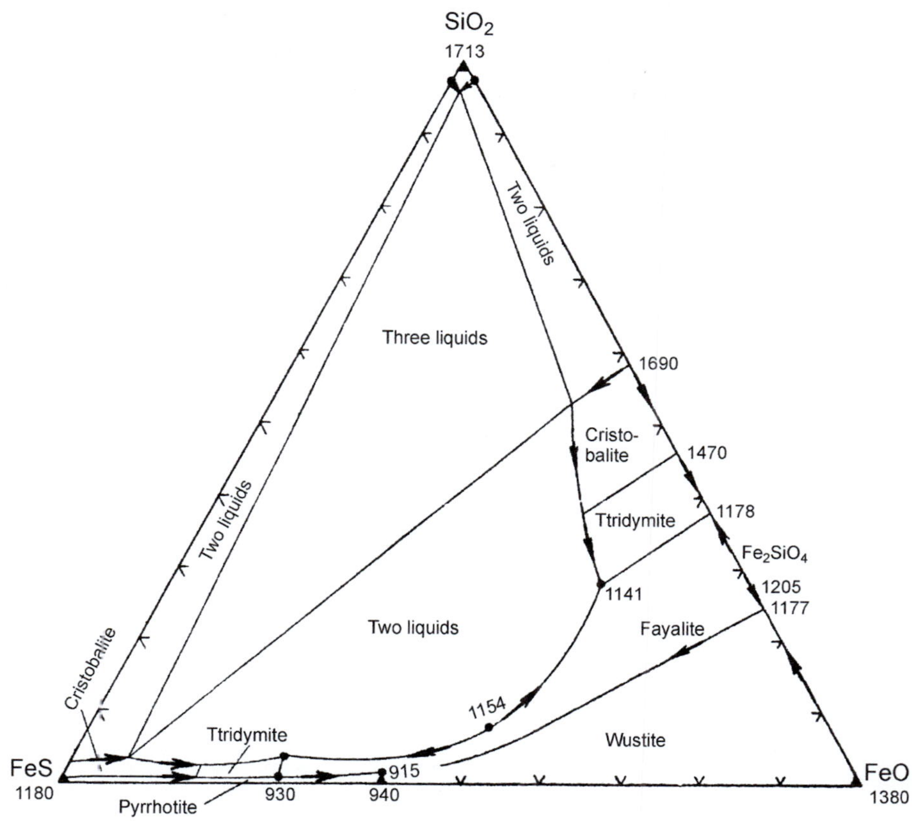

Fig. 1.9 Liquidus relationships in the FeS–FeO–SiO_2 system (From Skinner and Peck 1969). Figures show temperature in degrees Celsius

in detail by Yazawa and Kameda (1953) (Fig. 1.9). Naslund (1983) obtained the fractionation of an immiscible liquid in the $KAlSi_3O_8$–$NaAlSi_3O_8$–FeO–Fe_2O_3–SiO_2 system that was similar to natural magmas. In doing this, the fluid was differentiated into liquids enriched in iron and silicic acid in four regimes at fO_2 10^{-22}, 10^{-9}, 10^{-5}, and 10^{-7} atm. MacLean (1969) studied immiscibility phenomena in the FeS–FeO–Fe_3O_4–SiO_2 system, and Shimazaki and Clark (1973) studied these phenomena in the FeS–FeO–SiO_2–Na_2O system. Fractionation in ore liquids was studied by Newhouse (1927) in the pyrrhotite–pentlandite system and by Kullerud (1962) in the Fe–Ni–S system.

It is difficult to embrace the whole scope of available information about liquid immiscibility in such a short space because the number of publications devoted to this issue exceeds 100 per year. A review of these investigations concerning petrogenesis and ore genesis was given earlier (Ryabov 1989c).

It is known that the formation of immiscible liquids can occur in dry melts according to Greig (1928), enriched in Fe, Cr, Ti, and other elements (Babushkin and Zhunina 1983), as well as in melts saturated with volatile elements, in particular phosphorus (Fisher 1950) and fluorine (Grigor'ev 1935). From the final state following the initiation of a liquid-in-liquid emulsion, there is a close similarity between the theories of liquid immiscibility and fluid–magmatic differentiation (Marakushev 1976, 1979, 1981; Zharikov et al. 1986; Letnikov et al. 1988). Therefore, the term liquid immiscibility, that is, differentiation of immiscible liquids, is likely to be more extensional, and it includes all particular types of immiscible liquids formed with/without fluids in silicate, ore–silicate, ore, or salt systems.

In a number of publications based on experimental results, the role of single elements and compounds as active agents of liquid immiscibility such as HF, HCl, CH_4, H_2O, CO_2, H_2S, B_2O_3, P_2O_5, Cr, Ti, and others (Ryabov 1989b, c) has been shown. Many researchers note a profound effect of volatile elements on the ore-generating role of fluid–magmatic differentiation during petrogenesis and ore-forming processes (Marakushev 1976, 1981; Zharikov et al. 1986).

Volatile elements essentially influence the temperature and succession of mineral crystallization. The volatile elements dissolved in magmas are responsible for changes in physicochemical properties such as viscosity and degree of polymerization. If magma is saturated with volatile elements, the magma will differentiate into silicate–silicate and/or silicate–salt (ore, carbonate, phosphate, fluoride, etc.) melts. Magmatogenic fluid and salt melts separated from a magma are able to concentrate and transport many petrogenic and ore elements. The attention of geologists has long been drawn to the problems of fluid–magmatic

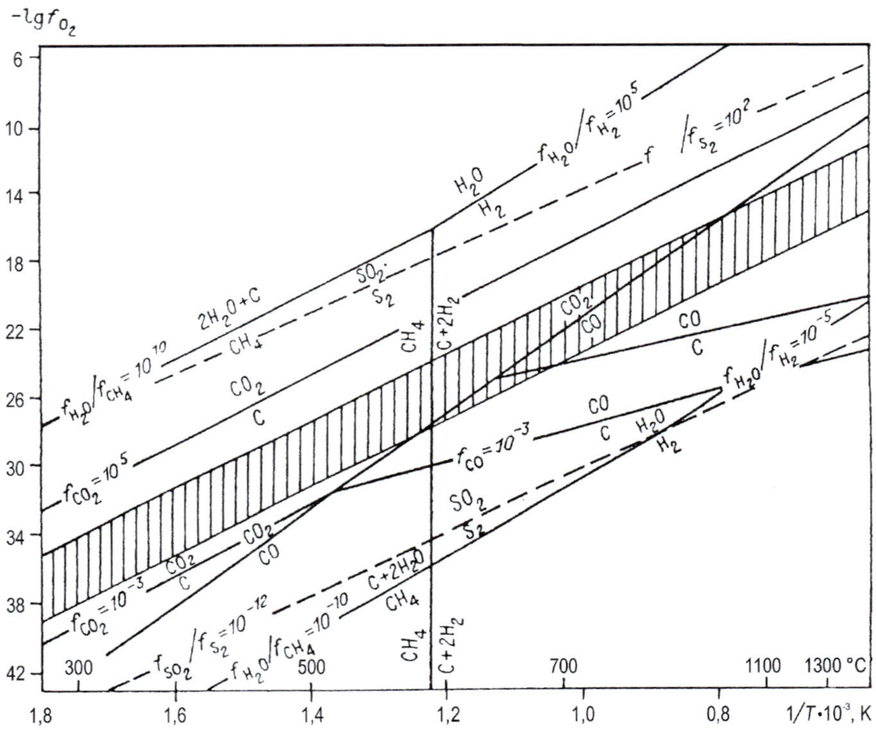

Fig. 1.10 Plot of fO_2 versus temperature for typical phase equilibriums in the C–O–S system. Crosshatched area indicates optimal values for component volatility and crystallization temperatures of Noril'sky-type ores (Nekrasov and Gorbachev 1978)

interactions. The importance of the fluid regime in magmatic process has been emphasized in works by Korzhinsky (1973) and Zavaritsky and Sobolev (1961).

In regard to Traps, data from experiments conducted for melts with volatile elements have demonstrated the possibility of melts to be differentiated into mafic–ultramafic and mafic–granophyric liquids (MacBirney and Nakamura 1974; Marakushev et al. 1982), as well as a trend of trachybasalt melts to liquid immiscibility (Gorbachev 1989).

Deep fluids are considered to have a reducing ability and represent a complex gas–salt mixture, which in some approximations has the form of the C— system (Marakushev and Perchyuk 1974; Nikol'sky 1978). For Traps and, in particular, for the Noril'sk intrusions and related ores, it is also proposed that the fluids arising from the melts, which are oxidized during the evolution of the magmatic system, were initially reducing (Nekrasov and Gorbachev 1978; Genkin et al. 1981; Ryabov 1989c).

In regard to Traps and ores associated with them, the Fe–S–O–SiO_2 system is one of the most extensively studied, individual units of which are known from various publications (Muan 1983; Olshansky 1951; Naldrett 1969; MacLean and Shimazaki 1976). The investigations performed by Nekrasov and Gorbachev (1978) allow the entire succession of formation of copper–nickel ores to be traced throughout the crystallization path of a sulfide melt. It has been found that the most important parameters controlling ore-forming process are temperature and the fugacity of O_2, S_2, and H_2S; the lowering of the fugacity in any of these, in the ore-forming process, leads to a change in the oxide–sulfide equilibriums and an extension of the stability fields of sulfides. Occurrences of graphite and carbides in sulfide ores suggest that the latter are in equilibrium with H_2O, CO_2, CO, and CH_4 (Fig. 1.10). The calculations have shown that the environment during the ore-forming process had reducing nature, f_{H_2O}:f_{CH_4} was unchanged, f_{CO_2} decreased from 10 to 10^{-1} atm, and fSO_2:fS_2 was 10^{-6}–10^{-8}.

During the ore-forming process, fO_2 changed from 10^{-12} (at 1,300°C) to 10^{-37} atm (at 1,300°C) and fS_2 changed from 10^{-3} to 10^{-16} atm. With increasing temperature, the crystallization of a sulfide-rich liquid was buffered by the pyrrhotite–magnetite association with or without olivine, by bornite–magnetite association, and others (Fig. 1.11), and high-temperature monosulfides under saturated in sulfur were replaced by multi-sulfides and disulfides against a background of decreasing fS_2.

Experiments associated with sulfide systems containing sulfur are of great interest in understanding the processes of sulfide–ore formation. Investigations conducted into these systems show that when solid ultramafic rocks, olivine, or a silicate-rich melt are exposed to solutions and gases containing sulfur, Fe and Ni are actively extracted from the silicate phase with the formation of sulfides (Kullerud and Yoder 1965; Petrenko et al. 1974; Mutanen 1957; Naldrett et al. 1984; Arutyunyan et al. 1988). It is remarkable that in the sulfurization process of a basaltic magma, the concentration

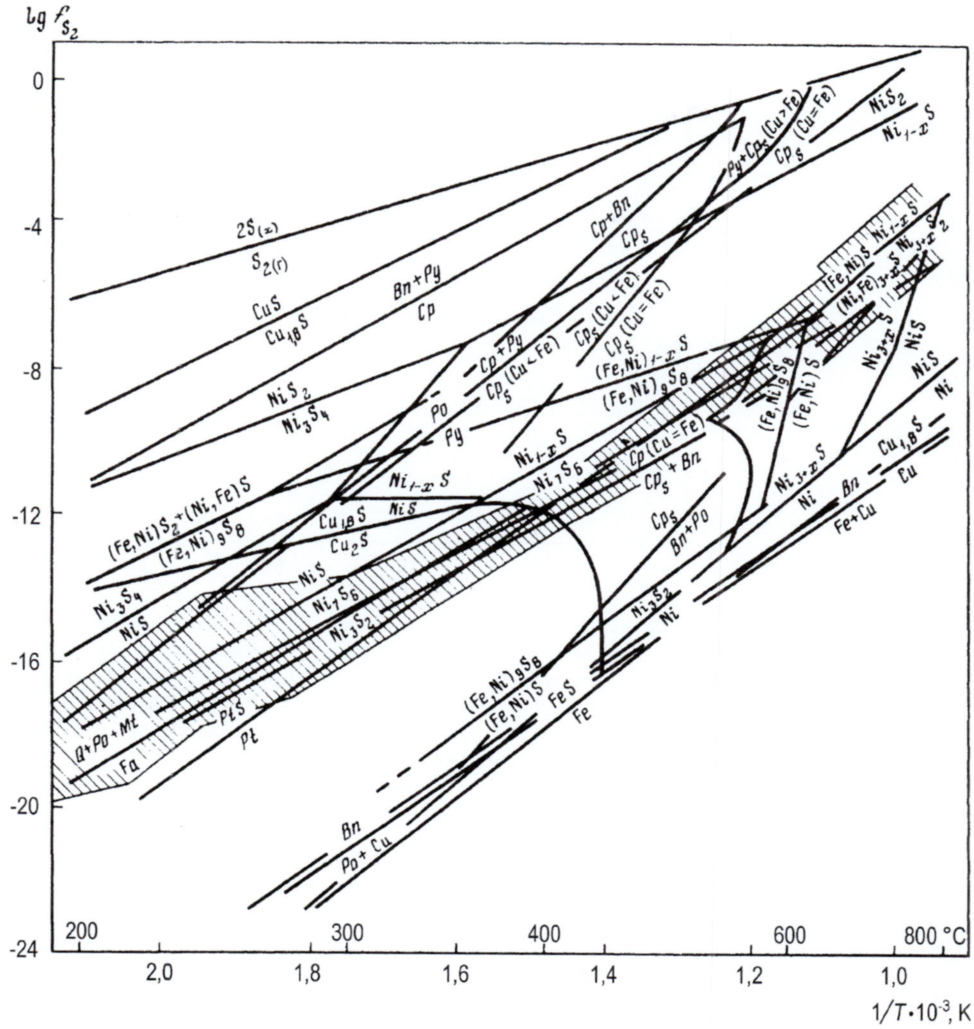

Fig. 1.11 Plot of lg fS_2 versus temperature for principal sulfide equilibriums of copper–nickel ore deposits. Bn—bornite, Cp—chalcopyrite, chalcopyrite solid solution, Po—pyrrhotite, Py—pyrite. Crosshatched area indicates Noril'sk ore crystallization (Nekrasov and Gorbachev 1978)

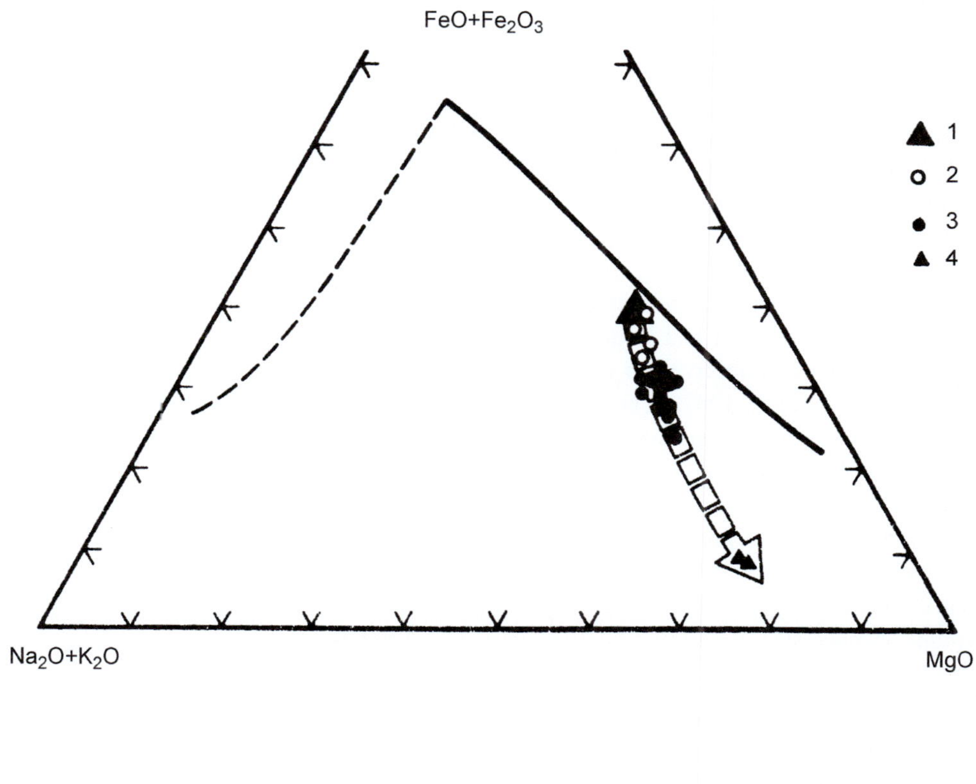

Fig. 1.12 Variation in composition (mol%) of basaltic magma in experiments on sulfurization (Almukhamedov and Medvedev 1982). *1* initial composition (sample TR-2), *2–4* interaction with pyrrhotite, *3* pyrite, *4* $CaSO_4$. The *wide arrow* indicates the direction of magma differentiation

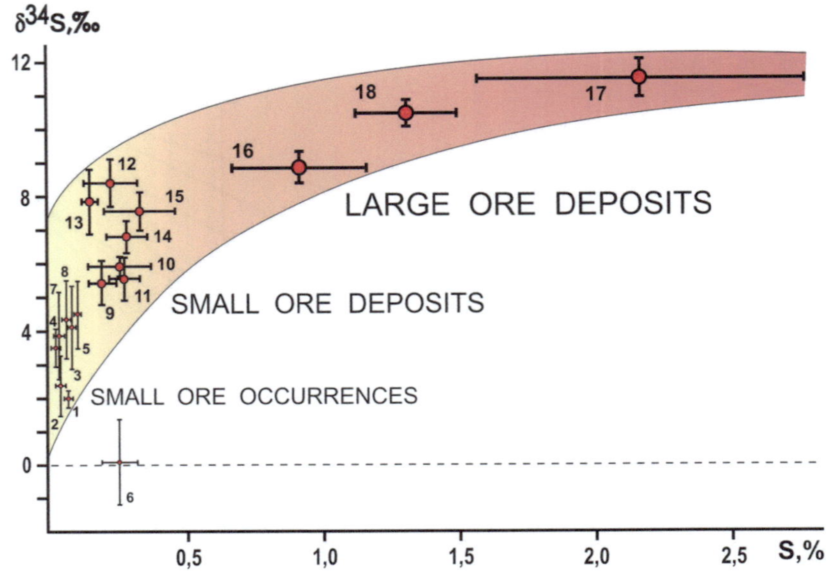

Fig. 1.13 Content and isotopic composition of sulfur from gabbrodolerites intrusions (Grinenko 1987). *Straight-line* segments indicate confidence intervals for average sulfur contents and $\delta^{34}S$ values. Their intersections correspond to average sulfur contents and average $\delta^{34}S$ values. The following intrusions are indicated: 1—F-81; 2—PE-3; 3—F-88; 4—Daldykanskaya; 5—the first rapid of the Kureyka River; 6—Khyukta; 7—Putanaya; 8—Nizhny-II; 9—Nakokhozskaya; 10—Zelenaya Griva; 11—Pyasino-Vologochanskaya; 12—Lower Talnakh; 13—Manturovskaya; 14—Imangda; 15—Mt. Chernaya; 16—Noril'sk-I; 17, 18—Upper Talnakh, Northeastern (17) and Northwestern (18) branches

of Fe decreases in direct proportion to amount of excess sulfur, and subsequently a proportional increase in the MgO content of the magma takes place (Fig. 1.12).

Studies of sulfur isotopes are of great importance in the resolution of a source of ore-forming material for sulfide deposits, since sulfur is in essence the only available parameter to map the deposits geochemical history. Broad analytical data on isotopes of sulfur from rocks and ores of the Siberian Platform have been gathered (Godlevsky and Grinenko 1963; Grinenko and Grinenko 1974; Goryainov and Yalovoy 1972; Gorbachev and Grinenko 1973; Nesterenko and Almukhamedov 1973; Kovalenker et al. 1974; Grinenko et al. 1976; Kuz'min and Tuganova 1977; Avgustinchik 1981; Grinenko 1985, 1987). The associated studies show that there is an increase in heavy sulfur isotopes and an associated increase in the overall sulfur content with increasing ore mineral abundance as the size of the ore deposit increases, from small ore occurrences in rock to small and large deposits (Fig. 1.13). A similar tendency is observed in the Pechenga ore-bearing intrusions (Grinenko 1987). The increase in the contribution of heavy sulfur isotope into ores of Noril'sk deposits is assumed to be associated with the assimilation of sulfur or sulfates from sedimentary rocks (Godlevsky and Grinenko 1963). In fact, data reported in reports by many geologists indicate significant variations in the $^{32}S/^{34}S$ ratio within an individual deposit, and even within each individual sulfide phenocryst. The magnitude of these variations depends on many of physicochemical factors including fO_2, fS_2, T, and P (Likhachev and Strizhev 1974; Ohmoto and Rye 1979; Grinenko and Grinenko 1974), and only detailed investigations provide the possibility of a solution to the problem of the source of the sulfur, mechanisms of its fractionation, and the genesis of Noril'sk deposits (Ovchinnikov and Ryabova 1975, 1977).

References

Almukhamedov AI, Medvedev AY (1982) Geochemistry of sulfur in evolution processes in mafic magmas (ed Chernyshov LV). Nauka, Moscow, 148 p (in Russian)

Andreev NS, Masurin OV, Poray-Koshits EA, Roskova GP, Filippovich VP (1974) Immiscibility phenomena in glasses. Nauka, Leningrad, 220 p (in Russian)

Anfilogov VN (1975) The nature and petrochemical criteria of immiscibility in magmatic melts. Geokhimiya 7:1035–1042 (in Russian)

Ariskin AA (1999) Phase equilibriums and dynamics of fractionation of basalt magma. Unpublished Doctor Sc thesis, GEOKHI RAS, Moscow, 52 p (in Russian)

Arkhipova AI (1975) Geochemical characteristic of intrusive traps of the Noril'sk plateau. Trudy NIIGA (ed Kavardin GI). Nedra, Leningrad, 135 p (in Russian)

Arkhipova AI, Natorkhin IA (1968) Toward the problem of genesis of "taxitic gabbrodolerites" from Noril'sk differentiated intrusions. In: Urvantsev NN (ed) Geology of mineral resources of the Noril'sk mining region. NMMC Noril'sk, pp 132–134 (in Russian)

Arutyunyan LA, Sargsyan GO, Smolkin VF (1988) On nickeliferous olivines crystallizing in silicate and sulfide-silicate systems. In: Mitrophanov FP, Gorbunov GI (eds) Nickeliferous mafic-ultramafic complexes of Ukraina, Siberia and Far East. Apatity, pp 78–81 (in Russian)

Avgustinchik IA (1981) On composition of sulfide mineralization of the Lower Talnakh intrusion. In: Likhachev AP (ed) Genesis and localization conditions of copper-nickel mineralization, vol 162. Trydy TSNIGRI, Moscow (in Russian)
Babushkin OS, Zhunina LA (1983) On immiscibility in chrome-bearing glass of the pyroxene system. Steklo, sitally i silikaty 12:26–27 (in Russian)
Bondar' IA, Toropov NA (1963) Ascertainment of immiscibility in fluorine-bearing slaggy and rare metal silicate systems. In: Poray-Koshits EA (ed) Glass-like state, issue 1, AS USSR, Leningrad, pp 39–41 (in Russian)
Borisov VI (1972) The role of plagioclatization of rocks in generation of the Talnakh ore-bearing intrusion. In: Urvantsev NN (ed) Copper-nickel ores of the Talnakh ore junction. Leningrad, pp 113–118 (in Russian)
Bowen NL (1928) The evolution of igneous rocks. Princeton University Press, Princeton, 334 p
Bulgakova EI (1971) Temperature conditions of formation of Noril'sk differentiated intrusions. Unpublished Ph.D. thesis, Institute of Geology and Geophysics, Novosibirsk, 30 p (in Russian)
Cameron EN (1970) Composition of certain co-existing phases in the eastern part of Bushveld Complex: symposium on the Bushveld Igneous Complex and other layered intrusions. Spec Publ Geol Soc S Afr 1:46–58
Carstens H (1979) Liquid immiscibility in mafic alkaline magmas. Chem Geol 27(4):297–307
Cawthorn RG (ed-in-chief) (1996) Layered intrusions (developments in petrology), vol 15, Elsevier, Amsterdam, 544 p
Chernyshev NM (1971) Sulfide copper-nickel deposits of the south-eastern part of the Voronezhsky crystalline massif. Voronezhsky Univ Pub, Voronezh, 298 p (in Russian)
Daly R (1933) Igneous rocks and depths of the Earth. McGraw-Hill, New York/London, 533 p
Distler VV, Duzhikov OA (1988) Formations of sulfide copper-nickel deposits. In: Obolensky AA, Sotnikov VI, Sharapov VN (eds) Ore formation and genetic models of endogenous ore formations. Nauka, Novosibirsk, pp 166–172 (in Russian)
Distler VV, Smirnov LV, Grokhovskaya TL, Filimonova AA, Muravitskaya GN (1979) Stratification, invisible layering of trap intrusions and conditions of formation of sulfide mineralization. In: Smirnov VI (ed) Conditions of formation of magmatic ore deposits. Nauka, Moscow, pp 211–269 (in Russian)
Dmitriev AN, Zolotukhin VV, Vasil'ev YR (1968) The experience of application of discrete mathematical processing of the data on differentiated trap intrusions at the northwest of the Siberian platform. Sovetskaya Geologiya 12:98–108 (in Russian)
Dneprovskaya MB, Frenkel MYa, Ariskin AA (1985) Modeling of layering of the Talnakh ore-bearing intrusion as a method for study of its formation mechanism. In: Thermodynamics in geology, vol 1. The first All-Union Symposium, Suzdal, March 12–14 1985. Chernogolovka, pp 228–229 (in Russian)
Dodin DA (1967) Petrology of traps of the East Khraelakh (Noril'sk region). Unpublished Ph.D. thesis, VSEGEI, Leningrad, 31 p (in Russian)
Dodin DA (1982) Nickeliferous magmatic complexes of the north-western part of the Siberian platform. Unpublished Doctor Sc thesis, VSEGEI, Leningrad, 34 p (in Russian)
Dodin DA, Batuev BN (1971) Geology and petrology of Talnakh differentiated intrusions and their metamorphic aureole. In: Urvantsev NN (ed) Petrology and Metallogeny of Talnakh and Noril'sk differentiated intrusions. Nedra, Leningrad, pp 31–100 (in Russian)
Dodin DA, Sadikov MA (1967) Some aspects of trap differentiation by the example of Kharaelakh mountains. In: Urvantsev NN (ed) Petrology of traps of the Siberian platform. Nedra, Leningrad, pp 141–152 (in Russian)
Dodin DA, Batuev BN, Mitenkov GA, Izoitko VM (1971) Atlas of rocks and ores of Noril'sk copper-nickel deposits. Nedra, Leningrad, 560 p (in Russian)
Dodin DA, Dodina TS, Fridman AI (1987) Fluid regime of magmatism and formation of the Noril'sk region. In: Petrology of fluid-silicate systems. Nauka, Novosibirsk, pp 121–140 (in Russian)
Durocher J (1857) Liquid fractionation. Part I. Ann des Mines, Series 5, 11, pp 217–239
Duzhikov OA, Distler VV, Sherman MK, Sluzhenikin SF (1988) Sulfide copper-nickel ore formation. In: Geology and Metallogeny of Sulfide Deposits of Noril'sk region USSR. Nauka Moscow, pp 77–172 (in Russian) (refer to English translation of 1988 Russian edition by Duzikov OA, Distler VV, Strunin VM, Mkrtychyan AK, Sherman ML, Sluzhenikin SF, Lurje AM. Soc Econ Geol Spec Publ, No. 1, 242 p)
Edwards AV (1942) Differentiation of the dolerites of Tasmania. J Geol 50(6):451–480
Eliseev EN (1959) Geochemistry of the main sulfide copper-nickel provinces of USSR. In: Lazarenko EK (ed) Problems of geochemistry, issue 1. Lviv Univer Pub, Lviv, pp 5–184 (in Russian)
Esin OA, Gel'd PV (1966) Physical chemistry of fire processes. Part II. Metallurgiya, Moscow, 103 p (in Russian)
Feoktistov GD (1978) Petrology and conditions for generation of trap sills (ed Zolotukhin VV). Nauka, Novosibirsk, 168 p (in Russian)
Feoktistov GD (1984) Computer calculation program for crystallization of silicate melt. Application of mathematical methods in geology and geophysics. Earth's Crust Institute SB AS USSR Irkutsk, pp 160–178 (in Russian)
Feoktistov GD, Ushchapovskaya ZF, Vasil'ev VK, Lakhno TA, Elizar'eva TI (1975) Mineralogy of traps in the South of the Siberian platform (ed-in-chief Feoktistov GD). Nauka, Novosibirsk, 87 p (in Russian)
Fersman A (1940) Pegmatites, vol I. AS USSR, Moscow/Leningrad, 712 p (in Russian)
Filippovich VN, Dmitriev DD (1971) Statistic immiscibility model of three-component glasses: Vitreous state (ed Poray-Koshits EA). Nauka, Leningrad, pp 60–62 (in Russian)
Fisher R (1950) Entmichungen in Schmelzen aus Schwermetallxyden, Silikaten und Phospaten, Ihre geochemische end lagerstattenkundliche Bedeutung. Neues Jahrb, fur Min, 81, No 3, pp 315–364
Frenkel MYa, Yaroshevsky , Ariskin AA, Barmina GS, Koptev-Dvornikov EV, Kireev BS (1988) Dynamics of interchamber differentiation of mafic magmas (ed-in-chief Frenkel MYa). Nauka, Moscow, 216 p (in Russian)
Genkin AD, Distler VV, Gladyshev GD, Filimonova AA, Evstigneeva TL, Kovalenker VA, Laputina IP, Smirnov VA, Grokhovskaya TL (1981) Copper-nickel sulfide ores from Noril'sk ore deposits (ed Shadlun TL). Nauka, Moscow, 234 p (in Russian)
Godlevsky MN (1959) Traps and ore-bearing intrusions of the Noril'sk region. Gosgeoltekhizdat, Moscow, 68 p (in Russian)
Godlevsky MN (1961) Review of hypotheses concerning origin of sulfide copper-nickel deposits. Trudy, VSEGEI, Leningrad, New Series, issue 45: pp 71–82 (in Russian)
Godlevsky MN (1967) Sources of mineralization associated with traps of the Siberian platform. In: Urvantsev NN (ed) Petrology of traps of Siberian platform. Nedra, Leningrad, pp 173–189 (in Russian)
Godlevsky MN (1968) Magmatic ore deposits. In: Genesis of endogenous ore deposits. Nedra, Moscow, pp 8–73 (in Russian)
Godlevsky MN (1971) The proportions between sulfide and silicate parts in a coarse of evolution of mafic magma. In: Abstracts of I Internat Geochem Congress, vol 1. Moscow, pp 65–77 (in Russian)
Godlevsky MN, Grinenko LN (1963) Some data on sulfur isotope compositions from sulfides of the Noril'sk deposits. Geokhimiya 6:491–499 (in Russian)

Godlevsky MN, Likhachev AP (1977) On perspectives and trends of research and exploration and prospecting works for nickel over East Siberia. Trudy, TSNIGRI, Moscow, issue 127, pp 3–8 (in Russian)
Godlevsky MN, Likhachev AP (1979) Conditions for initiation of crystallization of ore magmas forming copper-nickel deposits. In: Kuznetsov VA (ed) The main parameters of processes of endogenous mineralization, vol 1. Nauka, Novosibirsk, pp 109–118 (in Russian)
Gorbachev NS (1989) Fluid-magmatic interaction in sulfide-silicate systems (ed Zyryanov VN). Nauka, Moscow, 126 p (in Russian)
Gorbachev NS, Grinenko LN (1973) Sulfur isotope compositions from sulfides and sulfates of the Oktyabrsky nickel-copper deposits (Noril'sk region) and some aspects of their genesis. Geokhimiya 8:1127–1136 (in Russian)
Gorbunov GI (1968) Geology and genesis of sulfide copper-nickel deposits of Pechenga (ed Smirnov VI). Nedra, Moscow, 352 p (in Russian)
Goryainov IN (1969) On genesis of layered intrusions by example of the Talnakh massif. In: Zaridze GM (ed) Magmatism, metamorphism and metasomatism. Central Committee Communist Party of Georgia Pub, Tbilisi, pp 47–61 (in Russian)
Goryainov IN (1971) About genesis of the Talnakh ore deposit. In: Urvantsev NN (ed) Petrology and mineralization of Talnakh and Noril'sk intrusions. Nedra, Leningrad, pp 182–196 (in Russian)
Goryainov IN, Yalovoy AA (1972) Sulfur isotope fractionation of sulfides of the Talnakh ore deposit. In: Urvantsev NN (ed) Copper-nickel ores of the Talnakhsky ore junction. NIIGA, Leningrad, pp 58–62 (in Russian)
Greig JW (1928) On the evidence which has been presented for liquid silicate immiscibility in the laboratory and in the rocks of Agate Point. Am J Sci 215:375–402
Grigor'ev DP (1935) Immiscibility of silicate systems compositionally close to natural rocks. Zapiski VMO 64(1):250–267 (in Russian)
Grigor'ev DP (1946) An Experience in systematics and terminology of elementary processes of magmatic differentiation. In: Honor of Academician DS Belyankin. AS USSR Publ, Moscow, pp 184–188 (in Russian)
Grigor'ev DP, Iskyul EV (1937) Differentiation of some silicate melts as a result of formation of two immiscible liquids. AS USSR, Geol Series 1:77–106 (in Russian)
Grinenko LN (1985) Sources of sulfur of the nickeliferous and barren gabbrodolerites intrusions in the northwest part of the Siberian platform. Geologiya rudnykh mestorozhdeniy 1:3–15 (in Russian)
Grinenko LN (1987) Genetic model of formation of copper-nickel deposits based on isotope-geochemical data. In: Sotnikov VI (ed) Construction of models of ore-forming systems. Nauka, Novosibirsk, pp 119–128 (in Russian)
Grinenko VA, Grinenko LN (1974) Geochemistry of sulfur isotopes (ed Vinogradov AP). Nauka, Moscow, 274 p (in Russian)
Grinenko LN, Staritsyna GN, Goryainov IN (1976) Peculiarities of copper-nickel mineralization at the northeast of the Tunguskaya syneclise (based on sulfur isotopic composition of ores and rocks). Geokhimiya 11:1662–1672 (in Russian)
Gulin SA, Sukhov LG (1973) Some remarks on liquation-magmatic hypothesis concerning formation of the Norlsk-type copper-nickel deposits. Sovetskaya Geologiya 2:24–35 (in Russian)
Gulin SA, Sukhov LG (1974) Criteria for prediction of copper-nickel deposits based on their metasomatic origin conceptions. In: Lind EN (ed) The state and trends of investigations into metallogeny of traps. Krasnoyarsk TGU, Krasnoyarsk, pp 67–69 (in Russian)
Gurulev SA (1965) Geology and formation conditions of Yoko-Dovyrensky gabbro-peridotite massif. Nauka, Moscow, 122 p (in Russian)
Hill RET, Roeder PL (1967) Stability of spinel in basaltic melts. In: Geological Society of America, annual meeting, Abstract, pp 96–97
Ivanov MK, Ivanova TK, Shatkov VA (1971a) Some facial peculiarities of formation of intrusive and effusive picrites. In: Urvantsev NN (ed) Geology and mineral resources of the Noril'sk region, NIIGA, Leningrad, pp 83–84 (in Russian)
Ivanov MK, Ivanova TK, Tarasov AV, Shatkov VA (1971b) Peculiarities of petrology and mineralization of differentiated intrusions of the Noril'sk ore junction (Noril'sk-I, Noril'sk-II and Mt. Chernaya deposits). In: Urvantsev NN (ed) Petrology and metallogeny of the Talnakh and Noril'sk differentiated intrusions. Nedra, Leningrad, pp 197–304 (in Russian)
Karbyshev VD (1986) A study of rock composition and estimate of metallogeny of the Noril'sk region applying mathematical methods: Unpublished Ph.D. thesis, Institute of Geology and Geophysics SB AS USSR, 17 p (in Russian)
Karbyshev VD, Zolotukhin VV (1978) Estimate of self-descriptiveness of characteristic features in prediction of trap intrusion productivity. Nauka, Novosibirsk, 167 p (in Russian). In: Sobolev VS (ed) Criteria for Exploration of the Noril'sk-type Sulfide Ores. Nauka, Novosibirsk, pp 134–149 (in Russian)
Karbyshev VD, Bishaev AA, Zolotukhin VV, Dmitriev AN, Vasil'ev YR, Ryabov VV (1978) Perspectives of metallogeny of differentiated intrusions at the north of the Siberian Platform (on the base of logic-mathematical methods of processing geological information). In: Zolotukhin VV, Vilensky AM (eds) Petrology and perspectives of metallogeny of traps at the north of the Siberian platform. Nauka, Novosibirsk, pp 216–257 (in Russian)
Kavardin GI, Mitenkov GA (1971) Copper-nickel ores of the Talnakh deposit. In: Urvantsev NN (ed) Geology and Metallogeny of the Talnakh and Noril'sk Differentiated Intrusions. Nedra, Leningrad, pp 123–181 (in Russian)
Kavardin GI, Staritsyna GN, Golubkov VS, Goryainov IN, Kravtsova LI (1968) Traps of the Yenisey ore province (ed Urvantsev NN). Nedra, Leningrad, 208 p (in Russian)
Kennedy GS (1948) Equilibrium between volatiles and iron oxides in igneous rocks. Am J Sci 246:529–549
Kennedy GC (1955) Some aspects of the role of water in igneous melts. Geol Soc Am Spec Pap 62:489–503
Koptev-Dvornikov EV, Barmina GS, Frenkel MY, Yaroshevsky AA (1976) Geological structure of differentiated trap intrusion at the Velminsky rapid (the Podkamennaya Tunguska River). Vestnik Moskovskokgo Universiteta: Geologiya 4:50–56 (in Russian)
Korovyakov IA (1960) Behavior of nickel in Siberian traps. In: Mineral'noe Syr'e, issue 1. VIMS, Moscow, pp 169–183 (in Russian)
Korovyakov IA, Nelyubin AE, Raikova ZA, Khortova LK (1963) Origin of the Noril'sk trap intrusions hosting sulfide copper-nickel ores. In: Trudy VIMS (ed Gon'shakova VI), New Series, issue 9, Gosgeoltekhizdat, Moscow, 102 p (in Russian)
Korzhinsky DS (1973) Metamagmatic processes. Izv Akad Nauk SSSR Geol Ser 12:3–6 (in Russian)
Kotul'sky VK (1948) The state-of-art problem of genesis of copper-nickel sulfide deposits. Sov Geol 29:11–24 (in Russian)
Kovalenker VA, Gladyshev GD, Nosik LP (1974) Isotope composition of sulfur from sulfides from deposits of the Talnakh ore junction with special reference to their selenium content. Izv Akad Nauk SSSR Geol Ser 2:80–91 (in Russian)
Kozlov EK (1973) Natural series of rocks of nickeliferous intrusions and their metallogeny (by example of Kola Peninsula) (ed Ivanova TN). Nauka, Leningrad, 288 p (in Russian)
Krivenko AP, Ponomarchuk VA, Skripchenko VA (1984) Distribution of rare-earth elements during immiscible differentiation of mafic melt with high phosphorus content. Doklady Akad Nauk SSSR 275 (2):468–471 (in Russian)
Kullerud GA (1962) The Fe–Ni–S system. Year book, vol 62. Carnegie Inst. Washington, Washington, DC, pp 175–179
Kullerud GA, Yoder HS (1965) Sulfide-silicate relation and their bearing on ore formation under magmatic, postmagmatic and

metamorphic condition. In: Proceedings of symposium: problems of postmagmatic ore deposition, vol 2, Prague, pp 327–331
Kutolin VA (1972) Problems of petrochemistry and petrology of basalts (ed Kuznetsov YuA). Nauka, Moscow, 208 p (in Russian)
Kuz'min VK, Tuganova EV (1977) New data on the isotopic composition of the sulfur in the copper–nickel sulfide ores of the northwestern part of the Siberian platform. Sov Geol Geophys 18 (4):98–100 (in Russian)
Lebedev AP (1955) Trap formation of the central part of the Tunguska basin. AS USSR, Moscow, 197 p (in Russian)
Lebedev AP (1957) On differentiation types in traps of the Siberian platform. Izv Akad Nauk SSSR Geol Ser 2:55–74 (in Russian)
Lebedev AP (1958) Problems of study of basalt magma. Izv Akad Nauk SSSR Geol Ser 12:30–45 (in Russian)
Letnikov FA, Feoktistov GD, Vilor NV, Grudinin MI, Gantimurova TP, Gantimurov AA, Dorogokupets PI, Zhatnuev NS, Menaker IG, Men'shagin YV, Savel'eva VB, Bal'shev SO (1988) Petrology and fluid regime of continental lithosphere (ed Karpov IK). Nauka, Novosibirsk, 185 p (in Russian)
Likhachev AP (1965) Role of leucocratic gabbro in the generation of Noril'sk differentiated intrusions. Izv Akad Nauk SSSR Geol Ser 10:75–89 (in Russian)
Likhachev AP (1975) On generation of Bushveld intrusive complex and related ore deposits. Izv Akad Nauk SSSR Geol Ser 12:50–66 (in Russian)
Likhachev AP (1977) On crystallization conditions of trap magmas of the northwestern Siberian platform. Zapiski VMO 5:594–606 (in Russian)
Likhachev AP (1978) On conditions of generation of ore-bearing and barren mafic-ultramafic magmas. Doklady Akad Nauk SSSR 238 (2):447–450 (in Russian)
Likhachev AP (1988) Genetic models of sulfide-nikeliferous formations and their relation to other endogenous formations. In: Obolensky AA, Sotnikov VI, Sharapov VN (eds) Ore formation and genetic models of endogenous ore formations. Nauka, Novosibirsk, pp 158–165 (in Russian)
Likhachev AP (2006) Platinum-copper-nickel and platinum deposits. Eslan, Moscow, p 496 (in Russian)
Likhachev AP, Strizhev VP (1974) About a change of sulfur isotope composition in sulfide ores of copper-nickel deposits of the Siberian platform. Zapiski VMO 103(3):305–313 (in Russian)
Loewinson-Lessing FYu (1933) Petrography. ONTI, Moscow, 462 p (in Russian)
Loewinson-Lessing, FYu (1935) Four coryphaei of petrography: Washington, Dupark, Sederholm, and Vogt. Izv Akad Nauk SSSR, Series VII, No. 3 (in Russian)
Loewinson-Lessing FYu, Struve EA (1963) Petrographic dictionary (eds Afanas'ev GD, Petrov VP, Ustiev EK). Gosnauchtekhizdat, 3rd edn. Moscow, 447 p (in Russian)
Lurje ML, Masaitis VL, Polunina LA (1962) Intrusive traps of the western margin of the Siberian platform. In: Afanas'ev GD (ed) The petrography of the Eastern Siberia, vol 1. AS USSR, Moscow, pp 5–70 (in Russian)
Lurje ML, Ledneva VP, Selivanovskaya TV, Semenov LS, Tuganova EV, Ryabchenko AA, Komarova MZ, Staritsyna GN, Tomanovskaya YuI (1976) Structures of traps of the Siberian platform (ed ML Lurje). Trudy VSEGEI, New Series, vol 235. Nedra, Leningrad, 171 p (in Russian)
Lyul'ko VA, Nesterovsky VS, Goverdovskaya TG (1972) Magmatogenic breccias of nickeliferous trap intrusions. In: Urvantsev NN (ed) Copper-nickel ores of the Talnakh ore junction. Nedra, Leningrad, pp 123–127 (in Russian)
MacLean WN (1969) Liquidus phase relations in the $FeS–FeO–Fe_3O_4–SiO_2$ system and their application in geology. Econ Geol 64:865–884
MacLean WN, Shimazaki H (1976) The partitioning of Co, Ni, Cu and Zn between sulfide and silicate liquids. Econ Geol 71:1049–1057
Malich NS, Staritskiy YG, Tuganova EV (1965) Regularities in distribution of nickeliferous trap intrusions of the Siberian platform. VIEMS, Moscow, 23 p (in Russian)
Malich NS, Masaitis VL, Surkov VS (eds) (1987) Geological structure of USSR and regularities in distribution of mineral resources, vol 4 (Siberian platform). Nedra, Leningrad, 448 p (in Russian)
Marakushev AA (ed) (1976) Petrography: Part I. Moscow Univ Pub, Moscow, 384 p (in Russian)
Marakushev AA (1978) Some aspects of petrogenesis in the light of the theory of fluid-magma interaction. In: Kuznetsov VA (ed) Problems of petrology of Earth's crust and upper mantle. Nauka, Novosibirsk, pp 65–83 (in Russian)
Marakushev AA (1979) Petrogenesis and ore formation (geochemical aspects (ed Korzhinsky DS)). Nauka, Moscow, 261 p (in Russian)
Marakushev AA (ed) (1981) Petrography: Part II. Moscow Univ Pub, Moscow, 328 p (in Russian)
Marakushev AA, Perchyuk LL (1974) Thermodynamic model of the Earth's fluid regime. Ocherki Fisiko-khimicheskoy petrologii, issue 4. Nauka, Moscow, pp 102–130 (in Russian)
Marakushev AA, Fenogenov AN, Emel'yanenko PF, Duzhikov OA, Skripnichenko VA (1982) Genesis of the Noril'sk-type layered intrusions. Vestnik MGU, Geol Ser 1:3–19 (in Russian)
Masaitis VL (1958) Petrology of the Alamdzhakh trap intrusion (the Vilyuy River basin) (ed Lurje ML). Trudy VSEGEI Leningrad, New Series, No 22, 133 p (in Russian)
Masaitis VL (1973) Mafic magmatic formations of the Siberian platform. In: Problems of magmatic geology. Nauka, Novosibirsk, pp 159–171 (in Russian)
Masaitis VL (1974) Magmatic formations. In: Malich NS, Masaitis VL, Staritsky YuG (eds) Geological formations of Pre-Cenozoic cover of the Siberian platform and their metallogeny. Nedra, Leningrad, pp 125–144 (in Russian)
McBirney AR, Nakamura Y (1974) Immiscibility in late-stage magmas of the Skaergaard intrusion. In: Carnegie Institution Washington, Yearbook 73, pp 348–352
Mitenkov GA, Khineiko AL, Sishkin NN (1977) Peculiarities of the taxitic horizon structure of the Talnakh ore-hosting intrusion and their genetic significance. Doklady Akad Nauk SSSR 237(1): 191–194 (in Russian)
Morse SA, Lindsley DH, Williams RJ (1980) Concerning intensive parameters in the Skaergaard intrusion. Amer J Sci 280A:159–170
Muan A (1955) Phase equilibria in system $FeO–Fe_2O_3–SiO_2$. Trans AIME 203:965–976
Muan A (1983) Crystallization in silicate systems. In: Yoder HS Jr (ed) The evolution of igneous rocks: fiftieth anniversary perspectives. Princeton University Press, Princeton
Mutanen T (1957) Emaksiseen magmatismiin liittjvasulfidimimuodostus. Eripainos Geologi–Lehdesta 2:89–105
Naldrett AJ (1969) A portion of the system Fe–S–O and its application to sulfide ore magmas. J Petrol 10:171–201
Naldrett AJ (2004) Magmatic Sulfide deposits: Geology, Geochemistry and Exploration. Springer, Berlin/Heidelberg/New York, 727 p
Naldrett AJ, Cabri LJ (1976) Ultramafic and related mafic rocks: Their classification and genesis with special reference to the concentration of nickel sulfides and platinum-group elements. Econ Geol 71:1131–1158
Naldrett AJ, Duke JM, Lightfoot PC, Thompson JFH (1984) Quantitative modeling of the segregation of magmatic sulfides: an exploration guide. CIM Bull 77(864):46–56
Naslund HR (1983) The effect of oxygen fugacity on liquid immiscibility in iron-bearing silicate melts. Am J Sci 282(10):1034–1059
Natan HD, Van Kirck CK (1978) Model of magmatic crystallization. J Petrol 19(1):66–94

Natorkhin IA, Arkhipova AI, Batuev BN (1977) Petrology of Talnakh intrusions (ed Egorov LS). Nedra, Leningrad, 236 p (in Russian)

Nekrasov IY, Gorbachev NS (1978) Physical-chemical conditions of formation of the Noril'sk-type differentiated intrusions and copper-nickel ores. Ocherki fisiko-khimicheskoy petrologii 7:92–123 (in Russian)

Nesterenko GV, Almukhamedov AI (1973) Geochemistry of differentiated traps (Siberian platform) (ed Shcherbina VV). Nauka, Moscow, 198 p (in Russian)

Newhouse WH (1927) The equilibrium diagram of pyrrhotite and pentlandite and their relations in natural occurrences. Econ Geol vol 22:288–299

Niggli P (1920) Volatile components in magma. Preisschriften der Furstlich Jablonwkischen Gesellschaft, Leipzig, 272 p (in German)

Nikol'sky NS (1978) Thermodynamics of mineral equilibria of basites (ed Govorov IN). Nauka, Moscow, 177 p (in Russian)

Nitsan U (1974) Stability field with respect to oxidation and reduction. J Geophys Res 79(5):706–711

Ohmoto X, Rye RO (1979) Sulfur and carbon isotopes: geochemistry of hydrothermal ore deposits. A Wiley-Interscience, New York, pp 405–450

Oleinikov BV (1979) Geochemistry and ore genesis of platform basites (ed Bazhenov AI). Nauka, Novosibirsk, 264 p (in Russian)

Oleinikov BV (1981) Metallization of magmatic melts and its petrological and ore genetic consequences. In: Koval'sky VV (ed) Abstracts of native mineral formation in magmatic process. Yakutsk, pp 5–11 (in Russian)

Oleinikov BV, Tomshin MD (1976) Deep magma differentiation of platform basites. Doklady Akad Nauk SSSR 231(1):177–180 (in Russian)

Olshansky YI (1948) About significant flowability of sulfide melts and probable geological significance of this phenomenon. Doklady Akad Nauk USSR 63(2):187–190 (in Russian)

Olshansky YI (1951) The Fe–FeS–FeO–SiO_2 system. Izv Akad Nauk SSSR, Geol Ser 6:128–176

Osborn EF (1959) Role of oxygen pressure in the crystallization of basaltic magma. Am J Sci 257:609–647

Osborn EF (1964) Experimental study of oxygen pressure, water content and crystallization sequence of basalts and andesites. In: Chemistry of Earth's crust, vol II. Nauka, Moscow, pp 75–87 (Russian translation)

Osborn EF (1979) The reaction principle. In: Yoder HH (ed) Evolution of igneous rocks. Princeton Univ Press, Princeton

Ovchinnikov LN, Ryabova TV (1975) Statistical study of variations in sulfur isotope content. Geologiya rudnykh mestorozhdeniy 17(6): 3–19 (in Russian)

Ovchinnikov LN, Ryabova TV (1977) On application of sulfur isotope analysis in geological investigations. Geologiya rudnykh mestorozhdeniy 19(4):135–140 (in Russian)

Pankov VYu (1983) Microimmiscibility in melted inclusions in olivine. In: Lazebnik KA (ed) Bull NTI, Yakutsk Affiliated Branch SB RAS SSSR, pp 25–28 (in Russian)

Pankov VYu (1986) Early magmatic evolution of tholeiite-basalt melt on evidence derived from study of inclusions in mafic minerals of the Siberian platform. Unpublished PhD thesis, TGU of Tomsk, 17 p (in Russian)

Pavlov AL (1983) Genesis of magmatic magnetite deposits (ed Sharapov VN). Nauka, Novosibirsk, 220 p (in Russian)

Petrenko GV, Arutyunyan LA, Zhangurov AA, Mityunin YK, Predovsky AA (1974) About possibility of nickel removal from olivinites under hydrothermal conditions. Geokhimiya 8: 1185–1192 (in Russian)

Philpotts AR (1982) Compositions of immiscible liquids in volcanic rocks. Contr Miner Petrol 80(3):201–218

Reverdatto VV (1963) Petrology of Anakitsky differentiated trap massif and its structure. Geologiya i geofisika 10:79–92 (in Russin)

Roedder E (1951) Low-temperature liquid iimiscibility in the syatem K_2O–FeO–A_2O_3–SiO_2. Am Mineral 36:282–286

Roedder E (1979) Silicate liquid immiscibility in magmas. In: Jr Yoder HS (ed) The evolution of Igneous magmas. Princeton Univ Press, Princeton, NJ, pp 15–57

Roedder E, Weiblen PW (1971) Petrology of silicate melt inclusions, Apollo 11 and Apollo 12 and terrestrial equivalents. In: Proceedings of the second lunar science conference, vol 1, MIT Press, Cambridge, MA, pp 507–528

Roeder PL, Emslie RF (1970) Olivine-liquid equilibrium. Contr Mineral Petrol 29:275–289

Roeder PL, Osborn EF (1966) Experimental data for in the system MgO–FeO–Fe_2O_3–$CaAl_2Si_2O_8$–SiO_2 and their petrological implication. Am J Sci 264(6):428–480

Rogover GB (1959) The Noril'sk-I deposit. Gosgeoltekhizdat, 168 p (in Russian)

Rosenbush H (1887) Mikroskopishe Physiographie der massigen Geisteine (1887), vol 2, 2nd edn. E Schwezerbart'sche Verlagshandlung, Stuttgart, p 877

Ryabov VV (1969) About origin of taxitic gabbrodolerites, leucocratic gabbro, and magmatic breccia in the Kharaelakhsky branch of the Talnakh intrusion. Geologiya i Geofisika 2:51–58 (in Russian)

Ryabov VV (1983) Compositional peculiarities of endocontact zones of Noril'sk intrusions. In: Trap magmatism of the Siberian platform in connection with tectonics and exploration of mineral resources. Krasnoyarsk, pp 186–188 (in Russian)

Ryabov VV (1984a) On composition of upper contact zones of Noril'sk intrusions hosting chromite-rich mineralization. In: Polyakov GV (ed) Petrochemistry: criteria for mineralzation of magmatic complexes. IGG SB AS USSR, Novosibirsk, pp 124–142 (in Russian)

Ryabov VV (1984b) Peculiarities of mineralogy of magnesian basites in the Noril'sk region. In: Sobolev VS (ed-in-chief) Magnesian basites of the western Siberian platform and aspects of their nickel content. Nauka, Novosibirsk, pp 150–158 (in Russian)

Ryabov VV (1985) Peculiarities of native mineral formation at the roof of chambers of trap intrusion. In: Oleinikov BV (ed) Formation of native elements under endogenous conditions (abstracts). Yakutsk, pp 18–21 (in Russian)

Ryabov VV (1988a) Immiscible differentiation in traps. In: Goncharenko AI (ed) Actual problems of Siberian geology, TGU, Tomsk, vol 1. pp 132–134 (in Russian)

Ryabov VV. (1988b) Immiscible liquid in traps of Siberian platform. In: 2nd International Conference on Natural Glasses, Prague, 21–25 September 1987, pp 401–405

Ryabov VV (1989a) Genetic types of high-magnesian traps. In: Oleinikov BV (ed) Mafic magmatism of the Siberian platform and its metallogeny. Yakutsk, pp 78–79 (in Russian)

Ryabov VV (1989b) Chemical composition of immiscible liquids in natural glasses from traps. Geologiya i Geofisika 11:69–78 (in Russian)

Ryabov VV (1989c) Immiscibility in natural glasses (by example of traps) (ed Zolotukhin VV). Nauka, Novosibirsk, 223 p (in Russian)

Ryabov VV, Pavlov AL (1984) Physicochemical principles of the formation of magnesial scarn paragenesis in chambers of stratified intrusives of Noril'sk type. Sov Geol Geophys 25(3):51–56

Ryabov VV, Yakobi NYa (1981) Behavior of chrome during differentiation of trap magma. In: Sobolev VS (ed) Problems of genetic petrology. Nauka, Novosibirsk, pp 85–111 (in Russian)

Ryabov VV, Zolotukhin VV (1977) Minerals of differentiated traps (ed Sobolev VS). Nauka, Novosibirsk, 392 p (in Russian)

Ryabov VV, Zolotukhin VV (1978) Composition of rock-forming minerals of traps as an indicator of important petrologic processes. In: Abstracts of XI Congress of Internat Mineralogical Assoc, vol 1. Novosibirsk, pp 51–53 (in Russian)

Ryabov VV, Konenko VF, Khmel'nikova OS (1985a) Rock-forming minerals of picritic basalts of the Noril'sk region. Sov Geol Geophys 26(4):77–84

Ryabov VV, Konenko VF, Krasov NF (1985b) Immiscibility phenomena in glasses from ores of native iron of Khungtukun intrusion. Doklady Akad Nauk SSSR 285(4):982–987 (in Russian)

Ryabov VV, Pavlov AL, Lopatin GG (1985c) Native iron of Siberian traps (ed Zolotukhin VV). Nauka, Novosibirsk, 169 p (in Russian)

Scoon RN, Mitchell AA (1994) Discordant iron-rich ultramafic pegmatites in the Bushveld Complex and their relationship to iron-rich intercumulus and residual liquids. J Petrol 35:881–917

Shannon EV (1924) The mineralogy and petrology of intrusive Triassic diabase at Goose Creek Loundoun Country, Virginia. Proc US Natl Mus 66(art 2):1–84

Sharapov VN, Cherepanov AN (1986) Dynamics of magma differentiation (eds Polyakov GV, Kirdyashkin AG). Nauka, Novosibirsk, 186 p (in Russian)

Sharkov EV (1980) Petrology of layered intrusions (ed Shurkin KA). Nauka, Leningrad, 183 p (in Russian)

Shimazaki H, Clark LA (1973) Liquid relations in the FeS–Feo–SiO_2–Na_2O system and geological implications. Econ Geol 68:79–96

Skinner BJ, Peck DL (1969) An immiscible sulfide melt from Hawaii. Econ Geol Monogr 4(3):10–32

Smirnov MF (1966) Structure of the Noril'sk nickeliferous intrusions and sulfide ores. Nedra, Moscow, 58 p (in Russian)

Sobolev VS (1986) Petrology of traps: Izbrannye trudy. Nauka, Novosibirsk, 209 p (in Russian)

Sobolev VS, Kostyuk VP (eds) (1975) Magmatogene crystallization from data of study of fluid inclusions in melts. Nauka, Novosibirsk, 232 p (in Russian)

Staritsyna GN, Tolmachevsky YN, Kravtsova LI (1972) Intrusive traps of the northeastern side of the Tungusskaya sineclise (ed Egorov LS). Nedra, Leningrad, 210 p (in Russian)

Stepanov VK (1975) Rock-forming minerals of the Talnakh intrusion and analysis of their paragenesises with development of criteria of mineralization. Unpublished PhD thesis, Moscow, 25 p (in Russian)

Stepanov VK (1977) Olivines and hypersthenes of the Talnakh intrusion. In: Geology, petrology and geochemistry of copper-nickel deposits. Trudy TSNIGRI, vol 127. TSNIGRI, Moscow, pp 17–34 (in Russian)

Stepanov VK (1981) Dynamic model of emplacement, crystallization and mineralization of the Noril'sk ore-hosting intrusions. In: Genesis and localization conditions of copper-nickel mineralization. Trudy TSNIGRI vol 162. TSNIGRI, Moscow, pp 13–19 (in Russian)

Sukhov LG (1985) Petrochemical criteria for quantitative predictive estimate of nickel content of trap formations. In: Polyakov GV(ed) Petrochemistry, genesis and metallogeny of magmatic formations of Siberia. Nauka, Novosibirsk, pp 209–216 (in Russian)

Tarasov AV (1976) On formation mechanisms of the Noril'sk intrusions and related sulfide bodies. In: Pospelov GL (ed) Replacement and intrusion during magmatism and ore formation. Nauka, Novosibirsk, pp 123–276, in Russian

Tikhonenkov PI, Nesternko GV, Naumov VB (1985) Specific character of immiscibility of flood basalts of the Siberian platform. In: Abstracts of VII All-Union conference: thermometry and geochemistry of fluid ore formation. Lviv, pp 140–141 (in Russian)

Tomkeieff SJ (1939) A Contribution to Petrology of the Whin Sill. Min Mag 22:100–120

Tuganova EV (1988) Petrological peculiarities of formation of Noril'sk-type nickeliferous intrusions. In: Mitrofanov FP, Gorbunov GI (eds) Presence of nickel in mafic-ultramafic complexes. Apatites, pp 36–40 (in Russian)

Urvantsev NN (1970) Geologic-tectonic peculiarities of generation of Noril'sk copper-nickel ores. Uchenye Zapiski NIIGA, Regional'naya Geologiya, issue 18. pp 39–55 (in Russian)

Uspensky NM (1968) Nongranitic pegmatites. Nedra, Moscow, 344 p (in Russian)

Varshall BG (1981) Chemical aspects of liquation phenomena and silicate and alumosilicate melts. In: Studies of structure of magmatic melts. UNTs AS SSSR, Sverdlovsk, pp 41–51

Vilensky AM (1967) Petrology of intrusive traps at the north of the Siberian platform (ed Lebedev AP). Nauka, Moscow, 270 p (in Russian)

Vilensky AM (1978a) Intrusive magmatism. In: Zolotukhin VV, Vilensky AM (eds) Petrology and perspectives of metallogeny of traps at the north of the Siberian platform. Nauka, Novosibirsk, pp 61–121 (in Russian)

Vilensky AM (1978b) Peculiarities of interchamber differentiation of trap intrusions and some general features of magmatism of the Noril'sk region. In: Sobolev VS (ed) Criteria for Exploration of the Noril'sk-type sulfide ores. Nauka, Novosibirsk, pp 30–44 (in Russian)

Vilensky AM, Oleinikov BV (1970) The main factors of diversity and aspects of classification of Siberian platform traps. In: Vilensky AM (ed) Geology and petrology of intrusive traps of the Siberian platform. Nauka, Moscow, pp 5–25 (in Russian)

Vilensky AM, Kavardin GI, Kravtsova LI, Staritsina GN (1964) Petrology of trap intrusions on the right-bank of the Yenisey River lower reach (ed Kazakov AN). Nauka, Moscow, 237 p (in Russian)

Vogt JHL (1923) Nickel in igneous rocks. Econ Geol 18(4):307–353

Vorontsov AE, Razvozzhaeva EA, Syngaevskiy ED, Khlebnikova AA (1986) Geochemical investigation of carbonaceous matters from diatremes of the Siberian platform. Geokhimiya 2:226–235 (in Russian)

Vortsepnev VV (1978) Conditions of formation of the Talnakh ore-bearing intrusion on the base of data on inclusions. Geologiya rudnykh mestorozhdeniy 2:64–70 (in Russian)

Wager LR, Brown GM (1968) Layered igneous rocks. Oliver and Boyd, Edinburgh/London, 588 p

Wager LR, Deer WA (1939) Geological investigations in East Greenland. Part III. The petrology of the Skaergaard Intrusion, Kangerdlugssuaq East Greenland. Meddelelser om Grønland 105(4), 352 p

Walker F, Poldervaart A (1949) Karoo dolerites of the Union of South Africa. Geol Soc Am Bull 60(4):591–706

Warren BE, Pinkus AG (1940) Atomic consideration of immiscibility in glass systems. J Am Ceram 23(10):301–304

Watson TB (1976) Two-liquid partition coefficients, experimental data and geochemical implications. Contr Mineral Petrol 56:119–134

Yazawa A, Kameda M (1953) Fundamental studies on copper smelting I Partial liquidus diagram for FeS-FeO-SiO_2 system. Tech Rep Tohoku Univ 18:40–58

Yoder HS Jr, Tilley CE (1962) Origin of basalt magmas: An experimental study of natural and synthetic rock systems. J Petrol 3: 342–532

Zavaritsky AN (1947) On pegmatites as transitional rocks between igneous rocks and veins. Zap VMO 76:36–50 (in Russian)

Zavaritsky AN, Sobolev VS (1961) Physical-chemical grounds of petrography of igneous rocks. Gosgeolteckhizdat, Moscow, 383 p (in Russian)

Zharikov VA (1976) Grounds of physical-chemical petrology. Moscow Univ Pub, Moscow, 420 p (in Russian)

Zharikov VA, Gorbachev NS, Ishbulatov RA (1986) Fluid-magmatic differentiation of mafic magamas. Sov Geol Grophys 27(7):25–29

Zolotukhin VV (1964) Main regularities of prototectonics and aspects of origin of ore-bearing trap intrusions (ed Sobolev VS). Nauka, Moscow, 192 p (in Russian)

Zolotukhin VV (1965) On peculiarities in distribution of nickel in the Norilsk-I intrusion. Doklady Akad Nauk SSSR 162(6):1390–1393 (in Russian)

Zolotukhin VV (1971) On genesis of so called "liquation" of copper-nickel ores in the light of new data (about infiltration-autometasomatic hypothesis). Geologiya i Geofisika 9:12–22 (in Russian)

Zolotukhin VV (1978a) Differentiated ore-bearing trap intrusions of the Noril'sk and adjacent regions. In: Zolotukhin VV, Vilensky AM (eds) Petrology and perspectives of metallogeny of traps at the north of the Siberian platform. Nauka, Novosibirsk, pp 121–160 (in Russian)

Zolotukhin VV (1978b) Taxitic rocks and gabbro as an important criterion of metallogeny of differentiated trap intrusions. In: Sobolev VS (ed) Criteria for Exploration of the Noril'sk-type Sulfide Ores. Nauka, Novosibirsk, pp 52–64 (in Russian)

Zolotukhin VV (1979) The problem of genesis of sulfide-nickel mineralization in mafic-ultramafic complexes. In: Genesis of mineralization in basites and ultrabasites. Sverdlovsk, pp 48–57 (in Russian)

Zolotukhin VV (1981) On genesis of magnesian traps and their metallogeny on the Siberian platform. In: Aspects of genetic petrology, Nauka, Novosibirsk, pp 21–39 (in Russian)

Zolotukhin VV (1982) On the process of magma differentiation in certain intrusives of the Noril'sk-type. Sov Geol Geophys 23(6): 47–54

Zolotukhin VV (1984a) Petrology of magnesian basites of the Tunguska (Middle Yenisei) province. In: Sobolev VS (ed-in-chief) Magnesian basites of the western Siberian platform and aspects of their nickel content. Nauka, Novosibirsk, pp 48–149 (in Russian)

Zolotukhin VV (1984b) Magnesian basites and sulfide mineralization. In: Sobolev VS (ed-in-chief) Magnesian basites of the western Siberian platform and aspects of their nickel content. Nauka, Novosibirsk, pp 176–200 (in Russian)

Zolotukhin VV (1988) A generalized model of sulfide copper-nickel ore formation as a process of sulfurization. In: Obolensky AA, Sotnikov VI, Sharapov VN (eds) Mineralization and genetic models of endogenous ore formations. Nauka, Novosibirsk, pp 172–181 (in Russian)

Zolotukhin VV (1997) Mafic pegmatoids of the Noril'sk ore-bearing intrusions and the problem of genesis of the Platinum-copper-nickel mineralization of the Noril'sk type (ed Polyakov GV). SB RAN NITS OIGGM, Novosibirsk, 88 p (in Russian)

Zolotukhin VV, Laguta ON (1985) On fractionation of magnesian mafic melts and diversification of traps in the Siberian platform. Doklady AN USSR 280(4):967–972 (in Russian)

Zolotukhin VV, Podgornykh NM (1998) On probable overheat of Permo-Triassic trap melts of the Siberian platform. Doklady Akad Nauk SSSR 363(2):226–229 (in Russian)

Zolotukhin VV, Ryabov VV (1977) Problems of genesis of sulfide copper-nickel ores and principal parameters of the process. In: Kuznetsov VA, Berzina AP (eds) Principal parameters of processes of endogenous ore formation, vol 1. Nauka, Novosibirsk, pp 50–53 (in Russian)

Zolotukhin VV, Shchedrin NF (1977) Differentiated intrusions of the Imangdinsky ore junction (ed Sobolev VS). Nauka, Novosibirsk, 135 p (in Russian)

Zolotukhin VV, Vasil'ev YR (1967) Peculiarities of formation mechanism of ore-bearing trap intrusions at the northwest of Siberian platform (ed Sobolev VS). Nauka, Moscow, 231 p (in Russian)

Zolotukhin VV, Vasil'ev YR (1975) Problems of platform magmatism: the depths of the magma chambers in the upper mantle and their part in creating the variety of igneous rocks. Sov Geol Geophys 16(2): 1–6

Zolotukhin VV, Vasil'ev YR (1976) Problems of platform magmatism. Article 2: Differentiation as a cause of the diversity of magmas. Sov Geol Geophys 17(4):45–52

Zolotukhin VV, Ryabov VV, Vasil'ev YR, Shatkov VA (1975) Petrology of the Talnakh ore-bearing differentiated intrusion (ed Sobolev VS). Nauka, Novosibirsk, 432 p (in Russian)

Zolotukhin VV, Vilensky AM, Duzhikov OA (1986) Basalts of the Siberian platform (eds Sobolev VS, Sobolev NV). Nauka, Novosibirsk, 255 p (in Russian)

Zolotukhin VV, Kovyazin SV, Shcherbakova ZV, Vasil'ev YuR (1988) Features, composition and temperatures of hypogene mafic and ultramafic melts in the Siberian Platform. In: Mineral formation and ore genesis, Naukova Dumka, Kiev, pp 20–28 (in Russian)

Zotov IA (1979) Genesis of trap intrusions and metamorphic formations of the Talnakh (ed Korzhinsky DS). Nauka, Moscow, 155 p (in Russian)

Zotov IA (1982) Transmagmatic fluids in geology. Priroda 7:45–56 (in Russian)

Effusive and Explosive Complexes of the Noril'sk Region

2

Abstract

Among the "large igneous provinces" of the world, the Siberian Traps province had been, from the geological standpoint, one of the most extensively studied regions as long ago as by the 1970s of the last century. The principle of tectonomagmatic recurrence laid the foundation for the unified stratigraphic scheme of partioning the volcanogenic series. Such recurrence is represented by rhythmic alternation of pyroclastic and effusive rocks in the volcanogenic sequence. Five tectonomagmatic cycles (phases) are distinguished, all rocks derived from these cycles being of regional occurrence. The most complete volcanogenic sequence is in the Noril'sk Region. In Sects. 2.1, 2.2 and 2.13, the general characteristics of flood basalts, the structure of lava flows, the mineralogical and petrographic characteristic of flood basalts, and the mineralogical, petrographic, and petrochemical description of rocks from 11 suits of the volcanogenic sequence are given. Section 2.14 is concerned with the description of nine marker horizons of basalt flows, which are characterized by great thickness (up to 120 m) and can be traced over the entire basalt field of the Platform. Section 2.15 is aimed to description of the complex of "anomalous formations" in volcanogenic rocks. The formations include local manifestations of high-magnesium and subalkaline basalt lavas among flood basalts, exotic manifestations of flows of komatiite-like rocks and magnetite lavas and tuff paleovolcanoes, the belt of numerous dykes of varying composition, diatremes with sulfide and magnetite ores, calderas with beds of limestone and anhydrites, as well as manifestations of native copper and bitumen in volcanogenic rocks. In Sect. 2.16, the large-scale flood basalt eruptions of the Siberian Platform are discussed.

2.1 Characteristics of Flood Basalts

Volcanogenic sequences of the Siberian Traps, up to 3,500 m thick, are found in the Noril'sk Region (Fig. 2.1), comprising alternating lava flows (90%) and tuff horizons (10%) (Distler and Kunilov 1994). In the upper parts of the sequence in the northern part of the Siberian Platform, lava flows dominate, whereas in the south, at the Lower Tunguska latitude, basalts are replaced by a tuffaceous formation. Overall, more than 350 lava flows and about 120 tuff horizons are observed. The thickness of the lava flows varies from 1 to 100 m (average 15 m) and tuffs from 1–10 m to 100 m. Some marker horizons of tuff and lava can be traced for tens or even hundreds of kilometers (Figs. 2.2 and 2.3).

The volcanogenic sequence consists of 11 suites, some of which are further subdivided into subsuites and units (Table 2.1). Most lava is derived from a tholeiitic basalt parental magma, with smaller volumes derived from a subalkaline basalt and picritic basalt parental magma. The flood basalts have aphyric and porphyritic textures with poikilophitic, tholeiitic, and intersertal groundmass textures. The main rock-forming minerals in the sequences are plagioclase, clinopyroxene, olivine, and varying amounts of titanomagnetite, ilmenite, glass, and palagonite. The classification

V.V. Ryabov et al., *Trap Magmatism and Ore Formation in the Siberian Noril'sk Region: Volume 1. Trap Petrology; Volume 2. Atlas of Magmatic Rocks*, Modern Approaches in Solid Earth Sciences 3, DOI 10.1007/978-94-007-5022-7_2,

Fig. 2.1 Volcanogenic series of the west Putorana, the vicinity of the Khantayskoe Lake

and nomenclature of the rocks of the trap formations have been completed. Based on geochemical data features, basalts were assigned to various stratigraphic divisions and the average composition of effusive lava of the Siberian Platform (Table 2.2) has been calculated.

The age of the effusively erupted and intrusive traps, measured by a variety of methods and in various laboratories (Renne and Basu 1991; Dalrymple et al. 1991; Campbell et al. 1992), has shown that the initiation of flood basalt volcanism took place at the Permian–Triassic boundary (Distler and Kunilov 1994).

Paleomagnetic studies of the volcanogenic sequences have shown that the basalts of Ivakinsky Suite and the upper part of Samoedsky Suite are characterized by reverse polarity, whereas the rest of the extrusive section, from the Syverminsky to the Samoedsky Suites, have normal polarity (Lind and Shchekoturov 1991). On the basis of this data, the basalts of the Ivakinsky Suite were attributed to late Permian, and the overlying effusively erupted lava flows to early Triassic age. It is supposed that the rocks formed with normal polarity, representing about 90% (by volume) of the suite; in the early Triassic period, this lasted for 0.6 million years (Haag and Heller 1991), and the duration of 3,500-m-thick volcanogenic sequences of the Noril'sk Region did not exceed one million years (Distler and Kunilov 1994).

Structures observed throughout the volcanogenic section provide evidence that the majority of lava flowed onto non-eroded and non-weathered surfaces of earlier lava flows. This is indicated by traces of lava flows on upper surfaces, quenching crusts, and lava flows complete with an upper amygdaloidal zone. The eruptions were numerous, and each subsequent volcanic event produced lava that flowed over the surface of the (still hot) previous flow. The high

Fig. 2.2 The basalt cover of Khonnamakitsky Suite, the vicinity of the Namakan River

Fig. 2.3 Columnar structure in basalts of Khonnamakitsky Suite, in the vicinity of the Namakan River

frequency of eruptions promoted heat preservation in the earlier, buried flows, which maintained the high temperature and low viscosity of the lava flows and enabled better crystallization of melt.

2.1.1 Tectonomagmatic Recurrence

Studies on the stratigraphy and petrology of volcanogenic rocks in the north of the Siberian Platform have revealed distinguishing characteristics that can be used to identify the evolution of volcanic events and prove the existence of five basic tectonomagmatic cycles or phases in the magmatism of the Noril'sk Region (Nemenenok and Vilensky 1978; Zolotukhin et al. 1986). The founding principles for the partition of volcanogenic formations and their relationship to volcanic events were based on studies by Rozhkov, Kotul'sky, Godlevsky, and Urvantsev and have been further established by authors such as Golubkov, Dodin, Duzhikov, Ivanova, Ivanov, Lind, Lyul'ko, Maslov, Mitroshin, Nagaitseva, Nemenenok, Nesterovsky, Starosel'tsev, Pol'kin, Staritsyna, Sukhov, Tarakhovsky, Tomanovsky, and Fedorenko. While accumulating geological data, the volcanic classification scheme as well as the principles of partitioning volcanogenic series and allocation to cycles were corrected, supplemented, and improved and eventually took the form that is used nowadays by geologists for geological mapping and in various genetic reconstructions (Table 2.1).

It is supposed that the volcanogenic rocks of the Noril'sk Region were formed during five volcanic cycles or tectonomagmatic phases, each of which united one or several suites (Nemenenok and Vilensky 1978). In addition, the whole volcanogenic series is divided into early and late volcanic stages. Rocks derived from tectonomagmatic phases I and II (generated during the Upper Permian-Lower Triassic) are assigned to the early stage, whereas those from phases III–V (formed in the Triassic) are assigned to the late stage.

Each tectonomagmatic phase is represented by a polycyclic complex of explosive, effusive, and intrusive rocks. It is considered that the formation of each phase or cycle began with explosive activity that, during the course of increasing tectonic activity, grew into mass extrusions of melt from

Table 2.1 Stratigraphic division scheme for volcanogenic rocks of the Noril'sk Region

Phase (cycle)	Age, index	Suite	Thickness	The basic type of rocks
V	T_1sm	Samoedsky	>600(?)	Tholeitic, aphyric, and porphyritic basalt
	T_1km	Kumginsky	160–210	Tholeitic and glomeroporphyritic basalt
	T_1h	Kharaelakhsky	380–620	Tholeitic, subalkaline, porphyritic, and glomeroporphyritic basalt
IV	T_1mk	Mokulaevsky	400–690	Tholeitic, aphyric, and porphyritic basalt
	T_1mr	Morongovsky	240–700	Tholeitic, aphyric, and subalkaline basalt, and ankaramites
III	T_1nd	Nadezhdinsky	150–530	Tholeitic, porphyritic, and glomeroporphyritic basalt, and tuffs
	T_1tk	Tuklonsky	0–220	Tholeitic and picritic basalt
	T_1hk	Khakanchansky	15–260	Tuffs, tuffites, rare layers of basalt
II	T_1gd	Gudchikhinsky	0–250	Tholeitic and picritic basalt
	T_1sv	Syverminsky	0–195	Tholeitic basalt
I	P_2iv	Ivakinsky	0–330	Subalkaline basalt, trachybasalts, trachyandesitic basalts, and tuffs

Table 2.2 Chemical composition (wt%) of the Siberian Platform basalts

Component	1	2	3	4	5
SiO_2	48.43	48.41	48.70	48.25	48.24 ± 5.41
TiO_2	1.48	1.30	1.36	1.18	1.32 ± 0.14
Al_2O_3	15.35	15.31	14.80	15.47	14.71 ± 2.75
Fe_2O_3	3.59	4.23	3.31	4.84	
FeO	9.39	8.24	7.27	7.64	
FeO_{tot}	12.62	12.00	10.25	12.00	11.29 ± 1.32
MnO	0.19	0.19	0.18	0.18	0.19 ± 0.01
MgO	6.18	7.04	7.30	6.90	7.17 ± 5.74
CaO	10.69	10.68	9.68	10.78	10.79 ± 3.59
Na_2O	2.18	1.81	2.37	2.20	2.34 ± 0.54
K_2O	0.73	0.68	0.85	0.42	0.56 ± 0.43
P_2O_5	0.13	0.22	0.11	0.15	0.17 ± 0.02
LOI		2.04	4.00		2.92 ± 3.49
Total	98.34	100.15	99.93	98.01	99.7
Number of analyses	176	401	219	845	444

1—Average composition of the Siberian Platform basalts (Kutolin 1969); 2—effusive traps weight average composition of the northwest Siberian Platform (Zelenshchikov and Duzhikov 1974); 3—average composition of the Siberian Platform basalts (Nesterenko et al. 1964); 4—average composition of basalts of the Tungussky syneclise (Ledneva and Poroshin 1981); 5—weighted average composition of basalts of the northwest Siberian Platform (Almukhamedov et al. 1988)

associated fractures. A subsequent decrease in volcanic activity led to isolation of individual shield volcanoes and an associated termination of mass lava outpourings, and volcanic activity was accompanied by the accumulation of tuffaceous material and the formation of intrusions (Fadeev 1962; Kuznetsov 1965; Daragan-Sushchov 1989). Each phase is characterized by a unique melt composition and characteristics associated with its evolution in time and space (Godlevsky 1959; Fedorenko 1981). The periodicity of the tectonomagmatic activity, as well as the duration of each eruptive stage, completeness of its development, volume and the surface distribution of volcanic rocks, depends upon the geodynamic circumstance.

Nemenenok (1978) and Duzhikov (Zolotukhin et al. 1986) note that the majority of tectonomagmatic phases have a three-part structure: a tuffaceous layer, greenstone basalts sometimes with pillow lavas, and relatively fresh basalts. Greenstone basalts have a tholeiitic, poikilophitic, poikilophitic intersertal, or ophitic texture, whereas fresh basalts are characterized mainly by a porphyritic with microdoleritic and intersertal texture. Both authors underline the rhythmic repetition of greenstone basalts in a volcanic formation. It was supposed that the greenstone metamorphism of basalts was caused by autometamorphism and regional metamorphism, yet the leading role in this change, probably, belonged to the gases from numerous lavas intruding the effusive bodies (Godlevsky 1959). The accepted viewpoint adhered by Rozhkov, Moor, Korovyakov, and Maslov et al. (Nemenenok and Vilensky 1978) supposes the occurrence of greenstone basalts, spilites, and similar rocks is the result of lavas erupting underwater.

In the volcanogenic sequences of the Putorana Plateau, tectonomagmatic phases are manifest as a characteristic alternation of thick (up to 120 m) fine-grained basalt layers with microdoleritic texture and thin (not more than 20 m) layers of coarse-grained and anamesitic basalts with an ophito-taxitic and poikilophitic intersertal texture (Fadeev 1962; Kuznetsov 1965, 1977). According to the authors, rock crystallinity was governed by the content of volatile material in melt, which affected the degree of magma overcooling and crystallization. Therefore, despite their thickness, melts poor in volatile material would generate fine-grained basalts, whereas thinner layers that were enriched with volatile material would crystallize as coarse-grained rocks. It is believed that fine-grained basalts are the result of eruption from fractures or fissures, and coarse-grained ones were erupted from volcanoes.

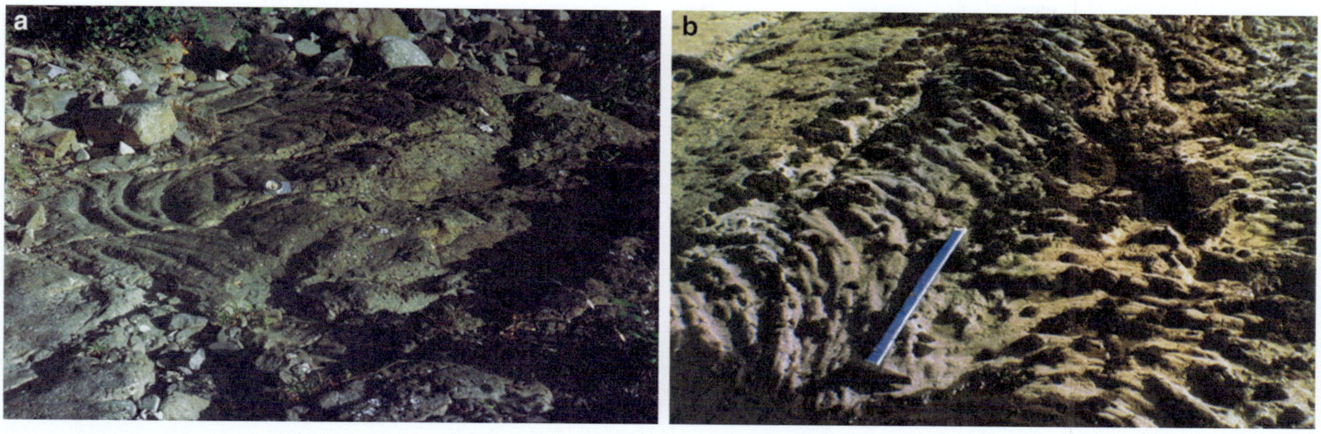

Fig. 2.4 Ropy surfaces of the lava covers of the Ivakinsky Suite in the vicinity of the Pyroxenovy Stream, inflow of the Mikchanda River (**a**), and of Pravoboyarskaya Suite at the Levaya Boyarka River (**b**)

2.1.2 Structure of Lava Flows

Basaltic lavas form very thin and thick flows that vary from 0.5–1 m to 100–140 m. Some thick flows possess characteristic features that enable them to be traced for hundreds of kilometers and have been used as marker beds (Starosel'tsev 1982). The lava flows have a regular structure that does not depend on their thickness or petrographic composition (Pol'kin 1961), consisting of three zones: bottom (amygdaloidal), central (massive), and top (amygdaloidal). Pol'kin (1961) also points at the existence of thin, glassy, etched crusts (1–2 cm) in the top parts of amygdaloidal basalts, which feature a smooth, glossy appearance and are usually flat or slightly wavy. The top surface of basalts with poikilophitic intersertal texture typically has a rough, "slaggy" shape. Pol'kin (1961) also observed lava flows that featured concentric flow "wrinkles" and surface "rope structures" on the southern shore of the western part of Lama Lake, in the valleys of the Vodopadny and Piritovy Streams, and between the Keta and Glubokoe Lakes. The "rope structures" of lava flows are also known from basalts of the Piroksenovy and Sirotsky Streams, on the Levaya Boyarka River, and various other localities (Fig. 2.4). Godlevsky (1959) noted that the viscosity of a tholeiitic basalt melt is close to pahoehoe-type lavas, and the viscosity of porphyritic lavas is close to -type lavas. In various parts of Khonnamakitsky Suite, around Lake Khantayskoe, on the surface of lava flows denuded by water, it is possible to see evidence of splitting of the top-etched surface.

The uniqueness of effusive lavas from the Siberian Platform, when compared with flood basalts from other platforms, is the high degree of crystallinity of a considerable part of the flows and the great volume of amygdaloidal zones, composing about 36 vol% of the lava flows in the Noril'sk Region, with the quantity of amygdales reaching 30–40 vol% (Distler and Kunilov 1994). In the lower parts of effusive flows, the volume of amygdaloidal basalts reaches 49 vol%, whereas at the top it reaches 31 vol%. Fine-grained and glassy basalts with abundant amygdales are usually called "mandelsteines," and rocks with fewer amygdales that retain original features are called amygdaloidal basalts. The quantity of amygdales and the thickness of amygdaloidal zones in separate flows vary widely. A detailed review of characteristics of basalt flow structures is given by Simanovich and Kudryavtsev (1981).

In thin flows with ophito-taxitic texture, amygdaloidal zones sometimes constitute 75 vol% of the section. For such rocks, the widespread development of palagonite I in the rock matrix and as amygdales is characteristic, and the presence of fresh glass is very rare. In some flows, differentiation within the flow is observed, which results in the formation of metadiorite (in Levinson Lessing's understanding) in the upper parts that is enriched by palagonite with large plagioclase and ilmenite crystals. Some lava flows are completely composed of greenstone basalts.

In thick flows of aphyric and plagioclase-phyric basalts, amygdaloidal zones constitute 10–25 vol% of the flow. At the top of a flow, the amygdaloidal zones are always thicker than at the base. Moving upward across an amygdaloidal zone, the diameter of amygdales increases from 2–3 mm to 7–12 mm, and they transform from spherical to ovoid shapes, and under the etch crusts, it is quite often possible to see large horizontally extended miarolitic cavities filled with calcite, quartz, and more rarely with zeolites. At the base of such flows, subvertical amygdale chains and tubular cavities 5–7 cm long, and rarely to 12 cm long, are observed with their ends frequently turned toward the direction of lava flow (Fig. 2.5).

Fig. 2.5 The contact of the basalt cover of Khonnamakitsky Suite with red knotty tuffs of the region of the Namakan River

In the central, denser parts of basalt flows, separate amygdales can be observed, but more frequent are accumulations of small rounded amygdales with sharp boundaries, the so-called gas bubbles. Amygdale accumulation can also form "gas channels." These are fusiform or slightly elongated (Fig. 2.6a, b) amygdales, 50–70 cm long, about 5 cm in diameter. Gas channels are distinct and formed through the accumulation (up to 50–60 vol%) of amygdales, arranged along columnar joints or bent in the direction of lava flow. In dense, aphyric basalts, chains of amygdales arranged along cooling joints are also observed.

In fact, the presence of "gas bubbles" in the basalt reflects the immiscibility of parts of the melt that are enriched with volatile material and the "dry" parts of the melt. In most cases, basalts with "gas bubbles" are simply underdeveloped pegmatoids, being poorly crystallized due to unfavorable conditions during formation, whereas "gas bubbles" containing considerable quantities of palagonite I represent glass enriched with volatile material, which had no opportunity to vitrify. An apparent exception here is the metadiorite formed from picritic basalt intrusions, whose coarse-grained texture still allows them to be classified as pegmatoids.

"Gas bubbles" are isolated portions of the melt enriched with volatile material formed at the top of the lava flow as a result of the accumulation of volatile material separating from the melt and rising upward. However, some gas bubbles, owing to the increased viscosity of the melt higher up, could not rise to the top of the lava flow and stopped at various hypsometric levels. These types of gas bubbles can be found throughout the flow. "Gas channels" were formed in more viscous melts, where volatile material could not rise upward and were consequently were pulled together along zones of weakness, such as cooling joints that correspond to the future columnar or sheet jointing of lava flows. The composition of the matrix contained in these "gas bubbles" differs from the enclosing basalts, featuring a greater abundance of plagioclase and palagonite, and rich ore dissemination or opacitizated glass.

As a rule, higher amygdale concentrations are observed in the top of the lava flow (see Fig. 2.6a, b), which gives the rock a pumice-like appearance. Usually, they have a spherical form and are up to 0.6–0.8 cm (though they can range higher) in size. They are typically filled with calcite, quartz, or zeolites or any combination of those minerals. Despite the abundance of amygdales, they maintain their spherical shape and seldom form large miarolitic cavities. Spherical amygdales are a characteristic feature of fine-grained basalts with microdoleritic, picritic, and intersertal texture. In coarse-grained basalts with ophito-taxitic, poikilophitic intersertal textures along with spherical amygdales, palagonite I segregations appear in the groundmass, frequently with calcite in their central parts. Besides these segregations, palagonite I also forms in the matrix along with glass or replacing it. Palagonite I is an equivalent of glass enriched with volatile material, whereas palagonite II is a secondary chlorite- or smectite-like product of rock replacement (Ryabov 1989b).

A prominent feature of coarse-grained basalts is the widespread formation of palagonite I instead of glass in the matrix, which most likely indicates the enrichment of the melt with volatile material. The higher volume of palagonite I in coarse-grained basalts is usually confined to the upper parts of a flow and frequently accompanied by metamorphosis of the rocks up to greenstone facies or metasomatism. In this way, the composition of rocks changes occasionally, and the differentiation of melt within a flow is foreshadowed. The phenomenon is most commonly expressed in picritic basalt lava flows.

Fig. 2.6 Amygdales in basalt covers. (**a**) Outcropping cf two basalt covers at the boundary of carbonaceous terrigenous sediments of the Tungussky series; in the roof of the lower cover, there are zones enriched with amygdales and large subhorizontally elongated miarolic cavities; the vicinity of the Mikchanda River; (**b**) palagonite-enriched gaseous channels aligned with columnar structure in basalts, the vicinity of the Khantayskoe Lake

Palagonite-enriched coarse-grained basalts in the lava flow alternate with the rocks of similar structures that contain opacitizated glass instead of palagonite. The former is more unstable and likely to weather (in comparison with the latter), and, as a result, it is possible to see alternating ridges of dense basalt that resist erosion and weathered vein-like hollows of friable basalts in exposures. These are described in the literature as "layered dolerites" (Dmitriev 1973; Simanovich and Kudryavtsev 1981). These formations are most clearly manifest in the Mokulaevsky Suite and its analogues—the Khonnamakitsky and Nidymsky Suites (Fig. 2.7). The thickness of "dense" basalts varies from 5 to 15 cm and "friable" 20 to 50 cm. In the latter, the presence of small (0.1–0.3 mm) pores (not more than 5 vol%), filled with calcite, was observed (Simanovich and Kudryavtsev 1981).

In the upper parts of the lava flows (of the above-mentioned suites), an alternating sequence of fine-grained glassy basalts with intersertal or poikilophitic intersertal textures is also observed. It forms a successive transition of fine-grained rock into coarse-grained rock within one section. We will discuss the features of these transitions in our description of the basalts of the Mokulaevsky Suite.

Despite the long-held concept of the relative "dryness" of trap magmas, in recent years, geologists have commonly recorded trap magmas enriched with volatiles. Other indirect evidence of against the concept of "dry" trap magmas includes high degree of crystallization of large parts of the basalts, thick amygdaloidal zones in lava flows, and, hence, a high amygdaloidal index, as well as high explosivity coefficient in the traps of the Siberian Platform.

Estimating the explosivity coefficient variations in different structure-formational zones and structure-facial zones in the northwest of the Platform is complex, and Nemenenok (1978) points at temporal and special variations in its value (Fig. 2.8). On average, the highest explosivity coefficient (70%) is from the Nizhnetungussky structure-facial zone, in Noril'sko-Vologochanky and Kharaelakh-Ikensky it is 11–15%, in Lama-Imangdinsky 21%, in Boyarsko-Del'kansky 14%, at the north of the Putoransky subprovince it is 21–24%, and in the south of the Putoransky subprovince it is 47%. Extreme explosive volcanism manifested during the formation of the Khakanchansky (Avamsky) Suite; explosivity coefficient almost everywhere reached a remarkable 90–98%. The author believes that a low explosivity coefficient testifies that fluids remained and accumulated in magma during intrusive formations, which later on promoted melt differentiation and mineralization. Accordingly, he considers that potentially mineralized are the structure-facial zones with low explosivity coefficient.

Fig. 2.7 Coarsely stratified basalts representing compact rocks pierced by chains of subhorizontal gaseous channels, enriched with palagonite; Khonnamakitsky Suite, the area of the middle reach of the Namakan River

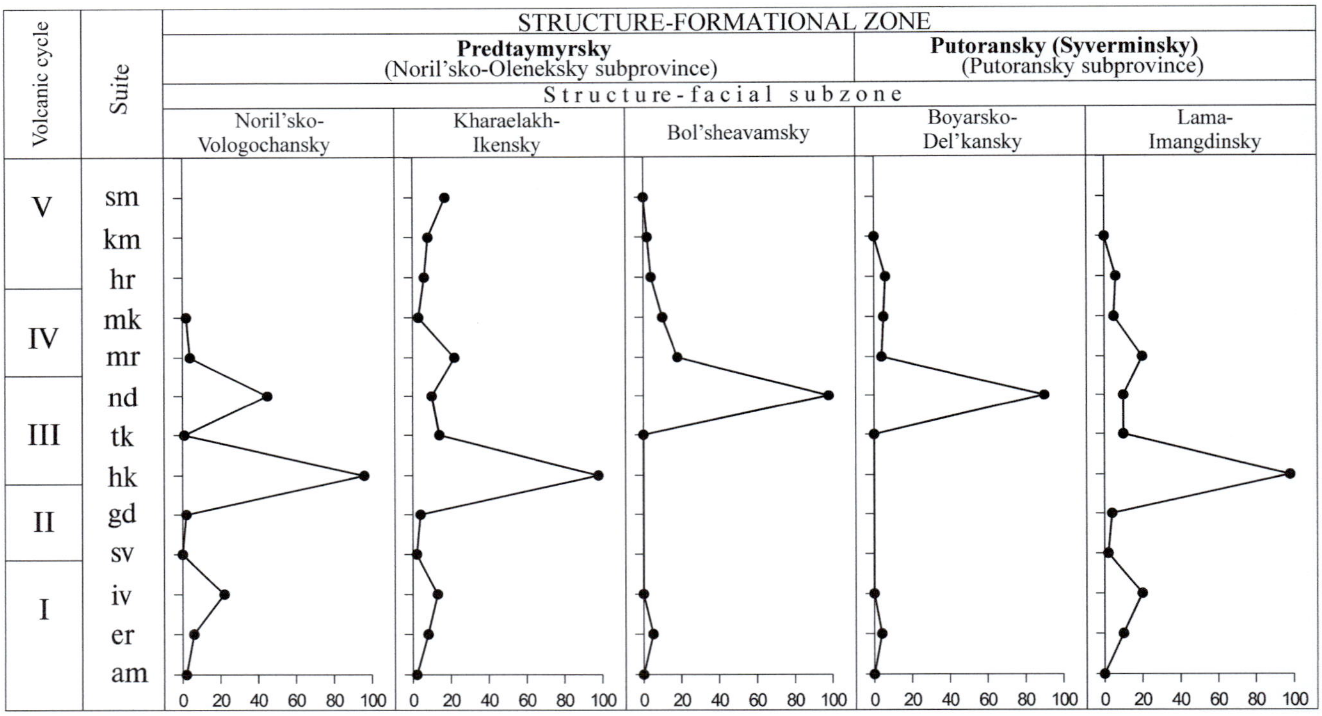

Fig. 2.8 Variation of explosivity index in section of volcanogenic series of the north of the Siberian Platform, according to T.I. Nemenenok (1978). The suites er—Ergalakhsky, am—Ambarninsky; for other suites see Table 2.1

2.2 Ivakinsky Suite (P_2iv)

In the northwest of the Siberian Platform, volcanic activity of the Ivakinsky Suite during the Late Permian was marked by the accumulation of tuffaceous material and lava flows that eroded terrigenous sediments of the Tungussky series. The thickness of the Ivakinsky Suite varies from 0 to 300 m and is divided into three subsuites: lower, middle, and upper (Distler and Kunilov 1994).

The lower subsuite consists of 1–13 trachybasalt lava flows, which features basal tuffs and tuff breccia. Thickness of the rocks of the subsuite varies from 0 to 240 m.

The middle subsuite consists of 1–5 subalkaline titanaugite basaltic lava flows with rare thin bands of tuff and tuff breccias. The thickness of the suite ranges from 0 to 100 m.

The lower and middle subsuites are located within the limits of Noril'sko-Vologochansky, Kharaelakh-Ikensky, and Lama-Imangdinsky structure-facial zones and are absent to the south of Keta Lake. The thickness of the lava sequence reaches 40 m. A complete flow features an upper and lower amygdaloidal zone with a central massive zone. Amygdaloidal zones constitute 15–25 vol% of the sequence, with the upper zones being three to six times thicker (Arkhipova and Dodin 1963).

The upper subsuite consists of two units, the lower composed of labradoritic andesitic–basaltic and trachyandesitic–basaltic rocks, and the upper composed of two-plagioclase andesitic–basaltic rock. Rocks of the upper subsuite occur on the main part of the Noril'sk Region and gradually thin out at its western, eastern, and southern borders. In the north of the Noril'sk Region and southwest of the Kharaelakhsky trough, labradoritic andesitic–basaltic rocks are exposed as one thick (from 25–30 to 60–70 m) lava flow and two-plagioclase andesitic–basaltic rocks as 1 to 4 lava flows (20–50 m) (Zolotukhin et al. 1986). Between these two units, the band of tuffaceous rocks and coaly argillites, sometimes with coal layers (up to 1–1.5 m thick), is observed. At the top of the Ivakinsky Suite, in the southern part of the Kharaelakhsky trough and in the western board of the Tungussky syneclise, there is a layer of coal 2–4 m thick called the Zametny formation.

2.2.1 Peculiarities in Rock Composition

Trachybasalts and subalkaline titanium-augite basalts, according to their chemical composition, are "high-alkaline basalts," and in the classification diagram (Na_2O + K_2O)–SiO_2, they are situated lower than the Kuno line, which divides the rocks of normal and high alkalinity. Trachybasalts are characterized by lower silica content in comparison with subalkaline titanium-augite basalts, with similar or higher content of alkali elements. Trachybasalts also feature higher concentrations of Fe_2O_3 + FeO, TiO_2, and P_2O_5 than subalkaline titanium-augite basalts.

According to the classification scheme of Bogatikov et al. (1981), both rock varieties correspond to a trachybasalt composition. Taking into account some peculiarities of rock extent and composition, observed by geologists in the Noril'sk Region, the authors have maintained the informal and regional names for these rocks, but have formally identified them as trachybasalts.

The texture of these rocks is either porphyritic, aphyric, poikilophitic, or ophito-tholeiitic with an intersertal, microdoleritic, tholeiitic groundmass. In amygdaloidal zones hyalopilitic, intersertal and tholeiitic textures are observed, whereas in central zones the rocks are more crystallized and sometimes feature a poikilophitic or ophitic texture. The rocks consist of pyroxene (15–25 vol%), plagioclase (45–60 vol%), olivine (0–10 vol%), apatite (0.5–2.5 vol%), glass and palagonite (15–20 vol%), and titanomagnetite and ilmenite (5–10 vol%). For trachydolerites, a higher concentration of oxide-ore minerals is characteristic. The most remarkable feature of these rocks is the distinct pleochroism in pinkish lilac titanaugite. Amygdales are filled with calcite, chalcedony, or a chlorite-like mineral.

Examples of the rocks from this suite and their texture and composition can be seen in the "Rock indications" derived from the analysis of rock samples, presented in the second volume of this book.

Subalkaline titanaugite dolerites correspond to $Wo_{37-43}En_{35-47}Fs_{16-24}$ (f = 25–41 wt%) and contain clinopyroxene that forms poikilocrysts with plagioclase intergrowths, small prismatic crystals, and granules. The mineral has a variable content of TiO_2 (0.76–2.22 wt%), Al_2O_3 (1.51–3.37 wt%), and Cr_2O_3 (0–0.46 wt%). High chromium content was noted in low ferriferous clinopyroxene, as well as a progression reduction in alumina concentration from the crystal center to margin (see Rock indications 1–3). Plagioclase in porphyry segregations has a labradorite ($Ab_{34-35}Or_{2-1}An_{65-64}$) composition, whereas laths in the groundmass and chadacrysts in clinopyroxene correspond to a labradorite–andesine composition ($Ab_{45-51}Or_{2-4}An_{53-46}$). In the interstices and in regions featuring plagioclase replacement, alkali feldspar $Ab_{52}Or_{47}An_1$ is observed. Titanomagnetite contains a high amount of V_2O_5 (0.69–0.86 wt%), and ilmenite is characteristically low in MgO content (0.67–1.38 wt%).

Labradoritic and two-plagioclase andesite–basalt and trachyandesite–basalt feature a porphyritic texture with a vitrophyric, trachytic texture to the groundmass and micro-ophitic, bostonitic, and hypidiomorphic granular elements. The rocks are characterized by the presence of two generations of plagioclase: partially melted crystals of An_{58-50} and An_{62-68} (Godlevsky 1959). The rocks consist of plagioclase (60–75 vol%), clinopyroxene (10–15 vol%), palagonite (5–15 vol%), rare olivine (up to 0.5 vol%), titanomagnetite and ilmenite (3–7 vol%), and apatite (0.5–1 vol%). Neocrystallizations of grossular are sometimes noted in palagonite in the matrix as result of autometamorphism (see Rock indication 4).

Labradorite trachyandesite–basalt of the Talnakh area consists of plagioclase ($Ab_{47-48}Or_3An_{50-49}$), which forms porphyritic segregations and prismatic crystals that form a seriate porphyritic texture and alkali oligoclase $Ab_{69}Or_{14}An_{17}$ in the matrix (see Rock indication 4). Clinopyroxene crystals are augite ($Wo_{39-40}En_{34-35}Fs_{23-26}$) and form small, sometimes twinned grains that contain TiO_2 (0.98–1.95 wt%) and Al_2O_3 (1.23–1.36 wt%). Olivine forms small ferrohortonolite granules (Fa_{71}) in the groundmass. Titanomagnetite and ilmenite have low MgO content, 0.12 and 0.38 wt%, respectively, and contain MnO (1.25–0.54 wt%) and V_2O_5 (0.59–0.26 wt%).

Two-plagioclase basalt consists of large and small plagioclase phenocrysts, with a groundmass characterized by intersertal, doleritic textures with ophitic and poikilophitic elements. Large phenocrysts have a zoned structure; their composition at their core is $Ab_{44}Or_3An_{53}$, at the edges $Ab_{54}Or_3An_{46}$ (see Rock indication 5). Small porphyritic segregations are $Ab_{53}Or_4An_{43}$, whereas microliths in the groundmass are $Ab_{64}Or_{11}An_{25}$. Clinopyroxene forms small isometric crystals or grains and oikocrysts. The mineral composition of augite is $Wo_{42-38}En_{37-35}Fs_{22-27}$, in which TiO_2 (1.58–4.03 wt%), Al_2O_3 (2.00–2.63 wt%), and Na_2O (0.37–0.40 wt%) were measured. The rock matrix contains fine grains of olivine (Fa_{67}), apatite (containing F 3.43–4.16 wt% and Cl 0.04–0.08 wt%), chlorite ($f = 89.5$ at%), titanomagnetite, and ilmenite. Titanomagnetite contains TiO_2 (19.57 wt%), Al_2O_3 (1.30 wt%), MnO (1.33 wt%), and V_2O_5 (0.45 wt%), and ilmenite contains MnO (0.62 wt%), MgO (1.28 wt%), and V_2O_5 (0.30 wt%).

According to the published analytical data (Godlevsky 1959; Zolotukhin and Vasil'ev 1967; etc.), the rocks of the upper Ivakinsky subsuite occupy the boundary region of basalt–andesite-basalt–alkaline basalt–trachyandesite-basalt in a ($Na_2O + K_2O$)–SiO_2 classification diagram. Results of the analysis of two-plagioclase andesite–basalt from the lava flow analyzed by Duzhikov (Zolotukhin et al. 1986) show that their composition varies to some extent: SiO_2 (49.90–51.40 wt%), MgO (4.40–6.69 wt%), CaO (4.00–8.24 wt%), and $Na_2O + K_2O$ (4.15–4.72 wt%). There is also a measurable increase in iron and magnesium content upward through the lava flow sequence.

2.3 Syverminsky Suite (T_1sv)

Rocks of the Syverminsky Suite present as a typically homogeneous stack of up to 20 tholeiitic and poikilophitic basalt lava flows, ranging from 4–5 to 15–20 m thick, and rarely up to 50 m thick. The maximum thickness of the suite (180–240 m) was measured of the western edge of the Tungussky syneclise (the Putorana Plateau); in the central part of the Noril'sko-Kharaelakhsky depression, it decreases to 120–150 m and in the south of the Noril'sky trough 20–70 m. On the watershed of the Gorbiachin and Kureyka Rivers, the Syverminsky Suite thins out completely (Nemenenok and Vilensky 1978).

Basalts of Syverminsky Suite are characterized by a tholeiitic and poikilophitic texture that gives them a small pea-like appearance. Tholeiitic basalts comprise plagioclase (41–59 vol%), clinopyroxene (17–26 vol%), volcanic glass and alteration products (13–34 vol%), and opaque minerals (2–4 vol%) (Godlevsky 1959). Elongate masses in the rock are formed of isometric or slightly elongated oikocrysts of clinopyroxene ($Wo_{39-37}En_{46-47}Fs_{15-16}$) with numerous lath-like chadacrysts of plagioclase ($Ab_{32-34}Or_{1-2}An_{66-65}$) (see Rock indications 6, 7). The bulk of the rock is composed of palagonite or opacitizated glass with laths of plagioclase $Ab_{37}Or_2An_{61}$ and fine grains of clinopyroxene $Wo_{35-33}En_{43-44}Fs_{22-24}$. Locally, small grains of olivine are replaced with bowlingite, and anisotropic lamellar chlorite-like formations ($f = 42.4$ at %) are also present. Clinopyroxene contains TiO_2 (0.77–1.16 wt%), Al_2O_3 (1.20–2.02 wt%), MnO (0.21–0.34 wt%), Na_2O (0.23–0.29 wt%), and Cr_2O_3 (0.009–0.36 wt%). Ilmenite has variable quantities of impurities including up to MnO (5.14 wt%), MgO (1.96 wt%), Cr_2O_3 (0.36 wt%), and V_2O_5 (0.33 wt%).

One of the tholeiitic basalt flow sequences, studied by Duzhikov (Zolotukhin et al. 1986), has revealed regular changes in rock-forming oxide content from the base to the top of the body; these are generally an increase in MgO and a decrease in CaO and FeO content of the rocks. It is remarkable that the maximum content of MgO is found in the central part of the flow sequence, whereas trachybasalts at the top contain SiO_2 (47.14 wt%) and $Na_2O + K_2O$ (4.20 wt %) and in the main part of the section contain SiO_2 (43.98–47.46 wt%) and $Na_2O + K_2O$ (3.08–3.25 wt%). This fact points at the existence of differentiation of tholeiitic basalt melt within the lava flow.

2.4 Gudchikhinsky Suite (T_1gd)

The suite is divided into three subsuites, based on composition and thickness, which vary widely across the region.

The lower subsuite in the Noril'sky trough and the south of the Vologochansky trough is composed of 2–4 flows of porphyritic basalt, and in the north of the Vologochansky trough, the the subsuite is composed of tholeiitic and poikilophitic basalt flows (Zolotukhin et al. 1986). In the southwestern part of the Kharaelakhsky trough, in the east of the Vologochansky trough, and in the western part of the Tungussky syneclise, a marker bed of glomeroporphyritic basalt is found. The maximum thickness of the subsuite (about 160 m) is observed in the western part of the Kharaelakhsky trough. The subsuite typically overlies a band of tuffaceous rocks.

The middle subsuite consists exclusively of picritic basalt flows, varying from 1–2 to 25–27 individual flows and total thickness from 3–5 to 200 m. The maximum thickness of the picritic basalt flows is found in the north of the Kharaelakhsky trough, while in its southwestern part in the north of the Noril'sky trough, the thickness of the subsuite varies from 30–40 to 100–110 m (Ivanova 1975; Zolotukhin et al. 1986; etc.). Around the Lama and Glubokoye Lakes, the thickness of the picritic basalt flows reaches 3–7 m, and further to the east and southeast, the subsuite is absent. Picritic basalts were described in the Siberian Traps for the first time in 1943 by Kotul'sky and were subsequently studied by many geologists (Korovyakov 1948; Godlevsky 1959; Dodin 1962; Duzhikov 1971; Ryabov et al. 1985b; Zolotukhin et al. 1986; etc.).

The upper subsuite is composed of a series of plagioclase basalt flows, which are not as extensive as the lower subsuites and may be missing in some areas. At the base of the Gudchikhinsky Suite is a 20-m-thick tuffaceous horizon, which, along with picritic basalt, represents a distinct marker horizon for the effusive flows of the northwestern part of the Siberian Platform.

Glomeroporphyritic basalts have porphyritic, glomeroporphyritic texture and both micro-ophitic and intersertal groundmass texture (see Rock indication 8). In the upper part of the flow sequence, basalt has subalkaline composition (SiO_2 47.40 wt%; ($Na_2O + K_2O$) 4.42 wt%). Composition of plagioclase chadacrysts intergrowing clinopyroxene oikocrysts corresponds to labradorite ($Ab_{34}Or_1An_{65}$), small plagioclase prisms are andesine ($Ab_{52}An_{48}$), and of the grains in the rock matrix are oligoclase ($Ab_{69}Or_6An_{25}$) Fine porphyritic segregations of clinopyroxene ($Wo_{42-39}En_{46-44}Fs_{12-16}$) are characterized by a zoned structure; crystals in the matrix are $Wo_{38}En_{45}Fs_{17}$. Augite contains TiO_2 (0.52–1.15 wt%), Al_2O_3 (1.47–3.57 wt%), Na_2O (0.23–0.35 wt%), and Cr_2O_3 (0.02–1.07 wt%). The highest Al_2O_3, Cr_2O_3, CaO, and Na_2O contents were found in the core of zoned crystals.

In the lower part of the lava flow sequence, the composition of glomeroporphyritic basalt corresponds to tholeiite (see Rock indication 9). Porphyritic plagioclase segregations have a zonal structure: $Ab_{35-49}Or_{1-2}An_{64-49}$. Composition of chadacrysts ($Ab_{38}Or_1An_{61}$) corresponds to the central zones of phenocrysts, and the laths in the groundmass correspond to $Ab_{37-46}Or_2An_{61-52}$. Clinopyroxene composition corresponds to augite, oikocrysts are $Wo_{42-41}En_{44-45}Fs_{13-14}$, and fine grains in the rock matrix are $Wo_{41-36}En_{45-42}Fs_{13-21}$. Clinopyroxene was observed to contain TiO_2 (0.79–1.42 wt %), Al_2O_3 (1.46–3.62 wt%), Na_2O (0.23–0.32 wt%), and Cr_2O_3 (0.009–0.90 wt%). The highest TiO_2, Al_2O_3, and Cr_2O_3 contents were found in the core of oikocrysts.

Among the ore minerals in these basalt flows, ilmenite (MnO 0.28–0.66 wt%, MgO 1.90–2.97 wt%, V_2O_5 0.51 wt %) was noted, and at the base of the flow sequence, mineralization is present as chromium spinel and Fe-armalcolite. Chromium spinel contains TiO_2 (2.41 wt%), Al_2O_3 (20.89 wt%), MgO (6.47 wt%), and Cr_2O_3 (30.30 wt%), and Fe-armalcolite contains TiO_2 (71.75 wt%), FeO (23.87 wt%), and MgO (1.47 wt%).

The Stoichiometric formula for armalcolite is (Fe, Mg) Ti_2O_5. In the Traps of the Siberian Platform, the mineral was observed in the intrusive rocks of the Ozernaya Mountain (Oleinikov et al. 1991). The variety of the mineral found by the authors differs from those described in previous works, containing higher iron and lower magnesium content. A recalculation of the composition of the mineral has shown that its formula reads $(Fe_{0.764}Mg_{0.084}Ca_{0.008}Mn_{0.007}Al_{0.005}Cr_{0.001})_{0.869}Ti_{2.064}O_5$.

The picritic basalt lava flows of the Gudchikhinsky Suite have a stratified structure, and the degree of their differentiation varies widely from flow to flow in the section and along each body (Ryabov 1992b). The differentiation phenomena in picritic basalts of the same age as the Gudchikhinsky Suite have been described in the literature for various parts of the Noril'sk Region. Most well known are the Val'kovsky (Dodin 1971), Daldykansky (Duzhikov 1971), and Talnakhsky (Ryabov 1992a, b, c) differentiated lava flows. Other types of stratification and crystallization of a picritic basalt melt is described for the Mikchandinsky (Dodin and Golubkov 1971; Dodin 1967; Ryabov et al. 1977; Ryabov 1992b) and Keta-Irbinsky (Kavardin 1988) differentiated lava flows. As both groups are similar, their characterization will be based on one representative sample.

2.4.1 Talnakhsky Differentiated Picritic Basalt Flow Sequence

An idealized complete lava flow section can be represented by the following rock types (in order, from top to bottom): mandelsteine (amygdaloidal basalt), metadiorite, olivine-phyric and ankaramite basalt, picritic basalt, and ankaramite and olivine-phyric basalts and mandelsteine.

The flows vary in thickness and structure (Fig. 2.9). The Talnakhsky flow sequence consists of four flows of varying thickness and structure. The most marked differentiation can be seen in the most powerful (30 m) flow that formed during the last eruption.

Mandelsteines are also formed by olivine porphyrites. The rocks consist of olivine phenocryst (15–25 vol%) in a bubbly hyaline, microlitic, intersertal, or pilotaxitic groundmass. Amygdales constitute about 10–15 to 25 vol% of the rock. In the upper part of the flow, an etched crust 0.5–2.5 cm thick with rare small amygdales and contraction joints filled with tuffaceous material is formed. Olivine phenocrysts are observed as subidiomorphic, prismatic, skeletal crystals of various sizes (from 0.05 to 1 mm) and segregations of crystals completely replaced by bowlingite, chalcedony, calcite, and late magnetite. Crystallites and microlites of plagioclase ($Ab_{51-37}Or_{5-0}An_{44-61}$) frequently take paniculate, sheaf-like, and dendroid forms (see Rock indication 10). In pseudomorphs after olivine and in the groundmass, relics of small chromite crystals with characteristic high content of Cr_2O_3 (45.13–46.13 wt%) and low TiO_2 content (1.70–1.83 wt%) have formed. Rarer,

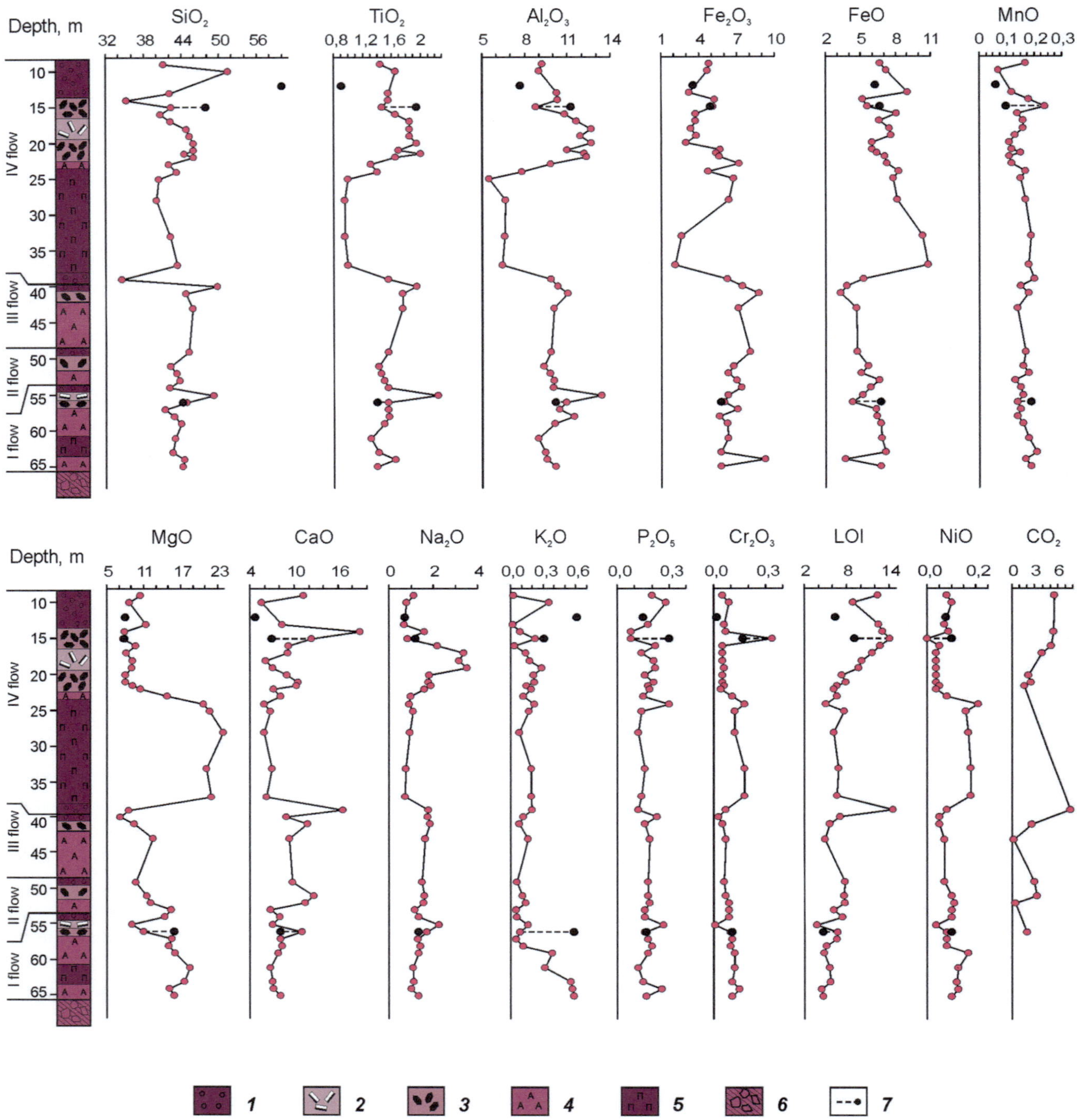

Fig. 2.9 Variation petrochemical diagram (wt%) of the Talnakhsky differentiated cover of picritic basalts: *1* mandelsteines, *2* metadiorites, *3* olivinophyric basalts, *4* ankaramites, *5* picritic basalts, *6* lava breccia, *7* probe purposely freed of amygdales

fine grains of amphibole with $FeO + Fe_2O_3$ content of 49 wt% and TiO_2 content of 7.31 wt% are also observed.

In view of the glassy character of mandelsteines and the presence of tiny granulated and skeletal crystals of olivine, it is possible to assume with confidence that olivine had crystallized during the eruption of lava and is not an intratelluric phase. In this way, it is possible to suppose that the picritic basalt melt was a homogeneous highly magnesian liquid rather than a product of deep fractional crystallization of mafic magmas in which accumulation of olivine protocrystals took place. Besides, the quantity of olivine in mandelsteines points to a possible initial composition of the melt, and the composition of chromite reflects the real composition of the mineral as products of crystallization of a picritic basalt magma that has not undergone chamber differentiation.

Metadiorites are porphyritic or pegmatoidal with an intersertal texture in the groundmass (see Rock indication 11). The composition of the rock corresponds to trachyandesite–basalt (SiO_2 52.00 wt%, ($Na_2O + K_2O$) 6.61 wt%). The rock is composed of large prisms of albite $Ab_{98}An_2$ and augite $Wo_{37-39}En_{52-54}Fs_{11-13}$. Its matrix consists of palagonite I in which fine grains of clinopyroxene $Wo_{35}En_{38}Fs_{27}$, segregations of alkali feldspar, and idiomorphic apatite crystals (F = 3.41, Cl = 0.09 wt%) are formed along with fine grains of ilmenite, titanomagnetite, and elongated crystals of an ore mineral, presumably Fe-armalcolite. Augite composition included TiO_2 (0.62–1.12 wt%), Al_2O_3 (1.81–2.33 wt%), and Cr_2O_3 (0–1.04 wt%). Higher Al_2O_3 and Cr_2O_3 content was determined in the cores of idiomorphic crystals. The composition of Fe-armalcolite, in the same way as those from glomeroporphyritic basalts, was calculated on 5 O. Mineral formulae are given in Rock indication 11.

In the Daldykansky lava flow sequence, there are dendritic crystals of clinopyroxene 5–8 cm long. Occasionally, in metadiorites fine olivine grains completely replaced with bowlingite are observed. Palagonite replacing glass or forming small amoeba-like aggregates in the matrix of the rock is widespread. The only ore minerals present are needles of ilmenite piercing plagioclase and clinopyroxene grains and ilmenite settled out in the rock matrix, as well as native copper disseminated in palagonite (most abundant in the rocks from the Khrebtovy Stream) and uncommon crystals of pyrite. On occasion, as a result of autometamorphic transformations, small idiomorphic crystals of garnet formed in palagonite in association with prehnite, calcite, and chlorite. The composition of garnet from metadiorites of the Daldykansky lava flow sequence was defined as SiO_2 (32.79 wt%), TiO_2 (0.12 wt%), Al_2O_3 (2.60 wt%), FeO (23.04 wt%), MnO (0.04 wt%), and CaO (32.80 wt%). Rarely, garnet has formed in bowlingite pseudomorphs after olivine.

Metadiorites have sharp contacts with mandelsteines and an indistinct contact with olivine-phyric basalts, which usually do not form independent bodies of constant thickness but rather form a transition zone between picritic and ankaramitic basalts.

Olivine basalts are porphyritic with an intersertal texture to the groundmass. Olivine phenocrysts (20–25%) are typically replaced with bowlingite. The groundmass is composed of plagioclase laths, fine grains of pyroxene, and needles of oxide ore minerals in a mass of devitrified glass. In these rocks, numerous miarole cavities filled with palagonite I are characteristic.

Ankaramites in some lava flows are the main rock type. Mandelsteines of such flows are composed of olivine porphyrites. Basalts are porphyritic with an intersertal, micro-ophitic groundmass and doleritic and segregational textural elements. The rocks feature numerous tiny amoeba-like and rounded miarole cavities. They are filled with either isotropic magmatogenic palagonite I, palagonite I with calcite, or palagonite I with calcite and an anisotropic chlorite-like mineral. Phenocrysts of olivine are fully replaced by iddingsite–bowlingite and clinopyroxene. The rocks are characterized by a high proportion of clinopyroxene as aggregations in the groundmass and as hypidiomorphic aggregates enclosing olivine. Under the microscope, olivine frequently reveals an hourglass structure, and its fine grains are characterized by deep pink coloring and distinct pleochroism.

Picritic basalts have a porphyritic, ophito-taxitic, microlitic, poikilitic, or poikilophitic texture. In various parts of the Noril'sk Region, outcrops of these rocks are rather similar. They are characterized by the presence of large quantities of olivine (from 20–40 vol% up to 60 vol%). Olivine, in thin sections, appears as large subidiomorphic skeletal crystals, sometimes with intergrowths of plagioclase and chromite, and tiny isometric grains (see Rock indications 12 and 13). Subidiomorphic olivine crystals, hexagonal in thin section, usually have perfect cleavage along (010). In picritic basalts, the composition of olivine has been observed to change from Fa_{18} to Fa_{29}. Large crystals usually contain less iron than smaller ones, though their iron content increases from the center to their margins from Fa_{19} to Fa_{22}. The most iron-rich minerals were found in thin parts of the flow (see Rock indication 12). The composition of olivine from picritic basalts is the most unique among all olivines from the traps, with an isomorphic impurity of NiO (0.39–0.45 wt%) (Ryabov 1992b), which indicates the initial concentration of silicate nickel in unmineralized traps. Olivine in picritic basalts are, as a rule, replaced by bowlingite–iddingsite or by talc with magnetite.

Pyroxenes in picritic basalts include augite, pigeonite, and enstatite. Augite ($Wo_{40-36}En_{50-52}Fs_{10-14}$) forms rather large and small subidiomorphic crystals and angular grains that contain TiO_2 (0.18–0.74 wt%), Al_2O_3 (1.41–2.50 wt%), MnO (0.13–0.35 wt%), Cr_2O_3 (0.40–1.07 wt%), and Na_2O (0.00–0.22 wt%). Besides the pyroxene compositions given in their Rock indication, one can find additional data in the work of Ryabov et al. (1985a, b, c). Pigeonite crystals ($Wo_7En_{70-69}Fs_{23-24}$) usually form growth rims on augite, preserving the augite crystal faces, and contain TiO_2 (0.3–0.32 wt%), Al_2O_3 (0.56–0.61 wt%), and Cr_2O_3 (0.06–0.09 wt%). Orthopyroxene ($Wo_4En_{72-71}Fs_{24-25}$) forms both separate crystals and growth rims on olivine and contains TiO_2 (0.66–0.82 wt%), Al_2O_3 (0.70–1.08 wt%), NiO (0.09–0.21 wt%), and Cr_2O_3 (0.18–0.19 wt%). As a whole,

Table 2.3 Chemical composition (wt%) of the Gudchikhinsky Suite picritic basalt differentiated lava flows

Component	1	2	3	4	5	6	7	8	9	10
SiO_2	44.47	47.47	56.84	45.01	43.75	43.08	44.22	43.87	46.22	43.49
TiO_2	1.57	2.00	0.20	1.56	1.49	1.15	1.54	1.53	1.61	1.44
Al_2O_3	9.86	12.71	25.71	11.11	9.94	7.38	10.36	10.35	10.26	9.10
Fe_2O_3	5.72	4.44	0.87	5.92	6.70	5.57	6.58	6.60	6.05	5.41
FeO	6.33	6.09	0.57	5.92	6.62	8.15	5.90	5.75	6.42	7.31
MnO	0.15	0.16	0.04	0.16	0.17	0.19	0.17	0.16	0.19	0.17
MgO	10.94	8.36	0.42	11.00	15.59	20.39	14.52	11.92	12.89	15.03
CaO	9.20	8.02	8.69	9.98	7.69	6.50	8.49	9.48	8.70	7.75
Na_2O	1.40	3.10	4.30	1.66	1.13	0.69	1.33	1.80	1.46	1.32
K_2O	0.14	0.95	1.23	0.24	0.24	0.23	0.28	0.21	0.18	0.22
P_2O_5	0.18	0.21	0.06	0.17	0.17	0.12	0.18	0.17	0.17	0.17
LOI	8.95	5.32	0.05	5.68	5.54	6.22	5.52	6.74	5.11	7.98
C.f.	52.41	55.74	77.42	51.84	46.07	40.22	46.22	50.89	49.17	45.84
Number of analyses	12	16	1	17	12	42	12	6	9	31

1—mandelsteine; 2—metadiorite; 3—labradorite; 4—olivine porphyritic basalt; 5—ankaramite; 6—picritic basalt; 7–10—average composition of the Talnakhsky picritic basalt lava flows (accordingly I-IV, drillhole Z-491)

pyroxenes in picritic basalts form along the augite–pigeonite–enstatite crystallization trend with a reduction in calcium content and increase in iron content in the minerals.

Plagioclase in picritic basalts forms prismatic crystals of various sizes and large xenomorphic laths. The latter are not zoned and are frequently twinned or contain hanging wedge-shaped twins. They are characterized by numerous poikilitic intergrowths of olivine and pyroxene. According to Fedorov's table, composition of plagioclase varies from anorthite An_{92} to labradorite An_{50}. According to microprobe analyses of large xenomorphic tabular crystals with poikilitic intergrowths, their composition corresponds to An_{67-65}, and the prisms correspond to An_{63-56} (see Rock indications 12, 13).

Chromite and ilmenite are common oxide ore minerals in picritic basalts. Chromite can be observed as an accessory mineral as small crystal intergrowths in all rock-forming minerals. The content of oxides in picritic basalts varies widely: TiO_2 (1.49–6.23 wt%), Al_2O_3 (3.04–14.46 wt%), Cr_2O_3 (18.20–48.12 wt%), Fe_2O_3 (6.73–36.16 wt%), FeO (21.85–29.29 wt%), MgO (2.27–8.9 wt%), and MnO (0.14–0.97 wt%) (Ryabov et al. 1985b). Peculiarities in the compositional variation of chromites from picritic basalts are clearly seen in the Al–Cr–(Fe^{3+}+2Ti) diagram (Ryabov 1999a, b), in which one can see that during the course of crystallization, a regular increase in titanomagnetic content takes place. Ilmenites of these rocks are characterized by high MgO contents (up to 5.46–6.56 wt%) and Cr_2O_3 (up to 0.54 wt%).

In addition, picritic basalts feature small irregular miarole cavities filled with a green chlorite-like mineral that is isotropic in the center and anisotropic at the border of each cavity. The cavities are non-uniformly distributed through the rock, dividing the enriched and depleted basalt. The quantity of miaroles varies from 1–3 to 10–15 vol%.

An indication of the chemical composition of various rocks found in picritic basalt lava flows can be obtained from Table 2.3, and the chemical composition for the complete section of the Talnakh differentiated lava flow sequence can be seen in Tables 1–3 (see the Appendix) and the variation diagram (Fig. 2.9). It follows that individual basalt flows in the flows of picritic basalts differ in chemical composition according to the quantitative ratios of main rock-forming minerals. To some extent, all the lava flows have a differentiated structure, which is most clearly manifest in the latest and thickest IV flow (see Table 2.3). Metadiorites here are characterized by high contents of TiO_2, Al_2O_3, and Na_2O, and picritic basalts are characterized by high MgO and FeO content and lower Fe_2O_3 and TiO_2 content. In picritic basalts of the IV flow, the content of rock-forming oxides is stable, higher concentrations of Cr_2O_3 and NiO are noted, and there is a trend for increasing oxidation of iron upward through the section.

2.4.1.1 Features of Petrology

Picritic lavas have the following characteristic features:

1. The overwhelming majority of picritic basalts flows have a differentiated structure.
2. Changing composition of lavas in various portions of the melt during the course of eruption.
3. Most magnesian and olivine-rich picritic lavas are from the final eruptions (Korovyakov 1948).
4. The first phases crystallized from picritic lavas were chromite and olivine.

5. The skeletal form and tiny size of olivine crystals in amygdaloidal zones points to its crystallization during eruption at the surface rather than in abyssal chambers.
6. The main rock types in differentiated lava flows are metadiorites and picritic basalts, with sharp contacts between them.
7. The picritic basalt horizon has a uniform distribution of rock-forming minerals and rock-forming oxides.
8. The most common association in picritic basalts is Fa_{19-23} + $Wo_{39-40}En_{50}Fs_{10-11}$ + $Wo_{6-8}En_{70-72}Fs_{21-23}$ + $Ab_{33-34}An_{67-56}$ + Cr-spinel.

2.4.1.2 Genetic Concepts

Among effusive bodies of various compositions, picritic basalts attract the special interest of geologists. They are used to aid in the determination of the parental magmas of Traps as a whole and for those of the Noril'sk mineralized intrusions in particular, as well as establishing mechanisms of differentiation in abyssal and hypabyssal chambers and existing ideas on comagmatism and petrogenesis. For many years, the interest of geologists in picritic lavas has remained strong, and the nature of these formations, despite being well-studied, remains in dispute. Wide use of outcrop relationships and the material composition of picritic basalts have helped in developing genetic concepts, including those connected with ore-bearing layered intrusions. Detailed study of these formations has enabled geologists to correct existing petrogenetic and ore-forming models.

Dodin (1967) explains the differentiated structure of the Val'kovsky sequence of picritic basalts by gravitational differentiation of olivine proto-segregations. Dodin (1967) additionally connects the formation of metadiorites in the upper parts of the sequence with enrichment of the melt with volatile material in the course of crystallization differentiation. A similar mechanism of gravitational separation of olivine, according to Duzhikov (1971), took place during formation of the Daldykansky flow sequence. In his opinion, metadiorites are only paragenetically connected with picrites in a single flow and they were formed due to late pneumohydrothermal processes, which took place million years after lava eruption synchronously with activation of the mobile Daldykansky zone. Later, the formation of fir-needlelike paniculate and spinifex-like crystals, along with the composition of the lavas themselves, was reason to classify these rocks as komatiites (Duzhikov and Distler 1981). The association of metadiorites with picritic basalts in differentiated lava flows was sufficient to propose the theory of abyssal mafic-ultramafic dissolution of magma into immiscible liquids (Marakushev et al. 1982). Simultaneously, the concept of within-flow crystallization–gravitational differentiation developed after observation of olivine separation and increased magnesium content from the upper to lower parts of the lava flow (Zolotukhin et al. 1986). Finally, the occurrence of large unzoned tabular plagioclase with intergrowths of olivine and pyroxene in picritic basalts allowed Likhachev (1997) to consider them relics of initial mantle material.

Taking into account the details mentioned above, and the available published data, one can suggest with confidence that the erupting picritic basalt magma was a homogenous, highly magnesian-rich liquid and did not contain olivine protosegregations. The first crystalline phases appeared with decreasing temperature as the melt began to solidify. The skeletal form and tiny size of early olivine and chromite crystals, along with the fast increasing viscosity of the melt (due to cooling), made it impossible for the gravitational accumulation of these minerals in the basal parts of the lava flows. Spinifex-like, dendritic crystals of minerals are evidence of fast crystallization under conditions of undercooling. The abundance of miarole cavities with palagonite I and, frequently, with calcite in all varieties of rocks of the lava flow sequence points at enrichment of the picritic basalt magma with volatile components. The splitting of the melt into metadiorite and picrite was due to fluid–magma differentiation, which occurred not in abyssal chambers but during its eruption. The same process created the high magnesium content of the later eruptions (flows) of picritic basalts.

2.4.2 Mikchandinsky Differentiated Basalt Flow

The lava flow sequence is situated in the vicinity of paleovolcano with the same name (Dodin and Golubkov 1971; Dodin 1967; Ryabov et al. 1977). It is unique in its subhorizontal alternation of large- and small-spherulitic pyroxenophyric basalts and horizons containing huge clinopyroxene dendrites (Fig. 2.10). Along the Piroksenovy Stream, the flow sequence is 13.5 m thick, and its structure is shown in Fig. 2.11. It is overlain by aphyric basalts. At the top of the Mikchandinsky flow sequence, picritic basalt flows have sharp, clearly visible contacts with aphyric basalt flows (see Rock indication 14). In the contact zone, aphyric basalt includes 5–6-cm-long dendritic clinopyroxene crystals that grow upward toward the picritic basalts, which are overgrown by olivine dendritic crystals 3–4 cm long (Fig. 2.12). In the lower parts of the section, the layered rock series comprises alternating horizons of large- and small-spherulite-rich pyroxene-phyri basalts and horizons with dendritic clinopyroxene megacrystals. The contacts of rocks containing spherulites with those containing dendritic clinopyroxene are always sharp.

Some idea of the content of the upper part of the Mikchandinsky flow sequence, shown in Fig. 2.12, can be obtained from Rock indications 14–18. Here is a brief

Fig. 2.10 Outcropping of the Mikchandinsky differentiated cover along the Pyroxenovy Stream, inflow of the Mikchanda River. The layered rock mass consists of altering spherolitic basalts and thin basalt layers (more *light-colored*) with dendrite-like clinopyroxene crystal

mineralogical and petrographic description of the rocks in the sequence.

Aphyric basalts overlying the rocks of Mikchandinsky flow sequence (see Fig. 2.12, zone) are an unaltered, uniform rock with a micro-ophitic texture and some elements of micropoikilophitic and microdoleritic texture (see Rock indication 14). Small clinopyroxene oikocrysts have a zoned appearance, and their composition from the center to the edges varies from $Wo_{36}En_{51}Fs_{13}$ to $Wo_{35}En_{40}Fs_{25}$. The basal rocks also include small angular grains of augite ($Wo_{35}En_{46}Fs_{19}$) and subidiomorphic zoned pigeonite crystals whose composition varies from $Wo_{11-10}En_{58-59}Fs_{31}$ to $Wo_{8-10}En_{51-49}Fs_{40-41}$. Augite contains TiO_2 (0.39–0.88wt %) and Al_2O_3 (1.40–1.87 wt%), and pigeonite contains TiO_2 (0.31–0.53 wt%) and Al_2O_3 (0.64–0.84 wt%). Prismatic plagioclase crystals correspond to bytownite $Ab_{28}Or_1An_{71}$, and microlites have a labradorite composition $Ab_{40}Or_2An_{58}$. Ilmenite is the sole ore mineral in the rock and has a composition that includes MnO (0.26–0.30 wt%) and MgO (1.27–1.50 wt%). Contact with the underlying zone is sharp and manifests itself as a decrease of rock grainsize with preservation of structural features.

Tholeiitic basalt with dendritic olivine (see Fig. 2.12, zone B) is a micro-ophitic rock that has some elements of intersertal and doleritic texture. In some areas, a trachytic texture was noted (see Rock indication 15). Dendritic olivine megacrystals grew through a fine-grained rock matrix, disturbing the trachytic texture and folding plagioclase laths into the olivine growth direction. Clinopyroxene is augite ($Wo_{39-36}En_{51-43}Fs_{11-20}$) and contains TiO_2 (0.32–1.26 wt%), Al_2O_3 (1.35–2.20 wt%), and Cr_2O_3 (0.17–0.61 wt%). The laths, and finer grains of plagioclase in the groundmass, have the composition $Ab_{27-30}Or_1An_{71-69}$. Ilmenite forms small prismatic and needlelike crystals in the rock matrix and contains MnO (0.38–0.41 wt%), MgO (3.68–3.84 wt%), and Cr_2O_3 (0.20–0.35 wt%).

Tholeiitic basalt with dendritic clinopyroxene has been studied in the upper part of the flow sequence (Fig. 2.12, zone C), at the boundary with the dendritic olivine zone in the lower part (Fig. 2.12, zone D), and at the contact with picritic basalts. The enclosing basalt has intersertal, doleritic, and micro-ophitic texture, and clinopyroxene megacrystals grow throughout the rock without affecting the rock fabric.

In zone C (see Rock indication 16), the matrix surrounding the dendritic clinopyroxene contains augite ($Wo_{40}En_{49}Fs_{11}$, with TiO_2 (0.31 wt%), Al_2O_3 (1.64 wt%), and Cr_2O_3 (0.33 wt%)), microlites of plagioclase $Ab_{27-30}Or_{1-2}An_{72-68}$, and ilmenite containing MnO (0.42 wt%), MgO (0.99–1.85 wt%), and Cr_2O_3 (0.16–0.17 wt%). Megacrysts of clinopyroxene, typically augite ($Wo_{40-36}En_{49-53}Fs_{11-12}$), contain TiO_2 (0.23–0.34 wt%), Al_2O_3 (1.35–1.99 wt%), and Cr_2O_3 (0.34–0.46 wt%). Sometimes, dendritic clinopyroxene is overgrown by long plagioclase prisms that changed composition as they grew, from ($Ab_{46}Or_3An_{51}$) to bytownite ($Ab_{25}Or_1An_{74}$).

In zone D (see Rock indication 17), the rock matrix is composed of small grains of augite ($Wo_{39}En_{50}Fs_{11}$), laths of labradorite ($Ab_{29}Or_2An_{69}$), titanomagnetite, and ilmenite. Clinopyroxene contains TiO_2 (0.30 wt%), Al_2O_3 (1.72 wt%), and Cr_2O_3 (0.45 wt%), and ilmenite contains MnO (0.24–0.40 wt%) and MgO (1.42–2.25 wt%). Dendritic clinopyroxene that grew between the contact of picritic and aphyric basalts has an augite composition ($Wo_{39-36}En_{50-52}Fs_{11-12}$) and contains TiO_2 (0.25–0.35 wt%), Al_2O_3 (1.29–2.18 wt%), and Cr_2O_3 (0.38–0.50 wt%). Laths of bytownite ($Ab_{23}Or_1An_{76}$) form intergrowths in dendritic clinopyroxene, and small subidiomorphic zonal crystals of pigeonite $Wo_8En_{63-56}Fs_{29-36}$ are also observed among the dendritic crystals.

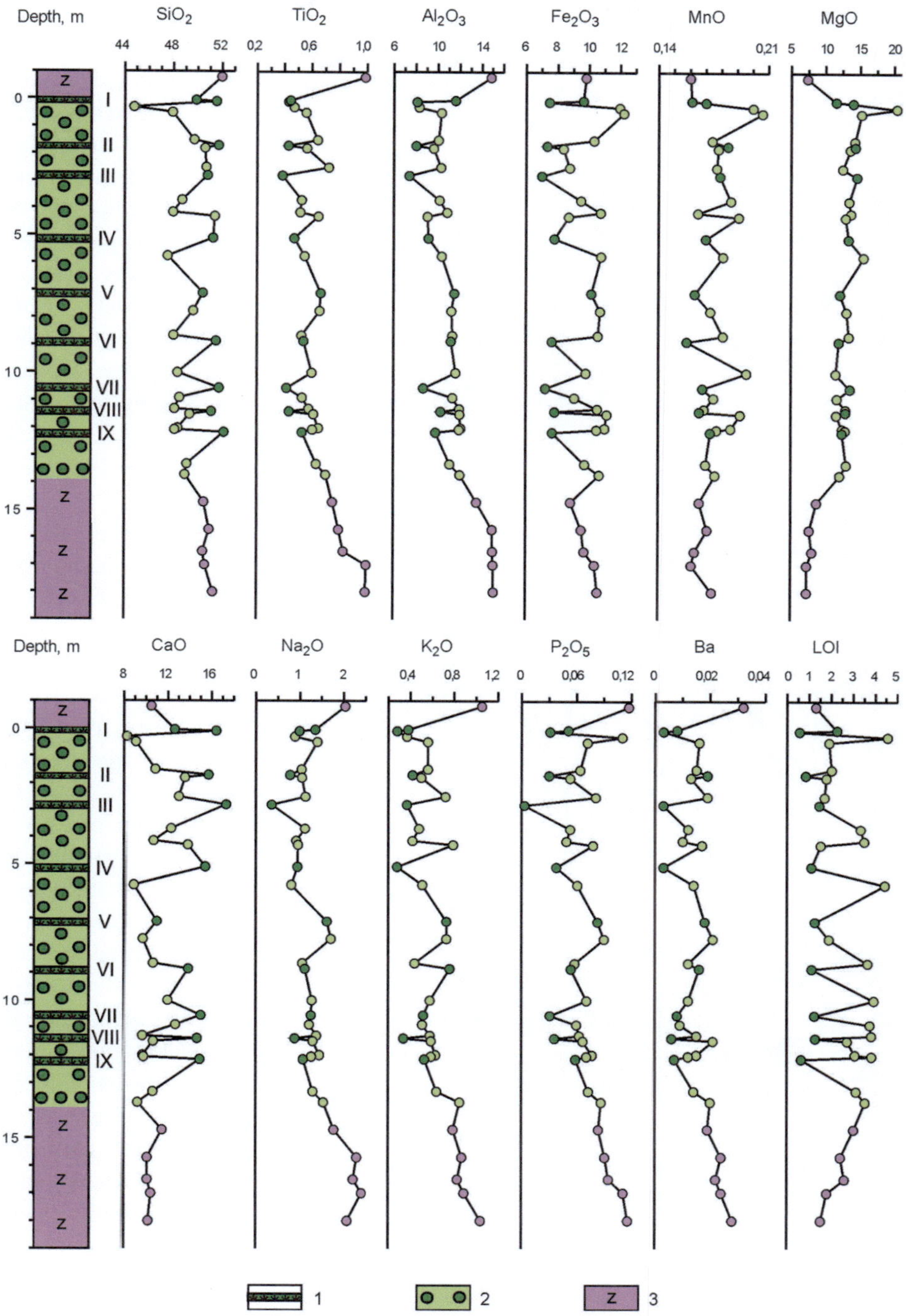

Fig. 2.11 Variation petrochemical diagram (wt%) for the Mikchandinsky sequence of lava flows along the Pyroxenovy Stream (inflow of the Mikchanda River). Painted circles correspond horizons with dendritic clinopyroxene megacrystals (for details, see the text of the book), unpainted—aphyric basalts and basalts containing clinopyroxene spherocrysts. *1* Interbeds with dendrite-like clinopyroxene crystals, *2* basalts with clinopyroxene spherocrysts, *3* aphyric basalts

Picritic basalt (see Rock indication 18), which underlies the layer with dendritic olivine and clinopyroxene, possesses seriate–porphyritic to porphyritic textures with micro-ophitic, doleritic, and poikilitic texture in the groundmass. Olivine forms crystals of various sizes: large subidiomorphic crystals with chromite intergrowths, and medium and small crystals that are isometric and slightly elongated and contain plagioclase inclusions. The composition of olivine changes from Fa_{23} to Fa_{28}, with NiO content from 0.12 to 0.06 wt%. Clinopyroxene ($Wo_{40–38}En_{50–49}Fs_{11–13}$) in the groundmass contains TiO_2 (0.18–0.37 wt%), Al_2O_3 (1.54–1.66 wt%), and Cr_2O_3 (0.28–0.63 wt%) and forms angular areas of subophitic intergrowth with plagioclase. Small grains from groundmass are augite ($Wo_{36}En_{49}Fs_{15}$), containing TiO_2 (0.57 wt%), Al_2O_3 (1.20 wt%), and Cr_2O_3 (0.06 wt%), and pigeonite ($Wo_8En_{69}Fs_{23}$). Pigeonite is also observed as overgrowths grown on olivine crystals. Olivine in picritic basalts is replaced by bowlingite, opaque minerals, and frequently clinopyroxene (Ryabov 1992a, b, c). The composition of clinopyroxene after olivine corresponds to $Wo_{39–40}En_{50}Fs_{11–10}$. Cr-spinel was determined to contain TiO_2 (1.87–2.12 wt%), Al_2O_3 (15.82–20.0 wt%), MgO

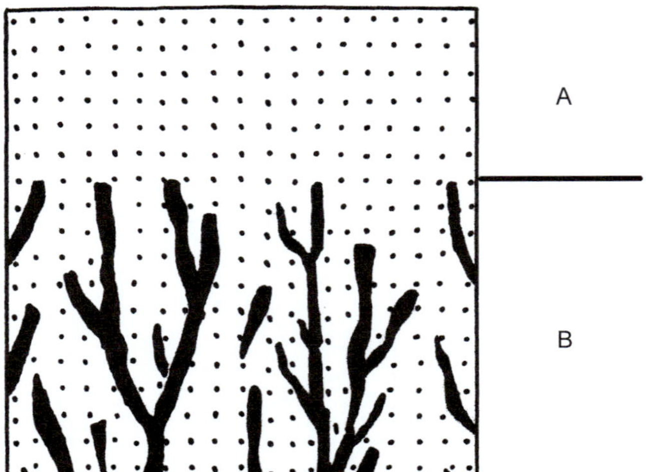

Fig. 2.12 Schematic view of the upper contact of the Mikchandinsky sequence of lava flows (sample MCH-60). *A* Aphyric tholeiitic basalt; *B* tholeiitic basalt with olivine dendrites; *C*, *D* (*E*) tholeiitic basalt with clinopyroxene dendrites; *E* picritic basalt. Composition of the rocks is given in Rock identification 14–18. For details, see the text of the book

(6.05–8.77 wt%), and Cr_2O_3 (36.90–38.90 wt%), and ilmenite contains MnO (0.35–0.39 wt%), MgO (5.22–5.25 wt%), and V_2O_5 (0.82 wt%).

Spherulitic pyroxenophyric basalts are composed of large spherical segregations of radially diverging prismatic clinopyroxene crystals and, as a rule, are found in picritic basalts (Fig. 2.13). In the center of the clinopyroxene spherocrysts, olivine seed crystals, typically replaced with bowlingite, are located. Disk-shaped subhorizontally elongated cavities (0.5 × 1.5 cm to 1.5 × 7.0 cm in size) with kelyphitic clinopyroxene rims about 0.5 cm wide are also rarely observed. These cavities contain palagonite, sometimes with calcite, or clinopyroxene–plagioclase intergrowths, forming miarolitic micropegmatites. The size of spherulites varies from 0.3 to 6.0 cm, though are typically 1.5–2.5 cm. Depending on the size, small (with diameters of spherocrysts 0.3–0.7 cm)- and large-spherulite (1.5–2.5 cm) pyroxenophyric basalts are distinguished, forming subhorizontal layers of constant thickness. Within each separate layer, the size and quantity of spherulites are similar; however, their quantity, across various layers, changes from 10–30 to 60–70 vol%. Less commonly, aggregates or continuous chains of large spherocrysts in the lower part of a layer are observed. In these cases, the spherical nature of the spherocrysts is locally disturbed by long dendrite-like clinopyroxene crystals (see Ryabov et al. 1977, Fig. 1.2c). The thickness of layers containing small- and large-spherulite basalts varies from 10–15 cm to 0.5–0.7 m.

In pyroxene-phyric basalts relatively poor in spherulites, relicts of unaltered picritic basalts are sometimes observed. In the matrix of small-spherulitic basalt, relicts of alternating picritic basalt are observed. The latter have a porphyritic texture, olivine (20–40 vol%) is replaced with bowlingite or opaque minerals, and the groundmass has a micro-ophitic and doleritic texture. The rock also features crystal aggregates and spherocrysts of brownish-green clinopyroxene ($Wo_{39}En_{50}Fs_{11}$), as well as tiny prisms and relict rims of orthopyroxene ($Wo_8En_{68}Fs_{24}$) around olivine (Ryabov et al. 1977). Palagonite is crystallized as amoeba-like segregations in the groundmass. In the upper part of the sequence of lava flows, among small-spherulite basalts, are lenses of metadiorite and olivine-phyric basalt (Ryabov et al. 1977; Kavardin 1988). In the same parts of the lava flow sequence, thin (7–15 cm) lenses of "bubbly" rock feature amygdales filled with zeolite and prehnite.

In large-spherulite basalts, the presence of translucent structures composed of picritic basalt in the matrix is rather rare. In most cases, the space between spherocrysts has a tholeiitic texture with palagonite instead of glass. Rarer are areas with bowlingite pseudomorphs after olivine phenocrysts in the groundmass.

Detailed studies of spherulitic pyroxene-phyric basalts were carried out on two samples (see Rock indications 19–20). They showed that spherocrysts are typically augite ($Wo_{40-35}En_{52-46}Fs_{10-19}$) that changes in composition from the center to the edges of the crystals in cross section and in the elongation direction in feather-like cases. The mineral contains TiO_2 (0.25–1.31 wt%), Al_2O_3 (1.32–2.23 wt%), and Cr_2O_3 (0.13–0.63 wt%). Olivines located in the centers of augite spherocrysts are Fa_{22} and contain NiO (0.10–0.14 wt%). Olivine crystals are frequently overgrown by kelyphitic rims of pigeonite ($Wo_8En_{73-61}Fs_{19-31}$) and contain tiny inclusions of Cr-spinel that contains TiO_2 (1.99–3.45 wt%), Al_2O_3 (12.32–15.51 wt%), MgO (6.82–8.03 wt%), and Cr_2O_3 (40.66–42.77 wt%).

Fig. 2.13 Spherolitic basalt. Clinopyroxene in the overgrowth rims on olivine crystals and miarolic cavities with micropegmatite, the Mikchandinsky sequence of lava flows. The sample is 10 × 22 cm in size

The matrix in the spherulitic basalts consists of phenocrysts of olivine (Fa_{27-29}, with NiO 0.07–0.08 wt%), small grains of augite ($Wo_{39-37}En_{50}Fs_{11-13}$), and prismatic crystals of zoned plagioclase ($Ab_{22-43}Or_{1-3}An_{77-55}$). In the rock, low quantities of titanomagnetite, ilmenite, and matrix, solitary apatite crystallites formed in palagonite. Augite in the rock mesostasis contains TiO_2 (0.34–0.41 wt%), Al_2O_3 (1.42–1.58 wt%), Cr_2O_3 (0.14–0.32 wt%), CeO_2 (0.39 wt%), F (2.89 wt%), and Cl (0.05 wt%); titanomagnetite contains TiO_2 (6.03 wt%), Al_2O_3 (0.77 wt%), MnO (0.10 wt%), and MgO (0.32 wt%); ilmenite contains MnO (0.26–0.37 wt%), MgO (1.08–4.87 wt %), and Cr_2O_3 (0.01–0.33 wt%).

Dendritic clinopyroxene crystals grew in the Mikchandinsky lava flow sequence, forming a series of subhorizontal layers 1–32 cm thick. In total, nine layers are defined in even the thinnest flow in the exposure. The layer has a poikilotaxitic texture, formed by dendrite-like megacrystals of light-green clinopyroxene that grew from the bottom of the layer to the top, in a micro-ophitic or microdoleritic groundmass of aphyric basalt. Saturation of the rocks with clinopyroxene varies from layer to layer. Thin layers consist entirely of prismatic clinopyroxene crystals (Fig. 2.14), and thicker layers are either composed of sprays or uniformly dendritic pyroxene. In all cases, the dendritic matrix is formed in unaltered aphyric basalt, and dendritic clinopyroxene is free from plagioclase intergrowths. In some cases, in the top layer of dendrites, long plagioclase prisms grow between diverging clinopyroxene megacrystals.

Dendrites have an augite composition which varies between that in the lower parts of the layer ($Wo_{39-40}En_{46-48}Fs_{13-14}$) and in the upper parts of the layer ($Wo_{38-40}En_{46-49}Fs_{11-16}$). Oxide content in the augite has been measured: TiO_2 (0.60–1.12 wt%), Al_2O_3 (3.34–5,21 wt%), Cr_2O_3 (0.30–0.62 wt%), MnO (0.06–0.17 wt%), and Na_2O (0.32–0.56 wt%) (Ryabov et al. 1977).

Thermobarometric–geochemical investigations performed by Bakumenko for melt inclusions in dendritic clinopyroxene crystals have shown that their homogenization occurred at 1,230–1,240° (Ryabov et al. 1977). Chromatographic analysis of clinopyroxene from dendrites (undertaken at IG UIGGM SB RAS by Konenko) points at the presence of a CH_4, CO, N_2, CO_2, H_2O, and O_2 gaseous mixture.

At the base of the Mikchandinsky sequence of lava flows is a layer of large-spherulitic pyroxene-phyric basalt containing a chain of clinopyroxene crystals along its lower contact. Down the section, about 0.3–0.5 m, in aphyric basalts, rare single clinopyroxene spherocrysts are noted, but farther down the section they are absent.

Geochemistry Typical variations in the chemical composition of rocks from the Mikchandinsky sequence of lava flows can be obtained from the variation diagram (see Fig. 2.11) and Table 4 (see the Appendix). Earlier work on the petrochemistry of the rocks of the lava flow sequence was discussed in Dodin and Golubkov (1971), Dodin (1967), and Ryabov et al. (1977). A study of the Keta-Irbinsky and Glubokoozersky lava flow sequences was published by Kavardin (1988).

As one can see in the variation diagram, the composition of rocks in the Mikchandinsky sequence of lava flows varies widely for all main rock-forming oxides, and these variation contents accord with changing petrographic features of rocks. The highest MgO concentrations take place in picritic basalts and spherulitic pyroxene-phyric basalts, in which picritic basalt relict textures were preserved. Enrichment of the rocks with clinopyroxene spherocrysts correlates with content, which reaches its maximum in layers containing dendritic clinopyroxene. Similar behavior is demonstrated by Cr_2O_3, which accumulates in pyroxene crystals. Cr-spinel are preserved only in unaltered relicts of picritic basalt in the top of the flow, whereas they are absent in spherulitic basalts. High concentrations of Al_2O_3 are characteristic for metadiorites enriched with feldspars, in which $Na_2O + K_2O$ content reaches 5.68 wt% (Ryabov et al. 1977). Low silica content is observed in rocks rich with clinopyroxene and olivine. Distribution of SiO_2 in the sequence of lava flows is controlled by the quantity of olivine and its alteration products, as well as by pyroxene and plagioclase. Content of SiO_2 is low when olivine content is high and high when the pyroxene and plagioclase content is high.

Aphyric basalts, overlying and underlying the Mikchandinsky sequence of lava flows and situated among

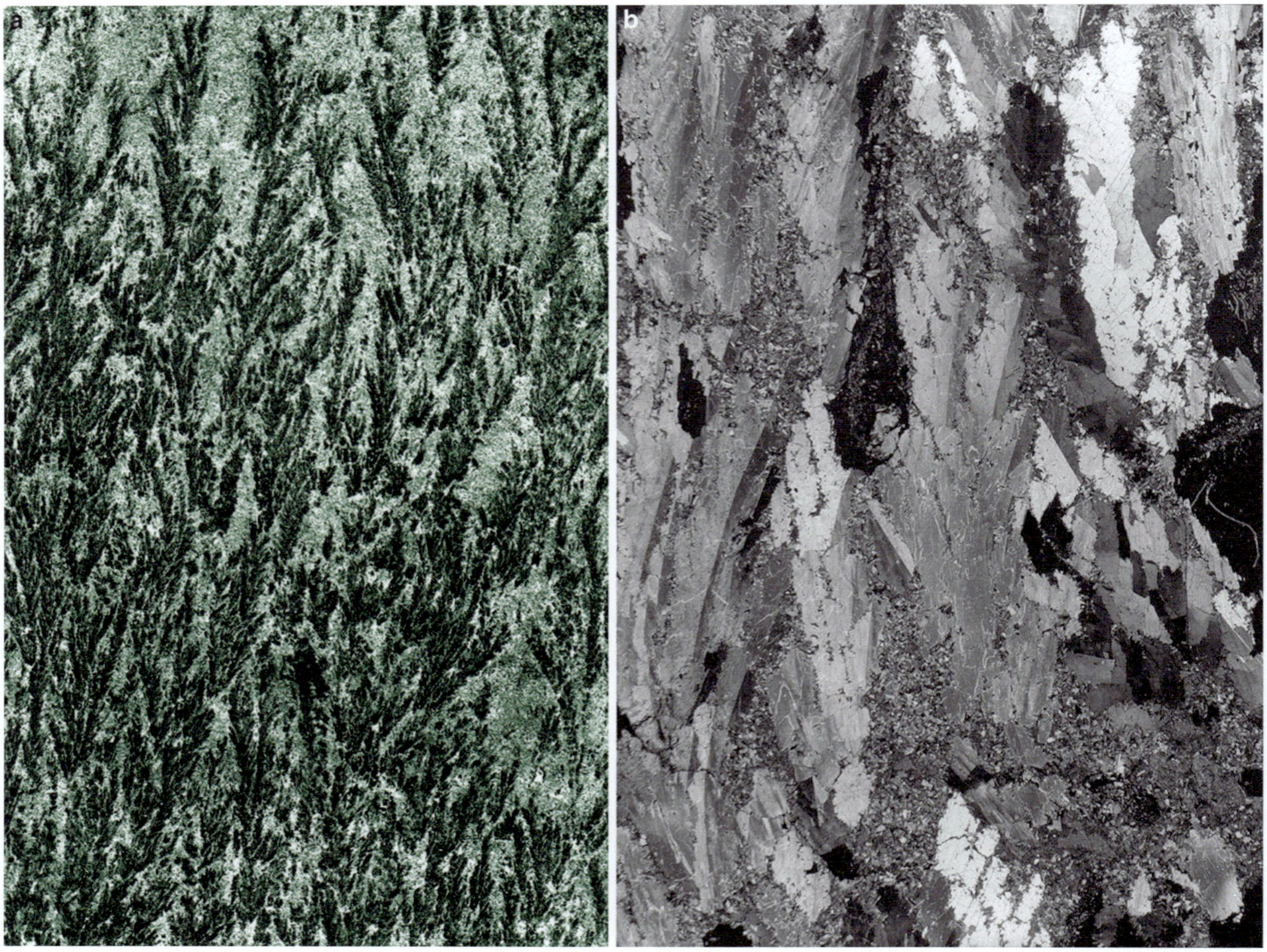

Fig. 2.14 Dendrite-like clinopyroxene megacrystals, the Mikchandinsky sequence of lava flows. (**a**) Clinopyroxene dendrites in aphyric basalt (natural scale), (**b**) a detail of the structure, the variety of clinopyroxene forms in megacrystals (polarized light, double amplification)

dendritic clinopyroxene, have similar composition and correspond to tholeiitic flood basalts.

2.4.2.1 Petrological Characteristics

Listed below are characteristic features of rocks of the Mikchandinsky sequence of lava flows that are important for gaining an understanding of its genesis:

1. The Mikchandinsky sequence of lava flows and the similar Keta-Irbinsky and Glubokoozersky sequence of lava flows (Kavardin 1988) have an expansive areal range, allowing us to regard them as lava lakes, situated in vicinity of a center of major volcanism.
2. The lack of temporal or erosional relations between the rocks of the Mikchandinsky sequence of lava flows and the enclosing aphyric basaltic rocks serves as conclusive evidence of the simultaneous eruption or intrusion of immiscible picritic and tholeiitic magmatic melts.
3. The lava flow sequence has a layered texture, resulting from the alternation of large- and small-spherulite basalts with aphyric basalts containing dendritic clinopyroxene.
4. There are relict fragments of picritic basalt, with their composition close to that of cognominal rocks of the Gudchikhinsky Suite.
5. The morphology of clinopyroxene crystals points to their formation after eruption of magma and crystallization during conditions of undercooling.
6. The relict fragments of alternating picritic basalt at the base of spherulitic pyroxene-phyric basalts are an evidence of heteromorphic crystallization of the picritic basalt melt.
7. The formation of spherocrysts took place around olivine crystals and gas bubbles that played the role of seeds, while dendritic crystals grew in tholeiitic (aphyric) basalts vertically from the contact with picritic basalts.

2.4.2.2 Current Genetic Ideas

Geologists have presented many hypotheses on the genesis of the Mikchandinsky sequence of lava flows. Dodin and Golubkov (1971) suppose eruption of a melt enriched with clinopyroxene and partially enriched with olivine protocrysts, due to abyssal crystallization–gravitational differentiation. During the movement of this heterogeneous melt, dynamic differentiation took place in feeder pipes. The authors believe this predetermined the layered structure of the sequence as well as the variation in clinopyroxene content in the rocks. Other groups of geologists regard the rocks of the Mikchandinsky sequence of lava flows and above-mentioned sequences as komatiites or komatiite-like (Duzhikov and Distler 1981; Zolotukhin et al. 1986; Kavardin 1988). The main arguments for this interpretation are the presence of spinifex texture and the rock composition, which is close to the published data on Australian and Canadian komatiites. Ryabov et al. (1977) stated that the initial melt composition of the Mikchandinsky sequence of lava flows had a composition close to that of the Gudchikhinsky picritic basalts, and the differences between the two rock groups were due to specific conditions of magma crystallization that took place near the erupting volcano—the erupting melt was layered and possessed high fluidity. Abrupt changes in thermodynamic conditions caused the unusual character observed in the crystallizing melt, creating a distinct heteromorphism.

Spinifex-like clinopyroxene and dendritic olivine are known from a number of other trap intrusions (Ryabov et al. 1977). Their presence points to quick crystallization in conditions of undercooling, and their orientation indicates the cooling front. Crystallization of dendritic clinopyroxene in layers of tholeiitic basalt within fluidized picritic basalt is probably connected to the gaseous–gravitational redistribution of mineral-forming components. Picritic basalts underlying tholeiitic basalts served as the base from which dendritic crystal growth began.

Preferential crystallization of clinopyroxene in nonequilibrium conditions results from its lower activation energy and, hence, wider range of crystallization on undercooling (340°c) in comparison with plagioclase, for example (Rashin 1963). This was discussed in detail by Ryabov et al. (1977). It is important to note that the growth of clinopyroxene spherocrysts was not from a homogenous liquid, but rather took place on olivine seeds. The source of the chemical elements necessary for the formation of clinopyroxene spherocrysts was the picritic basalt magma itself, in which crystallization of olivine and chromite had already begun. Under the conditions of undercooling, early crystal phases became unstable and underwent alteration and dissolution. As a result, the melt was enriched with magnesium and chromium, which later entered clinopyroxene crystals. It is worthy to note that the formation of pyroxene after olivine, as a result of high-temperature oxidation of olivine in the magmatic state, was noted in picritic basalts of the Hawaiian Islands and in dolerites of Karoo (Ryabov 1992b).

2.5 Khakanchansky Suite (T_1hk)

The rocks of the Khakanchansky Suite comprise tuffaceous deposits with sporadic, thin flows of greenstone-facies poikilophitic basalts. Lithologically, the rocks of the suite vary widely, both vertically and horizontally (Nemenenok and Vilensky 1978). Within the main part of the Noril'sky, Vologochansky, and Kharaelakhsky troughs, the rocks of the suite, also known as "the second Rozhkovsky horizon," have a roughly constant thickness (15–22 m) and represent alternating psammitic and aleuritic tuffs, tuffites, tuffaceous argillites, and tuffaceous sandstones. On the northern shore of the Khantayskoe Lake, distinctive ripple marks are observed in the tuffs as well as other textural features that have been attributed to rain drop imprints. In the vicinity of the Fokina River and the Lama Lake, the thickness of the suite is 100–150 m, and near the Keta and Khantayskoe Lakes, it extends to 250–370 m in thickness. The thickest sequences of tuffaceous rocks are confined to areas defined as paleovolcanoes.

In the Burkan Mountains (near the Irbo River) among tuff breccias, small-scale lenses of pillow lava (Fig. 2.15) 7–10 m thick are exposed.

2.6 Tuklonsky Suite (T_1tk)

Rocks of the Tuklonsky Suite comprise greenstone-facies poikilophitic intersertal and tholeiitic basalts (up to 8–10 lava flows) with subordinate olivine-rich, picritic (up to 3 lava flows), and porphyritic basalts with rare, thin tuff layers. The suite is located in the eastern part of the Noril'sk Region and is traced almost without interruption from the northern margin of the Kharaelakhsky trough to upstream of the Kulyumbe River. Its thickness within the Noril'sky trough reaches 170 m, in the northeast of the Kharaelakhsky trough 250 m, and in the northwest of the Tungussky syneclise 320 m (Ivanova 1975; Nemenenok and Vilensky 1978; Zolotukhin et al. 1986). Picritic and olivine-phyric basalts narrow locally in the basal parts of the thickest sections of the suite. Thickness of picritic basalts in the vicinity of the Lama and Glubokoe Lakes approaches 60–65 m (Distler and Kunilov 1994).

Fig. 2.15 Pillow lavas of Khakanchansky Suite at the Burkan Mountain

Rocks of the suite are characterized as high-magnesium basalts, which have been named elsewhere as picritic or olivine-phyric basalts. The rocks have a porphyritic, poikilophitic texture with a micro-ophitic groundmass and consist of olivine (10–20 vol%), plagioclase (25–40 vol%), clinopyroxene (35–40 vol%), palagonite and chlorite-like minerals (5–15 vol%), and oxide ore minerals (5–10 vol%). Olivine ($Fa_{17–20}$) forms rare porphyritic segregations of subidiomorphic crystals and isometric grains of various sizes in groundmass. Plagioclase ($Ab_{23}Or_1An_{76}$) forms prismatic crystals, whereas clinopyroxene forms small oikocryst ($Wo_{42}En_{47}Fs_{11}$) and tiny grains ($Wo_{38}En_{47}Fs_{15}$) in the groundmass (see Rock indication 21). An increase in iron content in the minerals during its crystallization was accompanied by increase of TiO_2 from 0.49 to 0.68 wt% and a decrease of Cr_2O_3 from 0.94 to 0.18 wt% and Al_2O_3 from 3.08 to 2.17 wt%.

According to the data published by Duzhikov (Zolotukhin et al. 1989), the main mineral associations of the Tuklonsky picritic (olivine-phyric) basalts are Ol ($Fa_{22–32}$) + CPx($Fs_{12–32}$) + OPx($Fs_{23–32}$) + Pl($An_{54–58}$) + Cr-Sp. Unlike the Gudchikhinsky picrites, Tuklonsky picrites are characterized by higher TiO_2, Fe_2O_3 + FeO, K_2O, and P_2O_5 content and lower Al_2O_3 and MgO content—MgO content shown as 20.39 wt% for the Gudchikhinsky picrites and 14.07 wt% for the Tuklonsky picrites (Table 2.4) (Zolotukhin et al. 1986). According to the present criteria (Bogatikov et al. 1981; Le Maitre 1989), picrites are characterized by a magnesium content higher than 18.0 wt%, so the magnesium-rich basalts of the Tuklonsky Suite correspond to olivine-phyric basalts rather than picrites.

Table 2.4 Magnesian effusive composition (wt%) of the northern part of the Siberian Platform

	gd	tk	mr	an	mk	mm	ar
	1	2	3	4	5	6	7
SiO_2	43.20	44.97	46.35	44.24	44.77	40.75	38.86
TiO_2	1.20	0.82	2.16	2.88	3.58	2.61	1.41
Al_2O_3	7.47	12.74	10.61	7.01	6.95	4.66	6.79
Fe_2O_3	6.12	4.16	7.66	4.99	15.57[a]	7.22	4.43
FeO	8.06	6.80	6.50	9.01	–	6.09	8.37
MnO	0.15	0.17	0.19	0.20	0.22	0.18	0.24
MgO	20.20	14.07	10.18	18.16	16.22	19.47	20.69
CaO	6.78	9.76	10.09	9.65	8.32	10.85	13.86
Na_2O	0.36	1.20	1.79	1.39	1.14	0.36	0.56
K_2O	0.12	0.41	0.55	0.75	0.64	1.38	1.40
P_2O_5	0.07	0.09	0.26	0.18	0.37	0.63	0.06
LOI	6.37	4.60	3.64	1.77	2.42	4.40	2.06
Number of analyses	1	12	17	1	6	3	2

Data from Zolotukhin et al. (1986) with attachments from authors
1— picritic basalt of the Gudchikhinsky Suite; 2—picritic basalt of the Tuklonsky Suite; 3—ankaramite basalt of the Morongovsky Suite; 4—picritic basalt of the Ayansky Suite; 5—limburgite of the Mokulaevsky (Khonnamakitsky) Suite; 6—picritic porphyrite of the Maymechinsky Suite; 7—melilitic picritic porphyrite of the Arydzhangssky Suite
[a]Fe_2O_3 + FeO

Olivine-phyric (picritic) basalts with dendritic growth of olivine, plagioclase and partially dendritic clinopyroxene crystals of the Tuklonsky Suite have similar petrographic and chemical composition throughout the region, excluding olivine-phyric basalts of the Sunduk Mountain. Until now, information about this unique lava flow sequence had not been published.

2.6.1 Differentiated Lava Flow Sequence of Olivine-Phyric Basalts of the Sunduk Mountain

Rocks of the Tuklonsky Suite on Sunduk Mountain comprise (from the base up) tholeiitic basalts (2–4 lava flows), tuff breccias with coal fragments, agglomerate tuffs with the lenses of tholeiitic and olivine-phyric (picrite) basalts (1 flow), and again by tholeiitic basalts with eroded upper portions.

The olivinophyric (picritic) basalts occupy a total thickness of about 20 m; the upper (8 m) and lower (8 m) portions are uniform, whereas the central part (4 m) features subhorizontal alternating layers of picritic basalt and layers containing dendritic olivine and plagioclase crystals. Thickness of the layers containing dendritic crystals varies from 4 to 13 cm, up to 30 cm near the upper layer, and the thickness of picritic basalt layers ranges from 0.2 to 2.8 m.

At the base of the olivine-phyric (picritic) basalt lava flows is porphyritic basalt with a hyalopilitic, intersertal, and micro-ophitic groundmass texture. Porphyritic segregations are represented by olivine, plagioclase, and more rarely by clinopyroxene. Olivine (5–7 vol%) forms as tiny (0.3–1.0 mm) crystals and accumulations of grains completely replaced by bowlingite. Plagioclase forms individual prisms or their glomeric accumulations. The groundmass comprises plagioclase laths, small clinopyroxene grains, altered glass, and palagonite. In the lower part of the sequence of flows, there is a glassy etched zone (about 20 cm thick), with subvertical tube-like amygdales, replaced by thin amygdaloidal zone with cavities filled with green palagonite, a radial chlorite-like mineral, dendritic calcite, Iceland spar, and frequently by garnet.

Olivine-phyric (picritic) basalts have a porphyritic and ophitic–poikilophitic texture. The rocks comprise olivine (5–25 vol%), clinopyroxene (25–30 vol%), plagioclase (40–60 vol%), palagonite (5–15 vol%), and oxide ore minerals (3–7 vol%).

Olivine forms rare subidiomorphic porphyritic segregations and small isometric grains and aggregations, which formed in the groundmass and intergrew with poikilitic clinopyroxene. In the central part of the sequence, alternating rock layers have varying olivine content (from about 5 vol% up to 20 vol%). In the same part of the sequence lie picritic basalts that are richest in olivine content (up to 30–40 vol%). Olivine composition is Fa_{17-20}.

Clinopyroxene forms large poikilocrysts and fine-grains in the rock matrix. In the upper part of the sequence of lava flows, clinopyroxene oikocrysts are frequently twinned and contain small quantities of plagioclase and pyroxene chadacrysts. Composition of clinopyroxene is $Wo_{42-38}En_{47-46}Fs_{11-16}$ (see Rock indication 21). Plagioclase ($Ab_{23}Or_1An_{76}$) forms rare porphyritic segregations and fine prismatic crystals in the groundmass. Ore minerals form small grains at the base of the flows. Chrome spinel contains TiO_2 (1.22–6.25 wt%), Al_2O_3 (4.63–18.51 wt%), Fe_2O_3 (12.85–38.98 wt%), Cr_2O_3 (14.68–34.60 wt%), FeO (25.81–31.24 wt%), and MgO (4.56–7.07 wt%) (Zolotukhin et al. 1986). The most chromium-rich variety has high concentrations of Al_2O_3 and MgO and low concentrations of TiO_2, Fe_2O_3, and FeO.

2.6.1.1 Olivine-Phyric Basalts with Dendritic Olivine and Plagioclase Crystals

Layers of this rock have micro-ophitic, poikilophitic, tholeiitic, and phylotaxitic textures. The latter, like rocks of Mikchandinsky lava flow sequence, are characterized by a porphyritic texture and dendritic phenocrysts. Dendritic layers on Sunduk Mountain are typically only faintly dendritic, usually composed of inclusion-free olivine, plagioclase, and more rarely clinopyroxene (up to 4–5 cm long), growing from the bottom upward.

Olivine-phyric basalt enclosing dendritic layers consists of olivine (7–15 vol%), plagioclase (50–60 vol%), clinopyroxene (20–30 vol%), palagonite and chlorite-like minerals (5–7 vol%), and magnetite (1–3 vol%). The tribute of dendrites in the bulk of minerals is olivine (3–10 vol%), plagioclase (3–15 vol%), and clinopyroxene (0–3 vol%).

Dendritric olivine consists of prismatic crystals bifurcating sequentially during growth. Fine dendritic olivine crystals are also found within the rock groundmass. Dendritic olivine crystals have a composition of Fa_{29-30} and a NiO content of 0.14 wt%. As a rule, olivine is replaced with bowlingite and iddingsite (see Rock indication 22). The rock matrix consists of small angular grains of augite ($Wo_{38}En_{49}Fs_{13}$) that contain TiO_2 (0.47 wt%), Al_2O_3 (1.3 wt%), Na_2O (0.26 wt%), and Cr_2O_3 (0.12 wt%), and prisms of bowlingite ($Ab_{22}Or_1An_{77}$.) Ilmenite has a higher content of MgO (5.5 wt%), V_2O_5 (1.1 wt%), and Cr_2O_3 (0.13 wt%).

Viewed under a microscope, dendritic plagioclase is revealed by its curved shape and undulating extinction. Bifurcation of crystals takes place along twinning boundaries and cleavage planes. The mineral content corresponds to bytownite ($Ab_{25}Or_1An_{74}$) (see Rock indication 23). Dendritic crystals of clinopyroxene ($Wo_{40}En_{49}Fs_{11}$) are rare as are clinopyroxene crystals intergrowing with plagioclase dendrites. Fine grains of the clinopyroxene in the groundmass have the composition $Wo_9En_{65}Fs_{26}$. Dendritic clinopyroxene crystals contain TiO_2 (0.41 wt%), Al_2O_3 (1.59 wt%), and Cr_2O_3 (0.44 wt%). Ilmenite from this rock is also high in MgO (4.53–6.73 wt%) and V_2O_5 (1.01–1.04 wt%).

2.6.1.2 Genetic Concepts

Based on the presented materials, one can suppose that the differentiated structure of rocks of the Sunduk Mountain is due to the initially layered character of erupting magma, which formed streams of melt with varying magnesium content. Dendritic crystal forms presumably reflect undercooling of the melt, whereas the degree of restriction of dendritic plagioclase and olivine to different layers is connected to changes in composition between the layers, as well as (probably) with different regimes of undercooling.

2.7 Nadezhdinsky Suite (T_1nd)

Within the Noril'sk Region, basalts of this stratigraphic unit are pervasive. The rocks of the suite are represented by tholeiitic, porphyritic, glomeroporphyritic, and aphyric basalts with rare tuff layers. Its thickness decreases from the west to the east and amounts to 530–570 m in the Noril'sky and Vologochansky troughs and 150–250 m in the northeast of the Kharaelakhsky trough and in the western margin of the Tungussky syneclise. The suite is divided into three units.

The lower unit is built up of 10–14 lava flows of porphyritic (polyphyritic, plagioclase-phyric) and aphyric basalts. Porphyritic varieties of basalts prevail in the west of the Noril'sk Region and tholeiitic varieties in the east. At the top of the subsuite, it is common to find a series of tholeiitic basalts and a layer of tuffaceous rocks, and at the base of the subsuite in the vicinity of Talnakh, there is a series of polyphyritic trachybasalts.

The middle unit constitutes 10–13 flows of porphyritic and more rarely polyphyritic, glomero-oligophyric, tholeiitic, and poikilophitic intersertal basalt. At the top of the subsuite, aphyric basalts are sometimes noted, and at the bottom a flow of glomeroporphyritic basalts is underlain by a horizon of tuffaceous rocks.

The upper unit consists of 1–8 flows of glomeroporphyritic basalts with a horizon of tuffaceous rocks at its base.

In different structure-facial subzones, some specific features and facial changes were registered. In the Lama-Imangdinsky and Khantaysko-Dyupkunsky subzones in sections of Nadezhdinsky Suite, lava flows of aphyric basalts prevail. In the vicinity of the Keta and Khantayskoe Lakes, at the base of the suite, lies a unit of pillow lavas 15–50 m thick. At the northeast end of the Kharaelakhsky trough, the suite is not divided into subsuites, but rather is present as a monotonous layer of porphyritic basalt passing to glomeroporphyritic basalt (Distler and Kunilov 1994). The explosivity coefficient and the proportion of tuffaceous rocks within the suite increases from south to east, away from the pre-Taimyr and Putorana structure-formational zones (Nemenenok 1978).

Microscope studies show that the basalts have porphyritic, glomeroporphyritic, or ophitic texture, and the groundmass is characterized by tholeiitic, micro-ophitic, intersertal, ophito-taxitic, and poikilophitic intersertal textures. Porphyritic segregations are usually labradorite (An_{60-70}), reaching 12 vol% in content, more rare clinopyroxene phenocrysts (not more than 1–2 vol%) are observed, and olivine is sporadic and is frequently altered. The groundmass in the basalts comprises plagioclase (An_{45-65}; 40–60 vol%), clinopyroxene (30–45 vol%), palagonite (5–15 vol%), oxide ore minerals (1–3 vol%), olivine (up to 3 vol%), and secondary minerals (3–7 vol%).

The main phases in the tholeiitic basalts are augite ($Wo_{38}En_{45}Fs_{17}$), plagioclase ($Ab_{32}Or_2An_{66}$), and palagonite (see Rock indication 24). Clinopyroxene contains TiO_2 (0.36 wt%), Al_2O_3 (1.96 wt%), and Cr_2O_3 (0.18 wt%). In the rock, fine olivine grains are completely replaced by bowlingite. In polyphyritic trachybasalts, the phenocrysts are typically plagioclase, olivine, or clinopyroxene (see Rock indication 25). Porphyritic segregations of early plagioclase are frequently replaced by prehnite and palagonite II. Large olivine phenocrysts are usually represented by bowlingite pseudomorphs, and sometimes a granoblastic aggregate of clinopyroxene grains forms within box-like olivine crystals. These also form individual subidiomorphic growths of $Wo_{38-37}En_{51-50}Fs_{12}$, which contain TiO_2 (0.19–0.22 wt%), Al_2O_3 (1.74–1.94 wt%), and Cr_2O_3 (0.33–0.36 wt%). Plagioclase phenocrysts have a labradorite ($Ab_{45}Or_3An_{52}$) composition. The rock matrix contains olivine and potassium spar is present.

The main group of basalts from the Nadezhdinsky Suite fall into the tholeiite series (Zolotukhin et al. 1989; etc.). One exception is a lava flow of polyphyritic trachybasalts (see Rock indication 25) that contained 6.10 wt% Na_2O + K_2O and 51.60 wt% SiO_2. High alkalinity is a characteristic for the glomeroporphyritic basalts of the Loganchi Mountain, where alkaline content is 5.43–6.42 wt% and silica 47.06–48.68 wt%. The rocks feature sulfide chalcopyrite–pyrrhotite mineralization (Ryabov 1997).

2.8 Morongovsky Suite (T_1mr)

The suite is widespread over the whole of the Noril'sk Region and can be traced beyond that region. Its facial analogues are the *Ayansky, Dvuroginsky, Korvunchansky,* and *Kogotoksky* Suites. The rocks of the Morongovsky Suite lie on the eroded surface of the Nadezhdinsky basalts, and a typical section through the Morongovsky Suite is characterized by volumes of tuffaceous rocks. Total thickness of the suite in the northeast of the Kharaelakhsky trough reaches 700 m, in the Noril'sky trough 250 m, and in a north–south direction, there is general decrease in thickness coupled with the replacement of lava flows with the tuffaceous layers of the Dvuroginsky and Korvuchansky Suites. The Morongovsky Suite is divided into two subsuites.

The lower subsuite comprises a number of aphyric basalt flows with an intersertal, or more rarely, porphyritic or poikilophitic texture and tuff layers. The two subsuites are divided by a thick (100 m) tuff horizon containing fragments of granite, gneiss, and schist (Nemenenok and Vilensky 1978).

The upper subsuite is composed of aphyric and poikilophitic (and more rarely plagioclase-phyric) basalt layers with tuff layers. In the north of the Kharaelakhsky trough, in the vicinity of Lake Arylakh, the upper part of the subsuite contains a 50-m-thick unit of ankaramite with high magnesium content. In the vicinity of Lake Khantayskoe, the Yuryakhsky Unit is composed of 7 trachybasalt flows with a total thickness of 50–60 m up to 140 m. Within the Kamensky Province near the Kaltama River is a facial analogue of the Morongovsky Suite composed of flows of polyphyritic basalts and 2–4 flows of picritic basalts. Further to the east, in the Boyarsko-Del'kansky structure-facial subzone, another facial analogue of the Morongovsky Suite is found among olivine-phyric basalts and rare layers of melano-nephelinite (Nemenenok and Vilensky 1978).

Aphyric basalt represents a rock of microdoleritic, intersertal, micro-ophitic, or tholeiitic texture with rarer poikilophitic texture. The rock consists of plagioclase ($Ab_{32}Or_1An_{67}$) and clinopyroxene ($Wo_{36}En_{42}Fs_{22}$) (see Rock indication 26), which contain TiO_2 (0.64 wt%), Al_2O_3 (2.60 wt%), and Na_2O (0.23 wt%). In the rock matrix, small amounts of alkali feldspar and palagonite are observed, as well as an oxide ore mineral, titanomagnetite, that contains TiO_2 (13.0 wt%), Al_2O_3 (1.56 wt%), and V_2O_5 (0.92 wt%).

Glomeroporphyritic basalt has a tholeiitic, intersertal, or micro-ophitic texture to the groundmass. Porphyritic segregations are represented by intergrowths of plagioclase grains or by individual prismatic crystals. Phenocrysts have a zonal structure and a frequently skeletal habit with numerous intergrowths of clinopyroxene (see Rock indication 27). In the central parts of porphyrocrysts, plagioclase has the composition $Ab_{22}Or_1An_{77}$, whereas in microlites in the groundmass, it is $Ab_{45}Or_2An_{52}$. Clinopyroxene corresponds to augite ($Wo_{40}En_{38}Fs_{22}$), with TiO_2 (0.99 wt%), Al_2O_3 (1.62 wt%), Na_2O (0.20 wt%), and Cr_2O_3 (0.017 wt%).

The uniqueness of the upper subsuite of Morongovsky Suite and its facial analogues in regions adjoining Noril'sk is due to the presence of lava flows of ankaramite, picritic basalts, and trachybasalts unlike any other suites in the northwest of the Siberian Platform.

2.8.1 Ankaramites of Arylakh

Basalt in Arylakh is a porphyritic or polyphyritic rock with microlitic, intersertal, tholeiitic, doleritic, or micro-ophitic groundmass texture. It consists of (see Rock indication28) olivine (5–15 vol%), clinopyroxene (35–45 vol%), orthopyroxene (3–5 to 15 vol%), plagioclase (30–40 vol%), oxide ore minerals (5–7 vol%), and palagonite (5–10 vol%).

Porphyritic segregations are formed by large subidiomorphic olivine crystals, sometimes with skeletal forms, usually forming aggregations and frequently replaced by bowlingite and fine, disseminated ore minerals. Olivine composition corresponds to chrysolite Fa_{15-27} (Zolotukhin et al. 1986). Clinopyroxene phenocrysts form short prisms 2.5–4 mm in size and spherical crystals with inclusions of bowlingite as pseudomorphs after olivine. Clinopyroxene inclusions usually exhibits irregular crystallographic forms and polysynthetic twinning. They often occur in accumulations of altered olivine and enclose resorbed orthopyroxene grains. Clinopyroxene phenocrysts are $Wo_{40-41}En_{49-47}Fs_{21}$ in composition and contain TiO_2 (0.50–0.63 wt%), Al_2O_3 (2.52–2.66 wt%), Na_2O (0.19–0.23 wt%), and Cr_2O_3 (0.99–0.83 wt%). The fine-grained mineral in the groundmass has a composition of $Wo_{35}En_{44}Fs_{21}$ and contains TiO_2 (1.16 wt%), Al_2O_3 (1.67 wt%), Na_2O (0.28 wt%), and Cr_2O_3 (0.043 wt%) (see Rock indication 28).

Pigeonite forms prismatic crystals with twinned augite rims and resorbed relicts inside augite phenocrysts. Its composition is $Wo_7En_{65}Fs_{28}$ and contains TiO_2 (0.41 wt%), Al_2O_3 (0.55 wt%), Cr_2O_3 (0.03 wt%), and V_2O_5 (0.016 wt%) (see Rock indication 29).

Plagioclase is observed as prismatic crystals of labradorite $Ab_{38-44}Or_{1-2}An_{60-54}$.

The ore minerals in the ankaramite include chromite intergrowths in olivine and titanomagnetite, ilmenite, and rutile in the groundmass. Chromite contains TiO_2 (2.67 wt%), Al_2O_3 (15.37 wt%), MgO (5.59 wt%), Cr_2O_3 (37.09 wt%), and V_2O_5 (0.92 wt%). Titanomagnetite is characterized by a high TiO_2 content (25.05 wt%), rutile is characterized by a high V_2O_5 content (0.71 wt%), whereas ilmenite has low concentrations of trace elements and varying contents of MgO (0.60–2.33 wt%) and V_2O_5 (0.17–0.73 wt%).

Ankaramites are characterized by the widespread occurrence of palagonite in interstitices and also in tiny miaroles. This fact may indirectly indicate the enrichment of the melt with volatile material.

The studies allow the supposition that the tholeiitic melt, from which the ankaramites crystallized, had high magnesium content, and its crystallization began with the formation of olivine. The high fluidity of the melt apparently enabled high-temperature oxidation of olivine (Ryabov 1992b) with its transformation into pigeonite and augite, as more stable phases under existing - parameters.

2.8.2 Picritic Basalts of the Ayan River

The picritic basalt sequence consists of three flows with total thickness of 40–70 m that can be traced along the Ayan River for 10–15 km (Tomanovskaya 1961; Zolotukhin et al. 1989). The individual lava flows all have a similar thickness (10–15 m) and are divided from each other by discreet, thin horizons of tuffaceous rocks. In each flow, an upper and lower amygdaloidal zone and a massive central zone can be distinguished. In the upper portion of the sequence, Vasil'ev (Zolotukhin et al. 1989) notes there is a subhorizontal branched pegmatoid vein composed of a coarse-grained pyroxene–plagioclase aggregate.

Amygdaloidal basalts are composed of olivine porphyrites consisting of olivine crystals (10–12 vol%) in a glassy groundmass.

Picritic basalts have a porphyritic texture and a groundmass with a micro-ophitic, microlitic, or intersertal texture. The rock consists of olivine (20–50 vol%), clinopyroxene (20–45 vol%), plagioclase (15–60 vol%), oxide ore minerals (1–8 vol%), and palagonite and secondary minerals. The palagonite and secondary minerals are noted, on occasion, to constitute a significant volume of the rock. The rock also contains other minerals such as biotite, amphibole, apatite, albite, chlorite, calcite and hydrogarnet, chromite, and ilmenite.

Metadiorites (pegmatoidal gabbro, according to Vasil'ev) have a gabbroic ophitic and pegmatoidal texture. They comprise plagioclase (30–50 vol%), clinopyroxene (50–65 vol%), oxide ore minerals (5–6 vol%), and accessory biotite and apatite. Palagonite is also widely distributed.

The most detailed study of the lava sequence was by Vasil'ev (Zolotukhin et al. 1989), who noted the following specific features of the rock composition. Olivine forms large (>0.8 cm) crystals with chromite inclusions. The I generation has a composition $Fa_{10–16}$, smaller olivine grains of the II generation are $Fa_{14–20}$, and the III generation is $Fa_{18–22}$. In olivine, varying quantities of (0.05–0.35 wt%) and NiO (0.28–0.36 wt%) were determined. The first oxide has a direct, and the second an inverse, correlation with iron content. Clinopyroxene is observed as short prismatic crystals with a composition corresponding to augite and containing TiO_2 (1.54–2.16 wt%), Al_2O_3 (2.30–3.50 wt%), Cr_2O_3 (0.10–0.87 wt%), and Na_2O (0.27–0.40 wt%). The total iron content in the mineral varies from 17.4 to 23.7 vol%. Chromite is characterized by varying contents of TiO_2 (3.23–6.34 wt%), Al_2O_3 (9.35–11.06 wt%), Cr_2O_3 (34.32–42.17 wt%), FeO (33.62–40.44 wt%), MnO (0.58–0.59 wt%), and MgO (6.56–7.88 wt%). For ilmenite, a high MgO content (2.60–5.16 wt%) is characteristic.

2.8.3 Trachybasalts of the Yuryakhsky Unit and Magnetite Lava in the Vicinity of Khantayskoe Lake

Trachybasalts of the Yuryakhsky Unit are a part of the Ayansky Suite (Nagaitseva et al. 1967; Pavlov 1969). The rocks are locally spread between the Keta Lake at the north and the Khakancha River at the south (Fig. 2.16) and vary from 0 to 150 m in thickness. At the southern coast of the Khantayskoe Lake, near the Magnetitovy Stream, three flows of trachybasalt, each 7–12 m thick, are exposed (Figs. 2.17 and 2.18). In the upper parts of the flows, mandelsteine zones composing 15–20 vol% of the section and denser regions with rare amygdales and slender prismatic jointing are noted. The first and the third flows have a similar structure and composition and, in fact, are typical of the whole Yuryakhsky Unit (see Rock indications 30–32). The second flow is unusual, featuring titanomagnetite lava (Ryabov and Pavlov 1991).

Trachybasalts here form aphyric or, more rarely, porphyritic varieties with microplagioclase-phyric, seriate–porphyritic, polyphyric texture and a vitrophyric, intersertal, tholeiitic groundmass. They consist of plagioclase (45–70 vol%), clinopyroxene (15–25 vol%), olivine (1–5 vol%), titanomagnetite (3–5 vol%), and glass with apatite and palagonite inclusions (10–35 vol%). Crystallinity of the rocks increases toward central parts of the flows and manifests first as a decrease in glass content and an increase in rock granulity. Amygdales in trachybasalts are filled with palagonite, chalcedony, quartz, and calcite.

The second flow comprises a mandelsteine zone about 2–4 m thick, a bleached aphyric trachybasalt zone 1–1.5 m thick, a zone of agglomerate trachybasalt 1–2 m thick, mineralized trachybreccia and ore lava 5–10 m thick, agglomerate trachybasalts 1–2 m thick, and a lower mandelsteine zone 0.5–1 m thick.

Mandelsteines are glassy rocks with vitrophyric, hyaloporphyritic, and intersertal texture. The groundmass is composed of glass in different stages of opacitization, with microporphyritic segregations and microlites of plagioclase (20–30 vol%), small prismatic clinopyroxene crystals (3–5 vol%), and magnetite grains (see Rock indication 33). With increasing degree of crystallinity, the quantity of plagioclase and clinopyroxene increases, which include, in their

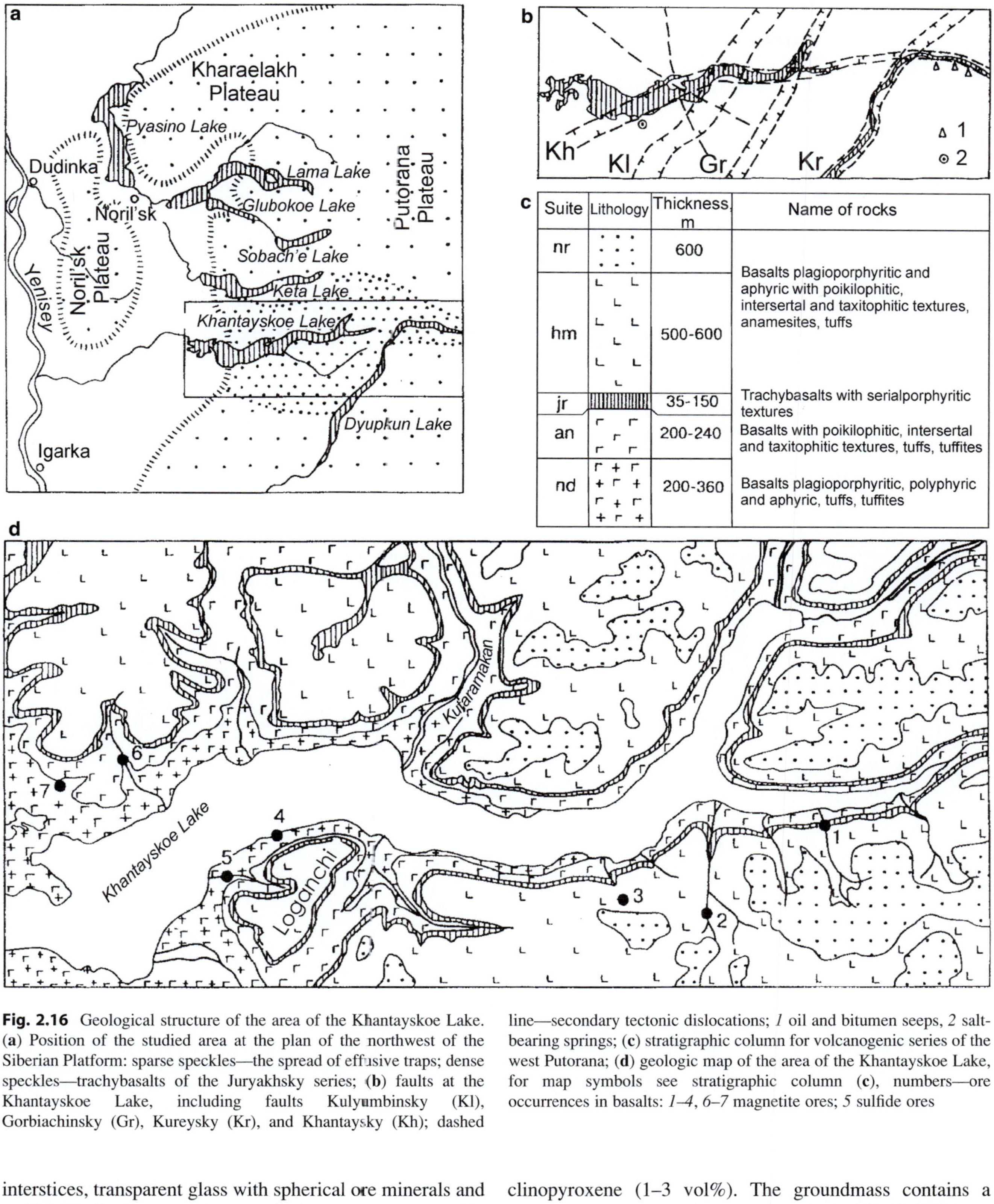

Fig. 2.16 Geological structure of the area of the Khantayskoe Lake. (**a**) Position of the studied area at the plan of the northwest of the Siberian Platform: sparse speckles—the spread of effusive traps; dense speckles—trachybasalts of the Juryakhsky series; (**b**) faults at the Khantayskoe Lake, including faults Kulyumbinsky (Kl), Gorbiachinsky (Gr), Kureysky (Kr), and Khantaysky (Kh); dashed line—secondary tectonic dislocations; *1* oil and bitumen seeps, *2* salt-bearing springs; (**c**) stratigraphic column for volcanogenic series of the west Putorana; (**d**) geologic map of the area of the Khantayskoe Lake, for map symbols see stratigraphic column (**c**), numbers—ore occurrences in basalts: *1–4*, *6–7* magnetite ores; *5* sulfide ores

interstices, transparent glass with spherical ore minerals and a "felt-like" shroud of apatite crystals (Ryabov 1989b; 1997). The lower mandelsteine zone has a porphyritic texture with intersertal and paniculate groundmass texture. Here, porphyritic segregations of olivine (3–5 vol%) are completely replaced by iddingsite–bowlingite and clinopyroxene (1–3 vol%). The groundmass contains a paniculate aggregate of long, prismatic plagioclase crystals, small clinopyroxene crystals, and magnetite and palagonite.

Bleached trachybasalts have an amygdaloidal texture, and microscopic study reveals an intersertal texture. The rock features widespread palagonite that, locally with glass,

Fig. 2.17 The inner structure of ore-bearing trachybasalt lava flow of the Yuryakhsky Unit in the vicinity of the Magnetitovy Stream. *1* Basalts of Khonnamakitsky Suite, *2* tuffites and tuff breccia, *3* trachybasalts, *4* bleached basalts and metabasites, *5* agglomerates, *6* magnetite lavas, *7* compact magnetites, *8* ore agglomerates

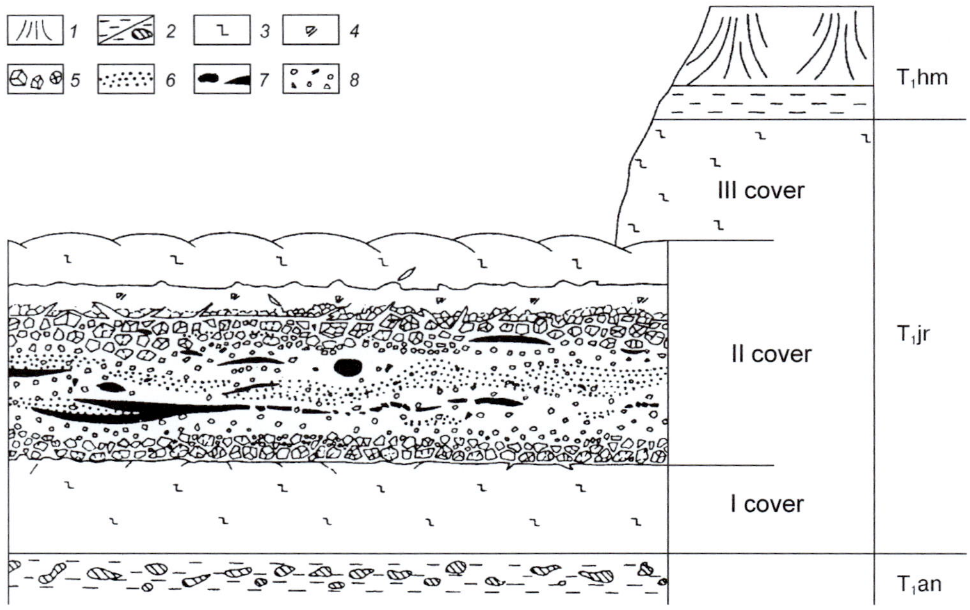

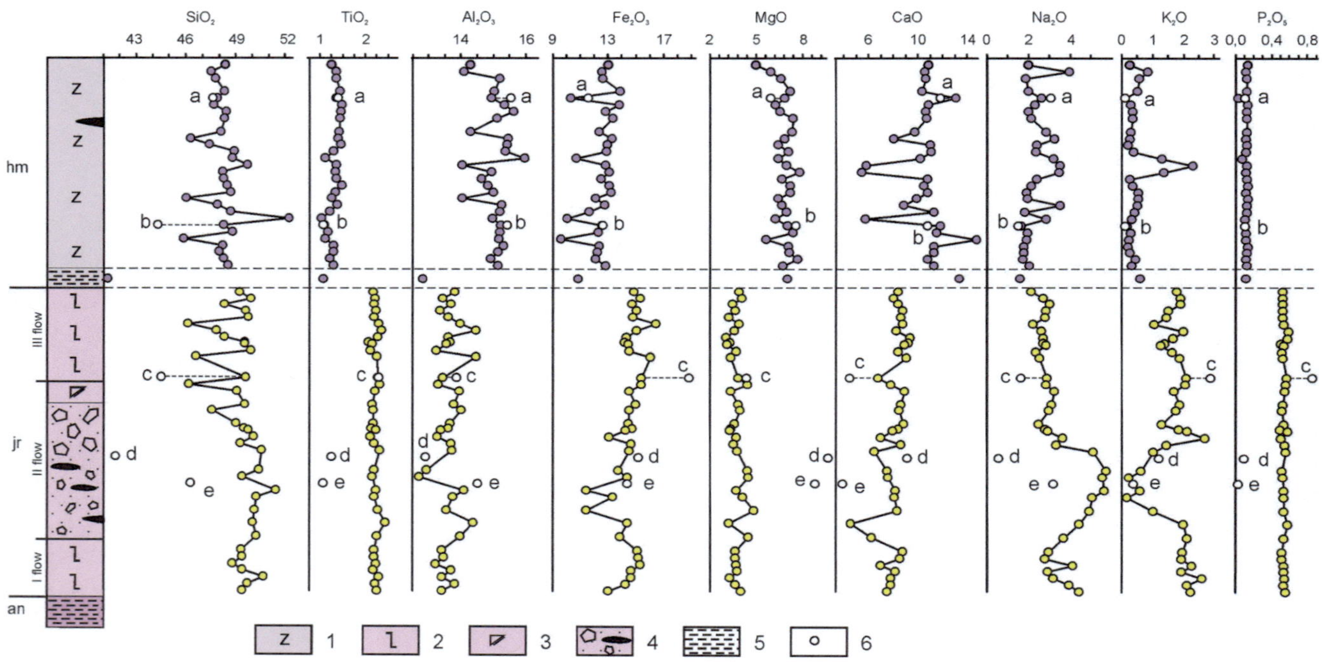

Fig. 2.18 Variation diagram of oxides (wt%) for a volcanogenic series of the western part of the Khantayskoe Lake. *1* Basalts of the Khonnamakitsky (hm) Suite; *2* trachybasalts of barren lava flows of the Yuryakhsky (jr) Unit; *3* bleached basalts and metabasites; *4* ore lavas and agglomerates containing lenses and veins of compact magnetite; *5* tuffaceous rocks; *6* composition of glasses (*a, b*), palagonite in basalts (*c*), palagonite from globules in magnetite lavas (*d, e*)

lies in interstices and forms "false amygdales," globules of palagonite I (1–7 mm in size) with orthopyroxene glass and palagonite I in the center. The palagonite also contains plagioclase microlites, box-like crystals of pink clinopyroxene, and fir-like skeletal magnetite crystals. Around globules of glass and palagonite, a tangential arrangement of plagioclase laths is often seen. Such manifestation of palagonite I leaves no doubt in its magmatic nature, in contrast with palagonite II, being a product of late palagonitization of glass and silicate minerals (Ryabov 1989b). Palagonite I also contains calcite globules, and the bleached basalt also features less common, black ore globules. Under microscope, they appear as a fine magnetite trellis containing palagonite.

Agglomerates of trachybasalt form a horizon of rocks composed of sharp angular fragments of bleached trachybasalt, close in composition to adjacent basalt. In the

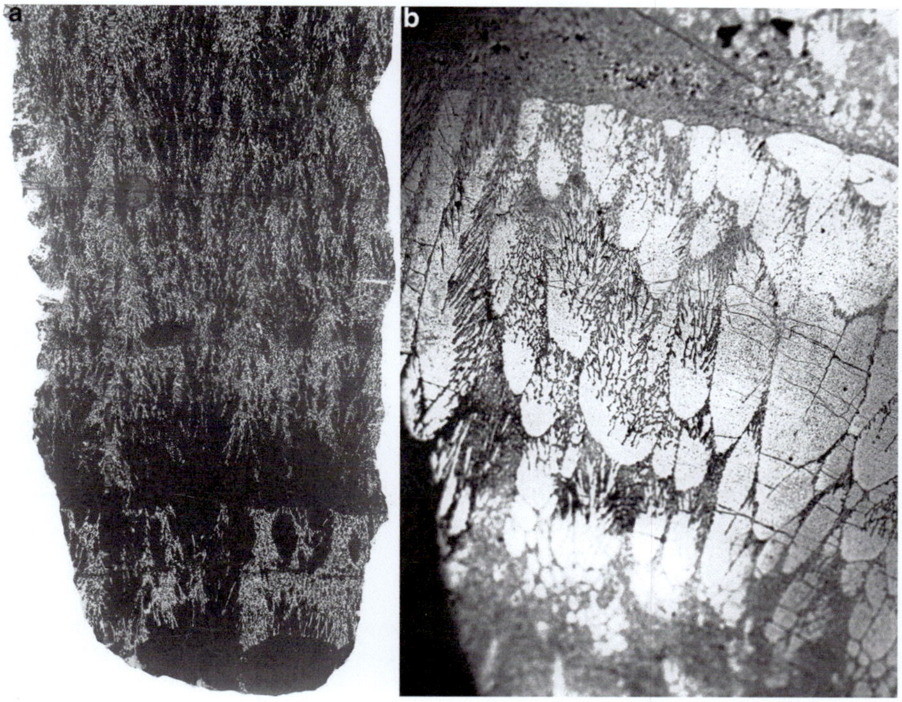

Fig. 2.19 Magnetite lava in trachybasalts of the Yuryakhsky Unit of the west Putorana, near Khantayskoe Lake. (**a**) Drop-like impregnation and dendrites of magnetite (*black* in incident light) in magnetite–palagonite bubbly matrix; (**b**) drop-like, dendrite-like, and spherolitic forms of magnetite from compact ores

lower part of the layer, magnetite and palagonite can be found in the material cementing the fragments.

Ore lava breccia represents the rock built up of various quantities (up to 60 vol%) of fragments cemented with magnetite-palagonite rock with a composition identical to that of the bleached trachybasalts. Large spherical bodies (up to 20 cm in diameter) of light-green palagonite with finely disseminated magnetite are also found in ore lava breccia. Analysis of these formations is shown in Fig. 2.18.

Ore lava has a bubbly consistency and contains tiny globules of partly crystallized magnetite grains in a palagonite or feldspar–palagonite matrix with numerous tiny hollows. Finely disseminated ore constitutes 30–60 vol% of the rock, and the total iron content (as Fe_2O_3) reaches 63 wt%. In the areas enriched with the ore phase, large magnetite droplets are observed, which accumulated as layers and nodules of massive magnetite due to gravitational settling and also feature dendritic growth patterns in the upper parts of the droplets. Agglutinated tiny magnetite globules sometimes form laced porous spheres 2 to 14 mm in diameter (Fig. 2.19).

Massive magnetite forms small flows and plate-like veins 1.3 m thick and 15–20 m long, lenses up to (10–15) × (70–90) cm, and spherical bodies up to 25 cm in diameter. In ore lava breccia and lava, sharp angular fragments of massive magnetite are observed, indicating solidification of the ore melt during lava flow movement. Massive bodies are composed of columnar and dendritic magnetite crystals up to 3–15 cm long. Small lenses frequently form hemispheres 7 to 40 cm in size with a radially fibrous and concentrically zoned structure, due to elongated forms of magnetite crystals and gas hollows.

2.8.3.1 Mineralogy and Geochemistry

In unmineralized flows of the Yuryakhsky Unit, the composition of plagioclase corresponds to albite ($Ab_{91-99}Or_{3-0}An_{6-1}$), and clinopyroxene corresponds to augite as idiomorphic porphyritic segregations ($Wo_{39-41}En_{41-37}Fs_{18-22}$) and as small grains and long prismatic crystals ($Wo_{43}En_{30-25}Fs_{27-32}$) (see Rock indications 30–31). Augite also contains TiO_2 (0.59–1.39 wt%) and Al_2O_3 (1.55–3.87 wt%) in the segregations and TiO_2 (2.50–3.19 wt%) and Al_2O_3 (3.66–3.75 wt%) in the prismatic crystals. Magnetite in trachybasalts is a titanium-poor variety containing TiO_2 (2.22–3.72 wt%), Al_2O_3 (0.49–0.85 wt%), MnO (0.23–0.58 wt%), and MgO (0.43–1.11 wt%).

In the ore-bearing lava flow, porphyritic segregations have the composition $Ab_{33-48}Or_{2-5}An_{60-47}$ (see Rock indications 32–34). Within the limits of one zoned crystal, the maficity of the mineral falls from its center toward margins by 2–7 units. Around prismatic labradorite crystals, a discontinuous margin of K–Na feldspar ($Ab_{60-57}Or_{33-37}An_{7-6}$) is sometimes observed. Clinopyroxene in the rock is augite whose composition appears to be rather stable in glassy varieties of the rock ($Wo_{42-39}En_{41-36}Fs_{17-24}$) and varies widely in devitrified ones ($Wo_{41-36}En_{41-27}Fs_{19-37}$). The mineral contains TiO_2 (0.75–1.59 wt%) and Al_2O_3 (0.85–3.41 wt%). Olivine from inclusions in pyroxene has the composition Fa_{48}. In groundmass, thin needles of apatite (CeO_2 (0.15–0.38 wt%), F (1.96–2.41 wt%), Cl (0.12–0.17 wt%)), titanomagnetite,

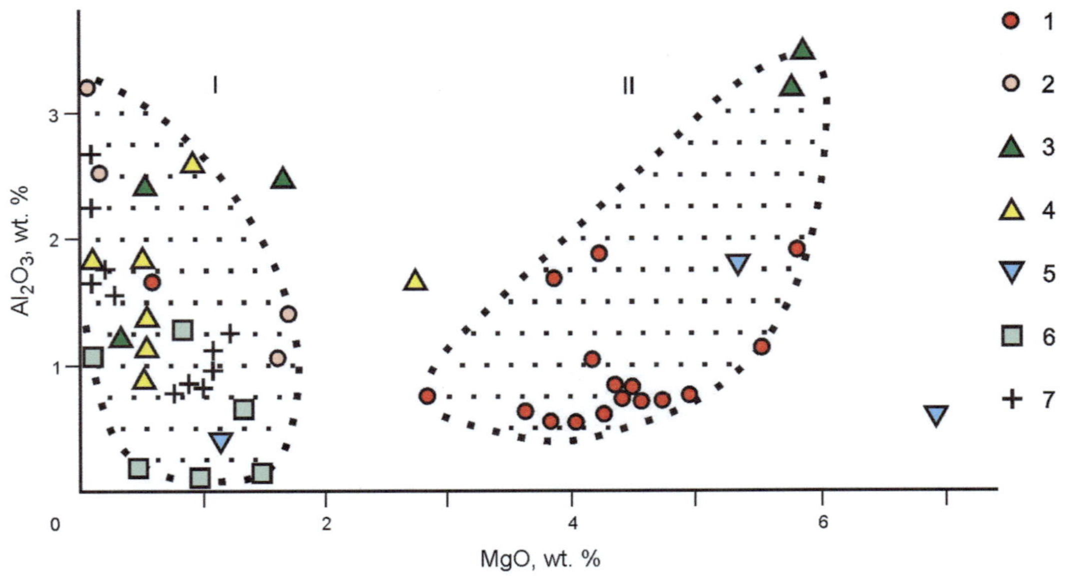

Fig. 2.20 Composition of magnetite from ore occurrences in the northwest of the Siberian Platform. *1* Magnetite lava from the Magnetitovy Stream, *2* diatreme of the Vetka Stream, *3* pyrrhotite and pyrrhotite–magnetite dykes of the Loganchi Mountain, *4* solid sulfide ores of the Talnakh deposit, *5* magnetite ores of the Anakitsky intrusion, *6* anhydrite–magnetite ores of the Talnakh, *7* rock-forming magnetites from basalts. I, II—The areas (for explanations, see the text of the book). Based on data from numerous authors (Ryabov 1997)

and ilmenite were determined. Titanomagnetite contains TiO_2 (24.52–24.92 wt%), Al_2O_3 (1.08–1.32 wt%), MnO (0.27–0.34 wt%), and MgO (0.40–0.67 wt%), and ilmenite contains MnO (0.43–0.58 wt%) and MgO (0.58–0.62 wt%). In microporphyritic trachybasalts, glasses with variable composition are noted, among which feldspar-normative, highly ferriferous, and basaltic varieties can be distinguished (see Rock indications 33–34). The glasses contain some F (0.23–0.57 wt%).

Composition of magnetite from ore lava of the second flow corresponds to magnesic-magnetites, containing Al_2O_3 (0.32–1.83 wt%), MgO (3.58–5.82 wt%), and MnO (0.31–0.38 wt%) (see Rock indication 35). The compositional peculiarities of magnetites from various iron deposits of the northwest of the Siberian Platform are given in Fig. 2.20. In accordance with their MgO content, magnetites are divided into two fields: the first, MgO-poor magnetites from a diatreme near the Vetka River and a magnetite deposit near the Makus River and from an anhydrite-–magnetite and sulfide ore of Talnakh; and the second, the magnesic-magnetites of ore lava from near the Magnetitovy Stream. Similar compositions characterize magnesic-magnetites from ore-rich rocks in the vicinity of the Severnaya, Kureyka, and Gorbiachin Rivers (Kavardin et al. 1968; Kavardin 1976).

Some understanding of the peculiarities in the variation of chemical composition of rocks in the trachybasalt lava flows of the Yuryakhsky Unit can be gained from Fig. 2.18 and Table 2.5. In the variation diagram, trachybasalts can be seen to differ markedly from the overlying basalts of the Khonnamakitsky Suite. Trachybasalts of the I and III flows have a relatively stable composition, except for a slight fluctuation in SiO_2 content. In the II ore-bearing flow, a sharp increase in Na_2O content and a decrease in K_2O content are noted. For palagonite segregations, higher concentrations of Fe_2O_3, K_2O, and P_2O_5 and lower concentrations of SiO_2, CaO, and Na_2O than in trachybasalts are characteristic. Similar palagonite formations from magnetite lava have higher concentrations of MgO and lower concentrations of SiO_2, TiO_2, Na_2O, and P_2O_5.

2.8.3.2 Genetic Concepts

The local occurrence of trachybasalt of the Yuryakhsky Unit in the vicinity of the Khantaysky Fault allowed Nagaitseva et al. (1967) to consider the fault as a magma feeder. Magnetite lava near the Magnetitovy Stream is the second known example of outpouring of ore magma to the Earth's surface. The first was found near the El Laco volcano in Chile. Magnetite mineralization in the vicinity of the Khantayskoe Lake along the Magnetitovy and Zhelezny Streams and the Khantay-Chopko River, as well as in some other places (Ryabov 1997), was observed during geological mapping as early as the beginning of the 1940s. They were studied by Kavardin (Kavardin et al. 1968; Kavardin 1976) who came to the conclusion that they were of a metasomatic–hydrothermal origin. Detailed petrologic studies allowed others to prove the magmatic origin of the magnetite mineralization (Ryabov and Pavlov 1991; Ryabov 1997) and to show that the ore melt occurred as a result of the interaction of basalt

Table 2.5 Average composition (wt%) of Honnamakitsky Suite basalts and Yuryakhsky Unit trachybasalts

	hm		jr, I		jr, II		jr, III	
Component	*X*	*δ*	*X*	*δ*	*X*	*δ*	*X*	*δ*
SiO_2	47.81	1.77	49.55	0.57	49.18	2.64	48.07	2.83
TiO_2	1.30	0.13	2.24	0.04	2.07	0.43	2.13	0.45
Al_2O_3	15.01	0.58	13.50	0.20	13.63	0.58	13.14	2.75
Fe_2O_3	12.21	1.42	14.64	0.81	13.74	1.22	14.94	1.34
MgO	6.89	0.61	3.71	0.22	4.94	2.08	3.56	0.64
CaO	10.68	2.12	7.99	0.61	7.18	1.61	9.25	4.31
Na_2O	2.31	0.59	3.50	0.67	4.26	1.48	2.59	0.62
K_2O	0.45	0.44	2.17	0.25	1.15	0.81	1.70	0.52
P_2O_5	0.12	0.03	0.54	0.02	0.46	0.19	0.54	0.12
Ba	0.03	0.05	0.12	0.05	0.07	0.05	0.08	0.02
LOI	3.07	2.22	2.01	0.50	3.28	1.96	3.85	3.75
Cu	0.0124	0.0049	0.0028	0.0002	0.0053	0.0051	0.0031	0.0011
Ni	0.0114	0.0036	0.0031	0.0015	0.0060	0.0069	0.0034	0.0039
Co	0.0043	0.0008	0.0032	0.0003	0.0032	0.0006	0.0032	0.0004
Cr	0.0209	0.0063	0 0102	0.0045	0.0110	0.0054	0.0067	0.0038
n	33		7		12		22	

Southern beach of the Khantayskoe Lake
X average, *δ* standard deviation, *n* number of analyses

magma with evaporite brines, which caused the subsequent intensive degradation of saliferous sediments (halites and dolomites), and the fluid-magmatic differentiation and ore liquid segregation. Similar ideas were proposed by Pavlov (1975) for the Angaro-Ilimsky iron-ore deposits.

2.9 Mokulaevsky Suite (T_1mk)

The suite is found in the Kharaelakhsky trough, where it overlies the eroded surface of the Morongovsky Suite, and is traced far to the east from the Noril'sk Region where a facial analogue is seen in the Khonnamakitsky Suite. The total thickness of Mokulaevsky Suite in the north of the region is 400–670 m and 150–250 m in the south. However, within the Noril'sky trough it is unknown. According to the order basalts are found and the different structure-textural characteristics, the Mokulaevsky Suite and its facial analogues are divided into two subsuites. Each subsuite, in turn, is further divided into two units: lower and upper. The lower unit of each subsuite on the Putorana Plateau contains 1–4 thick (20–60 m) flows of fine-grained aphyric, oligo-phyritic, and plagioclase-phyric basalt with a microdoleritic and intersertal groundmass texture. The upper unit is formed of a great number (10–50) of thin (1–3 m to 10 m) flows of crystalline basalt with an ophito-taxitic texture and anamesites (Fadeev 1962; Pavlov 1969). In section, the suite of lava flows dominates over volcano sedimentary formations (including some lenses of chalk). For the region as a whole, the following features of the suite are known.

2.9.1 Lower Mokulaevsky Subsuite

The lowermost unit in the region, also known as the Nadayansky horizon, is composed of a number (1–6) of flows of plagioclase-phyric, oligo-glomeroporphyritic, and aphyric basalts with a microdoleritic and intersertal groundmass texture. Basalts of the Nadayansky horizon have a characteristic radial, columnar, and fan-like jointing pattern (Fig. 2.21) that can be traced over large distances and, accordingly, serves as a marker bed (Starosel'tsev 1982). It varies in thickness from 20 to 100 m. At the bottom of the lower unit is a tuffaceous horizon that varies in thickness from 3–12 to 40–50 m. At the northern end of the Kharaelakhsky trough, the Arylakhsky native copper ore occurrence is connected with this horizon (Duzhikov 1988). The upper unit consists of 30–50 flows of relatively coarse-grained plagioclase-phyric basalts with a poikilo-phitic intersertal and ophito-taxitic groundmass texture. Thin (1–10 m) lava flows pinch out quickly and cannot be traced for long distances and do not correlate across different sections. About 50% of these rocks are composed of mandelsteines with voids filled with chalcedony-zeolite. Rare tuff layers and lenses of anamesite are noted locally. The thickness of the subsuite around the Putorana Plateau is 180–390 m.

2.9.2 Upper Mokulaevsky Subsuite

The base of the lower unit consists of aphyric, plagioclase-phyric, and oligo-glomeroporphyritic basalt with an

Fig. 2.21 Fan-like jointing in Khonnamakitsky Suite basalts, Khantayskoe Lake

intersertal and doleritic groundmass texture. In the western part of the Tungussky syneclise, the unit comprises 2–4 flows, 15–40 m thick. In the lower and central parts of these flows are basalts with a microdoleritic texture, whereas in the upper part are basalts with an ophito-taxitic texture. Mandelsteines compose 25–60% of the subsuite. Between the flows are thin horizons of tuffaceous rocks. The upper unit consists of 10–20 flows. The lower flows are aphyric and plagioclase-phyric basalt, and the upper flows are composed of poikilophitic intersertal, ophito-taxitic, and more rarely plagioclase-phyric basalt and anamesite. Tuffs are rare, and at the top of the tuffaceous rock horizon, Losev found the remnants of *Dicynodont Lystrosaurus Cope*, which dated them to the lower third of Lower Triassic (Distler and Kunilov 1994). The thickness of the subsuite in the region of the Putorana Plateau is 250–350 m.

In the Kamensky Province in the vicinity of the Namakan River and Bokovoe Lake, the top of the Khonnamakitsky Suite features localized flows limburgite typically 1.5–5 m thick (though they can reach 10 m thick) and alkaline mafic tuffs. The average composition of the rocks is given in Table 2.4.

Aphyric basalt has an intersertal, microdoleritic texture with micro-ophitic and micropoikilophitic elements (see Rock indication 36). The largest clinopyroxene crystals are zoned augite oikocrysts ($Wo_{41-39}En_{43}Fs_{16-18}$), ophitic intergrowths with plagioclase ($Wo_{39}En_{45}Fs_{16}$), and small granules ($Wo_{36}En_{45}Fs_{18}$). Comparison of the compositions at the center and at the margins of large grains and granules revealed decrease of the following components: TiO_2 (0.92–0.62 wt%), Al_2O_3 (3.24–1.62 wt%), Cr_2O_3 (0.182–0.045 wt%), and Na_2O (0.26–0.16 wt%). These variations are accompanied with some growth of MnO content, 0.21–0.25 wt%. The composition of plagioclase corresponds to $Ab_{28-34}Or_{1-3}An_{71-63}$. Ore minerals are represented by titanomagnetite and ilmenite. Titanomagnetite conatins TiO_2 (24.84–32.97 wt%), Al_2O_3 (1.21–1.43 wt%), MnO (1.29–3.92 wt%), and MgO (0.04–1.11 wt%), and ilmenite contains MnO (0.48 wt%) and MgO (1.44 wt%). The variety of the mineral most enriched with ulvospinel contained 1.20 wt% of V_2O_5.

Oligophyric basalt has an intersertal texture with microdoleritic, micro-ophitic, micropoikilophitic elements (see Rock indication 37). The rock contains small sporadic porphyritic segregations of zoned plagioclase ($Ab_{18-30}Or_{0-1}An_{82-69}$). Plagioclase chadacrysts that form intergrowths in clinopyroxene have the composition $Ab_{30}Or_1An_{69}$, and laths in the groundmass have the composition $Ab_{35}Or_2An_{63}$. Microporphyritic segregations of clinopyroxene also feature a zoned appearance; their cores correspond to augite ($Wo_{40-38}En_{46-45}Fs_{13-17}$) and their margins to pigeonite ($Wo_{18}En_{48}Fs_{33}$). Changes in the composition of minerals during crystallization in the augite–pigeonite series is accompanied by an increase of iron content, a decrease of calcium content, an increase in both TiO_2 (0.55–0.86 wt%) and MnO (0.22–0.48 wt%), as well as a decrease in Al_2O_3 (2.32–1.16 wt%) and Cr_2O_3 (0.54–0.00 wt%). The rock matrix contains fine grains of olivine–hortonolite (Fa_{52-56}) and prisms of amphibole ($f = 56$ at%). Ore minerals are titanomagnetite and ilmenite, with titanomagnetite containing TiO_2 (22.23 wt%), MnO (0.94 wt%), MgO (1.29 wt%), and V_2O_5 (1.48 wt %), and ilmenite MnO (1.47 wt%), MgO (1.31 wt%), and V_2O_5 (0.45 wt%).

Poikilophitic basalt is composed of large clinopyroxene oikocrysts, in a tholeiitic groundmass (see Rock indications 38, 39). In basalts in the vicinity of Lake Khantayskoe, rare glomeroporphyritic segregations of plagioclase are noted locally. Clinopyroxene oikocrysts have the composition $Wo_{38-37}En_{45-41}Fs_{17-22}$ and contain TiO_2 (0.90–1.08 wt%), Al_2O_3 (2.41–2.51 wt%), MnO (0.26–0.30 wt%), Na_2O (0.24 wt%), and Cr_2O_3 (0.01–0.25 wt%). Plagioclase chadacrysts and laths in the groundmass have the composition $Ab_{23-37}Or_1An_{76-62}$. Ore minerals, titanomagnetite and ilmenite, are confined to rock matrix. Titanomagnetite contains TiO_2 (22.61–23.84 wt%), Al_2O_3 (1.27–1.49 wt%), MnO (2.51–2.56 wt%), and V_2O_5 (0.95–1.34 wt%), and ilmenite contains MnO (0.46 wt%), MgO (1.80 wt%), and V_2O_5 (0.44 wt%).

The composition of basalts of the Mokulaevsky Suite and its facial analogue, the Khonnamakitsky Suite, corresponds to a tholeiite of the normal series (Fig. 2.18, Table 2.5) (Zolotukhin et al. 1986, 1989). One of the main features of the rocks in the suite is the numerous flood basalts that are anomalously depleted with K_2O (less than 0.3 wt%).

The results of detailed microprobe analysis of rock-forming minerals from poikilophitic basalt of Khonnamakitsky Suite are given in Sect. 4.1.2 (see Fig. 4.4, Rock indication 40).

2.10 Kharaelakhsky Suite (T_1hr)

The rocks of Kharaelakhsky Suite lie conformably over basalts of the Mokulaevsky Suite and are divided into two subsuites, each of which is further subdivided into two units. The total thickness of the suite is 520–600 m.

2.10.1 Lower Kharaelakhsky Subsuite

The lower unit consists of 1–4 flows of plagioclase-phyric basalt with an intersertal groundmass texture. At the base of the unit a horizon of tuffaceous rocks is noted, and at the top 1–2 flows of poikilophitic flood basalt are observed locally. The upper unit is composed of 1–4 (though up to 7 locally) flows of glomeroporphyritic basalt with a poikilophitic groundmass texture.

2.10.2 Upper Kharaelakhsky Subsuite

The lower unit consists of poikilophitic and aphyric basalt. In its central part is 1–2 flows of glomeroporphyritic basalts that locally feature 1–3 horizons of tuffaceous rocks with carbonate lenses. The upper unit is composed of alternating of basalt flows and tuffaceous horizons. These horizons underlie each flood basalt in some areas.

In section, the central part of the suite features aphyric and poikilophitic basalt at the base and plagioclase-phyric and glomeroporphyritic basalt at the top. However, in some sections in upper part of the suite, a flow of oligoclase andesite–basalt of varying thickness was observed.

In the Noril'sko–Vologochansky subzone, the Kharaelakhsky Suite and those overlying it are absent. Facial analogues of the Kharaelakhsky Suite within the borders of Kharaelakhsky-Ikensky subzone are the *Nerakarsky, Kochechumsky,* and *Maymechinsky Suites.* In the Central Putorana and Kamensky Provinces, the Nerakarsky Suite is composed of aphyric, poikilophitic, plagiophyritic, and glomeroporphyritic basalts with rare tuff layers (Fig. 2.22). At the bottom of the Nerakarsky Suite (in the western part of the Putorana Plateau, near Khantayskoe Lake) is a flow of aphyric oligoclase trachybasalts (see Rock indication 41). In the east Putorana Province the suite thins out, in the Khetsky Region it is absent, and in the Kureysko-Letninsky Region tuffaceous rocks constitute about half of the suite. The Kochechumsky Suite of the Tungussky Province consists of poikilophitic and porphyritic basalt with tuff layers that constitute about 1/3 of the suite in section. It is thought that the age-equivalent analogue of the Kharaelakhsky Suite in the Maymecha-–Kotuysky Province is the Maymechinsky Suite, composed of picritic porphyrites, meimechites, and tuff. It is about 800 m thick, and the average composition of the rocks is given in Table 2.4.

Aphyric trachybasalt has a microdoleritic texture with a bostonite fabric of plagioclase crystals (see Rock indication 41). The rock consists of fine grains of augite ($Wo_{43}En_{43}Fs_{14}$) and laths of oligoclase ($Ab_{84}Or_1An_{15}$). Clinopyroxene contains TiO_2 (0.78 wt%), Al_2O_3 (3.76 wt %), and Na_2O (0.59 wt%), and the ore phase in the rock is titanomagnetite that contains TiO_2 (9.42 wt%), Al_2O_3 (1.98 wt%), MnO (1.0 wt%), MgO (1.28 wt%), Cr_2O_3 (0.105 wt%), and V_2O_5 (0.47 wt%).

Aphyric basalt with low potassium content has a microdoleritic and micropoikilophitic texture (see Rock indication 42). Composition of the clinopyroxene micro-oikocrysts corresponds to augite ($Wo_{37}En_{42}Fs_{21}$) containing TiO_2 (0.92 wt%), Al_2O_3 (1.56 wt%), and Na_2O (0.28 wt%). Plagioclase forms laths and tiny prismatic crystals of composition $Ab_{33-42}Or_{1-2}An_{66-56}$. Ore minerals are

Fig. 2.22 A section of basalts from the Nerakarsky (*pink*) and Khonnamakitsky (*gray*) Suites, west branches of the Putorana Plateau

titanomagnetite and ilmenite; titanomagnetite contains TiO_2 (13.3 wt%), Al_2O_3 (1.12 wt%), MnO (0.51 wt%), MgO (0.21 wt%), Cr_2O_3 (0.12 wt%), and V_2O_5 (1.43 wt%), and ilmenite contains MnO (1.52 wt%), MgO (0.27 wt%), and V_2O_5 (0.38 wt%).

Aphyric andesite–basalt is a rock with a microdoleritic texture (see Rock indication 43). The composition of clinopyroxene in the rock corresponds to augite ($Wo_{38-33}En_{45-41}Fs_{16-26}$) and contains TiO_2 (0.61–0.94 wt%), Al_2O_3 (1.37–1.06 wt%), MnO (0.26–0.42 wt%), and Na_2O (0.23–0.26 wt%). Plagioclase laths have an albite and oligoclase composition ($Ab_{84-91}Or_{2-1}An_{14-8}$). The ore phase is titanomagnetite, which contains TiO_2 (22.48–23.84 wt%), Al_2O_3 (1.30–1.50 wt%), MnO (1.20–1.85 wt%), MgO (0.03–0.04 wt%), ZnO (0.45–0.54 wt%), and V_2O_5 (1.02–1.47 wt%).

Aphyric basalt has a microdoleritic groundmass texture with micro-ophitic and micropoikilophitic elements (see Rock indication 44). The rock locally contains microporphyritic plagioclase. Clinopyroxene ($Wo_{41-35}En_{45-40}Fs_{14-25}$) forms micro-oikocrysts and fine grains in doleritic parts of the rock, and an increase in iron content and decrease in calcium content are noted from the center of these crystals to their margins and as grainsize decreases. In the same way, a regular decrease in Al_2O_3 (2.74–1.8 wt%) and Cr_2O_3 (0.36–0.22 wt%) content takes place. Other impurities include TiO_2 (0.64–0.76 wt%), MnO (0.20–0.35 wt%), and Na_2O (0.22–0.29 wt%). In the rock matrix, fine grains of olivine Fa_{54} are observed.

Plagioclase forms laths in micro-ophitic intergrowths with pyroxene and microlites in parts with doleritic texture. The composition of plagioclase is $Ab_{30-31}Or_1An_{69-68}$. Ore minerals are titanomagnetite and ilmenite. Titanomagnetite contains TiO_2 (20.76–24.39 wt%), Al_2O_3 (1.23–1.58 wt%), MnO (0.40–1.04 wt%), and MgO (0.76–1.71 wt%), and ilmenite contains MnO (0.43–0.44 wt%) and MgO (1.67–1.84 wt%).

Glomeroporphyritic basalt has seriate-porphyritic texture (see Rock indication 45). Porphyritic segregations are formed by intergrowths of large crystals and singular plagioclase prisms of different sizes. The groundmass texture is intersertal, doleritic, and micro-ophitic. Plagioclase insets have a zoned structure, with composition varying from $Ab_{16}An_{84}$ in the center to $Ab_{28}Or_1An_{71}$ at the margins. In smaller phenocrysts, the composition changes from $Ab_{29}Or_1An_{70}$ to $Ab_{35}Or_1An_{64}$, and the composition of laths in the groundmass corresponds to $Ab_{51}Or_3An_{46}$.

Clinopyroxene forms small xenomorphic grains in micro-ophitic parts of the rock and granules intergrowing plagioclase. The mineral composition corresponds to a variable composition augite ($Wo_{39-26}En_{45-43}Fs_{16-31}$). Iron content has a direct correlation with TiO_2, which changes from 0.66 to 0.89 wt%, and a decrease in CaO in pyroxene is accompanied by decrease of Al_2O_3 from 2.18 to 1.24 wt% and Cr_2O_3 from 0.201 to 0.01 wt%.

Olivine forms fine grains of Fa_{54-56} in groundmass. The sole ore mineral is titanomagnetite containing TiO_2 (22.87–24.33 wt%), MnO (0.087–0.28 wt%), and MgO (0.43–1.34 wt%).

2.11 Kumginsky Suite (T_1km)

The suite is developed in the Noril'sko–Kharaelakhsky, Khantaysko–Rybninsky, and western part of the Kamensky Province. Total thickness of the rocks is 160–200 m. The suite is composed of 6–12 flows of glomeroporphyritic basalt with extremely rare tuff layers in the Khantaysko-Rybninsky Province and by poikilophitic basalt with tuff layers in the Tungussky Province. Thickness of the suite and its facial analogue in the Tungussky Province (the *Yambukansky Suite*) decreases to 80 m and 100–130 m, respectively (Zolotukhin et al. 1986).

Plagioclase-phyric, poikilophitic, and intersertal basalt from the lower part of the suite contains porphyritic segregations of plagioclase and commensurable small clinopyroxene poikilocrysts (approximately 25–30 vol%). Both are found in the groundmass of rocks with tholeiitic and intersertal textures (see Rock indication 46).

Glomeroporphyritic basalt from the central part of the section contains about 10–15 vol% of plagioclase in the groundmass with intersertal texture and poikilophitic and tholeiitic elements.

Glomeroporphyritic basalt from the upper part of the suite consists of plagioclase (5–12 vol%) in a poikilophitic intersertal groundmass. Fine clinopyroxene poikilocrysts (25–39 vol%) are found in the parts of the rock that have an intersertal or tholeiitic texture (see Rock indication 47) composed of opacitized glass and palagonite with plagioclase microlites and tiny titanomagnetite grains.

In glomeroporphyritic basalt, large plagioclase phenocrysts have the composition $Ab_{19-27}Or_{0-1}An_{80-72}$, whereas smaller prismatic crystals are $Ab_{28}Or_1An_{71}$, and laths in groundmass are $Ab_{36-41}Or_{1-2}An_{63-57}$ (see Rock indications 48–50). In "pea-like" basalts with a poikilophitic intersertal texture, a widening of plagioclase composition is marked: prismatic crystals similar in size to clinopyroxene oikocrysts are $Ab_{19}An_{81}$, laths are $Ab_{30}Or_1An_{69}$, whereas xenomorphic areas in the matrix are usually characterized by zoned crystals with a parquet-like extinction ($Ab_{36-45}Or_{1-2}An_{63-53}$) (see Rock indication 47). In aphyric basalts, plagioclase has a stable composition of $Ab_{37-40}Or_2An_{61-58}$ (see Rock indication 46).

Clinopyroxene from different varieties of basalt corresponds to augite ($Wo_{38-39}En_{47-42}Fs_{14-20}$). Iron content increases in the mineral along the following rock series: poikilophitic–glomero-porphyritic–aphyric basalt. Clinopyroxene contains TiO_2 (0.5–0.89 wt%), Al_2O_3 (1.42–2.01 wt%), and Cr_2O_3 (0.05–0.31 wt%). Clinopyroxene of oikocrysts contains minimal quantities of TiO_2 and Al_2O_3 and high quantities of Cr_2O_3.

Titanomagnetite contains TiO_2 (17.84–25.42 wt%), Al_2O_3 (1.06–1.86 wt%), MnO (0.85–3.75 wt%), MgO (0.04–0.22 wt%), ZnO (0.12–0.27 wt%), Cr_2O_3 (0.00–0.09 wt%), and V_2O_5 (0.89–1.38 wt%).

2.12 Samoedsky Suite (T_1sm)

The rocks of the suite are observed only in the Noril'sko–Kharaelakhsky Province and the northwestern part of the Kamensky Province. Maximum partial thickness (680 m) of the suite was determined in a section along the Verkhne-Talovaya River. The suite is divided into three subsuites: the lower subsuite is composed of plagioclase-porphyritic, aphyric, and more rarely poikilophitic and glomeroporphyritic basalt with numerous thin tuff layers; the central subsuite is composed of glomeroporphyritic and plagioclase-porphyritic basalt; and the upper subsuite is composed of aphyric, plagioclase-phyric, and poikilophitic basalt. From the base of the central subsuite to the top of the suite, six horizons of tuffaceous rocks are noted.

A sequence of basalts of the Samoedsky Suite was studied near a sharp turn in the Kumga River (Fig. 1.1). Here, two subsuites of the suite crop out. Studies were performed on lava flows of the lower and the upper subsuites on the right-hand river bank and the lower subsuite on the left-hand river bank.

Lower subsuite is composed of flows of oligophyritic and glomeroporphyritic basalt that are accompanied by alternating flows of large "pea-like" poikilophitic basalt and olivine-plagioclase-phyric and small "pea-like" poikilophitic basalt. Composition of the lavas corresponds to low-, medium-, and high-potassium tholeiites.

Oligophyric basalt (see Rock indication 51) is a rock with a porphyritic texture, typically with an intersertal and doleritic groundmass texture, and micro-ophitic and micropoikilophitic elements. Rare plagioclase insets have a zoned structure, corresponding to An_{75-70} in the center and labradorite (An_{60-59}) at the rim. The structural fabric of the rock matrix is determined by pyroxene, either as groups of fine grains among plagioclase microlites An_{66} or as individual angular grains resembling a fragment of ophitic or poikilophitic texture. The size of the fine pyroxene grains is 0.1–0.2 mm, and angular grains from micro-ophitic regions and micro-oikocrysts are 0.1–0.4 mm. Judging by interference colors, fine and large angular pyroxene grains are not uniform. Grains with smoothly changing coloring and zoning, as well as grains with sharp borders between differently colored parts, resembling disintegration textures, can be observed. The variations in the matrix and the unusual features of pyroxene in the matrix are characteristic for all basalts of Samoedsky Suite. Unfortunately, the composition of all coexisting pyroxene phases in different flows was determined for only a few of the basalts, yet the presented data give a good idea about the obvious disequilibrium in the magmatic system during crystallization. In oligophyric basalt, clinopyroxene micro-oikocrysts have the composition $Wo_{37-35}En_{41-37}Fs_{22-28}$, while fine grains represent a wide variety of compositions (see Rock indication 51): $Wo_{41}En_{45}Fs_{14}$, $Wo_{33}En_{41}Fs_{26}$, and $Wo_{12}En_{47}Fs_{41}$. Microscopic study shows that calcium-deprived pyroxenes form individual grains, rim onto high-calcium varieties, or form discrete rims around them. Decreasing calcium content in the mineral is accompanied by increasing iron content and less apparent variations in magnesium content. The content of other mineral-forming components are TiO_2 (0.56–1.62 wt%), Al_2O_3 (0.62–2.32 wt%), MnO (0.18–0.50 wt%), Na_2O (0.13–0.28 wt%), and Cr_2O_3 (0.0–0.33 wt%).

In the rock groundmass, fine grains of olivine Fa_{53-55} are observed, and ore minerals are titanomagnetite (TiO_2 21.55 wt% and MnO 4.83 wt%) and ilmenite (MnO 0.34 wt% and MgO 2.05 wt%).

Glomeroporphyritic basalt has an intersertal, doleritic, and micro-ophitic texture to the groundmass (see Rock indication 52). Porphyritic segregations contain intergrowths of large plagioclase crystals and microporphyritic olivine crystals. The composition of plagioclase insets corresponds to bytownite An_{81-80}. Chadacrysts in pyroxene and laths in glass correspond to labradorite: An_{66} and An_{63}, respectively. The composition of clinopyroxene in zoned micro-oikocrysts corresponds to $Wo_{42-33}En_{44-39}Fs_{14-28}$ and in fine grains to $Wo_{41}En_{46}Fs_{13}$. A decrease in calcium content in augite is accompanied by a sharp decline in Al_2O_3 (3.50–1.05 wt%) and Cr_2O_3 (0.87–0.001 wt%) content. The content of other oxides changes similarly: TiO_2 (0.97–0.79 wt%), MnO (0.20–0.37 wt%), and Na_2O (0.29–0.26 wt%). In fine clinopyroxene grains, oxide concentrations are TiO_2 (0.52 wt%), Al_2O_3 (2.16 wt%), MnO (0.17 wt%), Na_2O (0.26 wt%), and Cr_2O_3 (0.39 wt %). Composition of porphyritic segregations of olivine about 0.2 mm large corresponds to Fa_{38-40} (NiO, 0.076–0.102 wt%). Mineral granules in the rock matrix correspond to Fa_{48-60}. The ore phase in the rock is ilmenite, which contains MnO (0.35–0.38 wt%) and MgO (1.30–2.10 wt%).

Olivine-plagiophyritic or poliphyritic basalt has a porphyritic texture with a tholeiitic, doleritic, micro-ophitic, micropoikilophitic groundmass texture (see Rock indication 53). Porphyritic segregations are typically plagioclase and less common olivine. Olivine forms large (1 mm) subidiomorphic crystals of Fa_{37} and poikilocrysts (2–2.5 mm), totally replaced with red bowlingite. Fine olivine grains (0.4 mm) in the rock matrix are Fa_{53-54}, and granules (0.02 mm) are Fa_{58}.

Plagioclase phenocrysts have a multizoned structure with a gradual change in anorthite content 82–79–78–70–73 (vol%) from their center to margins. Clinopyroxene chadacrysts correspond to An_{65}, and plagioclase laths in groundmass correspond to An_{53}. In clinopyroxene micro-oikocrysts, a zoned structure was determined with a gradual change in composition from the center to the margins of $Wo_{42-35}En_{43-39}Fs_{15-26}$. Fine grains from groundmass have composition $Wo_{38}En_{44}Fs_{18}$. In titanomagnetite crystals, disintegration structures are clearly seen and form ilmenite and magnetite (see Rock indication 53). At the same time, individual prismatic ilmenite crystals are observed, which contain MnO (0.36 wt%) and MgO (1.61 wt%).

Poikilophitic basalts have a large and small "pea-like" texture, which are characterized by a microdoleritic groundmass or to a lesser degree by tholeiitic groundmass, in the large "pea-like" variety and by a tholeiitic groundmass in the small "pea-like" variety. The "pea-like" texture is due to clinopyroxene poikilocrysts, which vary from 2.5 to 3.5 mm in large "pea-like" (see Rock indication 54) varieties and 0.5–3.5 mm in small "pea-like" (see Rock indication 55) varieties. Clinopyroxene oikocrysts have a zoned structure. The central parts of the poikilocrysts from large and small "pea-like" basalts have a constant composition ($Wo_{39-42}En_{47-73}Fs_{14-15}$), whereas their composition is $Wo_{37-33}En_{43-35}Fs_{20-32}$ at the margin, which points to essential changes which occurred during crystallization. As one can see from the compositional formulas, crystal formation was accompanied by a decrease of Wo and En components and an increase of Fs. The tendency of calcium content to decrease from the center to the margin in zoned augite crystals can be seen in the formation of fine grains of pigeonite ($Wo_{14}En_{46}Fs_{40}$) in groundmass. The changes in concentration of main mineral-forming oxides is accompanied by the growth of TiO_2 content from 0.55 to 2.02 wt% and the lowering of Cr_2O_3 content from 0.80 to 0.004 wt%, from the center to the margins in zoned crystals. The central parts of oikocrysts, in comparison with their margins and with fine grains in the rock matrix, possess the highest concentrations of CaO and Al_2O_3.

Plagioclase forms prismatic crystals and microlites, either as chadacrysts in clinopyroxene oikocrysts or in the rock matrix. Composition of the chadacrysts in large "pea-like" basalts corresponds to An_{72} and in small "pea-like" basalts to An_{55}. The composition of prismatic crystals of different sizes and microlites from the groundmass varies from An_{68} to An_{60}.

Olivine in poikilophitic basalts forms fine grains in the rock matrix and, more rarely, intergrowths in clinopyroxene oikocrysts. The intergrowths usually contain less iron (Fa_{47})

than granules in the matrix ($Fa_{57–59}$). Ore minerals are titanomagnetite with disintegration structures, and ilmenite. Titanomagnetite contains TiO_2 (27.97 wt%), MnO (0.43 wt %), and MgO (2.12 wt%), and in lamellas of disintegration structures contain TiO_2 (3.73 wt%), Al_2O_3 (5.32 wt%), MnO (0.06 wt%), and MgO (1.58 wt%). Trace elements in ilmenite include Al_2O_3 (0.04–1.0 wt%), MnO (0.18–0.60 wt%), and MgO (1.15–1.39 wt%).

The central subsuite of basalts of Samoedsky Suite features a successive upward sequence of highly potassic (K_2O 1.25 wt%) aphyric and moderately potassic oligophyritic (K_2O 0.61 wt%) and aphyric (K_2O 0.31–0.32 wt%) basalt flows.

Aphyric basalt has a micro-ophitic texture with micropoikilo-ophitic elements (see Rock indication 56). The rock consists of plagioclase microlites of varying composition, in chadacrysts An_{68}, in prisms $An_{66–64}$, and in laths An_{59}. Clinopyroxene of micro-oikocrysts has a zoned structure, with a central composition of $Wo_{41–42}En_{47–46}Fs_{13}$ and composition at the margins of $Wo_{40–35}En_{45–43}Fs_{15–22}$. Fine pyroxene grains are augite ($Wo_{40}En_{47}Fs_{13}$) and pigeonite ($Wo_5En_{45}Fs_{50}$). From the center to margins of zoned augites, there is an increase in TiO_2 concentration from 0.54–0.56 to 0.69–0.83 wt% and a decrease in Al_2O_3 concentration from 2.31–2.38 to 1.28–1.98 wt% and Cr_2O_3 from 0.55–0.68 to 0.02–0.12 wt%. Olivine is fine-grained, similar in to clinopyroxene. Olivine composition varies from Fa_{47} as intergrowths in clinopyroxene to Fa_{58} in rock matrix. Ore minerals in basalt are titanomagnetite (TiO_2 13.85–21.12 wt %) and ilmenite. Titanomagnetite contains MnO (0.23–0.33 wt%) and MgO (1.19–1.28 wt%), and ilmenite contains MnO (0.38 wt%) and MgO (2.45 wt%).

Oligophyric basalt has a porphyritic texture with an intersertal groundmass texture (see Rock indication 57). The rock features elongate, dendrite-like clinopyroxene crystals, as well as regions that have a micro-ophitic texture and look like poikilocrysts in transmitted light. Sobolev (1986) proposed that the formation of this poikilophitic texture was the result of the assembly and recrystallization of angular pyroxene grains from micro-ophitic parts into one oikocryst. In the interstices of oligophyric basalt, palagonite and an unaltered light-brown glass with abundant emulsion of black glass globules are observed.

Porphyritic segregations in basalt are composed of zoned crystals of plagioclase $Ab_{36–61}Or_{2–7}An_{63–32}$. Composition of plagioclase microlites varies rather widely from $Ab_{43}Or_2An_{55}$ to $Ab_{56}Or_5An_{39}$. Composition of clinopyroxene corresponds to augite, with the composition of central parts of zoned crystals $Wo_{41–39}En_{38–42}Fs_{20–19}$, and at the margins $Wo_{40–37}En_{41–38}Fs_{19–25}$, and non-zoned fine grains correspond to $Wo_{41}En_{40}Fs_{19}$. Olivine from the rock matrix is hortonolite ($Fa_{60–61}$). In interstitional glasses containing immiscible liquids, only the composition of the glass matrix was determined, whereas the composition of other glass globules was not determined because of their tiny size. The glass matrix has a composition of an alkaline granite, and contains SiO_2 (71.46 wt%), Na_2O (2.19 wt%), and K_2O (6.73 wt%). Glasses with similar composition were described earlier (Ryabov 1989a, b). Oxide-ore minerals are titanomagnetite (TiO_2 23.17–29.55 wt%) and ilmenite. Trace element concentration in titanomagnetite was determined for MnO (0.86–1.20 wt%) and MgO (0.63–0.79 wt%) and in ilmenite: MnO (0.49 wt%) and MgO (1.84 wt%).

Aphyric moderately potassic basalt has an intersertal, doleritic, partly micro-ophitic texture (see Rock indications 58, 59). Plagioclase forms as microlites with a variable composition, An_{70}–An_{56}. The most mafic plagioclases occur as $An_{70–62}$ chadocrysts intergrown in clinopyroxene micro-oikocrysts and relatively large prismatic $An_{68–61}$ crystals. Small microlites in the rock matrix have the composition $An_{63–61}$ and in glass are An_{56}. Clinopyroxene grains from micro-oikocrystals and micro-ophitic parts of the rock correspond to augite ($Wo_{40–39}En_{43–44}Fs_{17–18}$). The mineral contains TiO_2 (0.75–1.43 wt%), Al_2O_3 (1.89–3.48 wt%), Na_2O (0.24–0.31 wt%), and Cr_2O_3 (0.01–0.37 wt%). Higher Al_2O_3 content is characteristic for cores of zoned crystals. In the rock matrix, olivine forms fine grains ($Fa_{52–60}$). Interstitional glass is more homogenous in its composition; SiO_2 was determined at 70.60–71.22 wt%. Oxide-ore minerals are titanomagnetite (TiO_2 25.36–26.81 wt%) and ilmenite. Titanomagnetite contains traces of MnO (1.52–1.64 wt%), and olivine contains MnO (0.43–0.46 wt %) and MgO (1.47–2.25 wt%).

2.13 Structural and Compositional Peculiarities of Volcanogenic Sequences of the Noril'sk Region

In last decade, precision geochemical analyses of Siberian Traps, undertaken in Canadian and US laboratories, have been published (Naldrett et al. 1992; Lightfoot et al. 1994). Based on these data, models were proposed explaining the variety of lava compositions by a plume source for melts (Campbell and Griffiths 1990; Sharma et al. 1992; Wooden et al. 1993) and by contamination, fractionation, and mixing of several types of mantle magmas (Lightfoot et al. 1990, 1993). However, the majority of Russian geologists in earlier and contemporary publications deny any influence of crustal contamination on magma composition (Almukhamedov et al. 1992; Gladkikh et al. 1994), explaining the magmatic variety by peculiarities in the basalt melt differentiation according to a local regime. Making no pretense to thorough discussion of the complicated problem of genesis of Siberian Traps, we will confine ourselves to a number of statements resulting from our investigations, essential for the development of a petrogenetic model of trap magmatism evolution.

Long-term research has allowed geologists to establish the oscillatory character of the volcanism, a continuity of violent explosions and quiescent lava outpouring, frequently followed by periods of rest and sedimentation. The discovered regularities, the cyclical nature of the magmatism, and the characteristic compositional features have created the basis for a scheme of differentiating effusive-explosive suites and intrusive complexes, as well as concepts for their comagmatic genesis.

Erupting lavas had varying fluid content and composition. Indirect evidence of the enrichment of magma with volatiles is the thick amygdaloidal zones, the high degree of rock crystallinity formed from portions of fluidized melt, and the evidence for volcanism with high explosivity coefficient. The presence of volatiles explains the phenomenon of fluid-magmatic differentiation in the flows into two-plagioclase and glomeroporphyritic basalts, most clearly expressed in the flows of picritic basalt of the Gudchikhinsky Suite.

Early magmas of effusive traps in the northwest of the Siberian Platform had a tholeiitic basalt, trachybasalt (subalkaline or alkaline olivine-basalt), and picritic basalt composition. The majority of effusive traps are represented by tholeiitic basalts (93 vol%), much more rarely subalkaline basalts and trachybasalts (6 vol%), as well as picritic basalts (1 vol%) (Gladkikh et al. 1994). These three main types of parental magma are reflected in a (Na_2O + K_2O)–SiO_2 diagram (Fig. 2.23). The diagram shows that the rock compositions in lava flows from the Syverminsky Suite to the Samoedsky Suite are predominantly tholeiitic basalt with some subalkaline varieties, and their compositions fall into the trachybasalt field. All three singled out fields form a chain in which tholeiites serve as a link between picritic basalts and trachybasalts. This allows us to assume with certainty that the genetic connection between picrite-, tholeiite- and trachybasalt melts, and their probable formation was the result of prechamber differentiation of a parental tholeiitic basalt magma.

Lavas with different compositions erupted within one volcanic cycle, as well as the recurrence of cycles with different combinations and proportions of these lavas, point at a genetic connection between tholeiitic and picritic basalts and trachybasalts, allowing us to suppose that their formation was the result of magmatic differentiation of a parental tholeiitic basalt melt.

Tholeiitic basalts are the main rock type of the Siberian Traps. However, the composition of the rocks varies for all geochemical indicators. The characteristic feature for these tholeiites is the sharp predominance of potassium depletion ($K_2O < 0.3$ wt%) with rare flows of moderately and highly potassic basalts (Zolotukhin and Almukhamedov 1991; Gladkikh et al. 1994).

The flows of subalkaline and alkaline basalt are observed repeatedly. They form the Ivakinsky Suite and Yuryakhsky Unit and sporadically form discontinuous flows in other regions, from Syverminsky to Kharaelakhsky, inclusively. Composition of the rocks within various volcanic depressions and suites varies, changing from early to late eruptions, as can be seen in the flows of Ivakinsky Suite (Fedorenko et al. 1984; Distler and Kunilov 1994).

Trachybasalts of the Yuryakhsky Unit uniquely feature magnetite lavas, serving as a proof of the existence of ore-magmas and ore-silicate liquid immiscibility in prechamber conditions.

Flows of highly magnesian lava occurred repeatedly in the Gudchikhinsky, Tuklonsky, Morongovsky, and Mokulaevsky Suites and are composed of picritic basalts of varying compositions, olivine-phyric basalt, ankaramite, and limburgite.

Petrographic studies of picritic basalt flows point to a lack of olivine gravitational separation. The small size and skeletal appearance of ilmenite crystals in mandelsteines of the picritic basalt flows exclude an intratelluric origin, instead pointing at nucleation and crystallization of mineral during the course of melt eruption. Peculiarities in the chemical composition and texture of picritic basalts do not give grounds to regard them as komatiites (Frolova and Kotorgin 1986).

The formation of picritic basalt differentiated lava flows (Talnakhsky, Daldykansky, and Val'kovsky) was the result of fluid-magmatic differentiation of a picrate-basalt melt in situ, which led to formation of metadiorites and picritic basalt with higher magnesium content.

Picritic basalt melts erupted at the surface as individual lavas or as an immiscible mixture with tholeiitic basalt melts. The uniqueness of the Mikchandinsky differentiated lava flow, composed of altering layers of dendrites and spherocrysts of clinopyroxene, is due to the crystallization of layered tholeiitic basalt and fluidized Picritic basalt liquids in undercooling conditions. The specific composition and the peculiarity in the crystallization of phenocrysts in the ankaramite of Arylakh allow us to theorize that their genesis may be the same as that of spherocrysts of the Mikchandinsky differentiated lava flow.

The structural peculiarities of effusive traps, and their crystal morphology and composition, allow us to claim there is an absence of mineral phases that can be regarded as intratelluric formations. Thus, the erupted magmatic melt was homogenous or split into immiscible liquids. Crystallization of Picritic basalt melts began in near-surface conditions and proceeded during the flow of lava across the surface.

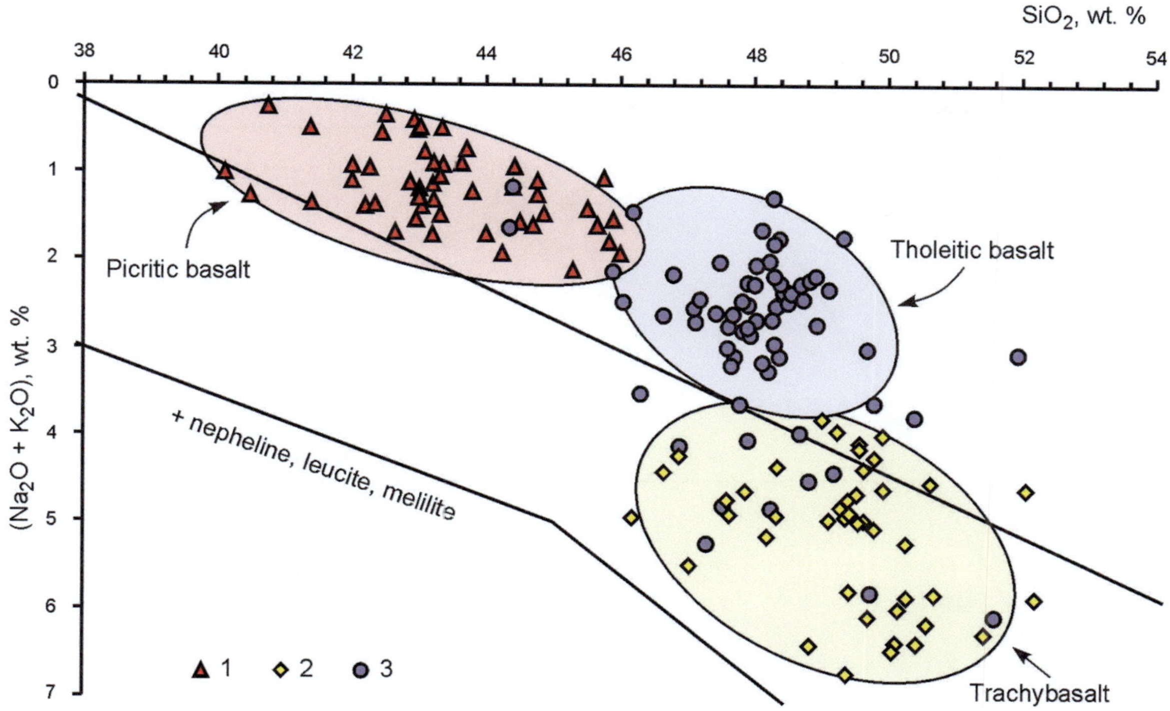

Fig. 2.23 ($Na_2O + K_2O$)—SiO_2 diagram for effusive rocks of the Noril'sk Region. *1* Picritic basalts of Gudchikhinsky and Tuklonsky Suites and ankaramites of Morongovsky Suite, *2* trachybasalts, trachyandesite-basalts, and subalkali basalts of Ivakinsky Suite and Yuryakhsky Unit, *3* basalts of the Syverminsky–Samoedsky Suites (for explanations, see the text of the book)

The main rock-forming minerals in effusive traps are plagioclase and pyroxene, in picritic basalt they are accompanied by olivine, and in trachybasalt by alkali feldspars. Ore minerals in picritic basalt are typically chrome spinel and in the other rocks are titanomagnetite and ilmenite.

The composition of plagioclase from the whole range of basalts lies in the range of An_{84}–An_1. As a rule, crystals have normal zonality. Changes in composition from the center to the margin within a single crystal can reach 35 units. Porphyritic segregations of plagioclase in tholeiitic basalts correspond to An_{84-49} and in trachybasalts to An_{65-49}. Microlites from the groundmass and chadacrysts include a range of compositions of plagioclase and K-Na feldspar: in tholeiites An_{72-46} and in trachybasalts An_{54-1} and $Ab_{59-52}Or_{33-47}An_{8-1}$. The composition of plagioclase from picritic basalt corresponds to bytownite An_{77-71}. More acid plagioclase is comparatively rare.

Pyroxenes of effusive traps form monoclinic and rhombic varieties. According to published pyroxene classification (Morimoto 1989), 93 vol% of 250 microprobe analyses of pyroxenes (160 given here, and 90 previously published) correspond to augite, 4 vol% to pigeonite, and 3 vol% to orthopyroxene. Augites are present in all varieties of traps: in tholeiitic basalts their composition is $Wo_{26-43}En_{31-51}Fs_{11-32}$, in picritic basalts $Wo_{32-42}En_{38-54}Fs_{10-27}$, and in trachybasalts $Wo_{36-43}En_{24-47}Fs_{16-37}$. Pigeonite formed in picritic and tholeiitic basalts. In picritic and olivine-phyric basalts and ankaramites of the Gudchikhinsky, Tuklonsky, and Morongovsky Suites, the composition of pigeonite is $Wo_{7-9}En_{65-71}Fs_{22-28}$. In tholeiitic basalts, pigeonite has been observed in the Nadezhdinsky, Mokulaevsky, and Samoedsky Suites. Crystals of pigeonite are most common in the Samoedsky Suite and are bound to final stage of crystallization of augites with varying calcium content. The composition of pigeonite in these suites corresponds to $Wo_{5-9}En_{45-49}Fs_{33-50}$.

Orthopyroxene is found only in picritic basalts of Gudchikhinsky Suite. Its composition corresponds to bronzite ($Wo_{3-4}En_{71-77}Fs_{20-25}$).

Pyroxenes of Mikchandinsky differentiated lava flow are augite ($Wo_{35-40}En_{40-53}Fs_{10-25}$) and pigeonite ($Wo_{8-11}En_{50-63}Fs_{29-41}$). Because this sequence of lava flows is so unique, analyses of clinopyroxene from the Mikchandinsky differentiated lava flow were not included in calculations of the mineral occurrence in basalt. It should be noted that the composition of augites from this lava flow sequence falls into the range of compositions of this mineral in picritic basalts, whereas the composition of Mikchandinsky pigeonites continues the compositional range of picritic basalts into the region of higher iron content.

Composition of olivines from picritic basalts is Fa_{19-30} and porphyritic segregations of olivine in tholeiitic basalts are Fa_{37-43}, whereas those of fine crystals in the rock matrix are Fa_{43-63}. Olivines in trachybasalts have a composition of Fa_{48} and in trachyandesite-basalts of Fa_{71}. The content of silicate nickel in olivine depends on its iron content. In picritic basalts, NiO content in olivine reaches 0.35–0.46 wt%, indicating the mineral is a probable source of sulfide mineralization of Noril'sk deposits.

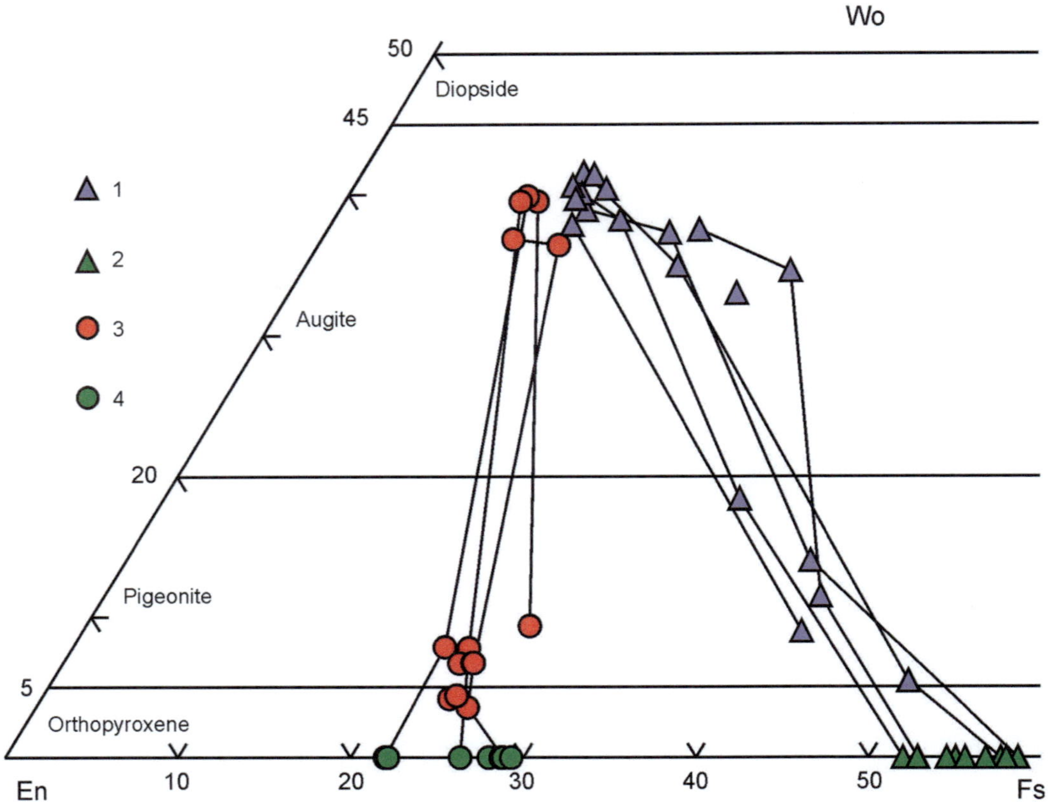

Fig. 2.24 A part of the Hess trapezium diagram for coexisting pyroxenes and olivines from tholeiitic and picritic basalts of the Noril'sk Region. *1–2* Tholeiitic basalts of Nadezhdinsky, Mokulaevsky, and Samoedsky Suites: *1* pyroxenes, *2* olivine (ferriferocity of olivine in Fa-component, %). *3–4* Picritic basalts of Gudchikhinsky and Tuklonsky Suites: *3* pyroxenes, *4* olivine (ferriferocity of olivine in Fa-component, %). Coarse lines and figures (%) mark boundary contents of Wo-component, dividing diopside, augite, pigeonite, and orthopyroxene (after Morimoto 1989)

General ideas of the evolutional peculiarities of pyroxenes in tholeiitic and picritic basalts are given in the Hess trapezium diagram (Fig. 2.24). Here, from the mass of existing analytical data, we have chosen a number of Rock indications, containing data for the center and margins of large clinopyroxene crystals and fine clinopyroxene and olivine grains from groundmass. Coexisting phases are connected by lines. The composition of olivine phenocrystals is not given in the diagram. According to the diagram, crystallization of pyroxene in picritic and tholeiitic basalts began with the formation of highly calcic augite. Further melt composition evolution led to a decrease in calcium content and an increase in iron content, leading to the formation of orthopyroxene and pigeonite in tholeiitic basalts. Crystallization of basalt melts was completed with the formation of olivine, which represents a link back to the uniform crystallization trend of mafic minerals.

Spinel group minerals in effusive traps are typically titanomagnetite in tholeiitic basalts and trachybasalts and by chrome spinels in picritic basalts. The ulvospinel component in titanomagnetites varies from 6 to 94 vol%. Mineral varieties close to stoichiometrically pure ulvospinel were detected in tholeiitic basalts, and titanomagnetites containing more than 50 vol% ulvospinel frequently occur in tholeiites and trachybasalts. The proportion of coulsonite in titanomagnetite reaches 2 vol%.

Chrome spinels, as a rule, form in picritic basalts. One exception is chromite in glomeroporphyritic basalt of Gudchikhinsky Suite. The content of chromite in chrome spinels varies from 48 to 61 vol%, the content of ulvospinel varies from 4 to 10 vol%, the content of Al-spinel varies from 23 to 39 vol%, and magnetite content varies from 8 to 15 vol%. The evolution of the composition of chrome spinels during crystallization proceeded toward a decrease in chromite component and an increase of ulvospinel and magnetite components. The most high-chromium spinels in the Traps are found in mandelsteine flows of picritic basalt.

2.14 Marker Horizons in Volcanogenic Units

The main part of the volcanogenic series consists of tholeiitic basalts of volcanic sequences III–V with a rather uniform elemental composition. Rhythmic alternation of pyroclastic and effusive rocks allows the volcanogenic series to be divided into suites and subsuites. Marker horizons represent one of the most important elements of stratigraphic separation (Mezhvilk 1962; Starosel'tsev 1989; Ryabov et al. 2009). They are situated in the base of suites and subsuites, are characterized by great thickness, and can be traced in each stratigraphic unit over the entire basalt field of the Platform. Upward through the volcanogenic series, nine reference flows are distinguished: the Kharpichsky flow and a "nameless" flow at the base of the middle subsuite in the Nadezhdinsky Suite, the Shadrinsky and Dyupkunsky (Anamsky) flows in the Ayansky Suite, the Nadayansky and Delochinsky flows in the Khonnamakitsky and Mokulaevsky Suites (the latter is similar in facies to the former), the Yaktaliysky and Agitkansky (Kaltaminsky) flows in the Nerakarsky Suite, and the Yambukansky flow in the Nerakarsky Suite (Fig. 2.25).

In the southern parts of the basalt field of the Platform, the names of some marker horizons are changed: Nadayansky to Nidymsky, Yaktaliysky to Kochechumsky, and Kaltamninsky to Agytkansky (Mezhvilk and Vasil'ev 1967). Outcrop is good, so contemporary erosion allows observation of a thickness of lava 500–600 m or more thick, with two to three marker horizons, traced over a large distance (Fig. 2.26). Marker horizons are situated at the base of suites and subsuites. As a rule, they overlay a surface of partly eroded horizon of tuffaceous rocks.

Paleomagnetic studies of more than 200 core samples in the Kharaelakhsky Suite confirmed the (already known) normal magnetization of Triassic lavas and reverse magnetization of Permian lavas (Mikhaltsov et al. 2008). In addition, it appears that within magnetic zones with prevailing polarity, there are zones of abrupt reversals in magnetic inclination (Fig. 2.27). In suites of tholeiitic basalt, the change of polarity occurs in marginal flows and adjacent tuffs at suites and subsuite boundaries. Polarity changes are also found at the boundaries between picritic and glomeroporphyritic tholeiitic basalts of Gudchikhinsky Suite, as well as the boundaries of trachybasalt subsuites of the Ivakinsky Suite. The confined nature of polarity changes at the boundaries of volcanogenic stratigraphic units formerly allocated by geologists is not well understood. Geoscientists who conducted this paleomagnetic research suppose the change of polarity to be due to the Earth's geomagnetic field excursions. The question is not definitively solved yet, but it is important as the coincidence of narrow intervals of reverse polarity with marker horizons gives an additional "magnetic" mark for the correlation of flood basalts of the region, and the zones with reverse polarity correspond to episodes of higher volcanic activity within the frames of each volcanic cycle.

The Nadayansky (150 m) and Yaktaliysky (135 m) marker horizons are the thickest. In outcrop, the area of the Yaktaliysky lava flows sequence is 292,500 km^2 and Delochinsky lava flow sequence is 237,500 km^2 (Starosel'tsev 1989). The original area of these and other marker horizons may have been much larger, for it is difficult to appraise the area that has been eroded and the unexposed area.

Marker horizons are traced over large distances, so we aimed to establish their petrographic, mineralogical, and geochemical features. To determine intraflow differentiation, samples were drawn from key sections through 5–10 m, depending on rock texture and flow thickness. To identify possible areal variation in rock composition, samples 1 m above the flow base were collected. The key marker horizon sections were studied for the Bol'sheavamsky, Ikensky, Kharaelakhsky, Vologochansky, Khantaysky, and Agatsky volcanic troughs of the Noril'sko–Kharaelakhsky depression and Tungussky syneclise. The Nadayansky lava flow sequence featured the best outcrop and was the least eroded and, hence, was studied in the most detail.

The Nadayansky lava flow sequence is about 100 × 170 km (Pol'kin 1962) to 270 × 250 km (Nesterenko et al. 1991) in size, based on outcrop. The thickness varies from 25 to 150 m, with an average thickness of 100 m. It spread over an area of about 70,000 km^2 and erupted more than 7,000 km^3 of magma.

In the most complete section of the Nadayansky lava flow sequence, basalts near the base (1–10 m) are characterized by thick columnar jointing (0.4–0.6 m diameter) (Fig. 2.26). Up through the sequence, the thick columns bifurcate abruptly, forming thin columns (0.15–0.3 m diameter). Thin columns of the jointing are vertical near the base and upward the sequence they form fan-like structures. The upper part of the sequence is an 8–15-m-thick amygdaloidal zone.

The basalts are composed of plagioclase (48–56%), clinopyroxene (37–42%), ore oxides (3–5%), olivine (+bowlingite) (0–5%), and palagonite (3–7%). Through the whole sequence, basalts have a porphyritic, serial porphyritic, or glomeroporphyritic texture. The size and quantity of phenocrysts increase in the middle part of the sequence. In the central part of the sequence, the groundmass texture is poikilophitic, in the upper and lower parts—doleritic with some elements of fluidal, trachytoid, tiny poikilophitic, and tholeiitic texture. Individual amygdales formed throughout the sequence, forming segregations in mandelsteines. Mandelsteines have a porphyric texture, with an intersertal groundmass.

Large phenocrysts are plagioclase An_{84-70} with sporadic ingrowths of olivine, smaller phoncrysts are An_{72-66}, and laths are An_{68-44}. Clinopyroxene forms oikocrysts of $Wo_{43-36}En_{36-49}Fs_{12-22}$ and small granules of $Wo_{41-30}En_{46-38}Fs_{13-32}$. Small olivine grains in the matrix are chrysolite Fa_{29}, though more frequently they are replaced with bowlingite. Ore oxides in basalts with a doleritic groundmass are typically homogenous titanomagnetite

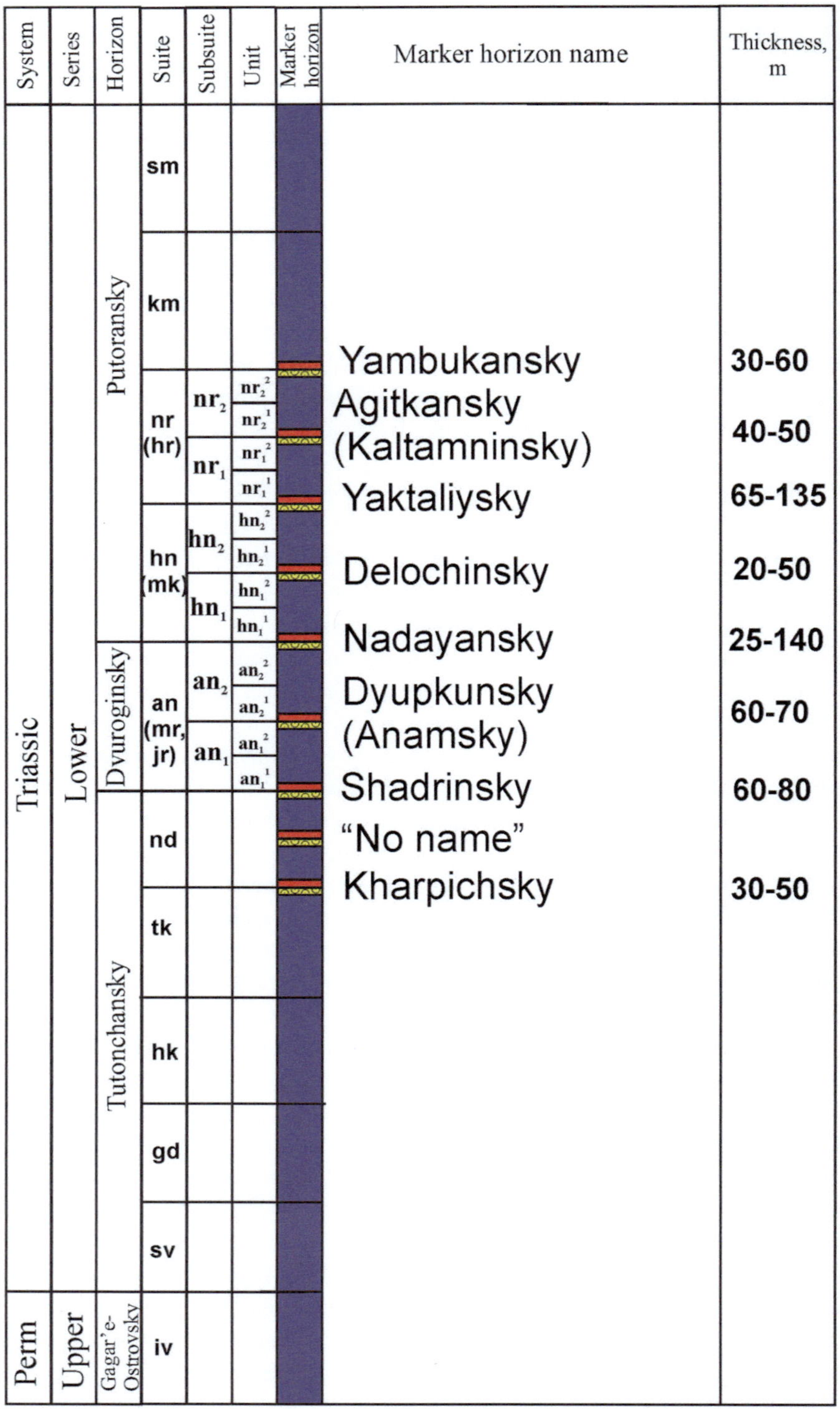

Fig. 2.25 Marker horizons of effusive traps of the Siberian Platform. Names of analogues are given in brackets. Names of the suites are given in Table 2.1

(TiO_2 16.7–28.28 wt%, MgO 0.11–1.49 wt%), whereas in basalts with a poikilophitic structure that are typically magnetite (TiO_2 1.55–3.98 wt%, MgO 1.23–1.87 wt%, MnO 0.19–2.18 wt%) and ilmenite (MgO 0.12–2.06; MnO 0.77–1.83 wt%).

The solidification thermal regime affected the degree of crystallinity and changed the texture of the basalts from contact zones to the center of the sequence. Crystallization differentiation manifested itself as changes of mineral composition, that is, in plagioclase deoxidation from An_{84} to An_{44}, an increase in iron content in clinopyroxene from Fs_{12} to Fs_{32}, and the breakdown in solid solution of homogenous titanomagnetite into magnetite and ilmenite.

Basalts of the Nadayansky lava flow sequence have a stable chemical composition. The contents of rock-forming oxides are (wt%): SiO_2 48.21–49.42, TiO_2 1.11–1.33, Al_2O_3

Fig. 2.26 Nadayansky marker horizon, the Uokhir' River

15.08–16.31, FeO_{tot} 10.65–12.21, MnO 0.16–0.19, MgO 6.24–7.49, CaO 11.24–12.03, Na_2O 1.91–2.24, K_2O 0.15–0.61, P_2O_5 0.12–0.14, and LOI is 0.56–2.54. Despite insignificant variations in oxide content upward through the sequence, some changes were noted. In the alkaline silica oxide diagram, the rock compositions form a local field coinciding with the field of rock compositions for the Khonnamakitsky Suite as well as the Mokulaevsky Suite (Fig. 2.28). The basalts vary from low to moderately potassic, which is reflected in variations in K_2O content. It is probably connected with the high mobility of potassium and magmatic as well as postmagmatic processes. This can also explain the behavior of rubidium, which mirrors potassium in course of all processes. Nonmobile elements such as Cr, V, Ni, Sc, as well as REE, (especially medium and heavy REE) reveal amazing stability in the diagrams; throughout the studied sequence, the content of these elements forms almost a straight line. Spidergrams show that the composition of basalts of the Nadayansky lava flow sequence practically coincides with the dots for effusive rocks of the flood stage of volcanism of the Siberian Platform (Almukhamedov et al. 2004). The stability of rock composition of the Nadayansky lava flow sequence demonstrates the same parental melt, fracture character of volcanism, and small formation time.

2.15 Anomalous Formations in Flood Basalts: Indicators of the Geodynamic Environment in Trap Magmatism

In the northwest of the Siberian Platform, among the overwhelming mass of tholeiitic basalt lava flows, a complex of formations offers the best insight into the peculiarities of volcanism on the Platform, despite their limited extent. These formations are uncommon, in comparison with flood basalts of uniform composition and textures, and are singled out as anomalous (Ryabov et al. 2005). These formations include local manifestations of high-magnesium and subalkaline basalt lavas among plateau basalts; exotic manifestations of flows of komatiite-like rocks, limburgites, picritic porphyries, and magnetite lavas; and tuff paleovolcanoes that were not flow generating, as well as numerous dykes of varying composition. In the same group are found the calderas with beds of limestone and anhydrites as well as dense bodies of native copper and bitumen in volcanogenic rocks and diatremes with sulfide and magnetite ores. Anomalous formations are characteristically confined to fracture zones, rift structures, and volcanic structures. The form, scale, and composition of the anomalous formations were predetermined by the depth of fracture formation, the maturity of the rift structures, and the history of tectonic development in the region preceding volcanic climaxes.

2.15.1 High-Magnesium Lavas in Traps

Despite their small volume, manifestations of high-magnesium effusive rocks (MgO $>$ 12 wt%) on the Platform are important for stratigraphic division of volcanogenic series, for correlation of sections, as well as for an understanding of the geodynamic situation during trap formation.

High-magnesium lava outpourings in the region took place repeatedly (Fig. 2.29); the largest area and thickness of lava flows enriched with olivine are in the Noril'sk (N) and Maymecha–Kotuysky (M–K) stratigraphic regions. In the Kamensky Province (K), lavas of picritic porphyries and limburgites are locally spread. In the Tungussky syneclise (TS), the flows of picritic and olivine-phyric basalt manifest in the western (WB) and the northern (NB) borders, being absent in the eastern border (EB) (Ryabov et al. 2003).

One of the features of manifestations of high-magnesium lavas on the Siberian Platform is their confined nature to a narrow chain of depressions along the northwestern and the north rims of the Platform and on the edge of riftogenic

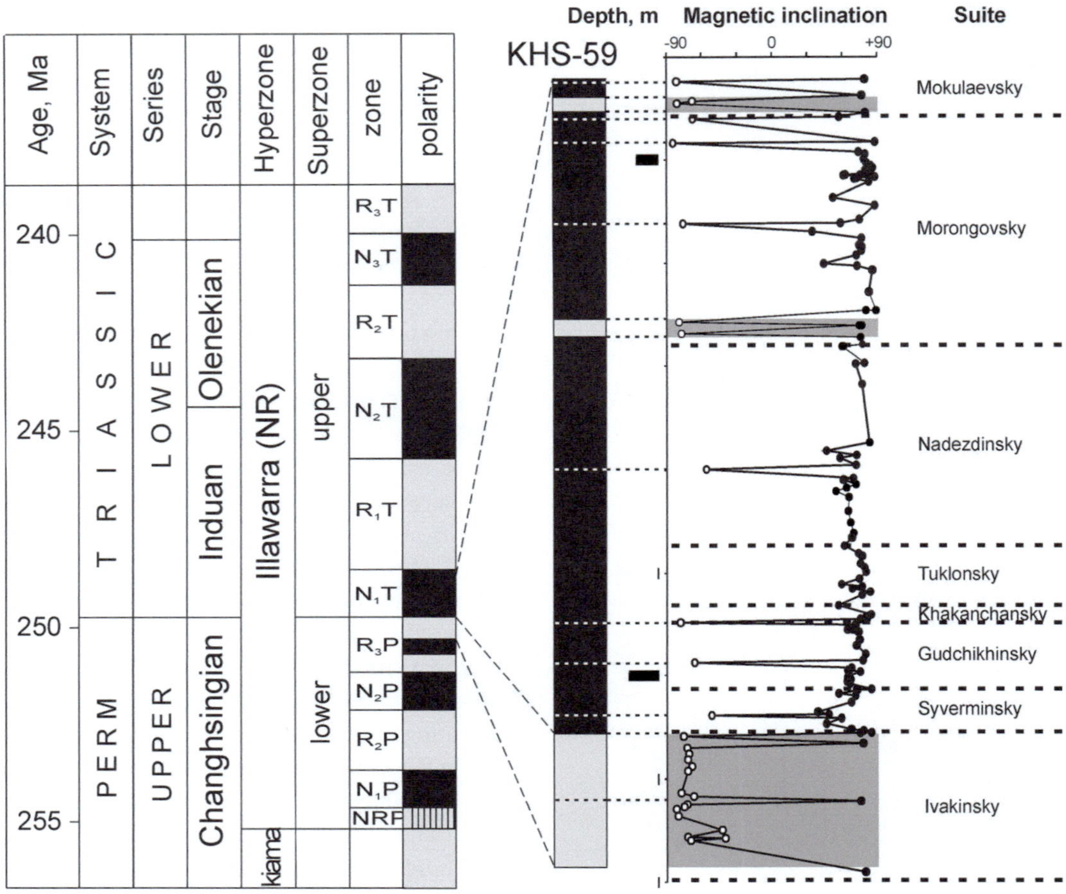

Fig. 2.27 Magnetostratigraphic section of volcanogenic formation according to the well No KhS-59 (the Kharaelakhsky depression) (Mikhaltsov et al. 2008)

structures (Fig. 2.30). Within the region, manifestations of high-magnesium lavas have a nonuniform distribution, being confined to the northwest of the lava field of the Platform in the Noril'sk Province and to the northeastern part in the Maymecha–Kotuysky Province.

At the northwest of the Siberian Platform, picritic and olivine-phyric basalts and ankaramites represent high-magnesium lavas. They are spread in volcanic troughs of the Noril'sko-Kharaelakhsky depression and Mikchanda-Khantaysky depression in the western limb of the Tungussky syneclise. Eruptions of high-magnesium lavas in this part of the Platform took place in three stages, reflected in the Gudchikhinsky, Tuklonsky, and Morongovsky Suites (see Figs. 2.29 and 2.30). The largest-scale outpouring of picritic basalt took place during formation of the Gudchikhinsky Suite. They are represented in all volcanic troughs of the Noril'sko-Kharaelakhsky depression, in the Noril'sky, Vologochansky, Kharaelakhsky, and Ikensky troughs, numbering 20 flows about 200 m thick. At the western border of the Tungussky syneclise, picritic basalt of the Gudchikhinsky Suite is noted in the Mikchandinsky, Lamsky, and Imangdinsky troughs. Here, the number of flow sequences varies from 1 to 3, with their total thickness from 3 to 15–20 m.

Isopachytes of picritic lava of the Gudchikhinsky Suite stretch along the Platform rim and open toward the Yenisey-Khatangsky riftogenic depression, whereas areas of maximum concentration lie close to the edges of the Platform (Zolotukhin et al. 1986). It is supposed that the elongated shape of isopachytes with maximum thickness of lava reflects magma-feeding channels.

Olivine-phyric and picritic basalt of Tuklonsky Suite in the northwest of the Platform were noted in the southern part of the Kharaelakhsky and Ikensky troughs and at the western border of the Tungussky syneclise in the Mikchandinsky, Lamsky, and Imangdinsky troughs. The number of high-magnesium lava flows varies from 1 to 3, with total thickness 5 to 15–25 m. In the northern part of the Ikensky trough, in the roof of Morongovsky Suite, a flow sequence of ankaramite and picritic basalt 2.5 to 12 m thick was observed.

In the northeast of the lava field in the Maymecha–Kotuysky Province, high-magnesium lavas in the

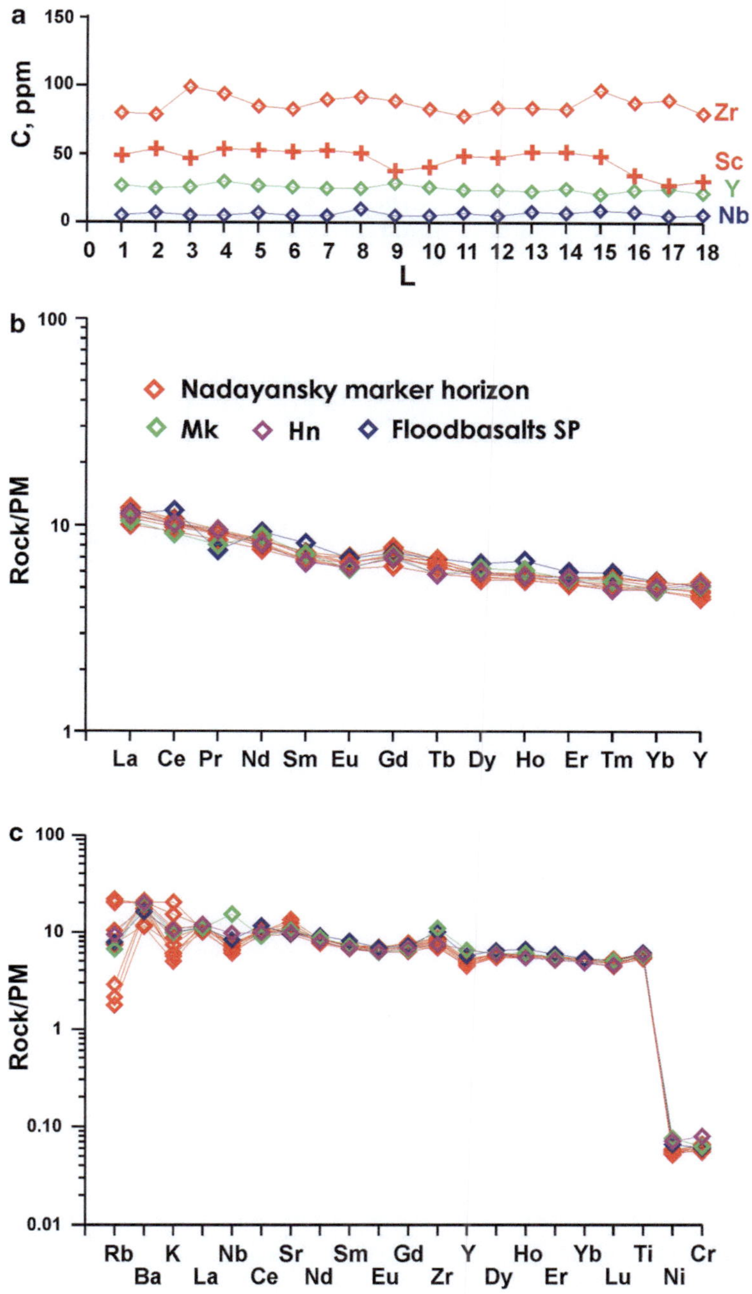

Fig. 2.28 Distribution of trace elements in basalts of the marking horizons. (**a**) Distribution of Zr, Sc, Y, and Nb in basalts of Nadayansky marker horizon, northwest of the Siberian Platform. *C* Concentration, *L* sample numbers. (**b**, **c**) Distribution of trace elements in the Nadayansky horizon, in comparison with that in the basalts of Mokulaevsky and Khonnamakitsky Suites and middle flood basalt of the Siberian Platform, after Almukhamedov et al. (2004)

Del'kansky trough form flows of picritic porphyries and limburgites in the Arydzhangsky, Kogotoksky, and Del'kansky Suites, and meimechites and picritic porphyries in the Maymechinsky Suite. The thickness of high-magnesium lavas in this region is 500–700 m. High-magnesium lavas of the province are characterized by an alkali-ultramafic composition.

Between the Noril'sk and Maymecha–Kotuysky Provinces, small-scale high-magnesium lava flows are found to the north of the Tungussky syneclise, at the confluence of the Kaltama and the Ayan Rivers (see Fig. 2.30). It represents a lava flow sequence of picritic porphyries enriched with alkali elements, about 17–25 m thick, confined to the upper part of Ayansky Suite, an analogue of the Morongovsky Suite.

2.15.2 Limburgite and Picritic Porphyrite Lava Flow Sequences

In the Bol'sheavamsky trough in the Kamensky Province of alkali-mafic rocks, the presence of high-magnesium lava was unknown until recently (see Fig. 2.30) (Ryabov 2003). The province is unique for the wide presence of small intrusions of alkali-mafic, alkali-ultramafic, and alkaline composition, close to that of similar magmatic formations in the Maymecha–Kotuysky Province. The small intrusions

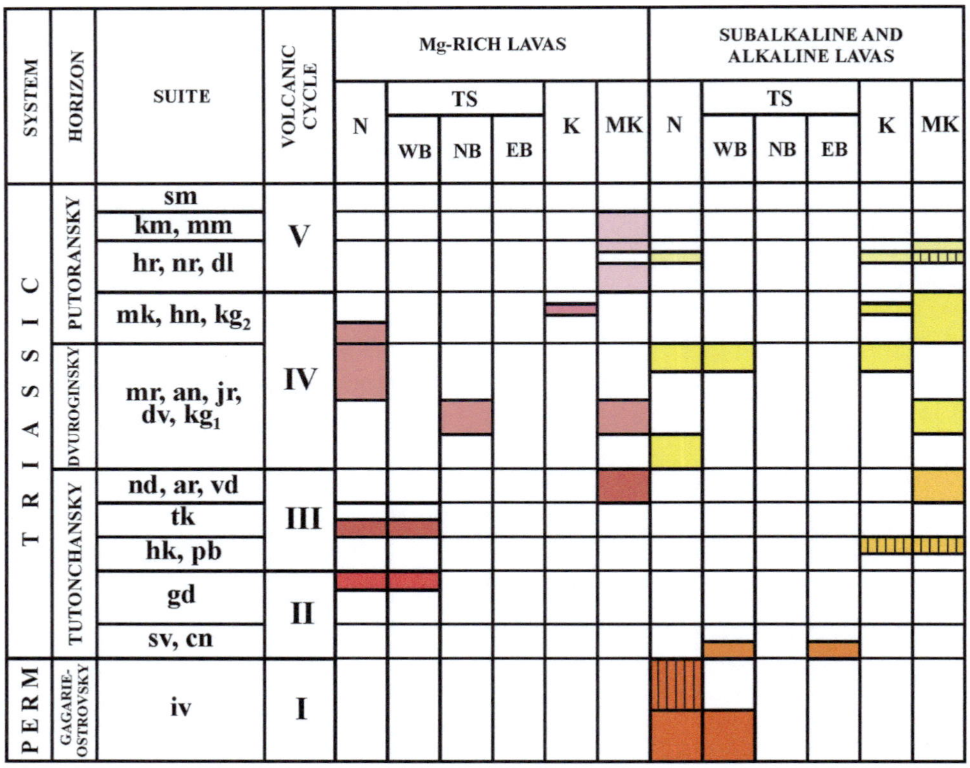

Fig. 2.29 Distribution of lava covers with anomalous compositions in the section of plateau basalts of the Siberian Platform. Suites: *ar* Arydzhangsky, *an* Ayansky, *cn* Changadinsky, *dl* Del'kansky, *dv* Dvuroginsky, *gd* Gudchikhinsky, *hk* Khakanchansky, *hr* Kharaelakhsky, *hn* Khonnamakitsky, *iv* Ivakinsky, *jr* Yuryakhsky, *kg* Kogotoksky, *km* Kumginsky, *mm* Maymechinsky, *mk* Mokulaevsky, *mr* Morongovsky, *nd* Nadezhdinsky, *nr* Nerakarsky, *pb* Pravoboyarsky, *sm* Samoedsky, *sv* Syverminsky, *tk* Tuklonsky, *vd* Vodopadny. The vertical hatching in the fields of distribution of subalkaline and alkaline lavas shows andesite–basalts and trachyandesite–basalts (for explanations, see the text of the book)

of the Kamensky Province have different compositions, and dykes of limburgites, picritic porphyries, and alkaline picrites were also noted. The high-magnesian lava flow found at the top of the first unit of the upper subsuite in the Khonnamakitsky (Mokulaevsky) Suite and traced over an area is of great interest. Along strike the composition of the lava flow rocks varies from limburgite to picritic porphyry. The limburgite and picritic porphyry are considered as a flow sequence due to the presence of textures formed at the surface (rope lava in the region of the Ketere River). In the Bokovoje Lake region, these rocks are represented only by limburgite. The rock outcrops in small areas at the tops of mountains, traced in a northeasterly direction. Other outcrops of high-magnesium lava are found on the left side of the Namakan River below the Uokhir' River branch and along the main inflow of the Ketere River. They form a sequence of picritic basalt flows 7–15 m thick, which is traced in outcrops for a number of kilometers. Some data on composition of the rocks of the lava in the three above-mentioned areas are given below.

Limburgite (BK-6) from the lava flow sequence in the region of Lake Bokovoje is a glassy rock with phenocrysts of olivine (25–35%) ($Fa_{19–20}$) and clinopyroxene (5–10%) ($Wo_{42–39}En_{43–49}Fs_{15–12}$). The groundmass consists of glass (30–40%) with small aggregations of clinopyroxene granules and fur-like titanomagnetite crystals. The glass color varies from black, opaque, to light-brown (most prevalent) and colorless. The glass composition varies widely. In clinopyroxene, in the row described as "center–rims–granule from groundmass," an increase in SiO_2 and MgO and a decrease in TiO_2, Al_2O_3, and FeO are noted, along with rather stable contents of CaO (Table 2.6, Fig. 2.31).

Picritic porphyry (N-137) from the lava flow sequence of the Namakan River has a seriate-porphyry texture with a granular, intersertal textured groundmass. Phenocrysts (30–35%) are olivine-forming large, subidiomorphic zoned crystals ($Fa_{14–20}$) and aggregates of small grains Fa_{21}, as well as single prisms of plagioclase ($Ab_{51}Or_4An_{45}$). The groundmass consists of granules of clinopyroxene ($Wo_{40}En_{46}Fs_{14}$), xenomorphic segregations of plagioclase ($Ab_{67-71}Or_{12}An_{21-17}$) and feldspar ($Ab_{37–57}Ort_{48–29}An_{15–14}$), glass, ilmenite prisms, rare titanomagnetite grains, and phlogopite. Beside interstitional glass, the rock contains segregations (5–10%) comparable in size with large phenocrysts. The color of the glass varies from light-brown to colorless. In the glass, long prisms of orthopyroxene

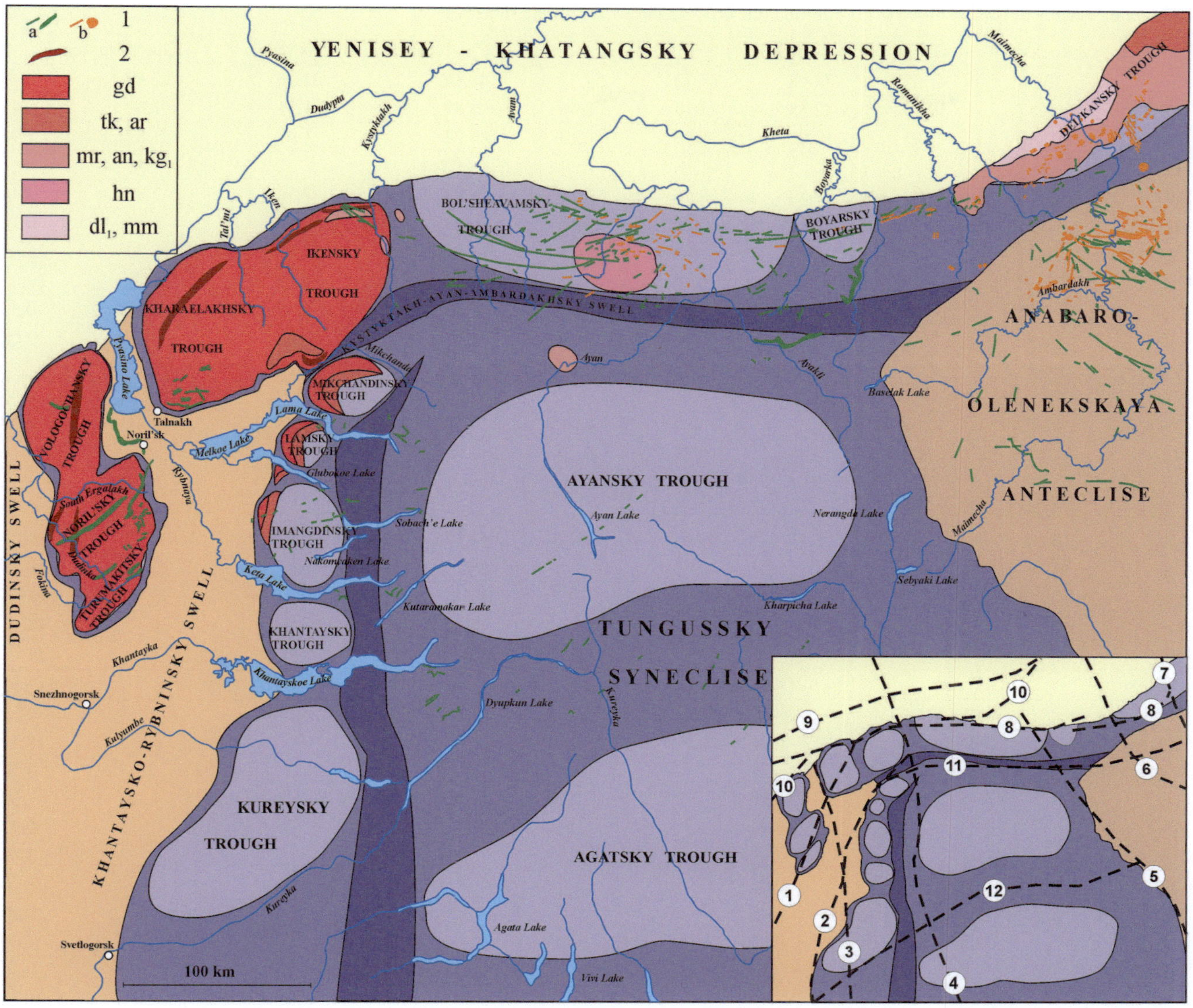

Fig. 2.30 Schematic map of distribution of high-magnesium lavas at the northwest of the Siberian Platform. *1* Dykes and small intrusions of mafic (*a*) and alkali-mafic–ultramafic (*b*) rocks; *2* areas of maximum density of picritic basalts of the Gudchikhinsky Suite. Various hatching shows distributive areas and ages (see Fig. 2.29) of high-magnesium lavas. Inset, the main abyssal faults are shown: *1* Noril'sko–Kharaelakhsky, *2* Imangdino–Letninsky, *3* Pyasinsky, *4* Kystyktakhsky, *5* Changaginsky, *6* Romanikhinsky, *7* Baykal–Taymyrsky, *8* Yenisey–Gulinsky, *9* Rassokhinsky, *10* Boganidsky, *11* Mikchanda–Boruryakhsky, *12* Kureysky

($Wo_2En_{77}Fs_{21}$) and phlogopite were observed. Composition of the glass is characterized with sharply increased SiO_2, Al_2O_3, and CaO (Table 2.7, Fig. 2.32).

Picritic porphyry (KE-35) from the lava flow sequence in the right branch of the Ketere River has seriate–porphyritic texture with a micro-ophitic, doleritic, and poikilitic groundmass. Phenocrysts are composed of subidiomorphic crystal and aggregations of olivine of different sizes (20–25%) (Fa_{28-33}) and prisms of clinopyroxene (3–5%) ($Wo_{42-39}En_{44-48}Fs_{14-13}$). The ground mass is feldspar prisms with intergrowths of clinopyroxene, tiny clinopyroxene prisms, as well as crystals of apatite, titanomagnetite, ilmenite, and zeolite (Table 2.8, Fig. 2.33).

The composition of high-magnesium lavas features an increase in rock alkalinity from west to east along the northern rim of the Platform. Within the Noril'sko–Kharaelakhsky depression and the western side of the Tungussky syneclise, the rocks are dominantly magnesium-rich basalt; in the northern side of the Tungussky syneclise and in the Kamensky Province, the rocks are subalkaline; and in the Maymecha–Kotuysky Province, the rocks are of subalkaline and alkali-ultramafic composition (Fedorenko and Czamanske 1997; Ryabov 2003; Ryabov et al. 2003).

Some idea on the compositional features of high-magnesium lavas in the north of the Siberian Platform can be gained from Fig. 2.34. For comparison, it shows the

Table 2.6 Chemical composition (wt%) of limburgite (BK-6) and rock-forming minerals

	Rock	Glass				Cpx (core)	Cpx (rim)	Clinopyroxene	
SiO_2	43.28	47.67	38.08	53.23	57.31	48.16	51.96	52.49	52.10
TiO_2	3.57	10.14	12.88	3.16	2.46	2.89	1.31	1.39	1.42
Al_2O_3	7.32	15.93	10.99	15.17	16.12	4.23	1.97	1.62	1.96
Fe_2O_3	4.68	–	–	–	–	–	–	–	–
FeO	10.61	10.04	19.54	8.56	7.26	8.93	7.53	7.87	7.14
MnO	0.29	0.14	0.12	0.16	0.12	0.14	0.14	0.16	0.16
MgO	17.21	0.39	4.63	2.04	2.15	14.51	17.37	17.29	16.82
CaO	7.99	7.18	7.68	6.30	6.45	20.16	19.62	19.07	20.05
Na_2O	1.21	0.47	0.49	0.59	0.65	0.42	0.35	0.27	0.33
K_2O	0.59	1.49	0.85	1.46	1.54	0.01	0.00	0.00	0.00
P_2O_5	0.43	–	–	–	–	–	–	–	–
NiO	0.10	–	–	–	–	–	–	–	–
Cr_2O_3	0.17	0.02	0.02	0.00	0.01	0.15	0.33	0.41	0.35
F	–	0.00	0.45	0.31	0.29	–	–	–	–
Cl	–	0.00	0.01	0.07	0.09	–	–	–	–
LOI	2.63	–	–	–	–	–	–	–	–
Total	100.08	93.47	95.74	91.05	94.45	99.60	100.58	100.57	100.33
C.f., f	47.0	–	–	–	–	25.7	19.6	20.3	19.2
Wo	–	–	–	–	–	42	39	38	41
En	–	–	–	–	–	43	49	49	48
Fs	–	–	–	–	–	15	12	13	11

	Ol (core)	Ol (rim)	Olivine		Mt	
SiO_2	38.46	38.82	38.78	39.08	–	–
TiO_2	0.02	0.04	0.04	0.09	19.25	18.53
Al_2O_3	–	–	–	–	3.87	4.36
FeO	17.45	17.29	18.33	18.53	66.42	67.00
MnO	0.23	0.24	0.23	0.26	0.63	0.37
MgO	42.41	42.09	40.77	40.89	2.05	3.83
CaO	0.22	0.35	0.31	0.37	–	–
NiO	0.39	0.46	0.34	0.33	0.18	0.20
Cr_2O_3	0.03	0.01	0.05	0.02	2.29	1.60
V_2O_5	–	–	–	–	0.45	0.46
Total	99.21	99.30	98.85	99.57	95.14	96.35
Fo	81	81	80	80	–	–
Fa	19	19	20	20	–	–

compositions of dykes of the Kamensky Province. From the diagram, one can see that olivine-phyric and picritic basalts and ankaramites of the northwest of the Platform are characterized by lower titanium content than picritic porphyries, limburgites, and meimechites from the flows in the northeast of the region and the northern parts of the Tungussky syneclise. The composition of limburgites and picritic porphyries of the Bol'sheavamsky trough is close to similar rocks of the Maymecha-Kotuysky Province, which is evidence for the eruption of the latter during the Mokulaevsky stage. The composition of dykes from the Kamensky Province lies within the field of compositions of high-magnesium lavas, so they can be considered possible comagmates.

2.15.3 Subalkaline Basalt Lavas in Traps

In the volcanogenic series of the Siberian Platform, rocks of increased alkalinity appear in various volcanic cycles and suites (see Fig. 2.29). They are represented mainly by trachybasalts, more rarely by andesite and trachyandesite-basalt, and still more rarely by trachytes and other rock types. The largest outpourings of these lavas are confined to the Noril'sk and Maymecha–Kotuysky Provinces (Fig. 2.35). In the Noril'sko–Kharaelakhsky depression, isopachytes of the trachybasalt lava flows of Ivakinsky Suite have a northwesterly orientation. Here, areas of maximum lava accumulation are found close to the border of the Siberian Platform. For subalkaline lavas, as well as picritic

Fig. 2.31 Limburgite (BK-6) from the lava flow sequence in the region of Lake Bokovoe

Table 2.7 Chemical composition (wt%) of picritic porphyry (N-137) and rock-forming minerals

	Rock	Plagioclase			Fsp		Phl		
SiO_2	44.08	56.15	60.96	62.62	63.47	62.74	40.04	40.28	39.52
TiO_2	4.00	–	–	–	–	–	5.07	5.70	6.30
Al_2O_3	7.04	26.75	22.62	22.17	16.44	21.10	10.86	11.50	11.85
Fe_2O_3	16.18	–	–	–	–	–	–	–	–
FeO	–	1.14	0.81	0.78	1.60	0.89	6.10	6.37	6.89
MnO	0.21	0.00	0.00	0.00	0.04	0.01	0.04	0.04	0.04
MgO	15.64	0.09	0.05	0.02	1.88	0.05	20.34	20.44	19.71
CaO	7.87	9.44	4.58	3.54	3.02	2.78	0.08	0.05	0.06
Na_2O	1.79	5.81	7.97	7.99	4.06	6.43	0.83	0.80	0.87
K_2O	0.99	0.64	2.19	2.01	8.01	4.92	8.70	9.14	9.09
P_2O_5	0.42	–	–	–	–	–	–	–	–
Cr_2O_3	–	0.00	0.01	0.02	0.02	0.01	0.10	0.04	0.06
Ba	0.05	–	–	–	–	–	–	–	–
BaO	–	0.08	0.37	0.46	0.23	0.59	–	–	–
SrO	–	0.32	0.09	0.08	0.03	0.10	–	–	–
F	–	–	–	–	–	–	4.07	4.27	4.23
Cl	–	–	–	–	–	–	0.09	0.06	0.07
LOI	2.22	–	–	–	–	–	–	–	–
Total	100.49	100.42	99.65	99.69	98.8	99.62	96.32	98.69	98.69
C.f., f	50.8	–	–	–	–	–	14.4	14.9	16.4
Ab	–	51	67	71	37	57	–	–	–
Or	–	4	12	12	48	29	–	–	–
An	–	45	21	17	15	14	–	–	–

	Glass		Ol (core)	Ol (rim)	Ol	Opx	Cpx	Mt	Ilm	
SiO_2	47.73	51.13	40.02	39.01	38.45	55.34	50.22	–	–	–
TiO_2	0.03	0.04	0.03	0.04	0.03		1.69	7.63	53.57	54.28
Al_2O_3	5.57	5.75	–	–	–	0.34	3.40	1.90	1.00	0.92
FeO	0.29	0.73	12.97	18.20	19.87	13.70	8.51	80.63	35.71	35.62
MnO	0.04	0.01	0.15	0.27	0.35	0.36	0.17	0.30	0.10	0.11
MgO	0.02	0.28	46.07	42.10	40.96	29.60	15.85	1.83	4.82	4.97
CaO	35.12	31.23	0.17	0.20	0.20	1.21	19.14	–	–	–
Na_2O	0.09	0.06	–	–	–	0.03	0.42	–	–	–
K_2O	0.06	0.08	–	–	–	0.01	0.00	–	–	–
NiO	–	–	0.50	0.43	0.29			0.42	0.03	0.01
Cr_2O_3	0.05	0.02	0.03	0.04	0.03	0.01	0.09	1.67	0.18	0.17
V_2O_5	–	–	–	–	–	–	–	0.51	0.32	0.26
F	0.00	0.00	–	–	–	–	–	–	–	–
Cl	0.00	0.00	–	–	–	–	–	–	–	–
Total	89.00	89.33	99.94	100.29	100.18	100.60	99.49	94.89	95.73	96.34
f	–	–	–	–	–	20.6	23.1	96.1	–	–
Fo	–	–	86	80	79	–	–	–	–	–
Fa	–	–	14	20	21	–	–	–	–	–
Wo	–	–	–	–	–	2	40	–	–	–
En	–	–	–	–	–	77	46	–	–	–
Fs	–	–	–	–	–	21	14	–	–	–

lavas, it is supposed that the isopachyte showing maximum lava thickness corresponds to the magma-feeding channel. In the Kamensky Province, there are only local occurrences of trachybasalt and subalkaline tuffs. In the Tungussky syneclise, trachybasalt flows of the Ivakinsky stage are confined to its western side to the Imangdino-Letninsky fracture zone. At the center of the syneclise, trachybasalt lava flow thickness decreases and then disappears from the sequence. More widespread in the Tungussky syneclise are trachybasalt lavas of the Yuryakhsky Unit. Their spread is controlled by the Khantaysko-Dyupkunsky rift traced in a sub-latitudinal direction between the Ayansky and Agatsky

Fig. 2.32 Picritic porphyry (N-137) from the lava flow sequence of the Namakan River

Table 2.8 Chemical composition (wt%) of picritic porphyry (KE-35) and rock-forming minerals

	Rock	Ol (core)	Ol (rim)	Ol	Cpx (core)	Cpx (rim)	Cpx	Mt	Ilm
SiO_2	44.78	37.90	36.07	35.89	49.39	51.63	50.80	–	–
TiO_2	3.37	0.02	0.02	0.03	2.04	1.60	1.90	18.18	49.69
Al_2O_3	8.19	–	–	–	4.05	1.88	2.86	0.57	0.04
Fe_2O_3	15.74	–	–	–	–	–	–	–	–
FeO	–	24.51	28.54	29.12	7.99	8.14	8.65	73.24	46.52
MnO	0.22	0.28	0.38	0.38	0.16	0.16	0.17	0.16	0.50
MgO	12.08	36.05	32.89	32.94	15.09	16.35	14.89	0.80	2.79
CaO	9.76	0.28	0.30	0.29	19.95	18.47	20.27	–	–
Na_2O	1.36	–	–	–	0.40	0.31	0.39	–	–
K_2O	1.21	–	–	–	0.00	0.00	0.02	–	–
P_2O_5	0.39	–	–	–	–	–	–	–	–
NiO	0.08	0.21	0.23	0.20	–	–	–	0.11	0.03
Cr_2O_3	–	0.00	0.01	0.01	0.73	0.42	0.11	0.59	0.12
V_2O_5	–	–	–	–	–	–	–	1.03	0.29
LOI	2.81	–	–	–	–	–	–	–	–
Total	99.99	99.25	98.44	98.86	99.80	98.96	100.06	94.68	99.98
C.f., f	56.6	–	–	–	22.9	21.8	24.6	–	–
Fo	–	72	67	67	–	–	–	–	–
Fa	–	28	33	33	–	–	–	–	–
Wo	–	–	–	–	42	39	42	–	–
En	–	–	–	–	44	48	43	–	–
Fs	–	–	–	–	14	13	15	–	–

	Potash feldspar					Zeolite	Apatite	
SiO_2	65.91	65.65	67.07	64.81	65.30	47.16	3.78	1.20
Al_2O_3	18.71	19.03	19.43	20.44	20.30	28.89	–	–
FeO	0.77	0.45	0.62	0.41	0.48	0.03	0.70	0.70
MnO	0.00	0.00	0.01	0.02	0.00	0.00	0.04	0.05
MgO	0.21	0.02	0.04	0.01	0.02	0.02	–	–
CaO	0.53	0.41	0.69	1.75	1.53	9.76	50.86	52.34
Na_2O	4.95	5.65	6.49	6.89	7.11	1.65	–	–
K_2O	7.82	8.32	5.68	5.73	5.48	0.00	–	–
P_2O_5	–	–	–	–	–	–	39.16	40.43
Cr_2O_3	0.03	0.00	0.01	0.00	0.00	0.00	–	–
BaO	0.10	0.06	0.03	0.53	0.26	–	–	–
SrO	0.02	0.00	0.00	0.02	0.00	–	0.33	0.36
CeO_2	–	–	–	–	–	–	0.58	0.48
F	–	–	–	–	–	–	3.95	4.76
Cl	–	–	–	–	–	–	0.19	0.19
	99.05	99.59	100.07	100.61	100.48	87.51	99.59	100.51
Ab	48	50	61	59	62	–	–	–
Or	49	48	35	33	31	–	–	–
An	3	2	4	8	7	–	–	–

troughs with its axis along the Khantayskoe and Dyupkun Lakes (Lipkov 1969). Along the axis of the trough, thenumber of trachybasalt lava flows decreases an d the thickness of the Yuryakhsky Unit also decreases until it pinches out entirely. The presence of trachybasalts of the Changadinsky Suite along the extension of the Khantaysko–Dyupkunsky rift in the eastern side of the Tungussky syneclise is also remarkable.

Detailed characteristics of the features of subalkali lavas of the region are given in the work (Ryabov et al. 2003). Alkali lavas are known only in the Maymecha–Kotuysky Province, and their characteristics are documented by Fedorenko and Czamanske (1997). By chemical composition, the lavas of the region correspond to the petrogenetic series of subalkaline and alkaline rocks. Their composition is shown on Fig. 2.36, and the compositions of trachydolerite from sills of the Ergalakhsky Suite and dykes of the Avamsky Complexes are also plotted. The former are comagmates of early outpourings of Ivakinsky Suite trachybasalt. The curves show the evolution of composition

Fig. 2.33 Picritic porphyry (KE-35) from the lava flow sequence in the *right* branch of the Ketere River

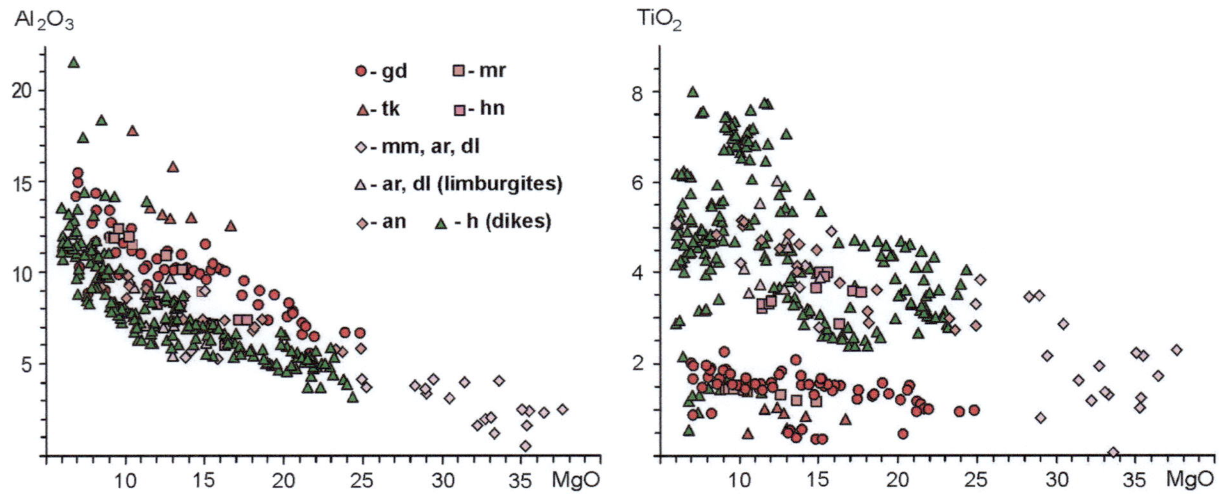

Fig. 2.34 Composition diagrams (wt%) of high-magnesium magmatic rocks of the northwest of the Siberian Platform. On the diagrams, the *dots* for compositions of small intrusions (h) of the Kamensky and Ust'-Del'kansky Complexes are plotted. For explanations, see text and Fig. 2.29

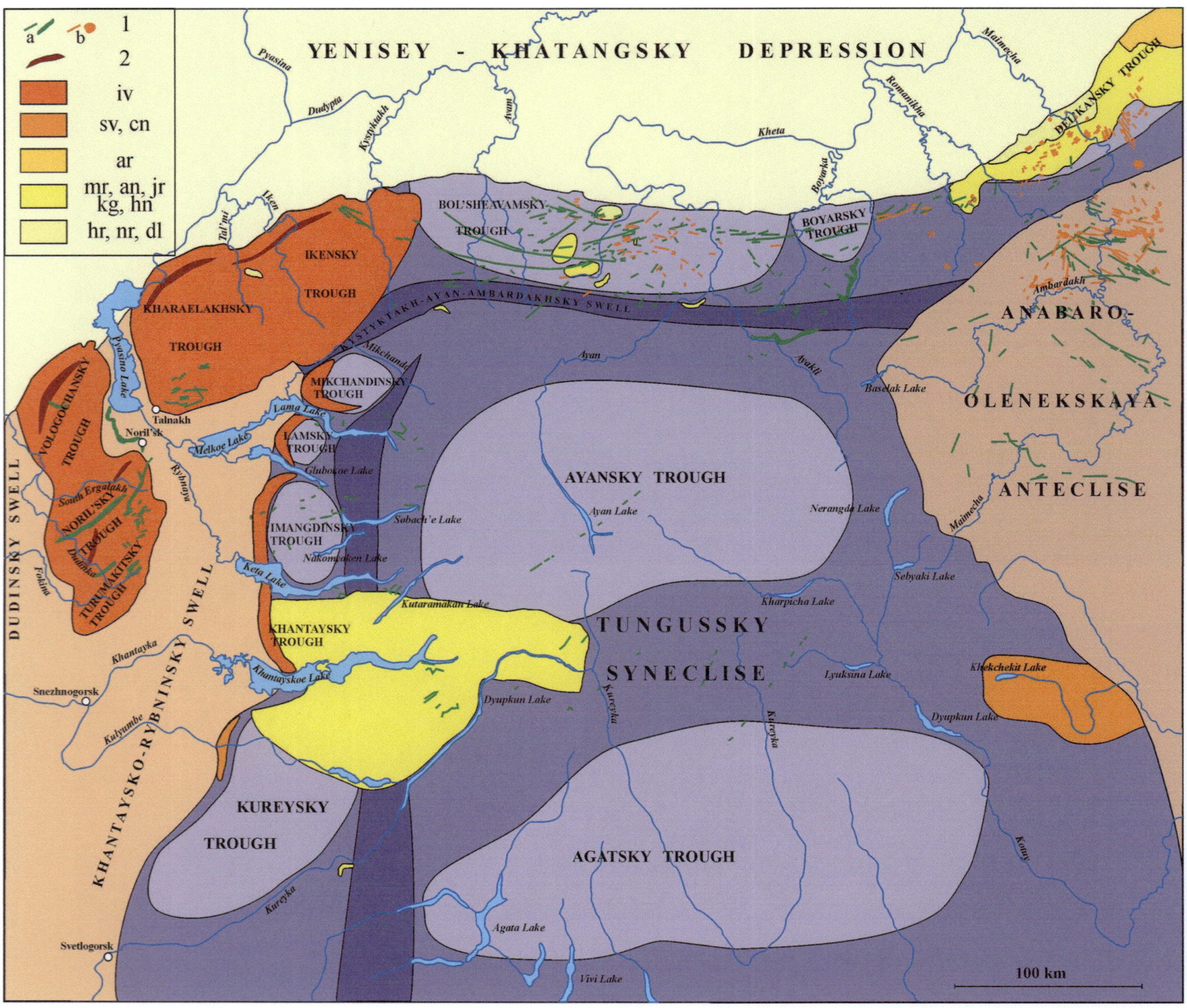

Fig. 2.35 Schematic map allocations of subalkali and alkali lavas at the northwest of the Siberian Platform. *1* Dykes and small intrusions of mafic (*a*) and alkali-mafic–ultramafic (*b*) rocks; *2* areas of maximum density of trachybasalts of the Ivakinsky Suite. Various hatching shows distributive areas and ages (see Fig. 2.29) of subalkali and alkali lavas

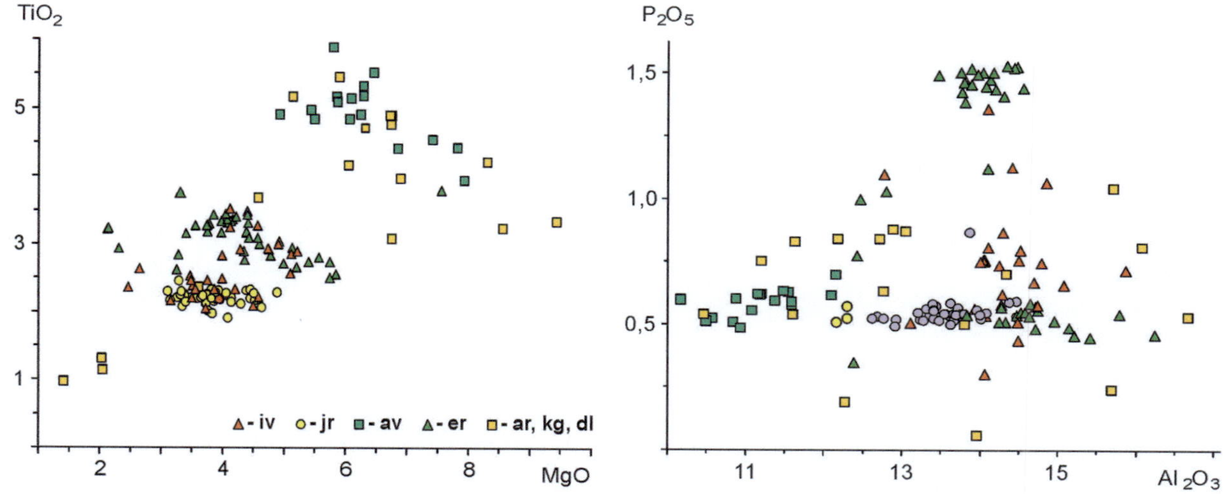

Fig. 2.36 Composition diagrams (wt%) of subalkali–alkali magmatic rocks of the northwest of the Siberian Platform. On the diagrams, the dots for compositions of small intrusions of Avamsky (av) Complex and sills of Ergalakhsky (er) Complex are plotted. For explanations, see text and Fig. 2.29

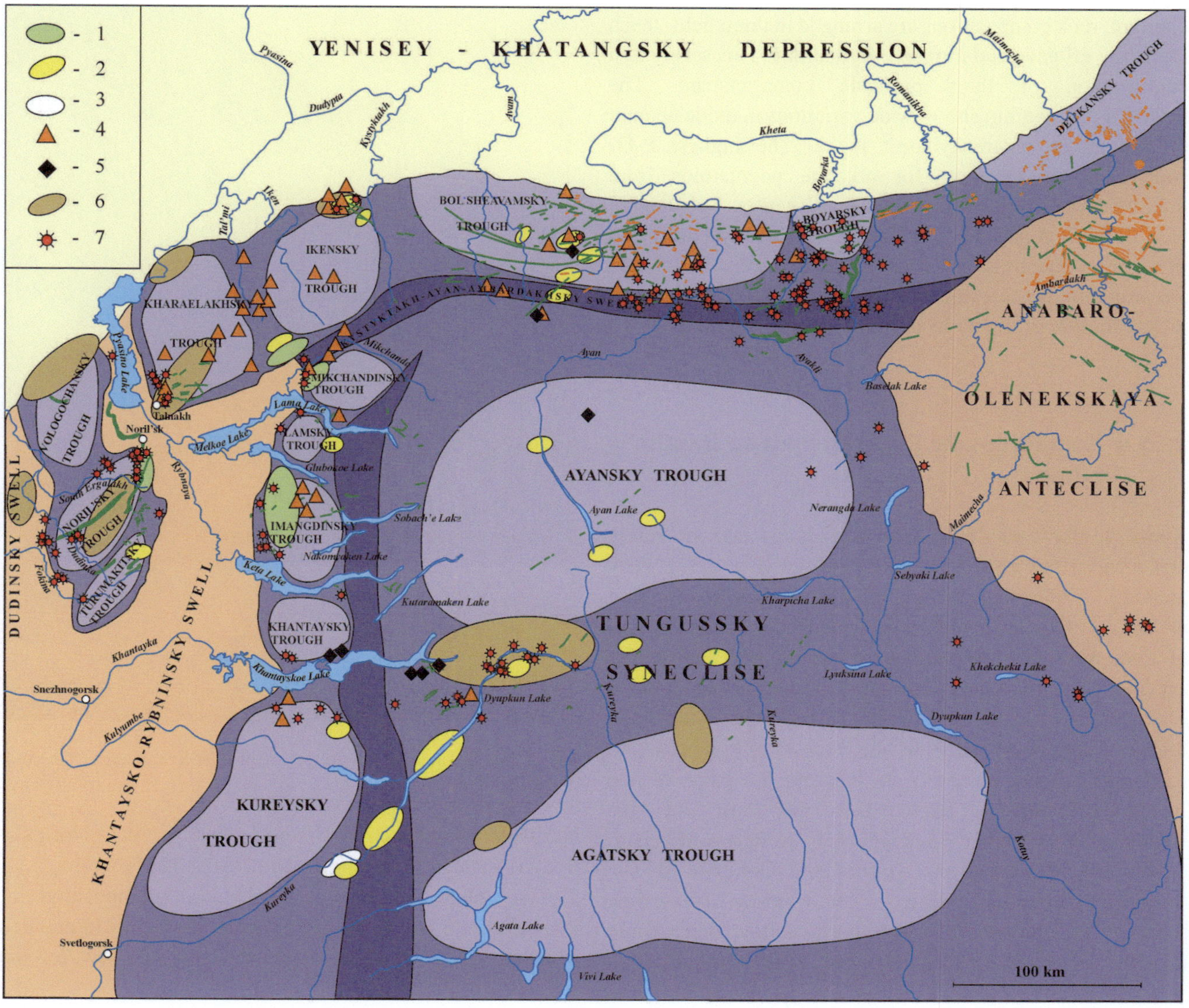

Fig. 2.37 Schematic map of allocations of some "abnormal formations" in basalts of the northwest of the Siberian Platform. *1* Komatiite-like rocks, *2* limestone lenses, *3* anhydrite lenses, *4* native copper, *5* magnetite lavas, *6* bitumen, *7* paleovolcanoes. The full legend is given in Fig. 2.30

Fig. 2.38 Limestone (*white and light-gray*) contact with mandelsteine (*dark-gray*), subsidence caldera of the Snezhny Stream, southern part of the Kharaelakhsky trough

of trachybasalt melts in time and in space. In the TiO_2–MgO diagram, rock compositions are arranged in three fields, each including effusive and intrusive formations. This serves as a confirmation of their probable comagmatism. The P_2O_5–Al_2O_3 diagram shows wide variations in oxide content in derivatives of trachybasalt melts. The highest P_2O_5 content is characteristic for alkaline pegmatoids from trachydolerite sills of the Ergalakhsky Complex. Subalkaline basaltic lava is spatially confined to abyssal fractures and riftogenic structures. Wide variations of rock composition are probably due to magmatic chamber depth and degree of differentiation of the melts in the presence of fluids.

2.15.4 Komatiite-Like Rocks in Plateau Basalts Flows

Komatiite-like rocks are exotic formations that are rarely observed in traps of the Siberian Platform (Fig. 2.37). They manifest within thick series of plateau basalts, forming uniformly thick horizons with dendrite-like megacrystals of clinopyroxene, more rarely of olivine and plagioclase crystals 1–2 up to 32 cm long. Dendrite-like mineral segregations resembling a spinifex texture was one reason for the description "komatiite-like." Flows of komatiite-like rocks have been observed in outcrop: in parts of the Imangdino-Letninsky fracture at the base of the Kharaelakhsky and Mikchandinsky troughs—Mikchandinsky lava flow sequence (Dodin and Golubkov 1971); in the Imangdinsky trough—Keta-Irbinsky lava flow sequence (Kavardin 1988) and the lava flow outcropping on the Sunduk Mountain; as well as in the Noril'sko–Kharaelakhsky Fault zone in the Noril'sky trough—unnamed (Duzhikov and Distler 1981).

Fig. 2.39 Outcrop of anhydrites of subsidence caldera in the Ande diatreme

The inner structure and material composition of komatiite-like rocks are best studied in the differentiated Mikchandinsky lava flow sequence and the flows of the

Sunduk Mountain (Ryabov et al. 1977; Ryabov 1992a, b, c). Both represent a simultaneous outpouring of various immiscible Picritic basalt and tholeiitic basalt melts. Detailed characteristics of komatiite-like rocks are given in parts 2.4 and 2.6. One can suppose the peculiar texture of komatiite-like rocks with dendrite-like and spherulitic megacrystals is due to the interaction of melts with varying composition and resorption of olivine and chrome spinel in picritic basalt in the presence of volatile elements (Ryabov et al. 1977; Ryabov 1992a, b, c).

2.15.5 Tuff Paleovolcanoes

The mechanism for the formation of enormous outpourings of flood basalts of the Siberian Platform has been discussed at length in geologic literature. Influenced by contemporary volcanism, some geoscientists developed models comprising one gigantic or a series of smaller shield volcanoes and stratovolcanoes of Hawaiian or Icelandic type that formed the entire volcanogenic series (Goryainov et al. 1976; Daragan-Sushchov 1984). Another theory describes a range of shield volcanoes situated along the rim of the Tungussky syneclise with fracture volcanoes within the Platform. The shield volcanoes had autonomic magmatic chambers that determined provincial features of magmatism. The volcanoes were polygenic and multichannel. Basalt floods formed as a result of lava outpourings from volcanoes of both a central and fracture type. Later, it was noted that the maps of isopachytes plotted for some suites, as well as for high-magnesium and subalkaline basalt lava flows (Zolotukhin et al. 1986), do not not support concentric lava accumulation around the assumed volcanic centers, but demonstrate a linearity that is complementary to the orientation of tectonic structures. This fact served as the basis for the definition of magma-feeding fractures and for rise of the role of fracture volcanoes in the formation of basalt lava flows.

In outcrop northwest of the Platform, 162 tuff paleo central-volcanoes were determined, partly shown on the schematic map (see Fig. 2.37). In the basalt field, volcanic structures are typically found in depression zones of rift structures and abyssal fractures confined to the limbs of volcanic troughs. Paleovolcanoes are quite randomly distributed over the territory. Typically, they are confined to tectonic stress zones adjoining the Yenisey–Khatangsky riftogenic depression. For example, the density of volcanoes in the Pravaya Boyarka River area reaches one per two km^2. A characteristic of paleovolcanoes is that crater facies are built up of pyroclastic material—tuffs, tuff breccias, and rare agglomerates and agglutinates. Central volcanoes that form lava flow sequences were not observed in the basalt field of the Platform.

Up to a latitude the same as the Nizhnyaja Tunguska River, the number of volcanic structures in the basalt field is about 500 (Daragan-Sushchov 1984). Geological maps only show those volcanic structures that are large enough to see at map scale. Small structures are not shown, so the number of volcanoes on maps is far smaller than in reality. In well-studied regions, such as the Noril'sky and Kharaelakhsky troughs, the existence of volcanic structures of the Ivakinsky, Syverminsky, Gudchikhinsky, Khakanchansky, and Morongovsky Suites was determined (Ivanova 1975; etc.). The structures functioned for a defined term, then ceased activity and were overlain by lavas of subsequent suites. Each volcanic cycle was characterized with its own paleovolcanoes that did not coincide with earlier volcanic structures. Due to the rather limited outcrop and bores, it is practically impossible to create a paleogeographic representation of the six cycles of volcanism, nor to count all volcanic structures. The number of small volcanic structures in the basalt field of the platform has been estimated in the thousands. It is known that the Laki volcanic fracture structure in Iceland (30 km long) features about 100 craters 30–50 m (and rarely up to 100 m) high (MacDonald 1972). It is feasible that the formation of 120 horizons of tuffaceous rock up to 200 m thick within lava flow sequences on the Siberian Platform was created by hundreds of volcanic structures.

Taking all this into account, it is possible that mass eruptions of lava took place during the gradual opening of a great quantity of fractures that functioned only once, then closed with no remaining melt in their magma-feeding channels. The central volcanoes apparently had no flow-forming role and functioned at final stages of the cycles of volcanism as tuff volcanoes. The largest was the Khakanchansky volcano with tuff thickness of up to 400 m in its aureole.

2.15.6 The Subsidence Calderas of Paleovolcanoes

Subsidence calderas were noted in the upper parts of the volcanogenic series, from the Morongovsky to Samoedsky Suites (inclusive), with the corresponding lava thickness more than 800 m. The calderas form separate bodies of various sizes or as a system of spatially contiguous basins confined to faults. As a rule they gather at tensile zones, such as faults and folds. For example, zones along the Kureysky abyssal fault and Kystyktakh–Ayan–Ambardakhsky swell north of the Tungussky syneclise. The calderas have a rounded or elongated oval form; their size varies from hundreds of meters to 7–10 km, rarely more. They are typically filled with tuffs and tuff breccias, more rarely with limestone, and occasionally with anhydrite. The thickness of carbonate and sulfate rocks reaches 20–30 m.

2.15.7 Limestone Lenses in Paleovolcano Calderas

Calderas filled with limestone are observed in different parts of the basalt field of the Platform (Ryabov et al. 2005; Ryabov and Lapkovsky 2009). In most long-lived calderas, thick horizons of fragmental pyroclastic and carbonate rocks alternate with tuff and solid limestone. In other cases, only limestone with smaller quantities of tuffaceous material is observed. Some carbonate-filled calderas feature bitumen and native copper, the latter sometimes forming ore-deposits.

Most frequently, limestone lenses and interbeds are found in calderas of the Ikensky, Bol'sheavamsky, and Kureysky troughs (see Fig. 2.37). In the central part of the Tungussky syneclise, a chain of carbonate- and sulfate-bearing calderas can be traced along the Kureysky Fault. Limestone forms small- to micro-grained layered rock, in which remnants of diatoms, spores, fern pollens, and coniferous plant sphenophytes were determined (Duzhikov 1988; Strunin et al. 1994). The thickness of limestone interbeds and lenses ranges from centimeters up to 3–6 m, and more rarely up to 30 m and can be traced for tens of meters, and more rarely for hundreds of meters (Pol'kin 1958). Along strike, limestones grade into tuffaceous rocks. In the upper contact of limestone interbeds, pillow lavas, and/or lava breccia, amygdaloidal and massive basalt is found, and downsection gives way to phreatic breccia and layered limestone. As a rule, limestone overlies tuffs, tuffites, or amygdaloidal basalts, usually with sharp contacts (Fig. 2.38).

It is believed that accumulation of limestone and tuff took place in shallow lakes and lagoons that formed after cessation of volcanic activity and local subsidence. The presence of limestone indicates that the formation of volcanogenic series up to the middle of the Morongovsky stage proceeded in subaqueous conditions (Duzhikov 1988). Yet geoscientists have doubted this traditional opinion on the nature of limestone interbeds and lenses in traps for a long time (Pol'kin 1958). The small size and lens-like form of limestone beds point at small-volume water basins in which sedimentation took place. The rate of sedimentation in seas in the process of limestone formation is estimated as about 1 cm per 100,000 years (oral communication Yu.I. Tesakov). Consequently, a 30-m-thick layer of limestone could be formed in three millions years, which is highly improbable given the absence of distinct erosion or weathering that would have manifest at the stratigraphic boundary. An alternative theory exists (Pol'kin 1958) that the source of calcium carbonate for the limestones was mineralized hydrotherms that led to formation of calcite and its accumulation in lake basins. In addition, it can be said that the form and the size of limestones indicate that maar lakes were formed in subsidence calderas in which highly mineralized fumaroles functioned. They produced calcium carbonate which formed limestone tuff and afterward, during denudation, became a limestone deposit.

This mechanism of formation of limestone lenses among basalt lava flows explains their small size and connection with volcanic structures. The interest in limestones in calderas of the Ikensky and Bol'sheavamsky troughs is the occurrence of higher native copper content in them and their concomitant tuffs (Duzhikov 1988). Native copper forms along stratification planes and small cross fractures in limestones, and as impregnations and veinlets in tuffs, whereas in the basalts underlying these copper-bearing limestones, the copper is found in amygdales.

2.15.8 Anhydrites of Subsidence Calderas

In volcanogenic series in the northwest of the Platform, only two calderas filled with gypsum and anhydrite are known: the Khuryuchi and Ende (Ryabov and Lapkovsky 2009). The Khuryuchi caldera is situated near the junction of the Khuryuchi spring, which flows into the Toutonchana River, and has a diameter of 100 m. The caldera is filled with uniform gypsum rock with an admixture of dolomite (2%) and calcite (2%). The visible thickness of gypsum layer is 10 m. It overlies the tuffs of Korvunchansky Suite of Lower Triassic age and is overlain by basalts of Khonnamakitsky Suite.

The Ende (or Kureysky) caldera is situated on the left bank of the Kureyka River, 17 km from the entry of the Ende River (see Fig. 2.37). It lies in the Kureysky abyssal fault zone and is confined to the tuffs of the Dvuroginsky Suite at the boundary with lavas of the Khonnamakitsky Suite of Lower Triassic age. The visible thickness of anhydrite layer reaches 25 m (Fig. 2.39). It is typically traced as a uniform body for 0.5 km and in some outcroppings for 6 km. The steep walls of the caldera's southern rim are exposed; the northern rim opens to the Kureyka River valley. In the central part of the caldera, round deep basins tens to hundreds of meters in diameter are situated, some filled with water. In the northern rim of the caldera, breccia anhydrite with veinlets of garnet skarn and a microdolerite dykes are exposed. The dyke crosses tuffs and overlying anhydrites and interfingers along their contact. The vertical wall of the southern rim about 20 m into its upper part is composed of fresh white and light-gray anhydrite and in the lower one by gray anhydrite with black thin (fractions of mm) layers of finely dispersed magnetite and carbonaceous matter. In flat areas, anhydrite underwent hydration, and in fractures secondary gypsum sinters can be seen. The isotopic composition is the following: anhydrite—$\delta^{34}S$ 23.2–23.7‰, $^{87}Sr/^{86}Sr$ 0.70824–0.70882‰, and $^{87}Rb/^{86}Sr$ 0.00169‰; gypsum—$\delta^{34}S$ 25.2‰, $^{87}Sr/^{86}Sr$ 0.70891‰, and $^{87}Rb/^{86}Sr$ 0.00265‰.

It can be supposed that fumaroles transported carbonate and sulfate material from Devonian sediments lying here 1 km down the section.

2.15.9 Diatremes with Pyrrhotite and Magnetite Ore

Diatremes containing magnetite, sulfide–magnetite, and pyrrhotite (pyrite) ores are found in the south Noril'sky and Kureysko–Gorbiachinsky volcanogenic structures (Ryabov and Lapkovsky 2009).

Diatremes filled with rich sulfide ores are rather rare on the Siberian Platform, represented by the Pyrrhotite diatreme, the fracture apparatus of the Canyon Stream (inflow of the Malaja Typtour-Orokta River), and ore veins of the Loganchi Mountain (The Khantayskoe Lake). The Pyrrhotite diatreme is an upward broadening cone-like structure filled with ore and metasomatically altered tuff breccia. The diatreme is cut by a sill of the Daldykansky Complex, evidence of the formation of the diatreme and the sulfide ores during the Daldykansky stage of tectonomagmatic activity in the region. Sulfides in diatremes are typically pyrrhotite and troilite, with small quantities cubanite, tetragonal chalcopyrite, pyrite, and more rarely sphalerite and galena. The isotopic composition of sulfur of compact sulfides, $\delta^{34}S$, is 12.2‰ in the Pyrrhotite diatreme, 16.3–16.4‰ at the Canyon Stream, and 19.7–21.5‰ in veins of Loganchi Mountain (Ryabov 1997; Ryabov and Lapkovsky 2009).

Magnetite ores are known in association with the diatremes of the Khikchele Stream (near the Malaja Typtour River) and the Vetka River. Compact magnetite–pyrrhotite ores were also determined in fault-line metasomatites of the Kureysko–Gorbiachinsky structures. The Vetkinsky magnetite deposit (the Magnetite diatreme) is situated in the south Noril'sky volcano-tectonic structure. Its morphology is close to that of the Angaro-Ilimsky magnetite deposits in the diatremes of the south of the Siberian Platform (Yudina et al. 1977). In plan view, the diatreme has elliptic section 400 × 800 m in size, and it has been drilled down to 732 m. The diatreme is composed of tuff breccia and fine clastic material (microdolerite and dolerite with varying degrees of crystallinity), a disseminated stockwork, compact magnetite ores, and metasomatites. The diatreme is crossed by a sill of gabbrodolerite of the Morongovsky Complex. At the contact of the diatreme, a thick metasomatite aureole is observed, with successive replacement of clinopyroxenite metasomatites by scapolitite and scapolite–albitite. Andradite–grossular garnet and epidote of calcite composition have formed after albitites and scapolitites, as well as gabbrodolerite. One of significant differences between ores of the Angaro-Ilimsky and Vetkinsky deposits is that manganomagnetite is one of the principal ore phases in the former, whereas almost pure magnetite is dominant ore in the latter. It is supposed that the depth of the volcanic structure was the main factor determining ore composition; in the Vetkinsky deposit, it was shallower than in the Angaro-Ilimsky deposit. The loss from erosion of the Vetkinsky diatreme is about 800 m (Yudina et al. 1977).

Basalt magma served as a source of iron for these magnetite ores. Extraction of iron from the melt and its transition and accumulation took place in the presence of chlorine-containing fluids, which were formed in the course of decomposition of halite in Devonian evaporite beds. In the later stages of ore formation, chlorine was fixed in scapolites and sodium was fixed in albitites and scapolitites (dipyres).

In tuffs of paleovolcanoes of the Siberian Platform, native iron forms metal balls in association with moissanite and diamonds (Goryainov et al. 1976). The occurrence of such phases, as well as bitumen in calderas and carbonaceous matter in pisolite tuffs, points to the migration of hydrocarbon fluids through the volcano-tectonic structures.

2.15.10 Native Copper in Volcanogenic Rocks and Diatremes

Accessory impregnation and accumulation of nature copper is confined to the upper half of the volcanogenic series, from the Tuklonsky to the Samoedsky Suite. Copper occurs in carbonate and tuffaceous rocks, amygdalous and compact basalt, dykes, and in differentiated flows of picritic basalt. In the northwest of the Siberian Platform, more than 50 ore manifestations and mineralized areas of native ore were determined (Strunin et al. 1994). The ore manifestations are characteristically confined to abyssal fault zones, for example, the chain of ore manifestations that form a ribbon along the Abagalakhsky Fault crossing the Kharaelakhsky and Ikensky troughs (Fig. 2.37). Increased tectonic deformation in the Bol'sheavamsky trough is reflected in increased copper concentration in the volcanogenic series in the region.

Copper aggregations are noted in limestone and carbonate tuff lenses confined to the Morongovsky and Mokulaevsky Suites. Within the Bol'sheavamsky trough, the largest ore manifestations are situated along the

Fig. 2.40 Occurrences of bitumen (*black*) in basalts of the Vologochansky trough

Ketere, Uokhir', Bol'shoj and Maly Avam, and Nerakachi Rivers (Rad'ko 2007; Savushkin 2008). The best studied Arylakhsky copper deposit is at the interfluve of the Arylakh–Samoedsky Rivers (Duzhikov 1988). The copper-bearing horizon was traced for 40 m. Upward, it is represented by a unit of tuffaceous rocks (1.5–12 m) and beds of brecciated and uniform limestone (0.7–6 m) with interbeds of volcanogenic rocks overlain by basalt and, locally by lava breccia (Strunin et al. 1994). The main ore minerals are native copper (samples reaches 24 kg) and chalcosine and more rarely cuprite, tenorite, bornite, covellite, pyrite, galena, sphalerite, magnetite, hematite, and gold. In the same association, calcite, chlorite, garnet, cerussite, zeolites, quarts, prehnite, adular, apatite, fluorite, malachite, and bitumen ($\delta^{13}C_{PDB}$ −9.7) were found.

Microprobe studies of disseminated copper in basalt have shown the presence of: As (0–410 g/t), S (10–1,410 g/t), Fe (810–8540 g/t), Ni (0–10 g/t), and Ag (130–560 g/t). Copper in limestone and tuff, including those of Arylakhsky deposit, contained: As (0–410 g/t), S (10–350 g/t), Fe (150–1,670 g/t), Ni (10–140 g/t), Ag (40–340 g/t), and Au (80–110 g/t). In cuprite and tenorite, the following impurities were observed: S (230–460 g/t), Fe (330–5380 g/t), and Ag (20–430 g/t). Chalcosine contains: As (up to 70 g/t), Ni (up to 230 g/t), Fe (1,570–16,400 g/t), and Ag (240–550 g/t).

Analysis of existing data points a connection between native copper in flood basalt and in limestone in subsidence calderas with endogenic fluids. Concentration took place in zones of excessive fissuring connected with abyssal faults and volcanic structures.

2.15.11 Flows of Magnetite Lava in Flood Basalt

Manifestations of iron-ore magma as lava flows are unique. Magnetite lava was mentioned in literature initially for the region surrounding the El Laco Volcano (Chile). The second find was made in the traps of the Siberian Platform in the outcrop along the Magnetytovy Stream, in the Khantayskoe Lake basin (Ryabov and Pavlov 1991; Ryabov 1997). Later, more magnetite lava in the Platform was discovered. Iron-ore lava flows were found and studied along the streams of the Sirota Mountain and Zhelezny Stream (the Khantayskoe Lake basin). Afterward, similar magnetite accumulations were discovered by V.A. Rad'ko at the head of the Uokhir' and Onkoyachin Rivers (Bol'sheavamsky trough) and by G.K. Shklyarik in outcrop at the Lake Duluk (Ayansky through) (Fig. 2.37). Manifestations of magnetite lava are confined to the upper part of the volcanogenic section, although a number of magnetite ore manifestations are known in basalt of the Kureysky and Khantaysky troughs. They have never been thoroughly studied, so are possible candidates into the list of iron-ore lava manifestations.

The iron-ore lava of the Magnetytovy Stream is connected to trachybasalt of the Yuryakhsky Unit, whereas all others lie in tholeiitic basalt of the upper part of the Mokulaevsky (Khonnamakitsky) Suite. The iron-ore lava forms flows of compact magnetite, basalt, and palagonite with rich in magnetite and also agglomerates with basalt fragments in hematite–magnetite cement. In compact ore, magnetite is observed as small spherulites (0.5–1.5 cm diameter), large hemispheres up to 12 cm diameter, and dendrite-like or columnar crystals 1–1.5 cm up to 10 cm long. Composition of magnetite is characterized by and increased MgO content of 3.58–5.82 wt% (Ryabov 1997), reaching 9.38–10.07 wt% in magnetite of Lake Douluk.

Manifestations of magnetite lava are confined to volcanic structures situated in abyssal fault zones. The connection of magnetite ore manifestations with fault tectonics is best revealed in the Khantaysko–Dyupkunsky paleorift zone and in the Kureysky trough crossed by the Kulyumbinsky, Gorbiachinsky, and Kureysky abyssal faults. Formation of iron-ore lava is explained as the result of interaction of basalt melt with chlorine-containing fluids in the course of magma movement in magma-feeding channels (Ryabov 1997).

2.15.12 Dyke Belt at the Northern Margin of the Tungussky Syneclise

Dyke formations are most widely spread in the northern margins of the Tungussky syneclise; on the schematic map (see Fig. 2.37) 580 dykes are plotted. Maximum concentration of dykes is confined to a depression along the northern rim of the Platform, including the Bol'sheavamsky, Boyarsky, and Del'kansky troughs and associated uplifted blocks. Local dykes are composed of dolerite and trachydolerites rocks of the Daldykansky and Avamsky Complexes, whereas smaller bodies are syenite–mela-nephelinite Kamensky Complex rock and lamprophyric Ust'-Del'kansky Complex rock. The dykes are frequently located along fault planes and form systems composed of a series of subparallel bodies or form accumulations in the aureoles around central-type intrusives (Bushkanets et al. 1970a, b; Zhabin and Surina 1970). Within the Tungussky syneclise, dykes are rare and are typically dolerite.

The widest variation in chemical composition is characteristic for dykes of the Kamensky and Ust'-Del'kansky Complexes (Ryabov and Grib 2002, 2005) and includes rocks of mafic alkali-mafic, alkali-ultramafic, and alkaline composition. Figures 2.34 and 2.36 present typical rock composition. The lack of age relationships between dykes is evidence of their almost simultaneous intrusion. Some dykes feature a dyke-in-dyke structure. The dykes cross the entire volcanogenic series as potential magma-feeding channels but did not erupt onto the surface. The only exception is presumably the above-described small flow of limburgite and picritic porphyry (Ryabov 2003). In the Kamensky Province, absolute Ar/Ar age of vogesite dykes is 238 million years, and the age of bostonite dykes is 239 million years. In the Noril'sk Province, the absolute age of minette dykes is 238 million years (Ryabov and Lapkovsky 2009; Ryabov et al. 2010; Travin et al. 2010).

The data show that the appearance of the dyke belt in the northern depression within the Tungussky syneclise took place with prestressed tectonic conditions at the border with the Yenisey-Khatangsky rift structure. The variety of dyke compositions is due to fluid-magmatic differentiation of trachybasalt melt as a result of its intrusion through a broken-up fluid-rich framework of rock (Ryabov and Grib 2005).

2.15.13 Bitumen in Flood Basalt

Recently, interest in the fluid regimes of ore-magmatic systems, including reductive fluids connected with plateau magmatism and formation of layered intrusions and ore deposits, has grown. Similarly, interest in manifestations of liquid and solid bitumen in traps has also risen.

Bitumen was observed in the volcanogenic series of the Noril'sky, Vologochansky, Kharaelakhsky, and Ikensky troughs adjacent to the Yenisey-Khatangsky rift and abyssal faults of the region (see Fig. 2.37). Within the northern part of the Tungussky syneclise, bitumen is known from areas near the Khantaysko–Dyupkunsky paleorift and abyssal faults in the vicinity of the Severnaja River, Lake Bel'douchan, and other local areas. Bitumen is observed in calderas as aggregates or impregnations in tuffaceous and carbonate rock. In the vicinity of the Erankan River (Lake Agata basin), an anthraxolite-rich fumarole was observed (Starosel'tsev and Starosel'tsev 1987). Bitumen fills hollows in amygdaloidal basalt in the Talnakh Region (the Skalny open-cast) and near the Listvianka River. In fractured basalts of the Boganidsky Fault zone, bitumen lies in steeply dipping veins that cross the whole volcanogenic series, from the

Ivakinsky to Mokulaevsky Suites inclusive (Fig. 2.40). Bitumen is associated with calcite, chalcedony, phrenite, chlorite, chalcopyrite, pirrhotine, and pyrite. Sporadically, in tuffs between basalt flows, subhorizontal lenses of asphaltite are observed. In the Arylakhsky deposit, bitumen is associated with native copper and copper sulfide (Duzhikov 1988). Isotopic composition $\delta^{13}C_{PDB}$ in bitumen and carbonaceous matter from volcanogenic formations varies from −19.4 to −27.6‰ (Ryabov et al. 2006).

2.15.14 Some Genetic Conclusions

Studies indicate the important role of abyssal faults and fracture tectonics in the build-up of "anomalous formations" (see Fig. 2.30) (Ryabov et al. 2005; Ryabov 2007; Ryabov and Lapkovsky 2009; etc.). The spatially confined nature of high-magnesium and subalkaline basalt lavas to rift structures and the linear elongation of isopachites of basalt flows (orientated with the main tectonic elements of the region) indicate lava eruption from fracture volcanic centers. Similar volcanic centers are indicated by the nature of komatiite-like and magnetite lava flows. Dyke belts in the north of the Siberian Platform are constrained by their position in tectonically stressed zone adjacent to the long-living Yenisey-Khatangsky rift structure. It opened numerous fractures and allowed the intrusion of melt during a period of tectonomagmatic activation. Intrusion of the melt through a fluid-rich zone of fissured rock initiated wide-scale fluid-magmatic differentiation and the formation of dykes with a great variety of composition.

The confined nature of paleo central-volcanoes sourced from abyssal faults and dykes, tuffaceous fill in crater zones of volcanic structures, and lack of lava flows connected to them point to the fact that they were not flow generating but appeared during the end of the volcanic cycle. The formation of subsidence calderas and the appearance of fumaroles and solfatares (with associated limestone and anhydrite in maare) are confined to the same stage. Formation of disseminated and aggregated native iron and copper, sulfides, and magnetite seems to be due to multicomponent fluids acting as a metal carrier.

Occurrence of decomposed rocks in rift zones played an important role in fractural tectonics. This led to migration of fluids from lithosphere sediments into zones of crustal weakness and formation there fluid-saturated zones. The zones of fissured rocks were favorable for the migration of volatiles as well as for intrusion of magma. This determined the fracture character of eruptions, the fluid-magmatic differentiation of melts, and the confined nature of anomalous formations. The various forms, scales, and composition of anomalous formations are due to various depths of fault location and degree of maturity of rift structures (Ryabov 2007; Ryabov and Lapkovsky 2009).

2.16 Characteristics of the Formation of Volcanogenic Units of the Siberian Platform

The mechanism for formation of flood basalt flows is still discussed in the geological literature, and unanimous consent has not been found (Dobretsov 1997; Dobretsov et al. 2001; Sharapov et al. 2008, etc.). Rock sample data remains the key for resolving the genesis of trap magmatism (Ryabov et al. 2006), and such data has been collected by geoscientists for more than 50 years. Materials collected in early investigations have not lost their significance but have often not been taken into account by creators of new genetic concepts.

A unified stratigraphic scheme of volcanogenic formations of the Siberian Platform has been created, based on the principle of tectonomagmatic periodicity. Tectonomagmatic periodicity manifests itself as a rhythmic alternation of pyroclastic and effusive rocks in the tuff lava series of the Platform. Six tectonomagmatic cycles, or phases, are separated (see Fig. 2.1) and divided from another by intervals, sometimes with angular displacement (Dodin and Goloubkov 1971; Fadeev 1962). Each tectonomagmatic phase is characterized by a complex of volcanogenic formations. In cross section, the cycles have a determinate variation in morphological and textural, as well as lithological, petrographic (Mezhvilk 1962; Kuznetsov 1965), and geochemical features (Nemenenok and Vilensky 1978; Fedorenko 1981; Zolotukhin et al. 1986). In the development of volcanogenic series, the quantity and character of tectonomagnetic phases does not stay constant. Areas with similar character are divided into different structure-facial or petrographic provinces. Synchronous manifestations of different facies, developed in different petrographic provinces, are grouped into horizons with stable stratigraphic position within the region (Pol'kin 1964; Daragan-Sushchov 1984). As was already stated, trap volcanism is divided into two main stages: early (up to the Khakanchansky Suite) and late (subsequent suites) (Duzhikov et al. 1971; Nemenenok 1978), or riftogenic (up to Nadezhdinsky Suite, inclusive) and flow-forming (from the Morongovsky to Samoedsky Suites) (Almukhamedov et al. 2004). The maximum thickness of the volcanogenic series is confined to the Noril'sko–Kharaelakhsky depression. In the Kharaelakhsky and Ikensky troughs, it is about 3.5 km, in the central part of the Tungussky syneclise about 2.5 km, and thins in southerly direction.

Various opinions exist on the mechanism for the formation of the volcanogenic series. Proponents of one of the older popular concepts supposed that lava outflow and formation of tuffaceous horizons in the troughs of the Noril'sko–Kharaelakhsky depression and in the basin of the Nizhnyaja Tunguska River took place from central volcanoes (Ivanova 1975; Mitroshin and Nevskaya 1975; Goryainov et al. 1976). Such volcanoes were divided into large shield volcanoes (up to 130 km diameter) and small stratovolcanoes (up to 60 km diameter). These geoscientists regarded the volcanic depressions or troughs within the fore deeps of the Platform as a reflection of abyssal magmatic chambers. In the Tungussky syneclise, only fracture volcanic structures were present. Volcanoes of the Hawaiian Islands and Iceland were considered as analogues of volcanoes of the Platform. The geologists that studied the Tungussky syneclise assigned fracture volcanic structures the main role in formation of the volcanogenic series. The formation of the basalt field of the Tungussky syneclise was believed to be due to one large volcano called Tungussky (Mezhvilk and Vasil'ev 1967). The volcano was situated within the Yangoda–Gorbitsky uplifted block in the Taymyr. The cone of the Tungussky volcano had undergone weathering during the Jurassic and Cretaceous transgressions. The remnants of the volcanic cone are raised over 4 km above its base and reach about 150 km in diameter in the upper part and more than 300 km in the lower part. The main eruption center was attended by smaller volcanoes. Subordinate volcanoes existed near the Goulinsky alkali-ultramafic intrusions. Lava flows of the Tungussky volcano moved southward for 1,300 km and reached the same latitude as the Nizhnyaja Tunguska River. Another theory states that the eruptive centers during the early (Permian) stage were central volcanoes and during the late (Triassic) stages were fracture type (Zolotukhin et al. 1986).

In aureoles surrounding central volcanoes, concentric hypsographic curves of basalt flow thicknesses around the apparent craters would be expected. Yet studies have shown that tube-like cones are absent, and hypsographic curves of the basalt flows' thicknesses are characterized by linear elongation (see Figs. 2.30 and 2.35) coinciding with abyssal fractures; a magma-feeding role was determined for these fractures (Zolotukhin et al. 1986; Fedorenko 1994). The issue of magma-feeding channels for flood basalts still remains unresolved. The only convincing evidence is the above-described transfer of the sill of titanaugitic dolerite of Ergalakhsky intrusive complex in the flow of titanaugite basalt of the Ivalinsky Suite. Within the lava field, any dykes that could be regarded as feeding channels for basalt flows are absent. On this basis, it is supposed that the fracture volcanic structures functioned only once, and then closed, leaving no trace, and large melt outflows that formed numerous flood basalts were the result of disseminated spreading and simultaneous opening of numerous fractures (Ryabov et al. 2005). At the same time, in various parts of the Platform in the vicinity of small volcanic structures, a number of local thin flows of basalt, limburgite, clastic lava, and agglomerate are known (Ryabov 2003). Within these volcanic structures, ring and radial dykes are frequently observed, as well as sill-like dolerite bodies that range in thickness from decimeters to tens of meters. They exist only within the volcanic structures (Sapronov 1986) and have no magma-feeding, flow-generating significance.

The probable power of the fracture volcanic structure is given by the spread of the Kamensky Province dyke belt over the Siberian Platform. The dykes cross the entire volcanogenic series and are not flow-forming. Their concentration as packs of subparallel dykes can reach 19 dykes per square km (Ryabov and Grib 2002). The dykes typically reach tens of kilometers long (up to 140 km) and are up to 100 m thick. The multiphase composition of some dykes with "dyke-in-dyke" structure is evidence of a pulsing melt intrusion (Bushkanets et al. 1970a, b; Mitroshin and Nevskaya 1975; Ryabov and Grib 2005; Ryabov et al. 2005).

Fracture volcanic structures are characterized by a complex of geologic and petrologic data. In a magma-feeding zone fracture volcano, basalt melts have maximum thickness and maximum number of flows within a single lithologic and stratigraphic horizon. Further from a fracture, the thickness of flows decreases and there are fewer in the horizon (Daragan-Sushchov 1989). An example of a volcanic body with these characteristics is the Yuryakhsky trachybasalt series connected with faults of the Khantaysko–Dyupkunsky paleorift.

The position close to magma-feeding fracture structures is indicated by specific layered and fluidal basalts. Layered basalts in outcrop have a characteristic look governed by the horizontal alternation of basalt with differing degrees of compactness and differing resistance to weathering. This gives outcrops a crinkled, slate-like appearance. Under microscope, the basalt has a porphyritic, oligophyric texture, with a microdoleritic groundmass in one part of the layers and tiny poikilo-ophitic in the other (with a higher content of palagonite and calcite-filled amygdales). Layers of basalt are classified according to their thickness: thin (0.5–3 cm), middle (5–15 up to 20–50 cm), and coarse (1.5–3 m). The thickness of the layers typically remains stable within an outcrop or gradually diminishes until it disappears upward through a flow and along of its lateral extent. Finally, along strike the layers split and compact varieties typically pinch out (Simanovich and Kudryavtsev 1981; Ryabov et al. 2006). Most frequently, such layered tholeiitic basalts are observed in Putoransky Horizon flows. Layered basalts from near-crater zones, basalts of Mikchandinsky, Keta-Irbinsky lava flow sequences, and the lava flows of the Sunduk Mountain alternate with picritic or olivine-phyric komatiite-like basalts with olivine-free tholeiitic varieties (Ryabov et al. 1977; Kavardin 1988).

Fluidal basalts are characteristic for near-crater zones of paleovolcanoes. They consist of alternating small- or medium-pisolitic poikilophitic basalt and aphyric or oligophyric anamesites with microdoleritic texture. The layers form flexed lenses and crescent-form bodies with characteristic swirl forms. Radius of curvature of the swirl structures varies from a few centimeters to 0.5–0.7 m. Basalt flows with different texture have sharp borders. In anamesites, areas with trachytoidal texture are frequently observed. Sometimes, a transition from fluidal basalts into parallel-layered varieties is observed. The thickness of fluidal basalts varies from 10 to 30 m. As a rule, they manifest strictly locally and are observed throughout the flow sequence or form in the roof of the thickest flows that feature poikilophitic basalts in lower part. Along the strike, fluidal basalts are replaced by massive poikilophitic basalts or anamesites. The texture of fluidal basalts points to turbulent melt flow that existed near magma-bearing channels. Fluidal basalts are noted in the Putoransky Horizon, in flows of the Vologochansky, Kharaelakhsky, and Bol'sheavamsky troughs and in the Tungussky syneclise.

Varying volumes of pillow lava are observed at the base of different suites and as horizons in tuff rocks (see Fig. 2.15). They are spread locally and usually form lens-like bodies and vary from 1 m to 10–15 m thick. Sometimes, for instance in the basin of the Nizhnyaja Tunguska River (at the base of the Khonnamakitsky (Nidymsky) and Nadezhdinsky (Tymersky) Suites), thick horizons of pillow lava (70–100 m thick) filled extensive trough-like depressions 25–30 km long and 15–20 km wide (Mitroshin and Sukhov 1969). Pillow lavas are underlain by volcanogenic-sedimentary rocks, such as tuff sandstones, tuff siltstones, and (in subsidence calderas of paleovolcanoes) limestones. Thick pillow lava horizons are composed of alternating big-and small-pillow lavas (10–15 layers) and hyaloclastites. Pinching out of lava and cross bedding are common and indicate the direction of lava flow. Formation of pillow lavas, by analogy with similar forms in spillites of geosynclinal areas, is supposed to take place during lava erupting into water (Mitroshin and Sukhov 1969; Simanovich and Kudryavtsev 1981). The existence of small water basins in the Platform in the early Triassic is proved by layered, sorted, redeposited volcanogenic-sedimentary rocks and lakes in subsidence calderas of paleovolcanoes, as well as by remnants of fresh-water fauna and flora. Pillow lavas are characterized by restricted fluidity and can flow only in the vicinity of magma-bearing channels, so they can serve as an indirect evidence of the close proximity of volcanic structures.

The upper surface of pillow lavas bears information about formation conditions, viscosity of the melt, volatile content, and distance from the eruptive center. Melt flow wrinkles and corded and ropy lava (pahoehoe type) indicate the distance to volcanic centers, and convex forms indicate the direction of melt flow. At some distance from the volcanic center, lava breccia and agglomerates were formed (aa type). Flow wrinkles were observed at the top of basalt flows of the Morongovsky and Syverminsky Suites (Pol'kin 1961), as well as the Khonnamakitsky, Gudchikhinsky, and Pravoboyarsky Suites (Fig. 2.4). All manifestations of wrinkled flow were found near faults producing magma extrusions, where the melt still retains its fluidity.

The existence of large paleo central-volcanoes—shield volcanoes and stratovolcanoes—is the basis of model constructions (Ivanova 1975; Mitroshin and Nevskaya 1975; Goryainov et al. 1976). It is supposed that these volcanoes played a lava flow-forming role, though the tube-like magma effluent channels, characteristic for such structures, were not found, and the concentric arrangement of lava flows around supposed crater zones is absent. On the Platform, small volcanic structures, diatremes, are widely spread and number more than 500. In plan view, the diatremes usually have a round form, and their size varies from meters to several hundred meters (more rarely up to a few kilometers). They frequently form chains along linear and ring faults, forming larger volcanic structures. Volcanic structures are often situated on isopachyte contours showing maximum basalt flow thickness or confined to faults producing magma extrusions. Their crater zones, as a rule, are composed of tuff and tuff breccia. More rarely, thick linearly elongated zones of tuff breccia controlled by fracture structures are met (Daragan-Sushchov 1989). Coarse-block breccias represent the near-crater fraction, and the size of clastic material in tuff breccia usually diminishes with distance from the crater. Clastic material in tuff breccias is usually trap rocks, though more rarely sandstone, siltstone, Tungussky series coal, or chemical-sedimentary rocks of different age (Tarakhovsky 1969; Sapronov 1986). Near-crater zones of paleovolcanoes are characterized by the presence of pisolitic and lapilli tuffs, sometimes forming layers up to 50–70 cm thick with 60–80% pisolites and lapillies.

Pyroclastic rocks characteristically show a great variety of petrographic compositions and manifestation forms in trap formations. The number of tuff-sedimentary formations in the volcanogenic series increases from north to south from 5–10% to 70–90% (Starosel'tsev 1989). Volcanoclastic rocks are divided into three main types according to their genesis: tuffs, tuffites and pyroclastic-sedimentary rocks, and volcanomictous rocks (Tarakhovsky 1969). Volcanic process began with explosive activity, continued with effusive lava, and ended with explosive activity again. Following eruption, weathering of tuff rocks occurred, and they were destroyed by water and wind, then redeposited resulting in the formation of volcanogenic-sedimentary horizons containing flora and fauna fossils. Pronounced powerful explosive activity, with volcanic structures observed in outcrop, took place during eruption of the Morongovsky (Ayansky, Korvunchansky) Suite (Sapronov 1986). This coincided with the formation of the Yuzhno-Ergalakhsky diatreme that contains detritus of crystalline schists, gneisses, and granites from the base of the Platform in its crater facies (Maslov and Nestyerovsky 1961).

The general scheme of the volcanic cycle and its products on the Siberian Platform has similarities with the eruption of fracture volcanoes Laki and Ogmundarraum in Iceland in 1783. These fracture structures were about 30 km long. During the eruption, the Laki fracture volcano produced 12.5 km^3 of melt. The volcanic paroxysm ended with formation of about 100 tuff cones, 10–15 m high and 30–50 m long, on each of the fractures (MacDonald 1972; Luchitsky 1985). It should be noted that the transformation of fracture channels into cylindrical channels with funnel-shaped outlets was noted long ago and experimentally modeled at the end of the nineteenth century by dropping gas pressure while passing through fractures in granite, limestone, steel, and glass (Moushketov 1929).

In a number of cases, the activity of fumaroles and solfatares was connected with volcanic structures. It promoted the formation of limestones and anhydrites in subsidence calderas up to 60 m thick. Diatremes are associated with the metasomatic and hydrothermal transformation of host rocks, as well as deposits and manifestations of native copper, sulfide and iron ore, optic quality calcite, bitumen, boron, and phosphorus (Ryabov et al. 2005).

Summing up, one could suggest that giant-scale flood basalts erupted through fracture magma channels. Tectonomagmatic cycles began with explosive activity through fracture structures that in the course of increasing tectonic activity were replaced by massive lava outpourings. A drop of pressure in the magma column led to closure of fractures and their transformation into central volcanoes with explosive activity. High density of simultaneously functioning fracture structures during tectonomagmatic cycles led to the formation of thick, wide-spread lava flows. Numerous explosive structures formed tuff cones and fields, whose erosion and redeposition led to loss of relief and formation of horizons of volcanogenic-sedimentary rocks during quiescent stages.

2.16.1 Comagmatism of Effusive and Intrusive Traps of the Siberian Platform

Trap formation in the northwest of the Siberian Platform is represented by effusive and intrusive facies. Effusive facies are divided into suites and intrusive facies into complexes. Formation of magmatic bodies took place during six phases of tectonomagmatic activation of the Platform. The accepted viewpoint is that for the first five phases started with lava flows and ended with intrusion of melts with the same composition, whereas during the sixth phase only sills and dykes of two intrusive complexes were formed. These cross the entire volcanogenic series and did not produce flows.

The main limitations for determination of effusive and intrusive comagmates lie in the similarity of their composition and formation within the same tectonomagmatic phase (Ryabov and Lapkovsky 2008). A confirmed case of comagmatism is seen in outcrops that show the transition of trachydolerite sills of the Ergalakhsky Complex into trachybasalt lava flows of the Ivakinsky Suite. Comagmatism of andesite–basalts of the upper unit of the Ivakinsky Suite and intrusions of the Pyasinsky Complex can be supposed on the basis of their close and specific mineralogical, petrographic, and chemical compositions. The subalkaline composition of the rocks of these comagmates is not typical for traps. This fact simplifies identification of the intrusive complexes.

Demonstrated comagmatism of products of picrite-basaltic melts is rare. It is supposed that comagmates of picritic basalts of the Gudchikhinsky Suite are found in high-magnesium intrusions of the Gudchikhinsky, Fokinsky, and Noril'sky Complexes. An important criterion of comagmatism of effusives and intrusives is their formation during the same volcano-plutonic cycle. According to this principle, all intrusions of picritic gabbrodolerites lying at a higher stratigraphic level are to be excluded from the list of comagmates of picritic basalts of the Gudchikhinsky Suite. This is also right for a number of undifferentiated intrusions as well as for layered intrusions of Noril'sky Complex, which are regarded by many geoscientists as comagmates of picritic basalts of the Gudchkhinsky Suite. Intrusions of the Noril'sk-tye cross volcanogenic series up to the Nadezhdinsky Suite and so cannot be comagmates of Gudchikhinsky Suite picritic basalts.

Most complicated is the determination of comagmates among lavas and intrusions of derivatives of tholeiitic-basaltic melts that make up the vast bulk (about 95%) of traps. Basalt flows lack channels, and dolerite sills are effectively bottomless intrusions. In the Noril'sk Region, there is one case of the transformation of a dolerite sill into a flow of tholeiitic basalt of the Syverminsky Suite. Here, the chemical composition of tholeiitic basalts and dolerite sills is constant enough for the determination of comagmates. It should be noted, however, that small variations in the content of petrogenic oxides and accessory elements in the rocks can be due to melt differentiation, redistribution in the presence of fluids, and by regional specificity of different parts of the Platform.

It is also interesting to note the existence of high-magnesium intrusions of Morongovsky age and Nizhnetalnakhsky type in the northwest of the Platform that have no compositional analogues among effusive formations. There are no known effusive comagmates of the layered intrusions, including those of Noril'sky, Dzhaltul'sky, and other types.

The problem of comagmatism in Platform magmatism is ongoing and requires further investigation.

References

Almukhamedov AI, Zolotukhin VV, Lozhkin VI, Gunicheva TN (1988) Average composition of basalts in the northwestern part of the Siberian Platform. Doklady Akad Nauk USSR 302(5): 1205–1208 (in Russian)

Almukhamedov AI, Plyusnin GS, Almukhamedov EA, Zolotukhin VV, Nikolaev VM, Kuznetsova SV, Sandomirova GP (1992) Permo-Triassic basalts of the Siberian Platform: 87Sr/86Sr isotopy. Russ Geol Geophys 33(7):38–47

Almukhamedov AI, Medvedev AY, Zolotukhin VV (2004) Spatial and temporal evolution of composition of Permo-Triassic basalts in the Siberian Platform. Petrologiya 12(4):339–353 (in Russian)

Arkhipova AI, Dodin DA (1963) New data on subalkaline trap magmatism in the northwestern part of the Siberian Platform (Khraelakh mountains). In: Datsenko VA (ed) Geology of north-west of the Siberian Platform. Gosgeoltekhizdat, Moscow, pp 153–167 (in Russian)

Bogatikov OA, Mikhailov NP, Gon'shakova VI (eds) (1981) Classification and nomenclature of igneous rocks. Nedra, Moscow, 160 p

Bushkanets YuS, Nevskaya AV, Belyakov GD (1970a) Dykes of traps in the interfluves of Kystyrtakh-Bol'shoy Avam Rivers. In: Urvantsev NN (ed) Geology and mineral resources of north-west of the Siberian Platform. Nedra, Leningrad, issue 2, pp 124–139 (in Russian)

Bushkanets YuS, Nevskaya AV, Belyakov GD (1970b) Kamenskaya magmatic province of alkalic rocks. In: Egorov LS (ed) Carbonatites and alkaline rocks of the northern of Siberia. Nedra, Leningrad, pp 157–169 (in Russian)

Campbell IH, Griffiths RW (1990) Implications of mantle plume structure for the evolution of flood basalts. Earth Planet Sci Lett 99:79–93

Campbell IH, Czamanske GK, Fedorenko VA, Hill RG, Stepanov VK (1992) Synchronism of the Siberian traps and the Permian–Triassic boundary. Science 358:1760–1763

Dalrymple GB, Czamanske GK, Lanphere MA, Stepanov VK, Fedorenko VA (1991) 40Ar/39Ar ages of samples from Norilsk–Talnakh ore-bearing intrusions and the Siberian flood basalts. EOS 72:570

Daragan-Sushchov YI (1984) About Late Paleozoic and Early Mesozoic volcanogenic rock mass of the Tunguska syneclise. Sovetskaya Geologiya 11:89–96 (in Russian)

Daragan-Sushchov YI (1989) Recognition of centers of volcanic eruptions by maximum amount of sheets in a sequence of flood basalt units. Doklady Akad Nauk USSR 307(5):1177–1180 (in Russian)

Distler VV, Kunilov VE (eds) (1994) Geology and ore deposits of the Noril'sk region. VII international platinum symposium guidebook. Moskovskiy Contact, Moscow, 67 p

Dmitriev YI (1973) Mesozoic trap volcanism in the central part and at periphery of the Tungusskaya sineclise. Izv AN SSSR Geol Ser 10:58–67 (in Russian)

Dobretsov NL (1997) Permian-Triassic magmatism and sedimentation in Eurasia as reflection of superplume. Doklady of Akad Nauk 354(2):220–223 (in Russian)

Dobretsov NL, Kirdyashkin AG, Kirdyashkin AA (2001) Depth geodynamics, SB RAS, Affiliated Branch GEO, Novosibirsk, 408 p (in Russian)

Dodin DA (1962) Oceanites and chrysotile-asbestos occurrences in the basin of the Upper Sukhoy Balek River (northwestern margin of the Siberian Platform). Inform Bull NIIGA. NIIGA, Leningrad, issue 29, pp 39–54 (in Russian)

Dodin DA (1967) Petrology of traps of the East Khraelakh (Noril'sk region). Unpublished Ph.D. thesis, VSEGEI, Leningrad, 31 p (in Russian)

Dodin DA (1971) About acid and subalkaline traps (by example of the northwestern part of Siberian Platform). In: Abstracts of Traps of the Siberian Platform and their metallogeny. Institute of Earth's Crust SB AS USSR, Irkutsk, pp 40–41 (in Russian)

Dodin DA, Golubkov VS (1971) On differentiated basalt sheet in the north-western part of the Siberian Platform (Noril'sk region). Doklady Akad Nauk USSR 108(2):403–406 (in Russian)

Duzhikov OA (1971) On picritic basalts of the Noril'sk region. Doklady Akad Nauk 197(6):1406–1410

Duzhikov OA (1988) Copper-zeolite ore formation. In: Geology and metalogeny of sulfide deposits of Noril'sk Region USSR. Nauka, Moscow, pp 173–179 (in Russian) [refer to English translation of 1988 Russian edition by Duzikov OA, Distler VV, Strunin VM, Mkrtychyan AK, Sherman ML, Sluzhenikin SF, Lurje AM. Soc Econ Geol Spec Pub, 1:242 p]

Duzhikov OA, Distler VV (1981) Komatiites of the Noril'sk ore region. Doklady Akad Nauk 261(5):1194–1198 (in Russian) (1992)

Duzhikov OA, Lind EN, Makeev MM, Iogin SM (1971) To the question on two-stage effusive trap magmatism in the northwest of the Siberian platform (in the instance of the Noril'sk-Kharaelakh ore field). In Traps of the Siberian platform and their metallogeny. Abstracts of II All-Union conference, Irkutsk, pp 32–33 (in Russian)

Fadeev VA (1962) Rhythmicity of volcanic processes as a base for subdivision of a volcanogenic unit. Informatsionnyi sbornik NIIGA, issue 31, pp 23 (in Russian)

Fedorenko VA (1981) Petrochemical series of extrusive rocks of the Noril'sk region. Sov Geol Geophys 22(6):66–74

Fedorenko VA (1994) Evolution of magmatism as reflected in the volcanic sequence of the Noril'sk region. Proc of the Sudbury-Noril'sk Symposium (eds PC Lightfoot, AJ Naldrett). Ontario Geol Surv Spec Pub 5:171–184

Fedorenko V, Czamanske G (1997) Results of new field and geochemical studies of the volcanic and intrusive rocks of the Maymecha-Kotuy area, Siberian Flood-Basalt Province, Russia. Int Geol Rev 39:479–531

Fedorenko VA, Stifeeva GT, Makeeva LV, Sukhareva MS, Kuznetsova NP (1984) Mafic and alkaline-mafic intrusions of the Noril'sk region in connection with their comagmatism to effusive formations. Geologiya i Geofisika 5:56–63 (in Russian)

Frolova TI, Kotorgin NF (1986) Toward the problem of classification of picrites and komatiites. Vestnik MGU, Geologiya, Ser 4, 1: pp 3–17

Gladkikh VS, Gusev GS, Peskov AI, Tarnovsky LL (1994) Geochemistry of volcanic rocks in the north-western part of Siberian trap province (Noril'sk region). Geokhimiya 3:367–381 (in Russian)

Godlevsky MN (1959) Traps and ore-bearing intrusions of the Noril'sk region. Gosgeoltekhizdat, Moscow, 68 p (in Russian)

Goryainov IN, Mitroshin MI, Leonova TS, Nevskaya AV, Ivanova TK, Ivanov MK (1976) "Meteorite" paragenesis – moissanite, native iron, (diamond?) in traps of the northwest of the Siberian Platform. Dokl Acad Nauk USSR (228)2:453–455 (in Russian)

Haag M, Heller H (1991) Late Permian to: Early Triassic magnetostratigraphy. Earth Planet Sci Lett 107:42–54

Ivanova TK (1975) Application of structural-facial analysis in volcanic reconstructions in the Noril'sk region, In: Urvantsev NN (ed) Copper-nickel ores at the northwest of the Siberian Platform. NIIGA, Leningrad, pp 52–75 (in Russian)

Kavardin GI (1976) Metallogeny of the northwestern part of the Siberian Platform (ed Urvantsev NN). Nedra, Leningrad, 158 p (in Russian)

Kavardin GI (1988) Komatiitic lavas of the Imangdinsky region. In: Dodin DA (ed) Geochemistry and mineralogy of ore formations of the Noril'sk region. Nauka, Leningrad, pp 83–87 (in Russian)

Kavardin GI, Staritsyna GN, Golubkov VS, Goryainov IN, Kravtsova LI (1968) Traps of the Yenisei ore province (ed Urvantsev NN). Nedra, Leningrad, 208 p (in Russian)

Korovyakov IA (1948) On picritic effusive traps of the northwestern part of the Siberian Platform. Doklady Akad Nauk USSR LXII (1):129–133 (in Russian)

Kutolin VA (1969) A statistic study of chemical peculiarities of basalts from different formations. Nauka, Novosibirsk, 140 p (in Russian)

Kuznetsov AA (1965) Some features of effusive traps at the northwest of the Siberian Platform in the light of current data. Uchenye Zapiski NIIGA, Regional'naya Geologiya 5:161–170 (in Russian)

Kuznetsov AA (1977) Tectonomagmatic process. Nedra, Leningrad, 118 p (in Russian)

Le Maitre (ed-in-chief) (1989) Classification of igneous rocks and glossary of terms. In: Recommendation of the international union of geological sciences subcommission on the systematics of igneous rocks. Blackwell Scientific Pub, Oxford/London/Edinburg/Melbourne, 248 p
Ledneva VP, Poroshin EE (1981) Petrochemical zoning of the basalts of the Tungusskaya sineclise. Sov Geol Geophys 22(11):119–122
Lightfoot PC, Hawkesworth CJ, Hergt J, Naldrett AJ, Gorbachev NS, Fedorenko VA, Doherty W (1993) Remobilisation of continental lithosphere by mantle plumes: Major trace element and Sr- Nd- and Pb-isotope evidence for picritic and tholeiitic lavas of the Noril'sk district Siberian Trap, Russia. Contrib Mineral Petrol 114:171–188
Lightfoot PC, Naldrett AJ, Gorbachev NS, Doherty W, Fedorenko VA (1990) Geochemistry of the Siberian trap of the Noril'sk area USSR, with implications for the relative contributions of crust and mantle to flood basalt magmatism. Contrib Mineral Petrol 104:631–644
Lightfoot PC, Naldrett AJ, Gorbachev NS, Fedorenko VA, Hawkesworth CJ, Hert J, Doherty W (1994) Chemostratigraphy of Siberian trap lavas, Noril'sk district Russia: implications and source of flood basalt magmas and their associated Ni-Cu mineralization. In: Lightfoot PC, Naldrett AJ (eds) Ontario Geological Survey special volume, Proceedings of the Sudbury – Noril'sk Symposium, Ministry of Northern Development and Mines, Toronto, vol 5, pp 283–312
Likhachev AP (1997) Trap magmatism and Pt–Cu–Ni ore formation in the Noril'sk region. Otechestv Geol 10:8–19 (in Russian)
Lind EN, Shchekoturov VV (1991) Paleomagnetic dating of trap volcanism of nortwestern part of the Siberian Platform. In: The fourth All-Union congress on paleomagnetism, Chapter II, CNIO SSSR, Vladimir–Suzdal, pp 79–80 (in Russian)
Lipkov LZ (1969) The tectonic structure and oil bearing prospects in the northeast part of the Dyupkun Lake basin. Uchenye Zapiski NIIGA, Regional'naya Geologiya 15:61–67 (in Russian)
Luchitsky IV (1985) Paleovolcanology (ed-in-chief Dymkin AM). Nauka, Moscow, 276 p (in Russian)
MacDonald GA (1972) Volcanoes. University of Hawaii Prentice-Hall Inc, Englewood Cliffs, 430 p
Marakushev AA, Fenogenov AN, Emel'yanenko PF, Duzhikov OA, Skripnichenko VA (1982) Genesis of the Noril'sk-type layered intrusions. Vestnik MGU, Geol Ser 1:3–19 (in Russian)
Maslov GD, Nesterovsky VS (1961) Eruptive rock fragments of the crystalline basement in Triassic tuffs. Geologiya i Geofizika 12:128–130 (in Russian)
Mezhvilk AA (1962) Reference horizons among effusive traps of the Siberian Platform. Geologiya I Geofisika 4:68–75 (in Russian)
Mezhvilk AA, Vasil'ev YuN (1967) Toward the history of formation of effusive traps in Tunguska syneclise/Petrology of traps in the Siberian Platform (editor-in-chief NN Urvantsev). Nedra, Leningrad, pp 66–77 (in Russian)
Mikhaltsov NE, Ryabov VV, Kazansky AYu, Shevko AYa, Kuprish OV (2008) Age and duration of Permian-Triassic magmatism in Siberia according to the paleomagnetic dating. In: The basement, framing structures of the West-Siberian Mesozoic-Cenozoic sedimentary basin, their geodynamic evolution and problems of oil-and-gas occurrence 29 September–2 October 2008. Proceedings of All-Russian scientific conference, Tyumen-Novosibirsk, pp 152–155 (in Russian)
Mitroshin MI, Nevskaya AV (1975) Morphological types of volcanostructures in the catchment area of the Severnaya River and their association with sulfide copper-nickel ores and datolite mineralization. In: Copper-nickel ores in the northwest of the Siberian platform (editor-in-chief NN Urvantsev). Leningrad, pp 38–51 (in Russian)
Mitroshin MI, Sukhov LG (1969) Features of the structure and formation of globular lavas and hyaloclastites at the west of syneclise. Uchenye Zapiski NIIGA, Regional'naya Geologiya, issue 16, pp 79–86 (in Russian)
Morimoto N (1989) Nomenclature of pyroxenes. Can Min 27:143–156
Mushketov DI (1929) The short course of General Geology. Nauchnoe khimiko-tekhynicheskoe izdatel'stvo, Leningrad, 368 p (in Russian)
Nagaitseva NN, Rogozhin VV, Ivanova TK (1967) On the composition and stratigraphic location of trachybasalts in the Yuryakhsky unit at the north-west of the Siberian Platform. Uchenye Zapiski NIIGA, Regional'naya Geologiya 10:202–207 (in Russian)
Naldrett AJ, Lightfoot PC, Fedorenko VA, Goebachev NS, Doherty W (1992) Geology and geochemistry of intrusions and flood basalts of the Noril'sk region USSR with implications for the origin of the Ni–Cu ores. Econ Geol 87(4):975–1004
Nemenenok TI (1978) Explosive magmatism. In: Zolotukhin VV, Vilensky AM (eds) Petrology and perspectives of metallogeny in traps at the north of the Siberian platform. Nauka, Novosibirsk, pp 8–43 (in Russian)
Nemenenok TI, Vilensky AM (1978) Effusive magmatism. In: Zolotukhin VV, Vilensky AM (eds) Petrology and perspectives of metallogeny of traps at the north of the Siberian platform. Nauka, Novosibirsk, pp 43–60 (in Russian)
Nesterenko GV, Amilova NS, Smirnova NP (1964) Rare elements in traps of the Siberian Platform. Geokhimiya 10:1015–1021 (in Russian)
Nesterenko GV, Tikhonenkov PI, Romashova TV (1991) Basalts of Putorana plateau. Geokhimiya 10:1419–1429 (in Russian)
Oleinikov BV, Kopylova AG, Pankov VYu (1991) Mineral associations of native iron in apodoleritic metasomatites of the Dzhaltaksky intrusion. In: Oleinikov BV (ed) Native mineral formation during magmatic process. Yakutsk Affiliated Branch AS USSR Pub, Yakutsk, pp 29–47 (in Russian)
Pavlov LG (1969) Regularities in formation of the volcanogenic complex of Keta and Dyupkun lakes (northwestern Siberian Platform). Uchenye Zapiski NIIGA, Reg Geol 15:97–101 (in Russian)
Pavlov DI (1975) Magnetite ore formation with exogenous chloride liquids involved. Nedra, Moscow, 246 p (in Russian)
Pol'kin YaI (1958) Toward the question on origin of limestones from effusive complex in the north of the Siberian Platform. In: Collected papers of Arctic region geology, Leningrad, issue 10, pp 18–23 (in Russian)
Pol'kin YaI (1961) Some peculiarities of surfaces of lava sheets in the northwest of the Siberian platform. In: Collected papers of Arctic region geology, Leningrad, issue 24, pp 31–34 (in Russian)
Pol'kin YaI (1964) Principles of the unified stratigraphic scheme of volcanic formations in the northwest of the Siberian Platform. Uchenye Zapiski NIIGA, Regional'naya Geologiya, issue 3, pp 5–26 (in Russian)
Rad'ko VA (2007) About indications of a large cuprum deposits at the north of the Putorana plateau. Geology and mineral deposits of the Krasnoyarsk Region, Krasnoyarsk, issue 8, pp 126–130 (in Russian)
Rashin GA (1963) Heteromorphosis and nonequilibrium states of mineral formation during crystallization of basalt melts. Izv AN SSSR Geol Ser 9:18–33 (in Russian)
Renne PR, Basu AR (1991) Rapid eruption of the Siberian Traps flood basalts at the Permo-Triassic boundary. Science 253:176–179
Ryabov VV (1989a) Chemical composition of immiscible liquids in natural glasses from traps. Geologiya i Geofisika 11:69–78 (in Russian)
Ryabov VV (1989b) Immiscibility in natural glasses (by example of traps) (ed Zolotukhin VV). Nauka, Novosibirsk, 223 p (in Russian)
Ryabov VV (1992a) Nontraditional types of platinum metal ores in the Noril'sk-type sulfide deposits. Russ Geol Geophys 33(11):79–88 (in Russian)
Ryabov VV (1992b) Olivines of Siberian traps as indicators of petrogenesis and ore formation (ed Zolotukhin VV). Nauka, Novosibirsk, 116 p (in Russian)
Ryabov VV (1992c) Types of low-sulfide of platinum-group metal ores in contact zones of trap intrusions. In: Distler VV, Evstigneeva TL, Kamshilina EM (eds) Geology and genesis of platinum metal deposits. IGEM Publication, Moscow, pp 40–41 (in Russian)
Ryabov VV (1997) Ore melts among basalts on the Putorana plateau (Siberian Platform). Russ Geol Geophys 38(12):1932–1949
Ryabov VV (1999a) Fluid regime of trap magmatism and ore formation (petrological aspect). Russ Geol Geophys 40(10):1437–1452

Ryabov VV (1999b) Ore geology of the Noril'sk region on the threshold of millennium. In: Simonov ON (ed) Abstracts of the regional symposium: trends in geological-exploration works on nickel, copper and platinoids in the Taimyr Autonomous Area in 2000–2005. Taimyrkomprirodresursy, Noril'sk, pp 25–26 (in Russian)
Ryabov VV (2003) The lava sheet of limburgites and picritic porphirites of the Mokulaevsky Suite as a new geodynamic border. Newslett Tomsk State Univ 3(III):182–185 (in Russian)
Ryabov VV (2007) Geodynamic control on trap magmatism and ore-formation (Siberian Platform). In: Large igneous provinces of Asia, mantle plumes and metallogeny. International symposium, Novosibirsk, pp 190–192
Ryabov VV, Grib DE (2002) The dyke belt of the spreading zone in the north of the Siberian Platform. In Petrology of magmatic and metamorphic complexes. In: Proceedings of All-Russian scientific conference, Tomsk State University Publication, Tomsk 3(1):208–214
Ryabov VV, Grib DE (2005) Multiphase dikes: signature of disperse springing in the northern Siberian craton. Russ Geol Geophys 46(5):471–485
Ryabov VV, Lapkovsky AA (2008) Problems of comagmatism of effusive and intrusive traps of the Siberian Platform. Petrology of lithosphere and origin of diamond. In Abstracts of international symposium marking the 100th anniversary of Academician VS Sobolev birth, Novosibirsk, 5–7 June 2008, Siberian Branch RAS, Novosibirsk, p 81 (in Russian)
Ryabov VV, Lapkovsky AA (2009) Volcano-tectonic structures in Siberian traps as areas of accumulation and discharge of formation fluids during the process of tectono-magmatic activation of the region. In: Volcanism and geodynamics: transactions of IV All-Russian symposium on volacanology and paleovocanology, Petropavlovsk-Kamchatsky, vol 2, pp 490–493 (in Russian)
Ryabov VV, Pavlov AL (1991) Magnetite lava in traps of the Siberian Platform. Doklady Akad Nauk SSSR 319(5):1193–1197 (in Russian)
Ryabov VV, Bakumenko IT, Fominykh IM (1977) Dendritic metacrysts in traps of the Noril'sk region and some aspects of their formation. In: Sobolev VS (ed) Proc on petrology and mineralogy. Nauka, Novosibirsk, pp 47–71 (in Russian)
Ryabov VV, Konenko VF, Krasov NF (1985a) Immiscibility phenomena in glasses from ores of native iron of Khungtukun intrusion. Doklady Akad Nauk SSSR 285(4):982–987 (in Russian)
Ryabov VV, Konenko VF, Khmel'nikova OS (1985b) Rock-forming minerals of picritic basalts of the Noril'sk region. Sov Geol Geophys 26(4):77–84
Ryabov VV, Pavlov AL, Lopatin GG (1985c) Native iron of Siberian traps (ed Zolotukhin VV). Nauka, Novosibirsk, 169 p (in Russian)
Ryabov VV, Shevko AYa, Zateeva SN (2003) The analysis of occurrence of high-magnesian and subalkaline lavas among flood basalts of the Siberian platform. Volcanism and geodynamics. In: Proceedings of II All-Russian symposium on volcanology and paleovolcanology, Ekaterinburg, pp 709–714 (in Russian)
Ryabov VV, Shevko AY, Zateeva SN (2005) "Abnormal formations" in traps of the Siberian Platform as indicators of geodynamic conditions of flood basalt formation. Litosfera 4:165–177 (in Russian)
Ryabov VV, Ponomarchuk VA, Talibova AG (2006) Bitumen, graphite and carbon-bearing matter in rocks and ores of the northwestern Siberian Platform: occurrence and carbon isotope composition. In: Obolensky AA (ed) Topical problems of ore formation and metallogeny. Abstracts of Internat meeting devoted to 100th anniversary of academician V.A. Kuznetsov. GEO, Novosibirsk, pp 189–190 (in Russian)
Ryabov VV, Medvedev AY, Napreev DE, Lapkovsky AA (2009) Marker Horizons in Lithostratigraphic sections of continental flood basalts of the Siberian platform. In: Large igneous provinces of Asia, mantle plumes and metallogeny. International symposium, Novosibirsk Russia, 6–9 August 2009, pp 268–270
Ryabov VV, Semenova DV, Ponomarchuk VA (2010) Carbon isotope composition in C-bearing species of the northwest of the Siberian platform/Contribution of international conference EDU-2009, Vienna, Austria, 2009. http://meetingorganizer.copernicus.org/EGU2010/poster_programme/2047 (EGU2010-8358)
Sapronov NL (1986) Ancient volcanic structures at the south of the Tunguska syneclise in Siberian Platform. Nauka, Novosibirsk, 103 p (in Russian)
Savushkin MP (2008) From the history of native cuprum occurrences in the west-northern Siberian platform. In: Geology and mineral deposits of the Krasnoyarsk Region. Krasnoyarsk (9):265–266 (in Russian)
Sharapov VN, Perepechko YV, Perepechko LN, Rakhmenkulova IF (2008) Mantle sources of Permian-Triassic traps (Western Siberian Plate and Siberian Craton). Russ Geol Geophys 49(7):492–502
Sharma M, Basu AR, Nesterenko GV (1992) Temporal Sr-, Nd-, Pb-isotopic variations in the Siberian flood basalts: implications for the plume-source characteristics. Earth Planet Sci Lett 113:365–381
Simanovich IM, Kudryavtsev DI (1981) Texture types of effusive basalts of the Tungusska sineclise (ed Peyve AV). Nauka, Moscow, 63 p (in Russian)
Sobolev VS (1986) In: Zolotukhin VV (ed) Petrology of traps: Izbrannye trudy. Nauka, Novosibirsk, 209 p (in Russian)
Starosel'tsev VS (1982) Tectonic grounds for prediction of oil-bearing sedimentary strata under basalt plateau of ancient platforms. Unpublished Doctor Sc thesis, IGG SB RAS, Novosibirsk, 32 p (in Russian)
Starosel'tsev VS (1989) Tectonics of basalt plateau and occurrence of oil-and-gas in underlying sediments. Nedra, Moscow, 256 p (in Russian)
Starosel'tsev KV, Starosel'tsev VS (1987) Fossil anthraxolie-zeolite fumarole in Triassic volcanites of the north Tunguska syneclise. Geologiya i Geofisika 3:113–117 (in Russian)
Strunin BM, Duzhikov OA, Barmina OA, Komarov VV (1994) Geological map of the Noril'sk region of 1:200,000 scale. In: Sherman ML, Mkrtychyan AK, Malich NS (eds) Explanatory note. AO Geoinformmark, Moscow, 118 p
Tarakhovsky AN (1969) Volcanogenic clastic rocks in the volcanogenic unit at the western edge of the Tunguska syneclise. Uchenye Zapiski NIIGA, Regional'naya Geologiya 16:87–91 (in Russian)
Tomanovskaya YI (1961) Picritic basalts of the Middle Ayan River (northwestern margin of the Siberian Platform). Inf Bull Inst Arct Reg Geol 24:12–16 (in Russian)
Travin AV, Ryabov VV, Ponomarchuk VA (2010) Siberian traps: what's new? Geochimica et Cosmochimica Acta 74(12):A1053
Wooden JL, Czamanske GK, Fedorenko VA, Arndt NT, Chauvel C, Bouse RM, King B-SW, Knight RJ, Siems DF (1993) Isotopic and trace element constraints on mantle and crustal contributions to characterization of Siberian continental flood basalts Noril'sk area Siberia. Geochim Cosmochim Acta 57:3677–3704
Yudina VV, Lyul'ko VA, Nemenenok TI (1977) The "Magnetitovaya" diatreme in the Noril'sk area. Sov Geol Geophys 18(6):73–81
Zelenshchikov GV, Duzhikov OA (1974) On comparative characteristcs of basalts of Voronezhsky anteclise and effusive traps in the north-western Siberian Platform. In: Abstracts of the third All-Union conference "The state and trends in investigations on trap metallogeny", KrTGU, Krasnoyarsk, pp 43–45 (in Russian)
Zhabin AG, Surina NP (1970) Petrology of dykes, sills and diatremes in the Maymecha-Kotuy province (ed VV Lyakhovich). Nauka, Moscow, 204 p (in Russian)
Zolotukhin VV, Almukhamedov AI (1991) Basalts of the Siberian Platform: Conditions of occurrences, composition and mechanism of formation. In: Traps of Siberia and Deccan, similarities and differences. Nauka, Novosibirsk, pp 7–38 (in Russian)
Zolotukhin VV, Vasil'ev YR (1967) Peculiarities of formation mechanism of ore-bearing trap intrusions at the northwest of Siberian platform (ed Sobolev VS). Nauka, Moscow, 231 p (in Russian)
Zolotukhin VV, Vilensky AM, Duzhikov OA (1986) Basalts of the Siberian platform (eds Sobolev VS, Sobolev NV). Nauka, Novosibirsk, 255 p (in Russian)
Zolotukhin VV, Vasil'ev YuR, Duzhikov OA (1989) Diversity of traps and initial magmas. Nauka, Novosibirsk, 247 p (in Russian)

Intrusive Complexes of the Noril'sk Region

3

Abstract

Trap intrusions of the Siberian Platform are subdivided into intrusive complexes, which were formed during six tectonomagmatic cycles. Ten intrusive complexes are recognized in the Noril'sk Region. The key parameters for subdivision of intrusive complexes are as follows: situation of intrusions in a geological structure of the region, shapes of intrusions, the degree of differentiation, mineralogical and chemical composition of rocks (normal, high magnesian, subalkaline, and silicate), features of the internal structure of intrusions, association of intrusions with deposits or manifestations of certain ore mineral resources, types of deposits or ore manifestations, age relationships to other intrusions, as well as features of aureoles of altered rocks associated with these intrusions. The most of intrusive complexes combine several types of intrusions, which possess some individual features apart from their common similarity.

Trap intrusions are subdivided into undifferentiated, weakly differentiated, and differentiated (layered). In Sects 3.1, 3.2, 3.3, 3.4, 3.5, 3.6, 3.7, 3.8, 3.9 and 3.10, the petrology of intrusions is described in which their own specific features are most clearly demonstrated. The most detailed petrology is given for the intrusions of Noril'sk Complex such as Upper Talnakh and Noril'sk-I, which are associated with sulfide Pt–Cu–Ni and Pt low-sulfide deposits.

Intrusive complexes of the Noril'sk Region have different morphologies, internal structure, composition, and ore content. The majority of the intrusions form sills. Dykes, batholiths, and gently crosscutting or sublayered, branched chonoliths and chimney-like intrusive bodies of complicated forms are less common. The intrusions vary in composition, and these are subdivided (according to their degree of differentiation) into undifferentiated (isotropic), weakly differentiated, and differentiated (layered). The first two types of intrusions are barren; ore manifestations, sulfide Pt–Cu–Ni ore deposits, and Pt low-sulfide ore deposits are only associated with layered intrusions.

Petrologic investigations carried out over many years have allowed geologists to develop the principles underpinning schemes for the subdivision of intrusions. The first schemes were proposed by Godlevsky (1959) and Lurje et al. (1962). Further schemes for subdivision were devised as new data were collected and, over time, a scheme developed that has provided the basis for the differentiation of all rocks in this region (Distler and Kunilov 1994).

At the present time, nine trap formation intrusive complexes are recognized in the Noril'sk Region (Table 3.1). Each intrusive complex combines magmatic formations, which possess closely allied mineralogical and chemical compositions; paleomagnetic properties, a similar appearance and differentiation extent; stable association of minerals; ore content; and an altered rock aureole. Within the borders of the Noril'sky Complex, four types of intrusions are distinguished that have their own specific features and are related to the Noril'sk ore-bearing batholith, and in a number of cases, these types of intrusions clearly demonstrate a genetic relationship with the Noril'sk Intrusion.

The age succession of intrusions in different complexes is controversial in many aspects, both in their relationship to each other and their relationship to effusive rocks. Effusive rocks and intrusions can be assumed, with a certain amount of confidence, to be comagmatic with formations of anomalous composition such as trachybasalt and, partly, picrite-basalt. In addition, the common occurrence of derivatives of tholeiitic basalt melts with other lavas and intrusions

V.V. Ryabov et al., *Trap Magmatism and Ore Formation in the Siberian Noril'sk Region: Volume 1. Trap Petrology; Volume 2. Atlas of Magmatic Rocks*, Modern Approaches in Solid Earth Sciences 3, DOI 10.1007/978-94-007-5022-7_3,

Table 3.1 Noril'sk region trap formation intrusive complexes

Complex	Type	Pt–Cu–Ni–sulfide and Pt low-sulfide deposits, ore occurrences	Intrusion petrotypes
Ergalakhsky		No	Sills of Listvianka River, of Noril'sk-I deposit region, of hole F-221 (the south of Noril'sk Plateau)
Pyasinsky		No	Intrusions of hole T-91, P-5, P-9 (the western and east beach Pyasino Lake)
Gudchikhinsky		No	Necks in region Mikchanda River (Intrusivny Stream) and in region Fokina River (Shiroky Log Stream), dyke of hole VP-1
Ogonersky		No	Ogoner Hill (hole SP-3)
Fokinsky		No	Lower Fokinsky, Magnitny Stream, Ergalachny (Pikritovy) Stream (hole E-14)
Noril'sky	Noril'sky	Noril'sk-I	Noril'sk-I (hole PE-35), Upper Talnakh (hole KZ-184, KZ-274, KZ-585, SV-16), Mountain Chernaya, Noril'sk-II, Imangdinsky
		Talnakhsky	
		Oktyabr'sky	
		Imangdinsky	
		Mountain Chernaya	
		Noril'sk-II	
	Lower Talnakhsky	Lower Talnakhsky Lower Noril'sky	Lower Talnakh (hole KZ-665, SV-16), Lower Noril'sk, Mountain Zelenaya Griva, Klukveny, Tulaek-Taas River
		North Kharaelakhsky (Klukveny)	
	Kruglogorsky	Olor River	Mountain Kruglaya, Lesnaya Ridge, Olor River (hole VKH-4)
		Ridge Lesnaya	
	Zubovsky	Mountain Zub-Marksheydersky	Zub-Marksheydersky (Zubovsky) (hole -28), Nakokhozsky, Manturovsky, South Pyasinsky (hole -6), Vologochansky (hole -8)
		Nakokhozsky	
Morongovsky		Mountain Morongo	Mountain Pegmatitovaya (hole E-28), Mountain Ruinnaya, Mountain Morongo, Mountain Putanaya (hole E-36), Mountain Picritovaya
		Mountain Ruinnaya	
		Mountain Pegmatitovaya	
Daldykansky		No	Daldykansky (hole NP-49), Mountain Zub, Middle Ergalakhsky, Mountain Gudchikha, Mountain Otdel'naya, Middle Dudinsky
Avamsky		No	Dykes of Bol'shoy and Maly Avam Rivers, of Ledyanaya River, of L'distaya River (LD/33–45)

impedes the establishment of comagmatic formations, especially for isotropic and weekly differentiated mafic sills. Detailed petrologic investigations of extensive layered intrusions show that their frontal, basal, and flanks have varying internal structures, extent of differentiation, composition, and ore content. This is known from the Upper Talnakh Intrusion (Zolotukhin et al. 1975; Sukhareva and Kuznetsova 1983; Likhachev 1996) and is manifested in the Mountain Pegmatitovaya and Mountain Putanaya Intrusions. The intrusions are clearly distinguished from each other, although they represent parts of a unified extensive magmatic body. Subsequent improvement of the subdivision scheme for the complexes is to aid in recognition of these types of intrusions or reference sections, which would characterize layered magmatic bodies (Ryabov 1999b).

The scheme for the subdivision of the intrusive complexes developed by Noril'sk geologists has been used as the basis for the presentation of data in this chapter. In the "intrusion petrotypes" column (see Table 3.1), our data supplement and extend characteristics of reference intrusions and are listed in addition to the formations recognized by Noril'sk geologists.

3.1 Ergalakhsky Intrusive Complex ($\tau\beta P_2er$)

Intrusions of the Ergalakhsky Complex are the most ancient magmatic formations in the region and differ in composition from the majority of other traps. The Complex is composed of moderately alkaline and alkaline mafic rocks with variable olivine content: from olivine-free to olivine-rich varieties with little quartz. These formations are reported in the literature as titanaugite dolerite and trachydolerite. According to earlier classifications (Marakushev 1976; Bogatikov et al. 1981), these formations fall into the field of subalkaline and alkaline magmatic rocks. Nevertheless, in what follows, we have kept names assigned by authors because they reflect important compositional features. The results of studies of alkaline mafic intrusions have been published in a number of works (Sobolev 1986; Godlevsky 1959; Lurje et al 1962; Arkhipova and Dodin 1963; Dodin 1971; Staritsyna 1971; Nemenenok 1982; Fedorenko et al. 1984).

Intrusions of the Ergalakhsky Complex are most abundant within the Noril'sky, Vologochansky, and Kharaelakhsky troughs and are less abundant on the sides of Rybninsky, Kulyumbe–Sukharikhinsky, and Kureysko–Letninsky uplifts and on the northwestern side of the Tungussky syneclise.

Intrusions of titanaugite dolerite and trachydolerite are sublayered bodies that vary in thickness from several meters and tens of meters up to 100–120 m, and rarely up to 200–240 m. The intrusions form thick sills or many-tiered systems of thin sublayered bodies, which can be traced along a strike for 25–50 km. In cross section, three types of sequences are recognized: (i) isotropic (undifferentiated); (ii) titanaugite dolerite intrusions with pegmatoid or rocks with increased alkalinity in the central part; and (iii) layered intrusions in which differentiation of olivine-free, olivine, and/or quartz bearing to olivine trachydolerite and small volumes of pegmatoid.

Isotropic intrusions are sills that vary in thickness from several meters to 5–10 m and rarely up to 25–30 m. The contact facies of the sills are composed of aphyric and porphyritic microdolerite 0.8–1.0 m thick. In the upper contact zone, a horizon of almond-shaped rocks is distinguished, and downward through the section, these rocks change to pea-shaped poikilo-ophitic trachydolerite. The central zone of sills consists of trachydolerite that is homogenous in structure and grain size.

Distinctive characteristics of the intrusions of the Ergalakhsky Complex include a poikilo-ophitic texture that creates a "pea-shaped" fabric and a bostonitic structure of plagioclase. The bostonitic structure is viewed as packets of suboriented laths with smoothly toothed edges rather than well-edged prisms (Stepanov and Zemskova 1988). In the transition from contact zones to the central zone, the size of pea-like clinopyroxene grains increases, and in the central zone section, it becomes uniform. In thin sills, the size of pea-like grains is about 0.3–0.5 cm and that in thick sills is as large as 0.6–0.8 cm.

Under microscope, clinopyroxene demonstrates clear pink or violet pleochroism and an hourglass shape. The clinopyroxene composition is $Wo_{39-47}En_{30-38}Fs_{21-24}$; the composition of the dominant plagioclase grains is $Ab_{49-53}Or_{2-4}An_{49-43}$. Titanomagnetite, ilmenite, and apatite are common in the rock, and analcime is rare, and palagonite and calcite are observed in the rock matrix and in amygdales (see Rock indications 60 and 61).

The titanaugite dolerite type of intrusion is observed in sills 20–60 m or more in thickness. Godlevsky (1959) considers this type of a sequence to be typical for intrusions of the Ergalakhsky Complex of the Noril'sky trough. Godlevsky (1959) details the internal structure of the sills. At the contact, a quench facies features a vitrophyric texture; inward from the contact, the facies changes to a porphyritic microdolerite with an interstitial texture and then to a microdolerite with a matrix and porphyritic phenocrysts of labradorite (An_{62-65}). The central parts of the sills feature porphyritic poikilo-ophitic titanaugite dolerite. The rock consists of two generations of plagioclase (30–45%) (the first generation is phenocrysts of An_{60-65} and the second generation is laths of An_{45}), large oikocrysts of brightly violet titanaugite (15–35%) with chadocrysts of andesine, and olivine (0–5%). Titanomagnetite and ilmenite (about 5%) and variable amounts of apatite, orthoclase, albite, sphene, analcime, biotite, and palagonite-I are found in interstices. Secondary minerals are iddingsite and bowlingite, chlorite, serpentine, and palagonite-II.

Dolerite and trachydolerite are observed in titanaugite dolerite as lens-like schlieren 0.2–4 m in size. Sometimes, the trachydolerite constitutes the entire central part of the sill. Trachydolerite possesses a porphyritic texture with a poikilo-ophitic, ophitic, or doleritic matrix texture. The composition of the rock is 40–60% plagioclase phenocrysts of An_{55-70} and laths of An_{35-45}, 25–40% violet titanaugite, 0–12% olivine, and 5–15% titanomagnetite and ilmenite. Other minerals include albite and orthoclase (up to 10%), apatite (up to 2–3%), sphene, analcime, biotite, zeolites, and chlorite. Zircon is accessory.

Compared to titanaugite dolerite, the characteristic compositional features of trachydolerite are increased clinopyroxene; ore minerals (oxides); apatite; higher titanium, iron, phosphorus content; and lower silica and alumina content.

Intrusions of trachydolerite of the Kharaelakhsky trough and the western side of the Tungussky syneclise have an internal structure similar to sills of the Noril'sky trough described above. In addition, attention is paid here to the fact that the structure of intrusions in thick bodies becomes inhomogeneous (Arkhipova and Dodin 1963). The rocks have a taxitic appearance due to the alternation of poikilo-ophitic and prismatic granular or panidiomorphic granular textures. The overall texture approaches taxito-ophitic, which is created by large titanaugite oikocrysts closely arranged with plagioclase chadocrysts, rounded olivine grains, as well as by tabular and prismatic plagioclase crystals. Trachydolerite contains plagioclase (50–52%); titanaugite (17–22%); olivine (3–8%); ore minerals (3–10%); palagonite, chlorite, biotite, analcime (8–15%); apatite (0.8–2.4%); and single orthoclase grains in interstices. The rocks have a porphyritic texture; phenocrysts are commonly composed of clinopyroxene and olivine.

The central part of the sills features lens-like coarse-grained leucocratic trachydolerite 5–15 m in size. Orthoclase and analcime content increases up to 17 and 10.6%, respectively, and the ore mineral and olivine content decreases (Arkhipova and Dodin 1963).

The third sequence type is layered intrusions. The intrusions are not as strongly differentiated as those of the Noril'sky type or from Mt. Pegmatitovaya, and some doubt exists as to whether these intrusions of the Ergalakhsky

Complex are typically isotropic bodies (Stepanov and Zemskova 1988). The differentiation extent of a melt changes along strike, and the types of sequences described above are observed in different sill intersections. Since data on layered trachydolerite sills on the northwest of the Siberian Platform are not available in the geologic literature, consideration of the petrology of this intrusion type is of interest.

3.1.1 Layered Intrusion of Trachydolerite, Drill Hole F-221

The intrusion is located in the southwestern part of the Noril'sky trough and forms a 237-m-thick subbedded body. The internal structure, from the top down, of the intrusion has been determined from drilling data:

Thickness, m	
Upper zone (UZ)	
Microdolerite, fine-grained pea-like olivine-free microdolerite and olivine-bearing trachydolerite	36
Central zone (CZ)	
Coarse-grained olivine trachydolerites with schlieren and lenses of pegmatoid	45
Olivine-bearing trachydolerite	33
Alternating olivine-bearing and olivine trachydolerite	40
Olivine trachydolerite	46
Lower zone (LZ)	
Olivine-bearing and olivine-free fine-grained pea-like trachydolerite, microdolerite, and porphyritic microdolerite	37

The results of X-ray fluorescence analysis of the rocks from the reference cross section in drill hole F-221 are given in the Appendix (Table A.5), and results of electron microprobe analyses of minerals from these rocks are given in Tables A.6–A.11.

3.1.1.1 Upper Zone

At the contact with the host rocks, a vitreous rock with an interstitial texture 10–15 cm thick has formed. Further from the contact, the vitreous rock is replaced by a 5-m-thick porphyritic microdolerite. The microdolerite grades into poikilo-ophitic trachydolerite with a pea-like texture. The pea-like grains gradually increase in size from 1.5–2 mm to 4–6 mm. The upper zone microdolerite contains about 5–10% amygdales (0.1–0.8 mm in diameter) that are filled with calcite and/or palagonite.

The microdolerite contains plagioclase ($Ab_{38-43}Or_{2-3}An_{60-55}$), clinopyroxene ($Wo_{43-42}En_{37-36}Fs_{20-22}$), apatite, and ore minerals such as titanomagnetite and ilmenite. In a number of cases, accumulations of palagonite are observed rather than clinopyroxene, commonly as a single stem-like crystal with altered olivine (see Rock indication 62).

In *poikilo-ophitic trachydolerite* in this zone, plagioclase has a clear bostonite appearance; clinopyroxene ($Wo_{43}En_{39}Fs_{18-19}$) forms elongate crystals, rarely intergrown with plagioclase ($Ab_{44-47}Or_{3-4}An_{53-50}$), which, as a rule, are confined to edges of oikocrysts (see Appendix, Table A.8). In these cases, the rock texture appears divergent (Lodochnikov 1946). Xenomorphic tabular plagioclase with wide twinned crystals in the matrix is andesine ($Ab_{65}Or_5An_{30}$). From the contact to the center of the sill, poikilocryst size increases, and they assume an isometric form. The volume of plagioclase intergrowths in poikilocrysts increases noticeably, and the plagioclase crystals are formed in different parts of oikocrysts. In the lower part of the UZ, small single isometric crystals and aggregates of olivine (up to 3–5%) are found in the matrix of trachydolerite between clinopyroxene poikilocrysts.

3.1.1.2 Central Zone

In section, three basic horizons in the intrusion are recognized: olivine trachydolerite (64 m thick), olivine-bearing trachydolerites (55 m), and pea-like coarse-grained olivine-bearing and olivine-free trachydolerite with schlieren and lenses of pegmatoids (up to 45 m thick). At the boundary between olivine-free trachydolerite and olivine-bearing trachydolerite, there is a zone where both of these rocks alternate. The texture of the rocks is seriate-porphyritic, poikilo-ophitic with a distinct tendency toward an increase in the size of poikilocrysts from the base upward through the section from olivine through olivine-bearing to pea-like coarse-grained trachydolerite.

Olivine trachydolerite consists of plagioclase $Ab_{38-43}Or_2An_{59-55}$ (35–50%), clinopyroxene $Wo_{44-42}En_{40-35}Fs_{17-22}$ (35–45%), and 7–17% olivine (Fa_{43-57}) as single subidiomorphic crystals and minor accumulations, as well as poikilocrysts and dendritic formations, and minor apatite (0.5–1%). Ore minerals are titanomagnetite and ilmenite (5–7%) (see Rock indications 63, 64).

Olivine-bearing trachydolerite is composed of plagioclase $Ab_{38-44}Or_{2-3}An_{60-53}$ (40–60%), clinopyroxene $Wo_{44-42}En_{37-31}Fs_{20-27}$ (25–40%), and olivine Fa_{48-58} (2–7%). The minerals are analogous in form to those described above, but the proportions vary, and the fraction of mafic minerals with higher Fe content increases. Xenomorphic aggregates of anorthoclase $Ab_{60}Or_{23}An_{17}$ (see Rock indication 65) are found in the matrix. Olivine more often forms palmate grains and/or poikilocrysts, and the volume of ore minerals (oxides) increases (up to 7–10%), and titanomagnetite often forms poikilocrysts with intergrown plagioclase.

In the layered intrusion sequence, a horizon of pea-like, coarse-grained poikilo-ophitic trachydolerite with isolated pegmatoids has been dislodged to the upper contact of the magmatic body. This demonstrates how the layered intrusions differ from the sills described above, in which pegmatoids are confined to central part.

Coarse-grained pea-like trachydolerite maintains a poikilo-ophitic texture, the oikocryst size sometimes as large as 1.5–1.7 cm. In addition, elements of ataxitic and pegmatitic texture are observed, which are caused by accumulations of rock-forming minerals of various sizes and intergrowths of plagioclase and pyroxene. The rock consists of plagioclase (50–60%), clinopyroxene (15–20%), olivine (0–3%), titanomagnetite and ilmenite (5–7%), apatite (0.5–1.5%), and K-feldspar (3–7%). Clinopyroxene poikilocrysts have isometric forms, and these form stem-like crystals with elongation (3–4):1 and small prismatic crystals. The poikilocrysts are composed of $Wo_{43-42}En_{36-31}Fs_{21-27}$ and small prisms of $Wo_{43}En_{18}Fs_{39}$ (see Rock indication 66). The plagioclase groundmass has a labradorite composition ($Ab_{39-41}Or_{3-2}An_{59-56}$), and in the matrix, xenomorphic oligoclase aggregates are observed. Large palmate crystals and olivine oikocryst are characterized by a lower Fe content (Fa_{48}) as compared to small grains in the rock matrix (Fa_{58}).

Three main types of pegmatitic trachydolerite are distinguished from the pegmatoid formations: quartz-bearing, titanomagnetite-bearing, and true pegmatitic trachydolerite. The first two types contain small or increased quartz or titanomagnetite content; the third type differs from common trachydolerite by their macrocrystalline prismatic texture, usually with elements of the pegmatitic texture, and more leucocratic composition.

Quartz-bearing pegmatitic trachydolerite contains clinopyroxene $Wo_{43}En_{30}Fs_{27}$ as intergrowths with plagioclase and markedly zoned idiomorphic crystals Wo_{43-41}-$En_{23-18}Fs_{34-41}$, in which the iron content within a single crystal can increase by 7 wt% from the crystal center to the edge (see Rock indication 67). *Titanomagnetite-bearing pegmatitic trachydolerite* has a heightened content (up to 10–15%) of ore oxide minerals. Titanomagnetite in these rocks forms large skeletal crystals with intergrown plagioclase and with a heightened ulvospinel content. Clinopyroxene ($Wo_{46-41}En_{35-30}Fs_{23-29}$) has a zoned structure. Plagioclase forms zoned andesine crystals that change in composition from $Ab_{51}Or_5An_{44}$ to $Ab_{60}Or_8An_{33}$ (center to rim). In pegmatoids, granules and dendrite-like ferrohortonolite Fa_{71} crystals (see Rock indications 68, 69) are not infrequent. *Proper pegmatitic trachydolerite* has the same phase content as common trachydolerites but contains more coarse grains, higher plagioclase (55–62%) and apatite (2.5–3%) content, and lower mafic mineral (15–20%) and oxide-ore mineral (5–7%) content.

A comparative analysis of trachydolerite and schlieren of pegmatitic trachydolerite derived from sills of the Syverma Plateau shows that both these rocks are similar in the chemical composition but have a noticeably different association of main rock-forming minerals. The trachydolerite consists of an association of $Wo_{46}En_{35}Fs_{19}$ and $Ab_{50}Or_1An_{49}$ (see Rock indication 70), and pegmatoid contains Fa_{71}, $Wo_{46}En_{31}Fs_{24}$, $Ab_{60-62}Or_1An_{36-29}$ (as prisms), and $Ab_{45}Or_{54}An_1$ (in the matrix) (see Rock indication 69). Titanomagnetite crystals in these rocks have a similar composition and feature a high ulvospinel content.

3.1.1.3 Lower Zone

Further away from the horizon of olivine trachydolerite in the CZ, the amount of olivine decreases to 3–5%, and there are layers of olivine-free rock. The size of poikilocrysts decreases, and the rock changes to a porphyritic microdolerite. A quench-texture crust has a vitrophyric structure, and the rocks contain small amygdales filled with calcite.

Olivine-bearing trachydolerite contains an association of olivine (Fa_{60-62}), clinopyroxene ($Wo_{43-42}En_{37}Fs_{20-21}$), and labradorite. Porphyritic microdolerite is olivine-free, plagioclase phenocrysts are labradorite ($Ab_{40-42}Or_2An_{57-56}$), laths in the matrix are andesine ($Ab_{53}Or_4An_{43}$), and small pyroxene grains are titanaugite ($Wo_{44-43}En_{33-32}Fs_{23-25}$) (see Appendix, Table A.7). Oxide-ore minerals in rocks of the LZ are titanomagnetite and ilmenite (7–10%), and minor amounts (0.5–0.8%) of apatite and, more rarely, K-feldspar are found, as well as secondary minerals palagonite-II and chlorite.

3.1.2 Rock-Forming Minerals

Clinopyroxene in rocks of the Ergalakhsky Complex forms poikilocrysts with variable intergrowth of plagioclase. Angular clinopyroxene in combination with plagioclase, stalk-like varieties, and star-shaped crystals forms an ophitic and divergent texture (according to Lodochnikov 1946). From the contact to the center of the sills, a transition from an interstitial texture to doleritic, microdoleritic, and, finally, to poikilo-ophitic texture is observed. In isotropic sills, the grain size increases from the contact to the center, and growth of clinopyroxene poikilocrysts in layered intrusions increases upward through the CZ and reaches a maximum (1–1.5 cm) in the pegmatite horizon.

Most of the clinopyroxene in the trachydolerite from drill hole F-221 fit a MgO-rich aluminum augite composition (Morimoto 1989). The Al_2O_3 content (wt%) in clinopyroxene changes from 2.29 to 3.27 in the CZ and reaches 4.06 and 5.08, respectively, in the UZ and LZ. This pyroxene is also characterized by an increased TiO_2 content that is not below 1.7 wt%.

In literature on the Noril'sk Region, as a rule, this mineral is referred to as titanaugite. The justification for the name is not only due to increased amounts of TiO_2 in the mineral but also the presence of Ti3+ in its lattice that, in contrast to Ti^{4+}, is chromophoric (Tsvetkov 1951). This is responsible for pink-violet color of the mineral and pleochroism of the mineral. The pink-violet color is characteristic for rocks of the Ergalakhsky Complex.

Rocks of the upper part of the CZ feature iron-rich and magnesium-rich augite, in which Al_2O_3 and TiO_2 contents are lower and seldom exceed 1.5 wt% each. In trachydolerites of the Khukoldysekit mountain layered sill, clinopyroxene is an aluminum diopside.

The composition of clinopyroxene in the complex is variable. The most stable composition is in isotropic bodies, and due to differentiation occurring in these bodies, an increase in variation of clinopyroxene composition is observed. The clinopyroxene composition in isotropic bodies is $Wo_{39–40}En_{37–38}Fs_{22–23}$, including 1.06–1.19 wt% TiO_2 and 1.67–1.8 wt% Al_2O_3 (see Rock indications 60 and 61). In layered sills, the range of clinopyroxene content is extended: $Wo_{41–44}En_{18–40}Fs_{17–41}$ (f from 30 to 70 wt%). Oxides include 0.63–3.88 wt% TiO_2, 0.28–5.08 wt% Al_2O_3, 0–0.22 wt% Cr_2O_3, 0.20–0.56 wt% MnO, and 0.4–0.55 wt% Na_2O. The highest concentrations of TiO_2 and Al_2O_3 are in the contact trachydolerites of the UZ and LZ, and highest Cr_2O_3 content is in olivine varieties of CZ rocks.

In contact facies of layered intrusions, the Fe content in pyroxene is close to that in isotropic sills and has a tendency to decrease from quenched crusts to inner part of the UZ and LZ. Upward over the CZ cross section, the Fe content of augite decreases and reaches a minimum (17.3 wt%) in the lower part of the olivine horizon. Upward through the cross section, pyroxenes with various Fe content are found in a horizon of alternating olivine and olivine-bearing trachydolerite. Clinopyroxene with varying Fe content, including those with *f* up to 68.4 wt%, is found in coarse-grained pegmatoidal pea-like trachydolerites. Significant variations in Fe content, including the highest values, are found in the horizon of pegmatoidal trachydolerites, varying from 36.2 to 69.6 wt%. Augites with a high Fe content are found in coarse-grained pea-like trachydolerites as second-generation crystals in the rock matrix. Augite of the same composition in pegmatoids forms large subidiomorphic prisms and graphic intergrowths with plagioclase. It is important to note that augites from quartz-bearing rocks are characterized by a higher total Fe content (52.6–69.6 wt% (see Rock indication 67)) compared to those from titanomagnetite pegmatoids (38.3–49.1 wt% (see Rock indication 68)); the content of other minerals in these rocks is similar.

A comparison of clinopyroxene composition from trachydolerite to those in enclosed schlieren of pegmatoidal trachydolerite from the Syverma Plateau shows that the former has a lower total Fe content (34.8 wt%) and increased content of (wt%): 2.88 TiO_2 and 4.41 Al_2O_3. For clinopyroxene from pegmatoid, these values are, respectively, 43.0 wt% f, 1.79 wt% TiO_2, and 2.8 wt% Al_2O_3.

Changes in the Fe content of pyroxene in the intrusion are shown in Fig. 3.1. The lowest Fe content clinopyroxene is from trachydolerites and early generations of the mineral from olivine-bearing trachydolerites. The Fe content increases upward over through a layered series of rocks in the CZ and toward the intrusion contact. The Fe content of clinopyroxene directly correlates with the fractionation coefficient of the rocks.

Olivine in isotropic intrusions is observed as small grains or aggregates commonly replaced by bowlingite. In layered intrusions, two olivine varieties are recognized: small isometric grains and large elongate stem-like and palmate-like crystals. In a number of cases, various olivine morphologies are present in the same rock. There is no relationship between the morphology of crystals and their compositions. In olivine-bearing trachydolerite of the CZ (see Rock indication 66), large crystals have a Fa_{48} composition, and small granules are Fa_{58}. In olivine trachydolerites, aggregates of small Fa_{49} grains and large $Fa_{43–50}$ and $Fa_{50–57}$-zoned crystals are present together, or relict Fa_{48} grains are found with new crystals of $Fa_{55–57}$, and later-formed large Fa_{54} crystals are found near Fa_{56} granules (see Appendix, Table A.6).

The olivine composition changes over the intrusion sequence from 43 to 62 mol% Fa component (see Fig. 3.1). Varieties of the mineral with a comparatively higher MgO content are confined to central parts of sills enriched in olivine. Toward the intrusion contacts, the Fe content in olivine increases, reaching the highest value (Fa_{62}) in contact dolerites of the LZ. In the upper part of the intrusion, olivine is completely replaced by bowlingite. The olivine Fe content in olivine trachydolerites varies from Fa_{43} to Fa_{58}, in olivine-bearing trachydolerites varies from Fa_{48} to Fa_{62}, and in pegmatites reaches Fa_{71} (see Rock indication 69). Zoning of large crystals ranges 2–7%, and the Fe content in small matrix grains and in rim zones of large crystals ranges 1–2%. A direct relationship between Fe and MnO contents in olivine is observed, varying from 0.49 to 0.80 wt%. The amount of CaO in olivine is comparatively invariable from 0.23 to 0.43 wt%, and the NiO content is less than 0.01–0.02 wt%.

Plagioclase forms porphyritic aggregates, laths, and packages of lath-like intergrowths with bostonitic fabric typical to trachydolerite (Stepanov and Zemskova 1988). The composition of plagioclase is comparatively invariable in isotropic and layered trachydolerite bodies. In the central parts of crystals, the composition is $An_{60–57}$ and more rarely up to An_{43}. At the crystal edges, the composition is $An_{56–43}$. The content of orthoclase ranges from 2 to 9 mol%. In the dolerite matrix, xenomorphic aggregates of K-oligoclase $Ab_{60–65}Or_{5–7}An_{30–33}$ are observed. Iron oxide (FeO; 0.43–1.09 wt%) is a typical impurity in plagioclase; other elements are variable and constitute less than 0.2–0.3 wt%. Increased BaO content of up to 0.55 wt% is confined to K-rich feldspars.

Spinels in intrusive rocks form small idiomorphic crystals which are frequently arranged as small aggregates and

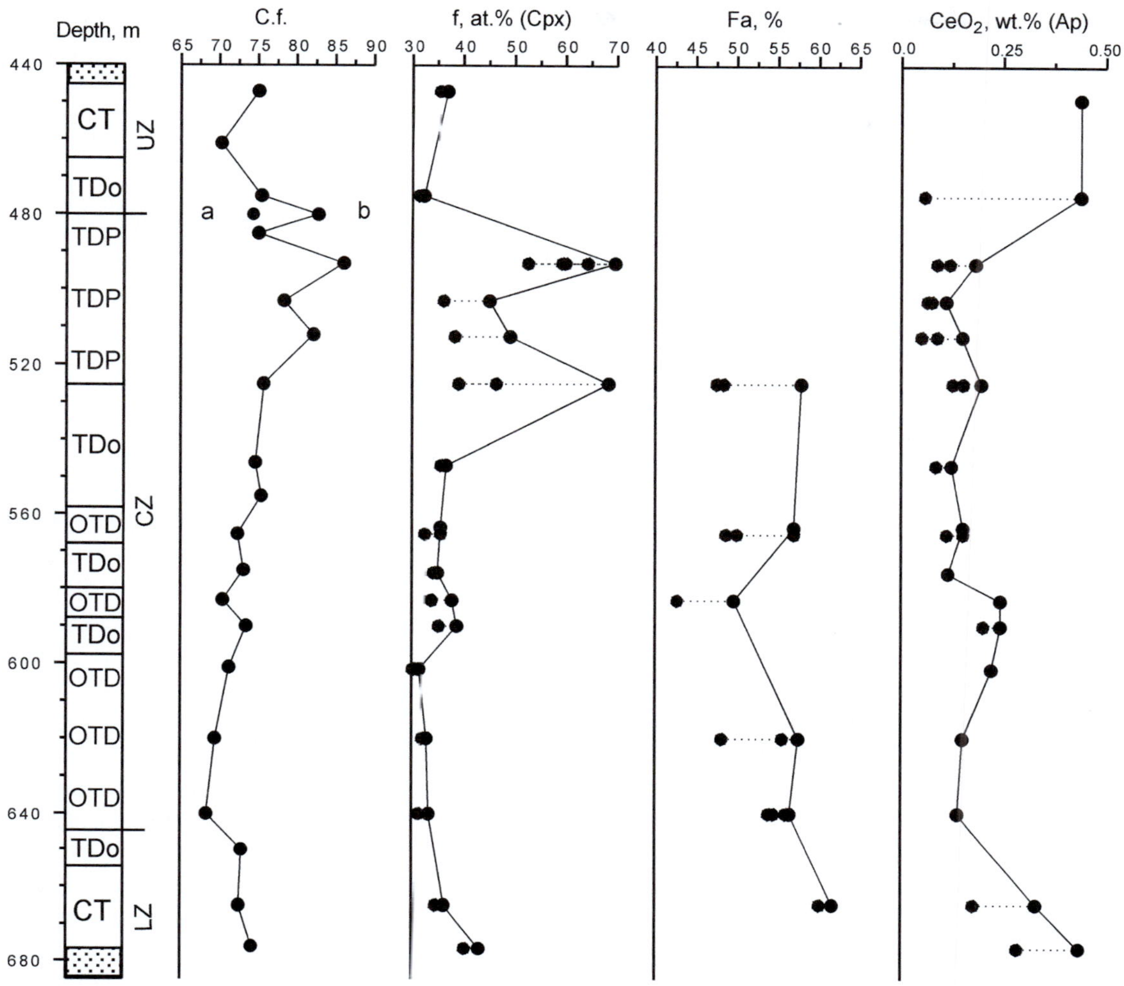

Fig. 3.1 Some indicators for differentiation in layered sill of trachydolerites, drill hole F-221. (**a**) Schlieren of pegmatoidal trachydolerite; (**b**) trachydolerite enclosing schlieren; dotted line connects points of analyses of mineral from a single sample (individual analyses are given in Rock indications)

Table 3.2 The component concentration in spinel in a stratified trachydolerite intrusion, hole F-221

Zone	Rock	Usp (mol%)	Al_2O_3 (wt%)	MnO (wt%)	MgO (wt%)	V_2O_5 (wt%)
UZ	Microdolerite	47	0.52	0.37	0.50	0.76
	Olivine-bearing trachydolerite	–	1.46	0.38	0.02	0.47
CZ	Pegmatoid trachydolerite	74–82	0.36–0.84	0.46–0.68	0.14–0.28	0.13–0.28
	Olivine-bearing trachydolerite	81–99	0.61–0.87	0.54–1.14	0.86–0.93	0.28–0.41
	Olivine trachydolerite	53–69	1.77–2.09	1.15–1.18	0.83–1.15	0.90–1.04
LZ	Olivine-bearing trachydolerite	59	1.64	1.23	0.08	0.68
	Microdolerite	60	1.00	0.65	0.04	0.72

relatively large skeletal crystals with intergrown plagioclase. The magnetite series from rocks of the Ergalakhsky Complex differs in composition from all other minerals and features an increased TiO_2 content (see Appendix, Table A.9). As a whole, the compositions of spinels from trachydolerite intrusions with varying degrees of differentiation are found to be similar. Variations in rock-forming mineral content reflect variations in spinel content in layered intrusions (Table 3.2). The TiO_2 content in spinel varies from 16.6 to 34.9 wt%, much higher than other mineral-forming oxides. The ulvospinel (Usp) content of the spinels varies from 47 to 99%. Decreased Usp component is observed in trachydolerites of isotropic intrusions and in contact facies of layered intrusions (see Rock indications 60 and 62), and a variety of mineral similar to Usp in stoichiometric composition is identified in olivine-bearing trachydolerite of the CZ.

Table 3.3 The component concentration in ilmenites of stratified trachydolerite intrusion, hole F-221

Zone	Rock	MnO (wt%)	MgO (wt%)	V_2O_5 (wt%)
UZ	Microdolerite	0.84	0.48	0.23
	Olivine-bearing trachydolerite	0.30–0.31	0.57–0.71	0.06–0.14
CZ	Pegmatoid trachydolerite	0.42	0.06	0.04
	Poikilo-ophitic trachydolerite	0.40	0.99	0.12
	Olivine-bearing trachydolerite	0.38–0.53	0.59–1.64	0.21–0.38
	Olivine trachydolerite	0.33–0.36	2.10–2.35	0.16–0.37
LZ	Olivine-bearing trachydolerite Microdolerite	0.35	2.29	0.27
		0.32	0.36	0.22

Under microscope in reflected light, the spinel exhibits a homogenous structure, and in some cases, a plate-like structure attributed to solid solution decomposition is observed. These plate-like structures are gray and light-gray phases that demonstrate insignificant variations in Usp component (see Rock indication 69).

Impurities in the spinels regularly change in the intrusion sequence. Olivine trachydolerites of the CZ exhibit maximum Al_2O_3, MgO, and V_2O_5, and the highest MnO is observed in olivine-bearing trachydolerites of the LZ. Upward from the base of the intrusion, the oxide content decreases through the rock series: olivine, olivine-bearing trachydolerites, and pegmatoids. This trend of changing spinel composition is only maintained by a continued decrease of MnO content in the UZ and LZ; the behavior of other elements is not regular. Where microdolerites change into trachydolerites in the UZ and LZ, an increase in Al_2O_3 content and decrease in V_2O_5 content are noted, and MgO content increases toward the roof of the body and decreases toward the base.

Ilmenite is found in all rocks of the complex as individual crystals in association with titanomagnetite. Impurities in ilmenite include MnO (0.30–0.84 wt%), MgO (0.36–2.35 wt%), and V_2O_5 (0.04–0.38 wt%) (see Appendix, Table A.10). Highest impurity content is confined to ilmenite in olivine and olivine-bearing trachydolerites of the CZ (Table 3.3). There is a tendency to decreasing content as contact margins are approached. Ilmenite in quartz-bearing pegmatoidal trachydolerite has anomalously low MgO (0.06 wt%). Lowest V_2O_5 content is confined to ilmenite from pegmatoids and adjacent olivine-bearing trachydolerite in the UZ.

Apatite is observed as hair-like or prismatic crystals restricted to plagioclase or to the rock matrix. Crystal size and quantity in pegmatoids are, as a rule, greater than in common trachydolerites. Apatite is typically fluorine-bearing (up to 3.4–4.0 wt% F), and increased Cl contents (up to 1.5 wt%) have been determined in pegmatoids (see Appendix, Table A.11). Apatite also contains up to 1.42 wt% SiO_2, 0.72 wt% FeO, and 0.44 wt% CeO_2. The distribution of cerium oxide in apatites decreases upward in cross section (see Fig. 3.1).

3.1.3 Geochemistry and Genesis

3.1.3.1 Petrochemistry

The chemical composition of rocks in layered trachydolerite sills can be visualized from Fig. 3.2 (see Appendix, Table A.5). Variation curves show fluctuations in content of virtually all rock-forming oxides, for example, 46.13–49.96 wt% SiO_2, 2.50–3.75 wt% TiO_2, 12.42–16.25 wt% Al_2O_3, 12.06–15.56 wt% Fe_2O_3, 0.16–0.27 wt% MnO, 2.31–5.84 wt% MgO, 6.26–8.81 wt% CaO, 3.08–4.66 wt% Na_2O, 1.21–2.58 wt% K_2O, 0.45–1.03 wt% P_2O_5, and 0.070–0.116 wt% Ba. Loss on ignition is in the range 1.08–5.77 wt%. The most significant variations in oxide content and anomalous rock composition are observed in the pegmatoid horizon. This is related to the occurrence of multicolored quartz-bearing and quartz-free rocks, which are enriched or depleted in titanium and/or alkali feldspar, clinopyroxene, and apatite.

The shape of variation diagrams indicates that the intrusion has a slightly differentiated structure and shows a clear trend of increasing TiO_2, Fe_2O_3, MnO, and Na_2O content and decreasing Al_2O_3, MgO, and CaO content. The behavior of other oxides is less significant. The trend in changing chemical composition across the intrusion becomes apparent in the transition from olivine trachydolerite to olivine-bearing trachydolerite, as well as through coarse-grained trachydolerite and, in a number of cases, pegmatoid. It should be noted that maximum MgO content is in the UZ.

The proportions of SiO_2 and ($Na_2O + K_2O$) in all rocks of the intrusion, with the exception of pegmatoids, are similar to that in trachydolerite; pegmatoids are considered alkaline gabbros. This will be explored later, but we note here that it can be assumed that fluid-magmatic differentiation of the trachybasalt melt resulted in the segregation of high-alkali liquid.

Among intrusions of the Ergalakhsky Complex, Fedorenko et al. (1984) recognize two groups corresponding to trachydolerite and titanaugite dolerite. Trachydolerite differs from the titanaugite dolerite by having a higher TiO_2 content (3.48 and 2.75 wt%), P_2O_5 content (1.25 and 0.62 wt%), FeO + Fe_2O_3 content (15.0 and 13.27 wt%), and

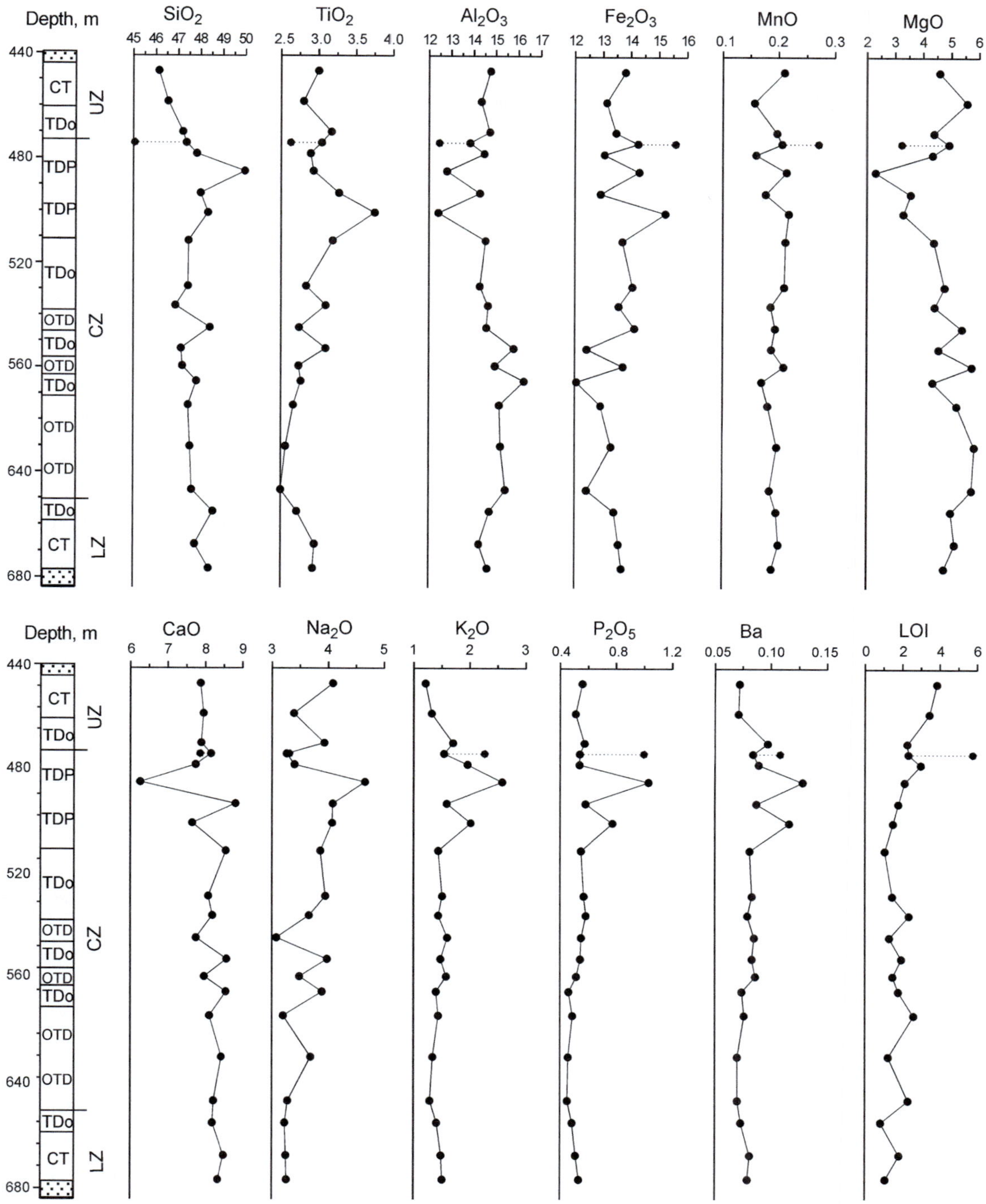

Fig. 3.2 Rock-forming oxide content of rocks in a layered trachydolerite sill, drill hole F-221. Compositions of pegmatoidal schlieren and trachydolerite enclosing the schlieren are shown for the depth 480 m. Trachydolerite and other rocks are connected by general variation curve

a lower SiO_2 content (45.0 and 47.47 wt%). These intrusions are developed in most of the Noril'sk Region except in the north and northwest.

On data from Noril'sk geologists, magmatic melts that formed sills of subalkaline and alkaline rocks of the Ergalakhsky Complex (Fig. 3.3) show some features common to rocks in different parts of the region. This is noted in the oxide content listed above.

3.1.3.2 Contact Alteration of Rocks

Intrusions of the Ergalakhsky Complex are characterized by a slight contact effect on host rocks. This effect is expressed in the formation of thin hornfels zones of the amphibole and muscovite–corneous temperature facies (Turovtsev 1979). At the contact with coaly argillites and coal, the trachydolerite becomes calcified, and coals are converted to graphite and rheomorphically intrude magmatic rocks through

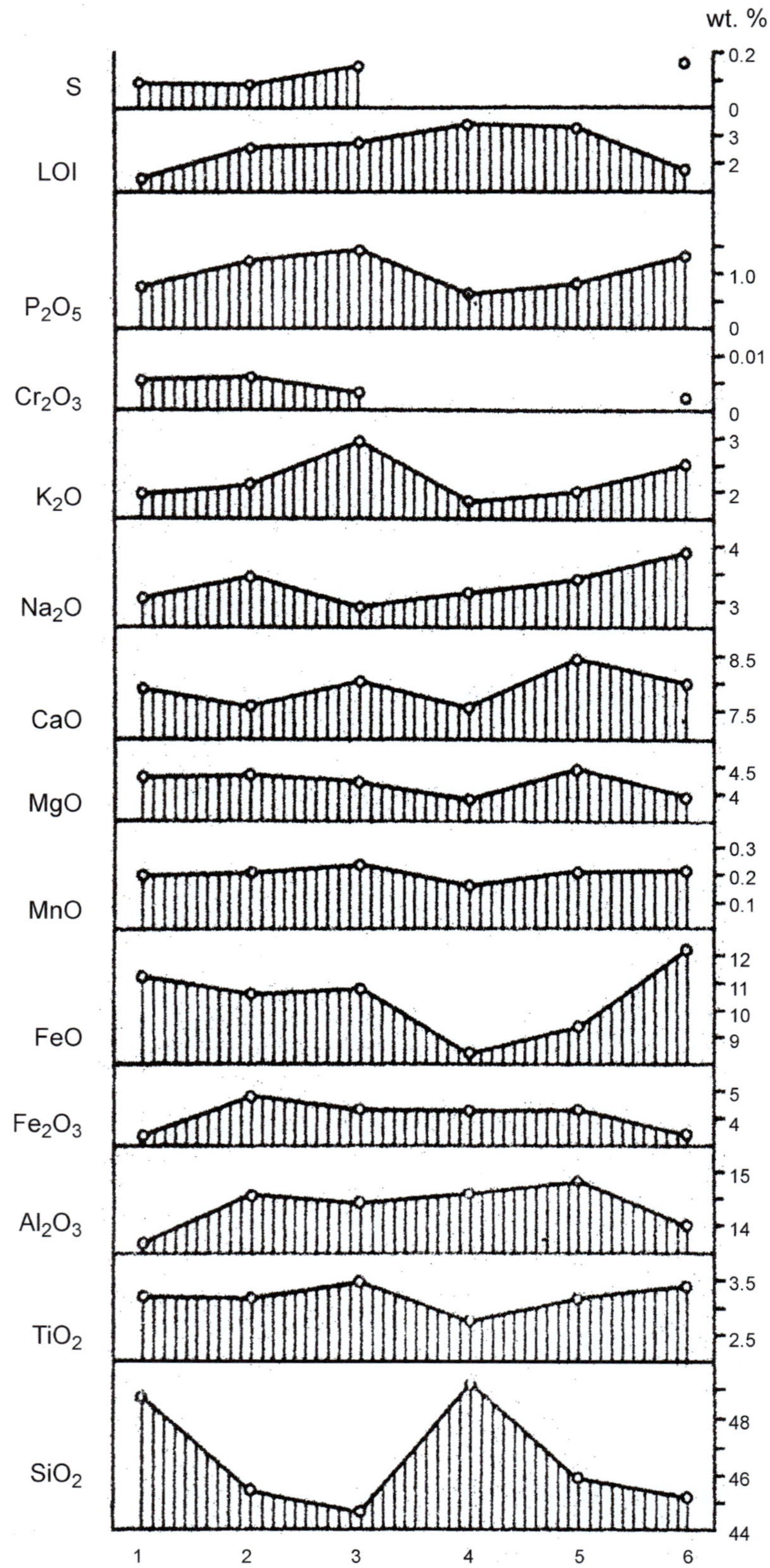

Fig. 3.3 Compositional features of trachydolerites and titanium augite dolerites in isotropic sills in the Noril'sk Region (from data of NKGRE 1975). 1—Noril'sky trough (southern part), 2—Noril'sky trough (northern part), 3—Kharaelakhsky trough, 4—Vologochansky trough (southeastern part), 5—Vologochansky trough (northeastern part), 6—Imagdinsky trough

contraction cracks. The upper and lower intrusion haloes of altered rocks are symmetric and increase in size as intrusion thickness increases. Data from Turovtsev indicates the value of the Ingersoll criterion, that is, the ratio of the altered rock thickness to intrusion thickness, varies from 0.05 to 0.20. Metasomatic crystal growth associated with these intrusions is rare.

Features of the Ergalakhsky Complex intrusions are listed below:

1. Chemical composition of the rocks is similar to trachydolerite and alkaline gabbro with variable SiO_2, TiO_2, Fe_2O_3 total, $Na_2O + K_2O$, and P_2O_5 content.
2. Some bodies are isotropic and have a slightly differentiated structure.
3. Changes in the composition of rock-forming oxides in the intrusion (e.g., drill hole F-221; an upward increase in TiO_2, Fe_2O_3, and Na_2O and decrease in Al_2O_3, MgO, and CaO) point to differentiation of magmatic melt in a magmatic chamber.
4. The main rock types are olivine trachydolerite, olivine-bearing trachydolerite and olivine-free trachydolerite, and pegmatitic trachydolerite.
5. The rocks feature a poikilo-ophitic structure, pink-violet clinopyroxene crystals, bostonitic plagioclase, and apatite and sphene (Komarova and Lyul'ko 1967; Stepanov and Zemskova 1988).
6. The rock-forming mineral association has variable composition: olivine trachydolerite Fa_{43-57}, Wo_{42-44}-$En_{40-35}Fs_{17-22}$, $Ab_{38-43}Or_2An_{60-55}$, and Usp_{53-69} and olivine-bearing trachydolerite Fa_{48-62}, $Wo_{42-44}En_{37-18}$-Fs_{20-39}, $Ab_{38-44}Or_{2-3}An_{60-53}$, $Ab_{60}Or_{23}An_{17}$, and Usp_{81-99}.
7. Pegmatoids are composed of Fa_{71}, $Wo_{41-43}En_{27-18}$-Fs_{30-41}, $Wo_{41-42}En_{35-30}Fs_{23-29}$, $Wo_{46}En_{31}Fs_{23}$, $Ab_{51-62}Or_{5-9}An_{44-49}$, $Ab_{45}Or_{54}An_1$, and Usp_{74-82} and have increased apatite and anomalously high alkali feldspar, titanomagnetite, and quartz content. Compositionally, the pegmatoids correspond to alkaline mafic rocks.
8. According to Morimoto (1989), clinopyroxenes of olivine-bearing and olivine trachydolerite are Al- and Mg-rich augite, and pegmatoids feature Fe-rich and Mg-rich augite and Al-rich diopside.
9. Feldspar in olivine trachydolerite is typically labradorite, feldspar in olivine-bearing trachydolerite is typically labradorite (I generation) and anorthoclase (II generation), and feldspar in pegmatoids is typically andesine (I generation) and sanidine (II generation).
10. Hidden stratification of the intrusions is manifest in a regular change in Fe content in olivine and clinopyroxene and CeO_2 content in apatite.
11. Typomorphic minerals are ulvospinel and titanomagnetite (containing more than 50% Usp).

3.1.3.3 Genetic Concepts

Investigations of titanaugite dolerite intrusions associated with trachydolerite were carried out by Godlevsky (1959). The investigations showed that these rocks are observed in deposits from the Cambrian to the Upper Tungussky series. Since these intrusions are not found in higher stratigraphic horizons, these intrusions are the earliest magmatic formations in the Noril'sk Region and are regarded as Upper Permian in age. Because of the similarity in composition of titanaugite basalt and trachydolerite, these effusive and intrusive rocks are regarded to be comagmatic. Later, Permian subalkaline magmatic rocks of the Ergalakhsky Complex were recognized (Lurje et al. 1962). More recently, Nemenenok (1982) described the direct transition of titanaugite dolerite sill into a lava sheet, which demonstrates these rocks are comagmatic.

Sobolev (1986) presented the first interpretation of a trachydolerite intrusion nature and considered that subalkaline rocks were associated with specific magma crystallization conditions or that they represent the "response of other, nontrap formation." Based on a study of sills in which the contact facies are composed of titanaugite dolerite and central sills are composed of trachydolerite, Godlevsky (1959) concluded that alkali mafic rocks were the result of crystallization differentiation. In this way, end crystallization stages are characterized by the accumulation of alkali elements (especially potassium), an increase in the FeO/MgO ratio, and a decrease in anorthite content.

Most geologists, as Staritsyna (1971) notes, share the view that enrichment in alkali elements at the edges of basalt magma bodies occurs during hypogeal differentiation. The author's view that there is an independent parental magma of subalkaline composition is less common. At the same time, it is apparent that differentiated mafic rocks in layered intrusions are genetically associated with trap formations.

In addition to the concept of hypogeal evolution of trap magma that formed the "alkalic" differentiation product (Sobolev 1986), other opinions have been expressed about the possibility of a multiplicity of original magma melts, including subalkaline melts (Masaitis 1964); about the leading role of hypogeal crystallization differentiation of Mg-bearing magma and alkalization of the melts by hypogenic fluids (Zolotukhin et al 1986); and, lastly, about various depths of magma melting and different scales of melting original substrate (Zolotukhin and Vasil'ev 1987).

Dodin (1971) considers trachydolerite and trachybasalt with titanaugite as derivatives of a subalkaline melt isolated from the basalt magma at the early stages of magmatic chamber development and enriched in alkali elements and Fe, Ti, Zr, Y, and P.

In summary, intrusions of titanaugite dolerite and trachydolerite are crystallization products of moderately alkalic trachybasalt melts associated with trap tholeiitic basalt magmas. The mineralogical–geochemical properties of trachydolerite are similar to those displayed in other mafic rocks. However, the origin of trachybasalt melts and the nature of their association with tholeiitic basalt magmas remain controversial. Petrologic data provide evidence that the progressive accumulation of alkali elements and the formation of rocks with increased alkalinity take place as a result of crystallization differentiation in sills of titanaugite dolerite and trachydolerite. Occurrences of high-alkali pegmatitic trachydolerite in these intrusions as schlieren, lenses, and structurally isolated horizons point to the fact that as volatile components (P_2O_5, Cl, F) are accumulated at the completion stages of magmatic process, a fluid–magmatic differentiation takes place that is isolated from "dry melt" poor in volatile components. The tendency of trachybasalt melts to undergo fluid–magmatic differentiation has been shown experimentally by Gorbachev (1989) as applied to sills of trachydolerite of the Noril'sk Region. Taking this fact, as well as the known cases of occurrences of high-alkalinity rocks in layered trap intrusions, into consideration, one can assume a genetic relationship between the trachybasalt and tholeiitic basalt melts.

3.2 Pyasinsky Intrusive Complex ($\tau\alpha\beta P_2$ps)

Intrusions of this complex are intersected by drilling on the eastern and western sides of Pyasino Lake. They are viewed as sheet-like bodies 7–8 m thick that are confined to Devonian halite–anhydrite–carbonate deposits of Vologochansky and the northwestern side of Kharaelakhsky trough.

The intrusions are isotropic bodies of porphyritic trachydolerite that correspond in composition and paleomagnetic properties to labradoritic and two-feldspar basalt and andesite–basalts of the Ivakinsky Suite and are comagmatic with both.

The rock texture is giant porphyritic with a microophitic groundmass texture. The rocks are noted for their large crystals of labradorite ($Ab_{42-46}Or_{3-4}An_{55-50}$) (2–5) × (25–35) mm in size. The groundmass includes subophitic intergrowths of anorthoclase $Ab_{62}Or_{15}An_{23}$ with brownish (f = 63 wt%) and green (f = 72 wt%) hornblende, single subidiomorphic clinopyroxene crystals, needle-like apatite crystals, xenomorphic K-feldspar grains, and quartz (see Rock indication 71). Clinopyroxene is augite ($Wo_{37}En_{42}Fs_{21}$) and contains 1.29–1.36 wt% TiO_2, 2.87 wt% Al_2O_3, 0.20 wt% MnO, and 0.34–0.38 wt % Na_2O. Amphiboles contain 1.36–1.53 wt% TiO_2, 1.47–1.82 wt% Na_2O, and 0.14–0.16 wt% V_2O_5. Apatite composition features increased SiO_2 (2.56–3.98 wt%) and is a F-bearing variety: 1.89–2.2 wt% F, 0.305–0.341 wt% CeO_2, and 0.634–0.642 wt% CeO_2. Ore minerals, titanomagnetite, and ilmenite are present. Titanomagnetite contains 16.68 wt% TiO_2, 2.79 wt% Al_2O_3, 0.57 wt% MnO, 0.04 wt% MgO, and 1.02 wt% V_2O_5. Ilmenite contains MnO (1.84–2.02 wt%) and MgO (0.06–0.23 wt%). Impurities constitute less than hundredth of 1%.

Contact alterations are limited to slight hornfelsing and compaction of host rocks. There are no indicator minerals associated with intrusion of the complex.

The origin, composition, and age of the intrusions are considered to indicate that the effusive trachyandesites of the Ivakinsky Suite are comagmatic with these intrusions.

3.3 Gudchikhinsky Intrusive Complex ($\omega\beta T_1$gd)

The complex is composed of picritic dolerite that forms a "neck" near the Mikchanda River (Intrusivny Stream), a neck or dyke in the Shiroky Log (the Fokina River), and a dyke intersected by drill hole VP-1 upstream of the Pyasina River.

The Mikchandinsky neck is situated in the vent zone of the Mikchandinsky paleovolcano confined to the junction of the western limb of the Tungussky syneclise and the Kharaelakhsky trough. Judging by the distribution of vent facies rocks and annular dykes, the size of the paleovolcano vent is 1.2 × 3 km (Nemenenok 1972). Volcanic masses of near-vent facies rock are composed of lumps and agglomerates of tuffs and xenotuffs, lava breccia, and clastic lavas injected by branching dykes of microdolerite and olivine dolerite. The author's data indicates the neck size is 60 × 20 m. On the left side of the Intrusivny Stream, a distinct contact between picritic dolerite and tuff agglomerate is observed. The neck is isotropic and is composed of picritic gabbrodolerite; under microscope, the rock displays a porphyritic segregation texture and a local poikilitic texture, a characteristic for these rocks. The rock consists of 40–50% olivine, 35–40% plagioclase, 10–15% pyroxene, and about 1–2% ore minerals (see Rock indication 72). Secondary minerals serpentine, bowlingite, saponite, and prehnite are also present.

The rock is noted for the wide variation in olivine size (0.2 to 2–3 mm). Large subidiomorphic olivine crystals impart a porphyritic texture to the rock. The composition of olivine changes with grain size (large to small) from Fa_{17} to Fa_{31}, and NiO content decreases (0.36 to 0.27–0.31 wt%) with increasing Fe content. Olivine encloses ore mineral "balls," small idiomorphic chrome spinel crystals, ilmenite, and large (up to 100 μm) rounded inclusions of altered silicate minerals. Plagioclase forms large, wide, polysynthetically twinned tabular grains ($Ab_{31}Or_1An_{66}$) with poikilitic intergrowths of olivine and clinopyroxene grains and zoned $Ab_{32-55}Or_{1-4}An_{67-41}$ prisms. Clinopyroxene forms relatively large subidiomorphic crystals, overgrowths on olivine, and small and needle-shaped crystals in the groundmass. Large clinopyroxene crystals are zoned and have the composition $Wo_{39-37}En_{50-40}Fs_{11-14}$; small crystals are $Wo_{35}En_{47}Fs_{18}$. A regular change in clinopyroxene composition, from core to rim in small grains, is manifest as a decrease in Al_2O_3 content (2.64 to 1.92 wt%), Cr_2O_3 content (0.82–0.096 wt%), and TiO_2 content (0.63–1.32 wt%). Pigeonite ($Wo_7En_{65}Fs_{28}$) forms small needle-shaped crystals near the base of the complex. Ore minerals are chrome spinel, chrome magnetite, and ilmenite. All these minerals form inclusions in olivine and in the matrix. Chrome spinel contains 1.39–2.02 wt% TiO_2, 12.38–13.58 wt% AL_2O_3, 6.77–8.05 wt% MgO, and 42.64–44.59 wt% Cr_2O_3. Chrome magnetite contains 7.49 wt% TiO_2, 3.17 wt% Al_2O_3, 3.53 wt% MgO, 14.3 wt% Cr_2O_3, and 1.94 wt% V_2O_5. Ilmenite forms inclusions in olivine and is present in the matrix. Its composition is stable, but one sample from the groundmass contains 12.19 wt% Cr_2O_3, 3.04 wt% Al_2O_3, 3.89 wt% MgO, and 0.65 wt% V_2O_5. Ilmenites with such a high content of Cr_2O_3 have not been noted in the literature (Chukhrov 1967); so this may be a unique type of ilmenite.

The neck or dyke of picritic dolerite from a vent facies of a paleovolcano in the Shiroky Log locality (Ivanova 1975) has a visible thickness of 4 m, and it is situated among lump and agglomerate tuffs and classic lavas. In the vent facies of the paleovolcano, there is a wide diversity of effusive rocks, including layers of bomb-like picritic dolerite.

Upper Pyasinsky Intrusion is a magmatic body intersected by drilling Devonian rocks in the north of the Kharaelakh Plateau. It has an isotropic structure and is composed of picritic gabbrodolerite with a porphyritic texture and a micro-ophitic groundmass. The body is 4 m thick (Ryabov 1992a).

The rock is composed of subidiomorphic olivine Fa_{23} crystals (50–55%) (often with cleavage), wide tabular labradorite (30–35%), and xenomorphic ortho- and clinopyroxene (10–15%) crystals. Unlike picritic rocks from other intrusions, there is an abundance of brownish biotite (up to 3–4%) and needle-shaped apatite aggregates (typically confined to biotite). In addition, large and small octahedral, skeletal titanomagnetite crystals, needle-shaped ilmenite, and rhomb-shaped chrome spinel are observed (see Rock indication 73). Secondary minerals include plate-like and fibrous serpentine and fine magnetite.

Contact alteration of rocks from the Mikchandinsky neck and Shiroky Log dyke is not observed, but in the Upper Pyasinsky, they are viewed as insignificant recrystallization of carbonate rock. Mineralization is not present in rocks of the Gudchikhinsky Complex.

Petrologic features are as follows:

1. Intrusions of the complex are holocrystalline.
2. Intrusions are composed of picritic gabbrodolerite that contains more than 20 wt% MgO and 0.15 wt% Cr_2O_3 and less than 8 wt% Al_2O_3 and 6 wt% CaO.
3. The Fe and NiO content in olivine fit the geochemical norm for olivine in picritic basalt magmas.
4. The presence of accessory chrome spinel with Cr_2O_3 content up to 14.3 wt% and V_2O_5 content up to 1.9 wt%.
5. Chrome ilmenite with up to 12.2 wt% Cr_2O_3.
6. A mineral association of Fa_{17-23}, $Wo_{4-7}En_{75-65}Fs_{21-28}$, $Wo_{39-35}En_{49-47}Fs_{12-18}$, $Ab_{32-55}Or_{1-4}An_{67-41}$, Cr–Sp, and Cr–Ilm.
7. Concentrations of element impurities in the rocks fit to Clarks in feldspar picrites.

3.3.1 Genetic Concepts

The compositional resemblance of the intrusions with picritic basalts of the Gudchikhinsky Suite and their position lower in the stratigraphic sequence indicates that these rocks are comagmatic, and the intrusive picritic gabbrodolerite can be considered feeders for the picritic basalt lavas.

3.4 Ogonersky Intrusive Complex ($\beta - \delta\nu\beta T_1og$)

This complex combines multiple intrusions of poikilo-ophitic "pea-like" dolerite and gabbrodolerite that form sheet-like bodies or, more rarely, crosscutting bodies situated in Silurian sedimentary deposits to tuff lavas of the Permian–Triassic age. These formations are widely developed and may be traced over many tens of kilometers. The thickness of the bodies changes from several m to 30–40 m and, more rarely, to 100 m. Visually, the intrusions have an isotropic structure; most thick bodies are slightly layered. Most sills are olivine-free; olivine-bearing dolerites are rare. Intrusions of the Ogonersky Complex are notable for their poikilo-ophitic structure and a plagioclase to clinopyroxene ratio close to 1:1; the content of olivine is less than 3%, magnetite and ilmenite content is low, and rare single apatite grains are observed (Stepanov and Zemskova 1988). The composition of clinopyroxene oikocrysts is augite ($Wo_{38}En_{47}Fs_{15}$), and plagioclase chadacrysts are labradorite ($Ab_{41}Or_2An_{57}$) (see Rock indication 74) and vary from An_{48} to An_{65}. Compositionally, these intrusions are related to tholeiites. In spite of the petrographical uniformity of the rocks, investigation results from rock samples show that differentiation and compositional changes take place vertically through the rocks (Fig. 3.4). Intrusions of the Ogonersky Complex have a positive remanent magnetization polarity. In the opinion of Lyul'ko et al. (1972), widespread lavas of porphyritic, glomeroporphyritic, and poikilo-ophitic (tholeiitic) basalts of the tuff lava Permian–Triassic series are comagmatic with effusive intrusions of the Ogonersky Complex.

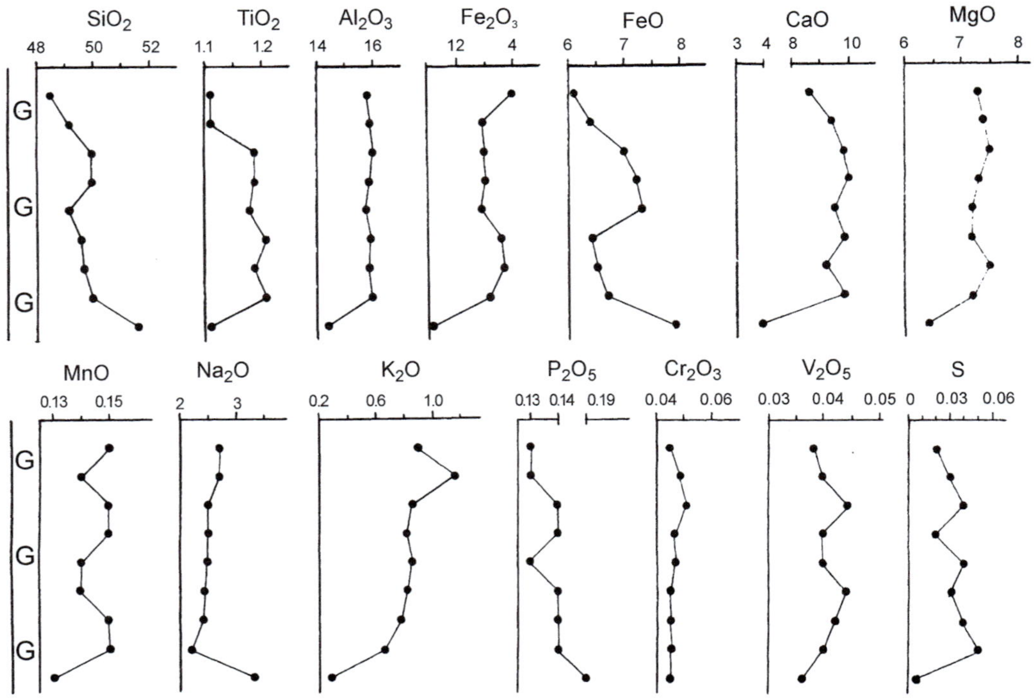

Fig. 3.4 Variation petrochemical diagram for a sequence of undifferentiated sills of poikilo-ophitic dolerites near the Fokina River. Oxide content and S, wt%. Thickness of sill is 3 m

3.4.1 Ogoner Hill Intrusion

The intrusion crops out and is intersected by drill holes. It is 30–40 m thick, and contact facies rocks are composed of olivine-bearing gabbrodolerite and quartz–olivine-bearing gabbrodolerite with coarse spot-like pegmatoid aggregates (same composition as the central zone).

In the Ogoner Hill Intrusion, the upper zone is eroded, and only the central zone crops out. The lower one is obscured by diluvium. The complete sequence is intersected by drill holes.

3.4.1.1 Mineralogy

Gabbrodolerite has a poikilo-ophitic texture with rarer poikilitic and micrographic texture. The rock comprises plagioclase (45–60%), clinopyroxene (35–50%), and olivine (3–5 to 7%, rarely to 10%), with a quartz, micrographophyre, magnetite, ilmenite, biotite, and palagonite (about 5–10%) matrix.

Plagioclase ($Ab_{19–45}Or_{1–3}An_{80–52}$) forms wide tabular grains and prisms with pronounced zoning (from core to rim), and plagioclase chadacrysts ($Ab_{82–48}Or_{1–3}An_{66–49}$ and $Ab_{28–36}Or_{1–2}An_{71–62}$) are found in clinopyroxene and olivine, respectively (see Rock indications 75 and 80). Large plagioclase grains enclose intergrowths of small plagioclase prisms.

Clinopyroxene poikilocrysts and large angular grains form an ophitic structure in combination with plagioclase. The mineral is augite $Wo_{31–38}En_{42–30}Fs_{23–33}$. Twinning is not uncommon in clinopyroxene, as well as lamellae of pigeonite $Wo_9En_{54}Fs_{37}$ (see Rock indication 76). Clinopyroxene adjoining palagonite is sometimes replaced by a dirty-green amphibole. Clinopyroxene from the Ogoner Hill Intrusion is noted for its low CaO content and insignificant variation in TiO_2 content (0.86–1.07 wt%) and Al_2O_3 content (1.38–2.52 wt%). The Cr_2O_3 content in augite is less than 0.067 wt%.

Olivine is observed as large palmate-shaped grains with intergrowths of plagioclase and isometric grains of various sizes. Composition varies from Fa_{36} to Fa_{67}, and in outcrops at Ogoner Hill, the composition of large olivine crystals in the upper part of the cross section is $Fa_{36–38}$, and small grains in augite are Fa_{45}, and grains in the matrix are of composition Fa_{42}. In the middle of the intrusion, the composition of large zoned crystals (from core to rim) is $Fa_{57–67}$, and that in the lower part of the outcrop is $Fa_{39–50}$ (see Rock indication 79–81). Olivine, in a number of cases, is situated directly at the boundary with quartz or micropegmatite. In the groundmass, small sheets of brownish biotite are common, and prismatic dirty-green amphibole crystals are rarer. The Fe content in biotites varies from 59 to 86 wt% and contains 0.76–4.48 wt% TiO_2, 0.56–1.07 wt% F, and 0.08–0.81 wt%

Cl. The Fe content in amphibole is 73–76 wt% and contains 1.3–1.4 wt% TiO_2, 1.74–1.75 wt% Na_2O, 0.89 wt% F, and 0.30 wt% Cl (see Rock indication 75).

Titanomagnetite forms large, skeletal crystals and contains 17.10–20.20 wt% TiO_2, 1.34–3.30 wt% Al_2O_3, 0.61–1.40 wt% MnO, 0.01–0.70 wt% MgO, 0.017–0.092 wt % Cr_2O_3, and 0.62–1.60 wt% V_2O_5. Prismatic ilmenite crystals in the groundmass contain 0.45–1.08 wt% MnO, 0.11–1.08 wt% MgO, and 0.22–1.01 wt% V_2O_5. Chlorite forms fine-scale aggregates confined to the matrix and palagonite aggregates. The Fe content in chlorite varies from 47 to 62 wt% and contains 0.56–0.73 wt% F.

Quartz is confined to the matrix as small grains and micrographic intergrowths with feldspar. Apatite forms needle-shaped crystals, which are intergrown in plagioclase and palagonite.

3.4.1.2 Geochemistry

The chemical composition of isotropic sills of "pea-like" dolerite is similar to tholeiitic basalts of the normal series and is close in content to the mean trap content calculated by Duzhikov for the northwest of the Siberian Platform (Zolotukhin et al. 1986).

In moderately thick sills of the Ogonersky Intrusive Complex, slightly pronounced layering of rocks is seen as a change in the rock-forming oxide content. The variation diagram (Fig. 3.5) demonstrates that the distribution of oxide contents over the body cross section is controlled mainly by its proximity to the margin and axis line of the sill. The trends toward increasing or decreasing oxide content from the sill margins to the center repeat the typical tendency for crystallization of a sheet-like body. Some deviation from this tendency and complication of variation diagrams seems to be associated with increased volatiles.

Metamorphism and mineralization as a result of the intrusion are observed as a slight hornfelsing of rocks at the contact and rare sulfide phenocrysts in gabbrodolerite that are found throughout the intrusion.

3.4.2 Petrologic Features

1. Most intrusions of the Ogonersky Intrusive Complex form isotropic sills of "pea-like" dolerite, including slightly differentiated bodies with some pegmatoid.
2. The chemical composition of isotropic bodies is similar to tholeiitic basalts of a normal series; in layered intrusions, redistribution of rock-forming oxides takes place with a tendency toward increased magnesium in the central parts of sills.
3. The rocks typically have a poikilo-ophitic texture with a plagioclase-porphyritic texture at the margins of sills.
4. Clinopyroxene and plagioclase are the dominant minerals in dolerite of the Ogonersky Complex and are present are in equal proportions.
5. Thick sills feature a variable amount of large olivine crystals (from 1–3% to 10%) and quartz in the matrix.
6. Principal mineral parageneses are $Wo_{38}En_{47}Fs_{15}$ + $Ab_{41}Or_2An_{57}$ in thin isotropic sills of "pea-like" dolerite and Fa_{45-50} + $Wo_{34}En_{41}Fs_{25}$ + $Ab_{35-50}An_{65-50}$ and quartz in thick sills.
7. Layering is manifest in the distribution of rock-forming oxides and impurities.
8. Petrographic and mineral–geochemical features of the rock are evidence that the intrusions of the complex were formed as a result of crystallization differentiation of a tholeiitic basalt melt, and solidification of the melt occurred from the margins to the central axis of the sills.

3.4.3 Genetic Concepts

Intrusions of the Ogonersky Complex maintain the original magma composition and did not undergo hypogeal differentiation; they are identical to tholeiitic basalt melt of normal series. Intrusions of this complex show no noticeable chamber differentiation. Therefore, the composition of these rocks is likely to be close to the original magma and similar to the composition of most flood basalt lavas in the region, from the Syverminsky and to Samoedsky Suites. Unfortunately, geologists studying Noril'sk note that reliable criteria that would allow confirmation of comagmatism and the genetic relationship between individual intrusions of the Ogonersky Complex are yet to be found. Therefore, for the present time, subdivision of the complex in more detail is not yet possible.

3.5 Fokinsky Intrusive Complex ($q\nu\beta - \omega\nu\beta$) T_1fk

Three intrusions comprise the complex: Lower Fokinsky, Magnitny Stream, and Ergalachny (Pikritovy) Stream. These are emplaced in the Noril'sky trough as sheet-shaped and slightly crosscutting bodies in the Permian–Carboniferous Tungussky series and lava tuffs of Triassic age up to the Morongovsky Suite. The intrusions are 115–150 m thick, and olivine-rich picritic and troctolitic gabbrodolerite constitute 80–85% of the intrusion, with olivine, olivine-bearing, and olivine-free quartz-bearing gabbrodolerite and pegmatoidal gabbro accounting for 15–20% (Fig. 3.6). The significant prevalence of olivine-rich gabbrodolerite over other varieties is reason to consider these intrusions slightly differentiated.

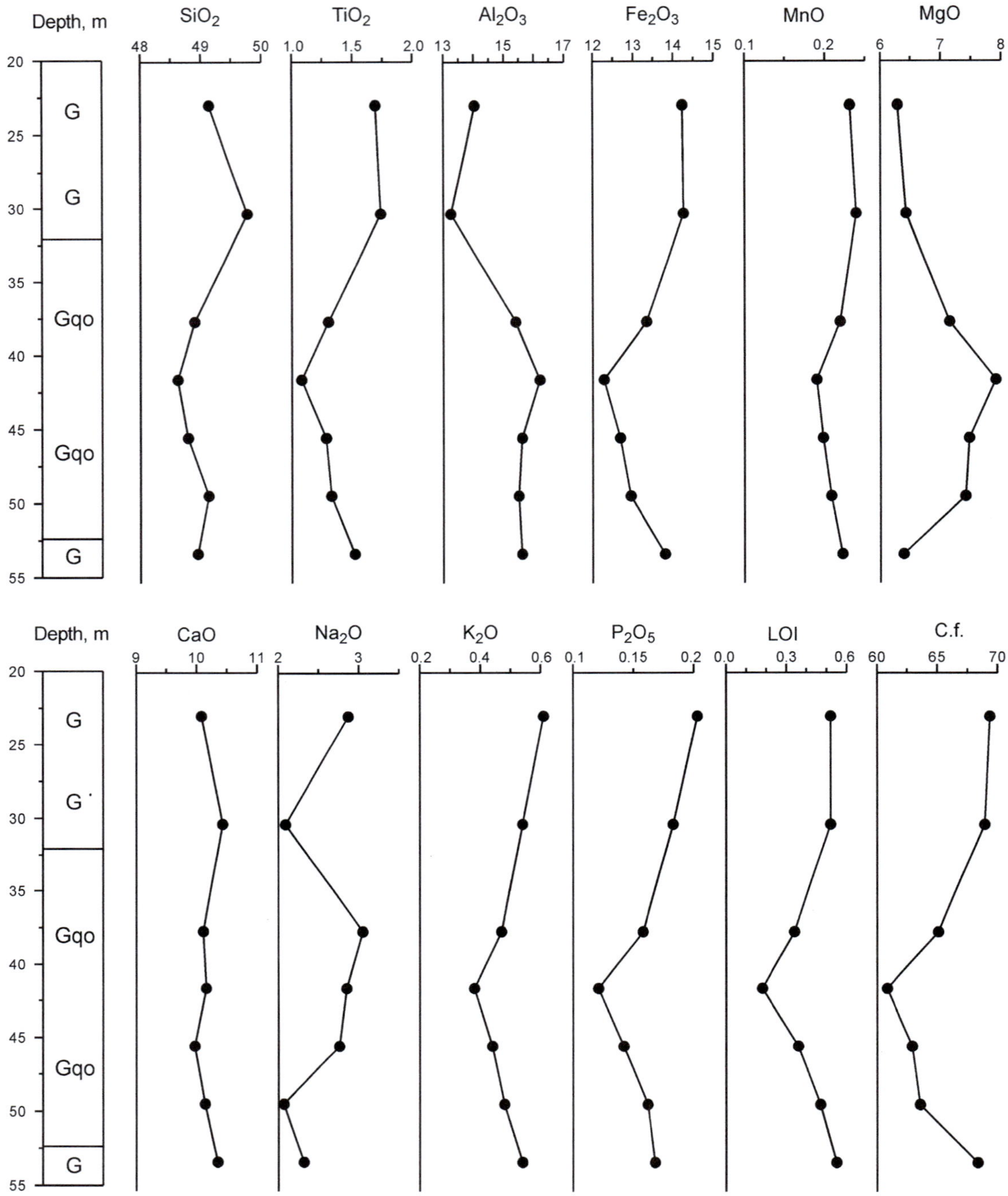

Fig. 3.5 Variation petrochemical diagram for the Ogoner Hill Intrusion sequence. Oxide content and loss on ignition, wt%

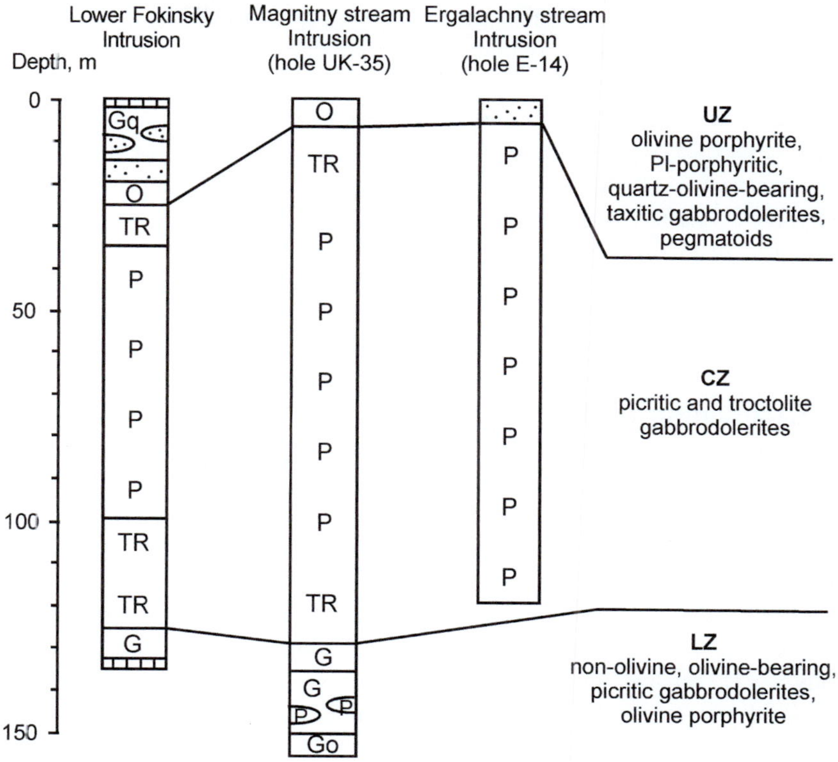

Fig. 3.6 Internal structure of the Fokinsky intrusive complex. Specks—pegmatoids, vertical hatching—terrigenous country rocks

3.5.1 Lower Fokinsky Intrusion

A study of the Lower Fokinsky intrusion was undertaken by NKGRE geologists, as well as by Arkhipova and Nachinkin (1964), Zolotukhin and Vasil'ev (1967), Oleinikov (1979), Duzhikov (Zolotukhin et al. 1989), and Ryabov (1989b, 1992a).

The intrusion is located in the southwestern part of the Noril'sky trough and extends toward the Fokinsko-Tangarlakhsky Fault where it is restricted to the hinge of the Lower Fokinsky uplift. It is a slightly crosscutting, sheet-like body 135 m thick, emplaced in terrigenous coal-bearing deposits of the Tungussky series. The intrusion crops out along the banks of the Fokina River and is intersected by drill hole F-12. In cross section, upper and lower zones are distinguished (Fig. 3.6).

Quenched contact facies of the intrusion are composed of amygdaloidal olivine porphyries and porphyritic olivine tachylites (see Rock indication 82), which grade into olivine-poor dolerite then to olivine-free gabbrodolerite. The quenched crusts are less than 10 cm thick, and contact facies rocks are 20–50 cm thick at the roof of the sill and up to 1.5 m thick at the base. A vitreous groundmass of quenched olivine tachylite becomes strongly calcitizated along the contact as a 2-cm-wide band, and away from the contact, the glass is opacitizated. Olivine porphyrocrysts are pseudomorphed by calcite with magnetite, bowlingite, serpentine, and talc. The olivine content in quenched rocks varies from 10–12% to 25–30%; olivine-rich and relatively olivine-poor sites are distributed as bands 1–2 cm thick along the contact (see Rock indication 82). Regardless of the statement about large olivine protocrystals (Zolotukhin et al. 1989), the olivine crystals in tachylites vary in size in the range from 0.1 to 1 mm, most of their accumulations form granules 0.1–0.3 mm in size. These crystals often stick together as aggregates. The largest crystals have a skeletal form, and idiomorphic crystals are infrequent. Small chrome spinel inclusions are observed in olivine pseudomorphs.

Olivine-rich vitreous rocks are rarely (if ever) found in traps. Their occurrence in quenched facies raises doubts about the independent existence of picrite-basalt magma; the first crystalline phases of which were chrome spinel and olivine (Ryabov 1989b, 1992a). Granule-like nucleations and skeletal forms of olivine crystals in vitreous quenched rocks are evidence that sharp cooling of a magnesium-rich melt took place rather than that there was an intratelluric nature that was peculiar to the melt, and there was a gravitational accumulation of large protocrystals in an olivine-rich melt.

3.5.1.1 Upper Zone

It includes olivine-free quartz-bearing pegmatoidal gabbrodolerite 10–12 m thick with lenses of pegmatoidal gabbrodolerite 0.5–6 m thick, which are emplaced at three levels in olivine-free gabbrodolerite, troctolitic gabbrodolerite, and at the troctolitic gabbrodolerite–picritic gabbrodolerite boundary. *Quartz-bearing varieties of gabbrodolerite* feature the wide development of star-shaped and graphic intergrowths of augite and plagioclase, long needle-shaped ilmenite crystals, and graphic and net-like intergrowths of quartz, albite, and K-feldspar with multiple apatite needles or palagonite with oval quartz inclusions in the matrix. Small olivine Fa_{55} crystals are also present (Ryabov 1992a).

The distinguishing feature of the *pegmatoidal gabbrodolerite* is the large (5–8 cm in length) light-green pinnate megacrysts of albite $Wo_{40}En_{50}Fs_{10}$ containing 1.07 wt% Cr_2O_3 (Vasil'ev 1970; Ryabov et al. 1977).

3.5.1.2 Central Zone

Central zone of the intrusion is composed of picritic gabbrodolerite 76.5 m in thickness. Above and below, the picritic core is framed by troctolitic gabbrodolerite 11.2 and 27.5 m in thickness, respectively (Arkhipova and Nachinkin 1964).

The *picritic* and *troctolitic gabbrodolerite* contain from 40–50% to 70% and 25 to 30% olivine, respectively. The olivine is situated in large fields of polysynthetically twinned plagioclase An_{56} (see Rock indications 83, 84) or as porphyritic crystals among ophitic pyroxene–plagioclase aggregates. The wide variation in olivine size (from 0.1–0.3 mm to 3–4 mm) is noteworthy, as well as the changes in olivine morphology (subidiomorphic crystals, skeletal forms, and granules are common), which are often observed in the same layer. Most olivine crystals in picritic gabbrodolerite have a stable composition Fa_{23-28}, but single elongate branches of olivine Fa_{34} are also found. In troctolitic gabbrodolerite, olivine of composition Fa_{32-34} is dominant.

Compositionally, clinopyroxenes and orthopyroxenes are Mg-rich augite ($Wo_{38-41}En_{49-48}Fs_{11-3}$) and bronzite, respectively. The characteristic feature of olivine-rich gabbrodolerite is the presence of accessory chrome spinel inclusions in olivine, pyroxene, and plagioclase.

3.5.1.3 Lower Zone

In cross section, the Lower Fokinsky Intrusion features olivine-free gabbrodolerite, which is located in the lower zone between a quenched crust composed of olivine porphyry below and troctolitic gabbrodolerite above. Gabbrodolerite consists of pegmatoidal intergrowths of plagioclase and clinopyroxene that grade into star-shaped, long, prismatic spherocrystals between which ophito-poikilitic aggregates of the same minerals are found. Clinopyroxene in the rock forms prismatic crystals ($Wo_{33-36}En_{53-50}Fs_{14}$) and angular grains ($Wo_{35-38}En_{49-45}Fs_{16-17}$), which differ in composition (see Rock indication 85). The former contain more Al_2O_3 (2.76–2.91 wt%) and Cr_2O_3 (0.33–0.63 wt%) compared to the latter containing 1.77–1.96 wt% and 0.01–0.08 wt%, respectively. Zoned plagioclase crystals are $Ab_{36-51}Or_{1-3}An_{63-46}$. In interstices, quartz and ilmenite are observed, and ilmenite contains 1.00–1.05 wt% MnO and 0.74–0.84 wt% V_2O_5.

In quench facies rocks of the LZ, fine pyrrhotite grains are disseminated and amount to 3–5%, which reflects the increased sulfur content of the rock (see Rock indication 82).

3.5.1.4 Mineralogy

Lower Fokinsky Intrusion rocks typically feature high olivine content, which is significantly higher in the quench facies. Olivine has a stable Fe content (Fa_{23-28}) and increased NiO content (0.22–0.36 wt%). Clinopyroxene content tends to increase slightly upward through the intrusion and is characterized by a high Fe content (Fs_{11-13}) and increased Cr_2O_3 content (0.90–0.98 wt%). Most plagioclase in quenched olivine porphyry (An_{67}) and picritic gabbrodolerite (An_{56}) is labradorite.

Chrome spinels have a variable composition, and the Cr_2O_3 content decreases from 23.2–36.7 wt% to 18.1–24.7 wt% when passing from picritic to troctolitic gabbrodolerite (Duzhikov et al. 1988).

High-chrome ($Cr_2O_3 = 44.9$ wt%) chrome spinel in a quench facies olivine porphyry (see Rock indication 82) is of great interest. The composition of this chrome spinel is close to those from picritic lavas of the Gudchikhinsky Suite and implies with certainty the initial phase composition of picrite-basalt melt. Crystallization of the melt with fluids resulted in variations in the chrome spinel composition that is observed throughout the intrusion.

3.5.1.5 Petrochemistry

Picritic and troctolitic gabbrodolerite of the Lower Fokinsky Intrusion has a relatively invariable composition, with insignificant variations in the content of principal petrogenic oxides: 43.1–46.5 wt% SiO_2, 0.9–1.15 wt% TiO_2, 6.7–8.0 wt% Al_2O_3, 1.9–2.4 wt% Fe_2O_3, 11.6–12.0 wt% FeO, 20.8–24.8 wt% MgO, 5.2–7.2 wt% CaO, 0.7–1.1 wt% Na_2O, 0.2–0.3 wt% K_2O, and 0.2–0.3 wt% Cr_2O_3. Gabbrodolerite visible in the UZ and LZ contains 49.6–51.1 wt% SiO_2, 2.0–2.4 wt% TiO_2, 13.5–13.9 wt% Al_2O_3, 1.1–3.0 wt% Fe_2O_3, 7.3–9.6 wt% FeO, 6.0–7.1 wt% MgO, 10.1–12.1 wt% CaO, 2.0–2.4 wt% Na_2O, and 0.5–0.7 wt% K_2O (Table 3.4). At the expense of an increase in clinopyroxene content in pegmatoids, an increase in MgO and CaO content and decrease in Al_2O_3 content take place. In quench facies of the intrusion, the MgO content is 12.65 wt% in the upper part and 14.11 wt% in the lower part; values for loss on ignition are, respectively, 2.57 and 3.84 wt%. These data, corrected for rock alteration, indicate the composition of the initial melt for the Lower Fokinsky Intrusion was picrite-basalt.

Table 3.4 The Fokinsky intrusive complex rocks' average content (wt%)

	Lower Fokinsky intrusion			Magnitny stream intrusion			Ergalachny stream intrusion		
Rock	P	G	PG	P	TR	PG	P	TR	PG
SiO_2	45.02	50.90	51.00	42.94	44.87	47.14	41.43	47.95	49.19
TiO_2	1.10	2.00	1.33	0.61	1.00	1.44	0.45	1.29	1.27
Al_2O_3	6.98	13.50	7.90	7.29	10.82	17.08	7.07	16.45	18.20
Fe_2O_3	2.62	1.07	1.61	2.96	3.48	4.02	4.39	2.38	2.10
FeO	11.24	9.56	7.55	9.80	8.80	7.55	9.78	9.29	7.24
MnO	0.18	0.16	0.17	0.20	0.20	0.19	0.18	0.17	0.15
MgO	22.28	5.96	11.26	25.90	16.56	7.29	26.31	7.04	4.73
CaO	6.17	10.86	14.91	5.01	7.40	10.97	4.97	10.01	10.83
Na_2O	0.86	2.17	2.32	0.88	1.38	1.62	0.61	2.00	2.78
K_2O	0.53	0.72	0.59	0.29	0.26	0.31	0.34	0.48	0.71
P_2O_5	0.12	0.19	0.10	0.05	0.07	0.08	0.07	0.15	0.16
LOI	3.15	2.46	2.04	3.17	3.93	1.24	5.57	2.17	2.16
Total	100.25	99.55	100.78	99.10	98.77	98.93	101.17	99.38	99.52

3.5.2 Magnitny Stream Intrusion

The magmatic body is located in the eastern part of the Noril'sk Plateau. It is exposed at the surface on the northern bank of the Chibichete River and is intersected by drill hole YuK-35. Partial data on the intrusion are reported by Urvantsev (1959) and Komarova and Lyul'ko (1967).

The intrusion is a sheet-shaped body 150 m thick, which slightly crosscuts basalts and tuffaceous rocks of the Morongovsky and Mokulaevsky Suites. The Magnitny Stream Intrusion is similar to the Lower Fokinsky Intrusion in internal structure and composition. The most complete cross section of the intrusion can be constructed by using data acquired from outcrop and drill cores.

The upper zone is composed of leucogabbro, olivine gabbro, taxitic gabbrodolerite, and quartz–olivine-bearing pegmatoid. Plagioclase is the principal mineral that occurs as large prisms, tabular crystals, and aggregates of small prisms providing a porphyritic texture to the rock. Clinopyroxene forms prismatic crystals and xenomorphic grains that contain structures indicative of solid solution breakdown. Olivine forms large (to 2 mm) isomorphic crystals and small grains, as well as stalked and palmate crystals, and varies from Fa_{39} to Fa_{47} (Ryabov 1992a). Quartz–olivine-bearing pegmatoidal gabbrodolerite has a pea-like fabric due to oikocrysts of clinopyroxene and olivine Fa_{61} about 0.5 cm in size enclosing chadocrysts of stubby prismatic plagioclase. Quartz is found in the matrix. In drill hole YuK-35, rocks of the UZ are olivine gabbrodolerite with porphyritic plagioclase. The transition to underlying rocks is abrupt.

The central zone is composed of picritic and troctolitic gabbrodolerite and is 125 m thick in drill hole YuK-35. The gabbrodolerite is commonly intersected by subbedded plagioclase veinlets 0.4 mm to 5 cm thick.

Olivine in the rock varies from 25 to 35% in *picritic gabbrodolerite* to 15–20% in *troctolitic gabbrodolerite*. It forms subidiomorphic, commonly elongate (2–3:1) crystals, and large palmate and small isometric grains (see Rock indication 86). Crystal size varies from 0.1 to 1.5 mm, and olivine composition is Fa_{18-21}. Chrome spinel, which forms inclusions in olivine, contains 6–14 wt% Cr_2O_3, 2.5–4 wt% Al_2O_3, 2.5–6 wt% MgO, 0.6–12 wt% TiO_2, and 0.1–0.4 wt% V_2O_5. Downward through the CZ, TiO_2 and Fe_2O_3 content increases, and Cr_2O_3, Al_2O_3, MgO and V_2O_5 content decreases. The increased MgO (3–5 wt%) content is unique to ilmenites of the CZ; other oxides include 0.5–1.2 wt% Cr_2O_3 and 0.4 wt% MnO.

The lower zone is composed of (from top downward) poikilo-ophitic gabbrodolerite (5 m), interlayered plagioclase-porphyritic and troctolitic gabbrodolerite (15 m), and olivine-bearing porphyry-like gabbrodolerite (2.5 m). The rocks sometimes have a taxitic-like fabric and often include plagioclase aggregates 3–5 cm in diameter that gives the rock a breccia-like structure.

Olivine content varies from sporadic in *poikilo-ophitic gabbrodolerite*, 3–5% in *olivine-bearing gabbrodolerite,* and up to 15–20% in *troctolitic gabbrodolerite*. Crystals are commonly isometric, but stalked crystals (elongation 1:5–15) and branched dendrites are also found. Olivine composition varies from Fa_{24} to Fa_{31}; in interlayers of troctolitic gabbrodolerite, the composition is Fa_{28} (0.15 wt% NiO).

At the margins of the intrusion, holocrystalline troctolitic gabbrodolerite with a pea-like fabric is developed from clinopyroxene ($Wo_{31}En_{43}Fs_{26}$) and olivine crystals framed by a fringe symplectite intergrowth with plagioclase (see Rock indication 87). In addition, dendritic olivine crystals 0.8–1.0 cm long that grow upward from the contact are observed.

Rare (1–2%) interstitial disseminated chalcopyrite and pyrrhotite are observed in rocks of LZ.

3.5.2.1 Mineralogy

The rocks are typically poor in iron and olivine (Fa_{18-21}) in picritic gabbrodolerite and contain 0.17–0.9 wt% NiO and chrome spinel. Chrome spinel in the CZ contains 6–41 wt% Cr_2O_3, 0.6–12 wt% TiO_2, 2.5–4 wt% Al_2O_3, 2.5–5.6 wt% MgO, and 0.1–1.4 wt% V_2O_5. Chrome spinel from the LZ contains 26–45 wt% Cr_2O_3, 0.1–2.0 wt% TiO_2, 6–10 wt% Al_2O_3, 2–6 wt% MgO, and 0–0.6 wt% V_2O_5. This wide variation in mineral composition is observed within a single specimen and points to nonequilibrium conditions during crystallization.

Ilmenite features increased MgO (3–5 wt%) and low MnO (0.4–0.5 wt%) content. The Cr_2O_3 content in ilmenite from rocks of the CZ is 0.5–1.2 wt% and is 0.1–0.2 wt% in the LZ.

3.5.2.2 Petrochemistry

The Magnitny Stream Intrusion contains three rock types: picritic, troctolitic, and olivine-poor gabbrodolerite. The composition of these rocks is controlled by the proportions of three mineral phases: olivine, pyroxene, and plagioclase. These proportions are exhibited in Al_2O_3, CaO, and MgO content. In picritic gabbrodolerite, the petrogenic oxide's composition is 42.5–43.4 wt% SiO_2, 0.6 wt% TiO_2, 6.6–8.0 wt% Al_2O_3, 2.9–3.0 wt% Fe_2O_3, 9.6–10.0 wt% FeO, 0.2 wt% MnO, 25.0–26.8 wt% MgO, 4.8–5.2 wt% CaO, 0.8–0.9 wt% Na_2O, 0.3 wt% K_2O, 0.04–0.05 wt % P_2O_5, and 0.33–0.34 wt% Cr_2O_3 (see Table 3.4). In troctolitic gabbrodolerite, the MgO and Cr_2O_3 content decreases to 16.0–17.1 wt% and 0.13 wt%, respectively, and Al_2O_3 content increases to 10.6–11.1 wt%. In olivine-poor and olivine-free gabbrodolerite, MgO content decreases to 6.9–0.9 wt%, and Al_2O_3 content increases to 13.7–16.3 wt%. The bulk of olivine-rich rocks is reflected in the high-magnesian weighted-mean composition of the Magnitny Stream Intrusion and turns out to be close to that of the Lower Fokinsky Intrusion.

3.5.3 Ergalachny (Picritovy) Stream Intrusion

This magmatic body is exposed at the surface at the head of the Ergalachny Stream (a tributary of the Sredniy Ergalakh River) and is intersected by drill hole E-14. The intrusion was studied by Tsyv'yan, and information is available in Komarova and Lyul'ko (1967), Komarova (1974), and Ryabov (1992a).

The Ergalachny Stream Intrusion is a steeply east-dipping magmatic body 116 m thick that crosscuts basalts and tuffs of the Morongovsky Suite. Investigations have shown that the bulk of the Ergalachny Stream Intrusion is composed of picritic and troctolitic gabbrodolerite, similar to the Lower Fokinsky and Magnitny Stream Intrusions, with a horizon of pegmatoid about 3 m thick in the roof of the intrusion (Komarova 1974).

The upper zone is composed of coarse-grained olivine gabbro and pegmatoidal quartz–olivine gabbro, which are emplaced in olivine-bearing gabbrodolerite. The quantity of olivine is less than 3–5%. Olivine in gabbro is Fa_{31-39}, and in pegmatoidal quartz–olivine, gabbro is Fa_{68}. In olivine gabbro, olivine and titanomagnetite form large (0.3–0.6 cm) oikocrysts with chadocrysts of plagioclase; in pegmatoidal gabbro, olivine has a palmate form, and quartz is observed in a matrix of granophyric micropegmatite together with needle-shaped apatite.

The central zone of the intrusion is composed of olivine-rich gabbrodolerite. *Picritic gabbrodolerite* is composed of olivine Fa_{14-26} (30–80%), orthopyroxene ($Wo_4En_{78}Fs_{18}$) and clinopyroxene (5–30%), plagioclase (10–15%), biotite (1–5%), chrome spinel, and magnetite (1–5%) (see Rock indication 88). As Komarova (1974) noted, some melanocratic rocks in the Ergalachny Stream Intrusion are completely devoid of plagioclase. A wide variety in the morphology and size of olivine crystals is characteristic of the Ergalachny Stream Intrusion, as in the Lower Fokinsky and Magnitny Stream Intrusions.

In *troctolitic gabbrodolerite*, the amount of plagioclase increases to 20–25%, and pyroxene decreases to 5–12%. Olivine forms large (0.6–0.8 cm) and, more rarely, small (0.1–0.3 mm) isometric grains and palmate and prismatic crystals (elongation 1:3–4). In large crystals of composition Fa_{32}, a cleavage is observed. The contact of the tractolitic gabbrodolerite with the overlying pegmatoids is abrupt.

3.5.3.1 Mineralogy

The Ergalachny Stream Intrusion contains olivine of two compositions: Fa_{31-68} in mafic rocks and Fa_{14-26} in ultramafic rocks; the NiO content in olivines from mafic rocks is 0.01–0.11 wt%. Orthopyroxene ($Wo_3En_{78}Fs_{18}$) in picritic gabbrodolerite contains 1.6 wt% Al_2O_3, 0.4 wt% Cr_2O_3, and 0.4 wt% TiO_2. Biotite in picritic gabbrodolerite has an Fe content of 21.8 wt%, as well as 7.2 wt% TiO_2, 0.74 wt% Cr_2O_3, 0.57 wt% F, and 0.08 wt% Cl. Amphibole is uncommon; its composition is $f = 18.8$ wt%, 0.62 wt% Cr_2O_3, and 1.2 wt% TiO_2. Chrome spinel contains 28.7–33.1 wt% Cr_2O_3, 5.3–7.8 wt% Al_2O_3, 3.6–5.3 wt% MgO, 5.3–7.8 wt% TiO_2, 0.5–0.7 wt% V_2O_5, and 0.14–0.18 wt% NiO and ilmenite contains 3.2 wt% MgO, 0.4 wt% MnO, 1.4 wt% Cr_2O_3, and 0.85 wt% V_2O_5.

3.5.3.2 Petrochemistry

Olivine-rich rocks of the Ergalachny Stream Intrusion have a relatively invariable chemical composition with insignificant variations of rock-forming oxide concentrations and impurities: 6.0–8.6 wt% Al_2O_3, 23.2–26.2 wt% MgO, 4.6–6.0 wt% CaO, 0.3–0.9 wt% Na_2O, 0.3–0.4 wt% K_2O, and 0.1–0.4 wt% Cr_2O_3. This evidences a picrite-basalt composition of the melt. In cross-section, some constituents sharply increase: Al_2O_3 (15.0–19.8 wt%), CaO (9.8–12.8 wt%), Na_2O (1.9–2.3 wt%), and K_2O (0.4–0.6 wt%), whereas MgO decreases to 3.2–7.0 wt% (see Table 3.4).

3.5.4 Petrology and Genesis

Petrological characteristics of the Fokinsky Intrusive Complex include:

1. Dominated by picritic and troctolitic gabbrodolerite (more than 80%).
2. Olivine-free and olivine (–quartz)-bearing gabbrodolerite in the UZ and LZ.
3. Pegmatoids have a mafic composition.
4. Chrome content in picritic gabbrodolerite corresponds to the abundance ratio in undifferentiated products of picrite-basalt melts.
5. Nickel content in olivine correlates with the unified geochemical trend for barren traps.
6. The presence of olivine porphyries in quench facies rocks is evidence for the existence of independent picrite-basalt magma.
7. The morphology of olivine crystals and their wide variations in their size with a relatively invariant composition, as well as the presence of inclusions of labradorite and symplectite intergrowths, point to crystallization in an intrusion chamber and the absence of gravitational differentiation of mineral phases.
8. Chromium-rich chrome spinel compositionally similar to chrome spinel in picritic lavas of the Gudchikhinsky Suite is evidence of the compositional resemblance of picrite-basalt lava to their initial magma.
9. Wide variations in the composition of chrome spinel are caused by fluids redistributing elements.

Contact alteration related to the Fokinsky Complex intrusions is limited to weak hornfelsing of sedimentary rocks.

Mineralization in intrusive rocks of the LZ is poor sporadic pyrrhotite mineralization with pyrrhotite content ranging up to 2%.

Genetic concepts for the Lower Fokinsky Intrusion presented by Zolotukhin and Vasil'ev (1967) highlighted the incompatibility between the weighted-mean composition of the intrusion and the contact facies. The calculated MgO content in the intrusion is 22.35 wt%, and the oxide content in contact rocks is 12.65 and 14.11 wt%. Taking this into consideration, Zolotukhin and Vasil'ev (1967) explain the formation of olivine-rich rocks by a high rate of crystallization in a heterogeneous melt with simultaneous settling of protocrysts and new olivine growth with new injections of melt.

Arkhipova (1975) and Oleinikov (1979) suggest the mechanism for the formation of the Lower Fokinsky Intrusion was the crystallization–gravitational differentiation of magma under interchamber conditions. They indicate the period of protocrystallization in an intermediate chamber is the critical phase in this magmatic process. Oleinikov (1979) considers that in the Lower Fokinsky Intrusion, olivine protocrysts had forsterite composition and now remain only as magnetite pseudomorphs in contact facies. Most of olivine crystals were crystallized from melt at lower temperatures and so have higher Fe contents.

Judging from the sequence of the intrusions that comprise the Fokinsky Intrusion, they were emplaced coeval with the Gudchikhinsky Suite through to the Mokulaevsky Suite. Based on the strong mineralogical–geochemical similarity between rocks of the Lower Fokinsky Intrusion and picritic basalts of the Gudchikhinsky Suite, these formations can be considered as comagmatic. The resemblance of the internal structure and mineralogical–geochemical characteristics of the Magnitny Stream and Ergalachny Stream Intrusions with the Lower Fokinsky Intrusion suggests that all these intrusions constitute a unified Fokinsky Complex. Moreover, emplacement of the Magnitny Stream Intrusion in basalts of the Morongovsky Suite and the Ergalachny Stream Intrusion into basalts of the Mokulaevsky Suite provides evidence: first, that picritic intrusions were emplaced post-Gudchikhinsky Suite; second, that there is an absence of effusive comagmatic with these intrusions in the Noril'sk Region; and third, of the mechanisms of development of the intrusions and the composition of picrite-like melts. The time of emplacement of intrusions of the Fokinsky Complex remains uncertain. It can only be said with assurance that picrite-basalt magmatism at the northwest of the Siberian Platform occurred repeatedly.

3.6 Noril'sky Intrusive Complex $(\nu\delta - \omega\nu\beta)$ T_1nr

The Noril'sky Intrusive Complex includes a number of closely spaced bodies, which are linked as a unified ore–magmatic system. However, the bodies differ in composition and extent of differentiation and mineralization. The complex is composed of differentiated Ni-bearing intrusions of the Noril'sky, Lower Talnakhsky, and Zubovsky type and slightly differentiated intrusions of the Kruglogorsky type. The emplacement of intrusions in a

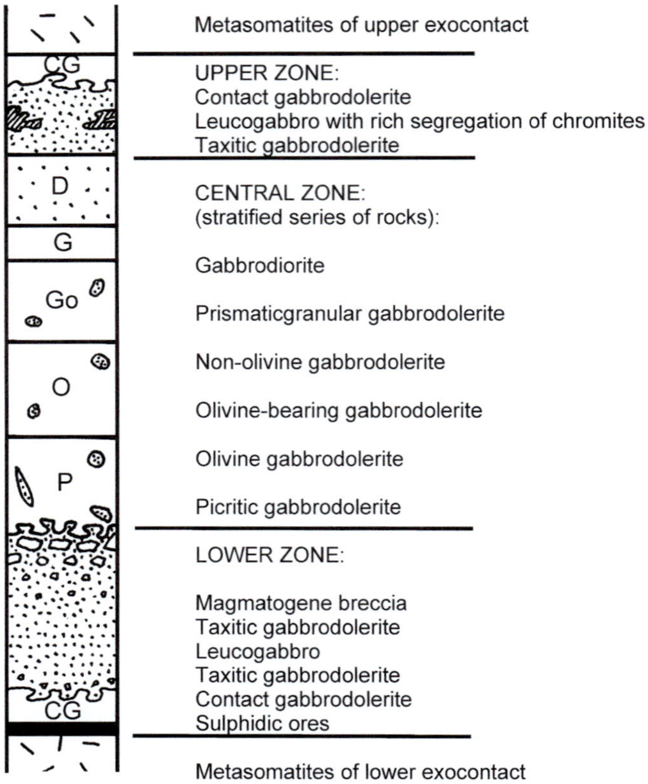

Fig. 3.7 Principal scheme for the structure of the Noril'sky-type intrusions. Specks designates pegmatoid formations: dense speckle—ore-bearing taxitic gabbrodolerite and leucogabbro, sparse speckle—barren gabbrodolerite, and prismatically granular gabbrodolerite

stratigraphic sequence, as well similarities in their composition with that of effusive rocks that are intersected by the intrusions, suggest the intrusions were comagmatic with the effusive rocks and were formed during the period between accumulation of Upper Gudchikhinsky and Upper Nadezhdinsky rocks.

3.6.1 Noril'sky-Type Intrusions $(\nu\delta - \omega\nu\beta)T_{1nr}^{nr}$

The Noril'sk-I, Upper Talnakh, Mt. Chernaya, Noril'sk-II, and Imagdinsky Intrusion are all Noril'sky-type intrusions. Studies of these intrusions, ores, and halos of altered rocks have been carried out over many years, and the results have been published in numerous papers and monographs: Godlevsky (1959), Rogover (1959), Smirnov (1966), Korovyakov et al. (1963), Zolotukhin (1964), Dodin and Batuev (1971), Zolotukhin et al. (1975), Tarasov (1976), Natorkhin et al.(1977), Zolotukhin and Shchedrin (1977), and Duzhikov et al. (1988).

Intrusions of the Noril'sky type are chonolites, ribbon-shaped and trough-shaped, slightly crosscutting bodies. The Upper Talnakh Intrusion extends more than 15 km, and its width and thickness are 0.8–2.0 km and about 100–200 m, respectively. The thickness of Noril'sk-I Intrusion reaches 320 m. The center of intrusions includes a three-member differentiated series of rocks such as picritic, olivine, and olivine-free gabbrodolerite, which grade at the margins into isotropic sills and dykes composed of gabbrodolerite or dolerite that are olivine-free or contain sporadic olivine.

In Noril'sky-type intrusions, the UZ, CZ, and LZ are distinguished (Fig. 3.7), and at the base, compact sulfide ores are found. The principal rock type of the UZ and LZ is pegmatoid: the upper horizons host low Pt–sulfide mineralization, and lower horizons are most enriched in sulfides. In the central zone, horizons of mafic and ultramafic rocks are distinguished. From the bottom upward, mafic rocks are layered into olivine, olivine-bearing, and olivine-free gabbrodolerite. In the upper part of the mafic rock horizon, prismatically granular gabbrodolerite and metadiorite are formed. The horizon of ultramafic rocks is composed of olivine-rich picritic and troctolitic gabbrodolerite. This differentiation of the intrusion sequence is occasionally violated by layers of olivine and picritic fine-grained gabbrodolerite, typically observed at boundaries between various layers.

At the base of the intrusion, pegmatoids commonly occupy the lower zone (see Fig. 3.7), but in some places, these are interlayered with picritic gabbrodolerite and are viewed as blind veinlets extending from a taxitic horizon or of a system of veinlets that give the rock a breccia-like fabric. In the leading section of the Kharaelakhsky branch of the Upper Talnakh Intrusion, pegmatoids form a fairly steeply dipping zone of magmatogenic breccias that cuts the layered gabbrodolerite rock series from the base of the intrusion to the top.

It has been shown that intrusions of the Noril'sky Complex are very similar to each other. Therefore, it is possible to extend the characteristics of the Upper Talnakh Intrusion listed below to other intrusions. However, in order to gain a thorough understanding of these formations, it is necessary to extract important genetic information from other intrusions.

3.6.1.1 Upper Talnakh Intrusion

The *Upper Talnakh Intrusion* is restricted to the intersection zone of the Kharaelakhsky trough and the Noril'sko–Kharaelakhsky deep fault and has a sheet-like form with swells and pinches and extends to a fan-shaped arrangement of intrusive fingers. The magmatic body gently dips from the bottom of Permian–Triassic tuff lava through Carboniferous–Permian terrigenous coal-bearing deposits of the Tungussky series and into Devonian carbonate, sulfate-bearing argillaceous carbonates, and argillaceous deposits. The Southwestern and Northeastern intrusive branches are situated in terrigenous deposits of the Tungussky Series, and the Northwestern and Kharaelakhsky intrusive branches lie in Devonian evaporite deposits

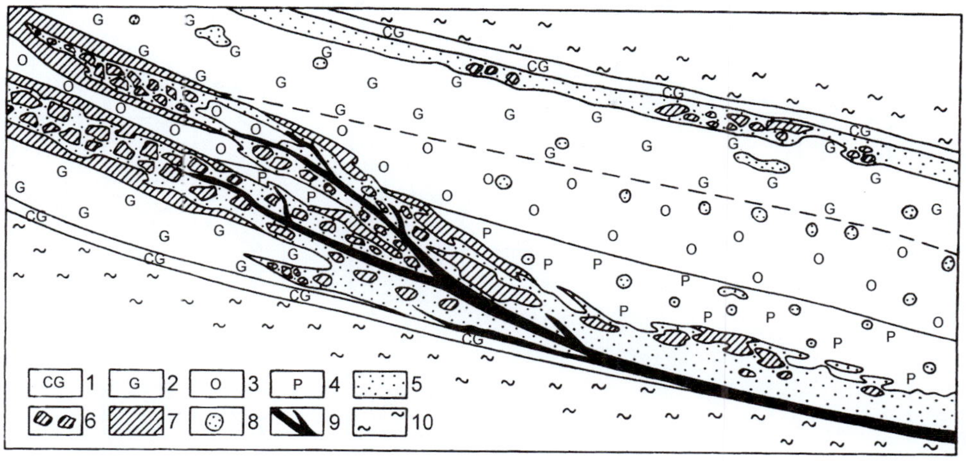

Fig. 3.8 Internal structure of the Kharaelakhsky branch of the Upper Talnakh Intrusion (east–west cross section constructed from drilling data). *1* contact granodiorite, *2* olivine-free and olivine-bearing gabbro, *3* olivine gabbro, *4* picritic gabbro, *5* taxitic gabbro and leucogabbro, *6* magmatogenic breccia, *7* taxitized melanocratic rocks, *8* "erratic boulders" of leucogabbro, *9* sulfide veins, *10* country rock

(Fig. 3.8) (Dodin and Batuev 1971; Zolotukhin et al. 1975; and Natorkhin et al. 1977). The Talnakhsky Pt–Cu–Ni sulfide deposit is associated with the first two intrusive branches; the Oktyabr'sky deposit is associated with the second two branches.

The intrusive branches form swells up to 220–250 m in thickness that are connected to each other by bridges composed of undifferentiated olivine-bearing gabbrodolerite. The swells in the intrusive branches are the most differentiated. On the flanks, the layering disappears and the sequence is composed of poikilo-ophitic gabbrodolerite with rare olivine.

The thickness of differentiated layers along different intrusive branches and down the sequence varies (Dodin and Batuev 1971), but it is relatively constant for ore-bearing differentiated layers and significantly varies for barren mafic horizons (Likhachev 1998).

The layered structure of the Upper Talnakh Intrusion is most distinct in the leading parts of Southwestern and Northeastern intrusive branches. Here, a classic sequence is observed (in order downward): gabbrodiorite (metadiorite), olivine-free, olivine bearing, olivine, olivine–biotite, picritic, taxitic, and contact gabbrodolerite and a sulfide-rich layer. This succession features a regular decrease in silicic acid and alkalinity, as well as an increase in the magnesium content. From the roof downward gabbrodolerite, olivine content increases, plagioclase becomes more mafic, and there is an increase in the ore content due to the presence of disseminated to compact masses of ore. These sequences have previously been demonstrated to have been generated by crystallization differentiation with gravitational fractionation of early mineral phases and accumulation at the base of magmatic chambers.

The flanks of the Southwestern and Northeastern intrusive branches are steep. Along the branch, different layers thin out in a sequence. In the Upper Talnakh Intrusion, picritic gabbrodolerite first thins out, then olivine poikilo-ophitic gabbrodolerite thins out. Taxitic gabbrodolerite then grades into taxito-ophitic, porphyry-like gabbrodolerite, and further along, they merge into a contact ophitic-poikilo-ophitic gabbrodolerite with sporadic olivine. This is similar to intrusions of the Noril'sky ore junction (Korovyakov et al. 1963).

In the leading parts of the Northwestern and Kharaelakhsky branches, breccia-like rocks are abundant, containing rock fragments ranging from several centimeters to blocks 1–3 m in size. The rocks contain vitreous aphanite–tachylite, dolerite, barren and mineralized gabbrodolerite, and various metasomatized rocks. The rock matrix is a metasomatized rock and/or compact sulfide-rich material. Expansion of the breccia in the intrusion was bounded and resembles a closed explosion at the front of a magmatic wedge. Geologists have not reached a consensus regarding the conditions under which this breccia was formed. Taking into consideration both the peculiarities of the occurrence and composition of the breccia, geologists have regarded it as "explosive" and "metasomatic" (Stetsenko 1966; Zotov 1976), a "magmatic (intrusive) pseudobreccia" (Ryabov 1969; Tarasov 1976), an "intrusive (eruptive) penetration breccia," a "pseudobreccia of pseudotachylites" (Ryabov and Zolotukhin 1970), and a "magmatogenic breccia" (Lyul'ko et al. 1972). Another type of breccia is formed of fragments of magmatic rocks cemented by a variable composition pegmatoid. Peculiarities of the occurrence and description of these rocks are given later.

The results of chemical analysis of rocks from the reference sections are listed in the Appendix: Tables A.12–A.17 for the Southwestern branch, Tables A.18–A.22 for the Northeastern branch, and Tables A.23–A.26 for the Kharaelakhsky branch.

Upper Zone

Contact facies rocks are strongly altered, taxitized, or metasomatized. Unaltered rock remains only in the peripheral parts of the zone. The UZ comprises gabbrodolerite; olivine-free, olivine-bearing, olivine, picritic, troctolitic, and

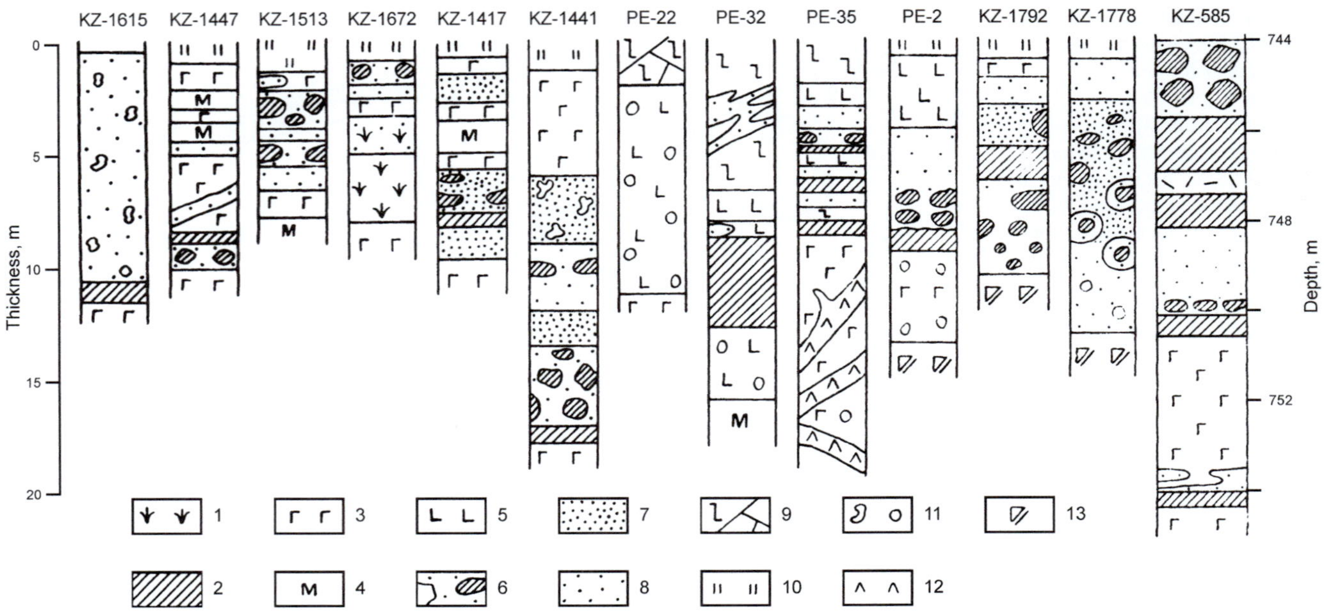

Fig. 3.9 Lithologic columns from drill holes showing the structure of the upper zones of the Upper Talnakh and Upper Noril'sk Intrusions (with drill hole marks); after drilling data from NKGRE. *1* dendrite-like pyroxene and olivine megacrystals, *2* gabbro enriched in chromite and/or olivine, *3* olivine-free and olivine-bearing gabbro, *4* prismatically granular gabbro, *5* dolerite, *6* magmatogenic breccia, *7* taxitic gabbro, *8* leucogabbro, *9* fractured basalt, *10* contact altered rocks, *11* miarolitic voids, *12* veins of vitreous aphanite, *13* metadiorite

taxitic gabbrodolerite; leucogabbro; olivine, graphitic, and palagonitic gabbro; and magmatogenic breccia. Most of these rocks, except leucogabbro, cannot be traced as layers for long distances.

Leucogabbro is the most widespread variety of UZ rock in the Noril'sky Intrusive Complex. The results of detailed investigations of these rocks in intrusions of the Noril'sky ore junction are reported in Likhachev (1965). In some places, the thickness of the leucogabbro horizon reaches several tens of meters, and these rocks extend deep into the flanks of the intrusion, cutting through olivine-free and contact facies gabbrodolerite creating a sharp (but not quench-textured) fine-grained contact (Likhachev 1965; Tarasov 1976). In the flanks of the Noril'sk-I and Upper Talnakh Intrusions, the upper and lower taxites merge (Korovyakov et al. 1963; Zolotukhin and Vasil'ev 1967; and Ryabov 1969).

In leucogabbro, small melanocratic aggregates are composed of altered and (rare) unaltered olivine. As the quantity of these aggregates increases, the rock assumes a taxitic- or breccia-like fabric. Aggregates as accumulations of intratelluric olivine in rocks of the UZ and olivine-rich gabbrodolerite near the roof of the Noril'sk-I Intrusion were first discussed by Rogover (1959) and Korovyakov et al. (1963). The most comprehensive information about the upper picrite is reported in Ivanov et al. (1971b), Tarasov (1976), and Ryabov (1982, 1984a, b).

The upper picrite and troctolite are known in the Noril'sk-I, Noril'sk-II, and Upper Talnakh Intrusions. The thickness of the rocks ranges from 1.5–4.5 m up to 50 m (Ivanov et al. 1971b). Data has shown that the upper picrite lies directly below an eruptive breccia at the contact with the host basalts and is surrounded by taxites and leucogabbro. Picrite forms elongate lenses that contain rounded leucogabbro schlieren, and small aggregates of large plagioclase crystals are not uncommon.

In addition to picritic gabbrodolerite, leucocratic gabbro in the UZ also contains magmatogenic breccia and single melanocratic aggregates. The magmatogenic breccia is composed of melanocratic fragments bound by a leucocratic gabbro. The fragments are several centimeters in size and have an angular form. Their boundaries are sharp but without quench texture. The fragments are often surrounded by plagioclase rims up to 1 cm thick, which grades into taxitic and troctolitic gabbrodolerite and then leucogabbro. The composition of fragments is that of picritic and troctolitic gabbrodolerite, olivinite (with or without plagioclase, spinel, and clinopyroxene), or pyroxenite. The fragments are commonly subjected to strong alteration with widespread serpentine, bowlingite, amphibole, mica, and chlorite. Fragments are not uncommonly surrounded by a zone enriched in coaly matter. The characteristic features of the magmatic breccia are the distinct fragment-like texture, the leucogabbro cement, and the plagioclase cement when fragments have a corroded, blurred form; the proportion of dark-colored minerals increases and texture becomes ataxitic. It is possible to trace the transition from shadow breccia-like textures into taxitic gabbrodolerite and leucogabbro.

The structure of the UZ of the Upper Talnakh Intrusion is shown in Fig. 3.9. The figure shows the variation in the UZ thickness and rock types; the contact facies of dolerite and gabbrodolerite is sporadic, and where it is absent, the rocks of UZ are immediately contiguous with metasomatized sedimentary rocks. Rocks of the UZ are underlain by gabbroic diorite or prismatically granular gabbrodolerite. Olivine-rich rocks (see Fig. 3.9, shaded regions) are located in different parts of the sequence, and these are permanently accompanied by taxites and leucogabbro. In the rocks, small rounded amygdale-like or miarolitic voids are filled with palagonite-I, biotite, apatite, amphibole, anhydrite, gypsum, calcite, and sulfides. In a number of cases, dendrite-like olivine or plagioclase (and clinopyroxene) crystals 3–6 cm in length are observed, as well as variolites, which are intergrowths of plagioclase and clinopyroxene in spherocrystals 2–3 cm in diameter.

Petrography

Most of the rocks in the UZ demonstrate significant resemblance to other varieties of gabbrodolerite in the intrusion sequence. However, a number of characteristics are recognized that are specific only to rocks of this horizon and are important in order to understand the petrogenesis of pegmatoids and the formation of Pt-poor sulfide ores restricted to the UZ.

Leucocratic gabbro is the main variety of UZ rocks. The rock has porphyritic-like, gabbroic, gabbro-ophitic texture with elements of ophitic, poikilo-ophitic, pegmatoidal, and ataxitic textures. Plagioclase (65–90%) and clinopyroxene (5–30%) are the dominant minerals. Other minerals, which often govern the rock texture, include palagonite (3–35%), olivine (0–10%), spinel minerals (1–20%), biotite (to 10%), apatite (to 5%), amphiboles (to 7%), sulfides (from 3–5% to 10%), quartz, and micropegmatite, as well as ilmenite, sphene, serpentine, bowlingite, chlorite, prehnite, calcite, gypsum, and zeolite. The leucogabbro has a relatively stable composition. The mineralogical–petrographic description of these minerals was reported by many authors (e.g., Likhachev 1965; Zolotukhin et al. 1975; Natorkhin et al. 1977). Changes in the proportions of pyroxene and plagioclase in leucogabbro are associated with an increase in dark-colored minerals and are commonly accompanied by appearance of the ataxitic texture.

The dominant texture of the leucogabbro is formed by the intergrowth of large plagioclase and clinopyroxene crystals. In interstices, fine-grained aggregates of the plagioclase and clinopyroxene are found with palagonite, ore minerals, biotite, apatite, calcite, chlorite, quartz, and micropegmatite. Amphibole forms poikilocrysts, pseudomorphs after clinopyroxene, and fine-grained aggregates in the matrix. Large clinopyroxene crystals are diopside ($Wo_{45-47}En_{43-41}Fs_{12-23}$), and fine crystals are augite ($Wo_{39}En_{45}Fs_{16}$) ($f = 25$ at%) and amphibole ($f = 23$–24 at%). Plagioclase forms large tabular bytownite ($Ab_{22-20}An_{79-78}$) crystals and labradorite ($Ab_{37-39}Or_1An_{62-60}$) prisms (see Rock indications 89 and 90). The size of angular interstices in leucogabbro varies from fractions of a millimeter to 0.1–0.3 cm, and their total volume is typically less than 3–5%. However, varieties with increased matrix volume have been observed.

Palagonite globular gabbro is a rock with ophitopoikilo-ophitic texture and a wide variety of felsic and mafic components. Large tabular plagioclase crystals are commonly altered, and smaller prisms have the composition $Ab_{47}Or_2An_{51}$. Pyroxene ($Wo_{39}En_{48}Fs_{13}$) crystals (see Rock indication 91) are usually unaltered. Blades of brownish biotite ($f = 19$ at%), chlorite, sphene, and leucoxene are present, and ilmenite forms large grains and small intergrowths in leucogabbro. The key feature of these rocks is the globular palagonite grains, which vary from 10–15% to 25–30% of the rock. Aggregates of palagonite have an irregular isometric or elongate oval or angular shape resembling miarolitic formations. The size of aggregates varies from 0.3–0.5 cm to 2–3 cm. Globules have sharp boundaries and are situated in the feldspar matrix, and grouped, large (5 × 7 cm) accumulations of clinopyroxene have a heteroblastic structure. The sizes of globules and accumulations of clinopyroxene are inversely related.

Under microscope, a deep-brownish, homogenous palagonite is observed in the globules or forming an alveolar structure around the globules, which appears as light-brownish and dark-brownish loops. It is also possible to trace the change from continuous palagonite into small drop-like minerals resembling bowlingite. Between these droplets, plagioclase ($Ab_{53}Or_2An_{45}$) grains are observed (see Rock indication 91). In some cases, among "drop-like" palagonites, olivine Fa_{31} is observed. Not infrequently, laths of plagioclase with Y-shaped ends and long-prism clinopyroxenes are observed in palagonite. Palagonite in continuous masses, as well as in alveolar and drop-like occurrences, may be anisotropic.

Graphite–palagonite globular gabbro is very similar in all structural and petrographic features to palagonite gabbro described above. The main distinctive feature of this rock is the increased graphite content of 5–7%; one or two layers of graphite line the walls of palagonite globules. Plagioclase laths are also situated tangentially to the surface of globules. On the outside of globules, graphite often forms plane hexagonal crystals, and on the inside surface, druses of needle-shaped crystals are found (see Rock indication 92). Other than these graphite "hedgehogs," rims of early plagioclase overgrowth, intergrowths of plagioclase with pyroxene, as well as circular and ring-like crosscuts that appear to be associated with small gas-filled bubbles are observed in the palagonite mass. Needle-like graphite crystals with a high fluorine content (6.9–10.4 wt%) are indicative of growth of the mineral from a gaseous phase. In addition, fluorine-rich apatite crystals (3.2–3.6 wt% F and 0.3–0.5 wt% Cl) and small skeletal clinopyroxene crystals are found in palagonite.

Large clinopyroxene crystals in gabbro are augite ($Wo_{45-44}En_{42-43}Fs_{13-12}$), and smaller crystals in palagonite are diopside ($Wo_{47-8}En_{25-29}Fs_{29-24}$). In addition, diopside has higher CaO, FeO, MnO, and Na_2O content and lower MgO, Al_2O_3, and Cr_2O_3 content. The trend for the zoned crystals to change composition from core to rim is similar for both clinopyroxenes.

Judging from the peculiarities of occurrence and composition of palagonite globules, it may be assumed that they originated through the accumulation of magmatogenic liquid, which was hampered in crystallization due to an abrupt decrease in the temperature. As a result, palagonite-I was formed, which is analogous to a glass enriched in volatiles. This palagonite should not be confused with palagonite-II, which is formed under low-temperature hydrothermal alteration of mafic rocks. Concepts concerning the magmatogenic nature of palagonite-I and hydrothermal origination of palagonite have been published by Belyankin, Serdyuchenko, Gon'shakova, and Lyakhovich et al. (Arkhipova and Natorkhin 1971). And it has been considered in detail by Ryabov (1989a).

Taxitic gabbrodolerite has a specific ataxitic texture due to a combination of rocks of varying compositions, grain sizes, and structural peculiarities to their intergrowths. The principal minerals of taxites are plagioclase and clinopyroxene, as well as olivine, oxides, and sulfides, variable amounts of which are uniformly present in the rocks. Other minerals including amphibole, biotite, apatite, and calcite are observed locally in significant volumes. Chlorite is also formed (see Rock indications 93–100).

In the UZ, one characteristic of pegmatoids in intrusions of the Noril'sk Region is the presence of chromite-enriched rock alongside chromite-free rock. Both varieties contain sulfides, but the sulfide content may be higher (10–15% or more) in chromite-free pegmatoids than that in chromite-bearing pegmatoids.

Chromite-free taxitic gabbrodolerite has a significant resemblance to rocks in the LZ. In upper taxites, fine-grained aggregates composed of clinopyroxene with palagonite or plagioclase are frequent; aggregates with small-granulated olivines that are specific for lower taxites are less frequent. Palagonite in taxitic gabbrodolerite no longer forms individual aggregates as those in palagonite globular gabbro. Palagonite content is lower than 3–5% and is found in a matrix of fine-grained minerals.

Microcrystalline intergrowths of clinopyroxene and plagioclase in taxitic gabbrodolerite are augite ($Wo_{44-49}En_{41-46}Fs_{9-11}$) and labradorite ($Ab_{42-44}Or_{2-3}An_{55-54}$), respectively, and rims of zoned plagioclase are andesine ($Ab_{63}Or_4An_{33}$) (see Rock indication 93). Clinopyroxene from fine-grained parts of the rock is diopside ($Wo_{45-50}En_{41-48}Fs_{7-10}$). The association of clinopyroxene and plagioclase enclosed in palagonite is also formed by diopside ($Wo_{50-52}En_{42-34}Fs_{8-15}$) and oligoclase ($Ab_{74}Or_4An_{22}$) (see Rock indication 94). The composition of diopside enclosed in palagonite from fine-grained areas in taxitic gabbrodolerite is found to have a higher Al_2O_3 content (3.48–6.04 wt%) than those from crystallized areas that have an Al_2O_3 content of 1.46–3.43 wt%.

Minerals (amphibole, biotite, apatite, calcite, gypsum, and sulfides) containing volatile components are uniformly distributed in taxitic gabbrodolerite. Small accumulations are usually found in the matrix, but not infrequently they form accumulations in association with plagioclase resembling globular palagonites in form and size. In such areas, the composition of rock-forming minerals shows significant fluctuations.

In taxitic gabbrodolerite enriched in sulfides (see Rock indications 95 and 96), large clinopyroxene prisms are augite ($Wo_{42-43}En_{47-48}Fs_{10}$), early plagioclase is altered, and small plagioclase prisms surrounding sulfide inclusions are $Ab_{82}Or_6An_{12}$ or pure albite (Ab_{100}). An increase in sulfide content is accompanied by a change in composition of clinopyroxene to diopside $Wo_{47-48}En_{46-51}Fs_{7-1}$ with a distinct decrease in Fe content from the core to rims of zoned crystals (see Rock indication 97). In these rocks, F-bearing biotite and apatite are present, and the fraction of chlorine and fluorine to total halogens is lower in the former than in the latter.

Apatite in taxitic gabbrodolerite sometimes reaches 3–5%. Large accumulations of sulfides, spherules of calcite, calcite together with apatite and albite, and albite with sphene are restricted to apatite-rich sites, where clinopyroxene and plagioclase are often altered. Short prismatic apatite crystals contain numerous fluid inclusions. An attempt at homogenization failed because the inclusions exploded on heating. In apatites from pegmatoids of the UZ, wide variations in F (0.21–5.0 wt%) and Cl (0–3.90 wt%) have been established. Most of the apatites are F-apatite.

Olivine crystals in the upper taxitic gabbrodolerite vary widely, producing a diversity of rocks from olivine-free to olivine-rich varieties. In most of the rocks, olivine forms small aggregates of small granules or singular large palmate crystals. Olivine content in these rocks is less than 3–5%, and it commonly is replaced by bowlingite. Fragments of *magmatogenic breccia* bound by leucogabbro are most enriched in olivine. A gradual transition is sometimes traced between magmatogenic breccia and taxitic gabbrodolerite. This is caused by variations in the extent of corrosion of melanocratic material (Ryabov 1984a, b). In taxitic gabbrodolerite, hornfelsed and granoblastic shadow relict textures of fine-grained fragments as well as of constituent minerals such as green spinel and forsterite, usually common in fragments of magmatogenic breccia, are sometimes preserved in gabbrodolerite. The magmatogenic breccia of the UZ is very similar to a breccia from the Kharaelakhsky branch of the Upper Talnakh Intrusion where it is diverse and extensive; because of this, these rocks will be discussed in detail later.

Fragments of magmatogenic breccia in the UZ are composed of a granoblastic aggregate of spinel, plagioclase, and olivine (Fa_{10}). Chrome spinel (27.31–31.02 wt% Cr_2O_3) is present as small inclusions in olivine and plagioclase, and green spinel (0.54–1.36 wt% Cr_2O_3) forms intergrowths in olivine and plagioclase, as well as individual crystals in the rock matrix. Olivine in the leucocratic gabbro cement of magmatogenic breccia is chrysolite (Fa_{14}) (see Rock indications 101 and 102).

Picritic and troctolitic gabbrodolerites of the UZ feature segregation, poikilo-ophitic and poikilitic textures. Olivine ranges from 25–30% to 40% and forms idiomorphic short prismatic crystals, small rounded and subidiomorphic intergrowths in clinopyroxene and plagioclase, and large crystals with skeletal forms. The skeletal forms feature perfect cleavage and common, rounded biotite–serpentine inclusions. Plagioclase and clinopyroxene often form large crystals with poikilitic olivine intergrowths. Clinopyroxene in picritic gabbrodolerite reaches 25–30% and in troctolitic gabbrodolerite is below 10–15%. One of the main distinctions between rocks is the presence of chrome spinel in picritic gabbrodolerite (see Rock indication 103) and the absence of chrome spinel in troctolitic gabbrodolerite (see Rock indication 104). Olivine in these rocks possesses similar Fe contents (Fa_{23-24} and Fa_{20-21}, respectively), but in troctolitic gabbrodolerite, generations of olivine depleted in Ni silicate appear. Clinopyroxene of picritic gabbrodolerite is diopside ($Wo_{45-50}En_{45-43}Fs_{10-7}$) exhibiting the tendency for decrease the Fe and increase Cr_2O_3 content (0.02 to 0.18 wt%) from the core to rims of zoned crystals. In troctolitic gabbrodolerite, amphibole ($f = 21.5$ at%) and biotite ($f = 17–18$ at%) contain more chlorine than fluorine. In picritic gabbrodolerite, drop-like and interstitial sulfide impregnation is present; the latter is virtually absent in troctolitic gabbrodolerite.

The amount of chrome spinel in picritic gabbrodolerite is commonly less than 3–5%. The mineral forms aggregates and streaks composed of small grains in clinopyroxene and plagioclase. Chrome spinel does not often form inside olivine crystals but is more commonly found in cracks. In addition, the amount of chrome spinel in olivine-rich rocks may reach 25–30% or more, but these rocks are commonly strongly altered, and the presence of olivine can only be judged from preserved textures, where olivine-free window spaces are rich in olivine microcrystalline growth. The absence of olivine provides a basis for the proposition that most ore oxides were formed after crystallization of olivine.

Chromite-bearing pegmatoids are rocks specific to the UZ and are only present in Noril'sky-type intrusions. The structural peculiarities and composition of the silicate component of the rocks are similar to those of chromite-free pegmatoids, and only the presence of increased chrome spinel combines these rocks into a group and distinguishes them from other pegmatoids.

The amount of spinel in pegmatoids ranges from 0.5% to 5–7%, sometimes to 20–30%, and more rarely to 40%. Increased spinel content is common in compositionally variable rocks, creating an ataxitic fabric (Ryabov 1984a, b). The spinel group of minerals is mainly represented by chrome spinels or chromites; green spinel is less common. Chlorites are distributed irregularly as small oval isometric sites (0.2–10 mm to 3–7 cm) with fine-grained (5–20 μm) ore cryptocrystallization. These areas, in turn, form small accumulations, larger schlieren, lenses, and systems of lenses, which vary in thickness and are complex in internal structure. Some view on the chromite-bearing horizon can be found in Sluzhenikin et al. (1994).

Areas dense with disseminated chromite have sharp contacts with barren rocks. Under microscope, this boundary often passes through a single crystal so that one half is rich in chromite intergrowths and the other is completely depleted. The disseminated chromite does not disrupt the rock fabric and looks the same as rocks that are chromite-free (see Rock indication 105). Chromite inclusions pierce clinopyroxene, most types of plagioclase, and biotite, and less commonly olivine crystals. However, chromite frequently forms overgrowths on olivine. In addition, chromite overgrows sulfide droplets, large tabular crystals of early plagioclase in leucogabbro, and rounded palagonite-I globules (see Rock indication 106). Distribution characteristics of chromite and its relationship with other mineral phases help determine the time of oxide formation during petrogenesis.

Blebby spinel forms in the matrix between large crystals of early plagioclase (see Rock indication 107) and is of particular interest. They form chains of grains on the inner surfaces of bubbles that contain biotite, chlorite, and relict-altered plagioclase. Spinel in the matrix can form compact chromite grains with rounded inclusions depleted in chromite. The blebby texture of rock is maintained in regions free of chromite accumulations. These regions feature a sharp decrease in spinel in the matrix, remaining only along bubble outlines, which become rarer. Interestingly, the composition of matrix spinel from spinel-rich rock areas in similar blebby formations is magnesioalumochromite. In areas depleted in spinel, it is green spinel (see Rock indication 107). The mineral changes abruptly, so spinel from neighboring bubbles may have different compositions.

The principal mafic mineral in chromite-rich rocks is light-green to colorless amphibole that sometimes forms continuous heterogranoblastic aggregates with inclusions of relict minerals or with persistent relict initial rock textures. Chromite is evenly disseminated in these rocks, and only small depleted areas are distinguished. The depleted areas represent relict early plagioclase and amphibole pseudomorphs (not infrequently, homoaxial) after olivine or bubbles, in which chromite forms chains of grains lining the vacuole surface with an inner chromite-free area composed of amphibole (see Rock indications 108 and 109).

The blebby and amygdaloidal texture and wide distribution of minerals containing volatile elements leave no doubt that volatile elements were involved into the petrogenesis and mineralization of pegmatoids in the UZ. As the magmatic body consolidated and temperature reduced, fluids did not disappear completely, but continued to take part in postmagmatic alteration of the rocks. In most high-temperature, ultramafic rocks, these fluids produce magnesian skarns, which are common in pegmatoids of the UZ, and these represent a link in the evolutionary chain of a fluid–magmatic system. A striking example of this is the high-chromium magnesian skarn in the leading part of the Northeastern branch of the Upper Talnakh Intrusion (Ryabov et al. 1996a).

Chromite-magnesian and magnesian-calcareous skarns consist of semitransparent, red-brownish spinel (30–80%), clinopyroxene (15–30%), bright-green garnet (5–20%), sulfides (1–5%), olivine, anhydrite, prehnite, chlorite, and coaly material (see Rock indication 110). Rocks are granoblastic and poikilitic. In magnesian skarns, relict high-chromium pegmatoid is sometimes preserved. The rock is unique due to the unusual paragenesis of Cr-rich minerals (Ryabov et al. 1996a). Unique garnet varieties contain up to 19.4 wt% Cr_2O_3 and up to 8.9 wt% TiO_2 (with a FeO total less than 0.4 wt%). Pyroxene is also unusually chrome rich; chrome-diopside and chrome-fassaite contain 0.8–3.9 and 0.6–2.0 wt% Cr_2O_3 and 3.1–5.2 wt% and 5.9–7.5 wt% Al_2O_3, respectively. These high Cr-content pyroxenes are a rare. Chrome spinel from skarns is magnesioalumochromite and magnesiochromite. They contain 0.3–3.2 wt% TiO_2, 10.1–35.0 wt% Al_2O_3, 30.8–68.6 wt% Cr_2O_3, 0.4–13.6 wt% FeO, 0.2–1.2 wt% MnO, and 15.6–24.8 wt% MgO. A comparative analysis of the composition of minerals from magnesian skarns and magmatic rocks will be given later in this work.

Another type of magnesian skarn formed in aggregate in leucocratic gabbro is spinel–monticellite rock with titanomagnetite and perovskite (see Rock indication 111). The spinel–monticellite rock contains 4.52–4.89 wt% FeO in monticellite, 0.91–3.62 wt% TiO_2 and 0.08–0.76 wt% Cr_2O_3 in spinel, and 8.34 wt% TiO_2, 1.10 wt% MnO, and 0.54 wt% V_2O_5 in titanomagnetite.

Central Zone

This part of the sequence gives an indication of the extent of differentiation in the magmatic bodies. The CZ of the Upper Talnakh Intrusion is composed of a three-member layered series of rocks: olivine-free, olivine-bearing, and picritic gabbrodolerite; this is indicative of a fully differentiated intrusion. The rocks of the CZ of Noril'sky Complex intrusions are petrographically very similar. This petrography has been described by Godlevsky (1959), Korovyakov et al. (1963), Zolotukhin (1964), Dodin and Batuev (1971), Zolotukhin et al. (1975), Natorkhin et al. (1977), and Distler et al. (1979), so only some textural features and compositional variations of the main types of gabbrodolerite from the base to top in the intrusion sequence will be discussed here.

The *horizon of olivine-free and olivine-bearing gabbrodolerite* is found in the upper part of the layered series. Grain size increases upward, and a succession of textural varieties is observed from poikilo-ophitic through ophitic to prismatically ophitic (prismatically granular).

The upper layers in the CZ feature a coarse-crystalline prismatically granular gabbrodolerite with a pegmatoid fabric with or without quartz (see Rock indication 112). The rock consists of plagioclase (40–60% $Ab_{50}Or_2An_{48}$), K-feldspar (1–3%), clinopyroxene (15–40% $Wo_{42–40}En_{47–45}Fs_{11–15}$) (ÊÇ-184/131.0, see Appendix, Table A.14), and palagonite (10–20%); amphibole ($f = 42$ at%) is consistently present (1–3%). At the front of the intrusion, amphibole content is markedly increased and becomes the main mafic mineral in metadiorites. Other minerals include biotite, apatite, calcite, prehnite, and chlorite, as well as ore minerals such as titanomagnetite and ilmenite (3–7%) and sulfides (0.5–5%).

Palagonite forms small aggregates in interstices that hydroxyl-containing minerals and ore minerals are restricted to. Increased palagonite content in the rock serves as an important petrogenetic feature that indicates volatile enrichment of the melt (Natorkhin et al. 1977). On the basis of experimental results from the homogenization of inclusions in quartz, the formation temperature for these rocks is in the range 1,060–1,070 °C (Vortsepnev 1978).

In the leading part of the Southwestern and Northeastern intrusive branches of the Upper Talnakh Intrusion, pink coarse-grained metadiorite or gabbroic diorite is widely developed. The rocks include plagioclase (40–50% $Ab_{50-70}Or_{2-4}An_{48-26}$ to $Ab_{98-100}Or_{0-2}$), K-feldspar (3%), clinopyroxene (10–5% $Wo_{40-35}En_{46-41}Fs_{13-23}$; 50–0% $Wo_{25}En_6Fs_{69}$), amphibole ($f = 92$ at%) (3–20%), and palagonite (10–30%). Prehnite, biotite, and ilmenite are less common (see Rock indications 113 and 114). The maximum thickness (to 110 m) of these rocks is observed in the leading parts of the magmatic body. At some intersections of the intrusion, metadiorites are common instead of olivine-free, olivine-bearing, and olivine gabbrodolerite (not present), and metadiorites are separated from the horizon of picritic gabbrodolerite by a layer of olivine–biotite gabbrodolerite 2–3 m thick. The change from prismatic, coarse gabbrodolerite to metadiorite toward the leading portion of a magmatic column, as well as the increase in metadiorite thickness, alteration of early magmatic minerals, and change in alkalinity of rocks, can be wholly explained by the migration of volatile elements upward through the magmatic body and accumulation in the magmatic column head.

Olivine-free, quartz, and olivine-bearing gabbrodolerite are medium-grained rocks with ophitic, gabbro-ophitic, and poikilo-ophitic (and rarely tholeiitic) textures. The rocks are composed of plagioclase (40–65% An_{79-48}), clinopyroxene (10–40% $Wo_{41-44}En_{49-42}Fs_{10-16}$), and variable amounts of olivine (0–5% Fa_{36-18}), quartz (0 to 4–6%), titanomagnetite and ilmenite (9–15%), palagonite (3–10%), K-feldspar (0–1.5%), amphibole (0–3%), biotite, and chlorite (see Rock indication 115; Appendix, Tables A.13–A.17). In olivine-free varieties, pseudomorphs of bowlingite after olivine are common.

The *horizon of olivine gabbrodolerite* is composed of rocks with a poikilo-ophitic, ophitic, and sometimes a panidiomorphic granular texture. Gabbrodolerite consists of plagioclase (30–75% $Ab_{13-24}Or_1An_{87-76}$), clinopyroxene (10–0% $Wo_{37-3}En_{45-47}Fs_{10-7}$), olivine (5–18% Fa_{25-35}), titanomagnetite, ilmenite, and sulfides (1–10%); orthopyroxene, biotite ($f = 26$–30 and 85 at%), apatite, and sphene are present in small amounts, and secondary minerals, prehnite, serpentine, bowlingite, and chlorite are observed (see Rock indications 116 and 118). Chrome spinel is also sometimes observed (see Rock indication 119). In addition, at the base of the horizon, some geologists recognize zones enriched in biotite, described as olivine–biotite gabbrodolerite.

Ivanov et al. (1971b) regarded the fine-grained olivine gabbrodolerite with poikilo-ophitic texture at the base of the horizon as a contact formation with underlying picritic gabbrodolerite. The relationship between olivine and picritic gabbrodolerite was explored by Godlevsky (1959), who presented the concept of a gradual transition between the rocks. This supported a mechanism of crystallization–gravitational differentiation of the magmatic melt under chamber condition. However, dense sampling has revealed the relatively smooth transition between olivine and picritic gabbrodolerite, and Smirnov (1966) identified a visually distinct contact between these rocks. Later studies by Ivanova et al. (Ivanov et al. 1971b) confirmed this. These studies revealed that the thickness of the transition zone was less than 10 cm. The sharp contact between these rocks is verified not only petrographically but by chemical analyses that show an inflection in concentration curves for petrogenic and oregenic elements (Ryabov and Yakobi 1981), which will be discussed later. It should be noted, in a number of cases, that interlayering of rocks or the presence of pegmatoidal taxitic and/or troctolitic dolerite at the boundary of mafic and ultramafic parts of the CZ sequence complicates or obscures the transition zone.

The *horizon of picritic gabbrodolerite* is typical for all Noril'sky-type intrusions. Its thickness across and along the section varies, but as a whole, it is similar in size to other horizons in the intrusions (80 m for the Upper Talnakh Intrusion and 120 m for Noril'sk-I-type intrusions). It is observed in recesses of the intrusion and is traced along the intrusion axis through to the intrusive branches as shown in isopach maps and in longitudinal sections (Dodin and Batuev 1971; Tarasov 1976; Zolotukhin et al. 1975; Natorkhin et al. 1977; Likhachev 1988). The sides of the intrusion are thinned and wedge out into lenses of rocks enriched in olivine and taxite-like gabbrodolerite, which are later replaced by ophitopoikilo-ophitic gabbrodolerite and dolerite.

Small rounded bodies of leucogabbro form inclusions in picritic gabbrodolerite, which increase in concentration at base of the horizon. The inclusions increase in size downward and are not infrequently lens-like or veinlet-like. At the contact with underlying rocks, leucocratic veinlets are so common that the rock assumes a pseudobreccia fabric that downward is replaced by taxitic gabbrodolerite.

Under microscope, picritic gabbrodolerite has segregation, hypidiomorphic, panidiomorphic granular, porphyritic, poikilo-ophitic, and ophitic with granular element textures. The rock consists of olivine (25–40%), plagioclase (10–30%), clino- and orthopyroxene (10–40%), and ore minerals (7–30%). Other minerals include biotite and chrome spinel, as well as serpentine, bowlingite, chlorite, prehnite, and late magnetite. The wide variation in rock-forming minerals of the picritic horizon means the rock can be classified as a troctolite, troctolitic gabbrodolerite, and melanocratic troctolite (Dodin and Batuev 1971). Rocks of this horizon contain up to 70% olivine and more are regarded by some researchers as plagioclase peridotites, plagioclase olivinites, and olivinites and are considered to be the end products of crystallization differentiation (Sukhanova 1968; Dodin and Batuev 1971). However, based on available data, these rocks cannot be considered as normal differentiates of a magmatic melt, as volatile elements are of primary importance in their formation. Characteristics of these rocks will be described in detail in the "Magmatogenic Breccia of Kharaelakh" section.

Quantitative mineralogical studies of the sequence of picrites in the Noril'sk-I Intrusion carried out by Sheshukova in 1959 have shown the existence of four accumulation maxima for olivine and sulfides, and Godlevsky has recognized six such maxima in the south of the deposit (Urvantsev 1959). The accumulation maxima for sulfides are always shifted with respect to olivine maxima; this reflects differences in the specific weight of gravitationally deposited phases. Smirnov (1966) noted that in the picrite horizon, there is an alternation of layers with different olivine and plagioclase content with relatively invariable pyroxene. Of six layers recognized by Smirnov, the upper layer and the second from the base are mostly enriched in olivine, and the most rich sulfide mineralization is observed in layers enriched in feldspars. The layer most increased in mineralization is restricted to the lower part of the picrite horizon at some distance from the base.

In the picritic horizon sequence of the Upper Talnakh Intrusion, an irregular distribution of sulfides is also observed, which is visible sometimes in a repeated alternation of layers with disseminated sulfide ores as large droplets (from 0.6–1.2 cm to 1.0–2.0 cm) and small droplets (from 0.2–0.4 cm to 0.6–0.8 cm). Disseminated ores near the bottom parts of intrusions, including picritic gabbrodolerite, have a zonal distribution of various types of sulfide ores (Genkin et al. 1981).

Olivine forms idiomorphic crystals and more rarely palmate and elongate grains. The size of grains may change by one order of magnitude within one thin section. In olivine crystals, idiomorphic and skeletal shapes are not infrequently combined. The perfect cleavage and solid solution breakdown structures are common as "dash-dot" intergrowths of an oxide-ore phase in the plane 010 or skeletal crystals in the plane 100. Olivine crystals have a preferred orientation, which is used by geologists to determine melt movement direction (Zolotukhin 1964) and to differentiate between turbulent or laminar flow (Likhachev 1996). These investigations assume that injection of olivine-rich melt is responsible for the orientation of olivine crystals. This is in contradiction with earlier assumptions on gravitational in situ separation of olivine crystals in chambers, for which rates of accumulations of these crystals have been calculated using the Stokes formula. Stepanov (1981) assigned a leading part in initiation of stratification to processes of rock fractionation under action of gravitational forces and considers that "undoubtedly, orientation of olivine grains is the result of a force vector impact during crystallization of olivine."

Olivine composition varies widely from Fa_{17} to Fa_{40} (Ryabov 1992a), and mineralogical studies have shown the presence of submicroscopic inclusions of Mg spinel in olivine (Stepanov 1977). Olivine is typically strongly altered, with sheet-like antigorite and magnetite in cracks. Often, only the outer parts of olivine are preserved; pseudomorphs of antigorite and bowlingite after olivine are common.

Both monoclinic and rhombic pyroxene are present. Clinopyroxene is usually zoned: green, chromous varieties are at the center and titanous, brownish varieties at the periphery. The general composition of the mineral is $Wo_{37-45}En_{46-50}Fs_{8-13}$. Orthopyroxene ($Wo_{2-4}En_{75-77}Fs_{19-22}$) forms individual crystals and overgrowth rims on olivine. The Fe content of orthopyroxene is, as a rule, higher than that of clinopyroxene. Plagioclase forms wide tabular grains ($Ab_{15-42}Or_{0-1}An_{85-58}$) with poikilo-ophitic intergrowths of olivine and pyroxenes and as prismatic crystals ($Ab_{48-64}Or_{1-5}An_{51-31}$). Other minerals include ilmenite, chrome spinel or magnetite, green spinel, fluorine–biotite ($f = 24.4$–28.6 at%), and apatite enriched in F and Cl (see Rock indications 120–126).

An important genetic feature of picritic gabbrodolerite is the presence of chrome spinel. This can serve as an indicator for rocks that belong to picrite-basalt melts of the Lower Fokinsky and Picrititovy Intrusions, as well as to picritic basalt lava of the Gudchikhinsky Suite. However, this mineral is extremely unevenly distributed in horizons of picritic gabbrodolerite of the Noril'sky-type intrusions. Increased concentrations of chromite up to 25–30% are restricted to the roof of the picrite horizon, whereas its base is completely devoid of chromite, and gabbrodolerites are also depleted in chromium (Ryabov and Yakobi 1981).

Two forms of chromite are distinguished: form I is represented by disseminated grains that are distributed relatively evenly in picritic gabbrodolerite (see Rock indication 127), and form 2 is represented by aggregates with globular textures. These types of disseminated chromite have been described by Genkin et al. (1979). They indicate that aggregates of chrome spinel are restricted to basal parts of the picrite horizon that, seemingly, shows agreement with the fact that chromite-rich and olivine rocks are situated at the bottom of layered intrusions (by analogy with classical examples known from the literature; Wager and Brown 1968). However, Kavardin and Mitenkov (1971) and Shatkov (Zolotukhin et al. 1975) noted that chromite accumulations are situated not at the bottom, but at the top of the picrite horizon. Investigations on field data show that

Fig. 3.10 Forms of sulfide disseminations in picritic gabbrodolerite

both aggregates and disseminated chrome spinel are restricted to the upper part of the picrite horizon, and no chrome spinel are found at the lower part of the horizon (Ryabov and Yakobi 1981).

Chromite forms oval, isometric, or elongate aggregates or system of aggregates that range in size from 1 cm to 1 m; in the aggregates, chromite reaches 50–60% (Genkin et al. 1979; Fenogenov 1987a, b). Chromite inclusions are observed in pyroxene and plagioclase, sometimes in olivine, but are not found in sulfides. The presence of inclusions is independent of structure and silicate arrangement so that plagioclase or clinopyroxene may be dominated by chromite inclusions in one area, but another almost identical area may be completely devoid of chromite. Chromite occurrence is invariably associated with globular formations of minerals such as prehnite, biotite, anhydrite, carbonate, zeolites, apatite, chlorite, and talc-containing volatiles (Genkin et al. 1979). These globules are usually found in the center of chromite masses. More information on the characteristics of chromite occurrence and compositions is given in the mineralogy section.

Disseminated sulfides are a characteristic feature of Noril'sky-type picritic gabbro. This dissemination may be responsible for compositional features of these ore-bearing intrusions (Ryabov 1989a, 1992a). Sulfides are observed as drop-like and interstitial disseminations and constitute 5 to 15% of the rocks, and intergrowths of olivine with plagioclase (and more rarely clinopyroxene) are set in continuous sulfide masses. Principal minerals include pyrrhotite, chalcopyrite, and pentlandite, as well as cubanite, troilite, talnakhite, and mooihoekite. Small drop-like inclusions often have a rounded form; larger ones also show flattened, ellipsoidal, and more rarely tube-like forms with subhorizontally elongated inclusions (Fig. 3.10). The drop-like disseminations are not infrequently stratified into chalcopyrite (upper) and pyrrhotite (lower) parts. In some cases, chalcopyrite surrounds pyrrhotite with a discontinuous rim or forms intergrowths with pyrrhotite. Above drop-like disseminations so-called reaction caps formed consisting of minerals containing volatile elements: biotite, serpentine, amphibole, anhydrite, calcite, and apatite. Godlevsky (1961) has attached important genetic significance to these drop-like sulfide minerals in picritic gabbrodolerite due to their immiscible origin. At the same time, "reaction caps" with hydrous minerals above drop-like sulfides have been considered as evidence in favor of a metasomatic (Zolotukhin 1964) or hydrothermal (Eliseev 1959) origin for disseminated sulfides or sulfide ores in general.

Characteristics of mineralization in picritic gabbrodolerite and ideas about the nature of the picritic gabbrodolerite and related ore minerals are presented later in this work.

Lower Zone

Rocks lying between picritic gabbrodolerite and the lower contact of the intrusion constitute this zone. Pegmatoid of various compositions formed along the contact with gabbrodolerite; here, the principal rock type is taxitic gabbrodolerite; troctolitic gabbrodolerite, troctolite, olivine gabbro, leucogabbro, and magmatogenic breccia are less developed. All these diverse rocks are regarded as "lower taxitic gabbrodolerite" as described by Godlevsky (1959), Zolotukhin (1964, 1997), Dodin and Batuev (1971), Ivanov et al. (1971a, b), Zolotukhin et al. (1975), Tarasov (1976), and Natorkhin et al. (1977).

Unaltered gabbrodolerite at the contact ("contact gabbrodolerite") is extremely rare since these rocks are often taxitized or metasomatized. The thickness of gabbrodolerite ranges from 0.6 to 5.8 m (2 m average) and rarely up to 10–14 m (commonly at the intrusion flanks).

The contact gabbrodolerite is a medium- and fine-grained rock with the poikilo-ophitic, ophitic, and doleritic texture. The rock contains plagioclase (45–55%), clinopyroxene (35–40%), titanomagnetite, and ilmenite (3–7%); olivine (Fa_{44}) or pseudomorphs of bowlingite are rare.

In the *contact gabbrodolerite*, at the boundary with taxitic gabbrodolerite, complex granular textures appear that are composed of clinopyroxene and plagioclase, or more rarely of olivine and plagioclase, as well as veinlets of coarse-grained plagioclase. Compact sulfides beneath the contact gabbrodolerite and related volatile elements actively affected magmatic rocks producing an aureole of wall-rock alteration, apodoleritic ore and barren metasomatites of biotite–feldspar composition, and pyroxenite, serpentinite, and anhydrite–pyroxene composition (Yudina 1965; Dodin and Batuev 1971; Zolotukhin et al. 1975; Turovtsev 1972).

Lower taxitic gabbrodolerite constitutes a uniformly thick rock horizon. In the leading parts of intrusive branches and in trough-shaped depressions at the intrusion base, the thickness of taxitic gabbrodolerite increases, and at flanks and in pinches between intrusive branches, the thickness decreases until the horizon completely wedges out. At the sides of the intrusion, both upper and lower taxitic gabbrodolerites merge into a single layer; this evidences their genetic relationship (Zolotukhin 1964; Vasil'ev 1966; Ryabov 1969). At the periphery of the intrusion, gabbro in sill-like apophyses sometimes constitutes the bulk of the body. Apophyses of leucogabbro and taxitic gabbrodolerite extend outward from the Noril'sk-I, Noril'sk-II, and Upper Talnakh Intrusions (Likhachev 1965; Ivanov et al. 1971a, b; Tarasov 1976).

Horizons of lower taxitic gabbrodolerite in different intrusions have a similar internal structure. The most complete sequence typically consists of picritic gabbrodolerite with leucocratic veinlets, magmatogenic breccia, melanocratic and taxitic gabbrodolerite with relict textures of olivine-rich rock, taxitic gabbrodolerite, olivine and leucocratic gabbro, leucocratic taxitic gabbrodolerite, and contact taxite-like gabbrodolerites (see Fig. 3.7). The presence of volatiles associated with the horizon of lower taxitic gabbrodolerite is responsible for the absence of chromites in lower parts of the picritic gabbrodolerite, which caused resorption of chromites and the outflow of chromium to the roof of the picritic horizon. When volatiles were present, recrystallization of picritic gabbrodolerite adjacent to taxites of picritic gabbrodolerite proceeded. Areas with granular olivine, aggregates and blind veinlets of coarse-grained plagioclase, and pyroxene–augite intergrowths formed in the picritic gabbrodolerite along with schlieren and veinlets of pegmatoidal gabbro responsible for then breccia-like fabric of picritic gabbrodolerite. In angular fragments of picritic gabbrodolerite, drop-like sulfide impregnations are often preserved. With increasing leucocratic cement, the amount and size of these fragments decrease. Downward from the base of the picritic horizon, an increase in leucocratic components continues; in the taxitic gabbrodolerite proper, fragments disappear leaving shadow relicts as fine-grained areas of rock. Downward through the sequence, a decrease in the mafic constituent of taxitic gabbrodolerite is observed, and the uneven-grained texture grades into even-grained pegmatoid—leucocratic gabbro is formed. Further toward the lower contact, the texture again becomes ataxitic, but the principal dark-colored mineral in fine-grained areas is usually clinopyroxene instead of olivine.

Zolotukhin et al. (1975) gives the most detailed description of lower taxitic gabbrodolerite of the Upper Talnakh Intrusion. Plagioclase (30–60%) forms wide tabular and prismatic grains with distinct zoning. The mineral composition varies from An_{100-70} to An_{60-35}–An_{30-25}. Clinopyroxene forms large crystals with sparse intergrowths of plagioclase and olivine. At the boundary with sulfides, clinopyroxene turns bright green.

Olivine is observed as small granules, subidiomorphic crystals, and palmate grains. The size of olivine crystals ranges from 0.02 to 2 mm or more. Small granules are scattered inside other minerals and form granoblastic aggregates elongated as narrow bands, chains of grains, and small areas of fine-grained olivine–plagioclase intergrowths.

Different theories about the crystal form and occurrence of olivine in taxitic gabbrodolerite have been proposed. Zolotukhin (1964) considers olivine granules as the result of the breakdown of idiomorphic olivine crystals from picritic gabbrodolerite and attributed palmate crystals to collective recrystallization of granules. He also considers the granules have higher Fe content than idiomorphic crystals and are in equilibrium with sulfides. Goryainov (1969) believes that the granule-like form of olivine is a characteristic feature of metasomatism. Lately, the leading role in the formation of granules is thought to be played by fluids that caused the recrystallization of early olivines, as well as promoted the redistribution of magnesium that formed the nuclei for olivine crystals from a magnesia-rich magma when the temperature fell (Ryabov 1989a, 1992a).

The composition of olivines in taxitic gabbrodolerite is Fa_{23} to Fa_{40}. In olivine-poor taxitic gabbrodolerite, small granules are characterized by higher Fe content as compared to subidiomorphic crystals, and most ferriferous are palmate olivines (see Rock indication 128). In olivine-rich taxitic gabbrodolerite, the composition of granules is magnesian chrisolite or forsterite.

The characteristic feature of lower taxitic gabbrodolerite is the absence of chrome spinel. Titanomagnetite or, more rarely, green spinel is the only spinel group minerals in the rock. Ilmenite is observed too. Genkin et al. (1979) believe that the absence of chromite in the lower taxitic gabbrodolerite is explained by their low fO_2 value, low melt temperature, and its leucocratic composition. However, these parameters are also valid for compositionally similar upper taxites, which contain rocks such as chromite anorthosite or plagioclase chromite (Ryabov 1984a, b, 1989a);

Fig 3.11 Magmatogenic breccia of the Upper Talnakh Intrusion. Fragments are appreciably olivine rocks with small amount of plagioclase, clinopyroxene, and green spinel; cement is pegmatoid gabbrodolerite. (**a**) UZ, samples KZ-1513/1486; (**b–d**) magmatogenic breccia of Kharaelakh, samples KZ-593/609.5, KZ-593/685, and KZ-862/855, respectively. Natural size

therefore, another explanation is needed. The rocks also contain biotite, serpentine, chlorite, and, more rarely, amphibole and anhydrite. Lower taxitic gabbrodolerite forms the horizon of Noril'sky-type intrusive rocks most enriched in sulfides. Based on this, Sobolev (1986) regards them as ore (sulfide) pegmatites. Leucocratic gabbro from the central part of this horizon is the most enriched in sulfides in the lower taxitic gabbrodolerite (Tarasov 1976).

Magmatogenic Breccia of Kharaelakh

Magmatogenic breccia is a rock that features fine-grained areas with distinct fragment-like forms, as opposed to taxitic gabbrodolerite that has fine-grained areas as blurred patches (Fig. 3.11). The fragment-like forms are melanocratic and are enclosed by medium- and coarse-grained leucocratic cement (see Rock indication 129). These breccia-like rocks and sparse fragments within taxitic gabbrodolerite are recorded from the Noril'sk-I, Noril'sk-II, and Talnakh Intrusions (Godlevsky 1959; Korovyakov et al. 1963; Zolotukhin 1964; Smirnov 1966; Ryabov 1969; Ivanov et al. 1971b; Lyul'ko et al. 1972; Tarasov 1976; Mitenkov et al. 1977; and Zotov and Krivoplyasov 1977). Rocks with fragments like these appear to have been described as ultramafic xenoliths by Vorontsov, Zontov, and Korovyakov in 1944 (Ivanov et al. 1971b). There is no consensus among geologists to the origin of the fragments. Some geologists regard them as hornfelsed picrites; others treat them as xenoliths of altered sedimentary rocks or hornfelsed and relict apodolomite or apopicritic skarns and so forth. In most cases, the composition of these breccia-like rocks and strongly altered leucogabbro fragments in taxites was not studied until the present time. The mineralogy of these rocks is not yet totally understood, and petrochemistry is represented only by three analyses of "outliers"—xenoliths in taxite from the Noril'sk-I Intrusion (Tarasov 1976).

Earlier, it has been shown that magmatogenic breccia and pseudobreccia are peculiar varieties of taxitic gabbrodolerite in which fine-grained areas of rock still retain their identity as fragments (Ryabov 1969). Where the fine-grained areas have been reincorporated into the rock, relict fine-grained areas in coarse-grained cement are preserved and are responsible for the ataxitic texture of the rock. Because the magmatogenic breccia is always associated with taxitic

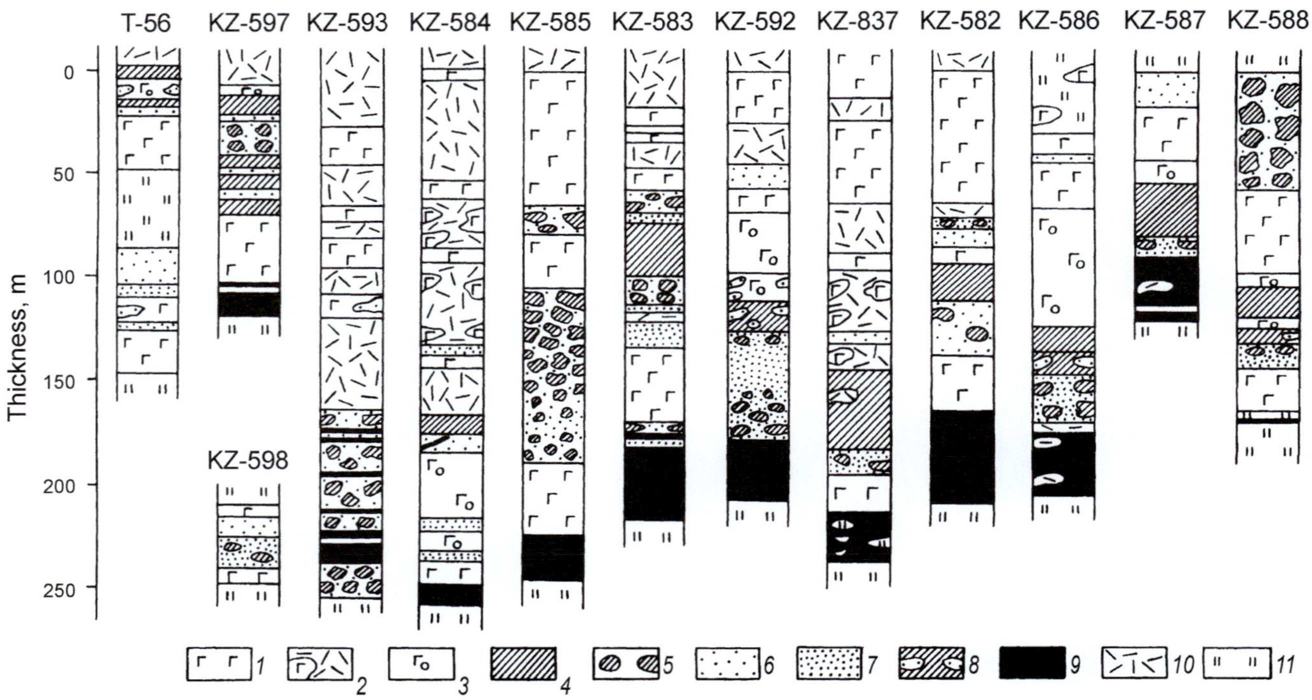

Fig. 3.12 Structure of magmatogenic breccia horizon of the Kharaelakhsky branch of the Upper Talnakh Intrusion (from drilling data of the Upper Talnakh Intrusion). *1* olivine-free gabbro, *2* metasomatite with relict gabbro, *3* olivine gabbro, *4* olivine-rich rock, *5* spinel–olivine rock and pyroxenite, *6* leucogabbro, *7* taxitic gabbro, *8* inclusions of leucogabbro in olivine-rich gabbro, *9* sulfide veins, *10* magnesian-calciferous metasomatites, *11* apoargillaceous metosomatites

horizons and there is a gradual transition from typical pseudobreccia to breccia-like rock with blurred outlines of fragments and finally to taxitic gabbrodolerite, it is reasonable to regard the magmatogenic breccia as a variety of taxite (Ryabov 1969).

The magmatogenic breccia is most pronounced in the Kharaelakhsky intrusive branch (see Fig. 3.8). Here, the magmatogenic breccia gently crosscuts a layered series of rocks in the CZ, gradually elevating from near the base to the leading part of the intrusion. In this direction (right to left in Fig. 3.8), the thickness of the LZ horizon increases, and typical gabbrodolerite grades into magmatogenic breccia. Compact sulfide ores at the base of the intrusion base continue into taxitic gabbrodolerite and then into magmatogenic breccia. Within the intrusion, magmatogenic breccia is observed instead of several differentiates of the CZ, which are absent here. On the left in the cross section, gabbrodolerite corresponding to different layered series members of the CZ is located between divergent vein-like tongues of magmatogenic breccia. The horizon of breccia is overlain and underlain by olivine-free gabbrodolerite. In the sequence, magmatogenic breccia in the UZ and LZ of the Upper Talnakh Intrusion described above is also shown. “Floating bodies” of leucogabbro are not to scale.

Figure 3.12 presents the position of magmatic breccia within the sequence in the leading part of the Kharaelakhsky intrusive branch, as determined from drill holes. The magmatogenic breccia is situated at the roof of the intrusion (drill holes T-56, KZ-597, 588), in the central area (drill holes KZ-585, 583, 582, 588), and near the base (drill holes KZ–593, 592, 837, 586, 587). Within the intrusion, the breccia may be repeated (drill holes KZ-585, 583, 582, 588), the fragment size and quantity in the rock decrease downward through the section (drill hole KZ-585), the horizon of taxitic gabbrodolerite is overlain and underlain by magmatogenic breccia (drill hole KZ-592), and lastly magmatogenic breccia is accompanied by a metasomatic breccia in a number of places. Judging from the drill hole, the volume of magmatogenic breccia often constitutes more than 50% of the intrusion sequence and in some cases reaches 70%. In drill holes T-56 and KZ-597, picritic gabbrodolerite, which is not observed lower in the section, is found at the roof of the intrusion, and in drill hole KZ-583, the picritic horizon is located in the middle of the sequence. These cases appeared to be what was described by geologists who wrote of inverted (reverse) differentiation and picritic horizons of the Kharaelakh intrusive branch (Sukhanova 1968; Natorkhin et al. 1977). In the Southwestern and

Northeastern intrusive branches, where ore-bearing rocks occupy only a limited part of the intrusion near the base, in the leading part of the Kharaelakhsky branch and partly in the Northwestern one, disseminated ore mineralization encompasses a significant part (up to 150 m) of the intrusion, and compact sulfide ores reach 30–45 m in thickness.

Sulfide mineralization is distributed in the magmatogenic breccia such that the cement is always richer in sulfides than fragments. The concentration of sulfides in taxitic gabbrodolerite may reach values similar to those in sulfide-rich ores. Fragments of magmatogenic breccia are poor in sulfide (less than 3%) and depleted in ore material.

Fragments of magmatogenic breccia are acute angled, range from centimeters to decimeters in size, and large blocks vary from 0.4–0.8 m up to 1.0–1.5 m and more. Proportions of fragments and cement vary widely, so that all transitions of pseudobreccia in which gabbrodolerite are intersected by numerous leucocratic veinlets to breccia when small fragments are set in the leucocratic cement can be traced. The alteration extent of fragments due to which these differ from gabbrodolerite increases with decreasing their sizes. The content of fragments in breccia varies in the wide range; for example, in drill hole KZ-585 (see Fig. 3.12), it reaches 85–90% of the rock bulk in the upper part of the horizon, and in the lower part, it does not exceed 30–50%.

In zones of reticular vein leucocratization of picritic gabbrodolerite, one can see initial stages of fragment alteration and a change in the cement composition. In the central parts of large fragments, the primary texture of gabbrodolerites is preserved with drop-like and interstitial sulfide inclusions, and in edge parts, virtually the continuous band of fine-grained olivine or its elongate crystals are observed transversely to the contact. Outward the contact of a fragment in most representative sections in a leucocratic veinlet, successive change of a pyroxenite zone into pyroxene–plagioclase and, in the central part of the veinlets, massive sulfides are observed. In places, spinel–olivine coronites appear at the fragment boundary. Small fragments are composed of granulated olivine and often are rimmed by plagioclase (see Rock indication 129), which change into pyroxene–plagioclase intergrowths. Characteristic feature of the magmatogenic breccia is a contrast between compositions of melanocratic fragments and leucocratic cement, the contrast being obliterated in taxitic gabbrodolerite.

In the Kharaelakhsky and Northwestern intrusive branches, metasomatites of various compositions are widely developed along with large volume of pegmatoids. The texture of original rock is strongly smoothed due to intense alteration of substrate with telescoping phenomena. This has generated different interpretations of the nature of metasomatites and metasomating solutions (Yudina 1965; Zolotukhin and Ryabov 1972; Zotov and Krivoplyasov 1977; Zotov 1979). Nevertheless, the available data allow one to consider that the main mass of metasomatites within outlines of the magmatic body is apointrusive, and their formation is associated with fluids separated from a melt, accumulated at the leading part of the magmatic column and developed with decreasing the temperature.

Petrography of Fragments from Magmatogenic Breccia

The diversity of breccia fragment composition is governed by variations of principal rock-forming minerals such as olivine, plagioclase, and spinel. The characteristic feature of magmatogenic breccia is the widespread occurrence of olivine-rich (up to olivinite) rock fragments and significant fluctuations in F content in olivine, as well as the presence of green spinel and fassaite clinopyroxene. The rock fragments are formed from several principal varieties.

Spinel–forsterite rock has coarse-grained granoblastic and poikilitic texture. Olivine constitutes 70–85% of the rock and includes colorless phlogopite, perovskite, anhydrite, and calcite (3–5% overall), as well as magnesioferrite (12.63% MgO), pyrrhotite (1–3%), brucite, and serpentine. Olivine (Fa_{0-2}) is observed as large (0.5–6 mm) elongate and isometric crystals with uneven rims and distinct cleavage. Spinel forms idiomorphic crystals and isometric grains between crystals of olivine and is observed as inclusions in olivine. The mineral is Al spinel with variable contents of Fe and Mg and has a green, brownish, and honey-yellow color. Magnesioferrite forms aggregates of small idiomorphic crystals intergrown with spinel (see Rock indication 130). The rocks are often strongly altered to brucite and serpentine.

Spinel–olivine rock shows both granoblastic and poikilitic textures. The rock differs from spinel–forsterite rock by the angular form of small isometric olivine crystals (70–80%) with a higher Fe content (Fa_{10-12}) and cleavage. Olivine grains are surrounded by serpentine that forms continuous loop-shaped aggregates. The amount of serpentine varies widely, and sometimes the rock is serpentinized completely. Green spinel (5–10%) as subidiomorphic crystals is observed between olivine grains and sometimes as submicroscopic inclusions in olivine. Spinel is rimmed by magnetite overgrowths. In addition, there are large single biotite leaflets with intergrown olivine and spinel and spinel with ilmenite prisms (see Rock indication 131).

Olivine rock is composed of heterogranoblastic, granoblastic, or granular aggregates of olivine (80–90%) that form the fabric of the rock. Other minerals include abundant serpentine and magnetite, and olivine is often completely replaced by serpentinite with relict granular olivine structures and textures. Lesser amounts of plagioclase, clinopyroxene, biotite, and sulfides are also found in the rock. Large olivine crystals are Fa_{12-14} (see Rock indication 132), and granules are Fa_{1} (see Appendix, Table A.25); these granules contain submicroscopic globular inclusions of ore minerals. At the boundary between olivinite and pegmatoid, symplectic intergrowths of forsterite and green spinel are observed.

Spinel–olivine–clinopyroxene rock is a granoblastic aggregate of polygonal clinopyroxene (50–65%), olivine (20–30%), and green spinel (from 5–7% to 12%). Olivine (Fa_{9-10}) forms accumulations of grains surrounded by loop-shaped spinel–clinopyroxene aggregate. Clinopyroxene is augite ($Wo_{29}En_{57}Fs_{14}$) and diopside ($Wo_{51}En_{42}Fs_{7}$) (see Rock indications 133 and 134). Spinel forms single idiomorphic crystals, intergrowths with fassaite, and small globules enclosed by olivine and clinopyroxene. Plagioclase (An_{100}) and ore minerals are rarer, not exceeding 5–10%.

Spinel–olivine–plagioclase rock has heterogranoblastic, granoblastic, and poikilitic textures. Principal minerals are plagioclase (30–60%), olivine Fa_{16-25} (40–70%), and green spinel (5–20%). Clinopyroxene, biotite, serpentine, magnetite, and sulfides are also observed in small amounts. Under microscope, there is a trend for decreasing spinel associated with increasing plagioclase. Olivine forms small isometric grains (with varying amounts of green spinel globules) or large subidiomorphic or palmate crystals (see Rock indications 135–137, Appendix, Table A.25). Plagioclase forms small crystals commensurable with olivine or forms large prisms with intergrown olivine granules or green spinel. Olivine granules also enclose spinel globules. Clinopyroxene forms small grains; poikilocrysts are rare.

Plagioclase–clinopyroxene–olivine rock is a granoblastic aggregate of polygonal olivine Fa_{20-21} grains that are observed as loops around small plagioclase and/or clinopyroxene clumps (see Rock indication 138). The loop-shaped nature of the olivine framework around plagioclase in fragments of magmatogenic breccia was noted by Lyul'ko et al. (1972). Olivine in these rocks is not infrequently replaced by bowlingite surrounded by isometric clinopyroxene grains.

Pyroxenites represent the most widely occurring type of rock fragment. The rock has a granoblastic and heterogranoblastic, hornfelsed, poikilitic, micro-ophitic texture and a homogenous, streaky, ataxitic fabric. The principal mineral is clinopyroxene, with small variable amounts of plagioclase, olivine, spinel, anhydrite, calcite, mica, and sulfides, along with bowlingite, chlorite, and prehnite. Elevated concentrations of one or another mineral give a ground for additional subdivision of the pyroxenite. One unusual type of pyroxenite is a heteroblastic aggregate of large augite ($Wo_{47}En_{43}Fs_{10}$) crystals with intergrowths of rounded plagioclase ($Ab_{10}An_{90}$) and pyroxene granules with fine granoblastic plagioclase aggregate (see Rock indication 139).

The petrographic characteristic of pyroxenites is reported in detail by Yudina (1965), Dodin and Batuev (1971), and Zolotukhin et al. (1975). Among pyroxenites, different genetic types are distinguished, but the absence of striking criteria to distinguish them creates difficulties in identification and formation environment.

Petrography of Magmatogenic Breccia Cement

The breccia cement is composed of taxitic gabbrodolerite and leucocratic, olivine, or pegmatoidal gabbro (see Rock indication 140).

Pegmatoidal gabbro (gabbropegmatite) has a coarse-grained ophitic, pegmatitic, poikilitic, and graphic texture. Principal minerals are plagioclase (60–70%) and clinopyroxene (10–30%), as well as olivine (0–5%), titanomagnetite (3–7%), and sulfides (3–15%), with lesser biotite and amphibole, as well as serpentine, bowlingite, and chlorite (see Rock indication 130). Gabbropegmatites are primarily developed in zones of veinlet-network leucocratinization. Where the intrusion thins out, pegmatoid veinlets often change into monomineral clinopyroxene or plagioclase veinlets.

Leucocratic and olivine gabbro have a coarse-grained gabbroic texture. The rock fabric is defined by short prismatic crystals of plagioclase (50–65%) and pyroxene (15–30%). Other minerals include large palmate olivine crystals (0–7%), amphibole, biotite, titanomagnetite (3–5%), spinel (less than 2%), serpentine, bowlingite, chlorite, and prehnite. The gabbro is characterized by an accumulation of skeletal crystals of green spinel. Likhachev (1965) and Stepanov (1977) pointed out that green spinel in leucogabbro plays an important role in determining rock petrogenesis.

Taxitic gabbrodolerite features coarse-grained areas composed of plagioclase, clinopyroxene, and olivine, and fine-grained areas composed of associations of minerals found in fragments of magmatogenic breccia. Olivines in taxitic gabbrodolerite are observed as large palmate crystals (Fa_{20-30}), clumps and narrow bands of small granules (Fa_{11}), and elongate cord-like aggregates of grains of various shapes and sizes (see Rock indication 138). Clinopyroxene ($Wo_{43}En_{49}Fs_{8}$) forms poikilocrysts with intergrowths of olivine and plagioclase. In the center of prismatic plagioclase crystals, aggregates of small green spinel grains are preserved.

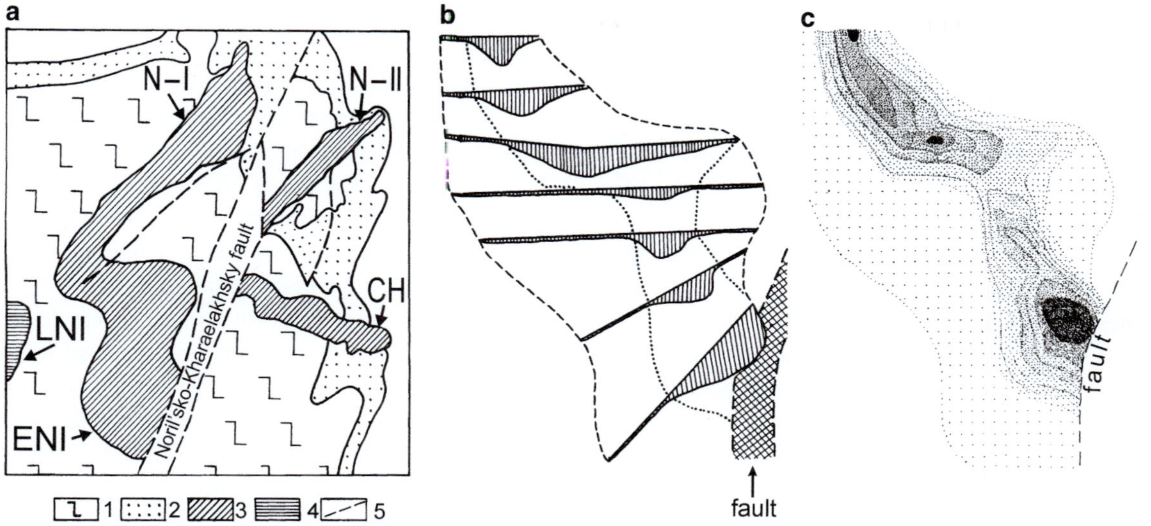

Fig. 3.13 Noril'sky ore junction. (**a**) Schematic geological map of the Noril'sky ore junction where locations and morphology of intrusions are shown: *N-I* Noril'sk-I, main body; *N-II* Noril'sk-II; *Ch* Mt. Chernaya Intrusion; *ENI* east Noril'sk intrusive branch of the Noril'sk-I Intrusion; *LNI* Lower Noril'sk Intrusion. *1* volcanic rock, *2* terrigenous rock, *3*, *4* ore-bearing and barren intrusions, *5* tectonic dislocations. (**b**), (**c**) morphology of east Noril'sk intrusive branch and isopach map (specks). From drilling data of NKGRE

3.6.1.2 Noril'sk-I Intrusion, Eastern Noril'sk Intrusive Branch

This part of the intrusion is located in the southern continuation of the Noril'sk-I Intrusion (Igrevskaya 1981; Distler and Kunilov 1994) and has a distinct stratified structure similar to other intrusions of the Noril'sky ore junction (Mt. Chernaya, Noril'sk-II) and fully differentiated Noril'sky-type intrusions.

In section, the magmatic body has a trough-like form with thin subbedded apophyses at the flanks (Fig. 3.13). The intrusion reaches 320 m thick, and the UZ, CZ, and LZ are all present. The petrography of rocks that resembles other Noril'sky-type intrusions is described in detail in Godlevsky (1959), Rogover (1959), Korovyakov et al. (1963), Zolotukhin (1964), Smirnov (1966), Zolotukhin and Vasil'ev (1967), Ivanov et al. (1971b), and Tarasov (1976). Hence, only the principal features, internal structure, and composition of rock of this intrusion will be described here.

Results of the chemical analysis of rocks from the reference cross section of the Eastern Noril'sk branch of the Noril'sk-I Intrusion and microprobe analyses of minerals from these rocks are given in the Appendix, Tables A.27–A.30.

Upper Zone

The upper zone is composed of contact gabbrodolerite and compositionally variable rocks that blend into the horizon of lower taxites. The contact facies is an olivine-bearing gabbrodolerite with poikilo-ophitic texture. Other rocks of the UZ include ophitic-poikilo-ophitic gabbrodolerite with variable olivine content, as well as picritic and taxitic gabbrodolerite, leucogabbro, pegmatoidal and palagonite gabbro, and magmatogenic breccia.

Upper picrites from intrusions of the Noril'sk ore junction have been described in Ivanov et al. (1971b) and Tarasov (1976). Picrites, as a rule, are associated with leuco and taxitic gabbrodolerite. The network of thin pegmatoid veinlets in picrite produces a pseudobreccia fabric, and with increasing pegmatoid bulk, the rock fabric becomes pseudobreccia-like (as in the Upper Talnakh Intrusion). Together with subidiomorphic crystals of olivine, small granules are observed, as well as palmate and dendrite crystals (1.5–2.0 cm in length). The composition of olivine is Fa_{20-27}; subidiomorphic crystals often have the dash-dot structures due to solid solution breakdown. Olivine is replaced by brownish bowlingite and a colorless chlorite-like mineral. Olivine grains among dense disseminated chrome spinel or pseudomorphs of bowlingite after chrome spinel are, as a rule, depleted in ore inclusions. Upper picritic gabbrodolerite of the Noril'sk-I and Upper Talnakh Intrusions shows a strong similarity to each other in mineralogical characteristics, texture, and composition (see Rock indication 141).

The amount of chromite in rocks of the Eastern Noril'sk branch ranges from 0 to 40%; elevated chromite content occurs not only in olivine-rich rocks but also in leucocratic gabbro devoid of olivine (see Rock indication 142). Chromite is unevenly distributed, forming oval and slightly elongate very fine-grained accumulations that commonly do not change the rock structure. In chromite-rich rocks, palagonite globules 0.1–0.3 cm and less in size are not uncommon. Biotite leafs, apatite, and amphibole are observed, together with cross-like and skeletal titanomagnetite crystals, clinopyroxene and box-like crystals, and laths of plagioclase. No chromite is observed in palagonite globules.

Inward-growing crystals of plagioclase and clinopyroxene at the boundaries of globules are depleted in chromite intergrowths but are otherwise rich in chromite. Chromite in gabbro is often restricted to palagonite interstices and concentrated near pseudomorphs of bowlingite after olivine.

Central Zone

The central zone consists of olivine-free, olivine-bearing, troctolitic, and picritic gabbrodolerite. In the olivine gabbrodolerite horizon, layering of gabbrodolerite lenses of uneven thickness with beds containing variable olivine content is observed: troctolitic and olivine-bearing gabbrodolerite are found in olivine dolerite, and olivine gabbrodolerite is found in troctolite and olivine-bearing gabbrodolerite (Igrevskaya 1981). The picritic gabbrodolerite horizon occupies a suspended position in the intrusion so that it overlies the thick horizon of olivine-bearing and olivine-free gabbrodolerite that overlies the horizon of lower taxites.

Some characteristic changes in rock textures up the sequence of the CZ may be viewed in a series of thin section microphotographs given in Rock indications 143–147.

Prismatically granular gabbrodolerites, metadiorites, and gabbroic diorites are characterized by a coarse-grained texture, increased palagonite in the matrix (7–12%), and variable quartz (0–5%) and amphibole (0–4%) content (see Rock indication 143). In addition, titanomagnetite, ilmenite, sphene, sulfides, and tourmaline are present in these rocks. Gabbrodiorite is most abundant at the roof of the intrusion, and the proportion of olivine-free rocks in the sequence increases southward, reaching 131 m thick.

Olivine-free and olivine-bearing gabbrodolerites have ophitic, poikilo-ophitic, and prismatically granular textures. The principal minerals include plagioclase (50–60%), clinopyroxene (25–30%), olivine (3–5%), and titanomagnetite (3–7%). Quartz, amphibole, palagonite, apatite, tourmaline, and sulfides are present, and serpentine, talc, bowlingite, chlorite, and calcite are also observed.

Olivine is found in virtually all the rock varieties, including quartz-bearing rocks. In the upper parts of the sequence, rare olivine grains are commonly replaced by bowlingite or serpentine–talc–magnetite aggregates. Unaltered large isometric crystals (0.3–0.5 cm) are found in olivine-bearing gabbrodolerite (see Rock indication 144). Quartz-bearing gabbrodolerite (with up to 10% quartz and/or micropegmatite) and gabbrodolerite enriched in titanomagnetite (up to 15%) are also olivine bearing. In the latter case, the titanomagnetite is sometimes a dark-color phase and forms large (to 0.6 cm) poikilocrysts with intergrown plagioclase.

Olivine gabbrodolerite has poikilo-ophitic and ophitic texture (see Rock indication 145). Olivine (7–12%) forms large isometric and elongate crystals 0.3–0.6 cm in size and composition Fa_{30-34} (rarely Fa_{25}). Plagioclase is observed as large prisms (An_{82-56}) and laths (An_{78-42}). Clinopyroxene ($Wo_{31-42}En_{45-49}Fs_{10-23}$) forms poikilocrysts with intergrowths of plagioclase laths and olivine granules (see Appendix, Tables A.28 and A.29).

In olivine gabbrodolerite at the flanks, some rocks contain drop-like ilmenite inclusions 0.3×0.5 cm and 0.6×1.2 cm in size, and oxides enclosing small troilite crystals (see Rock indication 146) are developed.

Picritic gabbrodolerite has a maximum thickness of 85–95 m, observed at the widest part of the intrusion; as the intrusion thins and wedges out, rock thickness decreases, and picritic gabbrodolerite is the first to disappear. In the picritic gabbrodolerite horizon, there are lenses of pegmatoids (olivine gabbrodolerite and taxitic gabbrodolerite up to several meters thick) and small isolated bodies of leucogabbro, which increase in frequency toward the lower taxite horizon.

Microscopic studies improved recognition of rocks with variable olivine content and different texture (see Rock indication 147) in the picritic gabbrodolerite horizon. Troctolite, troctolitic gabbrodolerite, plagioclase peridotite, and others are recognized by the proportion of principal minerals (Ivanov et al. 1971b; Igrevskaya 1981). Olivine (25–40%) forms subidiomorphic and isometric crystals, as well as small and palmate grains. Crystal size in the same thin section can differ by one order of magnitude. Olivine has the composition of Fa_{24-27}; clinopyroxene forms poikilocrysts and angular augite ($Wo_{37-43}En_{47-48}Fs_{10-15}$) grains, and plagioclase forms wide tabular crystals with poikilitic intergrowths of olivine or prismatic crystals.

The distribution of chromite in the picritic gabbrodolerite horizon sequence is similar to that in the Upper Talnakh Intrusion; accumulations of chromite up to 25% are restricted to the upper part of the horizon (see Rock indication 147), and chromite is either absent or sporadic in the lower parts of the sequence. A more complex chromite distribution character is observed in layering between picritic and taxitic gabbrodolerite.

Small erratic boulders of olivine gabbro in picritic gabbrodolerite consist of plagioclase (60–90%), clinopyroxene (3–20%), palagonite (0–10%), oxide-ore phase (3–7%), and granules of olivine and its pseudomorphs (0–3%). One such body is represented by granoblastic aggregate of anorthite (Ab_8An_{92}) with intergrowths of green spinel, pegmatite crystals of clinopyroxene, and ore–sulfide globules (see Rock indication 148).

Lower Zone

The lower zone of the intrusion consists of taxitic and olivine-bearing gabbrodolerite with ophitic and poikilo-ophitic texture. These rocks typically form a thick layer (tens of meters) between taxitic and troctolitic gabbrodolerite. In the rocks, lenses of taxitic and troctolitic gabbrodolerite, pegmatoidal gabbro nets, and areas of rocks rich in sulfides are observed.

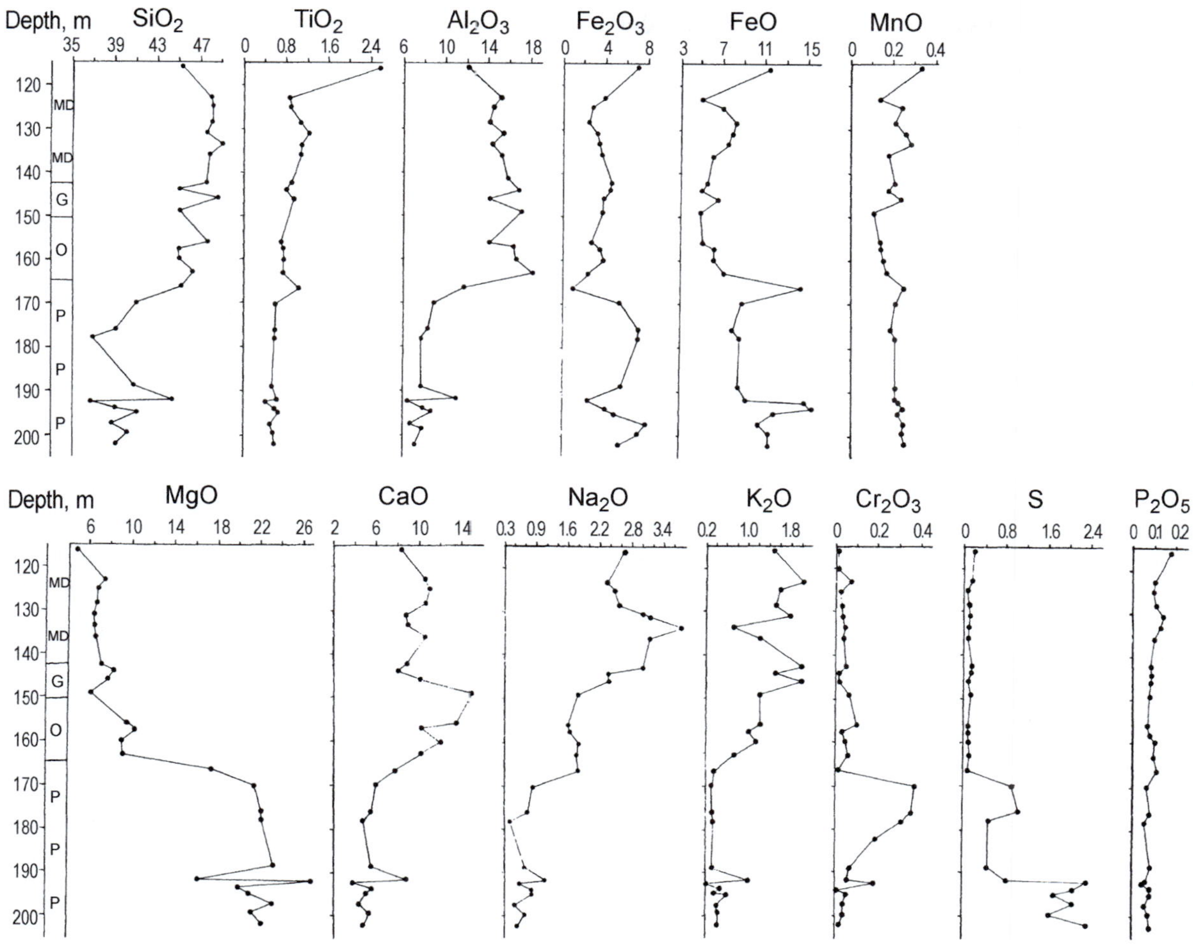

Fig. 3.14 Variation petrochemical diagram through a section of the Southwestern branch of the Upper Talnakh Intrusion; drill hole KZ-184. *1* gabbro, *2* olivine-free gabbro, *3* olivine-bearing and olivine gabbro, *4* picritic gabbro. Oxide and S content, wt%

Taxitic gabbrodolerite is thickest near the middle of the intrusion and thins along the flanks of the magmatic body. In the southern part of the intrusion, taxitic dolerite is absent; poikilo-ophitic gabbrodolerite with small leucocratic aggregates up to glomeroporphyritic accumulations of plagioclase is developed in its place (Igrevskaya 1981). In the upper part of the horizon, taxitic gabbrodolerite grades into picritic gabbrodolerite through magmatogenic breccia. The composition and structural features of the rocks are similar to taxitic gabbrodolerite of the Upper Talnakh Intrusion. The olivine content is Fa_{25-32}, and clinopyroxene is $Wo_{41}En_{48}Fs_{11}$ (see Rock indication 149).

Troctolitic gabbrodolerite is located in the transition zone between picritic and taxitic gabbrodolerite. The thickness of rocks is not over 2.5–6 m in the transition zones and less than 1.3–2 m in lenses among olivine-bearing gabbrodolerite. The rocks feature large palmate olivine crystals and low pyroxene content.

3.6.1.3 Compositional Features

Petrochemical peculiarities of rocks of the Noril'sky-type intrusions, including the Upper Talnakh Intrusion, have been repeatedly described in different papers and monographs (Dodin and Batuev 1971; Zolotukhin et al. 1975; Natorkhin et al. 1977; and Vilensky 1978a, b, 1984). These studies revealed the wide variation in rock composition and the average differentiate content and average weighted content. Petrochemical investigations undertaken by Vilensky (1967, 1974, 1978a) allow the Noril'sk-I and Upper Talnakh Intrusions to be assigned to the alkaline-earth branch of alkaline petrogenic series of intrusions. Crystallization of these intrusions proceeded with very low silicic acidity, increased potassium potential, relatively low alkalinity, and a high degree of maficity.

The petrochemical characteristics of the intrusion is reported in the references above, so only some reference cross sections of Southwestern, Northeastern, and Kharaelakhsky intrusive branches in the Upper Talnakh Intrusion and one cross section in the Eastern Noril'sk branch of the Noril'sk-I Intrusion will be considered in this volume.

The cross section through the leading part of the southwestern branch is observed in drill hole KZ-184 (Fig. 3.14). The behavior of principal rock-forming oxides demonstrates the layered structure of the intrusion. It can be seen from the variation diagram that the main change in behavior and content of oxides takes place at the boundary between mafic and ultramafic parts of the intrusions. In mafic layers, higher content of SiO_2, Al_2O_3, CaO, Na_2O, and K_2O is noted, and in ultramafic rocks, higher content of MgO, FeO, Fe_2O_3, Cr_2O_3, and S is typical. The content of TiO_2 and alkalis increases from the base to the roof of the magmatic body. The amount of MgO progressively increases downward through the mafic layer then sharply increases at the boundary between mafic and ultramafic rocks; MgO content is virtually invariable in ultramafic rocks. The variation in MgO values in the horizon is due to the presence of mafic pegmatoids or by taxitization of picrites. CaO and alkalis are accumulated in clinopyroxene gabbrodolerite and in rocks that contain a high quantity of plagioclase. Increased Cr_2O_3 and S concentration is attributed to chrome spinel and sulfides. The Cr_2O_3 content is highest at the roof, and it is less than 0.1 wt% in the lower part of the horizon (with low magnesium content). Maximum sulfur content is fixed at the bottom and roof of the picritic layer.

The section along drill hole KZ-274 (Fig. 3.15; see Appendix, Table A.18) represents the frontal part of the Northeastern branch of the Upper Talnakh Intrusion. Here, downward through the central zone, SiO_2, TiO_2, Fe_2O_3, MnO, Na_2O, K_2O, and P_2O_5 content decreases, and MgO, FeO, and S content increases. Content of Al_2O_3 and CaO first slightly increases then, at the boundary between olivine-bearing and olivine gabbrodolerite, begins to decrease. The transition between these rock types is remarkable due to variations in contents of other oxides. The most abrupt inflection in variation (e.g., bore hole KZ-184) is indicated by the change in FeO, MgO, and S content at the boundary between mafic and ultramafic rocks, and the sawtooth shape of plots is attributed to the presence of mafic pegmatoid.

The cross section along drill hole KZ-585 represents the leading part of the Kharaelakhsky branch of the Upper Talnakh Intrusion (Fig. 3.16; see Appendix, Tables A.23 and A.24). The remarkable peculiarity of its structure is the thick horizon of magmatogenic breccia overlain and underlain by olivine-free and olivine-bearing gabbrodolerite. As stated above, magmatogenic breccia consists of "fragments" of variable sizes both of unaltered and altered picrites and essentially olivine and spinel–olivine apopicritic rocks cemented by mafic pegmatoid. This collective occurrence of compositionally different rocks is responsible for the sawtooth shape of variation curves. No trend in changes in rock composition along the section is observed. It should be noted only that the fragments have elevated MgO content and lowered SiO_2, TiO_2, CaO, Na_2O, and K_2O content than those in the pegmatoid.

Insufficiently distinct behavior of Al_2O_3 seen in variation curves is attributed to the fact that both spinel and mafic plagioclase are present in variable amounts in fragments and in the magmatogenic breccia cement. Wide variations in Fe content are caused by uneven mineralization of rocks and, in part, by variable content of iron-rich minerals.

Attention is paid to the fact that variation diagrams for the cross section through the Eastern Noril'sk branch (Fig. 3.17) show no unified crystallization trend. The curves distinctly indicate three zones: UZ, CZ, and LZ. Allied absolute oxide contents and trends are observed only in a single horizon, and these experience a sudden change when mowing to another horizon. The most pronounced fluctuations in rock composition are observed in the UZ and LZ to which pegmatoid and mineralization are restricted.

The UZ is composed of two principal types of rocks: taxitic gabbrodolerite and leucogabbro. Taxitic gabbrodolerite is characterized by increased MgO and CaO content, and leucogabbro is enriched in Al_2O_3. This can be attributed to the fact that taxitic gabbrodolerite is enriched in olivine and clinopyroxene, and leucogabbro is enriched in plagioclase. Importantly, rocks of the UZ feature the maximum concentration of MgO, which in the cross section along drill hole PE-35 is caused by melanocratic taxites and in other cross sections of the Eastern Noril'sk branch is due to the presence of upper picrites.

In a cross section through the CZ, directional changes in content of each of rock-forming oxides have been outlined. In the horizon of pegmatoidal prismatically grained gabbrodolerite and metadiorites, high content of virtually all oxides is observed. In the lower series of olivine-free, olivine-bearing, and olivine gabbrodolerites, Al_2O_3, Fe_2O_3, and MgO content increases, whereas CaO and Na_2O content decreases; quantitative variations in other oxides are less explicit. At the boundaries between different horizons, variation curves show stepwise inflections. The sharpest contact and pronounced difference in the rock composition are observed at the boundary between mafic and ultramafic intrusion layers. The picritic horizon is characterized by higher FeO and MgO content and lower SiO_2, Al_2O_3, CaO, and Na_2O content compared to mafic rocks. In addition, the horizon of ultramafic rocks is not chemically uniform due to homogenous leucogabbro inclusions, veinlets of pegmatoidal gabbro, and interbedded layers of taxitic gabbrodolerite that produce wide variations in oxide contents and a sawtooth shape in variation curves.

The diversity in taxitic gabbrodolerite composition of the LZ is manifest as variations in oxide content, increasing SiO_2, TiO_2, Al_2O_3, Fe_2O_3, CaO, Na_2O, K_2O, and P_2O_5 content and decreasing FeO, MnO, and MgO content toward the lower contact (due to a decrease in olivine content).

3.6.1.4 Mineralogy

Olivine from the Upper Talnakh Intrusion falls within the range Fa_1 to Fa_{44}. Compositionally, the mineral widely ranges over the intrusion both within one rock type and a single sample.

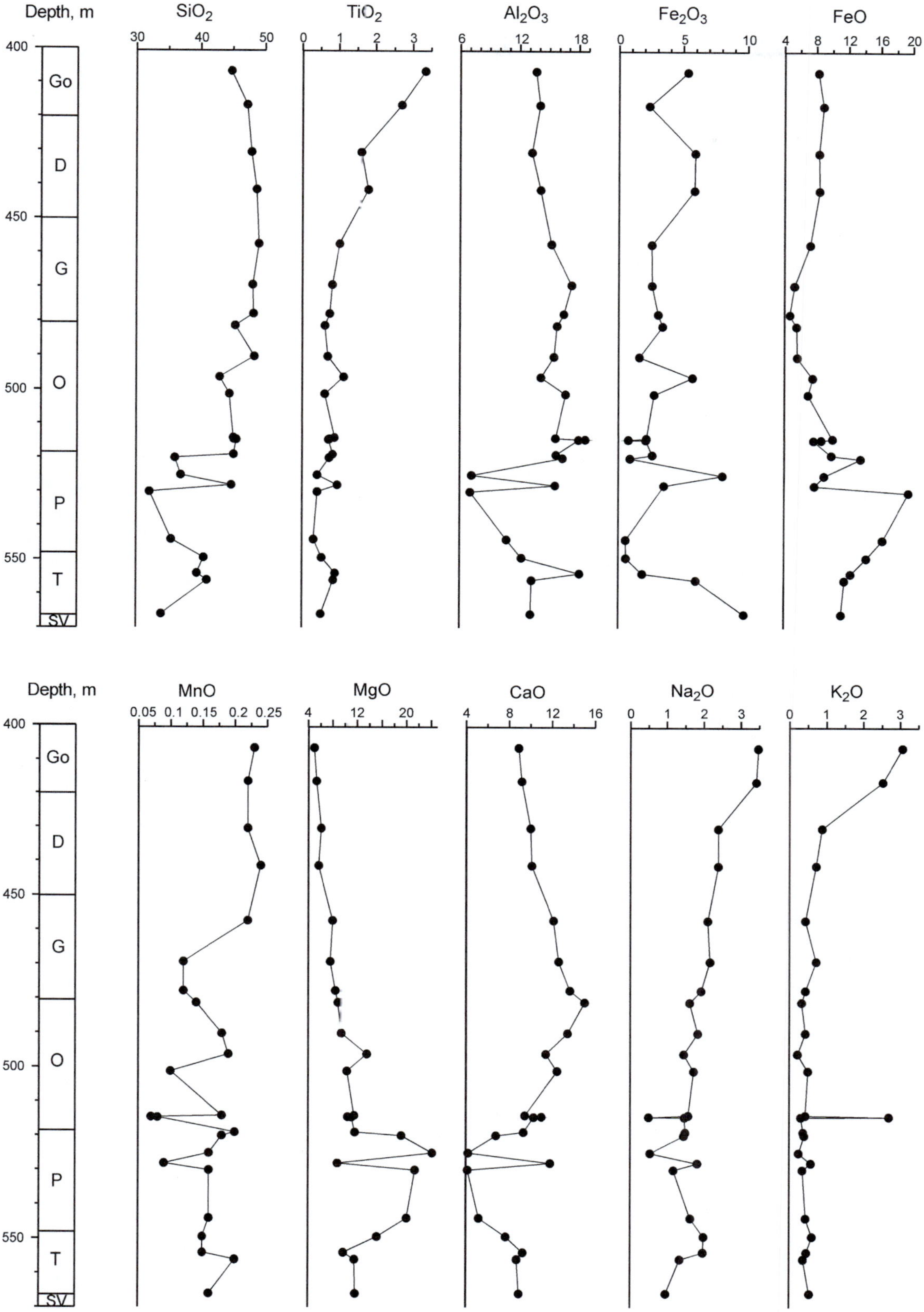

Fig. 3.15 Petrochemical variation diagram through the section in the leading part of the Northeastern branch of the Upper Talnakh Intrusion; drill hole KZ-274. Oxide and S content, wt%, element content, ppm

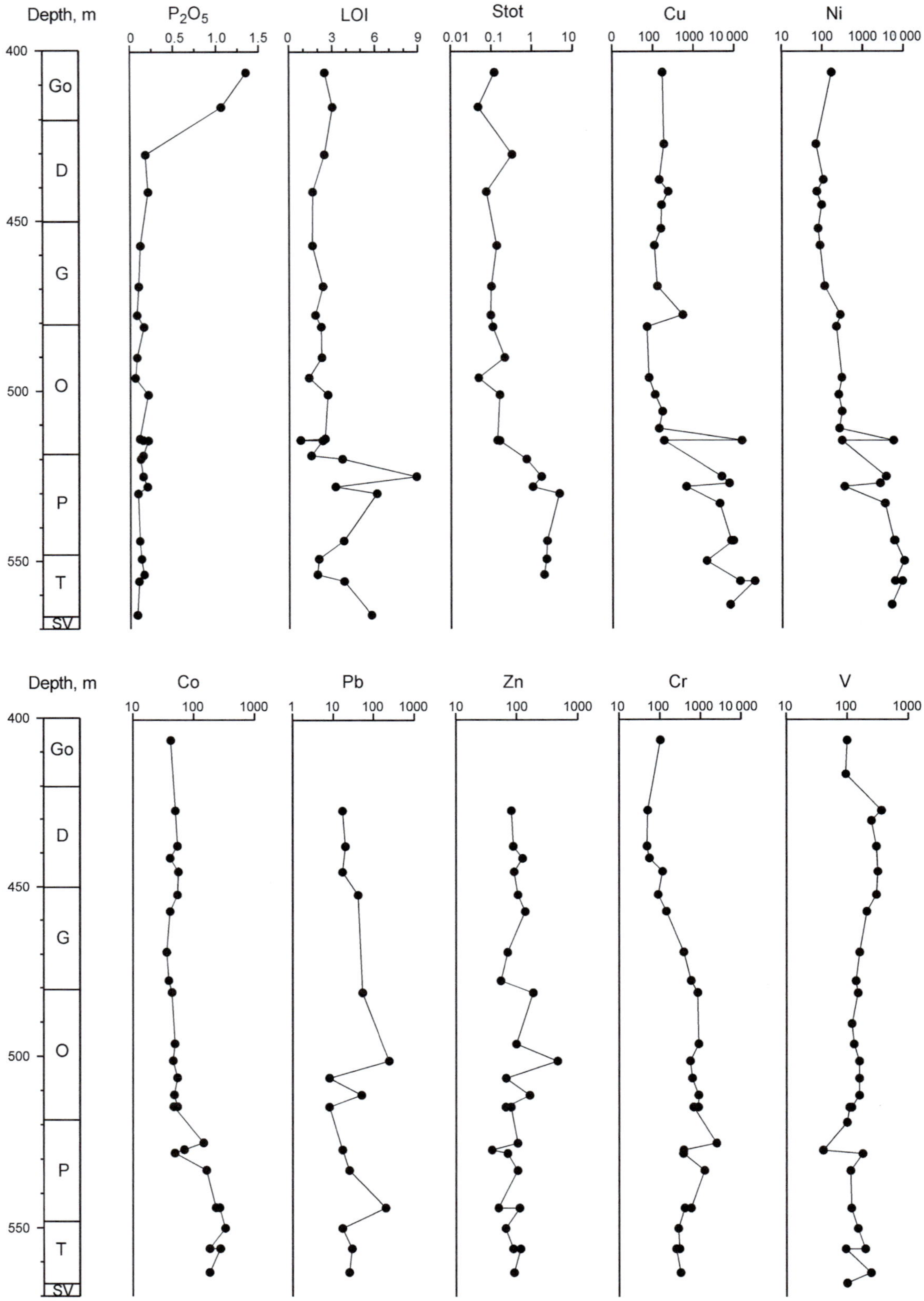

Fig. 3.15 (continued)

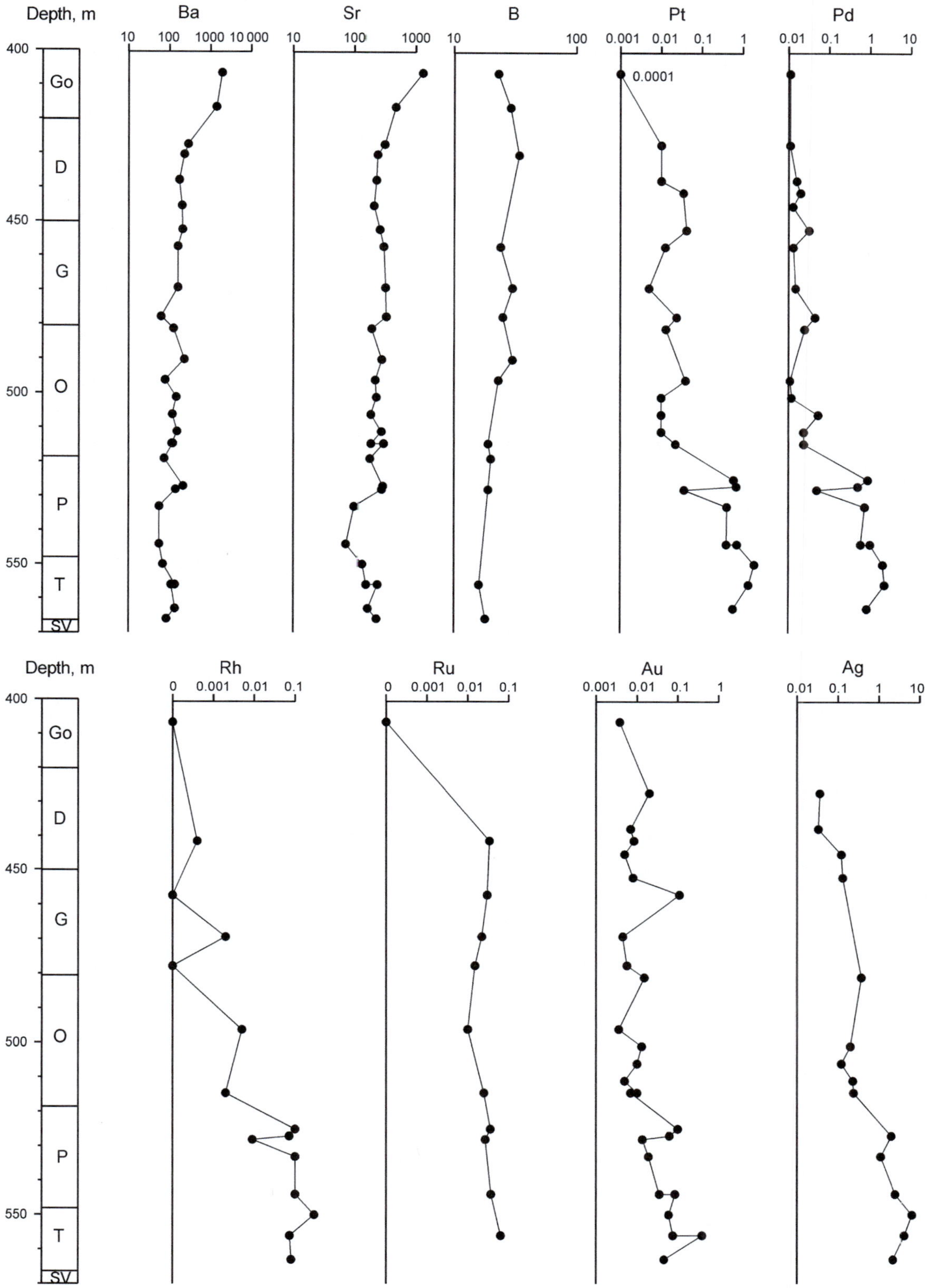

Fig. 3.15 (continued)

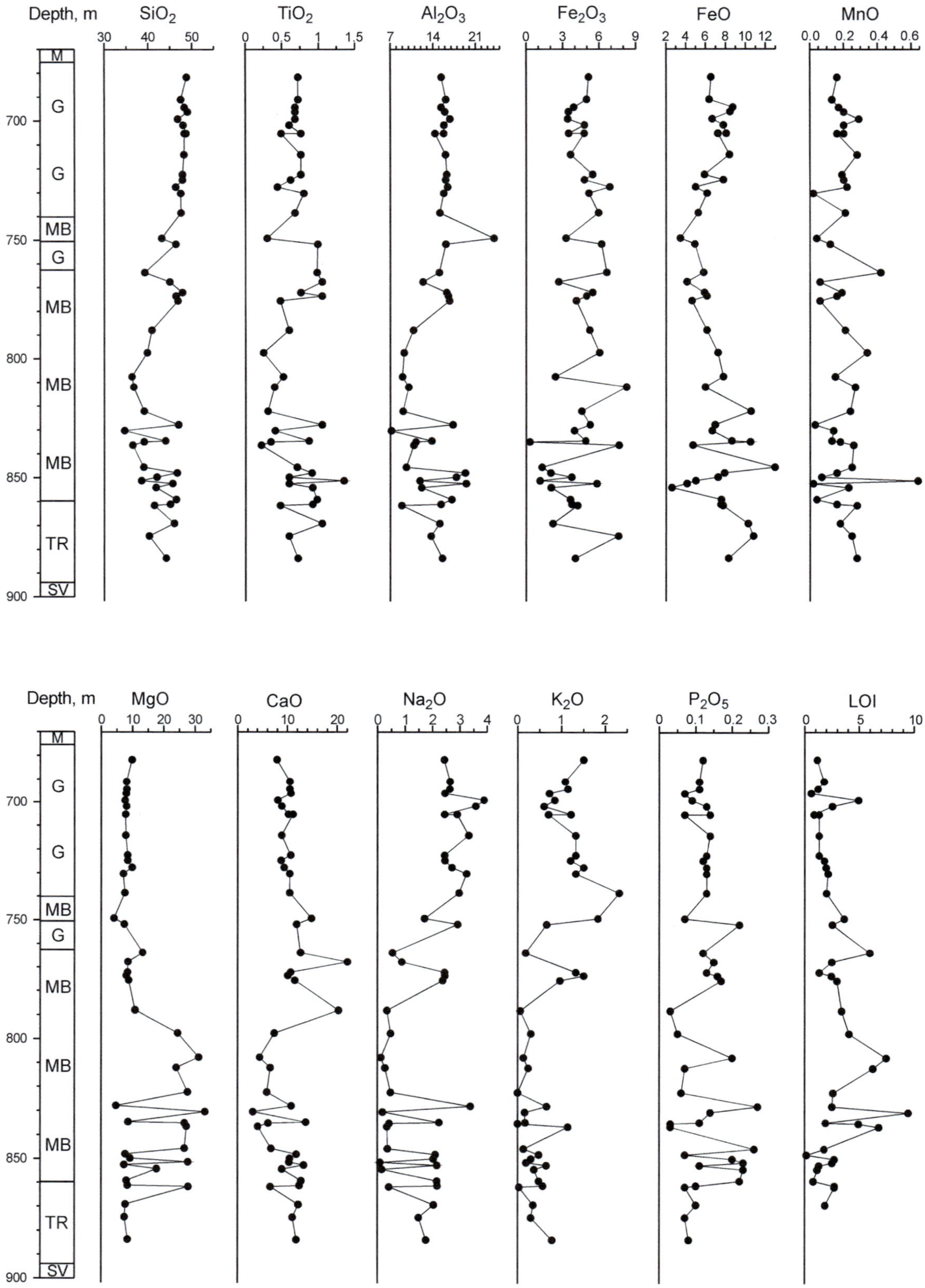

Fig. 3.16 Petrochemical variation diagram through the section in the leading part of the Kharaelakhsky branch of the Upper Talnakh Intrusion; drill hole KZ-585. Oxide and S content, wt%, element content, ppm

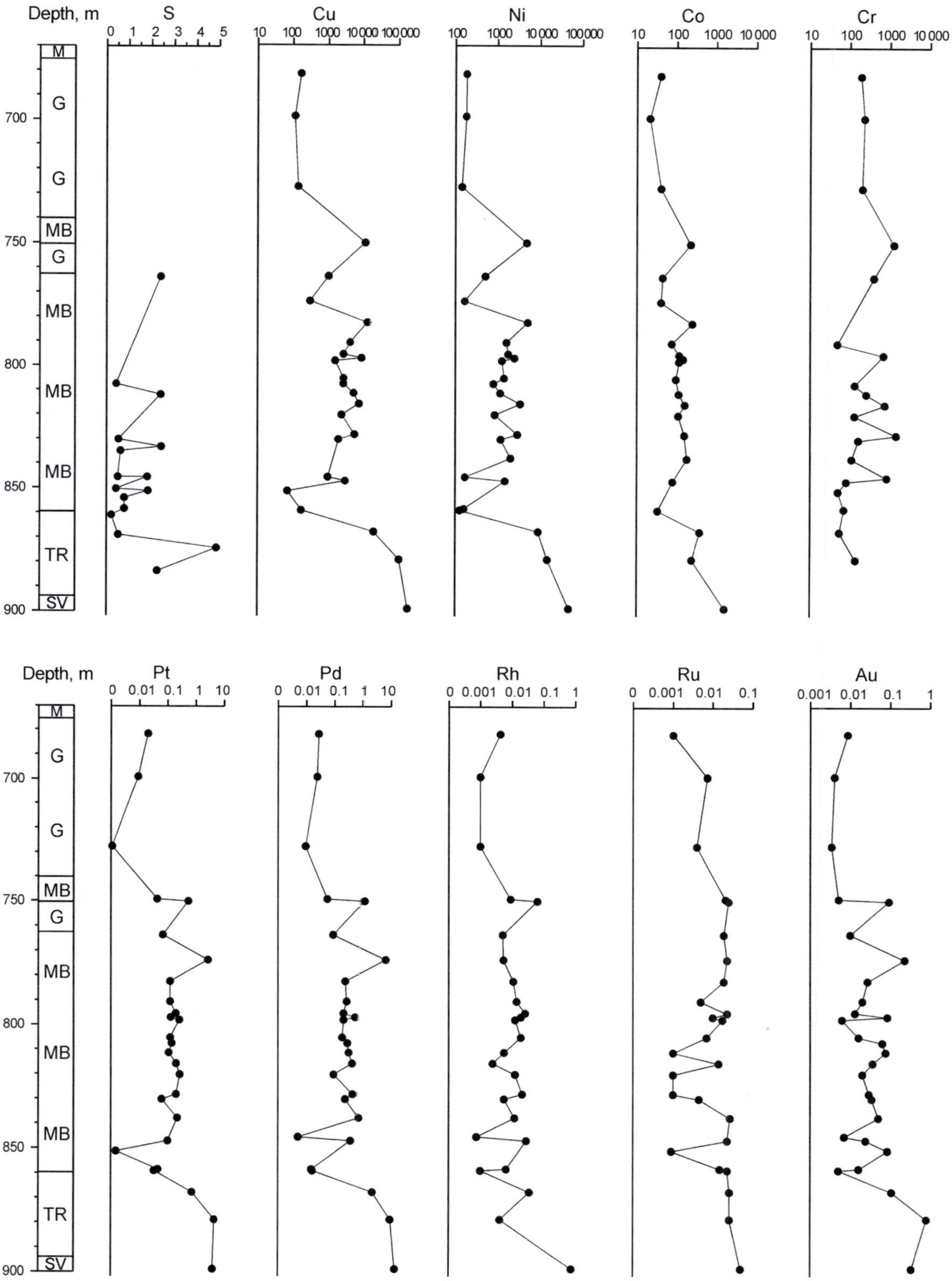

Fig. 3.16 (continued)

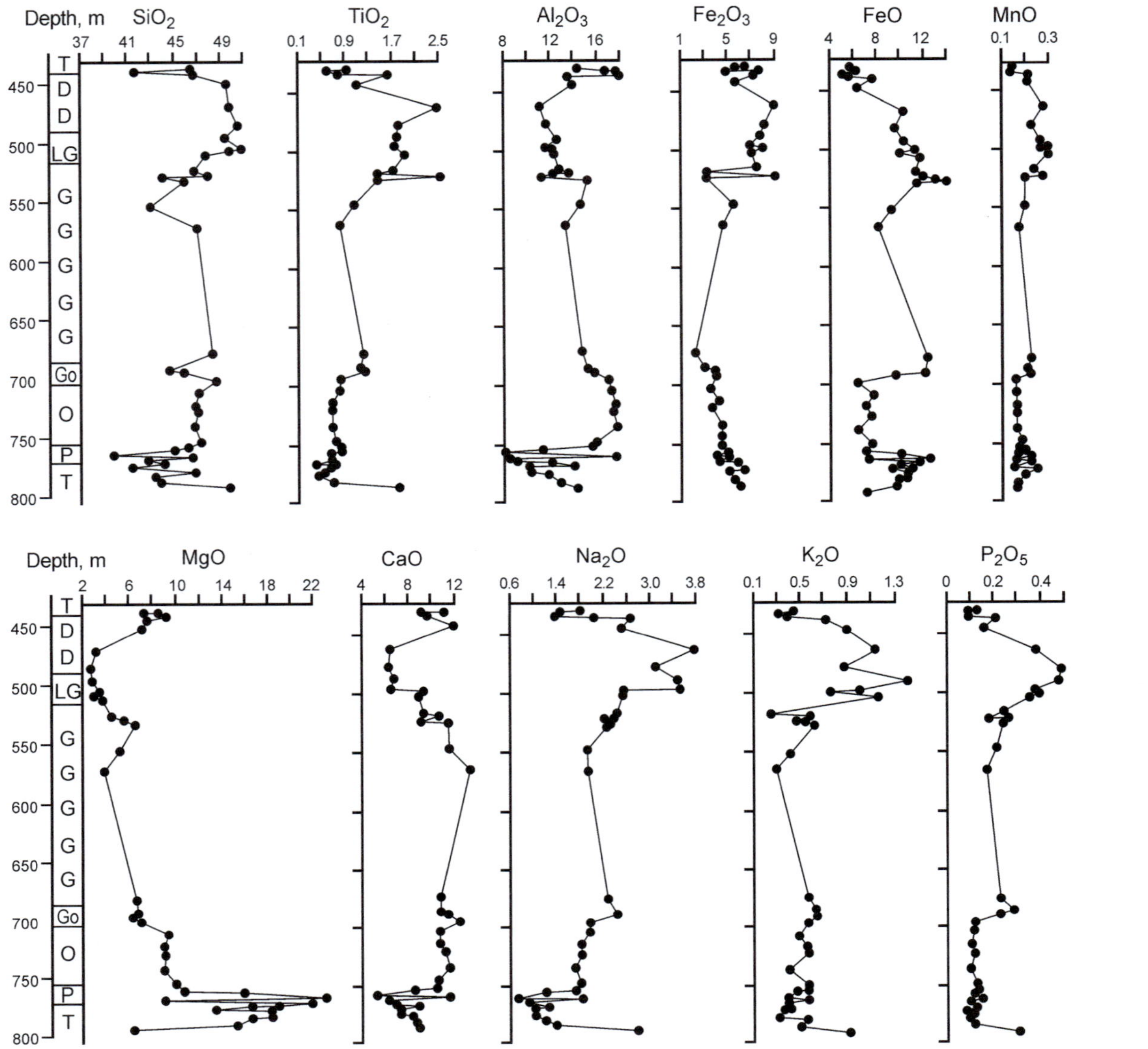

Fig. 3.17 Petrochemical variation diagram through the section in the Eastern Noril'sk branch of the Noril'sk-I Intrusion in drill hole PE-35. Oxide and S content, wt%, element content, ppm

Variation diagrams of olivine composition through the Northeastern branch of the Upper Talnakh Intrusion have a sawtooth shape (Fig. 3.18). The olivine compositions are Fa_{24-32}, Fa_{23-25}, and Fa_{28-39}, respectively, in the mafic, picritic, and taxitic horizons of the intrusion. The silicate nickel content in olivine (NiO_{Ol}) is 0.078–0.34 wt%; it is fairly stable in olivine-bearing and olivine gabbrodolerite and ranges from 0.151 to 0.215 wt% in picritic and taxitic gabbrodolerite.

Peculiarities in compositional variations of olivine and the behavior of some chemical elements can be seen in cross sections of picritic gabbrodolerite in the leading (KZ-184) and basal (ST-5) parts of the intrusion (Fig. 3.18). In the first case, the Fe content in olivine increases downward through the section, and mineral composition changes from Fa_{20} to Fa_{27}, and NiO_{Ol} content decreases downward from 0.25 to 0.007 wt%. The rocks exhibit a direct relationship between Fa–S and NiO_{Ol}–Cr and an inverse relationship between Fa–NiO_{Ol} and S–NiO. In the deepest part of the intrusion, the tendency for different components to change is complicated by the formation of pegmatoids in the picritic horizon. Here, the composition of olivine is Fa_{35}, changing to Fa_{35-40} in picritic gabbrodolerite, and at the boundary with olivine gabbrodolerite, Fe content decreases abruptly to Fa_{22}. It increases in the taxitic horizon from Fa_{22-31} to Fa_{36-43}, and NiO_{Ol} changes from 0.075 to 0.238 wt%.

Olivine from the Kharaelakhsky branch of the Upper Talnakh Intrusion features wide variations in Fe content (ranging from Fa_1 to Fa_{28}) as well as in NiO and CaO content

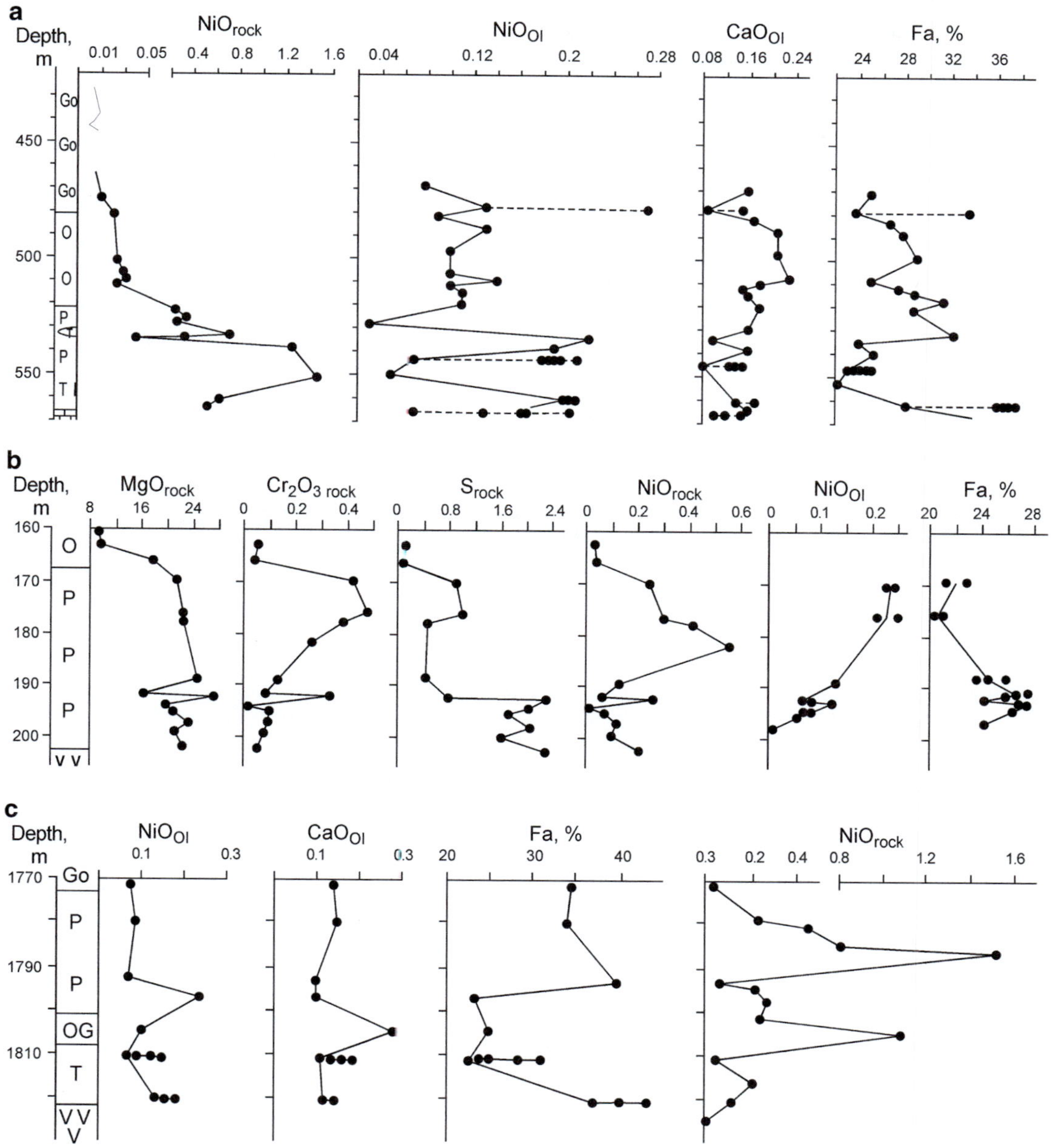

Fig. 3.18 Variation in olivine composition and chemical components of olivine-bearing rocks. (**a**) Cross section of the Northeastern branch of the Upper Talnakh Intrusion, drill hole KZ-274. Composition of olivine crystals from the same sample is connected by a dashed line. (**b**) Cross section of picritic gabbro in the western part of the Upper Talnakh Intrusion, drill hole KZ-184. (**c**) cross section of the hypogeal part of the Upper Talnakh Intrusion, drill hole ST-5. Altered sedimentary rocks and limestones are shown at the intrusion contact

(Fig. 3.19). Downward through this horizon, there is a tendency for Fe content to increase and NiO_{Ol} to decrease. Significant fluctuations in olivine content can be found within a single sample. Comparing the content of olivine from relict picritic gabbrodolerite and from olivine and spinel–olivine fragments in magmatogenic breccia reveals a distinct decrease in Fe NiO_{Ol} content from 0.197 wt% to 0 and an increase in CaO_{Ol} content from 0.04 to 0.80 wt%. The average olivine composition of magmatogenic breccia (excepting anomalous compositions) fits Fa_{10-11}; NiO and CaO content is 0.03–0.04 wt% and 0.11–0.23 wt%, respectively (Ryabov 1992a). The most magnesian olivine varieties from olivinites are Fa_1, with NiO and CaO content 0.01 and 0.53 wt%, respectively.

In the Noril'sk-I Intrusion sequence, olivine ranges from Fa_{23} to Fa_{40}. The average Fe content values in olivine decreases along the series: olivine-bearing gabbrodolerite, olivine gabbrodolerite, and picritic gabbrodolerite, although there is no tendency for Fe content to change over the cross section (Fig. 3.20). In addition, NiO content in olivine increases downward from 0.029 to 0.236 wt%, and CaO content decreases from 0.31 to 0.12 wt%.

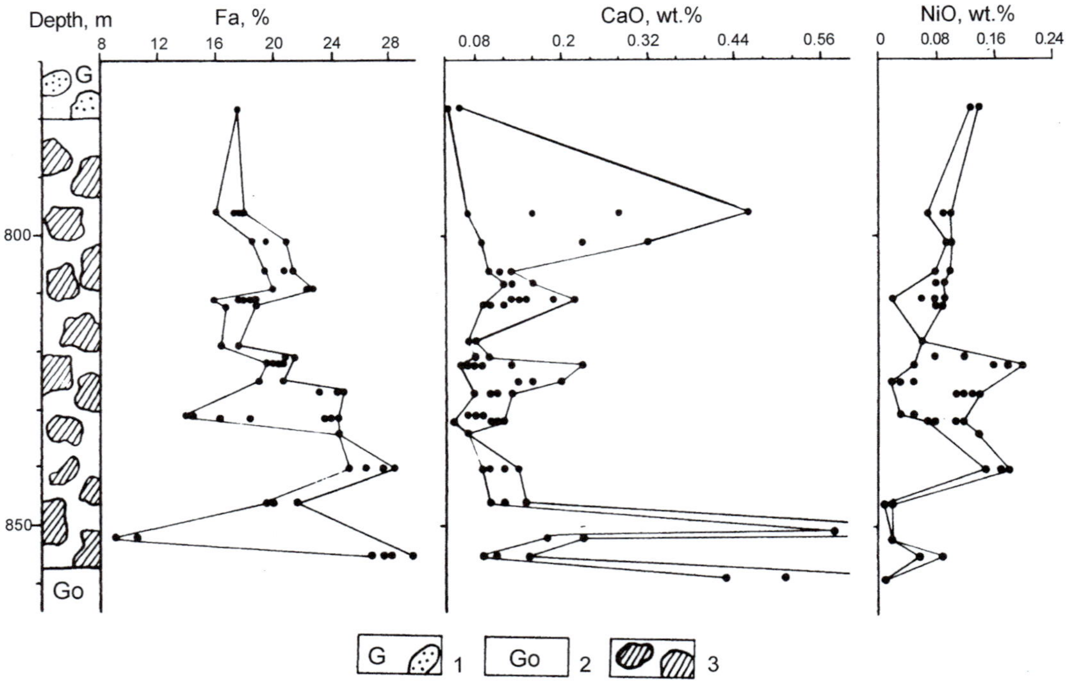

Fig. 3.19 Variation in olivine composition through a section of magmatogenic breccia of the Kharaelakhsky branch of the Upper Talnakh Intrusion, drill hole KZ-585. *1* schlieren of leucogabbro in olivine-free gabbro, *2* olivine-bearing gabbro, *3* magmatogenic breccia

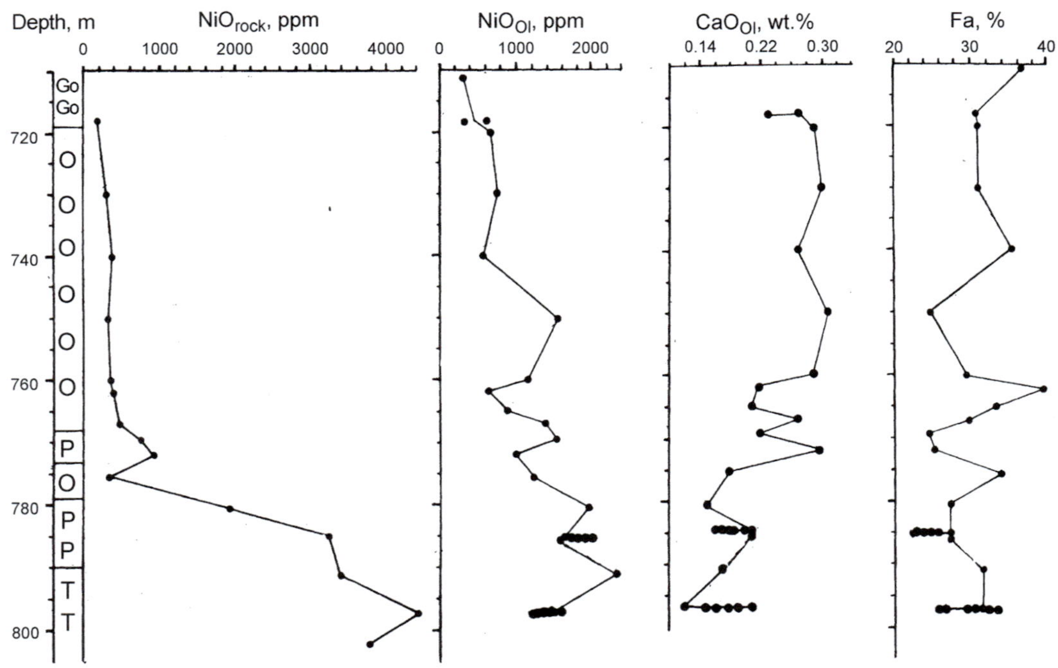

Fig. 3.20 Compositional variations of olivine in comparison with nickel oxide content through a section of the east Noril'sk branch of the Noril'sk-I Intrusion in drill hole PE-35. *1* olivine-bearing gabbro, *2* olivine gabbro, *3* picritic gabbro, *4* taxitic gabbro

Changes in olivine composition through the Upper Talnakh Intrusion are listed in Table 3.5. It can be seen that the upper picritic gabbrodolerite has a fairly magnesian composition and variable Fe content. In the layered series of rocks in the CZ, Fe content in olivine decreases downward through the intrusion despite variations in olivine composition. The Fe content in olivines from the LZ is commonly higher than in picritic gabbrodolerite. Olivine Fe content in magmatogenic breccia from different parts of the intrusion varies widely from $Fa_{1.3}$ to $Fa_{24.7}$, and NiO content is stable

Table 3.5 Olivine average content from the rocks' main varieties of the Noril'sky-type intrusions (Ryabov 1992a)

		Upper Talnakh intrusion		
Zone	Rock	Fa (%)	NiO (wt%)	CaO (wt%)
UZ	*Contact gabbrodolerite*	35.4	0.06	0.17
	Picritic gabbrodolerite I	20.1	0.05	0.19
	Picritic gabbrodolerite II	26.3	0.30	0.19
	Magmatogenic breccia	11.1	0.04	0.11
CZ	*Olivine-bearing gabbrodolerite*	32.8	0.08	0.18
	Olivine gabbrodolerite	29.4	0.11	0.18
	Picritic gabbrodolerite I	30.6	0.17	0.15
	Picritic gabbrodolerite II	19.3	0.10	0.20
	Magmatogenic breccia I[a]	9.6	0.03	0.23
	Magmatogenic breccia II	1.3	0.01	0.53
	Magmatogenic breccia III	16.6	0.06	0.10
LZ	Taxitic gabbrodolerite I	35.2	0.18	0.13
	Taxitic gabbrodolerite II	21.3	0.10	0.12
	Troctolite gabbrodolerite	26.6	0.02	0.12
	Magmatogenic breccia	24.7	0.09	0.09
	Contact gabbrodolerite	43.9	0.09	–

The Roman figures specify statistically steady compositions of minerals in one petrographic type of rocks
[a]Data from the Kharaelakhsky branch of the Upper Talnakh Intrusion where magmatogenic breccia cut the stratified series of rocks

and low from 0.01 to 0.09 wt%. In addition, olivine in picritic gabbrodolerite is $Fa_{24.1}$ and Fa_{26}, and the NiO_{Ol} contents are 0.18 wt% (average from 21 analyses) and 0.20 wt% (average from 33 analyses) in the leading and near-basal parts of the Upper Talnakh Intrusion, respectively (Ryabov 1992b).

Pyroxene is present as monoclinic and rhombic varieties in the Noril'sk Intrusions. The most abundant is clinopyroxene; orthopyroxene forms mainly in olivine-rich rocks (Godlevsky 1959; Stepanov 1977). Clinopyroxene in layered gabbrodolerite series is typically augite and in pegmatoids is typically augite and diopside. Bronzites and hypersthene are the most common orthopyroxenes.

Clinopyroxene in the Noril'sk Intrusions exhibits zoning: cores of crystals are chrome rich (Cr_2O_3 up to 1.25 wt%), and peripheral parts are enriched in titanium (TiO_2 up to 1.54 wt%). Maximum concentrations of these elements are associated with picritic gabbrodolerite.

The Upper Talnakh Intrusion features augite that varies in composition ($Wo_{35-45}En_{41-49}Fs_{8-23}$; Fig. 3.21; see Appendix, Table A.14). From the data given above, one can see that fluctuations in the total Fe content (*f*) within a single-zoned crystal may reach 8 wt%. Upward through the intrusion, *f* increases in the central parts of crystals from 14.2 to 33.3 wt% and from 17.1 to 35.3 wt% in rims.

The nature of variation of Ca and Mg (or Wo and En components) in clinopyroxenes through an intrusion is more complex than that of Fe or f. Smooth change in Wo and En is restricted to transitions of rock type and is noticeable in rims of crystals. The CaO content is relatively stable in brownish clinopyroxene in picritic gabbrodolerite with the Wo proportion about 40 (ranging from 37 to 44). The En content in brownish clinopyroxene from picritic gabbrodolerite is 49.0, and the minimum En content in metadiorite is 41.9.

In the core of clinopyroxene crystals (green in chrome-bearing varieties and brownish in chrome-free varieties), the Wo content slightly varies over the whole intrusion except in metadiorite where it sharply decreases approaching to the intrusion roof. The En content gradually deceases upward through the body with some fluctuations (from 50.1 to 41.4).

Additional information on variations in the clinopyroxene composition in the intrusion can be obtained from the Si/Al, (Ca + Na)/Al, Mg/Fe, and (Ca + Na)/Fe ratios. It should be noted that the fractionation coefficient (Kf) values of rocks at the boundary between mafic and ultramafic rocks exhibit a sharp jump that manifests itself in the Fe content of brownish clinopyroxenes, but it is not observed in green clinopyroxene.

A comparative analysis of variations in clinopyroxene and orthopyroxene compositions is shown in Fig. 3.22. Clinopyroxene exhibits stable principal rock-forming element content except one analysis which demonstrates a higher Fe, Ti, and Cr content compared to orthopyroxenes. Upward through the picritic horizon, Cr_2O_3 content increases and sharply decreases in olivine-bearing gabbrodolerite.

When examining the behavior of TiO_2 in cores and rims, attention is paid to the fact that cores of clinopyroxene crystals from the picritic horizon exhibit the higher TiO_2 content than those in the mafic horizon. The highest TiO_2 content in rims of the crystals is also restricted to the picritic horizon. It should be noted that the TiO_2 demonstrates individual behavior in each ultramafic and mafic horizon. Upward through each horizon, the tendency for TiO_2 to increase in cores of clinopyroxene crystals and for oxides to decrease in rims of the crystals is obvious. All variations in TiO_2 content reflect the tendency for an overall increase in TiO_2 in the rock.

The Cr_2O_3 content in the cores of crystals gradually decreases in the rock series: picritic gabbro-olivine gabbrodolerite-olivine-free gabbrodolerite-metadiorite. Along edge (or brownish) parts of clinopyroxene crystals, the amount of Cr_2O_3 remains increased in picritic gabbrodolerite but sharply decreases in overlying mafic rocks. The behavior of Cr_2O_3 in rocks is not associated with the Cr_2O_3 distribution in clinopyroxene.

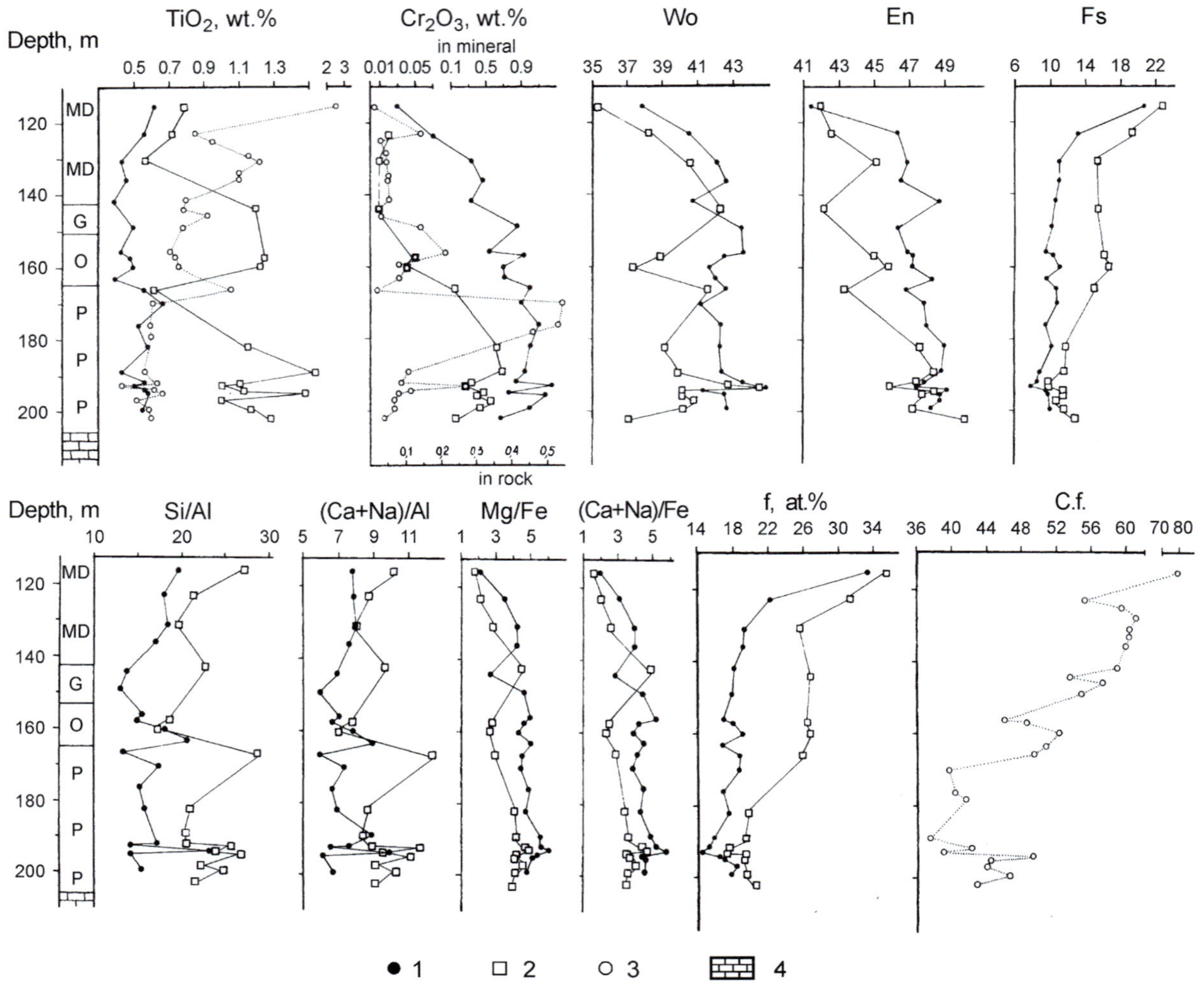

Fig. 3.21 Diagram showing variation in core and rim composition of zoned clinopyroxene crystals through a section of the Upper Talnakh Intrusion in drill hole KZ-184. *1* core, *2* rim of crystals, *3* rock, *4* limestones at the intrusion contact. For comparison, fractionation coefficient of rocks and titanium and chromium content are given

In the Noril'sk-I Intrusion, the variation trend for clinopyroxene crystals is, in general terms, similar to that described for the Upper Talnakh Intrusion. Here, zoning and the nature of distribution of both TiO_2 and Cr_2O_3 in cores and rims of crystals (Fig. 3.23) are clearly defined. It is seen on diagrams that the CaO content in rims of crystals decreases from $Wo_{39.5}$ to $Wo_{31.6}$ downward through the intrusion from the roof to the boundary between olivine-bearing and olivine gabbrodolerite and then increases to $Wo_{39.8}$ at the base of the intrusion. In the cores of crystals, the CaO content increases upward the intrusion, and it is always higher than in the crystal rims. From the intrusion roof to the boundary with olivine gabbro, the En component increases in cores of crystals and decreases in rims. Lower in the intrusion to the base, the En value remains constant. The Fe content in a single-zoned crystal reaches 19 wt%. In the cores of crystals, the total Fe content is about 8 wt% along the intrusion, and in rims of the crystals, it reaches 17 wt%.

The behavior of Cr_2O_3 and TiO_2 in clinopyroxene is fairly stable in olivine gabbrodolerite, crystal cores containing 0.60 wt% Cr_2O_3 and 0.32 wt% TiO_2, and rims 0.07 wt% Cr_2O_3 and 0.74 wt% TiO_2. From olivine gabbrodolerite to olivine-bearing gabbrodolerite, TiO_2 content in clinopyroxene cores increases from 0.28 to 0.49 wt% and decreases from 0.87–0.94 wt% to 0.53 wt% in rims. The Cr_2O_3 content in cores of crystals decreases from 0.79 to 0.04 wt% in the same direction but slightly increases from 0 to 0.03 wt% in rims of clinopyroxene grains. The behavior of principal pyroxene minals is unusual: fluctuations in values of Wo, En, and Fs components are within the range 31–41, 41–50, and 11–24, respectively. Differences in content are gradually equalized to $Wo_{40}En_{45}Fs_{15}$. Except in the UZ, which features clinopyroxenes that have their own specific character, it can be stated that insignificant but unidirectional variations along the intrusions' cross section take place in the layered series of rocks. Deflections from the

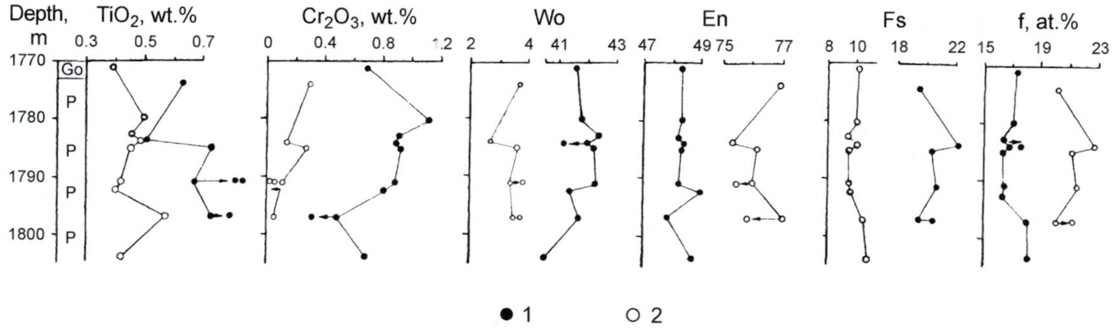

Fig. 3.22 Variation in composition of ortho- and clinopyroxene through a section of the picrite horizon of the Upper Talnakh Intrusion (drill hole ST-5). *1* clinopyroxene, *2* orthopyroxene

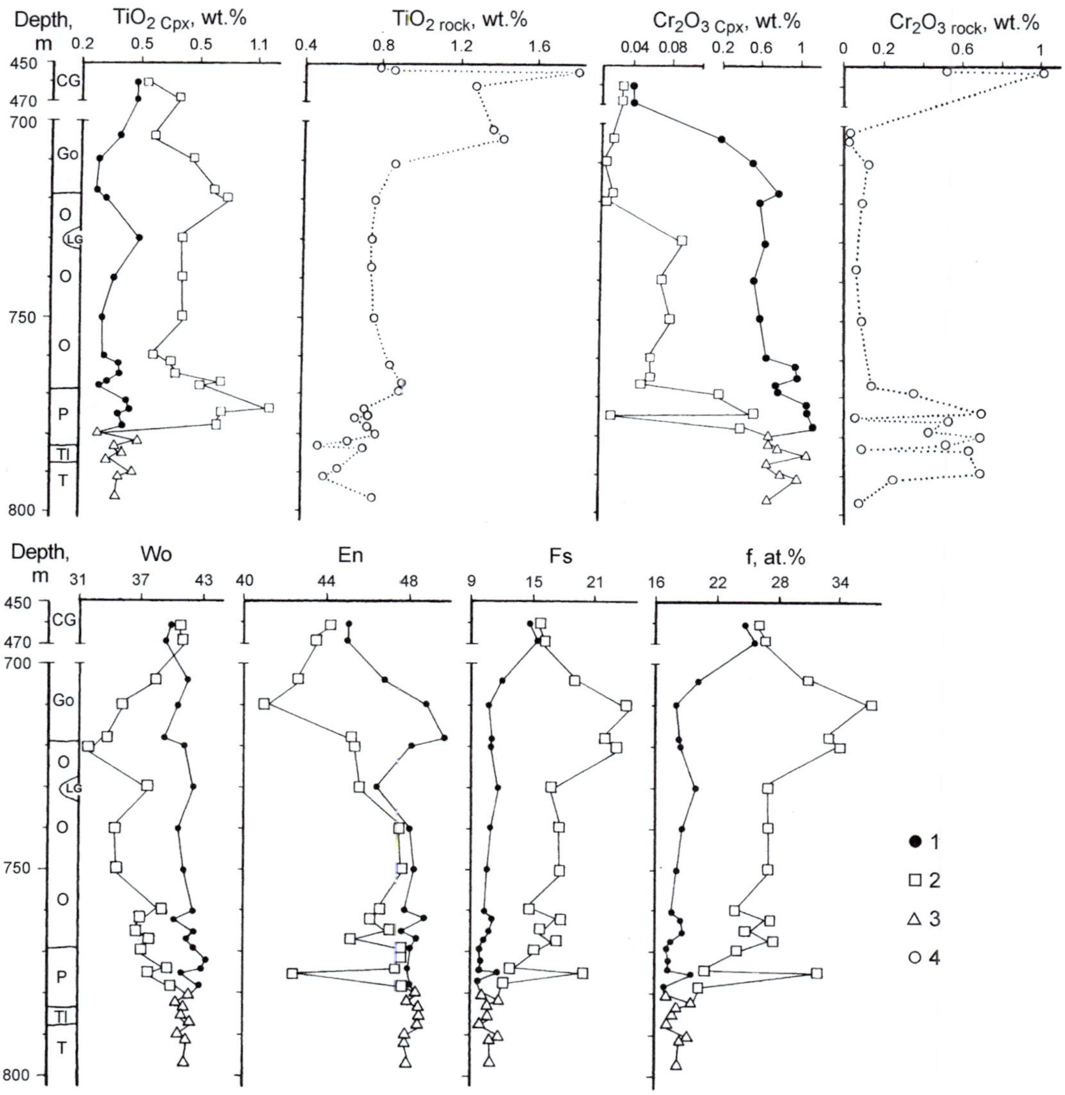

Fig. 3.23 Variation in composition of zoned clinopyroxene through a section of the east Noril'sk branch of the Noril'sk-I Intrusion (drill hole PE-35). *1* core, *2* rim of crystals, *3* angular grains, *4* rock

main trend on variation diagrams are associated with occurrences of small pegmatoid bodies, as well as with layering of different rocks or with transition of one gabbrodolerite to another.

Clinopyroxenes from pegmatoids of the both UZ and LZ are characterized by wide variations in all rock-forming oxides. In addition to augite, which is compositionally similar to other minerals of the layered series, varieties with anomalous composition are not infrequently found in pegmatoids. Here, diopsides with wide Al_2O_3 variations (up to 14.3 wt%), elevated Na_2O content (up to 2.44 wt%), and variable TiO_2 (0.26–1.67 wt%) and Cr_2O_3 content (0.01–082 wt%) occur. The range of the total Fe content of the mineral is within 7.6–34.8 wt%.

Compositional features of rock-forming minerals will be described in more detail later.

3.6.1.5 Contact Alteration

Country rocks enclosing Noril'sk Intrusions experienced significant metamorphic alteration, especially where a magmatic body was localized in sedimentary carbonate-clay sulfate-bearing rocks. Alteration was less intense in terrigenous rocks and much less intense in magmatogenic rocks. Most contact aureoles form polyfacial metamorphic complexes of hornfels and marble, which are most conspicuous adjacent to the Upper Talnakh Intrusion. Zoning of metamorphic complexes is exhibited in a facies change from spurrite–merwinite, pyroxene hornfels, biotite–amphibole hornfels, to chlorite–muscovite hornfels (Turovtsev 1972). The contact hornfels, as well as intrusive rocks, are replaced by alkaline, magnesian, and calciferous metasomatic rock with a widely varying composition, petrology, and texture (see Zolotukhin 1964; Yudina 1965; Turovtsev 1970, 1972; Dodin and Batuev 1971; Zolotukhin et al. 1975; Zotov 1979).

The most intense metamorphic alteration of country rocks and the intrusive rocks themselves is found in the leading portion of extended magmatic bodies; intensity of alteration decreasing strongly toward the root parts of bodies. The thickness of metasomatic rocks in the leading parts of intrusive branches of the Upper Talnakh Intrusion reaches 100–120 m. The temperature range of metasomatism determined from experiments on homogenization of inclusions in minerals is estimated as 820–140 °C (Bulgakova 1971). The available data allows one to link the thick aureole of altered rocks framing Noril'sky ore-bearing intrusions with a magmatic melt enriched with volatiles (Zotov 1979).

3.6.2 Lower Talnakhsky-Type Intrusions $(\nu\delta - \omega\nu\beta)T_{1nr}^{nt}$

During prospecting of the Talnakhsky ore deposit in the early 1960s, an additional magmatic body was discovered by drill holes beneath an ore-bearing intrusion. Considering the hypsometric location of these bodies, the upper body was referred to as Upper Talnakh Intrusion and the lower body as the Lower Talnakh Intrusion. It has been ascertained that the later intrusion includes all principal rock types specific for layered intrusions of the Noril'sky type, although most of the sequence consists of troctolitic gabbrodolerite. Due to this, the Lower Talnakh Intrusion has been assigned to the Morongovsky type of intrusions (Komarova and Lyul'ko 1967; Dodin and Sadikov 1967).

In the 1970s, on the southwest of Noril'sk-I, a magmatic body under an ore-bearing intrusion was intersected by drilling. This body turned out to be similar to the Lower Talnakh Intrusion in structure and composition, and it was referred to as the Lower Noril'sk Intrusion. In addition to these intrusions, the Zelenaya Griva Intrusion that also was discovered by drilling in the southern part of the Noril'sky trough has also been assigned to the Lower Talnakhsky-type intrusions.

3.6.2.1 Lower Talnakh Intrusion

The Lower Talnakh intrusion is a magmatic body with a complicated morphology; at the south of the Talnakhsky ore junction, the Lower Talnakh Intrusion is restricted to the core of an anticline and has a pipe-like shape; farther to the north, it changes into a plate-like bed with pinches and swells (Zolotukhin et al. 1975). In longitudinal cross section, both the Upper and Lower Talnakh Intrusions gently dip from the south to north, each body dipping within its own stratigraphic level. The intrusions gradually approach each other, and in one deep-seated locality, the Lower Talnakh Intrusion crosscuts and locally lies above the Upper Talnakh Intrusion before plunging beneath the ore-bearing intrusion. The thickness of the Lower Talnakh Intrusion ranges from 5–15 m to 380 m, and it is 60–100 m in sill-like apophyses. Contact interactions between the Upper and Lower Talnakh Intrusions are not observed. At the contact, no quenched crusts were found, and the transition simply resembles changes of rocks in other layered series.

Internal Intrusion Structure

The contact facies of the intrusions are composed of olivine-bearing gabbrodolerite, which is sometimes preserved along the sides and base of the intrusion. In other cases, rocks of a layered series are present at the boundary with country rocks: metadiorites at the roof and taxitic gabbrodolerite at the base (Fig. 3.24).

At the roof of the Lower Talnakh Intrusion, coarse-grained, greenish-pink, and greenish-gray *metadiorites, gabbroic diorites, and olivine-free, palagonite, prismatically granular gabbroic diorites* occur. The texture of these rocks is hypidiomorphic granular.

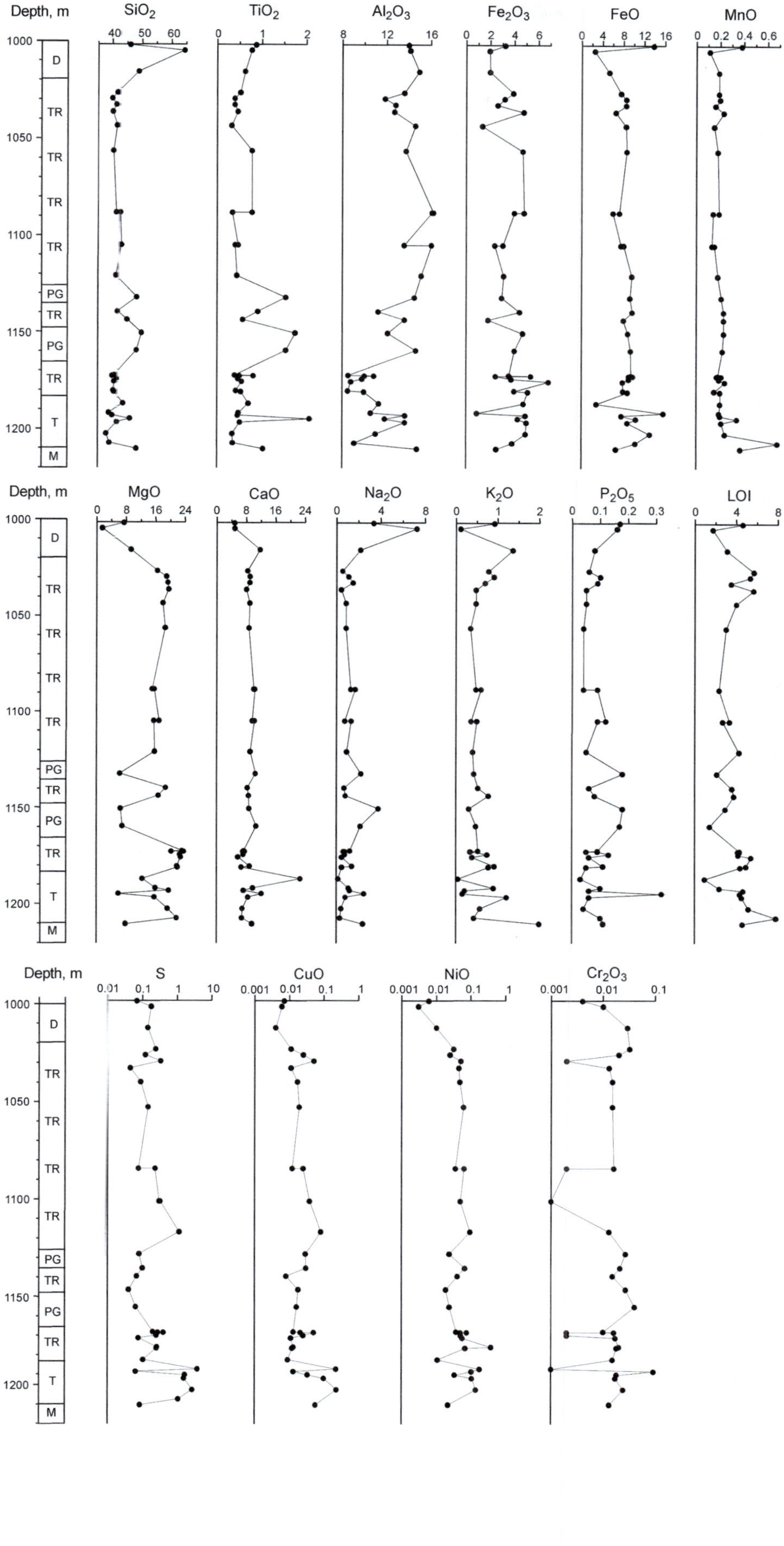

Fig. 3.24 Variation diagram of the Lower Talnakh Intrusion, drill hole KZ-665. Oxide and S content and loss on ignition, wt%

Principal minerals are K–Na feldspars, albite–oligoclase, amphibole, palagonite, quartz, and clinopyroxene. Sphene, apatite, titanomagnetite, ilmenite, biotite, sulfides, prehnite, and epidote are common. Downward, from head to root of the Lower Talnakh Intrusion, metadiorite changes into prismatically granular gabbrodolerite; clinopyroxene and intermediate plagioclase content increases and the portion of K–Na feldspar decreases. The transition of olivine-free gabbrodolerite into underlying troctolitic gabbrodolerite is relatively sharp.

Metadiorite from the Tulaek-Taas River Intrusion (see Rock indication 150) consists of prismatic albite $Ab_{99}An_1$ crystals, brownish-green amphibole ($f = 92.1$ wt%) crystals, palagonite, and sphene with plate-like ilmenite grains. Amphibole contains 0.66 wt% TiO_2, 4.08 wt% Al_2O_3, 1.08 wt% Na_2O, 0.11 wt% Cr_2O_3, and 0.32 wt% V_2O_5. Ilmenite contains 0.93 wt% MnO, and other impurity elements are below a few hundredths of a percent.

Troctolitic gabbrodolerite has porphyritic, ophitopoikilo-ophitic, panidiomorphic, hypidiomorphic granular, and poikilo-ophitic textures. Olivine ranges from 10–15% to 25–30% and usually forms subidiomorphic crystals with perfect cleavage, isometric grains, and more rarely palmate crystals and granules (see Rock indications 151 and 152). Olivine composition (Fa_{18-26}) is virtually unchanged over the whole troctolitic gabbrodolerite cross section. Plagioclase intergrowths and rounded inclusions composed of biotite, serpentine, chlorite, and magnetite are not infrequent in large olivine crystals. Plagioclase forms wide tabular crystals, large prisms, and rarely twinned prismatic crystals that impart a porphyritic texture to the rock. Prismatic crystals are also observed in the groundmass. The composition of porphyritic plagioclase is An_{88-85}, and inclusions in olivine are An_{72}. In small plagioclase prisms in the matrix, reaction rims of Ab_{100} are present. Poikilocrysts of clinopyroxene are Mg-rich augite $Wo_{45}En_{46}Fs_9$. Ore minerals include titanomagnetite and ilmenite (1–3%), sulfides are rare, and biotite is abundant.

Picritic gabbrodolerite in the intrusion sequences form interlayers in troctolitic gabbrodolerite. The texture of the rocks is segregated: poikilo-ophitic, ophitic, and poikilitic. Olivine accounts for 30–45% of the rock and has a Fa_{20-22} composition. Plagioclase (An_{86-83}) forms porphyric aggregates, prismatic crystals, and intergrowths in olivine An_{45}. Clinopyroxene is Mg-rich augite $Wo_{42}En_{49}Fs_9$. Other minerals in the rock include sparse biotite leaves, titanomagnetite and ilmenite (3–5%), and sulfides (0.5–3%) (see Rock indications 153 and 154). Sulfides form interstitial and drop-like disseminations. Near the base of the intrusion, thin interlayers of picritic gabbrodolerite with rich (up to 15–25%) sulfide dissemination occur. In these layers, the Fe content increases to Fa_{26-29}. There is an absence of chrome spinel in olivine-rich picritic and troctolitic gabbrodolerite of the Lower Talnakh Intrusion.

Taxitic gabbrodolerite is observed sporadically near the base of the intrusion. The rock horizon reaches 10 m thick and is composed of large subidiomorphic crystals and granules of olivine Fa_{27-29} and large clinopyroxene poikilocrysts with plagioclase intergrowths, as well as fine-grained ophitic pyroxene–plagioclase aggregate in the matrix. Areas composed entirely of serpentinized microgranoblastic aggregate and intercalations of olivine (Fa_{24}) and olivine gabbro are also observed. The volume of sulfide in the rocks ranges from 5 to 15%; in some cases, it reaches 25–30%.

Geochemistry and Mineralogy

The intrusion is noted for the large volume of rock enriched in magnesium, such as picritic and troctolitic gabbrodolerite (see Appendix, Tables A.31 and A.32). At the intrusion top, a horizon of metadiorite is formed up to 100 m thick (in the leading parts of the intrusion) accounting for 25% of the sequence. It decreases to 40 m and 10–20 m in middle and basal parts of the intrusion. Metadiorites feature a high SiO_2 content (52.24–56.64 wt%), high Na_2O content (0.51–7.70 wt%), and high K_2O content (1.55–5.08 wt%), with the total alkalinity reaching 5.59–9.49 wt%. The P_2O_5 content in these rocks reaches 2.08 wt%, and MgO does not exceed 3.16 wt% (Zolotukhin et al. 1975). Troctolitic varieties predominate among magnesian rocks, and only an insignificant portion of rocks is picritic gabbrodolerite. On the variation diagram (Fig. 3.24), a fairly sharp transition from metadiorites to troctolitic gabbrodolerite can be seen. The SiO_2 content in metadiorite reaches 64.4 wt%, and Na_2O reaches 7.25 wt% (See Appendix, Table A.31). The maximum K_2O content is 1.36 wt%, but in most of the sequence, it is significantly lower (1.0 wt%). In troctolitic gabbrodolerite, a high MgO content (15.07–20.29 wt%) and high K_2O content (10.86–16.24 wt%) are observed. In picritic gabbrodolerite, MgO content is 20.29–23.64 wt %, and K_2O content is 8.49–10.0 wt%. Other oxide contents are listed in Table 3.6.

In apophyses of the Lower Talnakh Intrusion intersected by drill holes in the Kharaelakh River and Kharaelakh Lake region, metadiorite and troctolitic gabbrodolerite are recognized in a sill-like body with a thickness of 40–60 m. Metadiorite retains high silicic acidity (53.28–59.4 wt% SiO_2) and high alkalinity (1.42–6.36 Na_2O, 0.14–6.8 wt% K_2O) (Zolotukhin et al. 1975).

Disseminated sulfide is observed in the upper pegmatoids of the Lower Talnakh Intrusion; these are also found sporadically in different parts of the intrusion (Avgustinchik 1981), and elevated sulfide concentrations are restricted to picritic and taxitic gabbrodolerite in near-basal parts. In some cases,

Table 3.6 The Lower Talnakhsky-type intrusion rocks' average content

	Lower Talnakh intrusion								
	Hole KZ-665				Hole SV-16				Lower Noril'sk intrusion
	TR		*P*		TR		*P*		*P*
Oxides	*X* (17)	δ	*X* (9)	δ	*X* (13)	δ	*X* (5)	δ	*X* (4)
SiO_2	41.27	1.66	40.71	1.26	42.68	1.14	38.36	1.50	40.84
TiO_2	0.50	0.52	0.15	0.44	0.03	0.39	0.07	0.63	0.63
Al_2O_3	13.53	1.74	9.64	1.01	15.34	1.66	7.52	2.12	8.73
Fe_2O_3	3.52	1.35	4.41	1.28	2.92	0.46	6.04	1.49	4.18
FeO	8.80	2.36	8.36	2.19	6.65	1.47	7.90	0.74	9.61
MnO	0.19	0.03	0.25	0.16	0.18	0.02	0.22	0.02	0.28
MgO	17.40	1.66	21.19	3.42	13.90	1.26	23.34	2.71	22.47
CaO	8.88	0.95	8.93	5.24	10.55	1.32	5.84	1.65	6.64
Na_2O	0.96	0.36	0.69	0.40	0.81	0.16	0.59	0.27	0.65
K_2O	0.60	0.25	0.50	0.27	0.80	0.41	0.45	0.12	0.80
P_2O_5	0.07	0.02	0.07	0.03	0.03	0.02	0.03	0.02	0.07
LOI	3.90	1.14	4.50	1.76	5.38	1.08	8.76	1.82	5.10[a]
Total	99.62		99.40		99.27		99.12		100.00

The notice. *X* the average (in brackets—number of analyses), δ standard deviation
[a]Calculated data

veinlets of sulfide are observed in gabbrodolerite, and copper ores are restricted to contact rocks. All these types of sulfide mineralization are observed in the described sequences of the Lower Talnakh Intrusion.

Additional information on the structure of the Lower Talnakh Intrusion can be extracted from drill hole SV-16, which penetrated the deep part of the magmatic body where the Lower Talnakh Intrusion overlies the Upper Talnakh Intrusion (Fig. 3.25). As can be seen in the figure, most of the Lower Talnakh Intrusion is composed of troctolitic gabbrodolerite, and the upper part features a horizon of olivine-bearing gabbrodolerite. Near the base of the intrusion, olivine gabbrodolerite changes successively into picritic and then taxitic gabbrodolerite. Downward through the intrusion, a decrease in SiO_2, Al_2O_3, Na_2O, and K_2O content and increase in MgO, Fe_2O_3, FeO, NiO, and CuO content are observed. Within the same rock horizon, quantitative variations of oxides are observed. Inclusions and interlayers of mafic pegmatoids impart a sawtooth shape to plots for rocks of the same horizon.

A comparison between the Upper and Lower Talnakh Intrusions intersected in the same drill hole allows one to see differences in elemental content and variations in chemical behavior. Notably, most of the Lower Talnakh Intrusion is composed of troctolitic gabbrodolerite, whereas the Upper Talnakh Intrusion is composed mostly of olivine-bearing gabbrodolerite. This manifests in the Upper Talnakh Intrusion as higher SiO_2, TiO_2, Na_2O, K_2O, P_2O_5, and V_2O_5 content; lower MgO content; and Al_2O_3, Fe_2O_3, FeO, CaO, and MnO content close to troctolitic gabbrodolerite. Transitions between horizons of mafic and ultramafic rocks are sharp in both intrusions. Trends in chemical element content downward through the cross section of the Upper and Lower Talnakh Intrusion are similar for some components and contrary for others. It is remarkable that the same trend is observed for some components in both magmatic bodies.

Changes in the internal structure of the Lower Talnakh Intrusion downward along dip can be viewed by comparing two intersections of the body (Figs. 3.24 and 3.25) and chemical analysis of the rocks (Table 3.6; Appendix, Tables A.31 and A.33). Judging from these data, MgO content decreases in troctolitic gabbrodolerite from the leading to basal parts of the body and increases in picritic gabbrodolerite.

3.6.2.2 Lower Noril'sk Intrusion

The Lower Noril'sk Intrusion features a large volume of olivine-free rocks such as troctolitic and picritic gabbrodolerite that account for about 80% of the sequence and the presence of a peculiar pegmatoid formation in the upper contact zone of the intrusion. The intrusion is about 150 m thick.

The intrusion roof contact is commonly altered due to alkalic–silicic acid transformations, among them relict olivine-bearing poikilo-ophitic gabbrodolerite with small isometric olivine grains (about 3%). At the intrusion base, in different intersections, rocks are composed of olivine-bearing (about 1%) or taxitic-like olivine (up to 20–30%) gabbrodolerite. The thickness of the contact zones is 0.5–3.0 m.

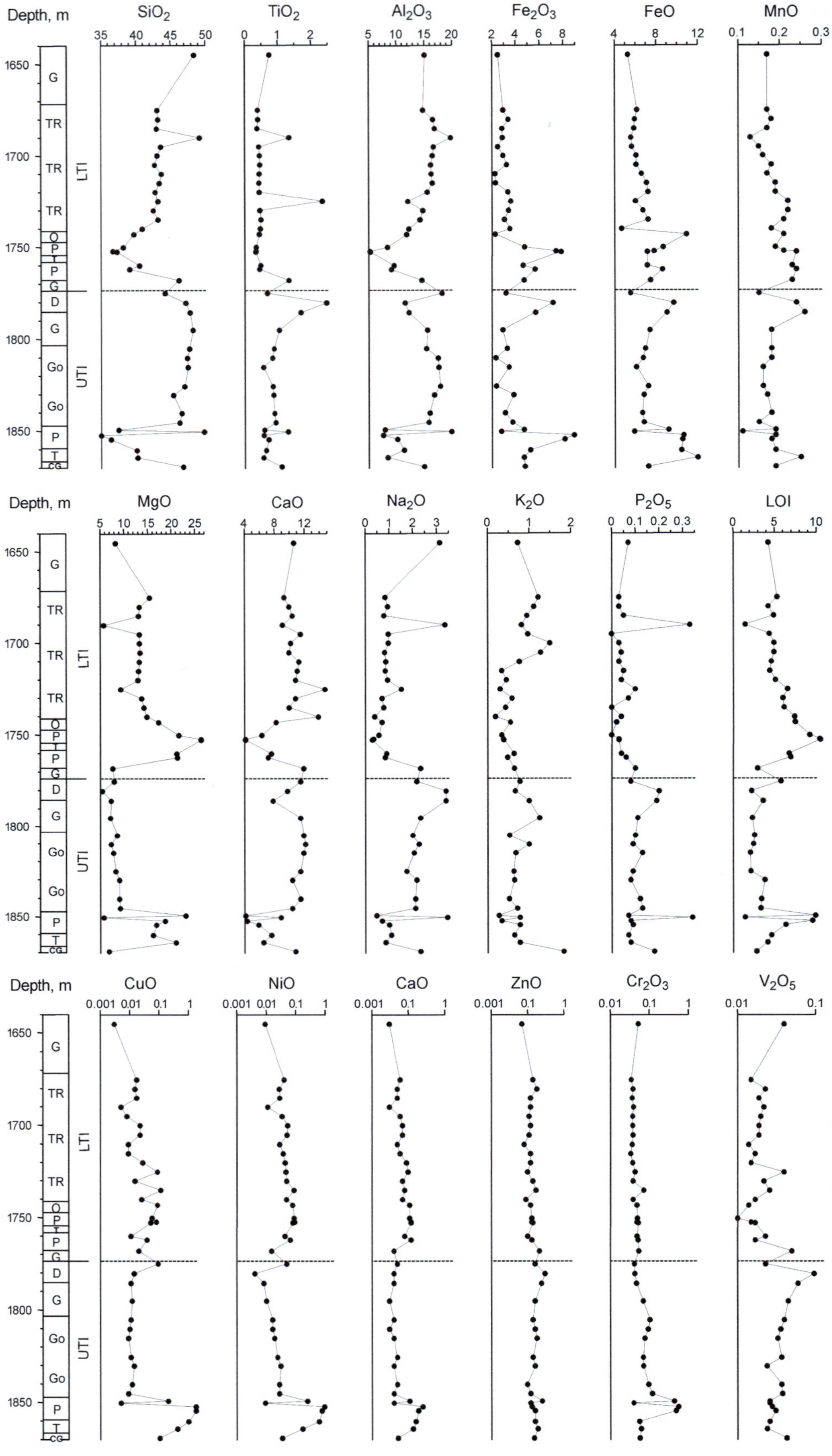

Fig. 3.25 Petrochemical variation diagram of a united section of the Lower and Upper Talnakh Intrusions, drill hole SV-16. Oxide content, wt%

Upper Zone

The upper zone of the intrusion consists of diorite, gabbroic diorite, monzonite, palagonite metadiorite, prismatically granular gabbrodolerite, and gabbropegmatites, which occur in different parts of the intrusion. Principal rock-forming minerals in these rocks are K–Na feldspars, quartz as individual grains (up to 5–7%), and, in micropegmatite, amphibole (up to 6–8%) and palagonite. Other minerals include (sometimes as aggregates) clinopyroxene, biotite, sphene, apatite, titanomagnetite, ilmenite, chlorite, and epidote. The volume of micropegmatite matrix sometimes is so significant that the rock can be classified as granophyre. In other cases, granophyre is absent. In some rocks, palagonite-I is common (up to 15–20%) and fills interstices and is observed as aggregates with intergrowths of plagioclase prisms resembling clinopyroxene oikocrysts (that are absent in this rock). Prismatically granular gabbrodolerite and gabbropegmatite are composed of plagioclase prisms, partly amphibolized clinopyroxene, skeletal titanomagnetite crystals, palmate olivine crystals commonly replaced by bowlingite, and palagonite in the matrix.

Between these coarse-grained silicate rocks and the underlying troctolitic gabbrodolerite, a transition zone composed of olivine-bearing and olivine gabbrodolerite with poikilo-ophitic texture is located. In these rocks, small olivine grains (3–10%) in the rock groundmass formed as intergrowths in other silicates.

Central Zone

Most intrusion rocks are olivine-rich troctolitic, picritic, and picrite-like gabbrodolerite (Zemskova 1981). The rocks have an ophitic, poikilo-ophitic, or a segregation or granular texture. The rock fabric is determined by olivine and plagioclase. Clinopyroxene is always present in minor amounts and orthopyroxene is observed sporadically.

The volumes of olivine troctolitic gabbrodolerite vary from 15 to 45% in troctolitic gabbrodolerite and from 50 to 65% in picritic gabbrodolerite (see Rock indication 155). Areas of picrite and troctolite with similar amounts of olivine exhibit gradual transitions. The maximum olivine content in the Lower Noril'sk Intrusion is restricted to the central part of the horizon (see Rock indication 156), repeating the olivine distribution in the Putanaya Intrusion. In some cases, the uniformity of the horizon of troctolitic gabbrodolerite is violated by thin lenses, veinlets, or rounded leucogabbro inclusions. In these cases, troctolite adjacent to leucogabbro is strongly enriched in olivine and changes into picritic gabbrodolerite.

Olivine crystals commonly have subidiomorphic outlines, but palmate, dendritic, and isometric grains with oval plagioclase intergrowths are also observed. Large olivine oikocrysts (6 mm) with plagioclase chadocrysts are less common. The olivine composition is Fa_{23-24}. Olivine crystals in troctolitic gabbrodolerite are, as a rule, large (3–6 mm), with perfect cleavage and occasional solid solution breakdown structures that form a dash-dot magnetite intergrowth, which preferentially form in olivine at the basal parts of the horizon. High olivine concentration produces the segregation texture. It is interesting to note the coexistence of olivine with similar composition (Fa_{20-25}) but sharply different in sizes: from 0.03–0.06 mm for granules enclosed by pyroxene to 1–3 mm for subidiomorphic crystals. The latter contains no inclusions, except biotite–serpentine globules, and often has cleavage, but no breakdown structures (Ryabov 1992a). The clinopyroxene composition is $Wo_{42}En_{49}Fs_9$ and that of orthopyroxene is $Wo_1En_{76}Fs_{23}$. The total Fe content in biotite is 20.6–27.2 wt%, and halogen content is 0.23–0.83 wt% F and 0.07–0.28 wt% Cl. In olivine-rich gabbrodolerite, minerals of the spinel group are mainly represented by magnetite poor in titanium, and, in a single case, chrome spinel (17.6 wt% Cr_2O_3) has been found as an inclusion in olivine (see Rock indication 155). The sulfide mineralization in the intrusion sequence is a poor interstitial dissemination, which is occasionally observed in different rocks, as well as thin interlayers in picritic gabbrodolerite with rich (up to 10–15%) droplet and interstitial mineralization.

Lower Zone

The lower zone consists of a variable thickness (0–3 m) interlayer of olivine-rich taxitic gabbrodolerite or a horizon of olivine-poor (3–7%) gabbrodolerite 8–10 m in thickness. For rocks of the LZ, as well as troctolitic gabbrodolerite, the coexistence of very small olivine granules and large palmate crystals is characteristic.

Internal features of the intrusion structure are reflected in their chemical composition (Fig. 3.26). On the variation diagram, two main contrasting horizons of rocks are distinguished: alkaline silicate pegmatoids and picritic gabbrodolerite. Pegmatoids contain 49.54–59.48 wt% SiO_2, 1.13–1.45 wt% TiO_2, 16.7–19.68 wt% Al_2O_3, 2.58–3.67 wt% Fe_2O_3, 5.45–6.93 wt% FeO, 2.19–6.96 wt% MgO, 2.44–7.11 wt% CaO, 2.67–3.60 wt% Na_2O, 1.56–2.75 wt% K_2O, and 0.13–0.33 wt% P_2O_5. For picritic gabbrodolerite, a high magnesium content (MgO = 18.38–28.70 wt%), low titanium content (TiO_2 = 0.44–0.77 wt%) and chromium content (Cr_2O_3 = 0.003–0.005 wt%), and an aluminum content of 5.24–9.27 wt% and alkali content of ($Na_2O + K_2O$) 0.52–1.98 wt% are characteristic. A comparison of gabbrodolerite compositions from the Lower Noril'sk and Lower Talnakh Intrusions (see Table 3.6) shows a strong similarity in rock-forming oxide content in the rocks. In addition, it is noted that iron in picritic gabbrodolerite of the Lower Noril'sk Intrusion is less oxidized, and potassium dominates over sodium.

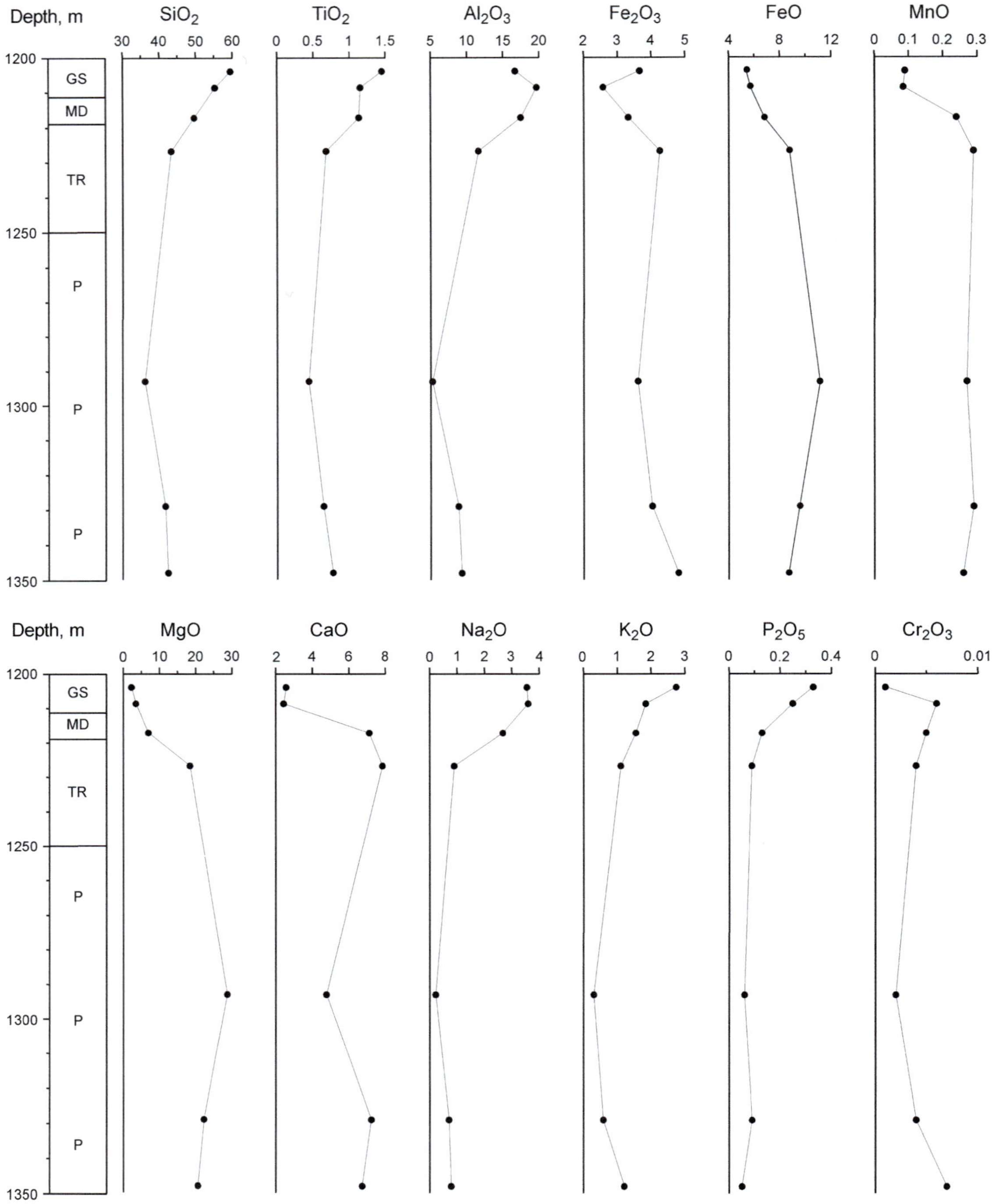

Fig. 3.26 Petrochemical variation diagram of a united section of the Lower Noril'sk Intrusion, drill hole NP-4. Oxide content, wt%

3.6.2.3 Petrology and Aspects of Genesis

The petrological characteristics of the Lower Talnakh type of intrusions are as follows:

1. A large volume of olivine-rich rocks (up to 80%) and the presence of alkaline mafic pegmatoids in the UZ.
2. The composition of picritic and troctolitic gabbrodolerite fits to chrysolite Fa_{20-28}, and NiO content (0.03–0.08 wt%) is evidence of depletion in silicate nickel as compared to the normal geochemical trend (Ryabov 1992b).
3. Picritic and troctolitic gabbrodolerites have the porphyritic texture. Plagioclase porphyrocrysts have a bytownite composition.
4. Subidiomorphic olivine crystals have a cleavage and contain lamellae of oxide-ore phase. Plagioclase of bytownite—labradorite composition is observed in these as inclusions.
5. The principal oxide-ore phases in olivine-rich rocks are magnetite and ilmenite as well as rarely chrome spinel (17.6 wt% Cr_2O_3) that is observed as inclusions in olivine.

6. Phlogopite ($f = 21$–27 wt%) is consistently found in the rocks, the greatest volume observed in gabbrodolerite of the Zelenaya Griva Intrusion (see Rock indication 157).
7. Picritic gabbrodolerite from a paragenesis comprising $Fa_{20-26} + Wo_{46}En_{47}Fs_{7} + Wo_{42-43}En_{49-48}Fs_{9} + Wo_{1}En_{76}Fs_{23} + Ab_{13-30}Or_{0-1}An_{86-69} + Ab_{52}Or_{3}An_{45} + Ilm$.
8. The intrusions are barren, although local rich disseminated segregations and veinlets of sulfides are found in gabbrodolerite, and copper ores are observed at the contact.
9. Intrusions have no effusive comagmates so they can be regarded as members of magmatogenic columns.

Genetic Concepts

The large volume of high-magnesium rocks in the Lower Talnakh Intrusion, as well as their porphyritic fabric (plagioclase insets), is the basis for their assignment as a Morongovsky-type intrusion (Komarova and Lyul'ko 1967; Dodin and Sadikov 1967). The investigations undertaken allowed geologists to recognize the independent Lower Talnakhsky type as a member of the Noril'sky intrusive complex. Sulfide mineralization in the UZ of the Lower Talnakh Intrusion was considered to be one of the main arguments in favor of a later age for this intrusion. The sulfide mineralization attracted the attention of geologists in sections located close to each other at the top of the Upper Talnakh and at the bottom of the Lower Talnakh Intrusions. This argument is being described in the literature, although geologists have long been aware that the upper contact zones in all layered intrusions, including the Lower Talnakh one, are characterized by increased sulfide content (Almukhamedov and Medvedev 1982; Ryabov 1989a). Therefore, the criterion of "superimposed" sulfide mineralization above the Upper Talnakh Intrusion is not, at least, perfect.

Study of the flanks of the Talnakhsky ore deposit revealed cases of injection of the Lower Talnakh Intrusion into the Upper Talnakh Intrusion, the conversion of xenolites of mineralized picritic gabbrodolerite from the Upper Talnakh Intrusion, and preservation of shadow relicts of Upper Talnakh Intrusion in rocks of the Lower Talnakh Intrusion (Sukhareva and Kuznetsova 1983). Based on these and other data present in the work, these authors have conclusions about successive intrusions close in time and the later generation of the Lower Talnakh Intrusion.

Some geologists compare the ore-bearing Noril'sk-I and Upper Talnakh Intrusions and the barren Lower Talnakh and Lower Noril'sk Intrusions and consider that the criteria for these intrusion to be recognized and assigned to ore-bearing or barren types are important (Zemskova 1981). In addition, there is the concept that barren high-magnesian bodies of the Lower Talnakh and Lower Noril'sk Intrusions and ore-bearing Noril'sk-I and Upper Talnakh Intrusions within the Noril'sky and Talnakhsky ore junctions are regularly adjacent. In this way, bodies of the Lower Talnakhsky type depleted in ore elements are considered to be indispensable constituents of the ore–magmatic system. These are not incidental magmatic formations but constituents of ore systems, and their observation can be considered as an exploration criterion for ore-bearing bodies (Ryabov 1980, 1992a, 1999b, 1984a, b). The Lower Talnakhsky-type bodies are assumed to represent the rear part of the magmatic column, the leading part containing ore-bearing intrusions of the Noril'sky type. Lower Talnakh and Lower Noril'sk Intrusions had primarily a picrite-basalt composition, but as a result of fluid movement, the intrusions are depleted in ore elements Ni, Cr, and platinum group elements that were accumulated in the ore-bearing horizons of the Noril'sky-type intrusions. Geochemistry of these elements will be considered in detail in a special.

3.6.3 Kruglogorsky Type Intrusions $(\pi\nu - \nu\beta)\ T_{1\ nr}^{kg}$

Intrusions of this type are known in the literature as sills of leucocratic gabbro (Korovyakov et al. 1963; Likhachev 1965). The intrusions form sheeted and gently cutting bodies 10–20 to 40 m in thickness that are separated from the ore-bearing Noril'sk-I and Upper Talnakh Intrusions as apophyses. Moreover, in the Noril'sky ore junction, such rocks constitute a bridge between the western (Ugol'ny Stream) and eastern (Medvezhy Stream) intrusion branches of the Noril'sk-I Intrusion. The subsheeted branch from the Mt. Chernaya Intrusion is referred to as the intrusion of Mt. Kruglaya (Kruglogorsky Intrusion). Such formations are also observed at the western and eastern flanks of the Upper Talnakh Intrusion where those were referred to as intrusions of the Lesnaya Ridge and the Olor River and in earlier works as the Gabbrovaya Intrusion.

The intrusions consist of porphyritic gabbrodolerite with variable olivine content (sometimes quartz bearing) and leucogabbro. The leucogabbro is characteristic by the great amount (from 40–60% to 70%) of large mafic plagioclase tablets, which produce a coarse-grained gabbroic fabric in the rock. Most observations of the Kruglogorsky Intrusion type are located near ore-bearing bodies. The thickness of leucogabbro horizon decreases as distance from ore-bearing bodies increases, the horizon becomes discontinuous, and it is interrupted by small bodies of leucocratic gabbro. Cross sections of the Kruglogorsky Intrusion type most distant from ore-bearing bodies form pea-like textured dolerite. The horizon of leucogabbro and coarse porphyric gabbrodolerite is commonly located in the axial part of a sill or in its UZ. Near the axial part of the sill, subhorizontal interlayers (3–5 cm in thickness) of rocks enriched in phenocrysts are not infrequent in the gabbrodolerite.

3.6.3.1 Intrusion of the Olor River

It is a subsheeted apophysis of the Upper Talnakh ore-bearing body. Its thickness varies from 15 to 60 m. In the intrusion sequence, upper, lower, and central contact zones are recognized (Fig. 3.27). The contact zones have a similar thickness from 4–6 m to 12 m depending on the total sill thickness. The thickness of the central zone varies from 6–10 m to 23 m.

The contact zones are composed of porphyritic and poikilo-ophitic dolerite with sparse amygdales. Phenocrysts (1–5%) are plagioclase $Ab_{24-35}Or_{1-2}An_{75-63}$. Clinopyroxene oikocrysts have the composition $Wo_{32}En_{41}Fs_{27}$, and chadocrysts of olivine have the composition $Ab_{21}Or_{1}An_{78}$ (see Rock indication 158). In the rock, leaflets of brownish biotite ($f = 55$ wt%) and ilmenite are observed. Farther away from the sill contact toward the center, the sizes of clinopyroxene poikilocrysts increase from 0.1–0.5 mm to 10 mm. Directly at the intrusion contact, a thin (5–15 cm) quenched crust composed of tachylites with trachytoidal arrangement of plagioclase laths (see Rock indication 159) and microporphyric olivine grains entirely replaced by bowlingite is observed.

Along the eastern flank of the Upper Talnakh Intrusion, clinopyroxene oikocrysts at the upper contact with poikilo-ophitic dolerite have the composition $Wo_{40-41}En_{44-39}Fs_{17-20}$; the composition of plagioclase enclosed by clinopyroxene is $Ab_{35}Or_{4}An_{61}$ (see Rock indication 160). Plagioclase laths in the rock matrix have a more acidic composition with relict crystals of $Ab_{31}Or_{18}An_{51}$. Brownish biotite ($f = 67.9$ wt%) is restricted to the matrix. Biotite in intergrowths with titanomagnetite is characterized by higher Fe content ($f = 73.4$ wt%). Ore minerals in dolerite include titanomagnetite and ilmenite (see Rock indications 159–160).

The *central zone* is a layered series of rocks and leucocratic gabbro. The layered series is composed of porphyritic, poikilo-ophitic dolerite and gabbrodolerite, which include olivine, olivine-bearing, and quartz-bearing varieties. Porphyritic plagioclase crystals are unevenly distributed throughout the intrusion, forming layers poor (from 1% to 3–5%) and enriched (25–30%) in phenocrysts. Overall, the sill is slightly layered; successive changes from olivine-rich through olivine-bearing rocks to quartz-bearing or quartz–olivine-bearing rocks (see Rock indication 161) are observed downward through the intrusion. In thin sills, the end members of the rock series are absent. In this case, the intrusion sequence is represented by contact dolerite or gabbrodolerite and in the central zone is composed of olivine-bearing dolerite or gabbrodolerite. The amount of olivine is commonly 3–7% in olivine-bearing gabbrodolerite; it increases to 10% in olivine gabbrodolerite and decreases to 1–3% in quartz–olivine-bearing rocks. Olivine forms small isometric crystals of Fa_{35-38} in olivine gabbrodolerite and Fa_{53} in olivine-bearing gabbrodolerite.

In the layered series, plagioclase forms porphyritic grains and intergrowths with clinopyroxene that create the poikilo-ophitic texture. Plagioclase is also found in the rock matrix. In olivine gabbrodolerite, phenocrysts and chadocrysts of plagioclase enclosed in clinopyroxene oikocrysts have similar compositions $Ab_{29-1}Or_{1}An_{70-7}$. In the matrix of these rocks, K–Na feldspar $Ab_{28}Or_{61}An_{11}$ is observed. In olivine-bearing gabbrodolerite, porphyritic plagioclase crystals have the composition $Ab_{22-31}Or_{1-2}An_{77-67}$ in crystal cores, and in crystals rims, plagioclase is deoxidized to $Ab_{59-60}Or_{9-14}An_{32-26}$. Plagioclase phenocrysts in quartz–olivine-bearing gabbrodolerite are characterized by the highest maficity $Ab_{17-28}Or_{1-2}An_{81-70}$. Inclusions of K-feldspar $Ab_{13}Or_{78}An_{9}$ found in these phenocrysts are unusual.

In rocks of the central zone, clinopyroxene forms oikocrysts with plagioclase intergrowths and small angular grains in the rock matrix. The clinopyroxene composition varies upward through the central zone from the axis of the sill to the roof. Clinopyroxene composition in olivine and olivine-bearing gabbrodolerite is $Wo_{37-36}En_{45-44}Fs_{18-20}$, and upward through the intrusion, its composition is $Wo_{42-37}En_{33-4}Fs_{25-29}$ and $Wo_{34}En_{35}Fs_{31}$ in olivine-bearing gabbrodolerite and in quartz–olivine-bearing gabbrodolerite, respectively. In the matrix of olivine-bearing gabbrodolerite, small pigeonite grains $Wo_{10}En_{49}Fs_{41}$ are found. Other minerals include biotite, chlorite, and more rarely amphibole, and ore minerals, titanomagnetite, ilmenite, and sporadic sulfides are also observed.

The *horizon of leucocratic gabbro* is located in the UZ at the boundary with contact dolerite and the layered series, or it extends along the axis of the sill. The contact between leucogabbro and their host rocks is abrupt (without traces of quenching), and a change from one to the other is marked by taxito-ophitic gabbrodolerite. Leucogabbros in the Olor River Intrusion, distant from the main body of the Upper Talnakh Intrusion, are outlined by small (several to tens centimeters) "erratic boulders." It is remarkable that these are sometimes observed with vitreous quenched facies.

The leucogabbros have a coarse-grained gabbroic texture. The taxito-ophitic variety of the densely porphyritic gabbrodolerite contains 60–70% phenocrysts, and the groundmass has an ophitic, prismatically ophitic, or gabbroic texture. The principal features of the rocks are the high plagioclase content (up to 80%), tabular or short-prismatical pattern of its larger crystals, and its highly mafic composition. Leucogabbro from the Olor River Intrusion consists of plagioclase $Ab_{23-26}Or_{2}An_{74-72}$, clinopyroxene $Wo_{40}En_{41}Fs_{19}$, biotite, and ilmenite (see Rock indication 166).

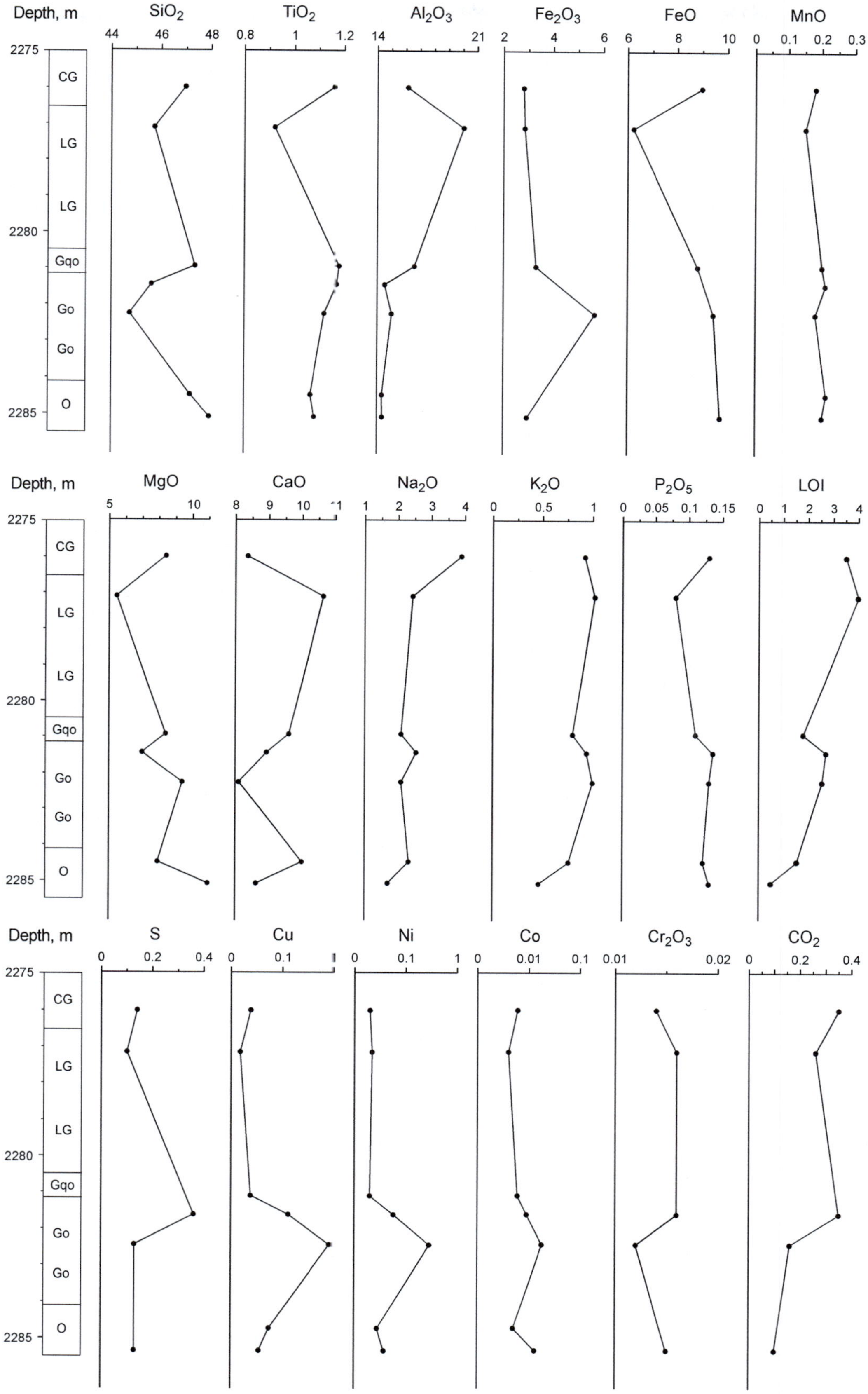

Fig. 3.27 Variation diagrams of oxides through a section of the Kruglogorsky-type intrusion, drill hole VKH-4. Oxide and S content and loss on ignition, wt%, element content, ppm

With regard to the extent of crystallization and grain sizes of the rocks in the intrusion, it should be noted there is a successive increase in sizes of "peas" of clinopyroxene poikilocrysts from the contacts of the sill to its axis. The regular increase in "peas" is most obvious in the contact zones of the intrusion (from 0.5 to 5 mm), and sizes of "peas" are maximal (8–12 mm in diameter) in the center of the sill, but not infrequently, coarse-grained pea-like gabbrodolerites are distinguished as a layer within overlying and underlying gabbrodolerite that have "peas" of smaller sizes (3–5 mm). The marked regularity is complicated in pegmatoids including leucogabbro. In this case, the horizon of coarse-grained rocks is located in the upper contact zone of the sill under the upper contact dolerite. The distribution of olivine through the sill is similar to that for pea-like grain sizes: maxima near the center decreasing to the contacts.

Schlieren of syenites and gabbroic syenite are found along the eastern flanks of the Upper Talnakh Intrusion. They are present in the UZ of sills as small (10–50 cm) homogeneous inclusions with sharp contacts without quenched zones. The characteristic feature of rocks in such intrusion sequences is the porphyritic texture with alternating bands enriched and poor in phenocrysts of feldspar. The rocks with schlieren-like inclusions have prismatically granular and hypidiomorphic granular textures. Gabbroic syenite is composed of large plagioclase $Ab_{83-87}Or_4An_{13-9}$ prisms, small laths of K–Na feldspar $Ab_{56}Or_{40}An_1$, and clinopyroxene $Wo_{46-40}En_{41}Fs_{13-19}$. Other minerals include sphene, quartz, chlorite, calcite, prehnite, and apatite; ore minerals are represented by ilmenite (see Rock indication 167). Syenites are composed of large albite $Ab_{90-93}Or_{5-3}An_{5-4}$ prisms, long prismatic crystals of altered amphibole, phlogopite ($f = 23$–28 wt%), chlorite ($f = 33$–43 wt%), quartz, calcite, palagonite, and small amount of apatite. From ore minerals, magnetite and ilmenite are present (see Rock indication 168).

Sulfide mineralization in the Kruglogorsky type of intrusions appears as a poor dissemination in different parts of the intrusion. Elevated concentrations of sulfide (to 3–5%, and in some cases to 10–15%) are commonly restricted to the UZ of layered bodies, axes of thin sills, or near-bottom zones. The distribution of sulfide in the rocks is uneven. Sulfide accumulations from 2–3 cm to 3–6 cm form patches 5–7 cm in size, which form layers 0.2 m thick. Disseminated sulfides are observed as interstices, small schlieren, and veinlets. The principal minerals are pyrrhotite, chalcopyrite and pyrite, and, more rarely, pentlandite. Increased concentrations of platinum group minerals (up to 3.3 g/t) are restricted to sulfide-rich rocks.

Metamorphic transformation of country rocks is typically observed as hornfelsing. The thickness of hornfelsed rocks is less than several meters. In some cases, at the contact with the intrusion, thin veinlets of calcareous skarns with related magnetite and pyrite mineralization are observed.

3.6.3.2 Mineralogy, Geochemistry, and Genetic Aspects

The following compositional features of rock-forming minerals in intrusions of the Kruglogorsky type should be emphasized.

The composition of plagioclase varies vertically through and along the intrusion cross section. The overall range of compositions covers a series from bytownite to albite. Cores of large phenocrysts $Ab_{17}Or_1An_{82}$–$Ab_{31}Or_2An_{67}$ are most mafic (see Rock indications 158, 161, 163–165), which are enclosed in quartz–olivine-bearing gabbrodolerite and contact dolerite. Zoning of porphyritic crystals is variable. In rock varieties with slightly pronounced zoning of plagioclase, K feldspars $Ab_{13-28}Or_{78-61}An_{9-11}$ occur. The composition of plagioclase chadocrysts exhibits slight variability ($Ab_{21-38}Or_{1-2}An_{78-60}$), and in a number of cases, this is the main generation of plagioclase in the rock (see Rock indications 158 and 162).

It has been ascertained in thin (5 m thick) subsheeted apophyses of the intrusion that porphyritic tabular crystals are $Ab_{39}Or_2An_{59}$ in olivine-bearing gabbrodolerite from the central part of the body, and chadocrysts have more mafic composition $Ab_{35}Or_2An_{63}$ (see Rock indication 162). At this intersection of the sill, microlites of anorthoclase $Ab_{24-18}Or_{30-49}An_{46-33}$ are observed together with laths of $Ab_{35}Or_6An_{59}$ (see Rock indication 159).

In cross section, the sill on the eastern flank of intrusion is a plagioclase-phyric dolerite that features gabbroic syenite, syenite, and leucogabbro in the UZ. The composition of large tabular plagioclase in leucogabbro varies, from core to rim, from $Ab_{26}Or_2An_{72}$ to $Ab_{38}Or_2An_{60}$. Prismatic crystals in the groundmass are $Ab_{24}Or_2An_{74}$, and chadocrysts in clinopyroxene are $Ab_{35}Or_2An_{63}$ (see Rock indication 166). Plagioclase of nearly the same composition ($Ab_{35}Or_4An_{61}$) forms chadocrysts in clinopyroxene oikocrysts in plagioclase-phyric dolerite in contact areas of the sill (see Rock indication 160). Gabbroic syenite contains plagioclase $Ab_{83-87}Or_4An_{13-9}$, K–Na feldspar $Ab_{56}Or_{40}An_4$, and orthoclase $Ab_{2-5}Or_{97-94}An_1$ (see Rock indication 167). The principal type of feldspar in syenite is albite $Ab_{93-90}Or_{3-5}An_{4-5}$ (see Rock indication 168).

The composition of *clinopyroxene* in contact facies of the sill is low Ca augite $Wo_{37-32}En_{40-1}Fs_{23-27}$ and contains of 0.71–1.07 wt% TiO_2 and 1.18–1.31 wt% Al_2O_3. In olivine gabbrodolerite and olivine-bearing gabbrodolerite in central part of sills, clinopyroxene has the lowest Fe content; its composition is $Wo_{36-36}En_{45-44}Fs_{18-20}$. Upward through the sill, Fe content in clinopyroxene increases in olivine-bearing gabbrodolerite, and its composition is $Wo_{42-37}En_{33-34}Fs_{25-29}$. In quartz–olivine-bearing gabbrodolerite, clinopyroxene has the composition $Wo_{34}En_{35}Fs_{31}$. Small pigeonite ($Wo_{10}En_{49}Fs_{41}$) grains are observed in the groundmass of olivine-bearing gabbrodolerite in association with augite. In augite from a layered series, TiO_2, Al_2O_3, and Cr_2O_3 contents are 0.60–1.06 wt%, 1.45–2.39 wt%, and 0.002–0.26 wt%.

Olivine gabbrodolerite features the highly magnesian *olivine* Fa_{35-38}. It contains 0.41–0.42 wt% MnO, 0.26–0.27 wt% CaO, and 0.099 wt% NiO. In olivine-bearing gabbrodolerite, olivine has the composition Fa_{53}, which contains 0.59 wt% MnO, 0.28 wt% CaO, and 0.043 wt% NiO.

Mica group minerals are present virtually in all intrusions as small lamellae. In syenite, the mica mineral is phlogopite; in other rocks, biotite is typical. The minimum total Fe content in the mineral is 23 wt% in syenite (see Rock indication 168) and 41–44 wt% in olivine gabbrodolerite (see Rock indication 161). The composition of mica from these rocks is characterized by increased TiO_2 (3.6 wt% and 1.65–4.87 wt %, respectively) and by prevalence of F over Cl: 2.98 wt% and 0.38 wt% in syenite and 0.60–0.93 wt% and 0.18–0.38 wt% in gabbrodolerite, respectively. Biotite in dolerite, in intrusion contact zones, has a higher total Fe content 55 wt%, TiO_2 content 4.21 wt%, F content 0.78 wt%, and Cl content 0.79 wt% (see Rock indication 158), and maximum Fe content of 68–87 wt% was determined in mica from olivine-bearing and quartz–olivine-bearing gabbrodolerite and plagioclase-phyric dolerite. The characteristic features of these iron-rich biotites are elevated concentrations of halogens and the prevalence of Cl (1.65–4.45 wt%) over F (0.57–0.69 wt%) (see Rock indications 160, 162, 164, and 165). The TiO_2 content in these micas varies from 0.27 to 3.75 wt%.

Amphibole forms small grains of brownish and green hornblende in olivine-bearing gabbrodolerite. The total Fe content of the mineral is 58 and 67 wt% in brownish and green hornblende, respectively (see Rock indication 163). In brownish hornblende, the Cl and F contents are 1.19 and 0.34 wt%, respectively.

Spinel group minerals include titanomagnetite and minor magnetite. Titanomagnetite contains 8.24 wt% TiO_2, 0.98–1.48 wt% Al_2O_3, 0.24–1.42 wt% MnO, 0.02–1.12 wt % MgO, 0.011–0.14 wt% NiO, 0.006–0.47 wt% Cr_2O_3, and 0.88–1.34 wt% V_2O_5. The content of impurity elements in magnetites does not exceed 1.0–7.5 wt%.

Ilmenite occurs virtually in all Kruglogorsky-type rocks. There are insignificant variations in impurity elements (wt%): 0.04–0.09 Al_2O_3, 0.62–2.50 MnO, 0.26–0.84 MgO, and 0.20–0.85 V_2O_5 with the exception of ilmenite in syenite in which the MnO content is as much as 4.0 wt% (see Rock indication 168).

Geochemistry

The distribution of rock-forming oxides throughout the intrusion is seen in Fig. 3.27. The contact facies of the intrusion is a plagioclase-phyric dolerite, and a thin leucogabbro layer is located at the boundary between the UZ and CZ of the intrusion. The CZ is composed of olivine-bearing gabbrodolerite with suspended interlayers of olivine gabbrodolerite and taxite-like gabbrodolerite, and quartz–olivine-bearing gabbrodolerite is located at the border with the taxite-like gabbrodolerite. In accordance with the mineralogical composition of the rocks, it can be seen on the diagram that plagioclase-rich leucogabbros have a high Al_2O_3 content, and gabbrodolerite enriched in olivine has increased MgO content. The behavior of oxides in the CZ is governed by the location of MgO-enriched rock layers, although the presence of volatile elements complicates the distribution of oxides. This is supported by sulfide mineralization, which amounts to 0.5–3% over the entire body thickness and to 7–8% in taxitic gabbrodolerite and two interlayers of gabbrodolerite (to which increased concentrations of platinum metals are restricted: 2.3 g/t in the upper and 1.7 g/t in lower interlayers) (see Rock indications 163 and 164).

The composition of rocks from intrusions of the Kruglogorsky type is marked on the classification diagram of ($Na_2O + K_2O$) versus SiO_2 (Marakushev 1976) and composition fields for rocks from the Upper Talnakh Intrusions and alkalic–silicic acid formations from the Zubovsky Intrusion (Godlevsky 1959), and the Lower Talnakh and Tulaek-Taas River Intrusions are shown for comparison (Fig. 3.28). The composition of the Kruglogorsky-type gabbroic intrusions falls within a compact field that provides evidence of their strong similarity to rocks of the Upper Talnakh Intrusion. These rocks feature significant fluctuations in $Na_2O + K_2O$ content (from 2 to 6 wt%) as compared to the relatively narrow range of variation in SiO_2 (from 45 to 50 wt%). Their analyses are therefore located in areas of the normal and subalkalic series on the plot and below the Kuno line separating these series. The sole exception is provided by syenite from schlieren on the eastern flanks of the Upper Talnakh Intrusion. These rocks with values of ($Na_2O + K_2O$) = 8.27 and SiO_2 = 58.92 wt% fall into the field that combines granite-, syenite-, and monzonite-like rocks at the roof parts of barren magnesian intrusions of the Noril'sky Complex. The variations in the content of rocks from Kruglogorsky-type intrusions, as well as the presence of syenite schlieren and K–Na feldspars and biotite in the gabbrodolerite matrix, indicate an alkaline trend in evolution of the melt and the possibility of fluid–magmatic separation of high-alkalic liquids from the melt.

Petrological Features

Characteristic features of the Kruglogorsky-type intrusions include the following:

1. Intrusions have a sheeted form and gently cut country rock with a thickness that ranges from 15–20 m to 40–50 m.
2. Intrusions are apophyses of Noril'sky-type intrusions that are located in their aureole.
3. Intrusions are slightly differentiated; layering is formed by relative enrichment in plagioclase forming leucogabbro

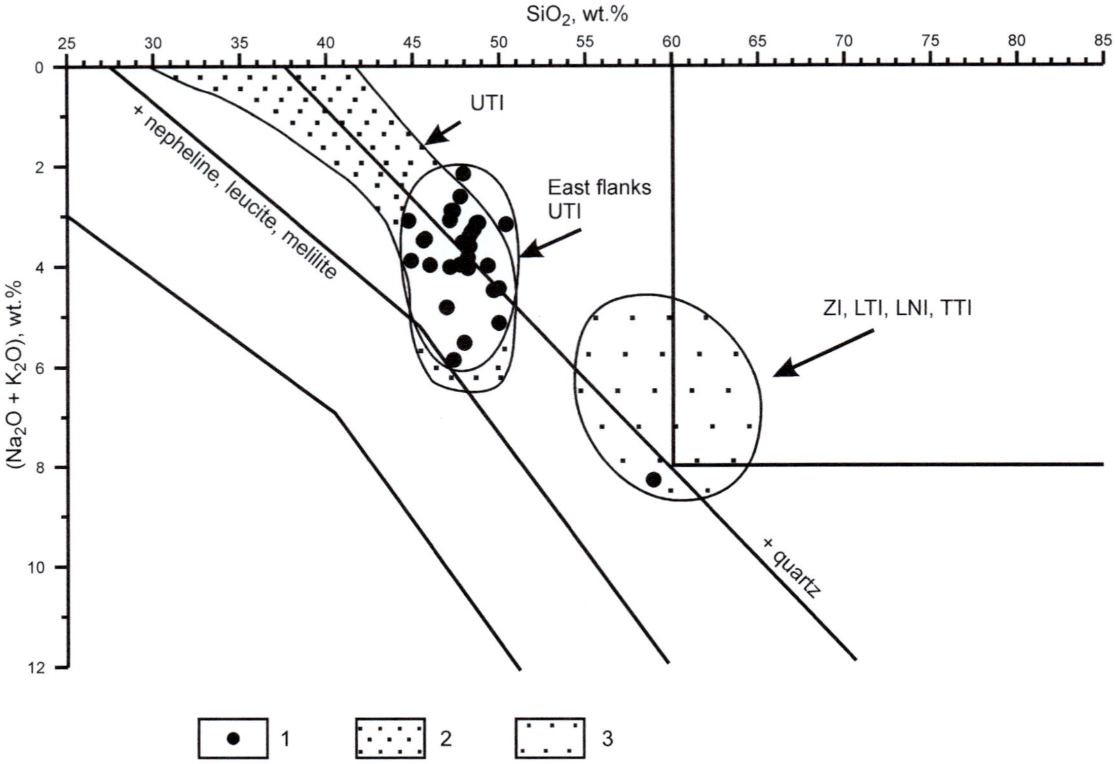

Fig. 3.28 Diagram of ($Na_2O + K_2O$) versus SiO_2 for some rocks of the Noril'sky complex intrusions. 1 rocks of Kruglogorsky-type intrusions from the eastern flanks of the Upper Talnakh Intrusion, 2 rocks of the Upper Talnakh Intrusion, 3 syenogranites, gabbrosyenites, syenites, and monazites from the Zubovsky Intrusion (from Godlevsky 1959), Lower Noril'sk, Lower Talnakh, and Tulaek-Taas River Intrusions. Analyses of rocks for reference intrusion sections are given in Rock indications and tables in Volume 2

or gabbrodolerite, as well as a vertical compositional variation in sills.

4. The groundmass is composed of olivine-free, olivine-bearing, and olivine dolerite and gabbrodolerite.
5. Leucogabbro and porphyritic gabbrodolerite feature bytownite plagioclase.
6. The evolution of the melt presents as variations in alkaline content; there is a wide range of variation in Na_2O + K_2O content in the rocks, and alkali feldspar locally forms in the matrix.
7. The final stage of differentiation of the intrusion is syenite, which forms schlieren.
8. Poor sulfide Cu–Ni mineralization and sporadic low-sulfide platinum mineralization are associated with the intrusions.
9. In the alteration aureole of the intrusion, there is evidence of thermal metamorphism and rarer calcareous skarns.

Genetic Concepts

The characteristic feature of the Kruglogorsky-type intrusions is the horizon of leoucocratic gabbro. Investigations by Likhachev (1965) have shown a genetic relationship between Noril'sky ore-bearing intrusions and sills of leucogabbro. Likhachev (1965) argues that there is a significant resemblance in the petrographic and mineralogical–geochemical features in these two types of intrusions. The origin of the leucogabbro is most controversial, considering that rocks from Kruglogorsky-type sills and rocks from Noril'sk Intrusions are identical. This raises further questions about the origin of taxitic gabbrodolerite and pegmatites in ore-bearing bodies.

Geologists have proposed numerous concepts for the origin of leucocratic gabbro. Some researchers believe that leucogabbro was formed before the layered series, whereas other geologists consider that leucogabbro formed after the layered series. The main factors in favor of early generation of leucogabbro in Noril'sk Intrusions are the presence of these rocks in different intrusions, their uniform composition, the large size of plagioclase crystals, their high maficity, and their accumulation as small "erratic boulders" that resemble hypogeal xenoliths (Likhachev 1965). Likhachev (1965) draws the attention to the disordered internal structure of plagioclase in leucogabbro as opposed to the regular structure in olivine and picritic gabbrodolerite and to the large crystal sizes that would require long crystallization times. The author believes that plagioclase of anorthite composition began to crystallize together with magnesian olivine at the early stages of magma moving into a hypabyssal chamber. Further increase in the size of olivine crystals led them to lagging behind the leading part of a magmatic column due to their increased weight. Conversely, plagioclase crystals, which would weigh less than the melt, accumulated at the leading part of a magmatic column forming plagioclase-rich rocks. Relict grains of forsterite (Fa_{10}) and green spines enclosed by anorthite (An_{92}) crystals provided the evidence in support of the model proposed by the author. A similar inference concerning early formation of leucogabbro as compared to other differentiates was drawn by Korovyakov et al. (1963) who considers leucogabbro as a protorock enriched in plagioclase. Oleinikov (1979) follows an analogous point of view and regards erratic leucogabbro boulders as xenoliths, which is evidence for an anorthosite trend of crystallization of melt in an intermediate chamber, and considers leucocratic or anorthitic gabbro to be representative of this trend.

The concept of later formation of leucogabbro with respect to other differentiates of the Noril'sk-I Intrusion has been proposed by Domarev, Kulichenko, and Sheshukova (Tarasov 1976), Rogover (1959), and Smirnov (1966). They regard leucogabbro as a product of crystallization of a residual magma or a separate magmatic phase. Zolotukhin has shown in a number of works that taxitic gabbrodolerite that formed after crystallization of the body from a second melt involving volatiles (Zolotukhin 1964; Zolotukhin and Vasil'ev 1967) has the same genetic nature as leucogabbro. He regards formation of "erratic boulders" of leucogabbro as a product of magmatic replacement (Zolotukhin and Vasil'ev 1967). A similar concept was stated by Tarasov (1976) who considered these rocks to be replaced by gabbro or metasomatically leucocratizated.

The information accumulated on pegmatoids of Noril'sky-type intrusions presented here suggests that extensive horizons of pegmatoids and "erratic" bodies of schlieren were formed as a result of fluid–magmatic interaction that were taking place in hypabyssal chambers at final stages of magmatic process involving transmagmatic fluids. Such a concept provides an explanation for the available data (Ryabov 1999a, b). These data were given in the description of pegmatoids of the Upper Talnakh Intrusion. The origin of leucocratic gabbro of Kruglogorsky-type intrusions is in close agreement with the mechanism of formation of pegmatoids from ore-bearing intrusions; the pegmatoids are apophyses of the ore-bearing intrusions. "Erratic boulders" of leucogabbro are not hypogeal formations in nature and are close to schlieren of pegmatoids in trap intrusions and in so-called gas bubbles that is rounded aggregates in basalt enriched with amygdales (Ryabov 1999b).

Schlieren-like gabbroic syenite and syenite in apophyses of the Upper Talnakh Intrusion are close in nature to "erratic boulders" of leucogabbro. The former was formed due to fluid–magmatic differentiation, which manifests as a separation of the melt into immiscible liquids with "dry" tholeiitic basalt liquid enriched in volatiles and alkalis. Minerals containing volatiles (including biotite, amphibole, apatite, chlorite, calcite, and sulfides) evidence that fluids were involved in petrogenesis. In addition, elevated concentrations of Cl and F in hydroxyl groups of silicates allow one to assume that halogens played an important role during the process of melt differentiation. Rocks enriched in alkali elements occur virtually in all types of intrusions in different magmatic complexes.

3.6.4 Zubovsky-Type Intrusions $(\nu\delta - \omega\nu\beta)T_{1nr}^{zb}$

This type of intrusion includes the Zub-Marksheydersky (Zubovsky), Pyasinsko-Vologochansky, Nakokhozsky, and Manturovsky Intrusions, which are subsheeted bodies with swells and pinches and feature separation into apophyses. This can be observed in cross section and in a plan view. The Zubovsky Intrusion is subdivided into the central (main) body, western, and eastern intrusive branches, and the Pyasinsko-Vologochansky Intrusion is subdivided into the South Pyasinsky and Vologochansky intrusive branches (Figs. 3.29 and 3.30). Both these intrusions, with their branches, are assumed to be one unified branched magmatic body (Goverdovskaya 1971a, b). An opinion exists that the Nakokhozsky and Manturovsky Intrusions are also branches of a single body. The thickness of the Zubovsky-type intrusions varies from 20–40 m to 300 m and average 100–120 m. The magmatic bodies are localized in carbonate–argillaceous sulfate-bearing sedimentary deposits of Lower and Middle Devonian age. In a sequence of the Zubovsky-type intrusions, the wide variability of rocks enables these rocks to be regarded as differentiated. But as opposed to Noril'sky type, layering of these rocks is less distinct; it changes laterally and along dip, transitions between olivine gabbrodolerite and picritic gabbrodolerite are often gradual, horizons of picritic and taxitic gabbrodolerite are not uniform in thickness, and sulfide mineralization is represented by poorly disseminated and sometimes by schlieren-like and interstitial ores.

The results of studies of these intrusions have been described by different researchers, for example, Godlevsky (1959), Korovyakov et al. (1963), Goverdovskaya (1971a, b), Zolotukhin and Shchedrin (1977), Duzhikov (1988), Duzhikov et al. (1988), and Strunin et al. (1988).

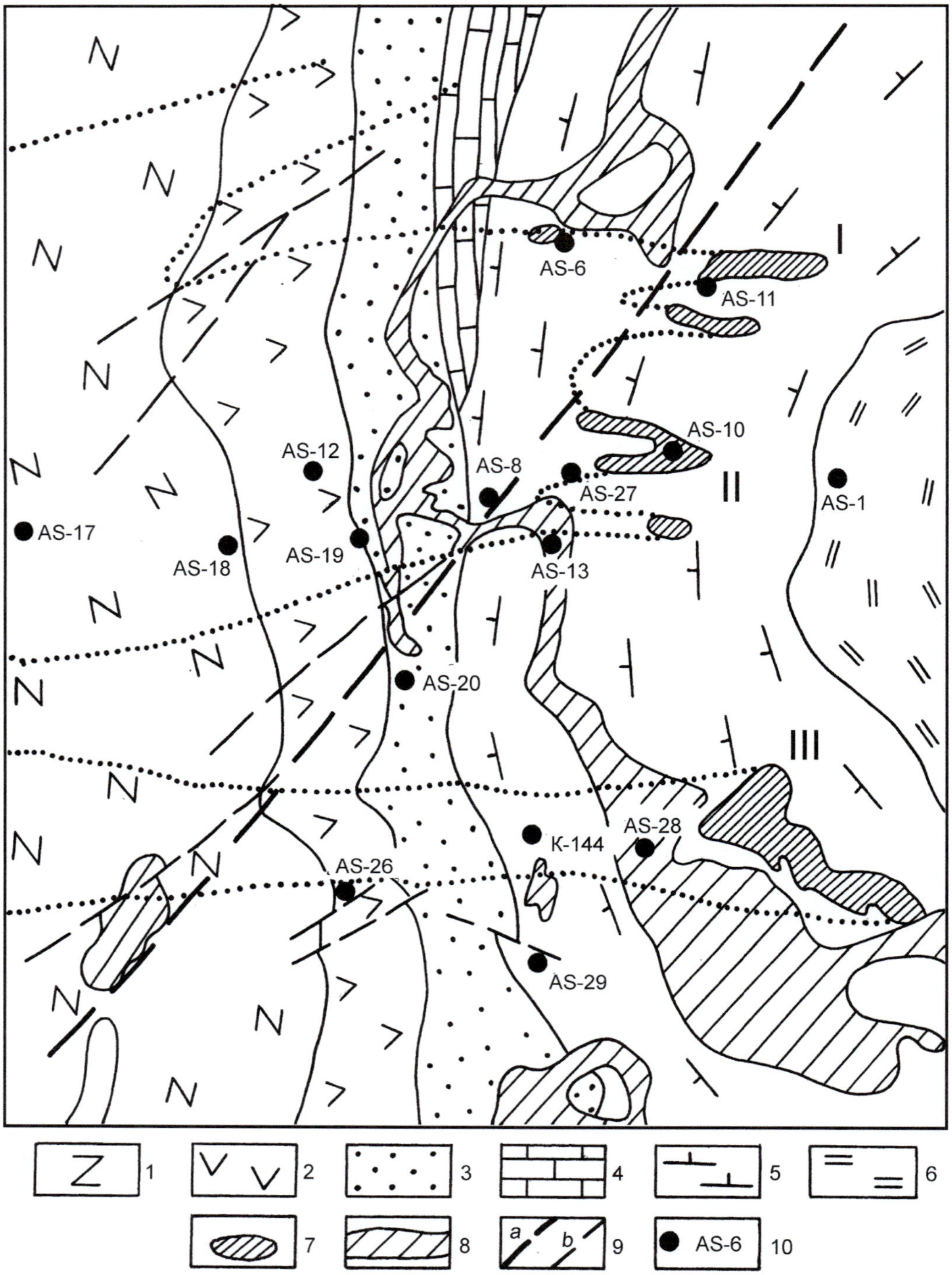

Fig. 3.29 Schematic geological map of the region of Pyasinsko-Vologochansky Intrusion. Constructed from data of NKGRE. *1* basalts of the Nadezhdinsky Suite, *2* basalts of the Gudchikhinsky, Syverminsky, and Ivakinsky Suites, *3* Tungusskaya series rocks, *4* Yuktinsky Suite carbonate rocks, *5* Devonian carbonate–argillaceous sulfate-bearing rocks, *6* Silurian carbonate rocks, *7, 8* differentiated and undifferentiated trap intrusions, *9* tectonic dislocations: (**a**) Tangaralakhsky deep fault, (**b**) subordinate faults; and *10* drill holes. Intrusive branches: I—South-Pyasinsky, II—Vologochansky, III—Zubovsky Intrusion

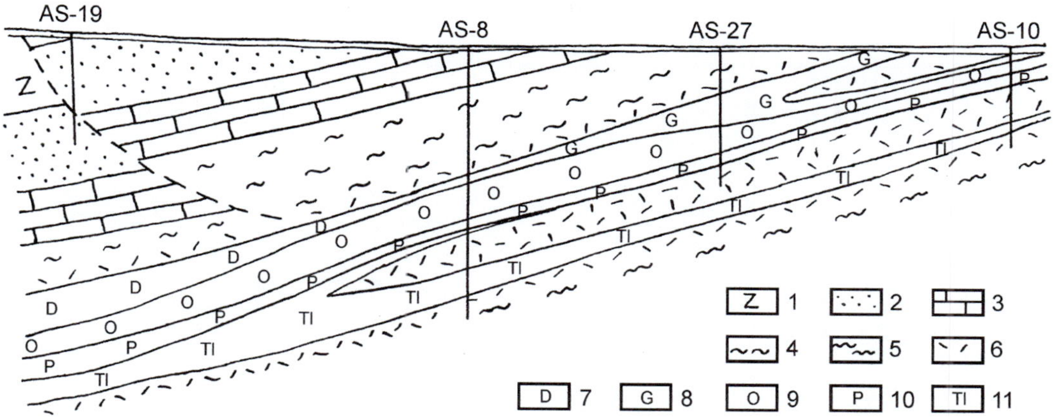

Fig. 3.30 Near east–west geologic cross section of the Vologochnsky branch of the Pyasinsko-Vologochansky Intrusion (constructed from drilling data of NKGRE). *1* Permian–Triassic effusive rocks, *2* Permo–Carboniferous terrigenous rocks, *3–5* Devonian rocks: limestones (*3*), marls and anhydrites (*4*), argillites (*5*), *6* metamorphic rocks, *7* gabbrodiorite, *8* olivine-free and olivine-bearing gabbrodolerite, *9* olivine gabbrodolerite, *10* picritic and troctolitic gabbrodolerite, *11* taxite-like gabbrodolerite

3.6.4.1 Zub-Marksheydersky Intrusion

Godlevsky (1959) and Korovyakov et al. (1963) recognize the following sequence downward through the intrusion: contaminated quartz diorite and acidic hybrid rocks (23%); quartz–olivine gabbrodiabase and gabbronorite (20%); olivine gabbrodiabase and noritediabase with interlayers of poikilo-ophitic gabbrodiabase with olivine (31%); interlayered picritic, taxitic, and olivine gabbrodiabase; and noritediabase (18%).

Picritic gabbrodolerite does not form thick, individual horizons. It commonly forms thin interlayers "suspended" in the horizon of olivine gabbrodolerite and (or) troctolitic gabbrodolerite. The rock consists of olivine Fa_{22-26} (30–45%), clinopyroxene (15–20%), orthopyroxene $Wo_4En_{74}Fs_{22}$ (1–5%), plagioclase $Ab_{27-34}Or_1An_{72-65}$ (25–35%), and biotite ($f = 24$–30 wt%); ore minerals include chrome spinel as inclusions in olivine, titanomagnetite, ilmenite, and variable amount of sulfide (from 1–3% to 10–15%) (see Rock indication 169; Tables A.36 and A.37 in Appendix)

Olivine, troctolitic, and olivine-bearing gabbrodolerites compose most of the intrusion. Olivine gabbrodolerite with poikilo-ophitic texture lies directly on picritic gabbrodolerite which is further changed into the olivine-bearing gabbrodolerite. The amount of olivine, clinopyroxene, plagioclase, and titanomagnetite is from 3–7% to 20%, 20–30%, 35–55%, and 5–7%, respectively; other minerals include orthopyroxene, quartz, biotite, ilmenite, and sulfide.

Quartz–olivine-bearing gabbrodolerite has an ophitic or prismatically ophitic texture. Quartz and olivine lie together and individually. As Godlevsky (1959) notes, micropegmatite is present in the rock along with quartz, especially in the presence of orthopyroxene. Orthopyroxene forms idiomorphic crystals of bronzite that are sometimes mantled by olivine and clinopyroxene, and xenomorphic hypersthene grains are noted in olivine-poor gabbrodolerite. However, the rocks enriched in orthopyroxene (such as norite) are developed locally, and these are not recognized in many drill holes into the Zubovsky Intrusion.

Contaminated quartz diorite and hybrid rock form a horizon of rock very variable in composition. Geologists have recognized rocks abnormally enriched in biotite, quartz, K-feldspar, and micropegmatite; thus, these rocks can be regarded as gabbroic syenogranite, syenite, and granophyre (Goverdovskaya 1971a).

Geochemistry

The compositional features of rocks from the lower part of the Zubovsky Intrusion can be viewed in Fig. 3.31. Here, magnesian rocks with interlayers of troctolitic and olivine gabbrodolerite, as well as rock abnormally enriched in olivine, are restricted to the picritic gabbrodolerite horizon. Troctolitic and olivine gabbrodolerite that overlie and underlie the horizon of picritic gabbrodolerite have similar Al_2O_3 contents but differ in MgO content. Elevated concentrations of Cu, Ni, and FeO are restricted to picritic gabbrodolerite.

3.6.4.2 Pyasinsko-Vologochansky Intrusion

The *Pyasinsko-Vologochansky Intrusion* is situated on the eastern flank of the Vologochansky trough and dips gently west to the center of the trough in conformity with the sedimentary rocks (Fig. 3.29). Morphologically, the body is represented by two narrow (300–400 m in width) and elongated (several kilometers) pipe-like bodies up to 300 m in thickness connected by thin (3–5 m) discontinuous sills. Single pipe-like bodies are regarded as individual South Pyasinsky and Vologochansky intrusive branches (Goverdovskaya 1971a). The structure of the Pyasinsko-Vologochansky

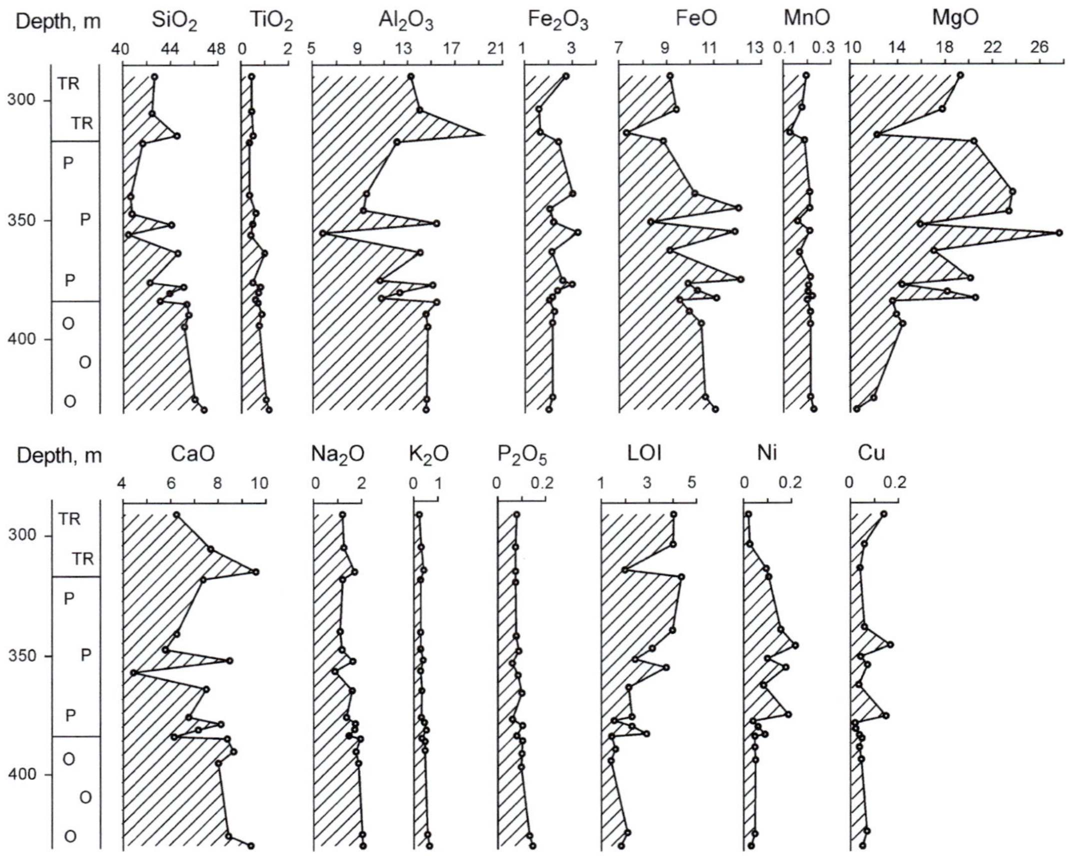

Fig. 3.31 Petrochemical variation diagram through the section of the Zubovsky Intrusion, drill hole AS-28 (from data of geologists of NKGRE and our data). Oxide content and loss on ignition, wt%, element content, ppm

is additionally complicated by the separation of the magmatic body not only into branches in the plan view but also into subsheets located at different hypsometric levels, which can be seen in the cross section (Fig. 3.30).

As opposed to the Zubovsky Intrusion, the horizon of contaminated and acidic hybrid rocks is absent in the sequence of the Pyasinsko-Vologochansky Intrusion. The most complete cross section of the Pyasinsko-Vologochansky Intrusion includes contact gabbrodolerite; metadiorite; olivine-free, quartz-, and olivine-bearing gabbrodolerite; olivine gabbrodolerite; and troctolitic, picritic, taxite-like, and lower contact gabbrodolerite. The rocks vary in thickness in cross section, and only a partial set of differentiates is observed in apophyses. The proportions of rocks throughout the intrusion appear to be olivine-free (33%), quartz–olivine-bearing (11%), troctolitic (32%), picritic (15%), and taxite-like gabbrodolerite (9%). In different parts of the intrusion, small schlieren, lenses, blind veinlets, and interlayers of leucogabbro, gabbroic pegmatite, and plagioclase are present. At the base of the intrusion, thin occurrences of magmatogenic breccia compositionally resemble those in the Kharaelakhsky branch of the Upper Talnakh Intrusion. By analogy with other intrusions, the three zones UZ, CZ, and LZ are recognized in the Pyasinsko-Vologochansky Intrusion.

The upper zone of the intrusion is composed of poikilo-ophitic and prismatically granular olivine-free gabbrodolerite with sporadic olivine and metadiorite. Gabbrodolerite is composed of plagioclase, clinopyroxene, micropegmatite, palagonite, and titanomagnetite. In metadiorite, elevated K-feldspar, quartz, amphibole, biotite, acidic plagioclase, and palagonite contents are observed.

The central zone is composed of ophitic and poikilo-ophitic gabbrodolerite with variable olivine content. In the CZ, quartz is associated with olivine. The proportion and distribution of olivine-rich and olivine-depleted rocks are variable through the intrusion. In a number of sections, picritic gabbrodolerite is absent, and olivine-rich rocks are troctolitic gabbrodolerites. Olivine-rich rocks are characteristically restricted to the axis of the subsheeted body, which is overlain and underlain by olivine gabbrodolerite and/or by olivine-bearing gabbrodolerite. Up dip, the picritic gabbrodolerite horizon appears to be shifted from the central part to the basal part of the magmatic body.

Quartz–olivine-bearing gabbrodolerite consists of plagioclase $Ab_{37-38}Or_{1-2}An_{61-51}$ (50–60%), clinopyroxene $Wo_{39-41}En_{45-41}Fs_{16-18}$ (25–35%), olivine Fa_{40-54} (1–5%), quartz (0.5–3%), titanomagnetite, and ilmenite (5–7%), as well as biotite and palagonite (see Rock indication 170). Olivine forms subidiomorphic crystals with perfect cleavage and intergrowth lamellae of magnetite. Aggregates of small plagioclase crystals occupy the cores of olivine crystals. In addition, dendrite-like crystals of late olivine are observed. The composition of the former is hyalosiderite, and the latter have the composition of hortonolite. Quartz is set in a matrix with palagonite and a chlorite-like mineral. Clinopyroxene forms large oikocrysts with chadocrysts of plagioclase and olivine, and short prismatic crystals in ophitic varieties of the rock. Plagioclase forms wide tabular crystals not infrequently with intergrowths of small titanomagnetite crystals, large and small prisms, intergrowths in clinopyroxene poikilocrysts, and small crystals in the rock matrix. Oxides form subidiomorphic crystals, often in association with biotite. Finely disseminated sulfides sporadically fill interstices.

Olivine gabbrodolerite consists of olivine Fa_{28} (5–7% to 10–15%), plagioclase $Ab_{59-51}Or_2An_{47-40}$ (35–55%), clinopyroxene $Wo_{40}En_{45}Fs_{15}$ (15–20%), titanomagnetite and ilmenite (5–7%), and biotite (0.5% to 1–3%) (see Rock indication 171). Olivine forms subidiomorphic crystals and poikilocrysts. It possesses a cleavage and forms rounded intergrowths with plagioclase, biotite, and magnetite. Clinopyroxene forms poikilocrysts and bands overgrowing olivine. Plagioclase forms large and small prismatic crystals. Biotite forms aggregates and single leaflets in the rock matrix, not infrequently in association with ore minerals. Titanomagnetite forms small idiomorphic crystals in olivine and early plagioclase.

Troctolitic gabbrodolerite contains olivine Fa_{27} (from 7–10% to 20–25%), plagioclase $Ab_{21-34}Or_1An_{79-65}$ (40–70%), clinopyroxene $Wo_{40-39}En_{45-47}Fs_{14-15}$ (3–10%), oxide-ore minerals (3–5%), and small amounts of biotite and sulfides (see Rock indication 172). Olivine forms subidiomorphic crystals (commonly with a cleavage and with plagioclase intergrowths in the cores), small granules, and poikilocrysts. Intergrowths of ore minerals in olivine are extremely rare. Plagioclase forms prismatic crystals of various sizes. Clinopyroxene forms oikocrysts with olivine and plagioclase chadocrysts and with olivine overgrowths.

Picritic gabbrodolerite has segregation, poikilitic, and poikilo-ophitic textures. The rock consists of olivine Fa_{22-23} and Fa_{28-31} (40–60%), plagioclase $Ab_{39-56}Or_{3-1}An_{59-41}$ (15–30%), clinopyroxene $Wo_{40}En_{46}Fs_{14}$ (5–7%), orthopyroxene (5–7%), oxide-ore minerals (5–7%), sulfides (0.5–15%), biotite, and serpentine (see Rock indications 173–175). Olivine forms subidiomorphic crystals with intergrowths of plagioclase and idiomorphic oxide crystals. Rounded grains of early olivine (sometimes pseudomorphed to antigorite) are observed as inclusions. Olivine is characterized by a cleavage, as well as a breakdown structure to magnetite lamellae. Clinopyroxene forms large poikilocrysts, and reaction rims surrounding olivine and orthopyroxene form subidiomorphic crystals with plagioclase intergrowths in rims. Plagioclase forms tabular crystals and large crystals (not twinned) with intergrowths of dark-colored and ore minerals. Oxides are observed as intergrowths in olivine and other minerals, as well as in the rock matrix. Sulfides fill small interstices and form large (0.6–1.2 cm in diameter) droplets, not infrequently with a biotite "cap."

The lower zone is composed of taxite-like gabbro with relicts of altered contact gabbro. In some parts of the intrusion, the LZ is represented by olivine gabbronorite with increased (up to 20%) sulfide and magmatogenic breccia.

Taxite-like gabbrodolerite is characterized by a pegmatoidal appearance and ataxitic texture due to large and small crystals of clinopyroxene and plagioclase. The rock features of olivine granules and other minerals include amphibole, serpentine, prehnite, calcite, titanomagnetite, and sulfides; ore minerals compose 3–7% of the rock. The magmatogenic breccia is composed of small (1–2 cm) fragments of spinel–clinopyroxene–olivine or spinel–olivine–plagioclase rock with microgranoblastic texture, and the fragments are cemented by olivine gabbro. Olivine granules have the composition Fa_9 (Ryabov 1992a).

Metamorphic rocks in the contact aureole of the Pyasinsko-Vologochansky Intrusion are appreciably conditioned by the carbonate–argillaceous composition of the country rocks. The most common altered rocks are pyroxenite, calciphyre (forsterite–phlogopite–green spinel–antigorite–calcite), serpentine–brucite and brucite marbles, and calcareous skarns (Goverdovskaya 1971a). Calcareous skarns developed after contact hornfelsing with gabbrodolerite of the intrusion. Skarns consist of olivine, diopside, plagioclase, garnet (grossular–andradite), calcite, and anhydrite.

3.6.4.3 Geochemistry and Genetic Aspects

Judging from the chemical composition of rocks, intrusions of the Zubovsky type are wholly differentiated. The extent of differentiation and the associated layered series vary in section and along the full extent of the magmatic body. Information about the distribution of chemical elements over cross sections through the Pyasinsko-Vologochansky Intrusion can be obtained from analytical results for rocks in the reference sections (see Appendix, Tables A.38 and A.40). It can be seen in the variation diagram that the successive change in rock through the intrusion section is accompanied by a variation in rock-forming oxides' content. The characteristic feature of the section through the South-Pyasinsky intrusive branch is the presence of two types of rock contrasting in

MgO content: olivine-rich and olivine-depleted and correspondingly enriched or depleted in MgO (Fig. 3.32). Troctolitic gabbro forms a horizon of rocks that are relatively stable in composition. In the upper part of the horizon, a thin interlayer of picritic gabbro is located. The increased MgO content of troctolitic gabbro is accompanied by a high alumina content that is typical for rocks of this type. The horizon of troctolitic gabbro is overlain and underlain by olivine gabbro that changes into taxite-like gabbro at the base of the intrusion. Elevated CaO content in olivine gabbro of the LZ is indicative of the rocks enrichment in clinopyroxene. Judging from the distribution of nonferrous metals in the intrusion section, sulfides are accumulated at the base of the body. Elevated concentrations of SiO_2, TiO_2, Na_2O, K_2O, and P_2O_5 in quartz–olivine-bearing gabbro testify that these rocks contain higher amounts of oxide-ore minerals, alkali feldspar, micropegmatite, quartz, and apatite. When comparing olivine-depleted rocks with olivine-rich rocks, it is noteworthy that chrome content is increased in olivine-depleted rocks. This is indirect evidence for the absence of chromites in magnesium-rich gabbro, and in the rocks poor in magnesium, minor chrome is accumulated in clinopyroxene at the early stage of its crystallization.

In the Vologochansky intrusive branch (Fig. 3.33), the horizon of picritic gabbro in the central part of the section is overlain and underlain by olivine gabbro. Accumulations of sulfides and increased concentrations of chrome, nonferrous metals, and platinoids (see Appendix, Table A.38) are also restricted to the same part of the section. The change from picritic gabbro into troctolitic gabbro is rather gradual and is related to an increase in plagioclase content.

The elevated concentrations of nonferrous and noble metals in rocks of the Pyasinsko-Vologochansky Intrusion are closely related to sulfide accumulations. The highest nonferrous metal content is observed in taxitic gabbro of the South Pyasinsky intrusive branch. These contents are 12,060 g/t Cu, 5,070 g/t Ni, 200 g/t Co, 0.4 g/t Pt, 0.61 g/t Pd, 0.036 g/t Rh, 0.0098 g/t Rh, and 0.14 g/t Au.

Petrologic Features

The following characteristic features of the Zubovsky-type intrusions are distinguished:

1. Intrusions have a complete set of differentiates specific for a Noril'sky-type layered series.
2. In section, clear layering according to the Noril'sky scheme is absent, and it is impossible to compare the thicknesses of different horizons of rocks
3. Horizons of picritic and taxitic gabbro are not uniform in thickness along strike.
4. Hybrid metasomatic and silicate rocks occur only locally.
5. High-magnesian olivine-rich gabbro occupies a hanging location in section and is restricted to the central part of sill-like bodies.
6. Horizons of olivine-rich rocks have sharp contacts with gabbro poor in olivine.
7. The main mass of olivine-rich rocks has a high ($> 30\%$) plagioclase content; thus, troctolitic gabbro is more abundant than picritic gabbro.
8. Gabbronorite and rock enriched in orthopyroxene are developed locally in Zubovsky-type intrusions.
9. Picritic and troctolitic gabbros have low chromium content as compared to Clarke for derivatives of picrite-basalt magmas.
10. Chromite sporadically forms relict crystals enclosed in early generation olivine in picritic gabbro.
11. The association of rock-forming minerals in picritic gabbro of the Zubovsky-type intrusions is Fa_{23} + $Wo_{38}En_{47}Fs_{15}$ + $Wo_4En_{74}Fs_{22}$ + $Ab_{27}Or_1An_{72}$ + titanomagnetite ± chrome spinel.
12. The association of rock-forming minerals in troctolitic gabbro of the Zubovsky-type intrusions is Fa_{27} + $Wo_{40}En_{45}Fs_{15}$ + $Ab_{21}An_{79}$ + ilmenite.
13. Sulfide mineralization occurs as not uniform horizons of disseminated ore and rarer massive ores with the low commodity content.

Genetic Aspects

The magnesian composition of the Zubovsky-type intrusions and layering peculiar to poor sulfide ores give geologists a ground for assigning these magmatic formations to the Noril'sky Complex. The presence of alkaline silicate rocks (granophyre, gabbroic syenite, syenitic diorite, diorite, gabbroic diorite, etc.) and their gradual transition into overlying eruptive breccia allowed Godlevsky (1959) to consider these rocks as hybrid or hybrid metasomatic, which originated as a result of the assimilation of roof rocks by a basaltic melt. Godlevsky believed that the increased rhombic pyroxene content in rocks of the Zubovsky Intrusion can be explained as a consequence of assimilation of sedimentary argillaceous rocks by an olivine gabbro melt.

Because basalt melts in hypabyssal chambers have a severely limited possibility for assimilation, other hypotheses have also been considered.

The origin of alkaline silicate rocks in intrusions was considered the result of crystallization (Sobolev 1986) or immiscible differentiation (Ryabov 1989a, b, c). Kavardin (1976) notes that the main trend for magma differentiation was due to enrichment in iron of a residual melt at the middle or late stages and in alkali elements during the final stages of crystallization. The composition of rocks in upper differentiates from layered intrusions, including the Zubovsky Intrusion, is often complicated due to metamagmatism, which is accompanied by introduction of alkali elements at late stages of the magmatic process. This leads to formation of a wide series of rocks with increasing

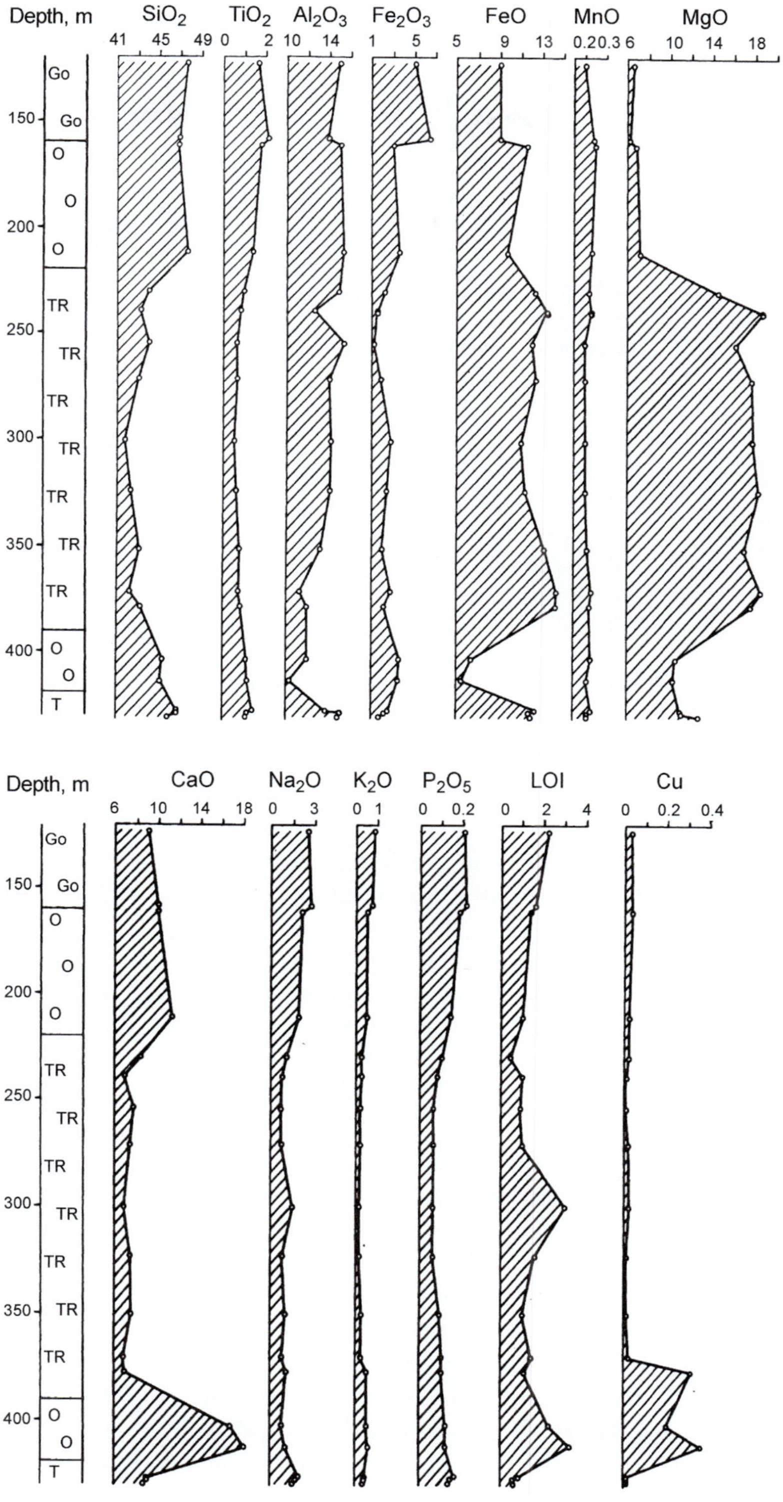

Fig. 3.32 Petrochemical variation diagram showing the behavior of chemical elements through a section of the South Pyasinsky branch, drill hole -6 (from data of both NKGRE and our). Oxide content and loss on ignition, wt%, element content, ppm

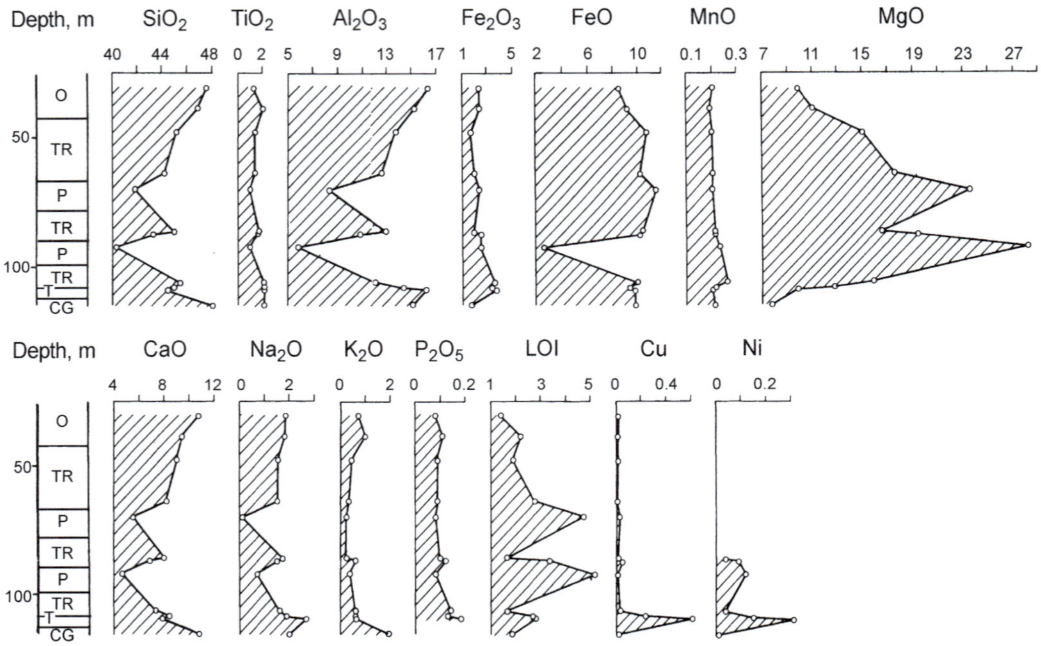

Fig. 3.33 Petrochemical variation diagram through a section of the Vologochnsky branch of the Pyasinasko-Vologochansky Intrusion, drill hole - 10 (from data of both NKGRE and our). Oxide content and loss on ignition, wt%, element content, ppm

alkalinity, the generation of which appreciably owes to fluid–magmatic interactions at the late stages of the magmatic process.

Taking into consideration the fact that basalt melts in hypabyssal chambers have a severely limited possibility for assimilation (Sharapov and Golubev 1976), the appearance of increased orthopyroxene concentration at the expense of dissolution of argillaceous sedimentary rocks in a melt appears to be improbable. Based on natural and experimental data, it has been suggested that orthopyroxene could be formed as a result of olivine oxidation (Ryabov 1992a). This is supported by the fact that traces of successive transformation of olivine revealed in rocks in many trap intrusions manifest themselves as a perfect cleavage of the mineral and loss of iron that forms magnetite lamellae oriented in particular crystallographic directions. As may be inferred from experimental data from Goode (1974), Haggerty and Barker (1967), Nitsan (1974), and Champness (1970) (cited by Ryabov 1992a), the generation of orthopyroxene after olivine at the magmatic stage proceeds with increasing fO_2. The thick horizon of hybrid rocks at the roof of the Zubovsky Intrusion, the wide spread of products of the high-temperature oxidation of olivine, and the replacement of olivine by orthopyroxene suggest that oxidized fluids enriched in alkali elements played an important role in formation of the intrusion.

3.7 Morongovsky Intrusive Complex ($\nu\delta - \omega\nu\beta$)$T_{1-2}$ mr

This complex unites high-magnesium mafic–ultramafic intrusions located in the central part of the Noril'sky trough along the Noril'sko-Kharaelakhsky Fault (Fig. 3.34). The Morongovsky group of intrusions includes the Mt. Picritovaya, Mt. Putanaya, Mt. Morongo, Mt. Ruinnaya, and Mt. Pegmatitovaya Intrusions, which are residual mountains that have been separated by erosion that are oriented in a south–north direction and cut volcanogenic rocks up to the Mokulaevsky Suite (Fig. 3.34b). The magmatic body has a complicated morphology and combines gently crosscutting subsheeted and dyke-like shapes, sometimes with subbedded apophyses. Judging from the basal jointing and banding in gabbro of the Mt. Ruinnaya and Mt. Morongo, the intrusion is subhorizontal or dips gently west. The banding in the Mt. Putanaya Intrusion is oriented at an angle of 45–55°, and the lower contact is oriented at 22–25°. The Morongovsky group of intrusions extends more than 10 km with 1.5–2.0 km in width. The thickness of the Mt. Putanaya, Mt. Morongo, Mt. Ruinnaya, and Mt. Pegmatitovaya in drill holes E-36, MK-1, M–1, and E-28 are 374 m, 67 m, 182 m, and 265 m, respectively.

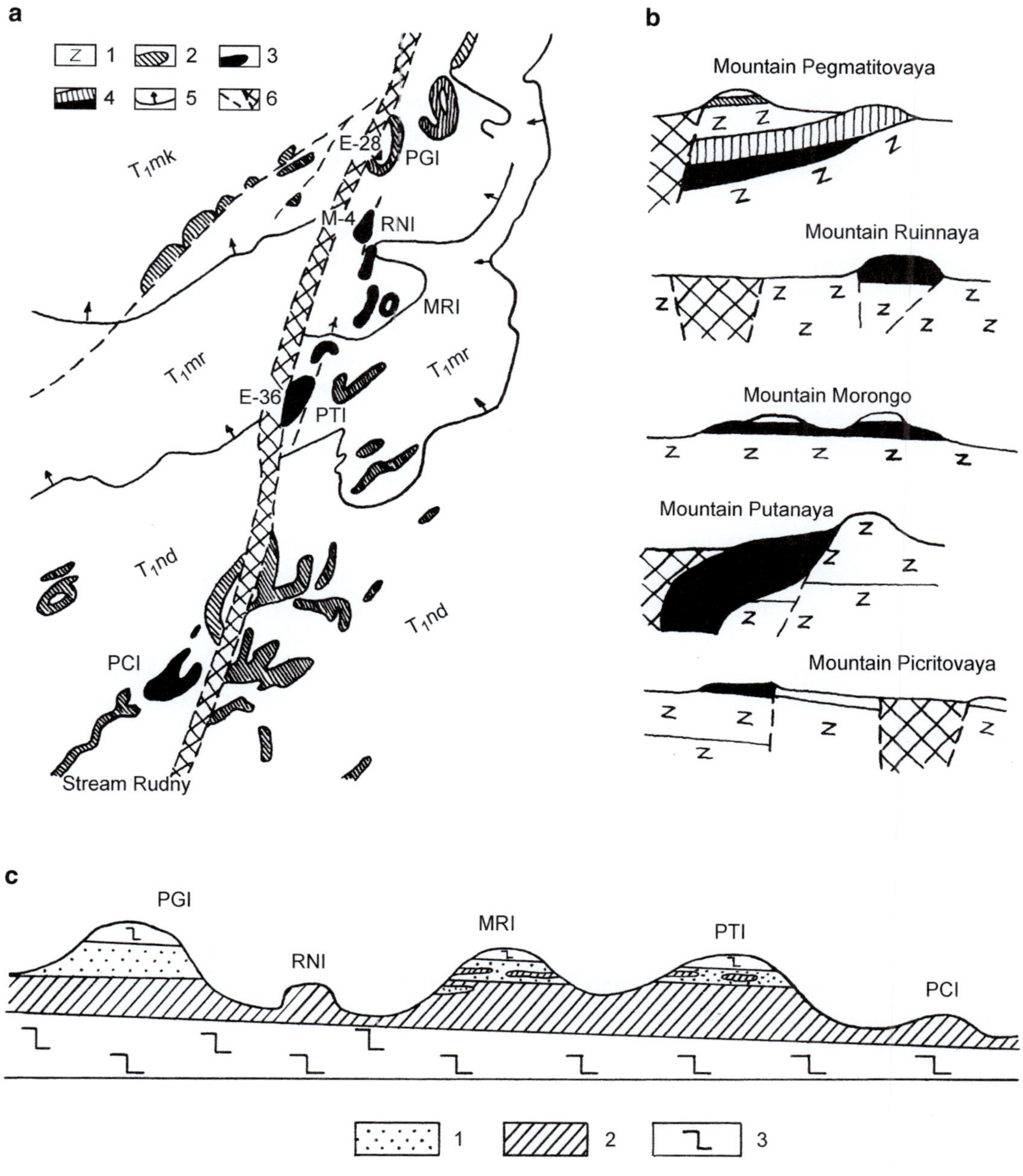

Fig. 3.34 Schematic geological map of the region around the Morongovsky Intrusions (**a**). Morphology of individual magmatic bodies (**b**), near north–south section through the Morongovsky group of intrusions (**c**). (**a**, **b**): *1* basalts (age, suites, and dips of rocks are indicated on the map), *2* undifferentiated mafic intrusions, *3* the Morongovsky-type magnesian intrusions, *4* mafic (vertical crosshatching) and essentially troctolitic layers in the Mt. Pegmatitovaya Intrusion cross section, *5* dips of volcanogenic rocks, *6* Noril'sko-Kharaelakhsky Fault with feathering fractures. (**c**): *1* mafic pegmatoids, *2* troctolitic gabbro, *3* basalts

The characteristic feature of the Morongovsky-type intrusions is the high-magnesium composition due to a large (about 80%) volume of olivine-rich rocks such as troctolitic gabbrodolerite and an insignificant proportion of gabbrodolerite poor in olivine. The sole exception is provided by the Mt. Pegmatitovaya Intrusion that is composed of mafic and ultramafic rocks in almost equal proportions.

Early petrological data on the Morongovsky Intrusions was published by Korovyakov et al. (1963). Later, these targets were studied by many other geologists, but at present, there are not available results of these investigations in the geologic literature. The unusual internal structure and composition of Morongovsky Intrusions are grounds to consider these as an individual complex, and they should be considered when constructing genetic models.

In addition, the leading and central parts of the extended magmatic body in the Morongovsky group of intrusions have their own features, and so two major sequence types for these intrusions should be considered.

The Mt. Picritovaya, Mt. Putanaya, Mt. Morongo, and Mt. Ruinnaya Intrusions represent the central part of an extended body and are very similar in internal structure and rock composition. It is these intrusions that geologists bear in mind when intrusions or complex of the Morongovsky type are concerned. Moreover, when the intrusion is considered from the north to south, some petrologic characteristics of this extended intrusion are traced upward through the intrusion.

The Mt. Picritovaya Intrusion is the most melanocratic of all Morongovsky Intrusions. It is composed of troctolitic gabbrodolerite with sparse thin (5–12 cm) lenses of gabbroic pegmatite enriched in apatite (to 3%) and subbedded veinlets of plagioclase 1–3 cm thick. At the roof of the Putanaya Intrusion, a horizon of mafic rocks accounting for 20% of the intrusion is observed, and in troctolitic gabbrodolerite, clinopyroxene increases as olivine decreases. In the central part of sequence, there is a thin (5–7 m) interlayer of picritic gabbrodolerite that is overlain and underlain by troctolitic gabbrodolerite. In the Mt. Morongo Intrusion, Morongovsky-type pegmatites are widely developed. Mafic rocks in the upper part of the Mt. Ruinnaya intrusion have been eroded, and only troctolitic gabbrodolerite has been preserved. Lastly, mafic rocks (145 m in thickness) in the Mt. Pegmatitovaya Intrusion dominate ultramafic rocks (120 m in thickness), and there is decreased olivine and MgO content.

Olivine gabbro that crops out at the top of the Mt. Morongo Intrusion has a gabbroic texture with some element of prismatically granular and ataxitic. The rock consists of olivine (3–5%), plagioclase (55–60%), clinopyroxene (25–35%), biotite, titanomagnetite, and ilmenite (7–10% overall) (see Rock indication 176). Olivine Fa_{47} (0.008 wt% NiO) forms large crystals and small grains in sites with the ataxitic texture. Plagioclase forms zoned crystals $Ab_{25-42}Or_{1-3}An_{74-4}$ and small prisms $Ab_{35}Or_1An_{64}$. Clinopyroxene forms subidiomorphic crystals with plagioclase intergrowths in rims, as well as small isometric grains. Compositionally, clinopyroxene is augite ($Wo_{38-37}En_{40-43}Fs_{21-20}$) and contains of 0.72–0.97 wt% TiO_2, 1.42–1.60 wt% Al_2O_3, and 0.32 wt% Na_2O. Brownish biotite has total Fe content of 35.6 wt%, and it contains 5.04 wt% TiO_2, 0.70 wt% F, and 0.20 wt% Cl. Titanomagnetite contains variable amounts of TiO_2 (6.89–12.53 wt%), Al_2O_3 (1.29–2.29 wt%), and MgO (0.07–1.05 wt%). Ilmenite contains 0.55 wt% MnO and 0.94 wt% MgO.

3.7.1 Mt. Putanaya Intrusion

In section, two zones (UZ and CZ) are distinctly recognized. Contact rocks have a holocrystalline texture and no quench textures were observed.

The upper zone is composed of pegmatoid with variable olivine content. Olivine and leucocratic gabbro and gabbroic pegmatite are most common here. Pegmatoids form layers and lenses from tens of centimeters to 1–3 m in thickness, which have an en-echelon arrangement and are separated by relict thin taxite-like troctolitic gabbrodolerite layers. Alternation of compositionally variable pegmatoid with troctolitic gabbrodolerite forms a bench of streaky gabbroidal rock in the UZ. In the streaky gabbros of the Mt. Morongo Intrusion, the leucocratic constituent is dominant, and the thickness of melanocratic interlayers does not exceed 1–5 cm.

The characteristic feature of olivine and leucocratic gabbro is the ophitic texture of the rocks, composed of large, short prisms (or tabular) of plagioclase and sharply subordinate clinopyroxene, sometimes with solid solution breakdown structures. Olivine forms aggregates of small grains and stalk-like crystals. Compositionally, clinopyroxene varies from Fa_{36} to Fa_{48} (Ryabov and Zolotukhin 1977). Both clinopyroxene and olivine are sometimes framed by symplectic intergrowths with plagioclase.

Gabbroic pegmatite has a prismatically granular texture formed by plagioclase An_{40-23} crystals, not infrequently with patchy albitization, and clinopyroxene. Small amounts of brownish and green hornblende and biotite are present, and ore minerals include titanomagnetite, ilmenite, and sulfides. Quartz, quartz–feldspar micropegmatite, and needle-shaped (up to 4 mm in length) apatite are found in the matrix.

The central zone is composed of troctolitic and picritic gabbrodolerite. Picritic gabbrodolerite comprises a thin (about 5 m) interlayer in the middle of the magmatic body (Fig. 3.35). The rock texture is porphyritic, poikilitic, and ophitic with elements of poikilo-ophitic. Plagioclase forms as porphyritic wide tabular crystals and prisms, as well as small prisms and glomerogranular aggregates in the groundmass. Clinopyroxene forms poikilocrysts $Wo_{39-41}En_{49-46}Fs_{12-13}$ with intergrowths of olivine and plagioclase (see Rock indication 177).

Quantitative mineralogical calculations undertaken by Raikova (Korovyakov et al. 1963) have shown that the rocks contain 44.9% olivine, 40.5% plagioclase, 11.4% pyroxene, 1% biotite, 1.7% ore minerals, and 0.5% secondary minerals. Raikova noted that the olivine content varies from 23 to 58%, and gravitational separation of the mineral is not observed. In picritic gabbrodolerite (as opposed to

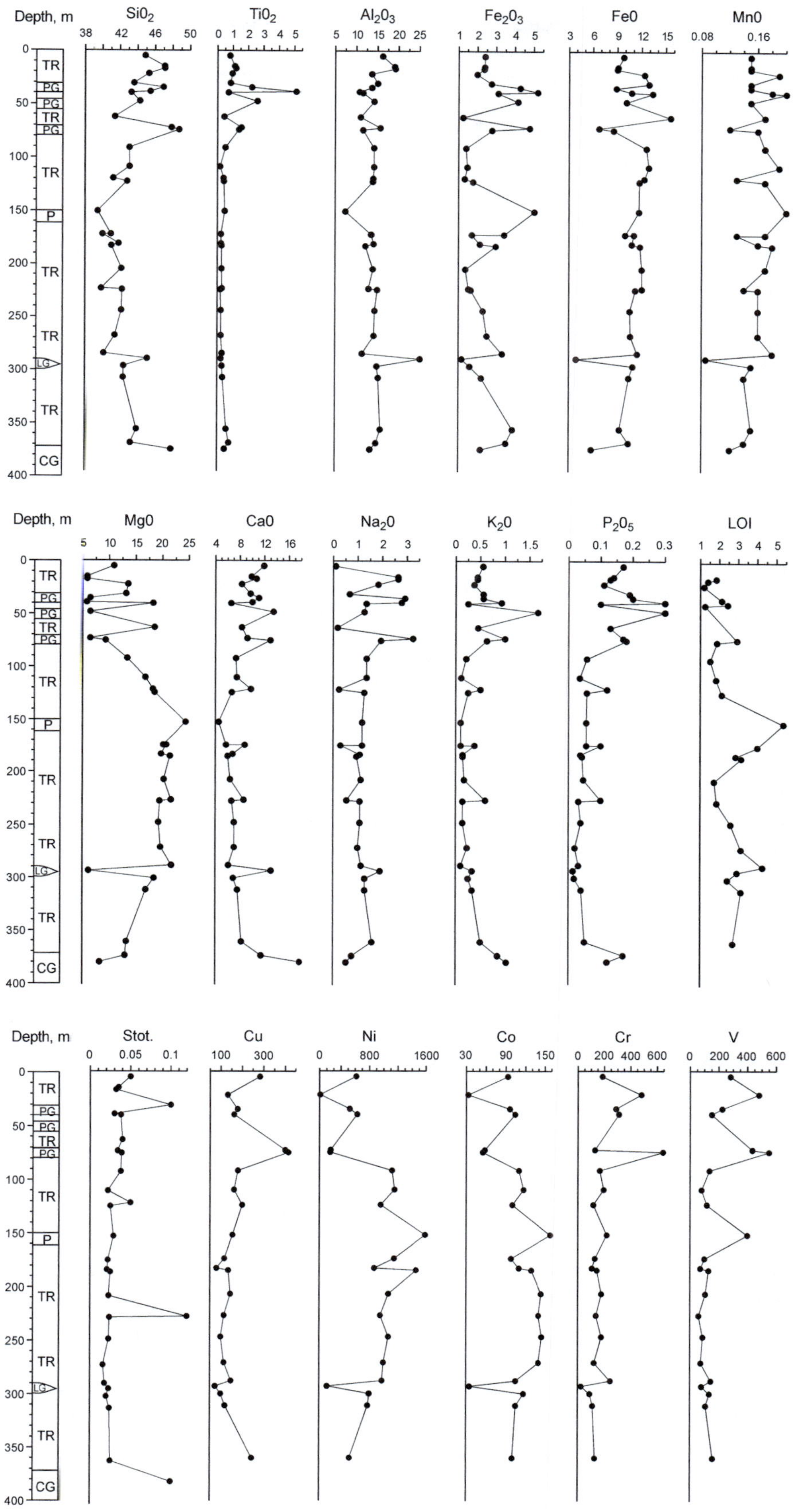

Fig. 3.35 Variation diagram for oxides through a section of the Mt. Putanaya Intrusion, hole E-36 (explanation in the text). Oxide and S content and loss on ignition, wt%, element content, ppm

troctolitic gabbrodolerite), olivine content increases to 55–60% at the expense of plagioclase.

The characteristic features of troctolitic gabbrodolerite of the Mt. Putanaya Intrusion and the Morongovsky Intrusions are porphyritic plagioclase crystals and banding due to plagioclase An_{71-72} veinlets (Ryabov and Zolotukhin 1977) along a basal protorock cleavage (thickness of several millimeters to 15–20 cm). Plagioclase often exhibits a trachytoid texture (see Rock indication 178). Olivine forms granules surrounded by clinopyroxene laminae. There is a characteristic increase in olivine content and therefore MgO content at the contact of plagioclase interlayers (see Rock indication 178); the thickness of olivine-rich troctolitic gabbrodolerite correlates with that of plagioclase interlayers.

Olivine Fa_{18-19} in troctolitic gabbrodolerite forms large subidiomorphic and skeletal, palmate, and stalk-like crystals of different sizes. It has perfect cleavage and often contains inclusions of oval and slightly elongated plagioclase An_{74} grains. Intergrowths of plagioclase in the cores of subidiomorphic olivine crystals are sometimes observed as symplectic, eutectoid intergrowths with the olivine. Porphyritic plagioclase crystals have the composition An_{78-64} (see Rock indication 178). In troctolitic gabbrodolerite, thin clinopyroxene laminae surrounding olivine grains are observed, and olivine is rarely observed as poikilocrysts. In palmate olivine, disseminated ore globules are observed. The olivine composition Fa_{25-31} in troctolitic gabbrodolerite throughout the entire CZ is fairly stable. Oxide-ore minerals are titanomagnetite, magnetite, and ilmenite.

Minerals contained in picritic gabbrodolerite are Fa_{28} + $Wo_{41}En_{46}Fs_{13}$ + $Ab_{35}Or_{2}An_{63}$ + Ti – Mt (see Rock indication 179).

The lower zone is virtually identical in composition and texture to overlying rocks. The only difference is an insignificant decrease in olivine content and increase in clinopyroxene content. This is reflected in an overall decrease of the MgO content.

3.7.1.1 Petrochemistry

Petrochemical features of the Mt. Putanaya Intrusion are revealed by the variation diagram of oxides for the magmatic body (Fig. 3.35). The saw-shaped plot distinctly distinguishes rocks of the UZ from rocks relatively stable in the contents of chemical components in the CZ. Alternation of compositionally varied rocks of the UZ with differing olivine content manifests itself as a variation in MgO content. Rocks of this zone feature large oxide variations, and highest TiO_2, Fe_2O_3, FeO, CaO, Na_2O, K_2O, P_2O_5, and S content is characteristic. The stable high MgO content with an increasing trend toward the central part of the intrusion and an extremely low Cr_2O_3 content of 0.2 wt% are remarkable in troctolitic gabbrodolerite of the CZ. This is continued with a rather low Cr content in highly magnesian rocks throughout the intrusion (Ryabov and Yakobi 1981). Compositions of plagioclase veinlets and adjacent rock in the lower part of the intrusion are plotted on the variation diagram.

3.7.1.2 Mineralogy

Olivine composition in the Mt. Putanaya Intrusion varies from Fa_{25} to Fa_{31}; iron-poorer and iron-richer varieties of the mineral are concentrated in the lower and upper parts of the sequence, respectively (Fig. 3.36). The NiO content in olivine correlates with its Fe content, and the highest NiO content in olivine of the Mt. Putanaya Intrusion is restricted to the central part of the sequence. Porphyritic plagioclase grains have a stable $Ab_{21}Or_{1}An_{78}$ composition throughout the entire body. Plagioclase composition in the groundmass is $Ab_{25-35}Or_{1}An_{63-74}$. Oxide-ore minerals include titanomagnetite (with the variable TiO_2 content) and ilmenite.

3.7.2 Mt. Pegmatitovaya Intrusion

Compared with other intrusions of the Siberian Platform, the Mt. Pegmatitovaya Intrusion features the most pronounced crystallization differentiation and invisible layering with distinct geochemical and crystallization trends.

The Mt. Pegmatitovaya Intrusion is the leading part of the Morongovsky group magmatic column. As opposed to other essentially ultramafic members of this group, the Mt. Pegmatitovaya Intrusion is a wholly differentiated body of mafic rock (Fig. 3.37). This is demonstrated by the results of quantitative mineralogical studies (Table 3.7) of rocks from the intrusion and the description of its sequence.

The upper zone is composed of contact microdolerite, eruptive breccia, and pegmatoid 7 m thick. The pegmatoid rocks have ataxitic and breccia-like textures. The breccia-like texture is created by fragments of fine-grained pyroxene–plagioclase aggregates cemented by coarse-crystalline amphibole–plagioclase aggregate (see Rock indication 180). Lower coarse granular pea-like poikilo-ophitic gabbrodolerite forms a 17-m-thick layer that is enriched in feldspar, amphibole, and apatite. Oval palagonite aggregates are found in the rock, and granophyre is observed in the matrix. The texture of the rock is created by poikilocrysts of clinopyroxene ($Wo_{38}En_{39}Fs_{23}$), olivine (Fa_{61-64}), and titanomagnetite, and prisms of plagioclase (An_{46}) (see Rock indication 181) are set in the groundmass as chadocrysts in mafic minerals. Further along the intrusion, this rock changes into giant granular gabbroic pegmatite (9 thick), which are composed of long prismatic plagioclase (An_{44-45}) crystals with hieroglyph-like palagonite inclusions in the cores of crystals and twinned

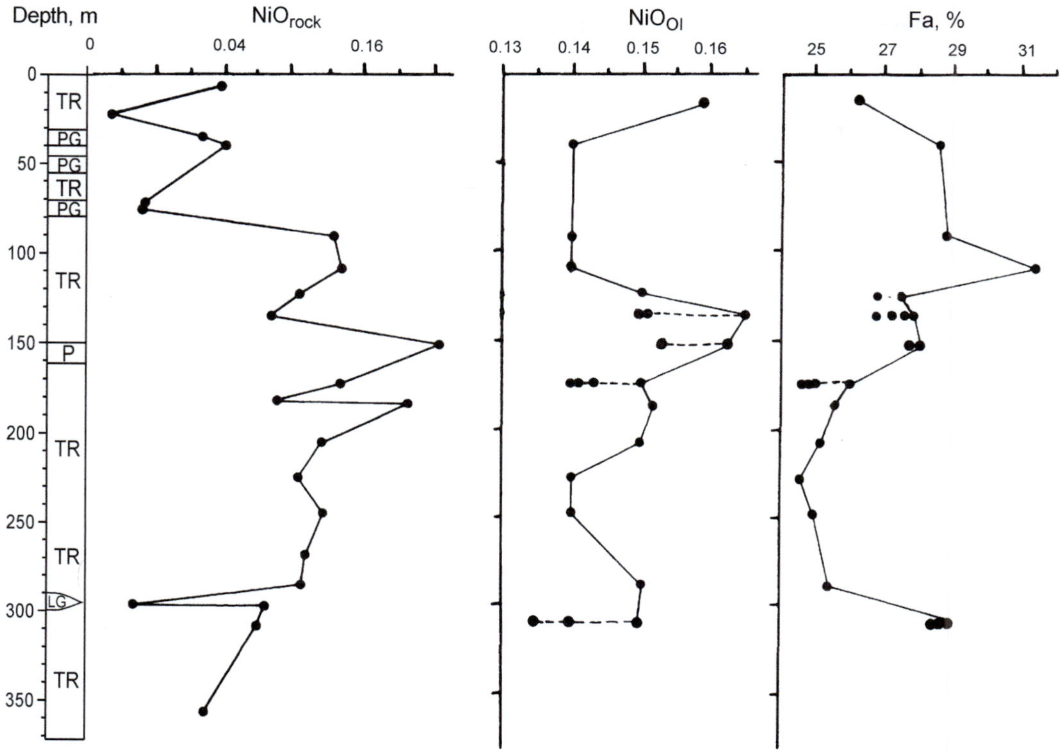

Fig. 3.36 Variation in NiO content compared to olivine composition in a section of the Mt. Putanaya Intrusion, drill hole E-36 (Ryabov 1992a). Oxide content, wt%

prisms of clinopyroxene ($Wo_{16-23}En_{25-34}Fs_{42-56}$) (see Rock indication 182) with solid solution breakdown structures and symplectic intergrowths with biotite. Palmate olivine (Fa_{70}) is present in minor amounts, and the matrix is characterized by elevated apatite concentrations. The contact with underlying rocks is sharp.

The central zone begins with coarse-grained gabbrodolerite (25 m thick) then grades into layered (10 thick) coarse- and medium-grained olivine-free and olivine-bearing gabbrodolerite (see Rock indications 183 and 184), 0.3–1.0 m thick. This gabbrodolerite grades into medium-grained pea-like olivine-bearing gabbrodolerite. Downward through the intrusion, "peas" decrease in size, and the porphyritic (pyroxene-aphyric) texture replaces the pronounced poikilo-ophitic texture (see Rock indication 185), for which single plagioclase intergrowths in phenocrysts of monolithic clinopyroxene are specific. "Peas" and phenocrysts decrease in size downward from 3.5–2.5 cm to 1.5–1.0 cm and 0.5–0.3 cm, respectively.

The rock texture changes from poikilo-ophitic into porphyritic. In coarse-grained gabbrodolerite, not only clinopyroxene but also olivine and titanomagnetite form poikilocrysts. As a rule, poikilocrysts abound with plagioclase intergrowths and not infrequently are observed as patches with skeletal eutectoid intergrowths of their constituent phases (see Rock indication 186). Moreover, clinopyroxene forms small angular grains in the matrix and as laminae surrounding grains of plagioclase and olivine. Porphyritic olivine varieties are often twinned. The most abundant forms of olivine are oval, isometric, and slightly elongated (less than 1:2) with imperfect or no cleavage; palmate forms are rare. Olivine is sometimes surrounded by thin, discrete laminae of orthopyroxene and clinopyroxene. The amount of olivine increases downward through the intrusion from 1–3% to 12–15%. Rounded plagioclase (An_{57}) grains are commonly intergrown with olivine. At the base of the layer, olivine with a cribriform surface is observed due to inclusions of plagioclase spherules.

At the base of the mafic layer about 35 m above the horizon of uniform troctolitic gabbrodolerite, melanocratic lenses and then interlayers of troctolitic gabbrodolerite within pea-like olivine gabbrodolerite begin to manifest. In the transition zone from olivine gabbrodolerite to troctolitic gabbrodolerite, perfect cleavage is observed in olivine, and clinopyroxene exhibits solid solution breakdown structures.

Troctolitic gabbrodolerite has a porphyritic texture (due to plagioclase) and a poikilo-ophitic texture to the groundmass. The upper boundary of the layer is distinguished by the disappearance of the distinct pea-like texture, although it can be seen under a microscope that the poikilo-ophitic texture is preserved throughout the layer. Oikocrysts become smaller, and the increase in olivine content to 25–30% changes the rock fabric (see Rock indication 187).

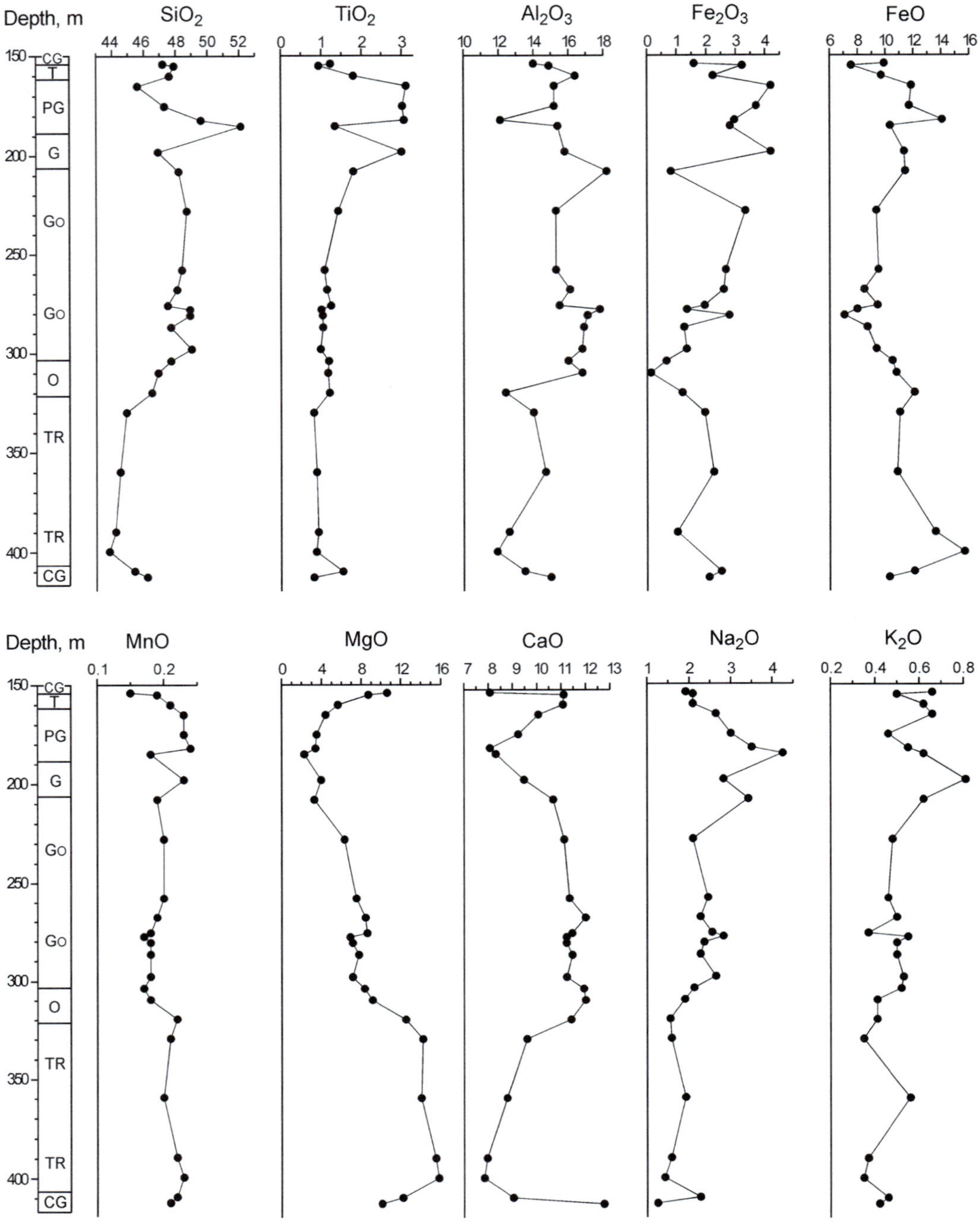

Fig. 3.37 Petrochemical variation diagram for the Mt. Pegmatitovaya Intrusion, drill hole, E-28. Oxide and S content and loss on ignition, wt%, Hg content, ppm

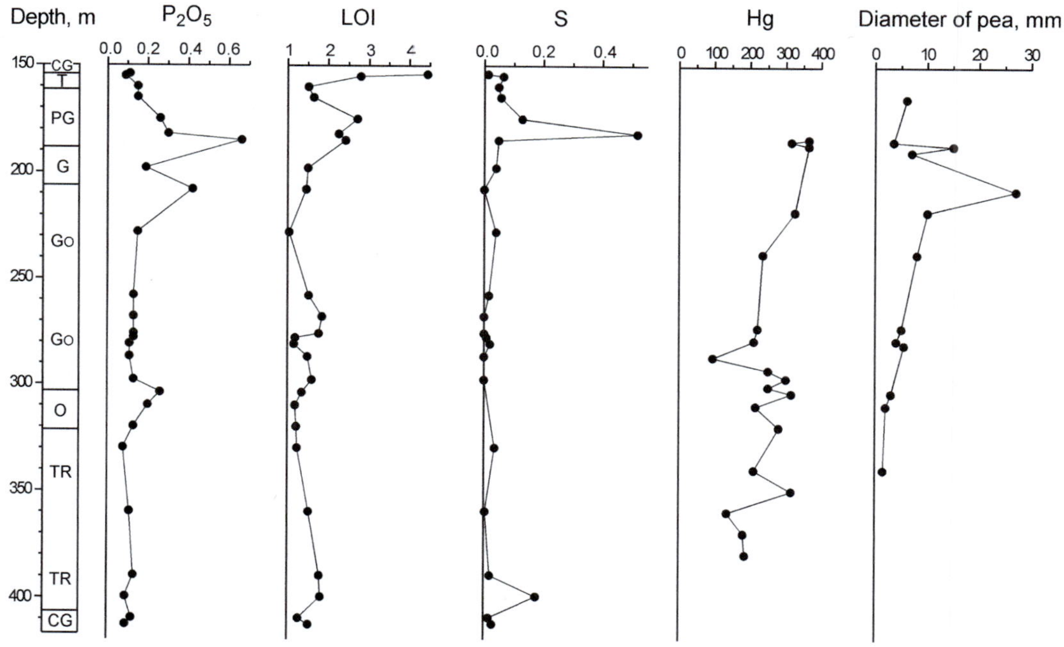

Fig. 3.37 (continued)

Table 3.7 Mineralogical composition of the Mountain Pegmatitovaya Intrusion (on hole E-28, Korovyakov et al. 1963)

Gabbrodolerite	Olivine	Plagioclase	Clinopyroxene	Orthopyroxene	Opaque	Others
Contact (upper)	0.6	57	22.8	–	0.6	19.0
Nonolivine	0.1	55	30.8	–	6.4	7.7
Olivine bearing	3.3	58.9	25.2	>0.1	3.5	9.0
Olivine	9.6	55.5	27.7	1.4	1.7	4.1
Troctolite	26.5	44.1	21.0	4.7	1.4	2.3
Contact (lower)	13.7	53	23.3	4.5	3.4	2.1

Olivine forms isometric, subidiomorphic crystals with smoothed vertices, and small elongate poikilocrysts with plagioclase intergrowths. In some olivine grains, the perfect cleavage is obvious. Olivine composition (Fa_{35-38}) is stable throughout the entire horizon. Clinopyroxene ($Wo_{37-42}En_{43-45}Fs_{18-14}$) is observed as small poikilocrysts and angular grains. Plagioclase forms polyzoned nontwinned short prisms, wide tabular crystals, and glomeroaggregates of narrow prisms. The composition of plagioclase is An_{74-56}. Ore minerals include titanomagnetite, ilmenite, and sparse sulfides.

The lower zone of the intrusion is composed of troctolitic gabbrodolerite with an obscure ataxitic texture, which can be seen in thin section due to small aggregates of rounded olivine grains. The zone is about 10 m thick.

3.7.2.1 Petrochemistry

Variation diagrams (Fig. 3.37; Appendix, Table A.44) show clear trends for distribution of chemical elements through the magmatic body. Variation in oxide contents is progressive without abrupt changes, except in pegmatoid of the UZ where accumulation of MgO up to 10.64% takes place. As has been shown earlier (Ryabov 1989a), the appearance of an extreme minimum in MgO content in the UZ is typical for several trap intrusions, and the MgO content may exceed that of oxides even from rock varieties in the intrusion sequence richest in magnesia. This appears to be due to the involvement of fluids.

The central intrusion zone is divided into two large groups of rocks: olivine-rich and olivine-poor mafic and ultramafic rocks. In mafic rocks, elevated SiO_2, TiO_2, Al_2O_3, Fe_2O_3, FeO, CaO, Na_2O, K_2O, and P_2O_5 contents are typical. These oxides reach their maximum contents in pegmatoid rocks. In ultramafic rocks, MgO and FeO are accumulated. The distribution of FeO through the intrusion varies between two distinct extrema, which are symmetrical about the interface between mafic and ultramafic rocks. It can be seen on variation diagrams that the transition zone between mafic and ultramafic rocks is recognized due to slight anomalies in the content of virtually all petrogenic oxides. This is evidence for a change in the crystallization nature at the boundary between melts of various compositions.

Variations in the mineralogical composition of rocks are reflected in variations of contents of rock-forming oxides. Increased calcium oxide content is manifest in clinopyroxene-rich gabbrodolerite, increased alumina and alkaline elements are noted in rocks enriched in feldspar, and the high magnesium content in olivine-rich rocks. The maximum P_2O_5 content is observed in coarse-grained pegmatoidal gabbrodolerite, there being the direct relationship between sizes of "peas" in clinopyroxene poikilocrysts and P_2O_5 content. The behavior of mercury in the intrusion is a sensitive indicator for a melt differentiation (Fig. 3.37; Appendix, Table A.45).

3.7.2.2 Mineralogy

One of the main indicators of interchamber differentiation of a melt is invisible layering that manifests itself as a directional change in the composition of rock-forming minerals. In the intrusion, the variation in clinopyroxene composition has a unified crystallization trend for both mafic and ultramafic rock horizons (Fig. 3.38). Most sensitive indicators of differentiation are the Fe and Cr_2O_3 content of a mineral. Upward through the intrusion, the composition of the mineral changes progressively from $Wo_{40}En_{46}Fs_{14}$ ($f = 22.9$ wt%) to $Wo_{16}En_{28}Fs_{56}$ ($f = 66.3$ wt%). An increase in Fe content correlates with the coefficient of rock fractionation that varies from 48 to 83%. The Cr_2O_3 content in the mineral varies from 0.61 to 0.01 wt% over the intrusion; concentrations of other components vary also, but these are less distinct.

Sensitive indicators for melt differentiation include the olivine composition, as well as the Fe content and impurities in the mineral (Fig. 3.39). Upward through the intrusion, the Fe content of olivine increases from Fa_{36} to Fa_{70}; CaO content also increases, whereas the NiO content decreases. The olivine composition is strongly related to the overall rock composition. The Fe content in mafic minerals versus the coefficient of rock fractionation is shown in Fig. 3.40, where the conjugate increase in f and K_f and the higher Fe content of olivine in comparison with its coexisting clinopyroxene are shown.

3.7.3 Geochemistry and Genetic Aspects

Contact alterations in connection with intrusions of the Morongovsky Complex manifest as a pyroxene–plagioclase hornfels with biotite, amphibole, and chlorite. The thickness of altered rocks does not exceed 5–10 m.

3.7.3.1 Mineralization

Rare scattered interstitial chalcopyrite–pyrrhotite ores, sometimes with pentlandite, are observed throughout the Morongovsky Complex intrusions. Large (0.6–1.2 cm) single sulfide inclusions form sporadically. In some cases, sulfide veinlets up to 0.5 cm thick are observed. Increased sulfide mineralization is restricted, as a rule, to rocks of the UZ and LZ (Fig. 3.37). These zones are also characterized by elevated concentrations of platinum group elements; in the UZ of the Mt. Pegmatitovaya Intrusion, the concentrations of Pt and Pd are up to 0.3 g/t and 1.7 g/t, respectively, and those in the LZ of the Mt. Ruinnaya Intrusion are, respectively, 0.12 g/t and 0.88 g/t (Dodin et al. 1994).

3.7.3.2 Magnetization of Rocks

Intrusions of the Morongovsky Complex are characterized by a slight magnetization (J_i and J_n) of the order $10–50 \times 10^{-5}$ CGS and significant scatter of declination (from 0 to 360°) and inclination (from $-$ 90° to +90°) of vector J_n. Magnetic cleaning of samples does not change reversed polarity of vector J_n, and normal polarity of vector J_n changes into the opposite (reversed) polarity. This evidences that the primary polarity of these samples was a reversed polarity.

3.7.3.3 Petrologic Peculiarities

The following features specific for the Morongovsky Complex intrusions should be emphasized:

1. The presence of a wide variety of rocks from olivine-free to picritic gabbrodolerite is indicative of the layered nature of the intrusion.
2. The vast volume (about 80%) of olivine-rich rocks in intrusions of Mt. Ruinnaya, Mt. Morongo, Mt. Putanaya, and Mt. Picritovaya is evidence for their slightly differentiated structure.
3. The leading part of the Morongovsky group of intrusions (Mt. Pegmatitovaya Intrusion) is a fully differentiated mafic rock intrusion.
4. The dominant rock type of intrusion is plagioclase-porphyritic troctolitic gabbrodolerite.
5. Olivine-rich gabbro of the Morongovsky Complex is distinguished from high-magnesian rocks of the Noril'sky and Fokinsky Complexes by a high (more than 40%) plagioclase and Al_2O_3 content.
6. The size and amount of olivine grains are larger than those of plagioclase (Stepanov and Zemskova 1988).
7. The olivine composition covers the range from Fa_{25} to Fa_{64}.
8. Olivine in picritic and troctolitic gabbrodolerite has the composition Fa_{25-37}.

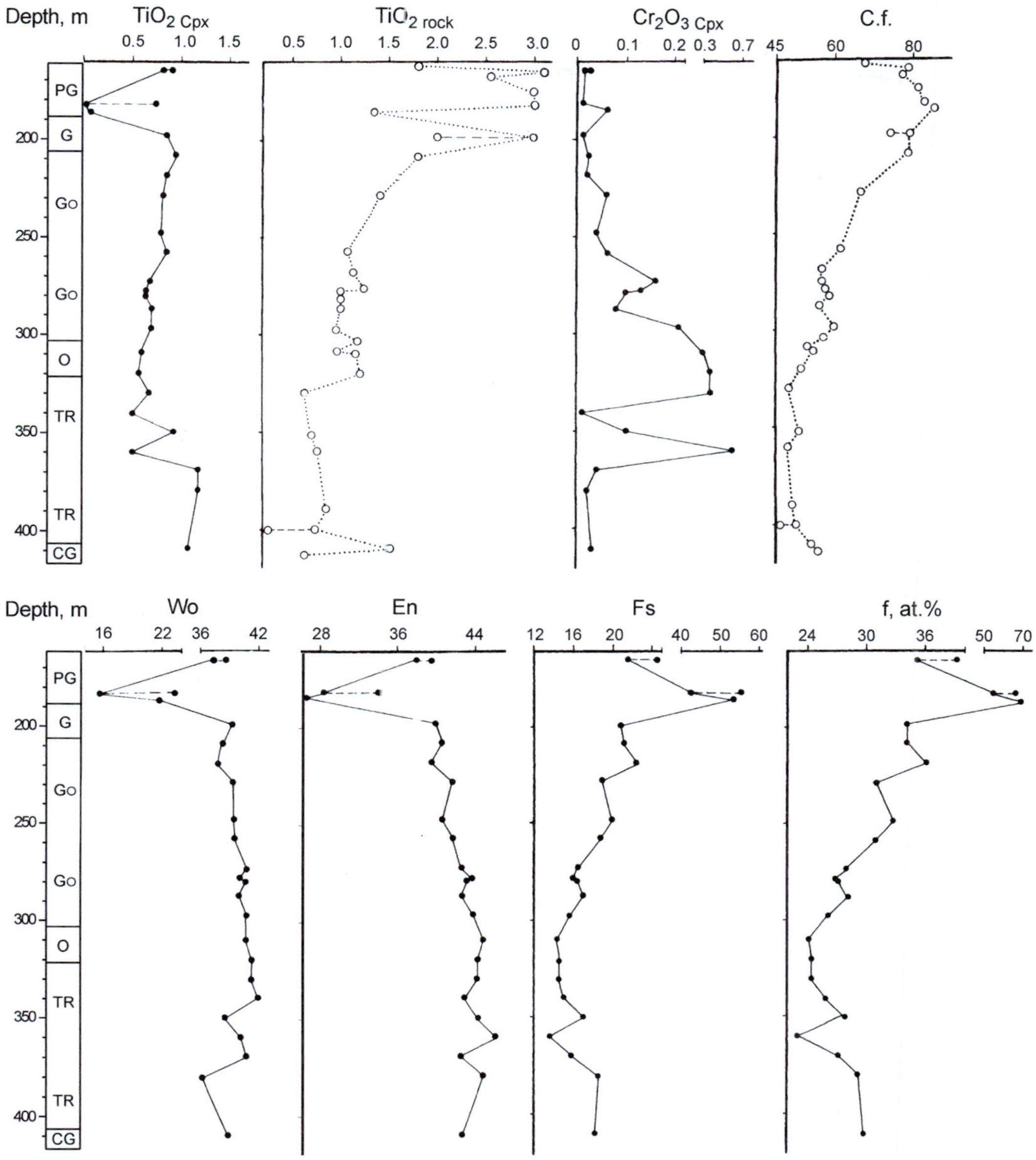

Fig. 3.38 Variation in clinopyroxene composition through a section of the Mt. Pegmatitovaya layered intrusion in drill hole E-28. Oxide content, wt%

9. The content of silicate nickel in olivine correlates with the Fe content in olivine, and judging from the compositional diagram, it is in accordance with the geochemical trend for traps.
10. The morphology of olivine crystals, wide variations in their size, and labradorite enclosed by olivine exclude an intratelluric nature of the mineral.
11. Olivine-rich picritic and troctolitic gabbrodolerites have low Cr_2O_3 content.
12. Chrome spinel are absent in gabbrodolerite of the complex, and minerals of the spinel group are represented by titanomagnetites containing 0.92–1.22 wt% V_2O_5.
13. Clinopyroxene in gabbrodolerite varies from $Wo_{42-37}En_{49-38}Fs_{12-25}$ to $Wo_{23-16}En_{34-24}Fs_{42-56}$.
14. Clinopyroxene in picritic and troctolitic gabbrodolerite is $Wo_{42-37}En_{49-42}Fs_{12-19}$.
15. The Cr_2O_3 content in clinopyroxene varies and reaches 0.61 wt%.
16. The Morongovsky Intrusions have no effusive comagmates.
17. There are no sulfide ores associated with intrusions of the Morongovsky Complex.
18. Accumulation of PGE up to 2 g/t takes place in intrusions of the UZ.

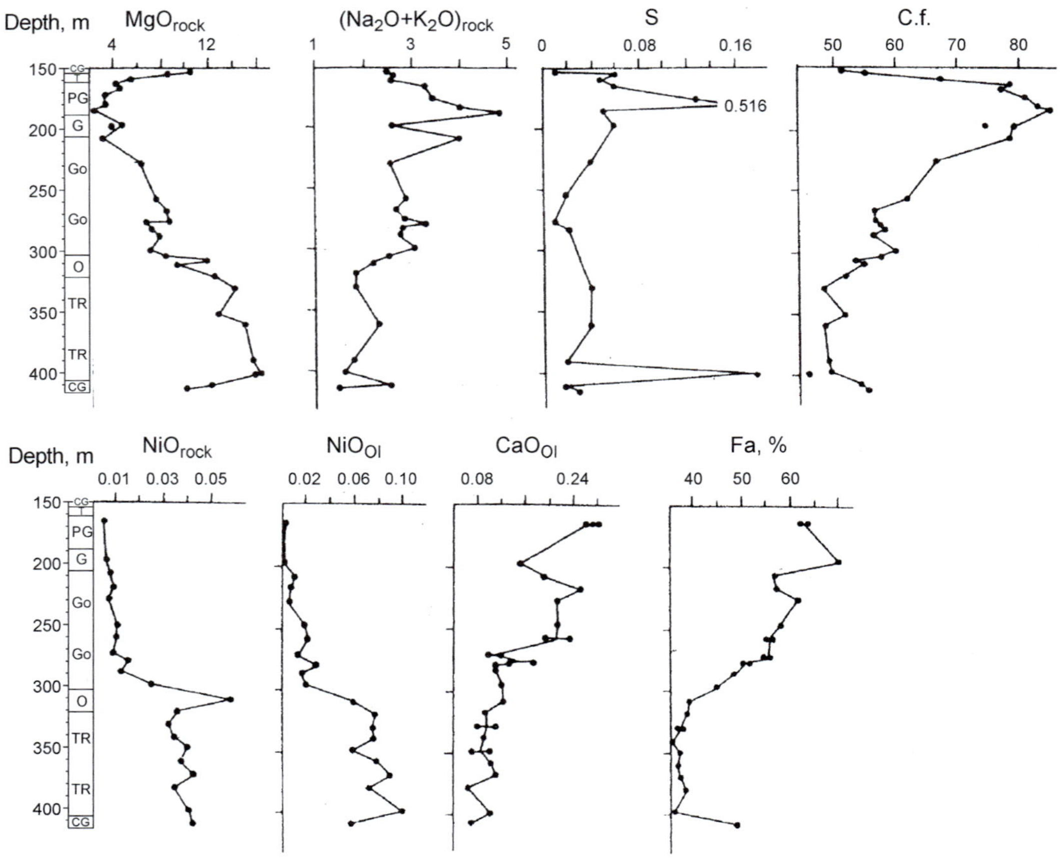

Fig. 3.39 Variation in olivine composition and content of chemical components in rocks through a section of the Mt. Pegmatitovaya Intrusion in drill hole E-28. Oxide content, wt%

3.7.3.4 Genetic Aspects

As it was mentioned above, an initial melt of mafic–ultramafic trap intrusions had an increased MgO content or picritic composition corresponding to the average-weighted composition of the Noril'sk-I Intrusion, the crystallization–gravitational differentiation of which was responsible for the variety of intrusive and effusive traps (Zolotukhin et al. 1986, 1989). Taking this into consideration, the composition of picritoid intrusions, including the Morongovsky ones, is similar to the initial melt, which had not experienced differentiation. Based on this, intrusions of the Morongovsky Complex and of the Upper Talnakhsky type have been considered as an integral group of melanocratic intrusions, which are distinctive, in comparison with other picritoid intrusions, by the absence of chromite (Duzhikov et al. 1988). Many geologists emphasize that rocks of these intrusions have the low nickel and chromium contents; however, they do not discuss the reasons of this phenomenon in the literature.

It is believed that young intrusions of the Morongovsky Complex in the region of the Mt. Pegmatitovaya cut through more ancient sills of the Middle Ergalakhsky type (Komarova and Lyul'ko 1967). This is based on information acquired in drill holes E-28 and E-29, in which dolerite of the Katangsky Complex cuts through troctolitic gabbrodolerite of the Morongovsky Complex. Rocks of these types have specific composition and texture: Morongovsky Complex rocks are essentially troctolitic and monolithic, and Katangsky Complex rocks are olivine poor and pea-like. The Mt. Pegmatitovaya Intrusion is composed of pea-like gabbrodolerite at the top and troctolitic gabbrodolerite at the base. The simple and distinct structure of the intrusion and strong visual similarity between rocks from one part with pea-like gabbrodolerite of the Middle Ergalakhsky Intrusion and rocks from another part with troctolitic gabbrodolerite of the Mt. Morongo Intrusion have been a reason for dividing the united body into two intrusions, one cutting the other.

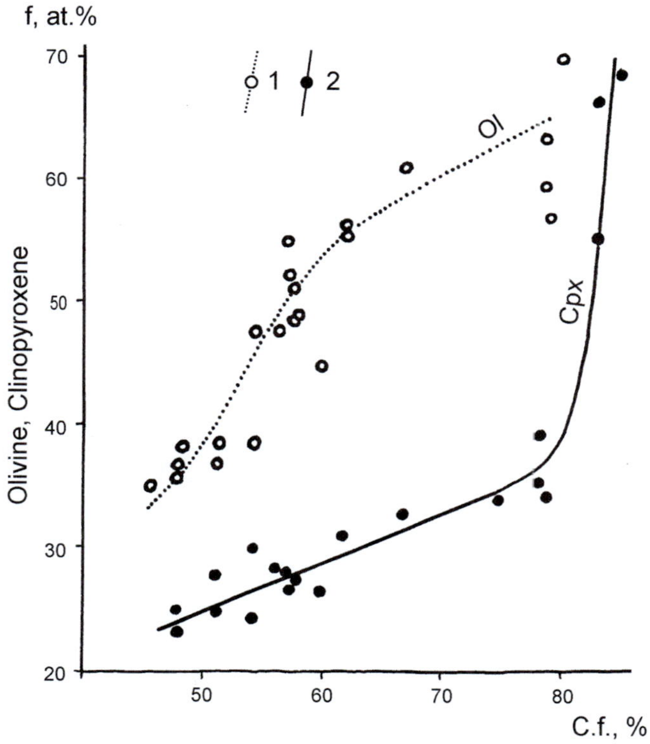

Fig. 3.40 The Fe content of iron-rich minerals versus the fractionation coefficient for rocks of the Mt. Pegmatitovaya Intrusion. *1* olivine, *2* clinopyroxene

Mineral–geochemical investigations and data reported by Korovyakov et al. (1963) show that the Mt. Pegmatitovaya Intrusion is a classical example of a layered mafic body with regular, stepped changes in rock composition, mineral association, and geochemical characteristics. Considering that the Morongovsky-type intrusions are fragments of a united body separated by erosion, it is meaningful to infer that the intrusion had an essentially troctolitic composition at the base and the proportion of mafic rock in the leading part of the rising body increased (Table 3.8). Ivanova (1975) has explained this phenomenon as the lagging of a more viscous ultramafic melt compared to a less viscous mafic melt in a unified magmatic channel. The capability for such mechanism to create a layered mafic–ultramafic melt prior to emplacement of a mafic melt is confirmed by the bimodal distribution of magnesium in intrusions of the Morongovsky Complex. This is evidence of the existence of two compositionally stable associations of mafic and ultramafic rock in intrusions of the Morongovsky Complex. Another confirmation of this phenomenon is the interlayering of olivine-rich and olivine-poor gabbrodolerites with sharp contacts, which are known from a number of trap intrusions. Thus, one can explain the mafic composition of the Mt. Pegmatitovaya Intrusion as the leading part of the magmatic column of the Morongovsky Complex intrusions. The absence of highly magnesian effusive comagmatic rocks in the Morongovsky-type intrusions can be explained similarly.

Based on investigations of the Morongovsky Complex intrusions, it is thought that the initial melt of these intrusions was of tholeiitic basalt composition. As a result of fluid–magmatic differentiation, the tholeiitic basalt was separated into immiscible mafic and ultramafic liquids, which were crystallized in accordance with their composition. The tholeiitic basalt composition of an initial magma explains the low concentrations of nickel and chromium in olivine-rich rocks, these concentrations being consistent with the abundance ratio of mafic rocks rather than to the abundance ratio of ultramafic rocks. The Fe content of olivine and the nickel content are typical for magnesian derivatives of tholeiitic melts rather than for picritic basalt. The morphology of olivine crystals and intergrowths of labradorite in olivine grains is evidence that their crystallization proceeded in hypabyssal intrusive chambers rather than in hypogeal hearths. The low abundance ratio of chrome in tholeiitic basalt melts in the Morongovsky Complex intrusions manifests itself as an absence of chrome spinel and accumulations of chrome in clinopyroxene.

3.8 Daldykansky Intrusive Complex $(\beta - \nu\beta)T_{1-2}$ dl

This intrusive complex includes, apart from the Daldykansky Intrusion itself, the Mt. Zub Intrusion, Middle Ergalakhsky, and Middle Dudinsky, as well as the Mt. Otdel'naya and Mt. Gudchikha Intrusions. Thick subbedded bodies of the Daldykansky Intrusion located at different hypsometric levels are subdivided into the Upper and Lower Daldykansky Intrusions. Intrusions of the Daldykansky Complex extend up to hundreds of square kilometers; the thickness of bodies varies from several meters to 100 m and more. The intrusions are magmatic bodies with complicated structure and form a system of sills with dyke-like bridges or sills that sharply change their hypsometric level, passing into a higher stratigraphic horizon as a steeply cutting body. Subbedded bodies are predominantly of Lower Devonian–Permo-Carboniferous age, and dykes cut through the entire volcanogenic unit including the youngest formations, such as basalts of the Samoedsky Suite.

Thin sills and dykes of the Daldykansky Complex have an isotropic structure and are composed of poikilo-ophitic pea-like olivine-free or olivine-bearing gabbro-dolerite, quartz-bearing dolerite, and plagioclase-porphyritic microdolerite. In the thickest subbedded bodies, slightly pronounced layering is observed with schlieren-like occurrences of gabbroic pegmatite. Stepanov and Zemskova (1988) note the uniformity in composition of rocks specific to the intrusions as well as the absence of an invisible layering.

Table 3.8 Chemical composition of mafic and ultramafic horizons and average weighted content of the Mountain Pegmatitovaya (PGI) and Mountain Putanaya (PTI) Intrusions (wt%)

	PGI			PTI		
Components	Mafic horizon (16)	Ultramafic horizon (6)	Average on intrusive (25)	Mafic horizon (7)	Ultramafic horizon (25)	Average on intrusive (33)
SiO_2	48.22	44.88	47.30	46.77	42.38	43.40
TiO_2	1.66	1.04	1.45	2.18	0.43	0.80
Al_2O_3	15.97	13.18	15.24	14.87	13.61	14.23
Fe_2O_3	2.29	1.78	2.21	3.71	2.35	2.61
FeO	10.11	12.54	10.58	9.01	11.34	10.62
MnO	0.20	0.22	0.20	0.15	0.16	0.16
MgO	6.12	14.04	8.26	6.58	17.57	14.89
CaO	10.67	9.10	10.41	11.04	8.08	8.86
Na_2O	2.69	1.72	2.35	2.49	0.98	1.33
K_2O	0.53	0.42	0.50	0.81	0.34	0.44
P_2O_5	0.22	0.11	0.18	0.20	0.08	0.10
LOI	1.63	1.48	1.63	2.10	2.66	2.53
H_2O^-	0.19	0.13	0.17			
Total	100.50	100.64	100.48	99.91	99.98	99.97
S_{tot}	0.062	0.050	0.057	0.034	0.039	0.037

3.8.1 Daldykansky Intrusion

In drill hole NP-49, three zones are distinguished: UZ (48 m thick), CZ (84 m thick), and LZ (22 m thick), which is typical for this intrusion.

The upper zone is composed of microdolerite, dolerite, and a horizon of schlieren-like gabbroic pegmatite, as well as rare ferrogabbro and granophyre, which are arranged among olivine-free and olivine-bearing gabbrodolerite.

Gabbrodolerite has plagioclase-porphyritic, seriate, porphyritic, ophitic, and poikilo-ophitic textures. The rock consists of plagioclase ($Ab_{32–46}Or_{1–2}An_{67–52}$; 55–65%), clinopyroxene ($Wo_{40}En_{44}Fs_{16}$; 25–30%), olivine (Fa_{62}; 0–7%), and apatite and titanomagnetite (5–10%) (see Rock indication 188). The following constituents have been determined in the rock: in clinopyroxene, 0.69 wt% TiO_2, 2.48 wt% Al_2O_3, 0.30 wt% Na_2O, and 0.25 wt% Cr_2O_3; in olivine 0.64–0.65 wt% MnO and 0.10–0.12 wt% CaO; in apatite 0.20–0.21 wt% SrO, 0.43–0.61 wt% CeO_2, 1.46–1.80 wt% F, and 0.71–0.77 wt% Cl; and in titanomagnetite 9.57 wt% TiO_2, 2.66 wt% Al_2O_3, 0.32 wt% MnO, 0.42 wt% MgO, 0.18 wt% Cr_2O_3, and 1.42 wt% V_2O_5.

Gabbroic pegmatite has coarse-grained and giant-grained textures. It is composed of plagioclase ($Ab_{62–64}Or_4An_{34–32}$; 35–70%), clinopyroxene ($Wo_{32–33}En_{30–32}Fs_{35–37}$; 15–25%), olivine ($Fa_{87–89}$; 1–2%), apatite (1–3%), and titanomagnetite and ilmenite (7–15%) (see Rock indication 189). Quartz is not infrequent in small amounts with olivine. In the rock matrix, granophyre micropegmatite and/or palagonite are common. Clinopyroxene contains 0.75–0.76 wt% TiO_2, 1.10–1.13 wt% Al_2O_3, and 0.23–0.25 wt% Na_2O; olivine contains 1.05–1.11 wt% MnO and 0.14–0.28 wt% CaO; apatite contains 0.21–0.25 wt% SrO, 0.13–0.18 wt% CeO_2, 2.25–2.86 wt% F, and 0.17–0.48 wt% Cl; titanomagnetite contains 11.27 wt% TiO_2, 1.62 wt% Al_2O_3, 0.36 wt% MnO, 0.09 wt% MgO, and 0.69 wt% V_2O_5; and ilmenite contains 0.55 wt% MnO, 0.40 wt% MgO, and 0.32 wt% V_2O_5.

The central zone is composed of coarse-grained pea-like olivine gabbrodolerite. The rock consists of 50–65% plagioclase, 25–30% clinopyroxene, 7–12% olivine, and 5–7% titanomagnetite and ilmenite. The rock features not only clinopyroxene but also olivine and titanomagnetite oikocrysts. The composition of porphyritic plagioclase is $Ab_{29–30}Or_1An_{70–69}$, plagioclase chadocrysts in oikocrysts of olivine are $Ab_{36}Or_2An_{62}$, and plagioclase in the rock groundmass is $Ab_{40–48}Or_{1–2}An_{59–49}$. Clinopyroxene is augite ($Wo_{37–39}En_{39–45}Fs_{16–23}$), although rare pigeonite ($Wo_{10}En_{52}Fs_{38}$) is also observed. Olivine oikocrysts have the composition $Fa_{58–61}$, and subidiomorphic and isometric grains are of composition $Fa_{49–50}$ (see Rock indications 190–192). The following constituents have been determined in the minerals: in clinopyroxene, 0.5–1.10 wt% TiO_2, 0.94–2.26 wt% Al_2O_3, 0.09–0.30 wt% Na_2O, and 0.06–0.25 wt% Cr_2O_3; in olivine, 0.54–0.63 wt% MnO and 0.08–0.16 wt% CaO; in apatite, 0.18–0.21 wt% SrO, 0.23–0.44 wt% CeO_2, 1.49–2.56 wt% F, and 0.64–0.87 wt% Cl; in titanomagnetite, 11.19–15.01 wt% TiO_2, 2.06–3.25 wt% Al_2O_3, 0.29–0.47 wt% MnO, 0.44–1.07 wt% MgO, 0.06–0.2 wt% Cr_2O_3, and 1.12–1.36 wt% V_2O_5; in ilmenite, 0.43 wt% MnO, 1.38 wt% MgO, and 0.34 wt% V_2O_5. The contact with underlying rocks is gradual.

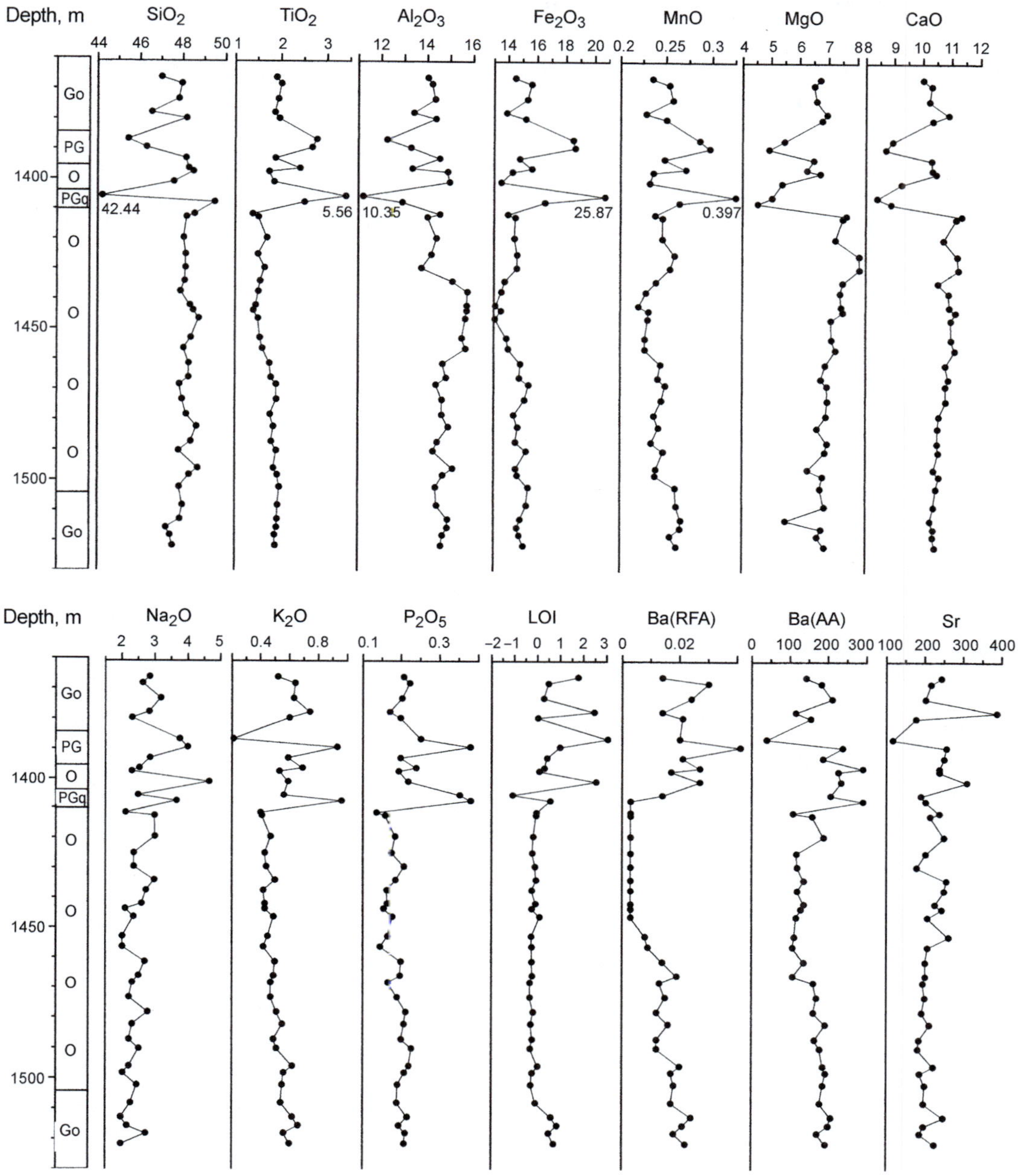

Fig. 3.41 Petrochemical variation diagram for the Daldykansky Intrusion, drill hole NP-49. Contents of oxides, Ba (XRF analysis), and loss on ignition, wt%, contents of elements, ppm

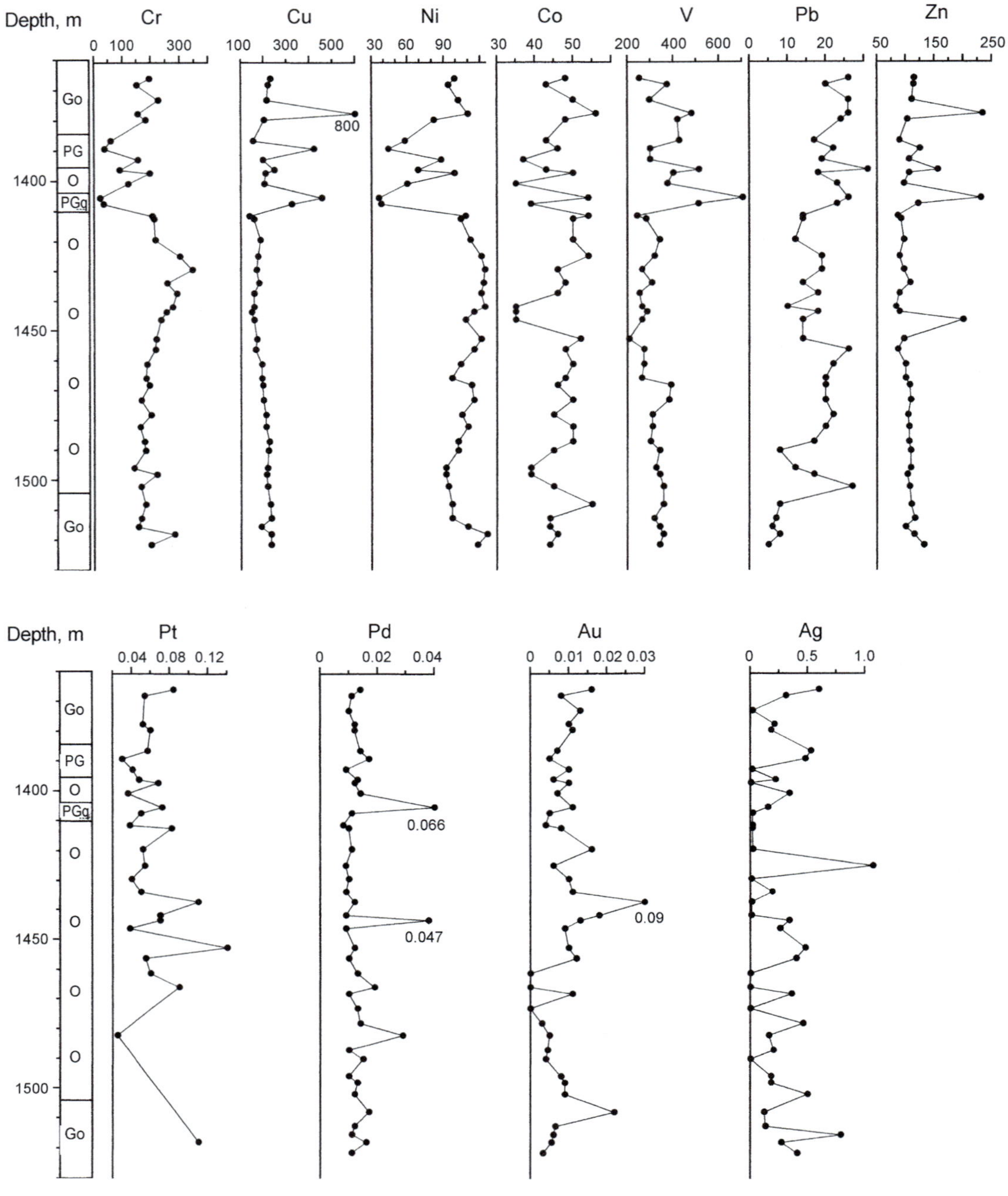

Fig. 3.41 (continued)

The lower zone of the intrusion is composed of poikilo-ophitic olivine-bearing gabbrodolerite, which when compared with overlying rocks show a decrease in olivine by up to 5–7% and a decrease in the size of clinopyroxene oikocrysts.

3.8.1.1 Petrochemistry

Compositional features of rocks from the Daldykansky Intrusion are given in Fig. 3.41 and Table A.48 (see Appendix). Rocks in the UZ are distinguished on the variation diagram by their unstable composition that manifests itself as saw-like plot shape. The extreme variation in oxide and element content is due to the redistribution of chemical elements by fluids in the pegmatoid and gabbrodolerite hosts.

In spite of insignificant variations in chemical components in the intrusion CZ, the concentrations of TiO_2, Fe_2O_3, MnO, K_2O, P_2O_5, Ba, Cu, V, and Zn distinctly increase, and those of MgO, CaO, Na_2O, Sr, Cr, and Pb decrease downward through the intrusion, whereas the behavior of other elements is not clear. It is remarkable that accumulation maxima of Al_2O_3, MgO, CaO, and Cr_2O_3 are restricted to the central part of the intrusion. This is conditioned by increased amounts of early magmatic high-magnesian plagioclase, olivine, and clinopyroxene and is related to the temperature gradient from the center to the edges of the magmatic body. Such temperature dependence in distribution of chemical elements is observed in a number of slightly differentiated intrusions poor in volatiles, the Mt. Putanaya Intrusion among them (Fig. 3.35).

3.8.1.2 Mineralogy

In rocks of the Daldykansky Intrusion, plagioclase composition covers the range $Ab_{29}Or_1An_{70}$–$Ab_{45}Or_2An_{53}$ in rocks of the layered series and $Ab_{32}Or_1An_{67}$–$Ab_{67}Or_4An_{34}$ in pegmatoid. Clinopyroxene is augite ($Wo_{39-32}En_{45-31}Fs_{16-37}$) and magnesian pigeonite ($Wo_{10}En_{53}Fs_{38}$). Total Fe content of augite is in the range 26–54 wt%; *f* varies by 3–7 wt% within a single grain from core to rim. Low concentrations of TiO_2 (0.55–1.10 wt%) and Al_2O_3 (0.94–2.48 wt%) are specific for clinopyroxene. Increased Cr_2O_3 (0.25–0.26 wt%) (with $f = 26$–27 wt%) concentration is characteristic for augite from the UZ and CZ. The Fe content in olivine is within the range Fa_{49-89}. The most iron-rich olivine, such as ferrohortonolite, is found in quartz–olivine-bearing gabbroic pegmatite. The Fe content in the mineral correlates with its MnO content which reaches 1.11 wt%. Apatite contains variable concentrations of F (1.49–2.86 wt%) and Cl (0.17–0.87 wt%). Ore minerals include titanomagnetite and ilmenite. Titanomagnetite contains 9.6–15.0 wt% TiO_2, 1.6–3.25 wt% Al_2O_3, 0.01–1.1 wt% MgO, and 0.7–1.4 wt% V_2O_5, and ilmenite contains 0.40–1.38 wt% MgO and 0.32 wt% V_2O_5.

3.8.2 Middle Dudinsky and Middle Ergalakhsky Intrusions

These two intrusions are parts of a unified body that extends north–south along the western side of the Noril'sky trough. The intrusions crop out in rapids of the Dudinka and Middle Ergalakh Rivers, as well as these are intersected by drill holes. The magmatic body is slightly differentiated; its thickness is 140–150 m. The body is situated in Permo-Carboniferous terrigenous deposits of the Tungussky series.

The upper zone of the intrusion is composed of olivine-free and olivine-bearing gabbrodolerite and magnetite ferrogabbro with schlieren of gabbroic pegmatite and streaky granophyre.

Ferrogabbro has an ophitic texture and trachytoid fabric. The rock consists of 45–60% plagioclase, 15–20% clinopyroxene, 1–5% amphibole, and 15–25% titanomagnetite and ilmenite. Olivine (Fa_{70}) occurs sporadically (Ryabov 1992a). The rock matrix consists of prisms of zoned plagioclase ($Ab_{50-66-76}Or_{2-3-8}An_{48-31-16}$) and augite ($Wo_{41}En_{28}Fs_{31}$) crystals ($f = 52$ wt%) and titanomagnetite (see Rock indication 193). Amphibole is a brownish hornblende ($f = 63.6$ wt%) that forms individual crystals and intergrowths with augite and blue–green amphibole ($f = 66.9$ wt%) developed after dark-colored minerals and is set in the rock matrix in association with oligoclase ($Ab_{74}Or_8An_{18}$) and apatite. Titanomagnetite features elevated V_2O_5 content (2.63 wt%) and low Cr_2O_3 content (0.12 wt%) and contains MgO (0.11 wt%), MnO (0,26 wt%), and Al_2O_3 (30.84 wt%). Ilmenite contains 0.20–0.02 wt% Cr_2O_3, 0.21–0.23 wt% MgO, and 0.42–0.43 wt% MnO. Apatite contains 0.20 wt% SrO, 0.46 wt% CeO_2, 1.84 wt% F, and 0.58 wt% Cl.

Gabbroic pegmatite is a coarse-grained to giant-grained rock composed of prismatic crystals of plagioclase and clinopyroxene. Amphibole, titanomagnetite, micropegmatite, quartz, apatite, biotite, calcite, and palagonite are present in variable amounts, sometimes in high concentrations. Dendrite-like megacrystals of titanium augite 12–17 cm in length (see Rock indication 194) are known in gabbroic pegmatite of the Middle Dudinsky and Middle Ergalakhsky Intrusions (Ryabov et al. 1977).

Granophyre forms a cement for magmatogenic breccia fragments of which are composed of fine-grained granoblastic aggregate of clinopyroxene (see Rock indication 195) that are intersected by thin graphite–calcite veinlets. In granophyre, plagioclase ($Ab_{7-11}Or_{88-93}An_1$) is observed in graphic intergrowths with quartz, and clinopyroxene is diopside ($Wo_{49}En_{26-29}Fs_{22-25}$).

The central zone of the intrusion is composed of pea-like quartz–olivine-bearing and olivine gabbrodolerite with poikilo-ophitic texture and elements of plagioclase-porphyritic and micro-ophitic textures. The size of "peas" (clinopyroxene poikilocrysts) decreases downward through the intrusion from 2–3 cm to 0.3–0.5 cm. In olivine gabbronorite, the amount of plagioclase chadocrysts in clinopyroxene oikocrysts decreases, and the rock looks like a pyroxene-aphyric gabbrodolerite.

In quartz–olivine-bearing gabbrodolerite (see Rock indication 196), large prisms of plagioclase have the composition $Ab_{27}Or_1An_{72}$ and are overgrown by laminae of $Ab_{73}Or_4An_{23}$. Chadocrysts of plagioclase in olivine have the composition $Ab_{31}Or_1An_{68}$, and those in clinopyroxene are consistent with $Ab_{37}Or_1An_{62}$. The Fe content of olivine is Fa, and the composition of clinopyroxene oikocrysts is $Wo_{37}En_{44}Fs_{19}$. In ilmenite, there is an elevated MnO content (1.13 wt%) and a low MgO and V_2O_5 content (0.14 wt% and 0.28 wt%, respectively).

In olivine pyroxene-aphyric gabbrodolerite (see Rock indication 197), large clinopyroxene porphyrocrysts have the composition $Wo_{41}En_{46}Fs_{13}$ and feature an elevated Cr_2O_3 content (0.84 wt%). Plagioclase chadocrysts in rims of grains have the composition $Ab_{26}Or_1An_{73}$. Olivine forms isometric grains of Fa_{53-56} both small and medium in size. Olivine intergrowths in large ilmenite crystals have the composition Fa_{40}. Discrete reaction rims of orthopyroxene $Wo_7En_{62}Fs_{31}$ are observed in some cases. The oxide-ore phase is ilmenite that forms large crystals and small needles. When comparing the former and the latter, a decrease in the MgO content from 1.64 to 0.43 wt% and in MnO content from 0.64 to 0.56 wt% is noted. The V_2O_5 content is 0.20–0.21 wt% and does not change significantly.

3.8.3 Mt. Gudchikha Intrusion

The intrusion varies from olivine-free to olivine gabbrodolerite. The MgO content ranges from 5.6 to 12 wt %, rarely up to 16.2 wt%; $Na_2O + K_2O$ content ranges from 1.3 to 4.4%. The Fe content of olivine in olivine gabbrodolerite is Fa_{30-31}. Clinopyroxene is augite ($Wo_{39-42}En_{46-48}Fs_{10-13}$) and contains 0.43–0.72 wt% TiO_2, 1.1–3.1 wt% Al_2O_3, and 0.1–0.73 wt% Cr_2O_3. Orthopyroxene ($Wo_4En_{70-71}Fs_{24-25}$) contains 2–2.2 wt% CaO, 1.1–1.3 wt% Al_2O_3, and 0.6 wt% TiO_2. Plagioclase crystals ($Ab_{28-20}Or_{0-1}An_{80-71}$) are porphyritic, in prisms in the groundmass have the composition $Ab_{35-39}Or_{1-2}An_{64-59}$, and grains in the matrix have the composition $Ab_{76-81}Or_{1-7}An_{21-12}$.

In outcrops of the UZ along the Kaskadny Stream, megacrystals of clinopyroxene up to 50–70 cm are observed in association of olivine Fa_{60-62} set in a rock groundmass (see Rock indication 198).

3.8.4 Geochemistry and Genetic Aspects

Contact alteration due to intrusion of the Daldykansky Complex manifests as a slight hornfelsing of rocks.

3.8.4.1 Mineralization

Sulfide mineralization in the Daldykansky Complex intrusions sporadically occurs either as poor interstitial disseminated ores at the bottom of magmatic bodies or as large (0.2–0.4 cm to 0.6–0.8 cm) single chalcopyrite-–pyrrhotite inclusions that are observed in different parts of the intrusion.

Elevated concentrations of ore metals in rocks of the Daldykansky Intrusion were studied in the UZ in drill hole NP-32, and these are as follows: 0.1 g/t Pt, 0.16 g/t Pd, 0,025 g/t Rh, and 2.96 g/t Ag. In the CZ, in drill hole VP-49, the metal content was 0.14 g/t Pt, 0.047 g/t Pd, 0.018 g/t Au, and 1.07 g/t Ag. Maximum Cu contents (800 g/t) have been determined in the UZ, and maximum Ni contents (up to 126 g/t) are in the LZ of the Daldykansky Intrusion.

3.8.4.2 Magnetization of Rocks

A negative remanent magnetization vector is a characteristic feature of Daldykansky Complex intrusions.

3.8.4.3 Petrological Features

The following features of the Daldykansky Complex intrusions should be noted:

1. The slightly differentiated internal structure of thick magmatic bodies and the isotropic structure of dykes.
2. The end members of the series of layered rocks in the CZ are olivine-free and/or quartz–olivine-bearing and olivine gabbrodolerites.
3. In the UZ of the intrusion, schlieren of gabbroic pegmatite, ferrogabbro, granophyres, and dendritic clinopyroxene megacrystals are observed.
4. Contact rock facies exhibit the plagioclase-porphyritic nature of the intrusions (Stepanov and Zemskova 1988).
5. The principal rock texture is poikilo-ophitic due to clinopyroxene and olivine oikocrysts.
6. The olivine content varies from single grains to 15% and rarely up to 20–25%.
7. The Fe content varies within the ranges Fa_{30}–Fa_{56} in the layered rock series and within Fa_{60}–Fa_{89} in the UZ rocks.
8. Clinopyroxene is Mg-rich augite with composition $Wo_{38-42}En_{39-48}Fs_{10-23}$ in the CZ and $Wo_{32-40}En_{31-44}Fs_{16-37}$ in pegmatoid.
9. The Cr_2O_3 content in clinopyroxene varies widely and may reach 0.84 wt% in zoned crystal cores.
10. The plagioclase composition in rocks of the layered series varies from $Ab_{20}An_{80}$ to $Ab_{39}Or_2An_{59}$, albite–oligoclase is common in the matrix, and the range of compositions $Ab_{32}Or_1An_{67}$–$Ab_{76}Or_8An_{16}$ is common for pegmatite.
11. Apatite contains variable halogen content: 1.49 wt% F and 0.17–0.87 wt% Cl.
12. Oxide-ore phases are represented by titanomagnetite and ilmenite; the former may contain abnormally high concentrations (up to 2.63 wt%) of V_2O_5 in magnetite ferrogabbro.
13. In the pegmatoid of the UZ, an accumulation of noble metals up to 0.26 g/t is noted.
14. Intrusions of the Daldykansky intrusive complex are posteffusive formations, and there are no lavas comagmatic with them.

3.8.4.4 Genetic Concepts

The chemical composition of rocks from isotropic sills and dykes of the Daldykansky Complex, as well as its slightly differentiated intrusions, is close to the composition of an "average trap" (Table 3.9) and is similar to that of the parental magma. The smooth variation in element content in the Daldykansky Intrusion is presumed to be due to

Table 3.9 The Daldykansky intrusive complex rocks' chemical composition (wt%)

Components	Average weighted content	Composition of dykes	Gabbropegmatite	Ferrogabbro	Granophyre	Mean trap (on Duzhikov[a])
SiO_2	48.02	47.93	47.52	44.25	57.2	48.41
TiO_2	1.75	2.441	2.44	5.08	0.8	1.30
Al_2O_3	14.70	13.14	13.26	10.61	13.1	15.31
Fe_2O_3	14.49	17.28	16.81	24.13	0.79	4.23
FeO	–	–	–	–	5.5	8.24
MnO	0.24	0.268	0.27	0.35	0.11	0.19
MgO	6.93	5.58	5.53	4.86	3.51	7.04
CaO	10.64	9.83	9.43	8.62	12.71	10.68
Na_2O	2.50	1.88	3.35	2.47	2	1.81
K_2O	0.52	0.79	0.68	0.67	2.53	0.68
P_2O_5	0.19	0.215	0.29	0.32	0.23	0.22
Ba	0.01	0.023	0.02	0.01	–	–
LOI	0.17	0.7	1.06	−0.80	1.24	2.04
Number of analyses	35	1	5	2	1	401

[a]Zelenshchikov and Duzhikov (1974)

differentiation under chamber conditions. Crystallization products of fluidized melts (pegmatoid, ferrogabbro, and granophyre) as thin veinlets, lenses, and scattered schlieren are evidence that the melt was depleted in volatile elements.

The poikilo-ophitic texture in most complex rocks is formed by oikocrysts not only of clinopyroxene but also olivine, as well as the presence of small olivine grains in the groundmass. This is evidence that crystallization was rapid, and there was no gravitational accumulation of olivine as an early crystalline phase. Phenocrysts of bytownite–labradorite composition are indicative of the plagioclase-porphyritic nature of the intrusions and the anorthositic trend for melt evolution. The pronounced zoning of minerals and association of olivine with quartz or granophyre micropegmatite, as well as the insignificant variations in mineral composition are indicative of abandoned equalization of the magmatic state. In spite of the small thickness of magmatic bodies, sometimes no more than 100 m, there is a slight layering to the Daldykansky Intrusion. It has been stated empirically that the extent and depth of differentiation correlate with the scale of pegmatoid occurrence and are related to a fluid regime in the magmatic system. Fluids are responsible for redistribution of elements producing directional crystallization and geochemical trends and stabilization of the system in general. In this way, we can assume that magmatic melts of the Daldykansky Complex were depleted in volatiles. Therefore, evolution of the magmatic melt under chamber conditions was limited by crystallization of porphyritic plagioclase and by the formation of porphyritic, pea-like gabbrodolerite. Nevertheless, an increase in the size of "peas" upward through the intrusion points an increase in volatile elements in this direction that is manifest as schlieren of pegmatoid. Prechamber differentiation is not typical for intrusions of this complex, although lenses of gabbrodolerite enriched in olivine and magnesium up to 11.3–16.2 wt% (Mt. Gudchikha Intrusion) that are found locally among olivine-poor rocks should be considered as a manifestation of prechamber differentiation.

Rocks of the Daldykansky Complex intrusion have remanent magnetization, and dyke formations of this complex cut through the entire volcanogenic section in the northwest of the Siberian Platform including the Samoedsky Suite. These data provide reasons to consider the Daldykansky intrusions as the youngest manifestation of tholeiite-basalt magmatism in the region, which have no effusive comagmates.

3.8.5 Dzhaltul Intrusion

3.8.5.1 General Characteristic of the Intrusion

The Dzhaltul Intrusion is located between Gorbiachin and Kureyka Rivers and represents a sheet-like body, which forms ellipsoidal exposures 25 × 70 km in plan view. The magmatic body was broken into blocks by faulting. These blocks were subsequently separated into individual units by erosional processes. Following initial exploration, they were assigned unique names: Mt. Dzhaltul, Lagernaya, Middle Gorbiachinsky, Mt. Ozernaya, Mt. Podkova, and Mt. Tiptur Intrusions (Fig. 3.42). Subsequent investigations have shown that all these intrusions are parts of the united extended magmatic body, but nevertheless, some parts have retained their former names in the geologic literature. The petrology of the Dzhaltul Intrusion has been described in a number of

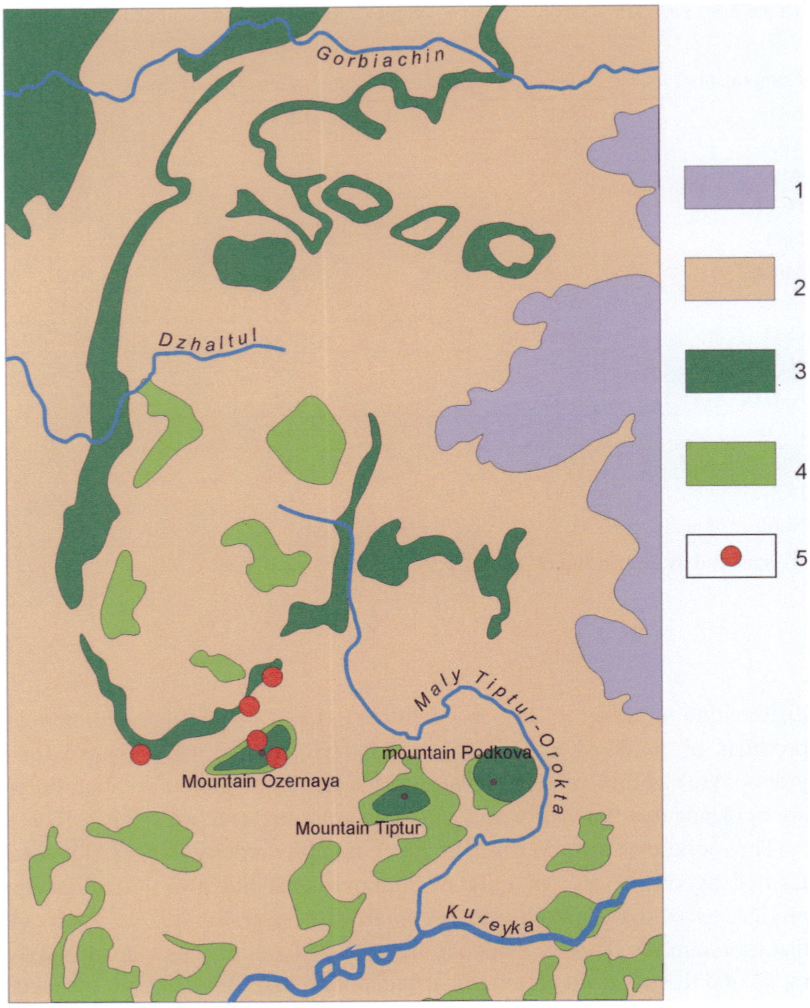

Fig. 3.42 Schematic geological map of the Dzhaltul Intrusion region. *1* Permo–Triassic volcanogenic unit, *2* coal-bearing deposits and limestones of Permo–Carboniferous age, *3* Dzhaltul differentiated intrusion, *4* intrusions of dolerite, olivine dolerite with pegmatoid, *5* occurrences of native iron

publications (Kravtsova 1958; Bazhenov et al. 1959; Lurje et al. 1962; Zolotukhin and Vasil'ev 1967; Kavardin 1976; Oleinikov et al. 1985; and Koroleva and Oleinikov 1998).

In early investigations, the Mt. Ozernaya, Mt. Dzhaltul, and Mt. Podkova Intrusions in the Kureyka, Gorbiachin, and Uchami river basin have been assigned to the Noril'sky intrusive complex (Kravtsova 1958; Lurje et al. 1962). This was due to the layering of the intrusions to picritic gabbrodolerite and the presence of sulfide Cu–Ni mineralization in taxitic gabbrodolerite. It is of interest that the authors related accumulations of sulfides in the horizon of taxitic gabbrodolerite to "xenoliths" of hornfelsed rocks. Later, Tuganova (Lurje et al. 1976) inferred that the Mt. Ozernaya, Mt. Podkova, and other intrusions on the Dzhaltul Plateau were vaguely differentiated. She combined them with the Kureysky type and recognized the Mt. Ozernaya subtype (Lurje et al. 1976). The Dzhaltulsky Intrusion shows a strong similarity in internal structure, composition, and native iron content to the Khungtukun, Khininda, and Maymechinsky Intrusions located in the Maymecha-Kotuysky Province. Based on the petrological–geochemical resemblance, these intrusions are regarded as individual Dzhaltul–Khungtukunsky-type intrusions (Ryabov and Anoshin 1999).

In the explanatory notes to the 1:200,000 geological map of the northwest Siberian Platform (published in 1999), Noril'sk geologists assigned the Dzhaltul Intrusion (with all other intrusions mentioned above) to the Kureysky intrusive complex. Characteristic petrological and mineralization features of the Dzhaltulsky Intrusion will be published in detail in separate publication.

3.8.5.2 Short Petrologic Characteristic

The Dzhaltul Intrusion lies in Middle Carboniferous–Permian terrigenous deposits. It varies in thickness from 30 m to over 600 m. The first petrographic description of the Dzhaltul Intrusion was made by Kravtsova (1958), observing the layered structure of the intrusion. The thickness of the intrusion part studied was 200–250 m and ranged from picritic to olivine-bearing gabbrodolerite. The middle part of the body is composed of olivine and olivine-free gabbrodolerite often with a pronounced trachytoid texture.

At the intrusion roof, dolerite and porphyritic microdolerite are developed. The picritic gabbrodolerite has a pronounced streaky appearance due to alternating bands of olivine–plagioclase composition and bands composed of olivine–plagioclase–orthopyroxene–clinopyroxene. The amount of olivine in the rocks is as much as 40%, and colored minerals, as a whole, account for 70%. Later Kavardin et al. (1968) noted melt flow indicators, such as streaks and lines due to oriented prismatic crystals. These observations become interesting in connection with aspects of a nature of layering a melt during intrusion, chamber differentiation, and the doubtful role of gravitational separation of olivine in situ. The petrology of the Dzhaltul Intrusion has been described in a number of works (Oleinikov et al. 1985; and Koroleva and Oleinikov 1998); therefore, we comment only principal internal structure and compositional features.

The magmatic body has a layered structure. Its lower part (about 400 m thick) is composed of gabbrodolerite alternating with rocks of different olivine content: from troctolitic gabbrodolerite to olivine-free rock. Both olivine and plagioclase are represented by two generations: Fa_{19-25} and Fa_{36-38} for olivine and An_{88-77} and An_{73-55} for plagioclase. Upward through the middle part of intrusion, the olivine content decreases, and olivine-free varieties become dominant. Quartz is commonly observed in the olivine-free rocks, and quartz-bearing gabbrodolerite becomes the main variety of rock. The composition of plagioclase in early generations is An_{81} and in late generations is An_{73-50}. For the middle part of the body, schlieren and lenses of gabbroic pegmatite and monzodiorites (metadiorites) are typical. The width of these bodies is several meters. In the upper part of the magmatic body, both gabbrodolerite and quartz-bearing gabbrodolerite become coarse-grained or pegmatoidal. Magnetite ferrogabbro up 35 m thick is also noted. Upward through the intrusion, in pegmatoidal gabbrodolerite, clinopyroxene and plagioclase content decreases, whereas quartz and K-feldspar content increases. Near the base of the intrusion, there are layers of gabbroic diorite and monzodiorite (metadiorite). These are composed of plagioclase An_{44-25} (14–37%), K-feldspar $Ab_{11-0}Or_{84-100}$Or Or_{5-0} (15–50%), quartz (14–37%), amphibole (3–25%), clinopyroxene (0–2%), and ore minerals (0.5–4%) (Koroleva and Oleinikov 1998). These authors suggest taxito-ophitic olivine gabbrodolerite to be dominant rocks in the Dzhaltul Intrusion.

In the Mt. Ozernaya Intrusion, the roof of the magmatic body is eroded, and the total thickness of the intrusion is about 350 m, 45 m of which is a horizon of taxitic and taxite-like gabbrodolerites with native iron in the UZ (Bazhenov et al. 1959). Inclusions of fine-grained rock are also restricted to this horizon. In the geologic literature, these inclusions have been described as xenoliths of hornfels (Zolotukhin and Vasil'ev 1967) or apodolerite metasomatite (Oleinikov et al. 1985). Most detailed investigations of these formations have shown that metasomatite has a dominantly pyroxene–plagioclase composition. Apodolerite metasomatite is subdivided into near-nodule, areal, and lens-like shapes. These evidence the presence of xenoliths from dolerites underlying the Dzhaltul Intrusion and aposedimentary metasomatites, which are similar to apodolerite ones in pyroxene–plagioclase and pyroxene composition. The layered texture that is observed for some xenoliths serves as a criterion for the sedimentary origin of the metasomatites. The authors note that xenoliths of dolerite experience only partial hornfelsing.

Metasomatic alteration of gabbrodolerite is observed as a granulation of rock-forming minerals (Oleinikov et al. 1991, p. 32) and the formation of granoblastic textures. These authors noted out that some features of metasomatic alteration of plagioclase from gabbrodolerite include, in particular, a sieve-like appearance of bytownite crystals due to numerous pyroxene granules and subsequent complete replacement of plagioclase by a granoblastic pyroxene aggregate. Similar insights into the nature of clinopyroxene granules and fine-grained rocks were provided earlier by Zolotukhin and Vasil'ev (1967). Near-nodule and areal metasomatites are typically found near occurrences of native iron, whereas lens-like metasomatite only sporadically contains disseminated sulfide (Oleinikov et al. 1991). These authors relate the formation of all these rocks to metasomatic transformation of intruding gabbrodolerite under the action of reducing fluids.

The diagram of ($Na_2O + K_2O$) versus SiO_2 (Fig. 3.43) gives an insight into the chemical composition of rocks from the Dzhaltul Intrusion. Judging from the composition fields on the diagram, one can propose with confidence that the initial magma of the Dzhaltul Intrusion had a tholeiitic basalt composition, and all the rock varieties, from troctolitic gabbrodolerite to monzonite, were formed by melt differentiation. Ore-free olivine–plagioclase-porphyritic gabbrodolerite with poorly disseminated native iron and sulfides show no significant differences in composition, and these fall into a compact field on the plot. However, fine-grained inclusions in gabbrodolerite exhibit significant compositional fluctuations that manifest as a scattering of points.

In spite of available geological information on the intrusion, aspects of the petrogenesis and mineralization have not become relevant until now. Geologists now have a renewed interest in the Dzhaltul–Khungtukunsky-type intrusions due to the connection between a concentration of native iron in traps and elevated PGE. In recent years, we gained a large amount of new information on the ore-bearing horizon of the Dzhaltul Intrusion. The investigation results will be published in a special work; we will only present some data on the petrographic composition of the rocks.

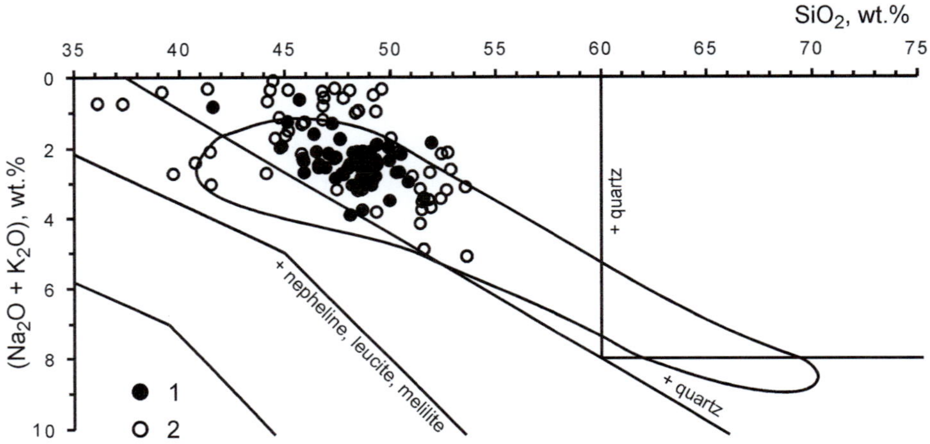

Fig. 3.43 Diagram of ($Na_2O + K_2O$) versus SiO_2 for some rocks of the Dzhaltul Intrusion. *1* gabbrodolerite of the southern part of the intrusion, *2* aphanites with variable ore saturation (native iron and sulfides) and barren. The field shows the variety of rocks in the massif (from literature)

Internal Structure of the UZ

The principal rock type of this zone is olivine–plagioclase-porphyritic gabbrodolerite, which includes nodules of native iron and inclusions of fine-grained rocks. Nodules have a rounded shape; their sizes vary from 5–7 cm to 20–30 cm, more rarely 50–70 cm in width. The nodules form separated "erratic boulders" throughout the UZ. Fine-grained rock (aphanite) form in association with nodules of native iron and single, rounded inclusions, or accumulations of dark-gray, light-gray, or pink angular fragments or aphanite aggregates in gabbrodolerite are more common. The sizes of fragments vary over a wide range from 0.5 × 0.7 cm, (3–5) × (10–15) cm up to (0.3–0.7) × (1–5) m. Large fragments form plates with a subhorizontal orientation, which sometimes feature a tail of small fragments. Accumulations of fragments commonly extend as a horizon. It is these rocks that have been described by previous researchers as xenoliths of hornfels and metasomatite. Taking into consideration the fine-grained nature of rocks, we assume that it is more likely to be correct to classify them as aphanites without attaching a genetic interpretation to their name.

Fragments of aphanite are cemented by gabbrodolerite. In the Ozernaya Intrusion, a large field of "composite" small-fragment breccia is formed. It is a compositionally varied ore-free aphanite cemented by porphyritic aphanite with disseminated droplets of native iron, and this breccia, in turn, is cut by a network of gabbrodolerite veinlets, which impart a breccia-like appearance to the rock. Contacts of fragments in the breccia are, as a rule, sharp, and in gabbrodolerite, collective recrystallization of small crystals into larger ones is observed around the rims of single inclusions.

Fragments of aphanites cemented by magmatic rocks constitute a magmatic breccia, in which fragments are autoliths, that is, fragments of earlier hardened products of the same magma, from which quenched facies of feeders of magmatic channels had been formed (Loewinson-Lessing and Struve 1963). In some cases, fragments of graphitized coal from country rocks found in breccia allow the magma to be considered eruptive.

In autoliths, the relict traxitoid texture of microdolerite, fiamme-like lenses of plagioclase laths in aphanite, or streakiness, which appear to be formed as a result of freezing of an overcooled melt on walls of a magmatic feeder conduit and its heteromorphic crystallization, are not infrequently preserved. Many autoliths have a zoned structure that manifests as a collective recrystallization of rock to a porphyritic texture at their contacts or as a variation of phase composition from core to rim of a fragment. The principal mineral in autoliths is clinopyroxene. Plagioclase is present in variable amounts, and olivine is observed in some rock types. Green spinel and biotite are sometimes observed in elevated amounts.

A silicate matrix with native iron nodules including sideronitic or spongy disseminated iron is commonly formed by olivine gabbrodolerite or taxitic gabbrodolerite. As well as this aphanite can form an oval body around a native iron nodule. The composition of the aphanite is, as a rule, pyroxene–plagioclase, and their formation is associated with fluid–magmatic differentiation of a melt.

Petrography of the UZ of the Intrusion

The Mt. Dzhaltul and Mt. Ozernaya Intrusions are composed of a wide range of rocks. In order to characterize these rocks briefly, a description of one reference section through gabbrodolerite of the UZ is given below; the composition of one of native iron nodule is considered in detail, and descriptions of principal types of fine-grained rocks and magmatogenic breccia are also given.

Reference Section of the UZ Gabbrodolerite

In the south part of the Mt. Dzhaltul Intrusion, at the eastern side of the Zirk location, a 65–70-m-thick gabbrodolerite section of the UZ has been studied with a sampling interval of 5 m. The intrusion roof in this part is eroded, and the UZ zone cross section is composed of olivine–plagioclase-porphyritic gabbrodolerite. Some textural features of these rocks can be viewed from Rock indications 199–204.

Olivine–plagioclase-porphyritic gabbrodolerite features porphyritic, glomeroporphyritic, seriate-porphyritic texture with a poikilo-ophitic and ophitic groundmass. The rock is composed of olivine (5–10%), plagioclase (50–60%), and clinopyroxene (30–40%); biotite, ilmenite, and sulfides are observed in small amounts (at most 1–3%). Porphyritic grains are typically plagioclase, less often clinopyroxene, and olivine is least common. The sizes of inset crystals vary from 2 to 6 mm. The latter are commonly in the form of glomeroporphyritic intergrowths from 0.4 to 2.0 cm in width.

Plagioclase forms large, medium, and small prisms and xenomorphic tabular crystals. Large prisms are commonly intergrown into glomeroporphyritic aggregates or are idiomorphic phenocrysts. Rounded olivine grains are commonly enclosed in rims of the phenocrysts. For insets, a polyzonal structure is typical. Small plagioclase prisms are observed as chadocrysts in clinopyroxene oikocrysts and are set in the groundmass as ophitic intergrowths with clinopyroxene and olivine. The composition of insets is approximately bytownite An_{79-73} in cores and labradorite An_{66-50} in rims. In some cases, zoned phenocrysts are observed, core and rims of which consist of Ab_9An_{91} and $Ab_{48}Or_3An_{49}$, respectively. The range of composition variation, from core to rim, is commonly about 18–25 numbers. The composition of plagioclase chadocrysts is An_{72}–An_{55}, and in prismatic crystals from the rock matrix, it is An_{66}–An_{46}.

Olivine forms crystals different in size and morphology. It forms subidiomorphic insets 2–2.5 mm in size or 3–8 grains are grown together. Phenocrysts feature solid solution breakdown structures as ingrown lamellae of magnetite, which form short streaks in the (010) plane and changes to skeletal in the (100) plane (see Rock indications 203 and 204). In addition, olivine forms palmate or dendrite-like crystals, medium-size isometric grains that sometimes have plagioclase intergrowths, and small granules that are observed in thin section. Olivine phenocrysts have the composition Fa_{37-48} (typically Fa_{42-46}), and medium and small grains consist of Fa_{43-51}. Olivine intergrowths in clinopyroxene oikocrysts have sometimes a low Fe content Fa_{37-41}, whereas the olivine in the cores of phenocrysts is Fa_{45-48} (see Rock indication 201). The composition of olivine crystals solid solution breakdown structures is Fa_{47-48}, and homogenous grains from this rocks are Fa_{44} (see Rock indication 204). The NiO content in olivine ranges from 0.06 to 0.13 wt%.

Clinopyroxene forms oikocrysts, small idiomorphic crystals, and angular grains in plagioclase interstices. Two types of oikocrysts have been recognized. Type I is represented by large isometric oikocrysts, which enclose a relatively small amount (about 30%) of small grains of plagioclase and more rarely olivine. Microscopic study in transmitted light shows that rims of theses oikocrysts have a gray color caused by submicroscopic magnetite lamellae as fine streaks. Type II oikocrysts are irregular aggregates that fill spaces between rounded oikocrysts of the type I. Type II oikocrysts typically feature multiple olivine and plagioclase intergrowths (up to 70–80%). At the edge of type II oikocrysts, spotted two-pyroxene solid solution breakdown texture is common and manifests as small orthopyroxene intergrowths in clinopyroxene (see Rock indication 200). Individual angular orthopyroxene grains are also associated with type II oikocrysts.

Clinopyroxene composition is $Wo_{42-34}En_{40-46}Fs_{12-22}$, and it typically contains 0.55–1.12 wt% TiO_2, 1.57–2.78 wt% Al_2O_3, and 0.003–0.54 wt% Cr_2O_3. Type I oikocryst cores have chrome concentrations of an order of magnitude greater than type II oikocrysts. Orthopyroxene from solid solution breakdown and individual grains in the rock matrix are hypersthene ($Wo_{2-3}En_{64-60}Fs_{34-37}$).

The rock contains minor amounts of biotite–phlogopite mica, commonly as intergrowths with ilmenite. Their Fe content ranges from 27 to 44 wt%. The composition of the micas includes 4.39–5.46 wt% TiO_2, 0.37–1.05 wt% F, and 0.14–0.73 wt% Cl. Ore minerals are represented by ilmenite and more rarely by titanomagnetite and sulfide. Ilmenite contains 0.92–2.53 wt% MgO and 0.31–0.47 wt% MnO; titanomagnetite contains 15.10 wt% TiO_2, 2.06 wt% Al_2O_3, 0.31 wt% MgO, and 0.26 wt% MnO. Sulfides form small interstices of pyrrhotite, more rarely as pyrrhotite with chalcopyrite, as well as sparse disseminated droplets or their aggregates.

A nodule of native iron is spongy iron-rich dissemination in a taxitic olivine gabbrodolerite. The examined nodule is 15–20 cm in size. In order to gain an understanding of the compositional features of the nodule and the ore-free rock enclosing it, three samples of the rock were analyzed: olivine gabbrodolerite with spongy ore (A), narrow (about 2–3 mm) transition zone (B), and ore-free gabbrodolerite enclosing the nodule (C) 2 cm from the contact with the nodule (see Rock indication 205).

Olivine gabbrodolerite with spongy dissemination of native iron (see Rock indication 205, A) has a coarse emulsion texture with elements of ataxitic and poikilitic textures. Ribbon-like intergrowths of silicate minerals with iron (ferrite) form the rock matrix. Silicate minerals include large olivine (Fa_{65}) crystals, pigeonite ($Wo_{9-12}En_{50-47}Fs_{41}$), quartz (as small nets and veinlets), orthoclase, and apatite. Spongy disseminations of ore aggregates are composed of ferrite, kamacite, ilmenite, and magnetite. The principal mineral in ore accumulations is ferrite surrounded by magnetite bands of oxidized metal. Kamacite is observed as plate-like breakdown structures in ferrite. Ilmenite forms individual prisms in the rock matrix, as well as in association with small magnetite grains and as intergrowths with ferrite. The olivine composition includes 0.32 wt% MnO. Pigeonite contains 0.76–0.80 wt% TiO_2 and 0.72–0.94 wt% Al_2O_3; apatite contains 0.19 wt% CeO_2, 2.23 wt% F, and 0.27 wt% Cl. In ferrite, the NiO content is 1.4–1.5 wt%, and in kamacite, NiO content is 5.45 wt%. Magnetite from bands around ferrite has the NiO content 1.96 wt%, and magnetite contains 0.06 wt% NiO.

Troctolitic gabbrodolerite from the transition zone between the ferrite nodule and ore-free rock has poikilo-ophitic, granulation, and ataxitic textures (see Rock indication 205 B). Olivine (Fa_{48}) forms large crystals, chadocrysts (Fa_{45}) in clinopyroxene, and small grains (Fa_{46}) in the rock matrix. The mineral contains 0.37–0.39 wt% MnO. Clinopyroxene forms large oikocrysts of augite ($Wo_{41}En_{45}Fs_{14}$) and small angular grains ($Wo_{34}En_{44}Fs_{22}$) in the groundmass. Discrete orthopyroxene ($Wo_3En_{59}Fs_{38}$) bands are observed around large olivine crystals. The composition of clinopyroxene includes 0.53–0.93 wt% TiO_2 and 1.52–2.22 wt% Al_2O_3. Plagioclase forms prismatic crystals in the rock matrix as well as intergrowths ($Ab_{35-34}Or_{1-2}An_{64}$) in olivine and clinopyroxene.

Olivine gabbrodolerite from the contact of the native iron nodule (see Rock indication 205, C) has serially porphyritic, poikilo-ophitic, and ataxitic textures. Olivine is observed in various crystal sizes, which exhibit a trend of increasing Fe content from Fa_{45} to Fa_{49} as grain size decreases. The mineral contains 0.44–0.50 wt% MnO. Clinopyroxene forms large oikocrysts ($Wo_{40}En_{46}Fs_{14}$) with chadocrysts of olivine and plagioclase. The rock matrix contains angular grains of augite ($Wo_{34}En_{44}Fs_{22}$) poor in calcium and displays two-pyroxene breakdown structures and individual small grains of pigeonite ($Wo_7En_{55}Fs_{38}$). Plagioclase forms idiomorphic prisms ($Ab_{24}Or_2An_{75}$), large tabular xenomorphic zoned crystals ($Ab_{29-43}Or_{1-2}An_{70-55}$), chadocrysts ($Ab_{29}Or_2An_{69}$) in olivine and chadocrysts ($Ab_{36}Or_2An_{62}$) in clinopyroxene, and small prisms ($Ab_{38-43}Or_3An_{59-53}$) in the rock matrix. In interstices, small grains of apatite and crystals of ilmenite are observed. Apatite contains 0.23 wt% CeO_2, 2.16 wt% F, and 1.37 wt% Cl; ilmenite contains 0.48–0.50 wt% MnO and 1.89–2.22 wt% MgO.

Fine-Grained Rocks from Inclusions in Gabbrodolerite

Characteristics of these rocks are given below as a description of typical varieties that form separate small autoliths or zones in large bodies.

Plagioclase–pyroxene aphanite forms the central part of a lens 40 cm thick and about 3 m long (see Rock indication 206). The texture of the rock is fine granular, heterogranoblastic, and microgabbroic. The rock is composed of an aggregate of short prismatic clinopyroxene crystals (55–60%), plagioclase (40–45%), and sparse small ilmenite grains with crystals varying in size from 0.02–0.04 mm to 0.08–0.1 mm. The rock resembles beerbachite in composition and structure.

Plagioclase has a stable anorthite ($Ab_{1-2}An_{99-98}$) composition. It encloses granules of clinopyroxene ($Wo_{49}En_{38}Fs_{13}$). Clinopyroxene in the granoblastic aggregate has the composition $Wo_{48}En_{37}Fs_{15}$, for which the Ca-tschermakite component (Ca-tscherm) accounts for 9–10%. The mineral contains 1.0–1.48 wt% TiO_2, 4.06–4.69 wt% Al_2O_3, 0.15–0.17 wt% MnO, 0.16–0.21 wt% Na_2O, and 0.0–0.02 wt% Cr_2O_3. Ilmenite contains 1.11 wt% MnO.

Microdolerite from an autolith 8 × 15 cm in size (see Rock indication 207) has a dolerite, granoblastic with elements of micropoikilo-ophitic texture under microscope. In some areas, the texture is microgranoblastic, and flows of traxitoid plagioclase lath aggregates are observed. Clinopyroxene is diopside ($Wo_{49-49}En_{39-32}Fs_{11-20}$) with a 2–4% Ca-tscherm component. The mineral contains 0.40–0.65 wt% TiO_2, 0.81–2.00 wt% Al_2O_3, and 0.04–0.10 wt% Cr_2O_3. Laths and granule of plagioclase have the composition $Ab_{8-11}An_{92-89}$, and relatively large microporphritic prisms have the composition $Ab_{17}An_{83}$.

Microdolerite in autoliths features a layered structure with different textures such as micropoikilo-ophitic, traxitoid with fiamme-like lenses of plagioclase laths, and plagioclase-porphyritic observed. The rock matrix in these cases retains a microdolerite texture. Clinopyroxene in small grains in microdolerite with small plagioclase lenses has the composition $Wo_{41-44}En_{43-41}Fs_{15-17}$ (0.14–0.73 wt% TiO_2 and 0.87–1.97 wt% Al_2O_3) (see Rock indication 208). Composition of plagioclase laths from fiamme-like lenses is bytownite ($Ab_{20}Or_1An_{79}$), and microliths and grains of plagioclase in the rock matrix are labradorite ($Ab_{35-41}Or_{1-2}An_{63-57}$). In the microdolerite with the poikilo-ophitic texture, spheroid oikocrysts of clinopyroxene have the composition $Wo_{44-41}En_{48-46}Fs_{9-13}$ (0.50–0.74 wt% TiO_2, 1.68–1.85 wt% Al_2O_3, 0.04–0.14 wt % Cr_2O_3). Plagioclase composition varies from core to rim from bytownite ($Ab_{22-47}Or_1$) to labradorite (An_{77-51}); chadocrysts in clinopyroxene and laths in the groundmass are labradorite ($Ab_{36-42}Or_{1-2}An_{62-56}$), and between laths, in interstices, grains of andesine $Ab_{50}Or_9An_{41}$ are not infrequent. In microdolerite, green amphibole ($f = 63$ wt%) is infrequent, and ore minerals are represented by ilmenite that contains 0.93–1.82 wt% MnO, 0.11–0.22 wt% MgO, and 0.02–0.43 wt% Cr_2O_3.

Spinel–olivine–pyroxene aphanite (see Rock indication 209) manifests in the central part of a plate-like autolith 0.7–2 m in size. At edge of this autolith, porphyritic plagioclase–pyroxene rock is located. Under microscope, the spinel–olivine–pyroxene rock has a granoblastic texture.

The rock consists of polygonal olivine (Fa_{26}) grains, clinopyroxene ($Wo_{52}En_{38}Fs_9$), Al spinel, and sulfide (ore minerals constitute 1–3%). The olivine contains 0.70 wt% MnO and 0.26 wt% CaO. Clinopyroxene contains 0.84–0.88 wt% TiO_2 and 8.93–9.24 wt% Al_2O_3, and a 19–20% Ca-tscherm component is calculated. Green spinel is observed as submicroscopic globules in silicates and pyrrhotite, as well as grains similar in size to the globules. The mineral contains 0.04–0.05 wt% TiO_2, 0.31 wt% MnO, and 0.04–0.06 wt% Cr_2O_3.

Spinel–plagioclase–olivine–pyroxene porphyritic aphanite (see Rock indication 210) is found in relatively large lenses and plate-like autoliths. The rock structure is porphyritic with the granoblastic and granulation textures in the groundmass. Under microscope, large prismatic crystals of disseminated plagioclase 1–3 mm in size and small porphyritic (0.3–0.5 mm) isometric olivine insets with intergrown green spinel are set in the granoblastic aggregate of plagioclase and pyroxene. The rock closely resembles the plagioclase–pyroxene aphanite described above. The main distinctive feature of the rock is the presence of olivine (5–7%) and spinel. Porphyritic plagioclase insets form skeletal box-like crystals, with cores composed of clinopyroxene granules cemented by plagioclase, and the band is a box with sparse intergrowths of clinopyroxene granules and, occasionally, spinel. The concentration of clinopyroxene granules varies in cores so that windows and fields of plagioclase are observed between the granules. Individual plagioclase insets have a relatively small poikilitic clinopyroxene intergrowth content. In some phenocrysts, "box in box"-type structures are observed that form areas that are poor in plagioclase intergrowths. The composition of plagioclase crystals morphologically different to porphyritic insets and in the groundmass is anorthite An_{98-99}. Olivine has a chrysolite composition Fa_{27-28}, and it contains 0.58–0.63 wt% MnO, 0.16–0.36 wt% CaO, and 0.01–0.02 wt% NiO. Clinopyroxene in granules inside plagioclase "boxes" and in granoblastic aggregates in the groundmass have a relatively stable composition ($Wo_{52-53}En_{40-37}Fs_{8-10}$) and incorporates a 16–20% Ca-tscherm component. The mineral contains 0.71–1.17 wt% TiO_2, 7.21–9.02 wt% Al_2O_3, 0.10–0.14 wt% MnO, and 0.04–0.07 wt% Na_2O. Spinel forms small green and black intergrowths in olivine and plagioclase. Its composition complies with pleonaste and contains 0.08–0.20 wt% TiO_2, 0.24–0.29 wt% MnO, and 0.05–0.24 wt% Cr_2O_3

Plagioclase–pyroxene aphanite (see Rock indication 211) is found in the lower contact zone of the large (0.3–0.4) × (5–7) m plate-like autolith in the middle part of the studied cross section in the UZ. The rock is composed of porphyritic plagioclase 1–3 mm in length enclosed in fine crystalline granoblastic plagioclase–pyroxene aggregate. The sizes of polygonal grains in the groundmass are 0.01–0.1 mm. Plagioclase insets are prismatic crystals with numerous intergrowths of small to submicroscopic clinopyroxene. In the cores of phenocrysts, clinopyroxene forms a granular aggregate with sparse "windows" of plagioclase, and in peripheral bands, the amount of clinopyroxene intergrowths in plagioclase decreases, and a box-like shape of phenocryst is clearly observed. Bands and "windows" of plagioclase inside a "box" have simultaneous extinction. The size of poikilo-ophitic clinopyroxene intergrowths is smaller inside phenocrysts than those in the groundmass. Cases of "box in box" texture are not infrequent in insets when inside a plagioclase box; one more plagioclase box is observed. The boxes are separated by a granoblastic aggregate of clinopyroxene. Clinopyroxene in granoblastic aggregate in the groundmass is fassaite ($Wo_{52-54}En_{36-34}Fs_{12}$) that contains 17–18% Ca-tscherm component. In the inner part of a plagioclase box, the composition of granules is $Wo_{53}En_{36}Fs_{11}$ (Ca-tscherm is 18%), and in granules enclosed in a plagioclase rim, it is $Wo_{53-52}En_{38-37}Fs_{9-11}$ (Ca-tscherm is 17–21%). Clinopyroxene contains 0.75–0.86 wt% TiO_2, 0.10–0.16 wt% MnO, 0.07–0.10 wt% Na_2O, and 0.01–0.04 wt% Cr_2O_3. Plagioclase enclosed into porphyritic insets and in bands of their boxes, as well as in the groundmass, complies with anorthite An_{98-99}. Ore minerals in the rock include aggregates of pyrrhotite of irregular shape from 0.5% to 3–5%.

The contact zone between aphanite and olivine–plagioclase-porphyritic gabbrodolerite has been studied in an autolith 40 cm thick and about 3 m long. The central zone of this autolith is composed of the homogenous, fine-grained, granoblastic plagioclase–pyroxene aggregate described above (see Rock indication 206), and the upper contact zone 1–2 cm thick comprises porphyritic aphanite. Structural and compositional peculiarities of this zone are shown in photographs (see Rock indication 212). The contact zone of the autolith is composed of porphyritic aphanite with enclosed plagioclase prisms 1.5–2.3 mm to 2.8–4.2 mm in size. Clinopyroxene granules 0.01–0.02 mm and smaller as accumulations with a regular geometric shape are enclosed into plagioclase prisms inside their margins creating the "box in box" texture. The groundmass is composed of a granoblastic aggregate of polygonal clinopyroxene grains 0.04–0.08 mm in size with small amounts of plagioclase between. Not infrequently, biotite oikocrysts and titanomagnetite insets with clinopyroxene chadocrysts are observed.

The composition of central parts of porphyritic plagioclase grains is anorthite ($Ab_{6-8}An_{94-92}$), and rims are bytownite ($Ab_{20}An_{80}$). The composition of clinopyroxene is $Wo_{46-42}En_{41-39}Fs_{12-17}$, in which Na-tscherm component is 4–11%. The mineral contains 0.51–0.97 wt% TiO_2, 2.00–4.87 wt% Al_2O_3, 0.12–0.24 wt% MnO, 0.20–0.35 wt%

Na_2O, and 0.00–0.02 wt% Cr_2O_3. Granules of clinopyroxene enclosed in plagioclase have the lowest FeO content and are enriched in Ca-tschermakite. Biotite has $f = 38.4$ wt% and contains 5.59 wt% TiO_2, 0.89 wt% F, and 0.46 wt% Cl. Titanomagnetite contains 8.33–12.46 wt% TiO_2, 2.25–2.56 wt% Al_2O_3, 0.25–0.27 wt% MnO, 0.26–0.74 wt% MgO, and 0.20–0.47 wt% Cr_2O_3.

In the contact zone of autolith, the rock has the ophitic-poikilo-ophitic texture: clinopyroxene and olivine form large oikocrysts, and idiomorphic plagioclase prisms have lost their porphyritic appearance but retain enclosed clinopyroxene granules as is observed in aphanite (see Rock indication 212). Amphibole, biotite, apatite, and titanomagnetite are present in the rock in small amounts.

Olivine is observed as large palmate crystals with break-down structures and homogeneous isometric grains smaller in size. Idiomorphic plagioclase prisms bound large crystals. The composition of olivine with magnetite lamellae is $Fa_{39–43}$, and homogeneous olivine has the composition $Fa_{45–47}$. The mineral contains 0.44–0.54 wt% MnO, 0.11–0.21 wt% CaO, and 0.05–0.07 wt% NiO. Olivine is sometimes surrounded by kelyphitic rims of orthopyroxene ($Wo_2En_{60}Fs_{38}$). Idiomorphic plagioclase prisms with poikilo-ophitic clinopyroxene intergrowths have a stable anorthite composition: $Ab_{7–8}An_{93–92}$ in the core of crystals and $Ab_{7–9}An_{93–91}$ in rims. Large tabular crystals are sometimes mantled by more acidic plagioclase $Ab_{36}Or_2An_{62}$, and small $Ab_{39}Or_2An_{60}$ grains form in the rock matrix. Clinopyroxene in small granules intergrown with plagioclase has a variable composition ($Wo_{48–44}En_{37–39}Fs1_{4–17}$). In addition to Al_2O_3 varieties (7.07–8.06 wt% Al_2O_3, 15–18% Ca-tscherm), grains relatively poor in Al_2O_3 (3.78 wt%, 8% Ca-tscherm) are not uncommon. The TiO_2 content is 0.52–1.05 wt%. The clinopyroxene composition in polygonal grains, which form a granoblastic aggregate, and in large angular aggregates is $Wo_{43–44}En_{40–39}Fs_{17}$, and the Ca-tscherm component decreases to 2–6%. The mineral contains 0.16–1.14 wt% TiO_2, 0.71–2.86 wt% Al_2O_3, 0.20–0.28 wt% MnO, 0.24–0.33 wt% Na_2O, and 0.01–0.03 wt% Cr_2O_3. The rock matrix is composed of amphibole and small prismatic crystals ($Wo_{47–46}En_{39}Fs_{14–15}$), which are characterized by the lowest content of Al_2O_3 (0.26–0.45 wt%, 0.6–1.0% Ca-tscherm) and TiO_2 (0.00–0.04 wt%). Amphibole forms large green crystals with clinopyroxene intergrowths and small prismatic crystals in the rock matrix. The total Fe content of the mineral is 50.9 wt%. Amphibole contains 1.33 wt% TiO_2, 2.03 Na_2O, 0.23 wt% Cr_2O_3, 0.42 wt% F, and 1.21 wt% Cl. Biotite forms large crystals with clinopyroxene intergrowths and not infrequent intergrowths with ore minerals. The total Fe content in the mineral is 36–45 wt %. Mica contains 4.64–7.41 wt% TiO_2, 0.54–0.84 wt% F, and 0.36–0.50 wt% Cl. The halogen content decreases as Fe content increases. Apatite forms intergrowths in mica and contains 0.52 wt% CeO_2, 1.00 wt% F, and 2.99 wt% Cl. Titanomagnetite contains 7.63 wt% TiO_2, 1.55 wt% Al_2O_3, 0.30 wt% MnO, and 0.22 wt% Cr_2O_3. Ilmenite contains 0.76–0.84 wt% MnO and 0.35–0.52 wt% MgO.

Collective recrystallization of the rock groundmass is observed in small autoliths (10–15 cm in width) (see Rock indication 213). As a result of recrystallization, the size of clinopyroxene grains increases, plagioclase crystals are more common, and the rock has a doleritic texture. The initial aphanite texture is retained in clinopyroxene granules preserved in plagioclase prisms.

Prismatic plagioclase crystals have a stable anorthite composition ($Ab_{6–9}An_{94–91}$) in different parts of the rock, and small grains and tabular crystals without pyroxene inclusions are bytownite ($Ab_{18}An_{82}$). Granules of poikilo-ophitic intergrowths and grains of pyroxene from the matrix in different parts of the rock also have a stable composition ($Wo_{44–42}En_{43–39}Fs_{13–19}$). Plagioclase features 6–10% Ca-tscherm component. The mineral contains 0.66–1.04 wt% TiO_2, 2.67–4.67 wt% Al_2O_3, 0.16–0.26 wt% MnO, 0.23–0.36 wt% Na_2O, and 0.07–0.16 wt% Cr_2O_3. Ore minerals in the rocks are titanomagnetite and ilmenite. Titanomagnetite contains 8.66–9.02 wt% TiO_2, 0.30–0.34 wt % MnO, 0.37–0.44 wt% MgO, and 0.63–1.61 wt% Cr_2O_3, and ilmenite contains 0.73 wt% MnO and 0.26 wt% MgO.

Magmatogenic Breccia

The rocks are composed of fragments of fine-grained rock cemented by medium- to coarse-grained gabbrodolerite. Small and large fragmental breccias are distinguished on fragment sizes. The composition and texture of the rocks vary widely, even within a single sample. Understating of mineralogical–petrographic peculiarities of magmatogenic breccia may be obtained from two descriptions below.

The first type of magmatogenic breccia consists of aphanite fragments of the pyroxene–plagioclase–olivine, plagioclase–pyroxene, and microgabbro composition cemented by picritic gabbrodolerite (see Rock indication 214). This magmatogenic breccia features poor (1–3% at least) sulfide mineralization of chalcopyrite–pyrrhotite composition, which is distributed both through the cement and in fragments, and the largest sulfide crystals are commonly restricted to fragment contacts.

Pyroxene–plagioclase olivine aphanite (see Rock indication 214 A) has granulation, granoblastic, and poikilo-ophitic textures. It is composed of small (0.02–0.1 mm) olivine grains (40–50%), plagioclase (35–50%), and pyroxene (10–15%) that also form sparse oikocrysts up to 1 mm wide. Olivine has a hyalosiderite (Fa_{39}) composition, plagioclase is bytownite ($Ab_{28}Or_1An_{71}$), and pyroxene is hypersthene ($Wo_{1–3}En_{61–59}Fs_{38}$).

Plagioclase–pyroxene aphanite (see Rock indication 214 B) with porphyritic texture is composed of large box-like plagioclase grains with intergrown plagioclase granules; the groundmass is a granoblastic aggregate of pyroxene with enclosed prismatic plagioclase crystals, which, in turn, enclose pyroxene granules. Compositionally, plagioclase crystals are anorthite, and pyroxene is diopside ($Wo_{51-45}En_{32-37}Fs_{17-18}$). The intergrowths in plagioclase contain an increased (16%) Ca-tscherm component (7.15 wt% Al_2O_3), and grains in the groundmass, inversely, contain an abnormally low (1%) Ca-tscherm component (0.48 wt% Al_2O_3). The TiO_2 contents in these clinopyroxenes are 0.88 and 0.10 wt%, respectively.

Microgabbro (see Rock indication 214 C) consists of clinopyroxene (40–45%) and plagioclase (55–60%). The rock has granoblastic, micro-ophitic, microgabbroic, and micropoikilo-ophitic textures and resembles beerbachite. The grain size is 0.02–0.06 mm. Plagioclase is labradorite ($Ab_{44}Or_2An_{54}$), and clinopyroxene is augite ($Wo_{41}En_{47}Fs_{12}$). The latter contains 0.24–0.26 wt% TiO_2, 1.86–1.91 wt% Al_2O_3, and 0.06–0.13 wt% Cr_2O_3.

Picritic gabbrodolerite (see Rock indication 214 D) from the magmatogenic breccia cement has porphyritic, ophitic, and poikilo-ophitic textures. Peculiar pegmatoid intergrowths of needle-like olivine, plagioclase, and pyroxene crystals are observed in the rock as rounded and lens-like insets 2–7 mm in size. Most olivine in the rock forms subidiomorphic homogenous crystals (Fa_{26-33}) that feature magnetite intergrowths—lamellae. Clinopyroxene oikocrysts are augite; the composition from cores to rims of crystals varies from $Wo_{42-41}En_{46}Fs_{12-13}$ to $Wo_{37}En_{44}Fs_{19}$. Pyroxene chadocrysts from oikocrysts of olivine are $Wo_{41-38}En_{47-48}Fs_{12-14}$. Interstices also contain angular hypersthene ($Wo_1En_{67}Fs_{32}$) grains. Clinopyroxene contains 0.56–1.24 wt% TiO_2, 1.89–2.39 wt% Al_2O_3, and 0.09–0.31 wt% Cr_2O_3, and the Ca-tscherm component is 4–5%. Plagioclase chadocrysts in oikocrysts of clinopyroxene are $Ab_{27}Or_0An_{73}$, and prismatic crystals from the rock groundmass are $Ab_{35}Or_1An_{64}$.

The second type of magmatogenic breccia (see Rock indication 215) is an ore-bearing breccia in porphyritic microdolerite that contains disseminated native iron, small fragments of plagioclase–olivine, plagioclase–pyroxene, and pyroxene aphanite, as well as fragments of dolerite and graphitized coal. In turn, fragments of microdolerite are intersected by a network of veinlets of olivine gabbrodolerite. This imparts a breccia-like appearance to rock, and the rock may also be classified as a magmatogenic breccia. The fragments and cement of magmatogenic breccia have a composition outlined below.

Plagioclase–olivine aphanite (see Rock indication 215 B) is a granoblastic or granulation aggregate of plagioclase and olivine (Fa_{42}) containing 0.51 wt% MnO and 0.04 wt% CaO. Disseminated native iron is observed in the rock as a fine rash.

Plagioclase–pyroxene aphanite (see Rock indication 215 C), porphyritic in texture, is similar to rocks with the same name described above. It is composed of large box-like plagioclase (Ab_5An_{95}) crystals with numerous clinopyroxene ($Wo_{52}En_{36}Fs_{12}$) granules (18% Ca-tscherm component), and a granoblastic clinopyroxene ($Wo_{50}En_{30}Fs_{20}$) aggregate (13% Ca-tscherm component) forms the rock groundmass. These clinopyroxenes contain 0.65 and 1.08 wt% TiO_2 and 8.13 and 5.81 wt% Al_2O_3, respectively.

Pyroxene aphanite (see Rock indication 215 D) has a fine crystalline texture. The rock consists of small (0.01 mm at least) clinopyroxene ($Wo_{46}En_{31}Fs_{23}$) granules and contains 0.11 wt% TiO_2 and 0.57 wt% Al_2O_3.

Porphyritic microdolerite (see Rock indication 215 A) with disseminated native iron cementing aphanite fragments described above has microporphyritic, microdoleritic, microgranoblastic, and granulation textures. Insets in the rock are prismatic plagioclase ($Ab_{24}Or_1An_{75}$) microphenocrysts. Clinopyroxene ($Wo_{36}En_{42}Fs_{22}$; 8% Ca-tscherm component) intergrowths are observed in the peripheral parts of insets. Clinopyroxene contains 1.74 wt% TiO_2 and 3.48 wt % Al_2O_3. The rock groundmass is composed of labradorite ($Ab_{49}Or_1An_{50}$) laths and small clinopyroxene ($Wo_{36-30}En_{40-36}Fs_{25-30}$) grains (1–3% Ca-tscherm component). Clinopyroxene contains 0.59–1.19 wt% TiO_2 and 0.55–1.32 wt% Al_2O_3. The distinctive feature of the rock is fine (0.03–1 mm) drop-like dissemination of native iron that constitutes about 3–7 vol%.

Olivine gabbrodolerite (see Rock indication 215 E) from the magmatogenic breccia cement has porphyritic, ophitic, and poikilo-ophitic textures. Olivine insets with intergrowths of magnetite lamellae are chrysolite (Fa_{28}) containing 0.34 wt% MnO and 0.28 wt% CaO. Clinopyroxene oikocrysts are $Wo_{37}En_{46}Fs_{17}$, and grains at the boundary with the autolith have the composition $Wo_{47}En_{35}Fs_{18}$. The Na-tscherm component in these clinopyroxenes are 4 and 3%, and Al_2O_3 and TiO_2 contents are 1.73 and 134 wt% and 0.89 and 0.35 wt%, respectively.

Mineralization in Rocks of the Intrusion UZ

Rocks of the UZ of the intrusion host ores of native iron and sulfide mineralization. Native iron forms compact accumulations as nodules in gabbrodolerite and as a fine dissemination in microdolerite.

Nodules of native iron have a globular or slightly flattened loaf shape and vary from 3–5 cm to 40–60 cm. During erosion, the nodules are isolated as nuggets from 100–200 g to 70–100 kg. The distribution of nodules throughout the intrusion and along strike is uneven. Two types of native iron occurrences in gabbrodolerite are recognized by their content and textural features: sideronitic type containing

20–35% ore phase and a spongy type containing 50–80% ore phase. In large nodules, the spongy ores sometimes change into sideronitic ore from core to rim. Poorly disseminated ores of native iron are virtually absent in gabbrodolerite. In section, spongy ores show a typical coarse-emulsion texture of ore and silicate liquids.

Another form of native iron occurrence is a thin emulsion of ore spheres in porphyritic inclusions of the second type of magmatogenic breccia. An important genetic feature of these rocks is the absence of ore rash in autoliths enclosed in microdolerite and gabbrodolerite. This small-droplet dissemination of native iron also occurs only in microdolerite. The ore dissemination manifests as an even distribution of small spherules 0.8–0.02 mm in diameter and less. The amount of the ore phase varies from 3% to 5–7% and commonly does not exceed 10%. The ore dissemination does not appreciably increase the iron content of the rock.

The principal ore phase in native iron ore is ferrite (a-Fe) depleted in impurities; Ni + Co content is at least 1.5–3 wt %. Other minerals include kamacite (3–9 wt% Ni and 1–2.3 wt% Co), ilmenite, cohenite, native copper, native lead, cube-octahedral graphite, taenite (32.9–44.4 wt% Ni), tetrataenite (45–50 wt% Ni and 2.8–3.6 wt% Co), armalcolite, schreibersite, and troilite (Oleinikov et al. 1985, 1991). Sulfides, as a rule, are not found in nodules of native iron though do occur in single small grains of troilite and occasionally as chalcopyrite restricted to silicate minerals. At nodule contacts, the phase composition of sulfides becomes more varied, and their content sharply increases; sometimes, they accumulate in the immediate vicinity of nodules.

Sulfide mineralization is found in all varieties of gabbrodolerite and aphanites in the UZ of the intrusion, although it is typically mutually exclusive with native iron occurrences. Few eutectoid troilite–ferrite–fayalite intergrowths have been described earlier (Ryabov 1989a). In gabbrodolerite of the UZ, sulfides form interstitial dissemination, small accumulations, and single drop-like insets from 0.4–0.6 cm to 0.8–1.3 cm in size. In some cases, blind sulfide veinlets up to 1 cm thick are present. Large sulfide insets, as it was noted, are not infrequently restricted to the aphanite and gabbrodolerite contact. In aphanite, typically no sulfides are present, although rare sparse interstitial dissemination that constitutes 1–3 vol% is observed. One of the autolith varieties contains rich (to 15 vol%) drop-like dissemination of troilite (Zolotukhin and Vasil'ev 1967).

In gabbrodolerite, sulfides are typically pyrrhotite and, more rarely, chalcopyrite, pentlandite, cubanite, bornite, and mackinawite (Oleinikov et al. 1985). Sulfides in aphanite are typically troilite and pyrrhotite.

Geochemical investigations undertaken earlier have shown that the UZ of intrusions is the most potentially ore bearing but most challenging for finding accumulations of PGE (Ryabov 1989c). This is inferred from Pt low-sulfide ores found in upper pegmatoid Noril'sk Intrusions, as well as from Pt occurrences in a series of other intrusions. The increased concentrations of platinum group elements in sulfide-free anorthosites in the Bootankaga Intrusion and graphite-bearing ores of native iron in the Khungtukun Intrusion are indicative of Pt–graphite type of mineralization (Ryabov 1992b, 1994). In the late 1990s, the distribution of noble metals in ores of native iron from Dzhaltul–Khungtukunsky-type intrusions was independently studied in Yakutsk and Novosibirsk. Mineralogical investigations have shown that native iron from the Dzhaltul Intrusion contains 0.01–12.90 (most geologists 2.42) g/t Pt, 0.33–42.0 (average 4.92) g/t Pd, 0.20–6.30 (average 1.51) g/t Ru, 0.10–5.25 (average 1.74) g/t Au, and 0.005–2.32 (average 0.29) g/t Ag (Kopylova and Oleinikov 1997). Studies of bulk ore samples have revealed that the concentrations of noble and nonferrous metals are of the same order but vary widely. The concentrations of elements in bulk samples from spongy ores of native iron reached 5.0 g/t Pt, 9.40 g/t Pd, 0.87 g/t Rh, 7.8 g/t Au, 1.97 g/t Ag, 500 g/t As, 8.4 wt% Ni, 0.9 wt% Cu, and 0.69 wt% Co. In dolerite containing a thin rash of native iron, the concentrations of these elements reached 0.03 g/t Pt, 0.02 g/t Pd, 0.008 g/t Rh, 0.31 g/t Au, 0.26 g/t Ag, 15 g/t As, 0.02 wt% Ni, 0.06 wt% Cu, and 0.01 wt% Co. The variation diagram (Fig. 3.44) exhibits the distribution of chemical elements in barren gabbrodolerite and in rocks containing disseminated native iron. As it is seen in the figure, the content of nonferrous metals is directly proportional to noble metals. The low PGE content and the high degree of chemical affinity of PGE with Fe and Ni are responsible for the fact that individual phases of platinum minerals have not been revealed. PGE appears to be isomorphically scattered but generally gravitates toward phases enriched in nickel. The positive correlation between concentrations of PGE and arsenic confirms that volatiles were involved in the formation of Pt–Fe metal ores (Ryabov and Anoshin 1999).

Petrologic Features

Based on investigations, it would seem possible to single out some basic data that are significant in the understanding of the formation process of the Dzhaltul Intrusion, crystallization of its melt, petrogenesis, and ore formation:

1. Parent magma of the Dzhaltul Intrusion had a tholeiitic basalt composition, and a conjugate increase in alkalinity and silica acidity of a melt was responsible for the wide range of rocks.
2. Intrusion of a melt by repetitive influxes appears to be the cause of quenched crusts formed at the walls of feeder conduits, stripping of these crusts, and the formation of magmatogenic breccia autolith of variable composition and size.

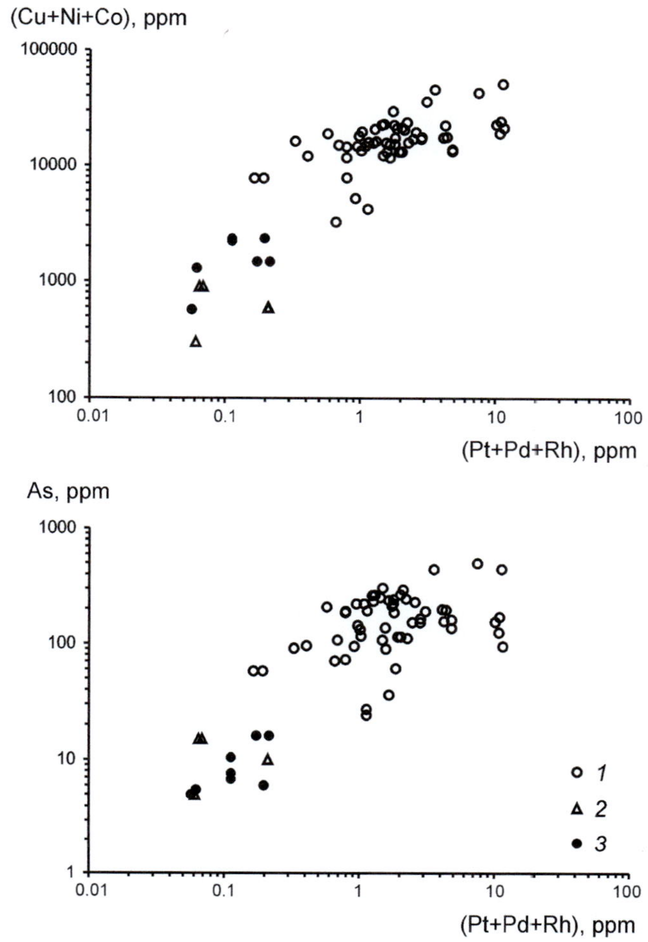

Fig. 3.44 Plots of (Cu + Ni + Co) versus (Pt + Pd + Ph) and As versus (Pt + Pd + Rh) for rocks of Dzhaltul Intrusion. *1* gabbrodolerite with native iron, *2* microdolerite with native iron, *3* gabbrodolerite with sulfide dissemination and barren dolerite

3. The principal rock type in the UZ in the southern part of the Mt. Dzaltul and Mt. Ozernaya Intrusions is olivine–plagioclase-porphyritic gabbrodolerite that contains elevated amounts of insets and glomeroporphyritic grains of plagioclase and olivine that impart a taxitic fabric to the rock.
4. Fluid–magmatic interaction during melt intrusion and in interchamber conditions was responsible for the diversity in composition and structural variation of autoliths, as well as the formation of platiniferous ores of native iron and sulfides.
5. Spinel–olivine–pyroxene paragenesis is typomorphic for the platiniferous UZ horizons of the Upper Talnakh and Khungtukun Intrusions and the Dzhaltul Intrusion.
6. The diversity of aphanites that form different size autoliths in gabbrodolerite of the Dzhaltul Intrusion is caused by wide variations in the amount and composition of clinopyroxene.
7. The characteristic feature of the clinopyroxene composition in aphanites in comparison with the mineral from gabbrodolerite is due to increased concentration of the tschermakite component $CaAl_2SiO_6$ reflected in the high CaO and Al_2O_3 content (Figs. 3.45 and 3.46). Such clinopyroxenes are known in ore-bearing bodies in magmatogenic breccia of the Kharaelakhsky intrusive branch and in upper pegmatoids of the Upper Talnakh and Khungtukun Intrusions. Crystallization of clinopyroxene enriched in CaO and Al_2O_3 proceeded from melt depleted in iron due to sulfide and iron metal formation.
8. Granulation and granoblastic textures of pyroxene aphanite and the high proportion of a tschermakite component in clinopyroxene composition suggest heterogenic crystallization of a melt under supercooling conditions, which caused the mass formation of granular nuclei. These granules contain up to 20–30% normative plagioclase.
9. Formation of native iron ores is associated with reducing volatiles that accompany intrusion of a melt; these extracted metals and formed a zone of fluidized melt, in which immiscible separation of metal liquid from silicate melt took place.
10. Nodules of native iron are schlieric ore pegmatites that form (by analogy with gas bubbles in basal sheets and erratic boulders of leucogabbro in Noril'sk Intrusions) due to isolation of fluidized melts and processes of ore–silicate separation (Ryabov 1989a).
11. Formation of Pt–Fe metal ores proceeded with involvement of reducing H–C–N fluids, which extracted valuable components from a melt and transported them to accumulation areas. Nickel-bearing iron played a role as a collector for noble metals (Ryabov and Anoshin 1999).
12. Observing sulfideless platinum ores in intrusions is evidence for the transfer and accumulation of metals without sulfur. This enhances the ore potential of intrusions and reveals new nontraditional types of platinum ore.

3.9 Avamsky Intrusive Complex ($q\beta T_2av$)

Dykes of trachydolerite that cut all volcanogenic units of the Noril'sk Region comprise the Avamsky Complex. Dykes in the eastern part of the Noril'sk Province in the region of the Kystyktakh and both Bol'shoy and Maly Avam Rivers (Bushkanets et al. 1970a), as well as the dyke of trachydolerite cutting through Lower Triassic deposits near Glubokoe Lake described by Ivanova (1975) are key examples of this complex.

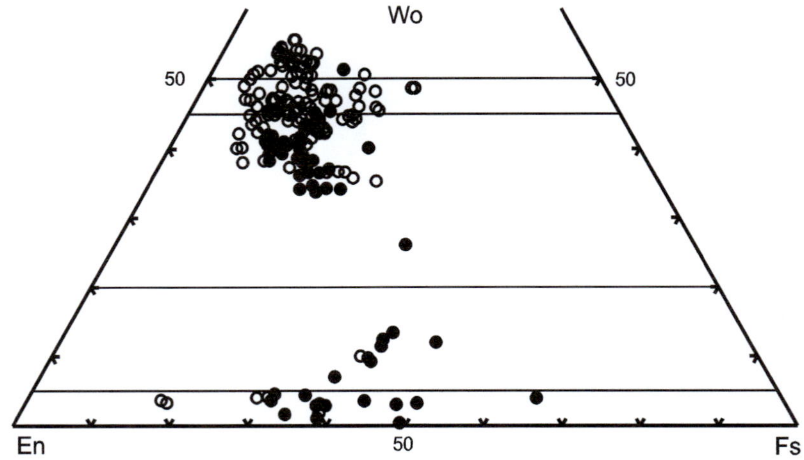

Fig. 3.45 Pyroxene compositions in rocks of Dzhaltul Intrusion. Horizontal lines show classification borders for pyroxenes of various compositions. Legend is given in Fig. 3.43

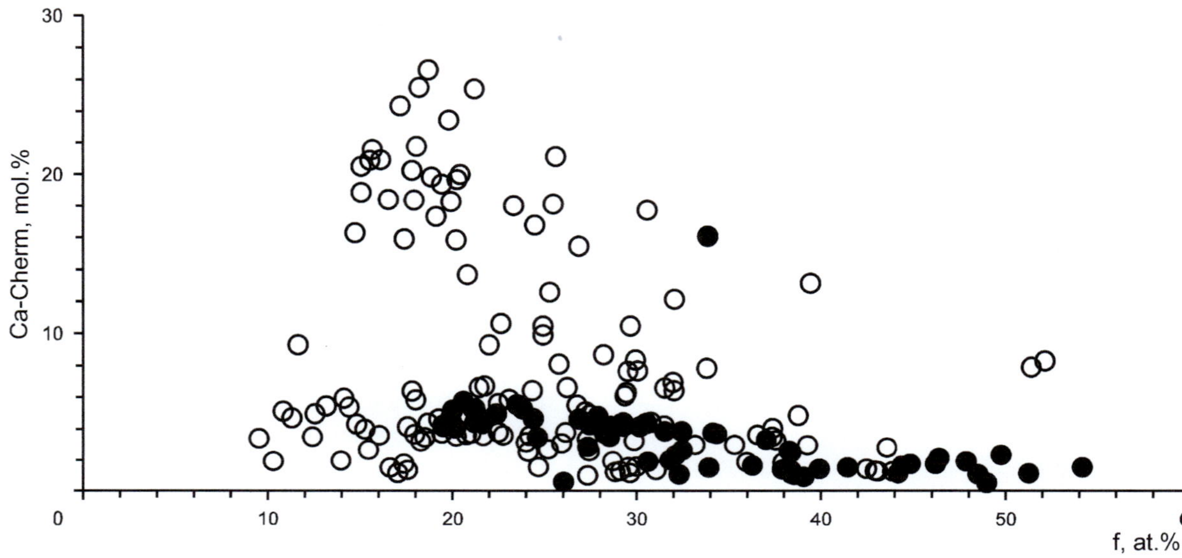

Fig. 3.46 Plot of Ca-tscherm versus *f* for pyroxenes of Dzhaltul Intrusion. Legend is given in Fig. 3.43

Dykes of trachydolerite are spread regionally. They extended along the northern margin of the Siberian Platform from Noril'sk to Maymecha–Kotuysky Provinces and have a predominantly east–west strike. The thickness of dykes varies from 0.5–1 m to 40–50 m, and they are extended over hundreds of meters to several kilometers (and more rarely more than 10 km) (Bushkanets et al. 1970b). There are isotropic (nondifferentiated) and slightly differentiated bodies, in which quartz-bearing, olivine-bearing, and olivine varieties are distinguished. Pegmatoids are also observed in the central parts of the dykes. Textural features of the dyke depend on the thickness.

Thin dykes are composed of fine-grained rocks with a trachytoid texture and large labradorite insets. Under microscope, the texture of rocks is seriate-porphyritic with plagioclase crystals ranging from 5–6 mm to 0.1 mm; the rock groundmass is doleritic and poikilitic. The rocks consist of plagioclase (30–45%), clinopyroxene (40–60%), olivine (0–2%), K-feldspar (to 2%), biotite (1–2%), ore minerals (8–10%), and apatite (0.5%); amphibole and chlorite are present in minor amounts; accessory minerals include tourmaline and zircon (Bushkanets et al 1970a).

Thick dykes have inhomogeneous symmetric structures. Toward the edge, porphyritic microdolerite is developed. Near the center of a dyke, the texture progressively changes to poikilitic and then to prismatically granular. In fine-grained edge facies rocks, a trachytoid fabric is observed. Porphyritic grains in the contact zones of trachydolerite are plagioclase and clinopyroxene that feature polysynthetic twins, anomalous interference colors, and often a sandglass texture under extinction. The poikilitic texture of trachydolerite (as opposed to the poikilo-ophitic texture of dolerite) features clinopyroxene intergrowths in large tabular plagioclase crystals. Judging from the composition of chadocrysts and oikocrysts, crystallization of clinopyroxene in trachydolerite began before plagioclase, and inversely, plagioclase in dolerite had crystallized earlier and was in turn enclosed in clinopyroxene poikilocrysts.

In thick dykes in the interfluve between the Bol'shoy and Maly Avam Rivers, the grain size of rocks increases, and the rocks become more leucocratic away from contact zones toward the center. The common variation range of rock-forming minerals in the dyke sequence is 32–70% plagioclase, 13–35% clinopyroxene, 2–20% for K-feldspar, 3–11% for biotite, 0.3–8% for hornblende, 0.1–3% for apatite, and 6–15% for ore minerals (Bushkanets et al. 1970a).

3.9.1 Dolerite–Trachydolerite Dykes LD/33-45

At the confluence of the Ledyanaya and Ldistaya Rivers, numerous, large, east–west striking dykes are exposed at the surface. These dykes are designated as trachydolerite–quartz trachydolerite. Our investigations have shown the dykes are similar in internal structure and composition, and therefore we will only discuss one in this volume.

Dyke LD/33-45 dips subvertically and is 28 m thick. The dyke is distinctly distinguished in relief by its profile, which is specific for this type of dyke (Fig. 3.47). It appears road-like or as a "gutter"; the central part of the dyke has a negative relief and is composed of plate-like leucocratic pegmatoid trachydolerite, whereas the fine-grained edge of the dolerite is more resistant to weathering. The magmatic body has a symmetric structure and consists of:

Fine-grained contact dolerite with quenched crust 0.1 m thick	4 m
Medium-grained dolerite	6 m
Coarse-grained trachydolerite leucocratic pegmatoid	8 m
Medium-grained dolerite	6 m
Fine-grained contact dolerite with quenched crust 0.1 m thick	4 m

3.9.1.1 Petrography

The quenched facies of the dyke is a black vitreous rock. Under microscope, it consists of a loop-like aggregate of small (0.02 mm) clinopyroxene granules, chains of which enclose eye-shaped cells of light-brownish cloudy glass. Glass insets are 1.2–1.5 mm in width (see Rock indication 216). Near the contact, a network of fine veins of oxide-ore mineral is observed in the glass. Away from the contact, the glass is ore-free and is observed as split dendrite-like or pinnate palmetto-like crystals, and the glass is anisotropic. The glass further from the contact changes into fan-shaped aggregates and bunches of thin plagioclase crystals (see Rock indication 217). Sparse small porphyritic clinopyroxene crystals with twins and abnormal interference colors, small quartz grains sometimes intersected by apatite needles, and small amphibole lenses are not infrequent in the quenched facies.

Microporphyritic clinopyroxene aggregates are augite ($Wo_{37–34}En_{53–51}Fs_{11–15}$), and small grains are $Wo_{38–43}En_{48–34}Fs_{13–23}$. The mineral contains 0.72–2.08 wt% TiO_2, 1.27–6.77 wt% Al_2O_3, 0.028–0.64 wt% Cr_2O_3, and 0.29–0.36 wt% Na_2O. Fan-shaped plagioclase aggregates are andesine ($Ab_{52–56}Or_{3–4}An_{44–40}$). The glass contains 55.13 wt% SiO_2, 18.51 wt% Al_2O_3, 7.25 wt% FeO, 2.38 wt% MgO, 4.79 wt % CaO, 3.16 wt% Na_2O, and 4.41 wt% K_2O. The anisotropic glass has been calculated for 8 (Î). The composition of the glass is very close to that of feldspar ($Ab_{36–47}Or_{37–15}An_{28–41}$) (see Rock indication 217). Ilmenite contains 0.07–0.73 wt% MnO, 0.20 wt% MgO, and 0.60 wt% MgO.

Fine-grained dolerite is dark gray; the rock has porphyritic, doleritic, and hypidiomorphic granular textures under microscope (see Rock indications 218 and 219). The rock appearance is specific due to accumulations of small polygonal clinopyroxene grains that are arranged and distributed as chains of grains bound together similar to those in the quenched facies. Plagioclase retains their fan-shaped appearance, but its crystals have, instead of thin laths, the form of split packs of intergrown long prisms. Sparse porphyritic grains of clinopyroxene, plagioclase, and olivine are present, the latter being commonly replaced by bowlingite. Ilmenite forms curved branchlets and bushes of intergrown crystals, as well as isometric and sheet-like ones. Xenomorphic K-feldspar, thin apatite needles, small lamellae of brownish biotite, chlorite, and pyrrhotite are observed in the rock matrix. Single amygdales are composed of calcite.

Porphyritic clinopyroxene grains are augite ($Wo_{34–37}En_{50–46}Fs_{15–17}$), and small grains are $Wo_{38–42}En_{48–39}Fs_{14–21}$. The mineral contains 1.16–2.64 wt % TiO2, 1.74–4.77 wt% Al_2O_3, 0.28–0.34 wt% Na_2O, and 0.01–0.45 wt% Cr_2O_3. Plagioclase of small porphyritic tabular crystals and laths are andesine ($Ab_{48–61}Or_{5–4}An_{47–35}$), and K-feldspar has the composition $Ab_{52}Or_{66}An_2$. The total Fe content of brownish biotite is 38–44 wt%. The mineral contains 4.14–4.25 wt% TiO2, 1.35–1.85 wt% TiO2, and 0.17–0.27 wt% Cl. Apatite contains 2.74 wt% F, 0.61 wt% Cl, and 0.58 wt% CeO_2.

Medium-grained dolerite has a gray color, doleritic, hypidiomorphic granular, poikilitic, and ataxitic textures (see Rock indications 220 and 221). Clinopyroxene forms aggregate of small grains, which are arranged as they are in the quenched facies as chains or are restricted to interstices. Plagioclase is observed as large prisms which are observed as fan-like intergrowths. The crystals become larger than those in fine-grained trachydolerite. Aggregates of small clinopyroxene grains retain in part a cellular shape, and plagioclase prisms surround such cells. A combination of small clinopyroxene grains with large plagioclase crystals is responsible for the ataxitic texture, and intergrowths of the former in the latter impart a poikilitic texture to the rock, pyroxene intergrowths commonly being restricted to rims of

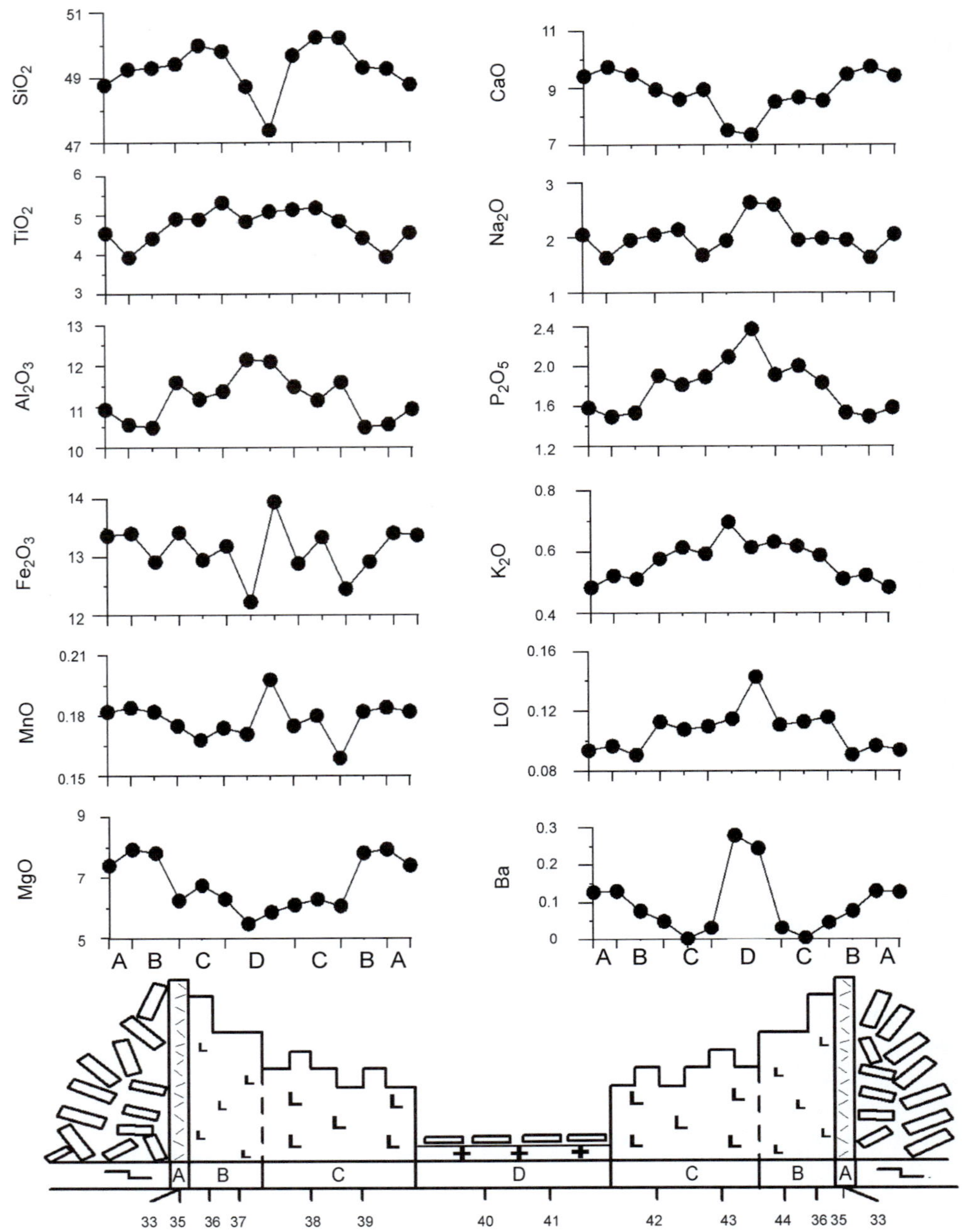

Fig. 3.47 Petrochemical variation diagram for oxides through a section of dyke LD/33–45, near the merging of the Ledyanaya and Ldistaya Rivers. At the *bottom*, the dyke profile and zones are shown schematically; *figures* indicate sample numbers. Zone A—quenched facies variolites, Zone B—fine-grained dolerite, Zone C—medium-grained dolerite, Zone D—trachydolerite. Country rocks are Ayansky Suite basalts. Explanations are given in the text

large plagioclase prisms. Ilmenite (up to 10–15%) is observed as skeletal isometric and prismatic crystals. Brownish biotite is not infrequent as an intergrowth with ore minerals. In the rock matrix, areas of K-feldspar with apatite needles and small prismatic crystals of brownish amphibole are observed.

Clinopyroxene is augite ($Wo_{36-39}En_{48-42}Fs_{14-21}$) and contains 0.99–1.71 wt% TiO_2, 1.98–2.81 wt% Al_2O_3, and 0.003–0.27 wt% Cr_2O_3. Plagioclase is andesine ($Ab_{46-50}Or_{3-4}An_{50-46}$), and K–Na feldspar in the matrix has the composition $Ab_{32-39}Or_{56-66}An_{5-2}$. Single K-feldspar grains are characterized by elevated BaO content (e.g., 1.80 wt%). Biotite contains 3.97 wt% TiO_2, 2.15 wt% F, and 0.43 wt Cl%, and the Fe content is 36.6 wt%. Apatite contains 2.02–3.70 wt% F, 0.161–0.27 wt% Cl, and 0.30–0.60 wt% CeO_2. Ilmenite is poor in mineral impurities and contains 1.02–1.34 wt% MgO; the other oxide contents do not exceed 100th of a percent.

Leucocratic trachydolerite is a coarse-grained, light-gray rock, sometimes with a pinkish tint. Under microscope, the rock exhibits prismatically granular, ophitic, poikilitic with elements of ataxitic and pegmatoid textures (see Rock indications 222 and 223). The rock consists of plagioclase (45–55%), K-feldspar (5–10%), clinopyroxene (20–25%), ore minerals (7–10%), biotite (3–5%), apatite (3–6%), chlorite (5–7%), and quartz (1–2%).

Plagioclase forms large (1.5–2.5 mm long) and small (0.12 mm) prismatic crystals. Large crystals exhibit slight, simple, wide twins and pronounced zoning and numerous intergrown apatite crystals. Large plagioclase prisms are labradorite ($Ab_{40}Or_2An_{58}$) in cores, and rims are andesine ($Ab_{60}Or_5An_{35}$) (see Rock indication 222). Small prisms have an andesine ($Ab_{50}Or_4An_{46}$) composition. K-feldspar ($Ab_{25}Or_{74}An_1$) is present with box-like albite crystals, and grains of quartz are found in interstices. In another sample from the central zone (see Rock indication 223), plagioclase prisms are all altered and feature wide tabular K-feldspar ($Ab_{39}Or_{59}An_2$) crystals, prisms of $Ab_{21}Or_{78}An_1$, and laths of $Ab_{47}Or_{50}An_3$. In rims of tabular K-feldspar and in the matrix, submicroscopic quartz–feldspar intergrowths are common.

Clinopyroxene in the rock is observed as large subidiomorphic crystals 0.3–0.5 mm and small grains (less than 0.1 mm) in the matrix. Large crystals have a stable composition $Wo_{42-39}En_{46-43}Fs_{14-17}$, and small grains are $Wo_{33}En_{37}Fs_{30}$. Pyroxene contains 0.53–1.63 wt% TiO_2, 1.56–3.43 wt% Al_2O_3, 024–0.38 wt% Na_2O, and 0.001–0.044 wt% Cr_2O_3.

Biotite forms prismatic crystals. The total Fe content of the mineral is 35–41 wt%, and it contains 5.05–5.75 wt% TiO_2, 1.41–1.92 wt% F, and 0.031–0.043 wt% Cl. In zoned crystals of mica, an increase in Cr_2O_3 content from 0.017 to 1.70 wt% (see Rock indication 223) is observed from core to rim.

Apatite is observed as short prismatic crystals, which are enclosed in clinopyroxene and feldspar, as well as in the chlorite-like rock matrix. Apatite is a fluorine-bearing variety containing 2.28–2.97 wt% F, 0.059–0.073 wt% Cl, and 0.114–0.143 wt% CeO_2.

Ore minerals in trachydolerite include titanomagnetite and ilmenite. Titanomagnetite forms large idiomorphic crystals that contain a fine rash, most likely of apatite, in the crystal core. In connection with this, the overall oxide content is reduced. Ilmenite features a wide variation in MgO from 1.19 to 5.50 wt%, the increased oxide content determined in trachydolerite with the elevated alkali content.

Mineralogical Peculiarities

The peculiarities in the composition of principal rock-forming minerals in the transverse section of the dyke are described below.

From the edge to the center of the dyke, there is a trend of increasing maficity in plagioclase from An_{28-44} to An_{36-58}. This is accompanied by the appearance of and subsequent increase in K-feldspar content that reaches a maximum in the central zone of the body. In the quenched facies of dyke, the composition of feldspars is unstable ($Ab_{36-47}Or_{37-15}An_{27-39}$). In medium-grained dolerite, plagioclase crystals with composition $Ab_{46-49}Or_4An_{50-47}$ are predominant, whereas K–Na feldspar $Ab_{39-32}Or_{56-66}An_{5-2}$ is found in the matrix. In trachydolerite, crystals of zoned plagioclase ($Ab_{40-60}Or_{2-5}An_{58-35}$) coexist with orthoclase ($Ab_{25}Or_{74}An_1$) crystals, or no plagioclase is found in the rock; the salic facies is represented by only K–Na feldspar ($Ab_{47-21}Or_{50-78}An_{3-1}$).

Clinopyroxene is a mineral phase that is most sensitive to crystallization conditions. Variation in mineral composition can be traced on diagrams (Fig. 3.48) constructed from data of Rock indications 216–223. Analytical data of pyroxenes are available only for the left part of the dyke. However, taking into consideration its symmetric structure, we believe that these data may be a reflection for the right part of the magmatic body. As it can be seen from the plots, clinopyroxene composition varies from the margin of the dyke to its center. The Fs component progressively increases from the quenched zone to the fine- and medium-grained zone, and then it sharply decreases in the central zone. The TiO_2 content in clinopyroxenes varies widely. In the contact zones of the dyke, high-titanium clinopyroxenes are present, which are more typical for trachydolerites, along with low-titanium varieties that are close to crystallization products of tholeiite-basalt melts. In the central zone, the upper limit of TiO_2 content in clinopyroxenes decreases, and low-titanium varieties appear. Clinopyroxenes of the quenched zone have an increased chrome content that is more characteristic for dolerite rather than for trachydolerite. The chrome content sharply decreases toward the center of the dyke.

Increased ilmenite and apatite is an important characteristic feature of these rocks, and the central zone is enriched in biotite.

Petrochemistry

The variation diagram indicates the composition of rocks in the dyke sequence (Fig. 3.47). In outcrops of dyke LD/33–45, its right-hand contact was buried under talus. Therefore, because of the symmetric structure of the magmatic body and in order to give a complete picture of its internal structure, we considered that it was possible to add analytical data acquired for the left contact of the data to the replace the missing data from its right contact. To this end, points 33, 35, and 36 were repeated in the right part. As it is seen from the plots, with increasing grain size, a decrease in SiO_2, MgO, and CaO content and an increase in Na_2O, K_2O, P_2O_5, and Ba and loss on ignition content are observed from the edge to the center of the dyke. The highest MgO and CaO content and the lowest TiO_2, Al_2O_3, K_2O, and P_2O_5 and loss on ignition content are typical for contact dolerites. Trends for SiO_2 and Ba observed at edge parts are reversed at the dyke center. In the central zone, increased Al_2O_3, Na_2O, K_2O, and P_2O_5 and loss on ignition content are characteristic due to the higher feldspar content and minerals containing volatile elements.

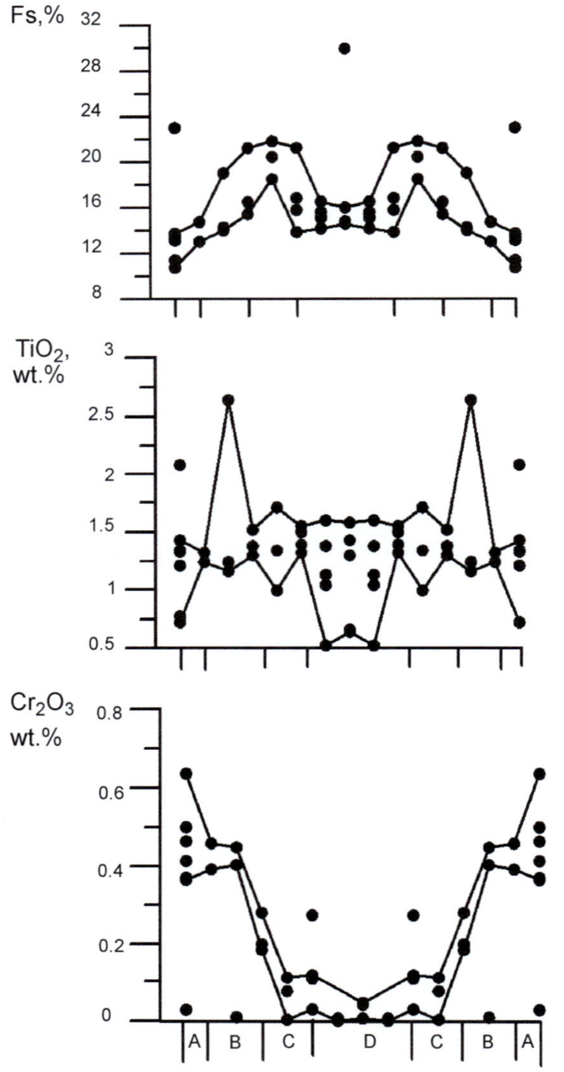

Fig. 3.48 Variation diagram for clinopyroxene content through a section of dyke LD/33–45. Zone A—quenched facies variolites, samples Ld-33 and Ld-35; Zone B—fine-grained dolerite, samples Ld-36 and Ld-37; Zone C—medium-grained dolerite, samples Ld-38 and Ld-39; Zone D—coarse-grained trachydolerite, samples Ld-40 and Ld-41. Explanations are given in the text

The characteristic feature of dykes of the Avamsky Complex is their zoned structure and the elevated TiO_2 (3.94–5.31 wt%) and P_2O_5 (0.48–0.62 wt%) content in dolerite contact zones and 4.84–5.09 wt% TiO_2 and 0.62–0.70 wt% P_2O_5 in the central zones of trachydolerite. Diagram of (Na_2O + K_2O) versus SiO_2 (Fig. 3.49) shows the composition of rocks of the Avamsky Complex in comparison with that of trachydolerites of the Ergalakhsky Complex. On the diagram, the Avamsky Complex is represented by dyke LD/33–45 and analyses of trachydolerite samples taken in the central zones of other dykes of the same region. The field of trachydolerites of the Ergalakhsky Complex is represented by sills at the northern end of the Noril'sk Plateau, drill hole F-221 (right group of points), and the Kharaelakh Plateau, drill hole NM-9 (left group of points). Judging from the distribution of fields and points, reducing the silica acidity of rocks is accompanied by an increase in their alkalinity; dyke formations of the Avamsky Complex are characterized by increased silica acidity. The Kuno line separating tholeiite and subalkaline basalts separates, in fact, contact dolerite and trachydolerite in the central zones of dykes. One of the most likely reasons for a layered structure of the Avamsky Complex is enrichment of the magmatic melt in volatiles. The composition of the volatiles can be indirectly judged by the composition of biotite and apatite, as well as of sulfide and carbonate found in the rock.

In the rock series, fine-grained dolerite–medium-grained dolerite–trachydolerite, in dyke LD-33–45, the F content in biotite and apatite increases from the edge to the center of the body and reaches a maximum (2.15 wt% in mica and 3.70 wt% in apatite) and then decreases from 1.41–1.92 wt% to 2.28–2.97 wt% in mica and apatites, respectively, in trachydolerite. Chlorine behavior is different to fluorine. Biotite and apatite in fine-grained dolerite are typically enriched in chlorine (0.27 and 0.61 wt%, respectively). The concentration of chlorine decreases in medium-grained dolerite, and in biotite and apatite from trachydolerite, these are 0.03–0.04 wt% and 0.07 wt%, respectively.

Additional information on the volatile content in rocks of dyke LD/33–45 is given in chromatographic analyses of the rocks (Table 3.10). For comparison, this table also lists data from trachydolerite of the Ergalakhsky Complex and trachybasalt from a dyke totally composed of glass. Judging from these data, CO and H_2 (the latter in the portion of analyses) are released from samples only on heating at high temperatures. The CO_2 and H_2 contents in samples vary widely.

Based on analytical data from samples taken from dyke LD/33–45 and heated to 600 °C, the gradual increase in the H_2O/CO_2 ratio from 11.9 to 14.5 and to 25.7, respectively, is observed, and the ratio value for samples of the same series heated to 1,000 °C decreases from 14.2 to 4.1 and to 0.88, respectively. With increased heating from 600 °C to 1,000 °C, an increase in the H_2O content in total (H_2O + CO_2) content is observed from contacts to the center of the dyke. The highest CO/CO_2 ratio has been determined for samples of contact variolites from dyke LD33-45 and a dyke of trachydolerite glass on heating to 1,000 °C.

Petrologic Features

Dykes of the Avamsky Complex have a symmetric layered structure: contact zones are composed of dolerite, and the central zone is trachydolerite. On the classification diagram of (Na_2O + K_2O) versus SiO_2, these are located in the tholeiite and subalkaline composition region, respectively (Fig. 3.49). The dolerite features elevated TiO_2 and P_2O_5

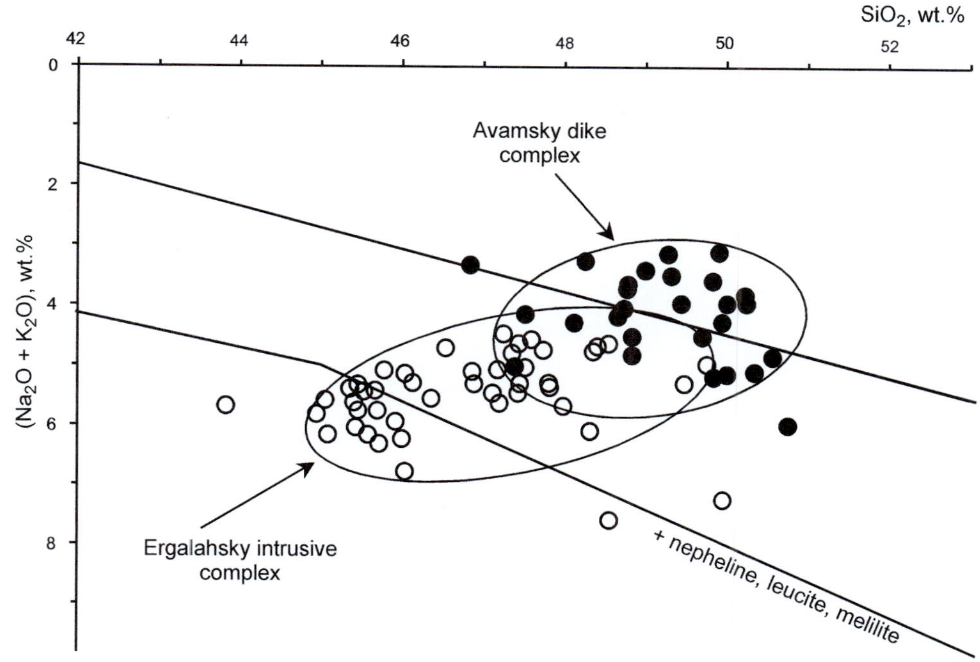

Fig. 3.49 Diagram of ($Na_2O + K_2O$) versus SiO_2 for rocks of the Avamsky and Ergalakhsky intrusive complexes

content, as distinct from dolerite from other petrographic complexes. Trachydolerite of the Avamsky Complex is similar to trachydolerite of the Ergalakhsky Complex, which is characterized by elevated TiO_2, P_2O_5, and alkali content.

Genetic Concepts

The problem of the origin of alkaline–mafic rocks in trap magmatism has been discussed in a literature, but it is largely unsettled. Most geologists consider that subalkaline basal melts originated as the result of crystallization differentiation of magmatic melts in deep-seated chambers. Some geologists explain the formation of subalkaline and tholeiite melts through differing magmatic chamber depths (Fedorenko 1994); others relate them to large mantle plumes (Campbell and Griffiths 1990) or suggest that the alkalization of mafic melts in intermediate chambers and feeder conduits proceeded under the action of mantle fluids (Zolotukhin 1991).

The presence of dolerite and trachydolerite dykes within a single magmatic body in the Avamsky Complex leaves no doubt that there is a genetic relationship between tholeiite and subalkaline basalt compositions and indicates that these were formed due to the differentiation of a unified magma. Using examples of layered intrusions from different complexes, described here, as well as other examples published elsewhere, it has been shown that during the process of chamber differentiation of a melt, an accumulation of volatiles and redistribution of alkali elements take place. During the final stages of a magmatic process, this leads to isolation of fluidized alkali melts that manifest as emplaced intrusions with elevated or high alkalinity. None of the researchers doubt that the basaltic or trachybasaltic magma was a source of alkali and volatile elements in schlieren-like pegmatoid bodies in these intrusions. Based on this, it is proposed that volatile elements played a crucial role during the basaltic magma differentiation process, and fluidized subalkaline melts may be separated from the magmas as immiscible liquids. It seems likely that immiscible separation of basalt magmas proceeded in a liquid phase under prechamber conditions similar to those observed in intrusions. As a result, compositionally variable fluidized liquids were formed (Ryabov 1989a, 1999b) including liquids enriched and depleted in alkali elements, with high K content or depleted in K, and these fit compositionally to lavas and intrusions of the region. The ability of a basaltic magma to undergo immiscible differentiation explains the diversity of daughter melts and means there is no need for simultaneous activation of various magmatic sources to explain the origin of subalkaline and tholeiitic melts. There is also no need to create a magma path through the Earth down to the lower mantle and core and allow subalkaline basaltic plumes to rise to the surface or to search for some other source of alkali and volatile elements for metasomatic alkalization of basaltic magma to form subalkaline basaltic melts.

Unfortunately, the available data on dykes of the Avamsky Complex are not enough to completely account for the mechanisms of basaltic melt differentiation. However, it is hopeful that results of studies from compositionally and structurally different small intrusions of the Kamensky Province, investigations of which were undertaken by the present authors during last years (Ryabov et al. 1996b), will clarify this matter.

Table 3.10 Chromatographic analytical results of rocks (ppm) from dyke section LD/33–45 in comparison with other trachybasalt melt derivates

Sample number	Heating temperature (°C)	CO_2	H_2O	H_2	CO	C_2H_2	H_2O/CO_2	CO/CO_2
LU-33	300	30	2,400	0.0	0.0	0.0	80	–
	600	160	1,900	0.0	10	2	11.9	0.06
	1,000	240	3,400	40	70	0.0	14.2	0.29
LD-38	300	20	1,600	0.0	0.0	0.0	80	–
	600	110	1,600	0.0	9	1	14.5	0.08
	1,000	630	2,600	0.0	40	0.0	4.1	0.06
LD-40	300	30	2,000	0.0	0.0	0.0	66.7	–
	600	210	5,400	0.0	10	0.0	25.7	0.05
	1,000	6,100	5,400	0.0	50	0.0	0.88	0.008
F-221/583.1	300	50	2,700	0.0	0.0	0.0	54	–
	600	220	4,100	10	20	10	18.6	0.09
	1,000	640	5,400	.	50	0.0	8.4	0.078
N-10	300	10	770	0.0	0.0	0.0	77	–
	600	60	540	0.0	5	.	9	0.08
	1,000	120	1,800	70	70	0.0	15	0.58

CH_4, N_2—not determined
Dyke LD/33–45: LD-35, contact variolite; LD-38, small-grained trachydolerite; LD-40, coarse-grained leucocratic trachydolerite; F-221/583.1, pegmatoid trachydolerite, Ergalakhsky intrusive complex; N-10, trachydolerite glass from dyke, middle course of the Namakan River

3.10 Bolgokhtokhsky Intrusive Complex and Lamprophyres of the Noril'sk Region ($\gamma - \gamma\delta\pi$)Mz

This complex consists of the syenogranite Bolgokhtokh Intrusion and related dyke rock complex, as well as dykes of lamprophyres, which occur in various genetic situations in the northern part of the Noril'sky trough. The dykes are distinguished as the Ostrogorsky type of intrusions (Distler and Kunilov 1994) because these were intersected by drill holes on the slope of the Mt. Ostraya.

3.10.1 Bolgokhtokh Granite–Syenite Intrusion

The Bolgokhtokh Intrusion of granitic porphyries and quartz–syenite porphyries is situated in the western part of the region (Fig. 1.1) at the junction of the Noril'sky and Vologochansky troughs and the Dudinsky uplift. The main disjunctive structure of the region is the Boganidsky Fault that trends nearly north–south and forms the western boundary of the Noril'sky and Vologochansky volcanic troughs. In the region surrounding the magmatic body, the vertical displacement of the fault is about 1,200 m. The intrusion is restricted to the core of anticline composed of carbonate–argillaceous and sulfate-bearing deposits of Ordovician, Silurian, and Devonian age. The roof of the intrusion is eroded or buried under a thick cover of unconsolidated Mesozoic–-Cenozoic deposits. According to drilling data and geophysical investigations, the intrusion has a cross section at the level of erosion of 2.1–2.3 km and a vertical thickness of more than 2.2 km.

The results of investigations into the Bolgokhtokh Intrusion have been published by Komarova (1968, 1983), Oleinikov and Pankov (1985), and Strunin et al. (1988). The intrusion is composed of granite, granodiorite, syenogranite, granite porphyry, diorite porphyry, quartz monzonite porphyry, and syenite porphyry. In the periphery of the intrusion, dykes and veins of aplite and lamprophyre are developed. Komarova (1968) noted that no pronounced differentiation vertically through the intrusion is observed, and veinlets of fine-grained syenite, aplite, and quartz cut the granitoids. A geological scheme, based on drilling data (Strunin et al. 1988), indicates the circular internal structure of the intrusion. At the center of the intrusion, granite porphyry is surrounded by other granite; continuing away from the center, the granite is surrounded by continuous metasomatically altered granitoids, and these are surrounded by granodiorite and granodiorite porphyry. Skarns have formed in the southern part of the intrusion.

Principal rock-forming minerals in rocks of the Bolgokhtokh Intrusion are K feldspar (25–50%), quartz (25–40%), and plagioclase (15–30%) (Komarova 1968). Mafic minerals include biotite and hornblende that constitute 1–3% to 10–15%, but clinopyroxene and muscovite are rare. Accessory minerals include sphene, zircon, rutile, and monazite, and secondary minerals include chlorite and sericite, and clay minerals are observed. The diversity of rocks in the intrusion is governed by the wide variations in mineral proportions, peculiarities in their structural interrelations, and a differing grain sizes including giant-, coarse-, and medium-grained rocks. Rocks of the intrusion are typically porphyritic. Phenocrysts are plagioclase, orthoclase, and less commonly quartz. The porphyritic component of the rocks ranges from 5–7% to 25–30%, in some cases up to 50–70%.

The rock groundmass has granoblastic, xenomorphic granular, and felsitic textures. It consists of alkali feldspars, quartz, and small aggregates of dark-colored minerals.

Porphyritic grains are typically feldspar and less common quartz. Plagioclase forms large twinned crystals that are commonly surrounded by bands of K-feldspar. The plagioclase composition ranges from andesine to albite. K-feldspar forms tabular crystals, granular masses, and granular fields irregular in shape. Compositionally, the mineral is within the range $Ab_{35-29}Or_{62-70}An_{1-3}$ (see Rock indication 224); orthoclase is more common than microcline. Quartz forms sparse, large, isometric crystals and aggregates of polygonal grains in the rock matrix and forms bands around crystals of feldspar and early quartz. Hornblende forms prismatic crystals that have brownish and green color and show pleochroism from green to yellow brownish. The total Fe content of amphibole is 30 wt%. Biotite forms individual dark-brown leaflets and "furry" aggregates that are not infrequently elongated along veinlets. Biotite replaces amphibole and develops in amphibole cleavage. The total Fe content of biotite is 35.4 wt%. Apatite forms prismatic crystals restricted to K-feldspar inclusions. Spinel group minerals in granodiorites are typically magnetite and contain 0.24–0.28 wt% Cr_2O_3, 0.36–0.40 wt% V_2O_5, and 0.10–0.29 wt% MnO.

Granitoids of the Bolgokhtokh Intrusion have experienced feldspatization, biotitization, and silification (Komarova 1968). The feldspatization manifests as an overgrowth of plagioclase and orthoclase by K-feldspar. One generation of syenite contains 9.47 wt% K_2O that appears to be due to later K-feldspatization (Oleinikov and Pankov 1985). Uneven distribution of biotite in granites and location of its scaly aggregates along rock fracturing confirm its metasomatic origin (Komarova 1968). Feldspatization and silification are also metasomatic in nature (Komarova 1968).

Dykes associated with the Bolgokhtokh Intrusion are thin (centimeter to several meters thick) vein-like bodies of syenite, syenogranite, aplitic granite, and lamprophyric syenite crosscutting the intrusion, based on data from Noril'sk geologists. Lamprophyric syenite has a xenomorphic granular aplite or felsite texture and consists of fine-grained orthoclase aggregate (65–80%), albite–oligoclase (5–15%), and quartz (up to 5%). Aplitic granite features a preserved texture and contains 20–30% quartz, 55–65% orthoclase, and 15–20% plagioclase.

Noril'sk geologists define the lamprophyres as vogesites and minettes, as well as biotite–hornblendite and pyroxene-bearing rocks. The rocks are lamprophyric with distinct idiomorphism of mafic minerals and xenomorphic granular feldspar aggregate. The rocks are composed of K-feldspar (40–55%), hornblende (up to 30%), or biotite (up to 15–20%). Other minerals include plagioclase (up to 15–20%), quartz (0–3%), apatite, sphene, and others. Information on ore minerals is not available.

Metasomatic alteration related to the Bolgokhtokh Intrusion includes hornfels, marble, and metasomatites. The temperature range of contact metamorphism covers pyroxene and amphibole hornfels facies. Among metasomatites, calcareous skarns that are developed after granites and their host rocks and quartz-feldspar rocks are mostly widespread. Skarns consist of garnet of a grossular andradite series, clinopyroxene, amphibole, epidote, calcite, anhydrite, scapolite, and hematite. Quartz–feldspar metasomatites are composed of quartz in variable amounts, orthoclase, albite, biotite, sericite, and chlorite. The composition of quartz–feldspar rocks ranges from the essentially feldspar to quartz–sericite, quartz–chlorite, and quartz–carbonate.

Ore mineralization is distinctly associated with hydrothermal metasomatic rocks. The mineralization is a complex of sulfides and oxides peculiar to traps; it also includes minerals peculiar to granitoids such as molybdenite, scheelite, fahlores, wolframite, galenite, and sphalerite (Strunin et al. 1988). Elevated concentrations of chalcopyrite and molybdenite in granites of the Bolgokhtokht Intrusion afford ground for recognition of the individual copper-porphyritic ore formation and copper–molybdenum occurrence in the Noril'sk Region.

3.10.1.1 Geochemistry

Alkaline, silicic, and acidic rocks of the Bolgokhtokh Intrusion have a relatively stable composition, which vary in biotite, amphibole, or quartz content (Table 3.11). For additional data on the composition of rocks, readers are referred to Komarova (1968) and Strunin et al. (1988). The rocks of the intrusion contain 63.73–67.20 wt% SiO_2, 0.17–0.59 wt% TiO_2, 14.07–15.40 wt% Al_2O_3, 0.42–3.10 wt% Fe_2O_3, 1.86–2.65 wt% FeO, 0.03–0.08 wt% MnO, 1.36–2.37 wt% MgO, 2.14–4.26 wt% CaO, 3.24–4.12 wt% Na_2O, and 4.04–5.75 wt% K_2O. On the diagram of ($Na_2O + K_2O$) versus SiO_2 (Fig. 3.11), these rocks form a group of points along the border zone of granodiorite–granite–syenite. Aplite dykes (Table 3.11) have a composition close to the principal varieties of rocks of the intrusion, and lamprophyres differ from aplites featuring a lower SiO_2 content and increased MgO, CaO, and FeO content.

3.10.1.2 Genetic Concepts

The emplacement of the Bolgokhtokh Intrusion appears to have occurred after the development of the Permo–Triassic volcanogenic unit. No direct contacts between rocks of the intrusion with trap lavas are observed, but the age of the granitoid can be argued indirectly from quartz veinlets cutting dolerite, as well as from intense sulfidization of basalts of the Nadezhdinsky and Morongovsky Suites and dolerite sills in the outer aureole of the intrusion (Komarova 1968). Absolute age determination of granitoid rocks (K–Ar method) is 183 ± 10 Ma and 244 Ma for K-feldspar (Strunin et al. 1988). Komarova (1968) and Strunin et al. (1988)

Table 3.11 The chemical composition of Bolgokhtokhsky intrusive complex rocks and Noril'sk Region lamprophyres (wt%)

	Granitoids				Lamprophyres				
Components	1	2	3	4	5	6	7	8	9
SiO_2	64.90	64.20	66.90	67.28	47.41	50.55	46.30	50.40	49.45
TiO_2	0.59	0.50	0.59	0.17	0.59	2.04	1.10	1.49	1.23
Al_2O_3	14.10	14.40	14.20	13.84	14.64	10.90	10.57	9.24	14.41
Fe_2O_3	3.87	3.27	2.58	4.20	6.71	7.29	7.90	7.89	7.90
MnO	0.06	0.04	0.07	0.04	0.11	0.11	0.26	0.10	0.13
MgO	2.23	2.37	1.99	0.96	9.38	6.79	6.03	9.28	8.26
CaO	3.08	2.98	4.26	2.95	9.24	8.25	8.11	7.31	6.73
Na_2O	3.67	3.48	3.81	4.35	2.00	1.50	0.97	0.30	2.54
K_2O	4.59	4.84	4.09	3.98	5.80	7.60	7.13	7.72	4.69
P_2O_5	0.49	0.35	0.38	–	–	1.52	1.67	2.52	1.12
LOI	0.98	1.74	0.89	–	–	3.05	12.10	3.14	3.04
Total	98.56	98.17	99.76	97.77	95.88	99.60	102.14	99.39	99.50

Bolgokhtokhsky stock: 1— amphibole granite (hole BG-26/125); 2—the same (hole BG-26/217); 3—granite porphyry (sample BG-2); 4—dyke aplite (Strunin et al. 1988); 5—dyke minette (hole B-8/370.6) (in the same place). Lamprophyres of Noril'sk ore junction: 6—dyke minette (hole UN-1/68.2–72.8) data of T.N. Lyul'ko and A.B. Dushatkin; 7—the same (hole NP-12/1508) (Distler and Kunilov 1994); 8—the same deposit Noril'sk-I (sample LU-1); 9— the same (average) (Daly 1933)

believe the posttrap age of the intrusion is supported by small intrusions of granite and syenite porphyry, presumably of the Mesozoic age, in the western Taymyr that crosscuts through volcanic rocks in the region. They are also compositionally very close to granitoid rocks of Bolgokhtokh Intrusion. The peculiarity of occurrence and the composition of copper–molybdenum mineralization within the intrusion aureole bear great resemblance to copper porphyry ores known from different parts of the world. The geochemical relationship of this mineralization with the granitoid composition of the intrusion is regular. There are different concepts concerning the question of what comprises a rock formation to which granitoids, and similar rocks in traps may be assigned and what is the nature of their occurrence. Some researchers argued that small intrusions of granitoid composition should be regarded as a separate formation at the northwest end of the Siberian Platform (Golubkov 1970) or as a formation of alkaline granite (Dodin et al. 1982). In addition, many geologists drew attention to the fact that alkaline silica acidic rocks in layered trap intrusions, including those of granitoid composition, are closely related to gabbrodolerite and can be considered as an evolutionary member of the magmatic system.

Based on a study of the Bolgokhtokh Intrusion, Strunin et al. (1988) have stated with categorical assurance that "...granitoid magmatism can be considered by no means as a derivative of mafic melt and it is the individual, later phase of magmatic activity."

In the opinion of these researchers, granitoid rocks may be considered as

> ...a consequence of evolution of the magmatic system during rifting.

Unfortunately, the authors have advanced insufficient arguments in favor of their views and have offered no explanations for the magmatic system itself, the composition of the initial magma, and characteristics of its evolution.

Granite-like rocks and other similar rocks on the Siberian Platform are known from different layered intrusions (Kavardin et al. 1968). Alkaline silica acidic magmatic rocks in a number of bodies have been described by different names. In intrusions of the Zub-Marksheydersky (Godlevsky 1959) and the Gorbiachin River (Zolotukhin and Oleinikov 1963), the formation of these rocks was associated with hybridization of a basaltic melt on dissolution of xenoliths of sedimentary rocks. In Noril'sk Intrusions, these rocks were considered as contaminates (Dodin and Batuev 1971), autometamorphic formations (of the metamafic type) (Ivanov et al. 1971b), or hybrid metasomatic rocks (Zolotukhin et al. 1975). In the Alamdzhakh Intrusion, granophyres are considered as end members of a typical differentiation series, and alkaline rocks are regarded as alkaline immiscibility products (Masaitis 1958). In the Dzhaltul Intrusion, these rocks are supposed to be formed as a result of the interaction of an acidic methane–hydrogen-bearing fluid with tholeiitic basalt magma under hypogeal conditions (Koroleva and Oleinikov 1998). The example set by trap intrusions and data on the composition of "glass-in-glass emulsion" validates the immiscible nature of a series of alkaline silica acidic formations (Ryabov 1989a). At a later time, this immiscibility was complicated due to fluid–magmatic interaction.

Thermodynamic investigations have shown that the ability of a basalt melt to assimilate in deep-seated chambers is insignificant (Sharapov and Golubev 1976). Therefore, the formation of alkaline silica acidic rocks due to contamination is improbable. Besides, occurrences of granite- and diorite-like rocks are known in intrusions lying not only in terrigenous sedimentary strata but also in saliferous argillaceous deposits and basalts. Petrologic investigations point to the fact that the composition and mechanism of formation of alkaline silica acidic rocks have their own peculiarities in different intrusions, but all these rocks are regular members of an evolvable magmatic system.

Peculiarities in rock composition variation during differentiation of tholeiitic basalt and trachydolerite are illustrated in the diagram of ($Na_2O + K_2O$) versus SiO_2 (see Fig. 3.50). Here, composition fields for rocks from the Dzhaltul-layered intrusion and for so-called metadiorite intrusions of the Noril'sky Complex including syenite-, granite–granosyenite-, and diorite-like rocks from intrusions of Zub-Marksheydersky, Upper Talnakh, Lower Talnakh, Lower Noril'sk, Tulaek-Taas River, and Gorbiachin River are shown, using our data and data from the literature. For comparison, there are given fields of compositions for rocks from trachydolerite sills of the Ergalakhsky Complex, analyses of which are given in the listed Rock indications and in the reference cross section table in Volume 2. The Bolgokhtokh Intrusion, in the opinion of some authors, is similar to the Taymyrsky Intrusions that include the layered Dyumtaley Intrusion (Ryabov et al. 1996c), intrusions of granitoid rocks (drill hole LP-1), and syenites (drill hole TP-42) of the Dikarabigaisky Complex.

On the basis of this diagram, one can suggest with assurance that the initial magma of the Dzhaltul Intrusion had a tholeiitic basalt composition, and magmatic crystallization differentiation was responsible for the wide variety of rocks including alkaline silica acidic rocks. The parent magma of the Dyumtaley Intrusion had a trachybasalt composition, and the diversity of rocks was created by differentiation. The syenite–monzodiorite field of the drill hole TP-42 intrusion is part of a continuation of the field of compositions for rocks from the Dyumtaley Intrusion. The fact that the rocks of these intrusions are in the same petrogenic series with overlapping composition fields implies that these rocks had similar parent magmas and that differentiation under prechamber conditions was responsible for these two types of intrusions. The characteristic feature of intrusion LP-1 is that its internal structure corresponds to a "dyke-in-dyke" type, with sharp variation in the rock composition. The main mass in the central zone has a syenogranite composition, and quenched crusts at contact zones are trachydolerite (see rhombs in Fig. 3.50). Such layered intrusions with different set of rocks are very common among small intrusions of the Kamensky Province. Petrologic characteristics and genetic conditions of these intrusions will be described in special publications by the authors; for now, we will state there is a genetic relation between granosyenites of the LP-1 intrusion and trachybasalt magma.

It is seen from the diagram (Fig. 3.50) that granitoid composition points fall into the LP-1 intrusion field, and the compositions turn out to be close to granite-like rocks and metadiorites of the Noril'sky Complex and Dzhaltul Intrusion. This resemblance, as well as the fact that these rocks are in the same petrogenetic series, suggests a genetic relationship between all these formations.

Summarizing the concepts stated above, it can be assumed that alkaline silica acidic rocks including granites and syenites had been formed as a result of chamber and prechamber differentiation of a tholeiite-basalt and trachybasalt magmas and emplacement of single differentiates as individual intrusions. Taking this fact into consideration, syenogranites of the Bolgokhtokhsky Intrusion should be considered regular members of the petrogenetic series of rocks related to trap magmatism.

3.10.2 Dykes of Lamprophyres of the Noril'sk Region

In the Noril'sk Region, lamprophyres are extremely uncommon. Lamprophyric vein rocks were noted by Moor[1] in 1957 in the core of an exploration drill hole in the Noril'sk I deposit. The rocks have been classified by the author as alkaline lamprophyres of the oachitite–alnoite type. Lamprophyres of the Bolgokhtokhsky Intrusion were later drilled, and in 1975, Voroshilov[1] revealed lamprophyre dykes in a core from drill holes Yu-1 and PE-51 at the north of the Noril'sky trough. The dykes cut basalts of the Ivakinsky, Tuklonsky, and Nadezhdinsky Suites at angle of 20–50° to the horizontal. The thickness of the bodies varied from 0.7 to 4.7 m. Their composition fitted to minettes–prowersites. Contact metamorphism of the dykes did not extend beyond a zone of slightly hornfelsed rock about 1 cm thick.

Investigations undertaken by Lyul'ko[1] and Dushatkin[1] have shown that the dykes are porphyritic medium-grained rocks with large quantity of mica crystals from 0.5–1 mm to 3–5 mm in size, and a preferred orientation imparts a trachytoid texture to the rock. In contact zones of dykes, the rock has a porphyritic texture with a cryptocrystalline groundmass. Phenocrysts include phlogopite (10.7%), apatite (6.7%), and diopside (2.8%). Away from the contact, the grain size of the rock increases, the trachytoid texture disappears, and the texture changes to prismatically granular. In the central, crystallized parts of the dyke, rocks consist of K-feldspar (46%), phlogopite (9–11%), diopside (8–9%), apatite (6.5–7%), amphibole (20–21%), aegirine–augite (5–6%), and ore minerals (2%). Carbonate and quartz are present in variable amounts.

Data of Dushatkin[1] indicates the absolute age of the lamprophyre from a dyke in drill hole PE-51 in phlogopite is 239 ± 8 Ma (determined using the K–Ar method).

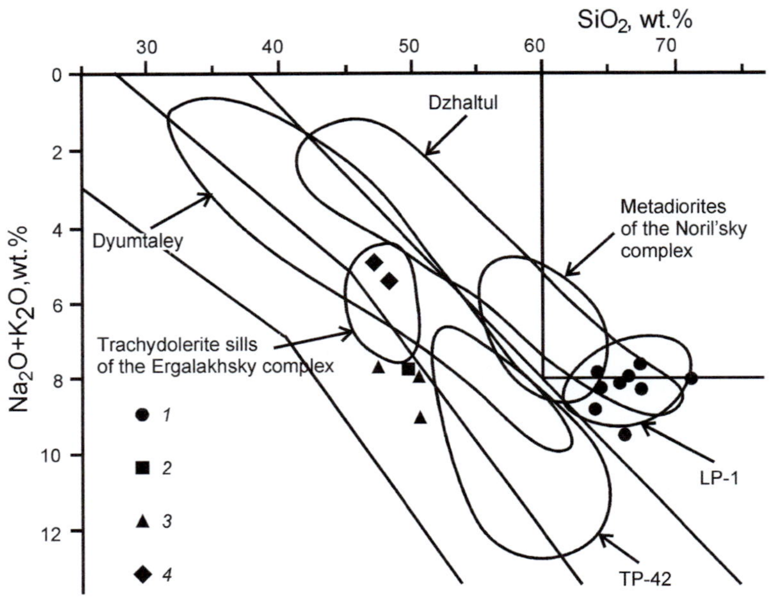

Fig. 3.50 Diagram of ($Na_2O + K_2O$) versus SiO_2 for rocks of normal subalkaline and alkaline series of the Siberian Platform and Taymyr Intrusions. *1*—syenogranites of the Bolgokhtokhsky Intrusion, *2–3* lamprophyres of the Bolgokhtokhsky Intrusion and the Noril'sk ore system, respectively, *4* LP-1 intrusion quench facies

One more lamprophyre dyke was discovered by Kulagov[1] in an underground working in the Zapolyarny mine of the Noril'sk-I deposit. The sample from the dyke has been given to V.A. Lyul'ko who kindly put it at our disposal for study. The rock is a minette. The rock texture is seriate porphyritic with a hypidiomorphic granular texture to the groundmass (see Rock indication 225). The rock is composed of phlogopite (30–60%), apatite (30–50%), calcite (5–7%), quartz (3–5%), K-feldspar (0.5–3%), amphibole (0.5%), and ilmenite (0.5–1%). Porphyritic insets are phlogopite that forms tabular crystals 2–3 mm in size, the oriented arrangement of which imparts a trachytoid appearance to the rock. The grain size gradually decreases from large phenocrysts to submicroscopic leaflets in the groundmass. Under microscope, phlogopite is brownish and shows bright pleochroism. Compositionally, the mineral is fluorine–phlogopite (2.00–2.17 wt% F). The total Fe content of large crystals is 8.4–9.1 at% and in small grains is 11.2–11.9 wt%. The content of other oxides is 1.34–1.36 wt% and 1.49–1.56 wt % TiO_2, 0.59–0.93 wt%, and 0.10–0.16 wt% Cr_2O_3.

Apatite forms prismatic crystals from 0.1 m in size to submicroscopic grains in the groundmass. Large intergrowths of apatite are found in rounded calcite grains. Compositionally, the mineral is a fluorine–apatite. Large crystals contain higher content of SiO_2 (0.37 wt%), MgO (0.58 wt%), and F (2.38 wt%) when compared to small grains. The small grains contain elevated concentrations of FeO (0.23 wt%) and CeO_2 (0.44 wt%).

Calcite forms small xenomorphic grains in the matrix and rounded aggregates to 1–1.5 mm in size, with apatite intergrowths not infrequently found in cores, and needle-like dirty-green amphibole crystals are commonly found in rims. The compositional feature of calcite is the increased SrO content (up to 2.04 wt%) and single grains of strontium calcite containing 18.45 wt% SrO and 39.07 wt% CaO.

In the rock matrix, small grains of diopside $Wo_{46-47}En_{47-46}Fs_7$, K-feldspar Ab_5Or_{95}, and sphene are observed, and ore minerals include ilmenite and pyrrhotite. The latter sometimes form rounded aggregates 1–3 mm in size with intergrown amphibole and apatite.

Table 3.11 shows that the composition of rocks varies for all rock-forming oxides, but in general, the rocks are compositionally close to minettes according to Daly (1933). On the diagram of ($Na_2O + K_2O$) versus SiO_2 (Fig. 3.50), lamprophyres of the Noril'sk Region (including those of the Bolgokhtokhsky Intrusion) are found along the continuation of the trachydolerites field from the Ergalakhsky Complex, and compositionally the lamprophyres are alkaline basalt theralites.

As it is known, the nomenclature and classification of lamprophyres are not yet completely developed (Bogatikov et al. 1981). Nevertheless, all dykes found in the Noril'sk Region are referred to as mica-rich lamprophyres, such as minettes. There is no consensus among geologists as to the origin of the lamprophyres and layered dykes and their involvement in trap magmatism. Hopefully, investigations of small intrusions of the Kamensky Province where lamprophyres are widely developed will make things clear in some way. The results of these investigations will be published in the near future. By analogy with the Kamensky Province, the formation of lamprophyres in the Noril'sk Region can be assumed to be due to fluid–magmatic differentiation of a trachybasalt melt and separation of a mafic liquid enriched in volatiles and alkalis from the melt (Ryabov et al. 1996b).

[1] "Designates that data of these researches have been taken from information of NKGRE."

References

Almukhamedov AI, Medvedev AY (1982) Geochemistry of sulfur in evolution processes in mafic magmas (ed Chernyshov LV). Nauka, Moscow, 148 p (in Russian)

Arkhipova AI (1975) Geochemical characteristic of intrusive traps of the Noril'sk plateau. Trudy NIIGA (ed Kavardin GI). Nedra, Leningrad, 135 p (in Russian)

Arkhipova AI, Dodin DA (1963) New data on subalkaline trap magmatism in the northwestern part of the Siberian Platform (Khraelakh mountains). In: Datsenko VA (ed) Geology of north-west of the Siberian Platform. Gosgeoltekhizdat, Moscow, pp 153–167 (in Russian)

Arkhipova AI, Nachinkin NG (1964) Lower Fokinsky differentiated intrusion at the western margin of the Noril'sk plateau. Uchenye Zapiski NIIGA, Regionalnaya Geologiya 3:51–69 (in Russian)

Arkhipova AI, Natorkhin IA (1971) Thermodynamic analysis of processes of formation of coal-bearing serpentine rocks in aureole of the Talnakh intrusion. In: Geology of mineral resources of the Noril'sk region. Nedra, Leningrad, pp 98–101 (in Russian)

Avgustinchik IA (1981) On composition of sulfide mineralization of the Lower Talnakh intrusion. In: Likhachev AP (ed) Genesis and localization conditions of copper-nickel mineralization, vol 162. Trydy TSNIGRI, Moscow (in Russian)

Bazhenov IK, Indukaev YN, Yakhno AV (1959) Native iron in gabbrodolerites of the Kureika River (Krasnoyarsk Territory). Zapiski VMO 88(2):180–184 (in Russian)

Bogatikov OA, Mikhailov NP, Gon'shakova VI (eds) (1981) Classification and nomenclature of igneous rocks. Nedra, Moscow, 160 p

Bulgakova EI (1971) Temperature conditions of formation of Noril'sk differentiated intrusions. Unpublished Ph.D. thesis. Institute of Geology and Geophysics, Novosibirsk, 30 p (in Russian)

Bushkanets YuS, Nevskaya AV, Belyakov GD (1970a) Dykes of traps in the interfluves of Kystyrtakh-Bol'shoy Avam Rivers. In: Urvantsev NN (ed) Geology and mineral resources of north-west of the Siberian Platform. Nedra, Leningrad, issue 2, pp 124–139 (in Russian)

Bushkanets YuS, Nevskaya AV, Belyakov GD (1970b) Kamenskaya magmatic province of alkalic rocks. In: Egorov LS (ed) Carbonatites and alkaline rocks of the northern of Siberia. Nedra, Leningrad, pp 157–169 (in Russian)

Campbell IH, Griffiths RW (1990) Implications of mantle plume structure for the evolution of flood basalts. Earth Planet Sci Lett 99:79–93

Champness PE (1970) Nucleation and growth of iron oxides in olivines, $(Mg, Fe)_2SiO_4$. Mineral Mag 37(291):790–800

Chukhrov FV, ed-in-chief (1967) Minerals: handbook, vol II, issue 3. Nauka, Moscow, 675 p (in Russian)

Daly R (1933) Igneous rocks and depths of the Earth. McGraw-Hill, New York/London, 533 p

Distler VV, Kunilov VE (eds) (1994) Geology and ore deposits of the Noril'sk Region. VII Internat Platinum Symposium Guidbook. Moscovskiy Contact Moscow, 67 p (in Russian)

Distler VV, Smirnov LV, Grokhovskaya TL, Filimonova AA, Muravitskaya GN (1979) Stratification, invisible layering of trap intrusions and conditions of formation of sulfide mineralization. In: Smirnov VI (ed) Conditions of formation of magmatic ore deposits. Nauka, Moscow, pp 211–269 (in Russian)

Dodin DA, Batuev BN, Mitenkov GA, Izoitko VM (1971) Atlas of rocks and ores of Noril'sk copper-nickel deposits. Nedra, Leningrad, 560 p (in Russian)

Dodin DA, Batuev BN (1971) Geology and petrology of Talnakh differentiated intrusions and their metamorphic aureole. In: Urvantsev NN (ed) Petrology and Metallogeny of Talnakh and Noril'sk differentiated intrusions. Nedra, Leningrad, pp 31–100 (in Russian)

Dodin DA, Sadikov MA (1967) Some aspects of trap differentiation by the example of Kharaelakh mountains. In: Urvantsev NN (ed) Petrology of traps of the Siberian platform. Nedra, Leningrad, pp 141–152 (in Russian)

Dodin DA, Sadikov MA, Shatkov VA (1982) Geochemical exploration criteria for copper-nickel deposits (ed Urvantsev NN). Nedra, Leningrad, 168 p (in Russian)

Dodin DA, Chernyshov NM, Polferov DV, Tarnovetsky LL (1994). Platinum metal deposits of the world. Vol 1: Low-sulfide platinum metal deposits in rhythmically-layered complexes. Geoinformmark, Moscow, 279 p (in Russian)

Duzhikov OA (1988) Copper-zeolite ore formation. In: Geology and metalogeny of sulfide deposits of Noril'sk Region USSR. Nauka, Moscow, pp 173–179 (in Russian) [refer to English translation of 1988 Russian edition by Duzikov OA, Distler VV, Strunin VM, Mkrtychyan AK, Sherman ML, Sluzhenikin SF, Lurje AM. Soc Econ Geol Spec Pub, 1:242 p]

Duzhikov OA, Distler VV, Sherman MK, Sluzhenikin SF (1988) Sulfide copper-nickel ore formation. In: Geology and Metallogeny of Sulfide Deposits of Noril'sk region USSR. Nauka Moscow, pp 77–172 (in Russian) (refer to English translation of 1988 Russian edition by Duzikov OA, Distler VV, Strunin VM, Mkrtychyan AK, Sherman ML, Sluzhenikin SF, Lurje AM. Soc Econ Geol Spec Publ, No. 1, 242 p)

Eliseev EN (1959) Geochemistry of the main sulfide copper-nickel provinces of USSR. In: Lazarenko EK (ed) Problems of geochemistry, issue 1. Lviv Univer Pub, Lviv, pp 5–184 (in Russian)

Fedorenko VA (1994) Evolution of magmatism as reflected in the volcanic sequence of the Noril'sk region. Proc of the Sudbury-Noril'sk Symposium (eds PC Lightfoot, AJ Naldrett). Ontario Geol Surv Spec Pub 5:171–184

Fedorenko VA, Stifeeva GT, Makeeva LV, Sukhareva MS, Kuznetsova NP (1984) Mafic and alkaline-mafic intrusions of the Noril'sk region in connection with their comagmatism to effusive formations. Geologiya i Geofisika 5:56–63 (in Russian)

Fenogenov AN (1987a) Petrological role of immiscibility in generation of igneous rocks. Izv vysshikh uchebnykh zavedeniy: Geologiya i Razvedka 2:25–39 (in Russian)

Fenogenov AN (1987b) Petrological role of immiscibility in generation of igneous rocks. Izv vysshikh uchebnykh zavedeniy: Geologiya i Razvedka 4:31–40 (in Russian)

Genkin AD, Distler VV, Laputina IP (1979) Chromite mineralization of differentiated trap intrusions. In: Conditions of formation of magamatic ore deposits. Nauka, Moscow, pp 105–127 (in Russian)

Genkin AD, Distler VV, Gladyshev GD, Filimonova AA, Evstigneeva TL, Kovalenker VA, Laputina IP, Smirnov VA, Grokhovskaya TL (1981) Copper-nickel sulfide ores from Noril'sk ore deposits (ed Shadlun TL). Nauka, Moscow, 234 p (in Russian)

Godlevsky MN (1959) Traps and ore-bearing intrusions of the Noril'sk region. Gosgeoltekhizdat, Moscow, 68 p (in Russian)

Godlevsky MN (1961) Review of hypotheses concerning origin of sulfide copper-nickel deposits. Trudy, VSEGEI, Leningrad, New Series, issue 45: pp 71–82 (in Russian)

Golubkov VS (1970) Mesozoic group of formations of near-fault metasomatism in the Yenisei ore province. Geology and mineral resources of north-western part of the Siberian Platform. Nedra, Leningrad, pp 23–49 (in Russian)

Goode ADT (1974) Oxidation of natural olivines. Nature 248 (5448):500–501

Gorbachev NS (1989) Fluid-magmatic interaction in sulfide-silicate systems (ed Zyryanov VN). Nauka, Moscow, 126 p (in Russian)

Goryainov IN (1969) On genesis of layered intrusions by example of the Talnakh massif. In: Zaridze GM (ed) Magmatism, metamorphism and metasomatism. Central Committee Communist Party of Georgia Pub, Tbilisi, pp 47–61 (in Russian)

Goverdovskaya TG (1971a) Pyasino-Vologochansky differentiated intrusion. In: Geology and mineral recourses of the Noril'sk region. Norilsk, pp 84–85 (in Russian)
Goverdovskaya TG (1971b) New data on the morphology of Zubovsky differentiated intrusion. In: Geology and natural recourses of the Noril'sk region. Noril'sk, pp 85–87 (in Russian)
Haggerty SE, Barker I (1967) The alteration of olivine in basaltic and associated lavas. Contrib Mineral Petrol 16(3):233–257
Igrevskaya LV (1981) New branch of the Noril'sk-I intrusion. In: Genesis and localization conditions of copper-nickel mineralization. TSNIGRI, Moscow, pp 79–85 (in Russian)
Ivanov MK, Ivanova TK, Shatkov VA (1971a) Some facial peculiarities of formation of intrusive and effusive picrites. In: Urvantsev NN (ed) Geology and mineral resources of the Noril'sk region, NIIGA, Leningrad, pp 83–84 (in Russian)
Ivanov MK, Ivanova TK, Tarasov AV, Shatkov VA (1971b) Peculiarities of petrology and mineralization of differentiated intrusions of the Noril'sk ore junction (Noril'sk-I, Noril'sk-II and Mt. Chernaya deposits). In: Urvantsev NN (ed) Petrology and metallogeny of the Talnakh and Noril'sk differentiated intrusions. Nedra, Leningrad, pp 197–304 (in Russian)
Ivanova TK (1975) Application of structural-facial analysis in volcanic reconstructions in the Noril'sk region, In: Urvantsev NN (ed) Copper-nickel ores at the northwest of the Siberian Platform. NIIGA, Leningrad, pp 52–75 (in Russian)
Kavardin GI (1976) Metallogeny of the northwestern part of the Siberian Platform (ed Urvantsev NN). Nedra, Leningrad, 158 p (in Russian)
Kavardin GI, Mitenkov GA (1971) Copper-nickel ores of the Talnakh deposit. In: Geology and Metallogeny of the Talnakh and Noril'sk Differentiated Intrusions. Nedra, Leningrad, pp 123–181 (in Russian)
Kavardin GI, Staritsyna GN, Golubkov VS, Goryainov IN, Kravtsova LI (1968) Traps of the Yenisei ore province (ed Urvantsev NN). Nedra, Leningrad, 208 p (in Russian)
Komarova MZ (1968) On granitoid intrusions in the Noril'sk region. Geologiya i Geofisika 5:104–108 (in Russian)
Komarova MZ (1974) Intrusive magmatism of the northern part of the Noril'sk plateau. Unpublished Ph.D. thesis, VSEGEI, Norilsk, 22 p (in Russian)
Komarova MZ (1983) Generation of small intrusions of sub-alkaline composition at the northwest of the Siberian Platform. In: Prediction and estimate of the nickel content in the north Siberian Platform. Leningrad, pp 83–94 (in Russian)
Komarova MZ, Lyul'ko TP (1967) On subdivision of trap intrusions of the Noril'sk region. In: Petrology of traps of the Siberian Platform. Nedra, Leningrad, pp 43–54 (in Russian)
Kopylova AG, Oleinikov BV (1997) Noble metals in native iron of continental basites. In: Geological structure and mineral resources of Sakha Republic (Yakutiya), vol 3. IG Yakutsk Affiliated Branch SB RAS, Yakutsk, pp 56–58 (in Russian)
Koroleva OV, Oleinikov BV (1998) Geochemistry and genesis of monzonitoids of the Dzaltul'sky trap intrusive. Geologiya i Geofisika 39(2):178–189 (in Russian)
Korovyakov IA, Nelyubin AE, Raikova ZA, Khortova LK (1963) Origin of the Noril'sk trap intrusions hosting sulfide copper-nickel ores. In: Trudy VIMS (ed Gon'shakova VI), New Series, issue 9, Gosgeoltekhizdat, Moscow, 102 p (in Russian)
Kravtsova LI (1958) Differentiated intrusion in the basin of the Dzhaltul River in the Yenisei River region. NIIGA Inf Bull 9:53–55 (in Russian)
Likhachev AP (1965) Role of leucocratic gabbro in the generation of Noril'sk differentiated intrusions. Izv Akad Nauk SSSR Geol Ser 10:75–89 (in Russian)
Likhachev AP (1988) Genetic models of sulfide-nikeliferous formations and their relation to other endogenous formations. In: Obolensky AA, Sotnikov VI, Sharapov VN (eds) Ore formation and genetic models of endogenous ore formations. Nauka, Novosibirsk, pp 158–165 (in Russian)
Likhachev AP (1996) Toward dynamics of emplacement of Talnakh ore-bearing intrusions and related platinum–copper–nickel deposits. Otechestv Geol 8:20–26 (in Russian)
Likhachev AP (1998) The Talnakh intrusion and its platinum–copper–nickel ores. Rudy i metally 1:36–46 (in Russian)
Lodochnikov VN (1946) About genetic significance of principal basalt and diabase structures, vol 7. Materials of All-Union Geological Institute (VSEGEI), Leningrad, pp 127–155 (in Russian)
Loewinson-Lessing FYu, Struve EA (1963) Petrographic dictionary (eds Afanas'ev GD, Petrov VP, Ustiev EK). Gosnauchtekhizdat, 3rd edn. Moscow, 447 p (in Russian)
Lurje ML, Masaitis VL, Polunina LA (1962) Intrusive traps of the western margin of the Siberian platform. In: Afanas'ev GD (ed) The petrography of the Eastern Siberia, vol 1. AS USSR, Moscow, pp 5–70 (in Russian)
Lurje ML, Ledneva VP, Selivanovskaya TV, Semenov LS, Tuganova EV, Ryabchenko AA, Komarova MZ, Staritsyna GN, Tomanovskaya YuI (1976) Structures of traps of the Siberian platform (ed ML Lurje). Trudy VSEGEI, New Series, vol 235. Nedra, Leningrad, 171 p (in Russian)
Lyul'ko VA, Nesterovsky VS, Goverdovskaya TG (1972) Magmatogenic breccias of nickeliferous trap intrusions. In: Urvantsev NN (ed) Copper-nickel ores of the Talnakh ore junction. Nedra, Leningrad, pp 123–127 (in Russian)
Marakushev AA (ed) (1976) Petrography: Part I. Moscow Univ Pub, Moscow, 384 p (in Russian)
Masaitis VL (1958) Petrology of the Alamdzhakh trap intrusion (the Vilyuy River basin) (ed Lurje ML). Trudy VSEGEI Leningrad, New Series, No 22, 133 p (in Russian)
Masaitis VL (1964) Magmatic trap subprovinces of the Siberian Platform. In: Transactions for geology of the Eastern Siberia. VSEGEI, Leningrad, pp 137–158 (in Russian)
Mitenkov GA, Khineiko AL, Sishkin NN (1977) Peculiarities of the taxitic horizon structure of the Talnakh ore-hosting intrusion and their genetic significance. Doklady Akad Nauk SSSR 237 (1):191–194 (in Russian)
Morimoto N (1989) Nomenclature of pyroxenes. Can Min 27:143–156
Natorkhin IA, Arkhipova AI, Batuev BN (1977) Petrology of Talnakh intrusions (ed Egorov LS). Nedra, Leningrad, 236 p (in Russian)
Nemenenok TI (1972) Some regularities of the spatial relationship between volcanoes of the central type and mafic-ultramafic nickeliferous massifs (Noril'sk region). Geologiya i Geofisika 5:135–139 (in Russian)
Nemenenok TI (1982) The roots and conduits of diatremes and lava sheets of the northwestern Siberian Platform. Sov Geol Geophys 23 (8):24–31
Nitsan U (1974) Stability field with respect to oxidation and reduction. J Geophys Res 79(5):706–711
Oleinikov BV (1979) Geochemistry and ore genesis of platform basites (ed Bazhenov AI). Nauka, Novosibirsk, 264 p (in Russian)
Oleinikov BV, Pankov VYu (1985) Melted inclusions and minerals-prisoners in products of basalt melt protocrystallization. In: Petrological–geochemical features of the deep evolution of a-substance from kimberlitic and mafic magmatic systems. Yakutsk Affiliated Branch AS USSR Pub, Yakutsk, pp 130–164 (in Russian)
Oleinikov BV, Okrugin AV, Tomshin MD, Levashov VK, Varganov AS, Kopylova AG, Pankov VYu (1985) Native metal formation in platform basites. Yakutsk Affiliated Branch AS USSR Pub, Yakutsk, 188 p (in Russian)
Oleinikov BV, Kopylova AG, Pankov VYu (1991) Mineral associations of native iron in apodoleritic metasomatites of the Dzhaltaksky intrusion. In: Oleinikov BV (ed) Native mineral formation during magmatic process. Yakutsk Affiliated Branch AS USSR Pub, Yakutsk, pp 29–47 (in Russian)

Rogover GB (1959) The Noril'sk-I deposit. Gosgeoltekhizdat, 168 p (in Russian)

Ryabov VV (1969) About origin of taxitic gabbrodolerites, leucocratic gabbro, and magmatic breccia in the Kharaelakh branch of the Talnakh intrusion. Geologiya i Geofisika 2:51–58 (in Russian)

Ryabov VV (1980) New prospective platinum-bearing horizon in the upper endocontact zone of Noril'sk intrusions (Funds of Krasnoyarsk VTS SB AS USSR. Program "Siberia", block 1.7.1.1). Krasnoyarsk, 3 p (in Russian)

Ryabov VV (1982) About concentration of chromium and platinoids at the roof of layered intrusions of the Noril'sk type. Doklady Akad Nauk SSSR 2(266):466–469 (in Russian)

Ryabov VV (1984b) Peculiarities of mineralogy of magnesian basites in the Noril'sk region. In: Sobolev VS (ed-in-chief) Magnesian basites of the western Siberian platform and aspects of their nickel content. Nauka, Novosibirsk, pp 150–158 (in Russian)

Ryabov VV (1984a) On composition of upper contact zones of Noril'sk intrusions hosting chromite-rich mineralization. In: Polyakov GV (ed) Petrochemistry: criteria for mineralzation of magmatic complexes. IGG SB AS USSR, Novosibirsk, pp 124–142 (in Russian)

Ryabov VV (1989a) Immiscibility in natural glasses (by example of traps) (ed Zolotukhin VV). Nauka, Novosibirsk, 223 p (in Russian)

Ryabov VV (1989b) Genetic types of high-magnesian traps. In: Oleinikov BV (ed) Mafic magmatism of the Siberian platform and its metallogeny. Yakutsk, pp 78–79 (in Russian)

Ryabov VV (1989c) Chemical composition of immiscible liquids in natural glasses from traps. Geologiya i Geofisika 11:69–78 (in Russian)

Ryabov VV (1992a) Olivines of Siberian traps as indicators of petrogenesis and ore formation (ed Zolotukhin VV). Nauka, Novosibirsk, 116 p (in Russian)

Ryabov VV (1992b) Types of low-sulfide of platinum-group metal ores in contact zones of trap intrusions. In: Distler VV, Evstigneeva TL, Kamshilina EM (eds) Geology and genesis of platinum metal deposits. IGEM Publication, Moscow, pp 40–41 (in Russian)

Ryabov VV (1992c) Nontraditional types of platinum metal ores in the Noril'sk-type sulfide deposits. Russ Geol Geophys 33(11):79–88 (in Russian)

Ryabov VV (1994) Platinum of Siberian traps. UIGGM SB RAS, Novosibirsk, 32 p (in Russian)

Ryabov VV (1999a) Ore geology of the Noril'sk region on the threshold of millennium. In: Simonov ON (ed) Abstracts of the regional symposium: trends in geological-exploration works on nickel, copper and platinoids in the Taimyr Autonomous Area in 2000–2005. Taimyrkomprirodresursy, Noril'sk, pp 25–26 (in Russian)

Ryabov VV (1999b) Fluid regime of trap magmatism and ore formation (petrological aspect). Russ Geol Geophys 40(10):1437–1452

Ryabov VV, Anoshin GN (1999) Platinum–iron mineralization in intrusive traps of the Siberian Platform. Russ Geol Geophys 40 (2):163–176

Ryabov VV, Yakobi NYa (1981) Behavior of chrome during differentiation of trap magma. In: Sobolev VS (ed) Problems of genetic petrology. Nauka, Novosibirsk, pp 85–111 (in Russian)

Ryabov VV, Zolotukhin VV (1970) About pseudotachylites of the Talnakh trap intrusion in the Siberian Platform. Geologiya i Geofisika 3:56–63 (in Russian)

Ryabov VV, Zolotukhin VV (1977) Minerals of differentiated traps (ed Sobolev VS). Nauka, Novosibirsk, 392 p (in Russian)

Ryabov VV, Bakumenko IT, Fominykh IM (1977) Dendritic metacrysts in traps of the Noril'sk region and some aspects of their formation. In: Sobolev VS (ed) Proc on petrology and mineralogy. Nauka, Novosibirsk, pp 47–71 (in Russian)

Ryabov VV, Shevko AY, Simonov ON, Anoshin GN (1996a) Composition of Pt-bearing Cr-rich skarns in Talnakh (Norrilsk region). Russ Geol Geophys 37(7):57–72 (in Russian)

Ryabov VV, Shevko AYa, Lyul'ko VA, Dushatkin AB (1996b) Compositional diversification of rocks in small intrusions as a consequence of magmatic differentiation of trachybasalt melt. In: Magmatism and geodynamics of Siberia. Abstracts of scientific conference in honor of 75 anniversary of MP Kortusov. Tomsk, pp 80–81 (in Russian)

Ryabov VV, Shevko AYa, Anoshin GN, Simonov ON, Krakovetsky YuK (1996c) Dyumtaley as a new classical example of a layered mafic intrusion with sulfide and magnetite mineralization. In: Abstracts of Ist internat symposium "Large and Unique Deposits of Rare and Noble Metals (problems of genesis and implementation)", St Petersburg, 8–11 Oct 1996, pp 77–79 (in Russian)

Sharapov VN, Golubev VS (1976) Dynamics of interaction between magma and rocks. Nauka, Novosibirsk, 237 p (in Russian)

Sluzhenikin SF, Distler VV, Duzhikov OA, Kravtsov VF, Kunilov VE, Laputina IP, Turovtsev DM (1994) Low-sulfide minertalization in the Noril'sk differentiated intrusions. Geologiya rudnykh mestorozhdeniy 36(3):195–217 (in Russian)

Smirnov MF (1966) Structure of the Noril'sk nickeliferous intrusions and sulfide ores. Nedra, Moscow, 58 p (in Russian)

Sobolev VS (1986) Petrology of traps: Izbrannye trudy. Nauka, Novosibirsk, 209 p (in Russian)

Staritsyna GN (1971) Acid and subalkaline differentiates of traps and their genesis. In: Traps of the Siberian Platform and their metallogeny (Abstracts), Irkutsk, pp 38–40 (in Russian)

Stepanov VK (1977) Olivines and hypersthenes of the Talnakh intrusion. In: Geology, petrology and geochemistry of copper-nickel deposits. Trudy TSNIGRI, vol 127. TSNIGRI, Moscow, pp 17–34 (in Russian)

Stepanov VK (1981) Dynamic model of emplacement, crystallization and mineralization of the Noril'sk ore-hosting intrusions. In: Genesis and localization conditions of copper-nickel mineralization. Trudy TSNIGRI vol 162. TSNIGRI, Moscow, pp 13–19 (in Russian)

Stepanov VK, Zemskova GV (1988) Petrographic types of trap intrusions of the northwestern Siberian Platform and their metallogeny. In: Nickeliferous mafic-ultramafic complexes of the Noril'sk region. Apatites, pp 54–60 (in Russian)

Stetsenko NS (1966) Metasomatic breccias associated with ore-hosting differentiated gabbrodolerite intrusions. In: Abstracts of 1Vth Krasnoyarsk Territory conference, Krasnoyarsk, pp 37–39 (in Russian)

Strunin BM, Distler VV, Mkrtychyan AK, (1988) Copper-porphyry ore formation. In: Geology and metalogeny of sulfide deposits of Noril'sk region USSR. Nauka, Moscow, pp 205–224 [refer to English translation of 1988 Russian edition by Duzikov OA, Distler VV, Strunin VM, Mkrtychyan AK, Sherman ML, Sluzhenikin SF, Lurje AM. Soc Econ Geol Spec Pub, 1:242 p]

Sukhanova EN (1968) About internal structure, chemical behavior and metalogeny of the northwestern (Kharaelakhsky) branch of the Talnakh intrusion. In: Geology and mineral deposits of the Noril'sk Mining region. Noril'sk, pp 112–114 (in Russian)

Sukhareva MS, Kuznetsova NP (1983) Toward a question of relation between differentiated intrusions of the Talnakh ore junction (by example of northern flanks). In: Abstracts of trap magmatism of the Siberian Platform with respect to its tectonics and exploration of mineral deposits. Krasnoyarskgeologiya Pub, Krasnoyarsk, pp 89–92 (in Russian)

Tarasov AV (1976) On formation mechanisms of the Noril'sk intrusions and related sulfide bodies. In: Pospelov GL (ed) Replacement and intrusion during magmatism and ore formation. Nauka, Novosibirsk, pp 123–276, in Russian

Tsvetkov NI (1951) Isomorphous substitutions in the group of alkalifree pyroxenes. Trudy Instita Geol Nauk AS SSSR, issue 138. AS SSSR Pub, Moscow, 108 p (in Russian)

Turovtsev DM (1970) Conditions of generation of contact hornfels formations in aureoles of differentiated trap intrusions in the

Talnakh ore deposit (Noril'sk region). In: Geology and petrology of intrusive traps of the Siberian Platform. Nauka, Moscow, pp 211–232 (in Russian)

Turovtsev DM (1972) Regularities of spatial location of metamorphic and metasomatic formations within contact aureoles of trap intrusions of the Talnakh ore junction. Unpublished PhD thesis, IGEM, Moscow 25 p (in Russian)

Turovtsev DM (1979) Metamorphism at contacts with traps of the Imagdinsky ore junction (Noril'sk region) in connection with development of metamorphic criteria for exploration. Trudy TSNIGRI 4:48–66 (in Russian)

Urvantsev NN (1959) The Yenisei ore field. Collected papers on Arctic region geology. Trudy NIIGA (ed Markov FG), vol 102, issue 10. Nedra, Leningrad, pp 28–48 (in Russian)

Vasil'ev YuR (1966) Petrology and structural features of complicate-differentiated Noril'sk-II trap intrusion. In: Geology and petrology of intrusive traps of the Siberian Platform. Nauka, Moscow, pp 101–116 (in Russian)

Vasil'ev YR (1970) Chromous clinopyroxene from Lower Fokinsky trap intrusion (north-west of the Siberian Platform). Geologuya I Geofisika 7:130–134 (in Russian)

Vilensky AM (1967) Petrology of intrusive traps at the north of the Siberian platform (ed Lebedev AP). Nauka, Moscow, 270 p (in Russian)

Vilensky AM (1974) Physicochemical conditions of deep-seated and interchamber petrogenesis of mafic trap melts. Sov Geol Geophys 15(3):27–32

Vilensky AM (1978a) Peculiarities of interchamber differentiation of trap intrusions and some general features of magmatism of the Noril'sk region. In: Sobolev VS (ed) Criteria for Exploration of the Noril'sk-type sulfide ores. Nauka, Novosibirsk, pp 30–44 (in Russian)

Vilensky AM (1978b) Intrusive magmatism. In: Zolotukhin VV, Vilensky AM (eds) Petrology and perspectives of metallogeny of traps at the north of the Siberian platform. Nauka, Novosibirsk, pp 61–121 (in Russian)

Vilensky AM (1984) Mafic formations of the Siberian Platform and Taimyr. In: Sobolev VS (ed-in-chief) Magnesian basites of the western Siberian Platform and aspects of their nickel content. Nauka, Novosibirsk, pp 17–47 (in Russian)

Vortsepnev VV (1978) Conditions of formation of the Talnakh ore-bearing intrusion on the base of data on inclusions. Geologiya rudnykh mestorozhdeniy 2:64–70 (in Russian)

Wager LR, Brown GM (1968) Layered igneous rocks. Oliver and Boyd, Edinburgh/London, 588 p

Yudina VV (1965) Some metamorphism and metasomatism phenomena associated with the Talnakh gabbrodolerite intrusion (Noril'sk region). In: The proportion between magmatism and metamorphism in genesis of ultrabasites. Nauka, Moscow, pp 112–175 (in Russian)

Zelenshchikov GV, Duzhikov OA (1974) On comparative characteristcs of basalts of Voronezhsky anteclise and effusive traps in the northwestern Siberian Platform. In: Abstracts of the third All-Union conference "The state and trends in investigations on trap metallogeny", KrTGU, Krasnoyarsk, pp 43–45 (in Russian)

Zemskova GV (1981) Petrographic characteristic of the "Lower Talnakh-type" intrusions (Noril'sk region). In: Genesis and conditions of localization of copper-nickel mineralization. TSNIGRI, Moscow, pp 28–34 (in Russian)

Zolotukhin VV (1964) Main regularities of prototectonics and aspects of origin of ore-bearing trap intrusions (ed Sobolev VS). Nauka, Moscow, 192 p (in Russian)

Zolotukhin VV (1991) Input of alkalis as a factor of diversity of natural picritic basalt melts. Geologiya i Geofisika 3:90–102 (in Russian)

Zolotukhin VV (1997) Mafic pegmatoids of the Noril'sk ore-bearing intrusions and the problem of genesis of the Platinum-copper-nickel mineralization of the Noril'sk type (ed Polyakov GV). SB RAN NITS OIGGM, Novosibirsk, 88 p (in Russian)

Zolotukhin VV, Oleinikov BV (1963) About acid hybrid rocks from the Gorbiachin River region (Siberian platform). In: Transactions on genetic and experimental mineralogy, vol 1. Nauka, Novosibitsk, pp 80–106 (in Russian)

Zolotukhin VV, Ryabov VV (1972) The main peculiarities of metamorphic and metasomatic alterations of the Talnakh deposit. In: Problems of petrology of mafic and ultramafic rocks. Nauka, Moscow, pp 218–269 (in Russian)

Zolotukhin VV, Shchedrin NF (1977) Differentiated intrusions of the Imangdinsky ore junction (ed Sobolev VS). Nauka, Novosibirsk, 135 p (in Russian)

Zolotukhin VV, Vasil'ev YR (1967) Peculiarities of formation mechanism of ore-bearing trap intrusions at the northwest of Siberian platform (ed Sobolev VS). Nauka, Moscow, 231 p (in Russian)

Zolotukhin VV, Vasil'ev YR (1987) Interrelations of ultramafic, mafic and alkaline-mafic magmatism (on the example of the Siberian Platform). Sov Geol Geophys 28(1):35–43

Zolotukhin VV, Ryabov VV, Vasil'ev YR, Shatkov VA (1975) Petrology of the Talnakh ore-bearing differentiated intrusion (ed Sobolev VS). Nauka, Novosibirsk, 432 p (in Russian)

Zolotukhin VV, Vilensky AM, Duzhikov OA (1986) Basalts of the Siberian platform (eds Sobolev VS, Sobolev NV). Nauka, Novosibirsk, 255 p (in Russian)

Zolotukhin VV, Vasil'ev YuR, Duzhikov OA (1989) Diversity of traps and initial magmas. Nauka, Novosibirsk, 247 p (in Russian)

Zotov IA (1976) Some peculiarities of interaction between trap magmas of Talnakh intrusions (Noril'sk region) and country rocks. In: Ocherki geneticheskoy petrologii. Nedra, Moscow, pp 250–260 (in Russian)

Zotov IA (1979) Genesis of trap intrusions and metamorphic formations of the Talnakh (ed Korzhinsky DS). Nauka, Moscow, 155 p (in Russian)

Zotov IA, Krivoplyasov GS (1977) Signs of two-phase nature of Talnakh gabbrodolerites intrusions (by example of Talnakh intrusive body). Doklady Akad Nauk SSSR 232(6):1383–1386 (in Russian)

4 Mineralogy and Geochemistry of Traps

Abstract

Variations in mineralogy and compositions of principal rock-forming minerals serve as an indicator of basaltic magma differentiation. Features of the composition of olivine, pyroxenes, plagioclase, spinel group minerals, micas, and ilmenite are described in Sect. 4.1. The compositions of olivine and plagioclase in layered intrusions vary from forsterite to fayalite and from albite to anorthite, respectively. Pyroxene is present as augite of the varying composition; pigeonite and orthopyroxene are less abundant. Minerals of the spinel group are represented by a series of spinel, chromite, and magnetite. Chromespinellides form accessory impregnations and accumulations in gabbrodolerites and pegmatoids. Micas occur as phlogopite–annite with the Fe/(Fe + Mg) ratio ranging from 1 to 90 at%.

In Sect. 4.2, features of the geochemistry of Cr, PGE, Ni, Ti, and S in basalts, barren, and ore-bearing intrusions with the different degree of differentiation are described. Features of the PGE and Cr distribution in sequences of ore-bearing horizons and in various rock types are considered. In the geochemistry of nickel, the main attention is drawn to its distribution in rocks and minerals, as well as to behavior of nickel in the process of sulfide formation. The Ti content in traps and its variations in the sequences of layered intrusions depend on the initial composition of a basaltic melt and the extent of its differentiation. In the geochemistry of S, the main attention is drawn to the sulfur isotopic composition in sulfides and sulfates of the Talnakhsky ore junction.

Results of investigations into the material composition of traps from the Noril'sk Region are reported in many articles and monographs, and a large body of petrochemical material has been accumulated concerning the variety of magmatic rocks from the Noril'sk Region. These data allowed geoscientists to establish the main compositional features of rocks from different volcanic and intrusive complexes, to extensively map them, to reveal regularities in compositional changes in layered intrusions, to establish petrochemical criteria for the degree of melt differentiation, and to outline the probable composition of parental magma for various magmatic complexes. These issues have been repeatedly discussed, and, therefore, this chapter will focus on theoccurrence and compositional features of major rock-forming minerals and geochemistry of chromium, platinumgroupelements, nickel, titanium, and sulfur, which are highly informative for solving genetic problems.

4.1 Major Rock-Forming Minerals of Traps: Indicators of Melt Differentiation, Petrogenesis, and Ore Formation

The variety of magmas in traps is the result of changes in the amount and composition of minerals of three series of solid solutions: olivine, pyroxene, and plagioclase. Traps also contain minor varying amounts of quartz, potassium feldspar, biotite, amphibole, titanomagnetite, chrome spinel, ilmenite, sulfide, apatite, biotite, calcite, zeolite, and also

V.V. Ryabov et al., *Trap Magmatism and Ore Formation in the Siberian Noril'sk Region: Volume 1. Trap Petrology; Volume 2. Atlas of Magmatic Rocks*, Modern Approaches in Solid Earth Sciences 3, DOI 10.1007/978-94-007-5022-7_4,

secondary minerals represented by bowlingite–iddingsite, serpentine, chlorite, talc, prehnite, and chlorite-like minerals (Ryabov and Zolotukhin 1977). Investigations of the genetic mineralogy of traps: morphogenesis of some minerals and mineral associations, morphology of crystals and structural intergrowths of different phases, and distribution of major and trace mineral-forming components provides valuable information for the solution of petrogenetic and ore formation problems.

Thorough genetic petrography and mineralogical–-geochemical studies related to the formation of trap rocks are scarce in the literature. Petrographic studies show that the vitreous textures of effusive rocks and intrusion contact facies result from rapid crystallization under decreasing magmatic melt temperature. A fine-grained doleritic texture is generated under supercooling, which leads to the origin of numerous nucleation centers. Porphyritic textures suggest that melt crystallization initiated with the formation of phases redundant relative to the eutectic. However, opinions as to the mechanism of formation of porphyry phenocrysts differ. The majority of geoscientists suggest that porphyry segregations and clusters are the products of deep-seated crystallization. There is also the opinion that phenocrysts were formed during in situ devitrification of the melt (Frikh-Khar 1977; Ashikhmina et al. 1987).

Some genetic concepts, based on the protocrystallization of plagioclase, olivine, or their intergrowths, relate trap variety (including layered inclusions) to crystallization differentiation of magmatic melts in deep chambers. The important role of protocrystallization of plagioclase as a characteristic feature indicating evolution of deep melts of potentially ore-forming intrusions was discussed by Likhachev (1965, 1997) and Oleinikov (1979). The deep-seated origin of olivine crystals in traps is confirmed for many scientists as it agrees with existing genetic concepts on melt differentiation. Therefore, this problem has not been discussed, even though the crystal morphology, composition of olivine, and its inclusions do not support the idea of its protocrystallization (Sobolev 1986; Goryainov 1969; Ryabov 1992a).

4.1.1 Olivine

The wide quantitative variations of olivine (referred to as Ol hereafter) in traps (from 0 to 100%) and its isomorphous series from fayalite to forsterite (in spite of the limited component composition) make it a highly informative indicator of abyssal magma formation and melt differentiation in present-day chambers. The specific features of olivine composition are also used for solving ore-genetic and forecasting tasks when studying and prospecting of the Noril'sky-type deposits (Zolotukhin 1965; Stepanov 1975; Likhachev 1977; Ryabov 1978b).

4.1.1.1 Features of Olivine in Traps

The quantity of olivine in traps and its distribution through magmatic bodies are important features for establishing the initial melt composition and degree of differentiation. The overwhelming majority of effusive traps are olivine-free or olivine-bearing varieties of basalt, whereas olivine-rich picritic and olivine-phyric rocks are extremely rare. Within one sheet of picritic lava, the olivine content is variable in different flows. The content of Ol phenocrysts in mandelsteines varies from 12 to 30% (averaging 22–25%). The mineral content in the dense, picritic parts of the flow ranges from 35–45 to 55%, with olivine content invariable from top to bottom within a single flow. This fact suggests the absence of gravitational separation of Ol in lavas.

Vertically through trap intrusions, Ol content varies considerably. According to the character of these variations, several types of bodies can be distinguished. In most mafic intrusions, the Ol content gradually increases from the top downward through the body and reaches a maximum of 8–15% in olivine gabbrodolerite (Middle Dudinsky, Middle Ergalakhsky, Khungtukun, and Maymechinsky Intrusions) or 35–40% in picrite and troctolite gabbrodolerite (Mount Pegmatitovaya, Mount Zub, and Pyasinsko–Vologachansky Intrusions). Similar intrusions were described in other parts of the Siberian Platform (Lebedev 1957; Masaitis 1958; Vilensky et al. 1964; Staritsyna et al. 1972; Feoktistov 1984; Zolotukhin 1984a).

A considerable part of the Lower Fokinsky, Lower Talnakh Intrusions, and Morongvsky group of intrusions consist of olivine-rich rocks. In the Lower Fokinsky Intrusion, they are picritic gabbrodolerite, and in the Lower Talnakh Intrusion and Morongvsky group of intrusions, troctolite and less often picritic gabbrodolerite. Researchers have shown that these rocks from the Morongvsky group of intrusions (Korovyakov et al. 1963), as well as from the Lower Talnakh and Lower Fokinsky Intrusions, lack any features consistent with in-chamber gravitational separation of Ol. The elevated Ol concentrations in picritic gabbrodolerite from the Lower Fokinsky, Yergalachny Stream, and other intrusions result from the initial picritoid melt, which are suggested by the Ol-rich glassy rocks of contact facies. In addition, in the Magnitny Stream Intrusion, picritic gabbrodolerite forms a single horizon with a sharp contact with Ol-poor gabbrodolerite rim rocks. The Morongvsky group of intrusions, Lower Talnakh, and Lower Noril'sk Intrusions are composed mostly of olivine-rich rock, and near-basal parts of extended magmatic bodies are composed of both mafic and the Noril'sky-type rocks.

In the Noril'sky-type of intrusion, the Ol content increases smoothly from single grains to 12–15% content from the top to bottom in the mafic part of the intrusions and then within a 10-cm interval abruptly increases to 45–60% content in the picritic gabbrodolerite horizon (Rogover 1959; Smirnov 1966; Ivanov et al. 1971a, b). In ataxite and equigranular mafic pegmatoids of the UZ and LZ of these intrusions, one can observe wide variations in Ol content from olivine-free to 45–50% in melanocratic troctolite to olivinite (Dodin and Batuev 1971; Ivanov et al. 1971a, b; Lyul'ko et al. 1972; Zolotukhin et al. 1975; Ryabov 1984a, b; Ryabov and Pavlov 1984). The contact facies of the Noril'sk Intrusions are composed only of olivine-free gabbrodolerite or with sporadic olivine, which is at odds with the concept of a troctolite composition for the initial melt.

In the Khungtukun Intrusion, rhythmic gravitational settling of Ol led to the formation of three horizons of Ol-enriched (10–20%) rock. Picritic gabbrodolerites richest in Ol (50–60%) are found in the UZ of the intrusions (Ryabov 1992b), which is at odds with the theory of fractional differentiation. It is worth noting that the enrichment of rocks with olivine in the UZ of trap intrusions was observed by A. P. Lebedev (1957) in a number of sills of the Tungussky syneclise, in the Ozernaya Mountain Intrusion (Zolotukhin and Vasil'ev 1967), and upper picrites were described in the Upper Talnakh and Eastern Noril'sk Intrusions (Ivanov et al. 1971a, b; Ryabov 1984a, b) and in the Pegmatitovaya Mountain Intrusion.

Crystal Morphology

Crystal morphology is of interest for understanding the crystallization conditions of rocks. Olivine in traps forms five morphological types of crystals: (1) isometric and slightly extended with regular crystallographic faces (the most widespread type in Ol-poor rocks); (2) poikilitic; (3) idiomorphic with skeletal form elements (typically in picrites); (4) skeletal, palmate, amoeboid, and dendroid (in equigranular pegmatoids); (5) granules and polygonal grains as granoblastic aggregates, accumulations, stripes, and single grains (in ataxitic pegmatoids and magmatogenic breccia).

The crystal morphology of olivine in different rocks and intrusions has been described in many works. Most researchers suggest that olivine crystals in olivine-rich rocks, including picritic gabbrodolerite, are of intertelluric origin. Olivine is believed to crystallize in deep and intermediate chambers and gravitationally accumulate in the basal parts of hypabyssal chambers, forming layers of picritic gabbrodolerite. Highly magnesium intrusions, for example, the Lower Fokinsky, Morongovsky, and Lower Talnakh Intrusions, are assumed to be intruded by a mass of intertelluric olivine phenocrysts.

The granular shapes of olivine crystals in taxitic gabbrodolerite are considered by Zolotukhin (1964) to result from the blocking of large crystals of early olivine under the effect of fluids during the magma replacement of picritic gabbrodolerite. Goryainov (1969), after examining all variations in olivine crystal occurrence, comes to the conclusion that the entire horizon of picritic gabbrodolerite from the Noril'sk Intrusions has a metasomatic origin.

Important information on the amount of olivine, morphology, and crystal size in effusive and intrusive picritic basalt can be obtained from mandelsteines of picritic lavas and quenched olivine porphyries from the Lower Fokinsky Intrusion (see Rock indication 10 and 21). They are evidence that the picritic basalt melt was initially homogeneous and did not contain any intertelluric crystals. Olivine crystals formed during the outflow of melt to the surface and intrusion into hypabyssal chambers as a result of magma supercooling, which is suggested from the skeletal form and small (0.1–0.5 mm) size of crystals. Further growth of crystals to large sizes took place in chamber conditions, the drastic drop in temperature being followed by the formation of skeletal shapes even in large crystals and trapping fragments of matrix inside grains.

The granular form of olivine results from the decomposition and recrystallization of previously formed crystals under the influence of fluids. Experimental data (Almukhamedov and Medvedev 1982; Gorbachev 1989) suggest that fluids promoted not only recrystallization of olivine but also the transfer of magnesium and formation of granules of olivine, increased crystal growth, and formation of olivinite segregations.

Breakdown Structures of Solid Solutions

Breakdown structures of solid solutions in olivine have been described in rocks from different formations, but they were not mentioned in the mineralogic–petrographic literature on traps of the Siberian Platform before Ryabov (1992a). Studies showed that breakdown structures of solid solutions in olivine form with varying frequency in different intrusions. In the Upper Talnakh Intrusion, lamellae of breakdown structures were observed in olivines from olivine-bearing and olivine gabbrodolerite and are absent in picrite (drill hole KZ-274). In another part of the intrusion, breakdown structures form in olivine gabbrodolerite and at the top of the picrite horizon (drill holes KZ-585, KZ-586). In drill hole SV-16, which intersected the Upper and Lower Talnakh Intrusions, olivine with breakdown structures were found only in picrites of the Lower Talnakh Intrusions. Breakdown structures of solid solutions are typically absent in taxitic gabbrodolerite.

Breakdown structures of solid solutions under microscope are tabular microinclusions lying in the *Np–Nm* (100) plane, and less often (001), and are skeletal. In the (010) plane, perpendicular to the plane of optical networks, these flat skeletal crystals have a dash-and-dot form. The thickness of plates is 0.1–0.5 (μm); the length is 60–80 (μm). One crystal may contain breakdown structures oriented in two directions. In some cases, breakdown structures form chains of dotted lines whose shape is nearly a solid line.

Some idea of the morphology of exsolution structures can be gained from microphotographs of gabbrodolerite.

Tabular crystals of breakdown structures are dark brownish to black in color, opaque, and similar to magnetite, though data for other regions in the literature evidence that these might be pyroxenes and other minerals as well. Breakdown structures are arranged in the center of olivine grains as zones or stripes that can be traced throughout the crystal but are typically concentrated in one area. Under microscope, it is clear that in olivine grains breakdown structures form only in some of them and less often trap all crystals.

Serpentinized olivine forms looped or continuous and up to homoaxial pseudomorphs of antigorite. Flat crystals are at first preserved, but during the recrystallization of serpentine, the ore phase constricts into balls or veinlets. Breakdown structures also occur in dense-brownish pseudomorphs of bowlingite after olivine and disappear after replacement by serpentine or talc.

The composition of olivine featuring breakdown structures appears to be more magnesium than minerals without these structures. Attempts to determine the composition of intergrowths in olivine failed due to their small size.

Cleavage

Cleavage in olivine formed in traps is a widespread phenomenon, especially among olivine-rich rocks, particularly troctolitic gabbrodolerite of the Morongovsky Complex. The presence of exsolution structures is often preceded or accompanied by the appearance of a perfect cleavage. This typically forms in subidiomorphic crystals from olivine-rich rocks. In palmate or granular grains, a cleavage is absent.

Alteration

One of the highest temperature transformations of olivine in traps is replacement by ortho- and clinopyroxene, as observed in ankaramites from Arylakh. Most often, this is observed in olivine-rich gabbrodolerite as pyroxene rims around olivine. Coronate or kelyphitic structures are often observed in the gabbronorite of Zubovsky and Mountain Putanaya Intrusions, where pyroxenes form interrupted or continuous film overgrowths on olivine, and grow to large poikilocrysts with rounded intergrowths of olivine, both large and abundant, or small and scarce. Resorption of olivine and its overgrowth by spherocrysts of clinopyroxene is observed in basalts of the Mikchandinsky flow sequence.

The most widespread replacement product of olivine with high iron content is bowlingite, whereas olivine with a high magnesium content is replaced by serpentine–magnetite (Ryabov and Zolotukhin 1977). In the mandelsteines of picritic lavas, olivine is replaced by chalcedony, calcite, serpentine, iddingsite, or is opacitized. The character of opacitization of olivine in picritic basalt is very similar to the products observed in experiments of high-temperature oxidation of olivine, described by Haggerty and Barker (1967). In intrusive traps, Ol is normally replaced by brownish-green bowlingite, serpentine, and (/or) talc–magnetite.

The chemical composition of the alteration products of Ol reported in Ryabov (1992a) shows that the main components of newly formed minerals are 32.5–54.7 SiO_2, 5.3–40.5wt% MgO, 1.6–35.2wt% FeO, and 2.2–1.2wt% Al_2O_3. Elevated iron content is typical of bowlingite, whereas high alumina content is typical of amphibole and chlorite-like phases.

4.1.1.2 Chemical Composition of Olivine

This chapter is based on the results of X-ray spectroscopic analyses of olivine from effusive and intrusive traps. These are our published materials (Ryabov 1992b), and new data reported in the Rock indications and tables of analyses for key sections. To calculate the average values and to plot diagrams, we also used some literature data (Ryabov and Zolotukhin 1977; Genkin et al. 1981; Oleinikov and Pankov 1985; Ryabov et al. 1985a, b, c; Zolotukhin et al. 1986, 1989). To compare olivine from traps with olivine from other formations, we used data from Vasil'ev and Zolotukhin (1975), Velinsky and Bannikov (1986), and Zolotukhin et al. (1986).

Olivine from Lava Flows

The iron content of Ol from lava ranges from Fa_{10} to Fa_{62}. The varieties of olivine with the highest Fe content were found in titanium-augite dolerite basalts (Fa_{62}), in olivine-phyric basalt ($Fa_{27–40}$), and in plagioclase-phyric basalt ($Fa_{38–45}$) (Table 4.1). Olivine with the lowest Fe content was observed in mantle limburgite from the Kamensky Province ($Fa_{10–14}$). Here, the iron content of the mineral is Fa_{11} in augites, $Fa_{10–20}$ in picrite of the Ayansky lava flow, and $Fa_{19–26}$ in picritic basalt of the Noril'sk Region.

The maximum nickel content on the platform was observed in olivine from limburgite flows (Fa_{12}) that contain 0.447wt% NiO. The average oxide value was 0.36wt% (Fa_{22}) in picritic basalt from the Noril'sk Region, and 0.30wt% (Fa_{15}), in the Ayansky lava flow sequence (Table 4.1).

The highest calcium content was found in olivine from subalkaline dolerite (0.74wt% CasO), and the lowest, in olivine-phyric basalt (0.10wt% CaO). The MnO content in olivine regularly increases with increasing iron content of the mineral: 0.16wt% in Fa_{15}, 0.32wt% in Fa_{21}, 0.61wt% in Fa_{45}, and 0.98wt% in Fa_{62} (Table. 4.1). The Cr_2O_3 content in olivine from picritic basalt ranges from 0.03 to 0.105wt%. To compare, in lunar samples, CaO, MnO, and Cr_2O_3 content reaches 0.21wt% (Haggerty and Barker 1967), and in terrestrial kimberlites, meimechites, and xenoliths from basaltoids, it varies from 0.03 to 0.11wt% (Velinsky and Bannikov 1986).

Table 4.1 Olivine composition of Siberian Platform effusive traps

Rock, location	Fa (%)	NiO (wt%)	CaO (wt%)	MnO (wt%)
Picritic basalt, Talnakh River	21.7(13)	0.36(8)	0.21(13)	0.32(6)
Picritic basalt, Mikchanda River	24.5(2)	0.09(2)	0.25(2)	–
Picritic basalt, Mount Sunduk	23.5(2)	0.37(2)	0.28(2)	–
Picritic basalt, Khrebtovy Stream	21.4(1)	0.36(1)	0.08(1)	–
Picritic basalt, Bistraya–Daldykan rivers	18.0(2)	0.34(2)	0.24(2)	–
Picritic basalt, Ayan River (Vasil'ev and Zolotukhin 1975)	15.1(9)	0.30(9)	0.28(9)	0.16(6)
Picritic basalt, Iken River (Zolotukhin et al. 1986)	22.2	–	–	–
Limburgite, Murukakit River	12.4(3)	0.447(3)	0.20(1)	–
Olivine-phyric basalt, Mount Sunduk	29.2(2)	0.14(2)	0.10(2)	–
Olivine-phyric basalt, Nidym River (Zolotukhin et al. 1985)	25.3;31.8	–	–	–
	43.7;49.6	–	–	–
	26.6(2)	–	0.35	
Olivine-phyric basalt, Lake Bokovoe (Zolotukhin et al. 1986)	26.5	–	0.35	–
Plagioclase-phyric basalt (Zolotukhin et al. 1986)	45.4	–	–	0.61
	38.0(2)	–	0.38	–
Ti-augite dolerite-basalt, Iken River	62.4(2)	–	0.74(2)	0.98(2)

Note: Brackets indicate the number of analyses used

Table 4.2 Olivine composition from trap intrusions contact facies

Intrusion	Sample number	Fa (%)	NiO (wt%)	CaO (wt%)	MnO (wt%)
Upper Talnakh[a]	ST-V upper	35.4	0.06	0.17	0.53
	731/940.1 lower	43.9	0.09	–	0.65
Lower Noril'sk[a]	NP-13/1284 lower	22.8	0.36	0.36	0.31
	–||–	18.1	0.07	0.39	0.27
Khungtukun[b]	Khung-3 upper	47.3	–	–	–
	Khung-7 upper	48.9	–	–	–
Mountain Pegmatitovaya	E-28/410.0 lower	47.6	0.055	0.07	–
Mountain Putanaya	E-36/311.0				
	Lower	27.3	0.14	–	–
	Lower	28.7	0.135	0.10	–
Magnitny Stream	Mg-36				
	Lower	60.7	0.022	0.24	–
	Lower	45.0	0.068	0.23	–

Notes: [a]Genkin et al. (1981),
[b]Ryabov (1981)
Upper and lower—accordingly upper and lower contact gabbrodolerite

Olivines in Intrusive Systems

Consideration is made of the alterations to Ol composition compared with petrochemical parameters in some layered intrusions. Results of average iron content values in Ol and trace element content in a variety of rocks from different intrusions are reported in Ryabov (1992a). Some notion about Ol composition from contact facies of trap intrusions can be gained from Table 4.2, which shows a discrepancy in the composition of the mineral from the upper and lower contact gabbro from the Upper Talnakh Intrusion and their elevated iron contact in typical traps. The rocks from the Lower Talnakh Intrusion have rather high magnesium content in Ol with low and unstable nickel content. The iron content of Ol from the lower contact gabbro from the Upper Talnakh Intrusion is close to that of the rocks from the Khungtukun Intrusion and the Pegmatitovaya Mountain Intrusion. Olivine with varying Fa components coexist in the contact facies of the Magnitny Stream Intrusion.

The Pegmatitovaya Mountain Intrusion is a classic example of cryptic layering after olivine crystallization, which is consistent with a regular change in magnesium content and C.f. of rocks in the body (see Fig. 3.39). The iron content in Ol from the top to bottom varies from Fa_{70} to Fa_{35}. A strong correlation exists in rocks and minerals, between MgO_{rock}, NiO_{rock}, and NiO_{Ol}, and between *Cf*, Ca_{Ol}, and Fa. A strong feedback exists between these two groups of components.

The Putanaya Mountain Intrusion is a low-Cr melanocratic intrusion of the Morongovsky Complex, which is composed mainly of troctolitic gabbrodolerite. The iron content in Ol ranges from Fa_{31} to Fa_{25} with a tendency for decreasing iron content downward through the intrusion (see Fig. 3.36).

In the *Upper Talnakh Intrusion*, iron content varies creating an Ol composition range from Fa_1 to Fa_{44}. The mineral composition also varies widely along the intrusion, and in some parts of the intrusion, it can vary even in one sample. In the mafic rock horizon of the Upper Talnakh Intrusion (see Fig. 3.18a), Ol composition ranges from Fa_{24} to Fa_{32}, the composition is more stable in the picrite layer (Fa_{23-25}), and in the taxite layer, the iron content increases and Olivine composition varies accordingly (Fa_{28-39}). Comparison of the variation in Ol composition in picrites of the leading and basal parts of the Upper Talnakh Intrusion (see Fig. 3.18b, c) shows that in the upper part of the intrusion, the iron content of the mineral increases and Ol composition changes from Fa_{20} to Fa_{27} with corresponding decreasing NiO content. In the illustrated cross section, one can see the relationship between the variations in Fa–S and NiO_{Ol}–Cr and the feedback behavior of Fa–NiO_{Ol} and S–NiO_{rock}. At the base of the intrusion, variations in the compared components are complicated by the presence of olivine gabbro and taxite gabbro. Nevertheless, the iron content of Ol in picrites is higher and increases from the top down (Fa_{35-40}), and at the border with the pegmatoid zone, it drastically decreases and Ol composition is Fa_{22}. In the taxite horizon, the iron content increases with depth, and Ol composition varies from Fa_{22-31} at the top to Fa_{36-43} at the base. The nickel content of Ol is stable in picrites; it increases at the border with olivine gabbro, then returns to the NiO content of picrites, and somewhat increases at the bottom of taxite layers. The calcium content of Ol varies slightly, except in olivine gabbro where it drastically increases to double the normal concentration. No relation between NiO_{rock} and Ol composition was observed.

The special features of Ol composition in the magmatogenic breccia are given in Fig. 3.19. All analyzed olivines were picked from olivine-rich rocks that form the breccia fragments. The breccia cement is represented mainly by olivine-free pegmatoidal gabbro. From the top down through the horizon of magmatogenic breccia, iron content of Ol tends to increase, and NiO content decreases. The main specific features of olivine are the pronounced differences in the compositions of coexisting minerals, abundant olivine ($<Fa_{20}$) with low-iron content, and a clear tendency for nickel depletion and calcium enrichment. The low iron content in this section creates an olivine range from Fa_1 to Fa_{28}. Tracing the variations in Ol composition from typical picrites, preserved in some fragments of magmatogenic breccia to olivinite restites, one can see an increase in the magnesium content depending on the degree of alteration of the rock and degree of taxitization. The NiO_{Ol} content in magmatogenic breccia varies from 0 to 0.197wt%, and CaO ranges from 0.04 to 0.80wt%. An elevated nickel content was found in relict picrites. Small polygonal grains of Ol in olivinites were found to be depleted in NiO and enriched in CaO.

As olivine contained in groundmass reveals a markedly stable depletion in nickel, it is worth noting that the magmatogenic breccia features poor sulfide mineralization with a rather stable Ni content and Cu/Ni ratio. Moreover, the Kharaelakhsky branch intrusion has the highest weighted average composition of sulfides (23.66%) compared to other branches of the Upper Talnakh Intrusion (Gulin and Sukhov 1973).

The base of the extended part of the Upper Talnakh Intrusion is observed in drill hole SV-16, where the Upper Talnakh and Lower Talnakh Intrusions are observed together (Fig. 4.1). Comparing them from the top down through the Upper Talnakh Intrusion, iron content in Ol decreases (Fa_{38} to Fa_{21}) from olivine-bearing gabbrodolerite to picritic gabbrodolerite and then somewhat increases (Fa_{27-31}) in taxite and gabbro. The decreasing iron content of olivine (av. Fa_{34}) in the mafic rock layer in no way influenced the magnesium content in the rock, which is virtually stable (MgO averages 9wt%) through the whole horizon and strongly increases only in picrites. The NiO_{Ol} and NiO_{rock} indices increase downward through the intrusion until the taxite zone after which both drastically decrease. CaO_{Ol} depends directly on the iron content of Ol and inversely on NiO_{Ol}.

A specific feature of Ol in intrusive traps is the coexistence of varying composition of the mineral with one sample (Table 4.3), which is mostly clearly manifest in ore-bearing intrusions. It is worth noting that earlier generations of olivine have a higher iron content than late olivine, magnesian varieties are lower in nickel than iron varieties, and finally, the rocks contain uncommon olivines that are rich or extremely depleted in NiO (Ryabov 1992a).

The Lower Talnakh Intrusion. The iron content of Ol in the intrusion creates a variation from Fa_{19} to Fa_{26}, with this range found within a single sample (Fig. 4.1). The content of NiO_{Ol} has a tendency to increase from bottom upward through the intrusion, whereas MgO_{rock} and NiO_{rock} decrease in this direction. This fact suggests that the latter two increase owing to the growth of Ol in the rocks.

Noril'sk-I Intrusion. In the Eastern Noril'sk Intrusive branch, the iron content of Ol tends to increase from the base upward through the intrusion with Ol varying from Fa_{23} to Fa_{40} and a corresponding decrease in CaO_{Ol} and NiO_{Ol} (see Fig. 3.20). In mineralized picrite and taxite gabbrodolerite, the composition of Ol is unstable, even within a single

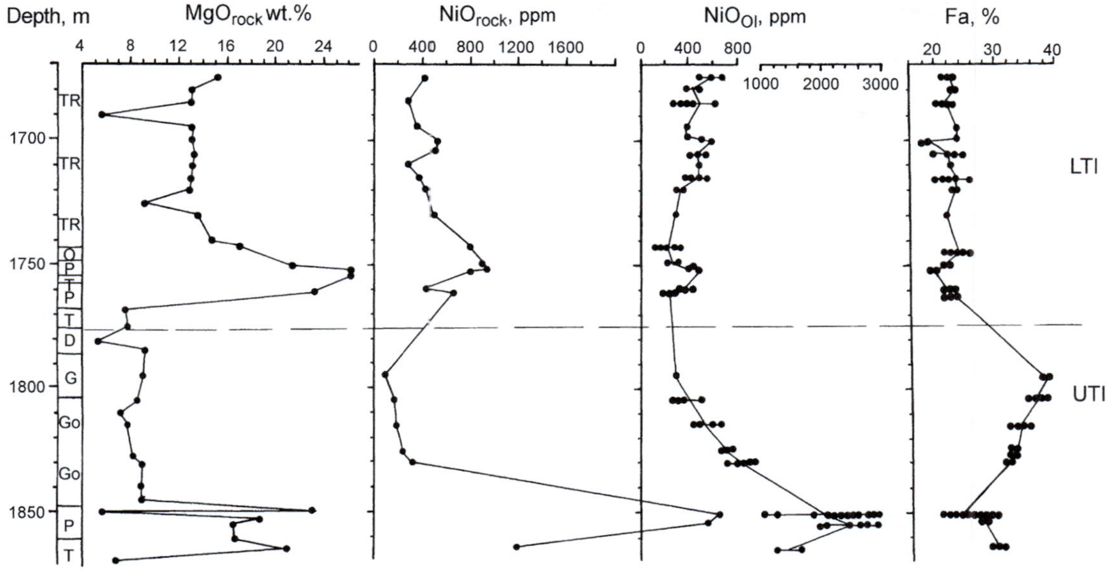

Fig. 4.1 Variations in olivine composition compared to the magnesium content in the combined section of the Lower Talnakh (LTI) and Upper Talnakh (UTI) Intrusions (drillhole SV-16). The intrusion boundary is shown by a *dashed line*

Table 4.3 Olivine composition in single samples

			1		2	
Sample number	Intrusion	Rock	Fa (%)	NiO (wt%)	Fa (%)	NiO (wt%)
SV-28/1674.8	Upper Talnakh	P	26.4	0.274	20.9	0.043
KZ-274/477.5	Upper Talnakh	Go	33.7	0.267	23.6	0.134
KZ-1551/1378.3	Upper Talnakh	P ore	22.8	0.208	16.0	0.169
KZ-586/949.2	Upper Talnakh	P ore	29.0	0.136	19.8	0.109
SV-16/1852	Upper Talnakh	P ore	30.1	0.105	22.2	0.282
ST-5/1811.1	Upper Talnakh	T ore	31.1	0.130	23.3	0.085
ST-5/1820	Upper Talnakh	T ore	43.8	0.143	37.6	0.148
KZ-585/831.8	Upper Talnakh	T	24.5	0.106	16.3	0.068
SV-16/1705	Lower Talnakhh	TR	24.5	0.041	19.9	0.051
SV-16/1715	Lower Talnakhh	TR	26.2	0.046	20.8	0.038
SV-16/1743	Lower Talnakhh	TR ore	26.4	0.236	23.0	0.020
PE-35/397	Noril'sk-I	Tl	32.0	0.136	24.7	0.147
NP-13/1465.4	Lower Noril'sk	TR	26.3	0.184	25.9	0.151
AS-6/211	Pyasinsko–Vologochansky	Go	53.8	–	27.0	–
AS-6/412	Pyasinsko–Vologochansky	Tl	42.8	–	28.5	–
MG-15	Magnitny Stream	Go	31.0	0.110	24.9	0.150
MG-23/24	Magnitny Stream	Tl	46.6	0.028	39.0	0.057
MG-36	Magnitny Stream	CG	60.7	0.022	45.0	0.068
KH-166	Khungtukun	Go ore	42.9	0.028	26.4	0.220
KH-182	Khungtukun	Go ore	41.7	0.043	23.5	0.204
KH-24/67.8	Khungtukun	P	47.4	–	40.4	–
KH-25/110	Khungtukun	TR	32.1	–	26.7	–
KH-25/170	Khungtukun	T	40.5	–	24.3	–
SB-27	Suhkaya Bakhta River (Zolotukhin 1984a, b)	O	58.4	–	37.7	–
SU-5	Lava flow	PB	25.5	0.350	21.3	0.389
MCH-60	Lava flow	PB	28.0	0.065	22.9	0.118

Note: Coexisting olivines: 1—ferrous, 2—magnesian

sample. The drastic growth of MgO and NiO in picritic gabbrodolerite is related to high Ol content in the rock and presence of sulfides.

Behavior of silicate nickel and iron content in olivines. To clearly demonstrate the behavior of NiO distribution in Ol with varying iron content, it is useful to plot NiO–FeO content in olivine.

Figure 4.2a shows the olivine composition of barren rocks from different trap intrusions of the northwestern Siberian Platform, including the Noril'sk Intrusions (Ryabov 1992a). One can see on the diagram that a decrease in NiO_{Ol} and an increase in FeO_{Ol} is followed by the appearance of a chain of compositional fields from picritic basalt, picritic dolerites, and gabbrodolerites of the Fokinsky Complex (Lower Fokinsky, Magnitny Stream, Upper Pyasinsky Intrusions) and troctolitic gabbrodolerite of the Morongovsky Complex (Mountain Morongo, Mountain Putanaya, Mountain Ruinnaya Intrusions and their apophyses in drill hole NP-13). This field also includes olivines from the Pyasinsko–Vologochansky and Zubovsky Intrusions, and further along the field olivines from troctolitic gabbro of the Mountain Pegmatitovaya Intrusion. The field separated from the chain includes olivine from picrite–troctolite gabbrodolerite from the Lower Talnakh and Lower Noril'sk Intrusions. The location of Ol in troctolitic gabbrodolerite from the Morongovsky Complex and Zubovsky-type intrusions within the field of olivine gabbrodolerite suggests a genetic relationship, whereas the isolated position of fields of olivine from the Morongovsky Complex and Lower Talnakh and Lower Noril'sk Intrusions suggests they have a different origin.

Olivine from picritic gabbrodolerite from the Noril'sk Intrusions (Upper Talnakh, Eastern Noril'sk, Noril'sk-I, Mountain Chernaya and Imangdinsky) forms an elongate field with a wide range of variations in NiO_{Ol} and minor variations in FeO_{Ol} (see Fig. 4.2b). The compositions of coexisting olivine may have different iron and nickel content (Ryabov 1992a). Olivine with the highest nickel content in the upper part of the field appeared to be similar to the mineral from picritic gabbrodolerite of the Fokinsky Complex and picritic basalt. This, on the one hand, seems to support the comagmatism of these rocks and, on the other, suggests the existence of processes leading to the depletion of the mineral in nickel ore-bearing gabbrodolerite.

It was shown that olivines from olivine and olivine-bearing gabbrodolerite from the Noril'sky-type intrusions have lower iron contents when compared to the ore-free intrusions, though their nickel contents are similar (Ryabov 1992b).

Olivines from mafic pegmatoids related to the Noril'sk Intrusions form a trail of points that intersects the geochemical trend formed by ore-free traps (see Fig. 4.2c). The trend is characterized by an inverse relationship between NiO and FeO, whereas in Ol from pegmatoids, the relationship is direct. The iron content gradually increases in Ol (Fa_1 to Fa_{34}) in the series of rocks formed by olivinites from magmatogenic breccia, melanocratic taxite (taxitized picrite), troctolite, taxite, and olivine gabbro. The Ol field from pegmatoids provides evidence that the mechanism controlling the distribution of NiO in the mineral of these rocks differs from those that form a normal geochemical trend. At the same time, the position of the olivine field from troctolite gabbrodolerite from the Lower Talnakh and Lower Noril'sk Intrusions within the outline of the Ol field from pegmatoids suggests that the mechanisms for their formation were similar.

Comparative analysis of Ol composition from olivine gabbro and picritic gabbrodolerite in the leading and deep parts of the intrusion showed that, though olivines from picritic gabbrodolerite have a similar iron content, the mineral from the leading part has a lower nickel content and a wide range of NiO contents (Ryabov 1992a). In olivine gabbrodolerite, olivines from the leading parts have higher nickel contents; both olivine compositions form a compact cluster of dots.

Some idea of the iron content of olivine from the Siberian Traps can be gained from Fig. 4.3, which indicates the range of variation in iron content in olivine. Olivine in the groundmass corresponds to Fa_{18-36}. Earlier studies (Ryabov 1992a) show that among olivine-bearing gabbrodolerite and olivine gabbro of the platform, olivine from the Noril'sk Intrusion has higher magnesium content than in other intrusions but they have similar NiO and CaO content. Compared to picrite and troctolitic gabbrodolerite from different intrusions, Noril'sk Intrusion olivine has a lower iron content and a considerably variable nickel content—in some cases differing from ore-free intrusions and in other cases similar. The distribution of CaO_{Ol} is similar.

Comparison of Ol composition from various rocks of the Noril'sky-type intrusions provides evidence that in the layered series from the CZ of the intrusion Ol from olivine gabbro and olivine-bearing gabbro has a higher iron content, Ol from picritic gabbrodolerite has a lower iron content, and their compositions overlap at about 22–31% Fa component. The widest range of iron content was found in mafic pegmatoids. These rocks contain two types of Ol, one with a higher and the other a lower iron content than in picrites. This is clear from the double-peak structure of the histogram reported in Ryabov (1992a). It is worth noting that Ol of troctolite gabbrodolerite from the Lower Talnakh Intrusion lies between these two peaks.

The nickel concentration in olivine is distributed so that low NiO content (0.01–0.08wt%) is found in troctolite gabbrodolerite from the Lower Talnakh and Lower Noril'sk Intrusions, whereas NiO content of 0.03–0.17wt% is found in olivine gabbrodolerite, 0.06–0.30wt% in picritic gabbrodolerite, and mafic pegmatoids show the widest

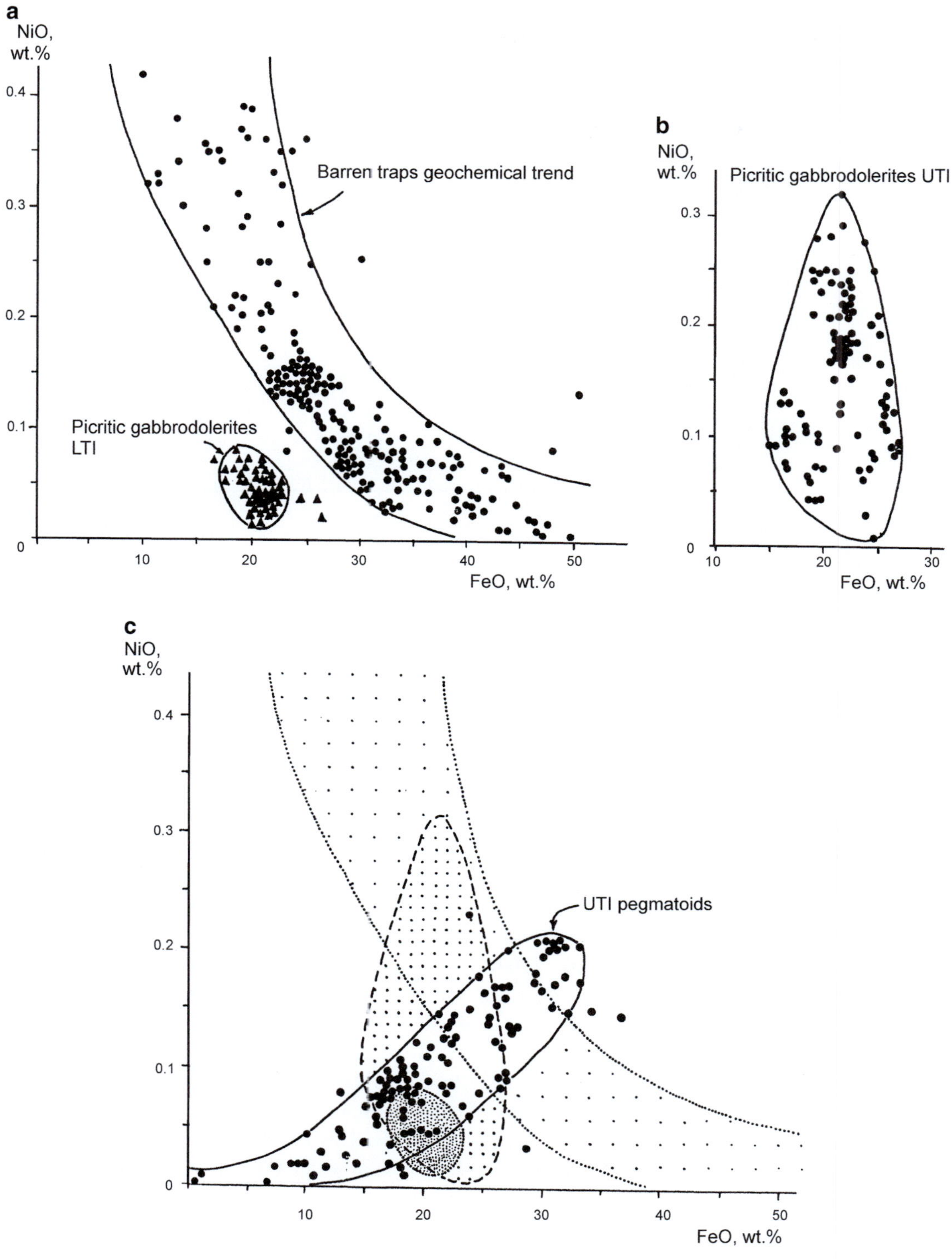

Fig. 4.2 Diagram of NiO versus FeO in olivines from traps of the Siberian Platform (Ryabov 1992a). *UTI* Upper Talnakh Intrusion, *LTI* Lower Talnakh Intrusion. Fields of analyses shown on plots A and B are indicated with a speckle pattern on plot C

variations from 0.01 to 0.30wt%. A comparison of olivine composition from high-magnesium lavas showed some specific features. Olivines from picritic basalt differ from those in troctolite gabbrodolerite from the Lower Talnakh Intrusion, having a lower iron content and significantly higher NiO and CaO content. Olivine from troctolitic gabbrodolerites from the Morongovsky Complex differ from similar rocks from intrusions of the Lower Talnakh Complex having a higher iron and nickel content but similar calcium content.

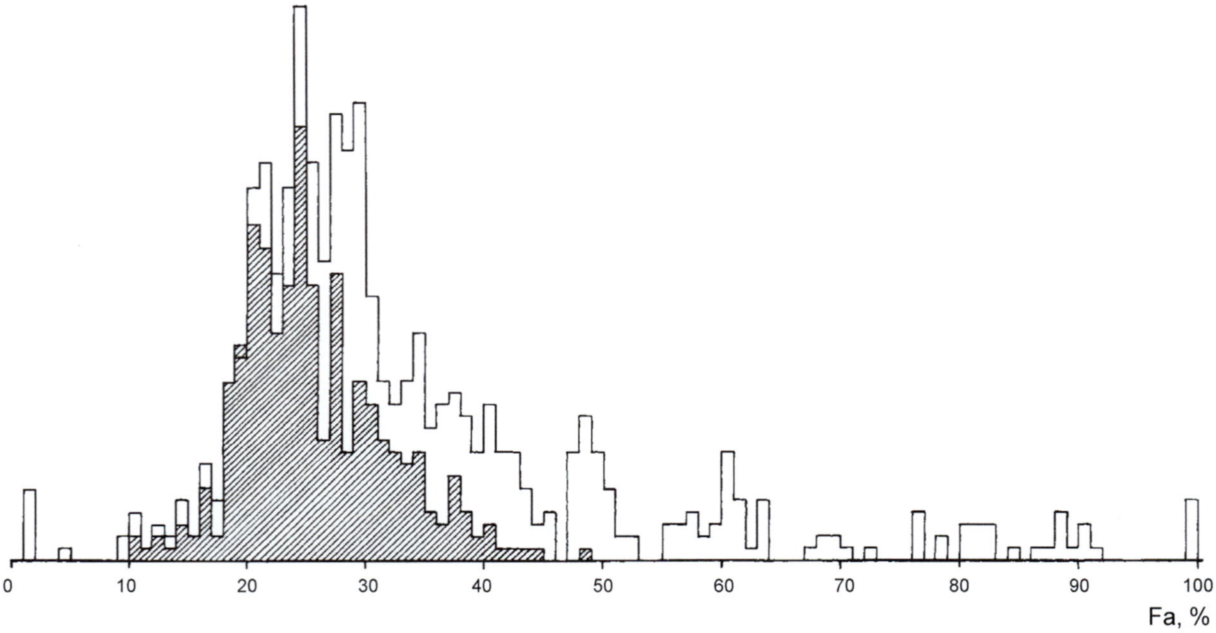

Fig. 4.3 Histogram of iron content distribution in olivines from traps of the Siberian Platform ($n = 605$). The field of composition of olivine from the Noril'sk Intrusions is shown by *shading*

4.1.2 Pyroxenes

Pyroxenes are the dominant and most widespread mafic mineral, and their multicomponent composition is a highly informative characteristic of crystallization conditions in rocks. Specific features of the evolution of pyroxene composition have, for a long time, been used to resolve the differentiation of magmatic melts. Studies in this field were carried out by many geoscientists (Sobolev 1986; Godlevsky 1959; Korovyakov et al. 1963; Vilensky et al. 1964; Zolotukhin 1964; Vilensky 1967).

4.1.2.1 Features of Clinopyroxene in Traps

It is known that most pyroxenes in traps are monoclinic varieties; rhombic varieties are less common. Clinopyroxene (Cpx) in most cases is augite with minor amounts of other varieties. The content of pyroxene in traps varies from 15–20 to 30–35wt%. In isotropic intrusions, the mineral content increases constantly, and with the increasing differentiation, the distribution of pyroxene becomes nonuniform. This is best observed in the Noril'sk Intrusion, in which the maximum pyroxene concentration was found in the middle part of the layered series of rocks. Mafic pegmatites contain variable pyroxene content (Godlevsky 1959; Zolotukhin 1964; Dodin and Batuev 1971; Ivanov et al. 1971a, b; Zolotukhin et al. 1975).

The morphology of pyroxene crystals in traps is rather diverse, and petrographic studies can provide important genetic information about the formation conditions of rocks. The description of Cpx crystals in traps and structures and their participation in the formation mechanism are available in Sobolev (1986), Dodin et al. (1971), Ryabov and Zolotukhin (1977), and Lurje et al. (1976).

Of great genetic interest is the poikilo-ophitic texture which creates the pea-shaped texture in rocks and is one of the most widespread textures of effusive and intrusive traps. This rather coarse-grained texture surprises geoscientists as it occurs not only in sills but even in thin basalt flows that are widespread over large areas. The occurrence of poikilo-ophitic texture in lavas and thin sills suggests that the texture is the result of rapid crystallization.

The size of pea-shaped poikilocrysts or oikocrysts varies widely in effusive and intrusive traps. Within one body, larger crystals are typically confined to the UZ of flow or intrusions. This is best observed in the Putanaya Mountain Intrusion, in which the size of Cpx pea-poikilocrysts gradually increases from the base upward through the body (from the boundary of olivine gabbrodolerite and troctolitic gabbrodolerite) from 0.2–0.4 cm to 2–3 cm in diameter (Fig. 3.37). The matrix between poikilocrysts is typically filled with pyroxene–plagioclase–ore, but there are cases when it is completely composed of tholeiitic glass or palagonite. This is seen on microphotographs of rocks of different effusive and intrusive complexes.

Different points of view exist concerning the formation mechanism of the poikilo-ophitic texture. Sobolev (1986) believes that it took place during slow crystallization, allowing clinopyroxene crystals to grow to large sizes and

Table 4.4 Separation coefficients of rock-forming oxides between mafic and sialic magmatic formations (proportion of oxides in coexisting formations, wt%)

	SiO_2	Al_2O_3	Na_2O	K_2O	TiO_2	FeO	MnO	MgO	CaO	P_2O_5
Mafic glass—sialic glass										
Liquation glass (Glg)	0.6	0.6	0.5	0.4	5.5	7.7	9.9	16.1	7.4	26.6
Pyroxene glass										
Glass globules with crystallite (Glc), GK-21	0.8	0.4	0.1	–	7.4	3.4	–	–	4.5	–
Glass globules with crystallite (Glc), GK-22	0.7	0.4	0.6	–	4.0	5.0	4.8	–	8.8	–
Aphanite	0.7	0.1	0.04	–	0.6	11.2	50.0	172.2	3.3	–
Diabase-pegmatite	0.7	0.3	0.2	–	1.9	5.6	6.4	20.2	4.2	5.0
Olivine-bearing gabbrodolerite	0.8	0.1	0.2	–	2.1	2.7	6.2	2.9	9.6	0.1
Coexisting rocks										
Ferrogabbro–granophyre	0.6	0.8	0.4	0.6	4.1	3.4	6.8	12.4	7.3	26.
Picritic basalt–metadiorite	0.7	0.7	0.6	0.2	1.7	8.8	1.4	1.8	0.5	1.4
Picritic gabbrodolerite–gabbrodolerite of stratified series	0.8	0.4	0.3	0.1	0.5	1.9	2.0	3.4	0.5	0.4
Picritic gabbrodolerite–leucogabbro	0.8	0.4	0.1	0.3	0.3	1.6	9.3	5.5	0.5	0.2

trap small laths of plagioclase. Vedernikova (1961) treats this texture as the result of selective recrystallization. Lebedev (1955) reports that the groundmass of most basalts crystallizes in accordance with the ophite scheme, but the arrangement of plagioclase chadacrysts in clinopyroxene oikocrysts points to the fact the former were separated prior to the latter. Godlevsky (1959) notes that the glassy edges of trap bodies typically contain plagioclase but little pyroxene, and that pyroxene and glass rarely occur together. Zolotukhin and Podgornykh (1998) believe that a good degree of crystallization in most basalts, including the coarse-grained poikilitic texture, and the elevated fluidity of lava sheets is explained by overheating of the melt, probably to 1,600°C. It is noteworthy that pigeonite, which is one of the criteria of high-temperature crystallization of basalt melt (Marakushev 1986), is rare to absent in effusive traps, and the assumption of the quoted authors is, at least, doubtful.

Some features of the formation of poikilo-ophitic textures in the rocks crystallized from high-alumina olivine-tholeiite magma were described by Mathison Charter (1987). Chadacrysts of plagioclase in oikocrysts of clinopyroxene are arranged rather chaotically; their shape is strongly elongated or, less often, rounded. Laths are held together at a wide angle, resembling skeleton crystals, have an inverse zonality, and contain 5–10% less An component than microliths of basalt which have an elongated shape and preferred orientation. The author suggests that chadacrysts of plagioclase formed during supercooling 25–40°C below equilibrium temperature.

Investigations into the crystallization forms of Cpx in glass from Khungtukun allow interpretation into the origin of morphologically different crystals and crystal aggregates that form different textures (Ryabov 1989b). During recrystallization of a mafic liquid continuous pyroxene fields, spherocrystals, panicullate, parquet-like, and other aggregates of pyroxene and regular pegmatoid intergrowths of Cpx and plagioclase and Cpx poikilocrysts with the intergrowths of plagioclase formed. Based on a study of glasses, it was assumed that the formation of poikilocrysts of Cpx was preceded by mafic liquid of a pyroxene-normative composition in the form of globules, which further crystallized pea-shaped Cpx (Ryabov 1988a, b, 1989b). An additional argument in favor of this assumption is the similarity of separation factors between some phases (Table 4.4).

In the Siberian Platform, traps, sills, and lava sheets feature a poikilo-ophitic texture in differing fashions. In some cases, the pea-like poikilocrysts of clinopyroxene are globular, whereas in others, they are short-prismatic or elongated dendroid crystals. The quantity of "peas" (oikocrysts with chadacrysts) varies from 5–7% to 35–50%, though is typically 30–45%. Within a single discrete rock type, the "peas" are distributed uniformly. There are rocks in which poikilocrysts intergrow each other, forming a solid clinopyroxene framework. The bulk of the rock enclosing poikilocrysts in effusive rocks is typically composed of palagonite-I containing plagioclase, magnetite, and less often olivine and other minerals which make up the intersertal texture. The matrix of intrusions also contains palagonite-I, but it is often crystallized. In different sheets and sills, the minimum size of poikilocrysts (about 0.5 mm) is observed near their contacts, and away from the contact, the size of "peas" increases and stabilizes. In thick sills, as was observed in the Pegmatitovaya Mountain Intrusion, the size of "peas" might grow to 3 cm in diameter. In some sheets, layered basalts are formed as sharply alternating layers of basalt with poikilo-ophitic and intersertal textures. The layers of basalts with an intersertal texture are from 2–3 to 5–7-mm thick. They form bands, often with weakly pronounced folds and wedging-out lenses. The laths of plagioclase exhibit trachytoid flow traces. The interlayers of poikilo-ophitic basalts are 2–6 to 30-mm thick. The thin layers can hold only 1–2 "peas." Wedging-out zones of these layers feature palagonite with microlaths of plagioclase. Thicker layers are persistent in length. In "pea"-containing layers, large-"pea" (3 mm in diameter) basalts with scarce "peas" (about 10–15%) transform into small-"pea" (0.5 mm in diameter) basalts with about 30% "peas" and, in places, into intersertal basalts.

Figure 4.4 shows a schematic sketch of the transition of the large-"pea" basalt into small-"pea" and homogeneous basalts, which was thoroughly studied using a microprobe. The total length of the section was 2 cm. The composition of minerals is given in Rock indication 40. A specific feature of the rock groundmass is the large quantity of palagonite in the large-"pea" basalt (25–30%), and a small amount in intersertal basalt (no more than 10%), which is characterized by opacitized glass and fragments with microdolerite texture.

Zone A is large-"pea" basalt. The rock texture is poikilo-ophitic and intersertal. The main fabric of the rock is formed by large (1.5–2 mm) oikocrysts of zoned clinopyroxene. Clinopyroxene composition changes from center to rim from $Wo_{38}En_{47}Fs_{15}$ to $Wo_{39}En_{42}Fs_{19}$. Chadacrysts of plagioclase consist of bytownite $Ab_{27}Or_1An_{72}$. The rock matrix is composed of microporphyritic segregations of plagioclase $Ab_{29}Or_1An_{69}$, single grains of olivine Fa_{55}, and titanomagnetite, which are found among microlaths of plagioclase $Ab_{29}Or_1An_{70}$ and small angular grains of clinopyroxene $Wo_{38}En_{31}Fs_{31}$, localized in palagonite-I with zones of opacitized glass.

Zone B has a poikilo-ophitic texture similar to the above-described basalt, although the oikocrysts of clinopyroxene are smaller (0.8–1.0 mm), and their composition is $Wo_{41}En_{45}Fs_{14}$. Chadacrysts are plagioclase ($Ab_{34}Or_1An_{65}$), and the microporphyritic segregations in groundmass are $Ab_{24}Or_1An_{75}$, and microliths are $Ab_{45}Or_2An_{53}$.

Zone C is composed of compactly arranged small (0.3–0.6 mm) poikilocrysts of clinopyroxene ($Wo_{41}En_{46}Fs_{13}$).

Zone D is basalt with microporphyritic segregations of olivine (Fa_{56}) and titanomagnetite and a groundmass of microliths of plagioclase ($Ab_{35}Or_1An_{64}$), opacitized glass, and rare palagonite-I.

Microprobe studies show that the central parts of clinopyroxene oikocrysts from zones A, B, C, and small grains from zone D have a similar composition ($Wo_{38-41}En_{44-47}Fs_{13-16}$). Variations in mineral composition during crystallization is observed by comparing the content of mineral-forming oxides in the following series: center–rim of oikocrysts—grains in the groundmass feature a decrease in SiO_2, Al_2O_3, MgO, and Cr_2O_3 content and an increase in TiO_2, FeO, Fe_2O_3, MnO, and V_2O_5 content.

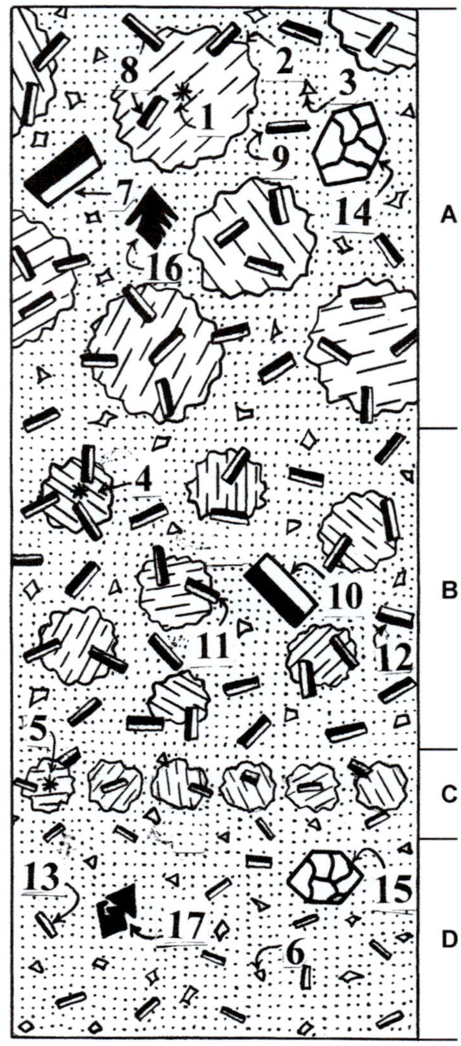

Fig. 4.4 Schematic sketch indicating the transition of large-pea-shaped poikilo-ophitic basalt into aphyric basalt with an interstitial texture (sample LD-73). *A* large-pea-shaped basalt, *B* small-pea-shaped basalt, *C* layer of small poikilocrysts, *D* basalt with interstitial texture. *1–6* clinopyroxene: zone: *A* core (*1*) and rims (*2*) of large oikocryst and small angular grains (*3*) from groundmass; zone *B* center of small oikocryst (*4*); zone *C* small oikocryst (*5*); zone *D* grain from groundmass (*6*); *7–13* plagioclase: zone: *A* phenocryst (*7*), chadacryst (*8*), microlite of groundmass (*9*); zone *B* phenocryst (*10*), chadacryst (*11*), microlite of groundmass (*12*); zone *D* microlite of groundmass (*13*); olivine microphenocrysts (*14*) and (*15*) from zones *A* and *D*, respectively; skeleton titanomagnetite crystals (*16*) and (*17*) from zones *A* and *D*, respectively. Numbers within the sketch correspond to analyses given in Rock Indication 40

The mineral composition in this sequence alters as $Wo_{38.1}En_{46.1}Fs_{15.2}$–$Wo_{39.4}En_{41.5}Fs_{19.1}$–$Wo_{37.7}En_{30.9}Fs_{31.4}$.

The central parts of clinopyroxene oikocrysts have elevated Cr_2O_3 (0.55–0.73wt%) content and decreased Al_2O_3 (2.55–0.85wt%) content. The maximum CaO (19.75wt%), Cr_2O_3, and Al_2O_3 content was found in clinopyroxene of small-"pea" basalt in zone B. Clinopyroxene from zones B, C, and D has increased CaO (19.37–19.75wt%) content and decreased CaO (17.66–18.75wt%) content in zone A. Clinopyroxene in the groundmass has a low Al_2O_3 (1.86–2.20wt%) content.

Plagioclase composition in layered basalt is variable, $Ab_{24-45}Or_{1-2}An_{75-53}$. Different zones exhibit specific features of compositional variations. In zone A, plagioclase of chadacrysts, porphyritic segregations, and microliths have a similar mafic nature (An_{69-72}). Zone B contains one generation of microliths of plagioclase (An_{64}), whereas zone C contains three generations of plagioclase: phenocrysts (An_{75}), chadacrysts (An_{65}), and microliths (An_{53}).

Some noteworthy features were observed by comparing the compositions of minerals from different zones:

1. Petrographic varieties of layered basalt feature unique variations in the composition of the main rock-forming minerals—clinopyroxene and plagioclase.
2. In coarse poikilo-ophitic basalt, one can observe wide variations in clinopyroxene composition, whereas in fine poikilo-ophitic, wide variations are observed in plagioclase composition.
3. The distribution of CaO and Al_2O_3 between chadacrysts of plagioclase and oikocrysts of clinopyroxene are inversely correlated, as a result plagioclase rich in An component is accompanied by a CaO-depleted clinopyroxene and vice versa.
4. The crystallization of pyroxene is accompanied by an increase in TiO_2, FeO, and MnO and a decrease in SiO_2, Al_2O_3, MgO, and Cr_2O_3.
5. An elevated concentration of Cr_2O_3 is confined to oikocrysts of clinopyroxene and has a direct relationship with Al_2O_3 content.

The crystal form and mineral composition suggest that the crystallization conditions were not the same in all areas. In zone A, melt crystallization initiated with the separation of plagioclase rich in An component, whose stable composition was found in chadacrysts, microprints, and microliths. Clinopyroxene formed after plagioclase from already CaO-depleted melt and was accompanied by a progressive decrease in CaO, Al_2O_3, and MgO content and increase in FeO content. In zones B and C, melt crystallization began with the formation of porphyritic segregations of bytownite, which later was changed by the joint crystallization of less mafic plagioclase (labradorite) and calcium-enriched clinopyroxene. At the final stage, acid labradorite crystallized in these zones. In zone D, a joint crystallization of labradorite and Ca-enriched clinopyroxene took place.

Differences in the structure of rocks and mineral compositions can, most likely, be explained by the varying degree of fluid saturation in the melt. Volatiles promoted liquation differentiation, which resulted in the formation of an emulsion of coexisting pyroxene- and plagioclase-normative liquids, giving rise to a poikilo-ophitic texture. A melt poor in volatile components on cooling resulted in an intersertal texture.

4.1.2.2 Chemical Composition of Pyroxenes

The problems of the chemical composition of pyroxenes have been discussed in many works, most of which consider the composition of pyroxenes from the Noril'sk Intrusions (e.g., Genkin et al. 1970; Godlevsky et al. 1971; Dodin and Shatkov 1971; Ivanov et al. 1971a, b; Ryabov and Zolotukhin 1977; Distler et al. 1976; Zolotukhin and Shchedrin 1977; Tuganova 1988). Some works are devoted to Cpx from intrusions of the eastern flank of the Tungussky syneclise (Masaitis 1958; Vilensky 1967; Staritsyna et al. 1972; Oleinikov et al. 1985), the middle course of the Yenisey River (Staritsyna et al. 1972; Zolotukhin 1984a), the northern Siberian Platform (Staritsyna et al. 1972; Oleinikov et al. 1985; Ryabov et al. 1985a; Ryabov 1989b), and platform sills (Vladimirov 1962; Feoktistov et al. 1975). The composition of pyroxenes from effusive traps was discussed by Poroshin and Ledneva (1977), Ryabov (1978a), Ryabov et al. (1985c), and Zolotukhin et al. (1986). A survey of the mineralogy of pyroxenes from different magmatic formations, including the Noril'sk Intrusions, was given by Ryabov and Zolotukhin (1977), and here, we will comment upon the results of the latest studies on this problem.

A specific feature of Cpx from traps is their zonal structure, which is clearly exhibited and well studied in the Noril'sk Intrusions (Genkin et al. 1970; Godlevsky et al. 1971; Dodin and Shatkov 1971; Ivanov et al. 1971a, b; Ryabov 1974; Ryabov and Zolotukhin 1977; Distler et al. 1976; Zolotukhin and Shchedrin 1977; Tuganova 1988).

Clinopyroxene crystals from the Noril'sk Intrusions have a pronounced color zoning: a green core and brownish (brown) periphery, which is due to the admixture of strong chromophores—chrome and titanium. The green variety forms in olivine-rich rocks: picrite, olivine gabbrodolerite, olivine-bearing gabbrodolerite, and troctolitic gabbrodolerite. The brownish variety is found in overgrowth rims surrounding green cores and forms independent crystals in olivine-free and olivine-poor varieties. The total iron content of green Cpx is typically lower than 21.9wt%, and brownish Cpx is lower (Ryabov 1978b).

The investigations carried out by Genkin et al. (1970) showed that the entire range of variations in the total iron content for Cpx from the Noril'sk-I Intrusion is about 10 wt%, and compositional variations approaching this value (7.1%) take place within one zoned crystal. The authors note that no increase in the iron content of Cpx from the base upward was observed in the intrusion and they believe that it does not reflect the cryptic layering of the intrusion. The same conclusion was made by other geoscientists (Godlevsky et al. 1971; Dodin and Shatkov 1971; Ivanov et al. 1971a, b). Indeed, the analyses of Cpx from the Noril'sk Intrusions, shown on the classification diagram (forming a small isometric group of dots), supports the opinion that Cpx does not reflect melt differentiation. An increasing number of analyses for the intrusions in the

Noril'sk Region and effusive traps (Ryabov et al. 1985c) has not changed the position and configuration of composition field on the diagram.

Our research showed that the main components and iron content of Cpx from random rock samples are low. We proposed to take (Ca + Na + K)/Al and Si/Al (Ryabov 1974), and then (Ca + Na + K)/Fe and g/Fe (Ryabov 1978a) as differentiation indices, that is, to normalize the major mineral-forming components to 1 and Fe. On the diagrams, the dots representing separate analyses of Cpx stretched along a line with minor dispersion, with CaO- and Al_2O_3-rich Cpx from picritic gabbrodolerite concentrated in one part of the diagram and Al_2O_3-poor and SiO_2-rich Cpx from olivine-free gabbrodolerite, in the other. The composition fields for Cpx from different rocks and minerals from one type of rock on the diagram overlap, but nevertheless, the dependence of Cpx composition on host rock and crystallization rock is clearly exhibited. A regular decrease in CaO and Al_2O_3 content and increase in SiO_2 content toward the center–periphery of zoned crystals is typical of Cpx from trap formations in contrast to (for example) layered intrusions of syenite–gabbro (Krivenko et al. 1974). This reflects the different tendencies for crystallization of clinopyroxene.

Variations in Clinopyroxene Composition in Intrusions

To gain a notion of the variations in Cpx composition in trap intrusions, let us consider the variational diagrams for the behavior of wollastonite (Wo), enstatite (En), and ferrosilite (Fs) components, the main admixtures (trace elements) and some other factors of Cpx composition in the Pegmatitovaya Mountain, Upper Talnakh, and Noril'sk-I Intrusions, and then return to the discussion of typical regularities of mineral composition in traps.

The Pegmatitovaya Mountain Intrusion is the first mafic massif on the Siberian Platform in which cryptic layering was studied in detail. In this intrusion (see Fig. 3.38), one can clearly observe a regular alteration of all mineral-forming components. The TiO_2 content in Cpx gradually increases from the base upward through the intrusion from 0.52 to 0.94wt%, and Cr_2O_3 content decreases from 0.61 to 0.01 wt%. The minimum concentration of chromium in clinopyroxene from contact facies is indicative of the low abundance ratio of the element in tholeiite-basalt melts. In addition, in the core of crystals from the central parts of the body, the chromium content was very high. Noteworthy is the maximum increase in the element content in Cpx from the UZ of intrusions. Calcium content (Wo) and magnesium content (En) show a tendency to decrease from the base upward through the section, and the iron content of Cpx (except in CaO-poor varieties) clearly increases from Fs_{14} ($f = 23$ at%) to Fs_{56} ($f = 69$ at%) and directly correlates with the *Cf* of the rocks.

In the Khungtukun and Maymechinsky Intrusions, studied earlier, we also observed a regular variation in Cpx composition in the series of rocks from troctolite and olivine gabbrodolerite to olivine-free pegmatoids (Ryabov et al. 1985a; Ryabov 1989b). The iron content varies from $Fs_{10–14}$ ($f = 18–23$ at%) to $Fs_{38–55}$ ($f = 57–62$ at%), respectively.

The Upper Talnakh Intrusion was studied by many geologists, but no detailed description concerning Cpx composition has been made. This intrusion is presented in Fig. 3.21. The diagram shows that the central parts of crystals are typically richer in chromium (Cr_2O_3 to 1.25wt %), whereas the peripheral parts are richer in titanium (TiO_2 to 1.54wt%). Interestingly, the maximum concentrations of these components were found in picritic gabbrodolerite. The content of TiO_2 in crystal cores from top down through the picrite and mafic rock horizons slightly decreases, and Cr_2O_3 gradually decreases from bottom to top of the body. For peripheral parts of crystals, one can observe a drastic decrease in the TiO_2 and Cr_2O_3 content from the base upward along the picrite layer. In the mafic rock horizon, TiO_2 accumulates in the rims of Cpx from gabbrodolerite and sharply decreases in metadiorite, and the Cr_2O_3 content decreases from gabbrodolerite to metadiorite.

The entire variety of Cpx compositions fits the range $Wo_{35–45}En_{41–49}Fs_{8–23}$. In spite of the significant variation in iron content within one crystal (to 8 at%), the variation diagram shows a clear growth of this parameter from the base upward through the intrusion. In the central parts of crystals, the iron content increases from 14.2 to 33.3 at%, and in the periphery, from 17.1 to 35.3 at%.

The behavior of calcium and magnesium in Cpx in the intrusion is more complicated than that of iron. The major variations in Wo and En components are confined to transitions from one type of rock to another and are most appreciable when comparing the central and peripheral parts of one crystal. The same can be said of the Si/Al and (Ca + Na)/Al ratios. The Mg/Fe and (Ca + Na)/Fe ratios drop distinctly from the base to the top of the section. One can see from the diagrams that all mineral-forming components reflect the process of crystallization differentiation.

Variations in the compositions of Cpx and orthopyroxene (Opx) in the picrite horizon are shown in Fig. 3.22. Clinopyroxene shows rather stable pyroxene end member content, ruling out the near-bottom analysis, and higher iron, titanium, and chromium content in Cpx than in Opx. Noteworthy is the Cr_2O_3 increase from the base upward through the picrite horizon both in Cpx and Opx.

The Noril'sk-I Intrusion. In the Eastern Noril'sk Intrusion branch, the variations in Cpx composition are typically similar to those described for the Upper Talnakh Intrusion. Here, one can observe a well-pronounced zonality and characteristic behavior of TiO_2 and Cr_2O_3 in the central and peripheral parts of clinopyroxene crystals (see

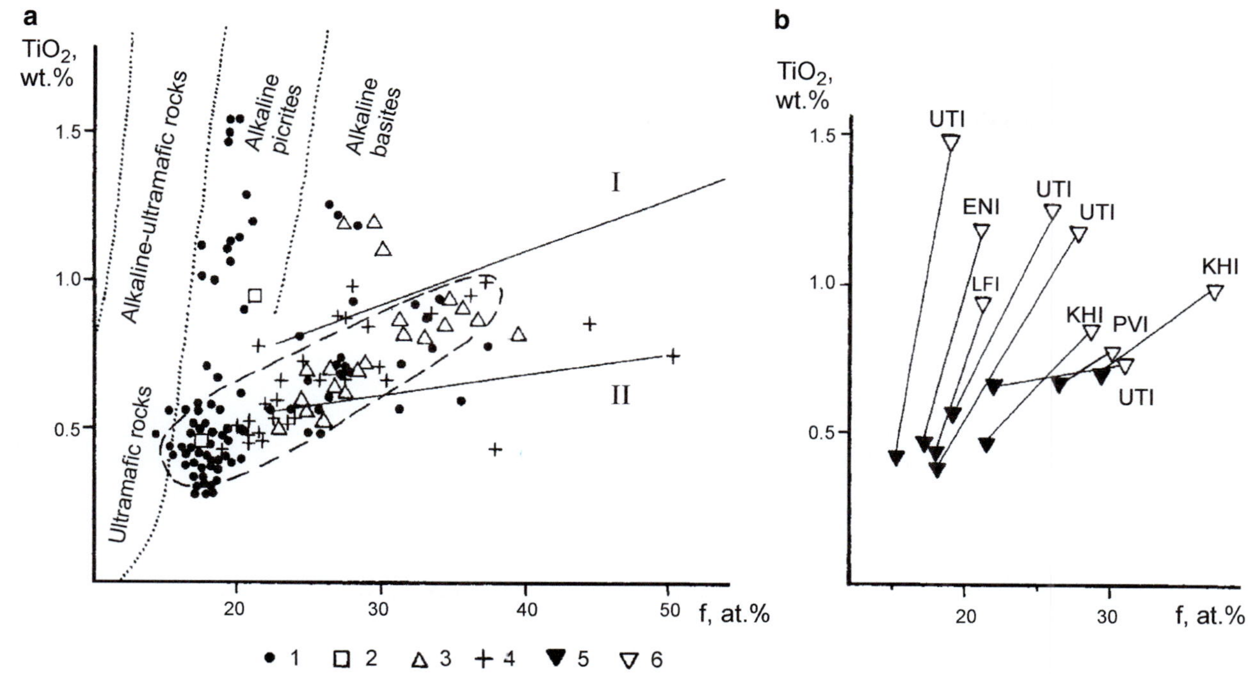

Fig. 4.5 Diagram TiO_2 versus f in clinopyroxenes from traps of the Siberian Platform (**a**) and behavior of oxides in zoned crystals (**b**). *1* rocks of the Upper Talnakh Intrusion (UTI); *2* picrite gabbrodolerite of the Lower Fokinsky Intrusion (LFI); *3* gabbrodolerites from the Mt. Pegmatitovaya Intrusion; *4* the same rocks from the Khungtukun Intrusion (KHI); *5, 6* core and rims of zoned crystals, respectively, of sampling grains from: *PVI* Pyasinsko–Vologochansky Intrusions, *ENI* east Noril'sk branch of the Noril'sk-I Intrusion. For comparison, the diagram shows the position of compositional fields of clinopyroxenes from rocks from various intrusions (Ryabov 1978a) and the trend for clinopyroxenes from sills of southern Siberian Platform (I) and intrusions of the eastern side of Tungussky syneclise (II) (Ryabov 1978b)

Fig. 3.23). The diagram shows that in the crystal cores unidirectional variations in the end member content take place from the base upward through the body: Wo decreases; En also decreases with some complication at the border of olivine and olivine-bearing gabbrodolerite, after which its value drops more abruptly; and Fs gradually increases. The composition of peripheral rims drastically changes at the boundary of the above-mentioned rocks. As a result, the Wo and En components in Cpx begin to increase dramatically, whereas Fs decreases.

The range of iron content in Cpx in one zoned crystal amounts to 19 at%. In separately taken Cpx cores, the whole range of total iron content in the intrusion section is 8 at%, whereas in peripheral rims, it is 7 at%. When considering the behavior of Cr_2O_3 and TiO_2 in Cpx, one can see that in olivine gabbrodolerite, clinopyroxenes have stable values of oxides both in the cores of crystals (0.60wt% Cr_2O_3 and 0.32wt% TiO_2) and in peripheral rims (0.07 and 0.74wt%), respectively. For the intrusion generally, the variations in Cpx composition in the body are evidence of the crystallization differentiation of melt.

Titanium in Clinopyroxenes

The behavior of TiO_2 and Cr_2O_3 in Cpx from rocks belonging to different formations, including traps, was earlier considered in Vilensky et al. (1970), Godlevsky et al. (1971), Dobretsov et al. (1971), Nesterenko and Almukhamedov (1973), Feoktistov et al. (1975), and Ryabov (1978a).

Vilensky et al. (1970), after revealing the relationship between the high-titanium pyroxenes and silicic acid undersaturated rocks with olivine and nepheline, come to the conclusion that the behavior of titanium in Cpx depends on the content of silicic acid rather than on alkalinity, that is, in rocks saturated with silicic acid Cpx must have a low titaninum content, whereas in silicic acid undersaturated rocks, Cpx has a high-titanium content.

Nesterenko and Almukhamedov (1973) compared the titanium content of Cpx from Skaergaard and draw the conclusion that the TiO_2 content in Cpx from traps must not exceed 1.0wt%. Feoktistov et al. (1975) consider these conclusions unfounded as they apparently did not establish the dependence of TiO_2 on iron content in Cpx and, moreover, did not take into account the high-iron Cpx found in Skaergaard but absent from traps. The problems of titanium content in Cpx were discussed by us earlier (Ryabov 1978a, b).

The arrangement of Cpx compositions fields from rocks with differing alkalinity (Fig. 4.5), which we plotted using materials from Dobretsov et al. (1971), implies that the titanium content of the mineral depends on rock alkalinity. Most analyses of Cpx from layered mafic rock intrusions on this diagram form a flat field with a minor increase in TiO_2 with increasing f. Lengthwise along the field, the

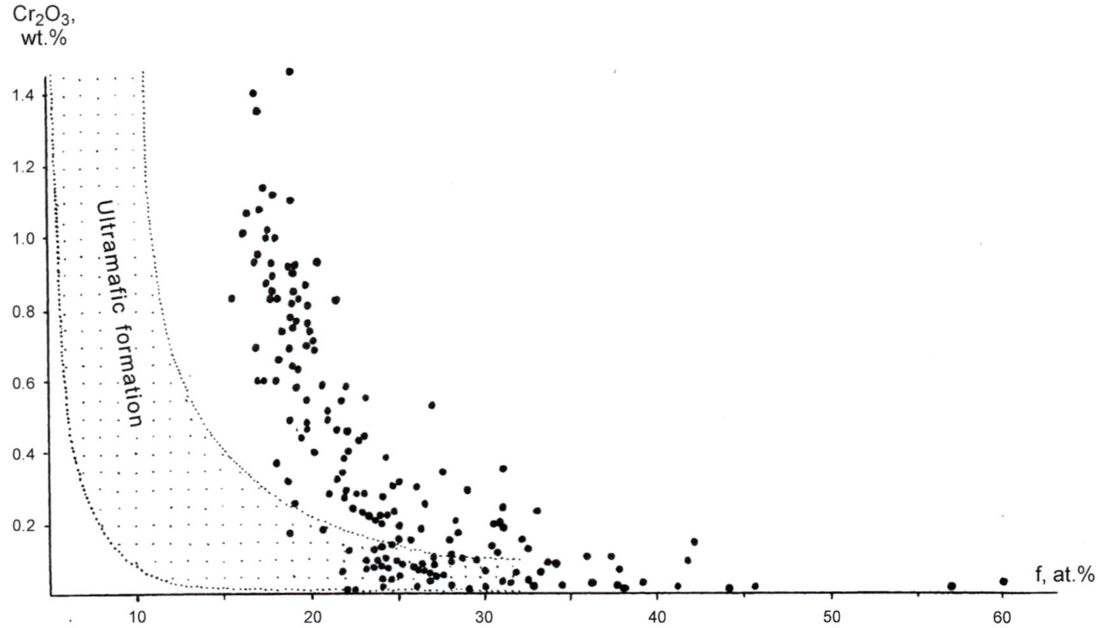

Fig. 4.6 Diagram of Cr_2O_3 versus f for clinopyroxenes from traps of the Siberian Platform (Ryabov 1978b). The compositional field of Cpx from ultramafic rocks from different intrusion is shown by speckles. Based on data from Dobretsov et al. (1971)

composition of the center and periphery of zoned crystals from different rocks of intrusions are localized. Clinopyroxene from the mafic rock parts of Noril'sky-type intrusions fall into the same field. A dramatic increase in titanium content was found in Cpx from picritic gabbrodolerite of the Upper Talnakh, Noril'sk-I, and Magnitny Stream Intrusions and from mafic pegmatoids of some intrusions, which fall into the field of alkaline picrite and mafic rocks. In such zoned crystals, TiO_2 dramatically increases from the center to the periphery (see Fig. 4.5b). This suggests an increase in melt alkalinity during the formation of frontal rims, which is consistent with the idea of the influence of elevated alkalinity and oxidation potential on the preferential incorporation of titanium into silicates rather than into oxides (Sobolev 1949).

The TiO_2 content of Cpx and alteration of the oxide in the center–periphery of zoned crystals in various intrusions differs significantly. This seems to be caused by the differences in the composition of the magmatic melt. An example of this is the geochemical trends of Cpx from the sills of the southern Siberian Platform and eastern flank of the Tungussky syneclise.

Chromium in Clinopyroxene

It was empirically established long ago that chromium is an indicator element in deep magnesium melts, but chromium content in basaltic magmas is negligible. In connection with this, detection of elevated chromium concentrations in Cpx from traps is of great interest as it aids in resolving the petrogenesis of the rock.

In 1970–1971, a few works were published concerning the behavior of chromium in pyroxenes from the Noril'sk-I, Upper Talnakh, and Lower Fokinsky Intrusions (Vasil'ev 1970; Genkin et al. 1970; Godlevsky et al. 1971; Dodin and Shatkov 1971; Ivanov et al. 1971a, b). Later, chromium in Cpx was discussed by Ryabov and Zolotukhin (1977), Ryabov (1978a, b), and Tuganova (1988). Chromium concentration in Cpx was found to vary widely, and its highest concentration was detected in the central parts of crystals. In different intrusions, the absolute value of Cr_2O_3 in Cpx varies considerably. The maximum Cr_2O_3 value of 1.27wt% was established in Cpx in picritic gabbrodolerite from the Upper Talnakh Intrusion. In spite of the fact that chromium is a trace element, its distribution in green Cpx obeys the normal distribution law and in brownish Cpx follows the law typical for these trace elements. This is related to the fact that in green varieties, chromium is one of the first to occupy the M2 position in pyroxene structure, which is similar to major components (Godlevsky et al. 1971).

The distribution of Cr_2O_3 in Cpx in traps, depending on the total iron content, is parabolic (Fig. 4.6). The plot shows that accumulation of chromium occurs in a narrow range of iron content. The minerals richest in chromium are green clinopyroxenes from picritic basalt and gabbrodolerite from the Lower Fokinsky, Magnitny Stream, Upper Talnakh, and Noril'sk-I Intrusions. Decreasing chromium content in Cpx from traps appears associated with decreasing magnesium content.

Chromium is considered an alien element for trap magmas, its abundance ratio in mafic rocks, according to

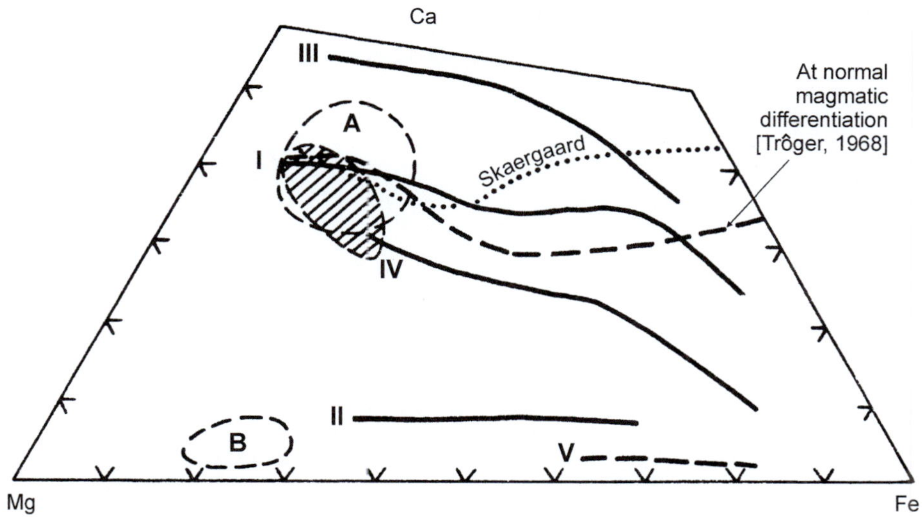

Fig. 4.7 Cumulative diagram of Ca–Mg–Fe illustrating crystallization trends of pyroxenes in Siberian Platform traps (Ryabov 1989b). Clinopyroxene compositional fields are outlined from the layered series of rocks (*shaded*), from traps and orthopyroxenes of the Noril'sk Region (*A*) and (*B*), respectively. Explanations are given in the text

Vinogradov, is 0.02%, which is an order lower than in ultramafic rocks (Godlevsky 1967). In spite of this fact, elevated concentrations of Cr_2O_3 have been determined recently in Cpx of some poorly differentiated intrusions: 0.55wt% Khuringda River Intrusion, 0.53wt% Sukhaya Bakhta Intrusion, 0.96wt% Nizhnii Imbak Intrusion (Zolotukhin 1984a), 0.74wt% Khungtukun Intrusion (Ryabov et al. 1985c), 0.84wt% Ust'-Khann'ya (Oleinikov et al. 1985), and 0.44wt% Tolstomysovsky sill (Feoktistov et al. 1975). These data provide evidence that chromium-rich Cpx are not a prerogative of the Noril'sk Intrusions. Taking into consideration the low solubility of chromium in tholeiite-basalt melts, the concentration of this element is not sufficient to form independent chromium-bearing phases (spinel), and for this reason, the element accumulates in the earliest high-temperature phase, that is, Cpx (Ryabov and Yakobi 1981).

The elevated chromium content of Cpx is typical of high-magnesium traps containing chromite. The compositional similarity of Cpx from picritic basalt and picritic gabbrodolerite from the Noril'sk Intrusions and picrites from the Fokinsky complex are an argument in favor of the comagmatism of these rocks. In addition, high-chromium Cpx in olivine-free rocks associated with picrite, such as metadiorite from differentiated picrite lavas, dendrites of Cpx from the Mikchadinsky lava flow sequence, and stalky crystals in gabbroic pegmatite from the Lower Fokinsky Intrusion, suggest chromium mobility at the magmatic stage.

Some problems of chromium distribution in Cpx from the Noril'sk Intrusions are discussed by Tuganova (1988). Based on the elevated chromium content of Cpx from Noril'sky-type picritic gabbrodolerite, similar to minerals from mantle-derived xenoliths and kimberlites, the author identifies their formation conditions (ignoring crystal morphology), which suggests the chamber crystallization of Cpx from the Noril'sk Intrusions.

Evolution of Pyroxene Composition During Crystallization Differentiation

The Hess trapezium diagram (Fig. 4.7) shows the compositional fields of groundmass clino- and orthopyroxene from traps and four crystallization trends evident in some bodies (Ryabov 1989b). Trend line I reflects the major crystallization trend of clinopyroxene from poorly differentiated intrusions. In the initial stages of crystallization, the trend has a similar direction to trends of the Skaergaard Intrusion and normal magmatic differentiation, but during the final stages has a higher iron content. Trend II forms pigeonite in poorly differentiated intrusions. Trend III forms crystals of the diopside–hedenbergite clinopyroxene series that are found mainly in mafic pegmatoid related to layered intrusions. Trend IV forms clinopyroxene of moderate and low calcium content, the latter being rare to absent. Trend V is formed by orthopyroxene with an increased iron content ($Fs_{63–82}$), which are followed by field-B bronzites of the Noril'sk Intrusions (see Fig. 4.7)

The main mineral-forming components of Cpx from the Noril'sk Intrusions were earlier assumed not to reflect the crystallization differentiation that took place (Genkin et al. 1971; Godlevsky et al. 1971; Dodin and Shatkov 1971). As mentioned above, our studies showed that normalization of Si, Ca, and Mg to 1 and (or) Fe components results in the

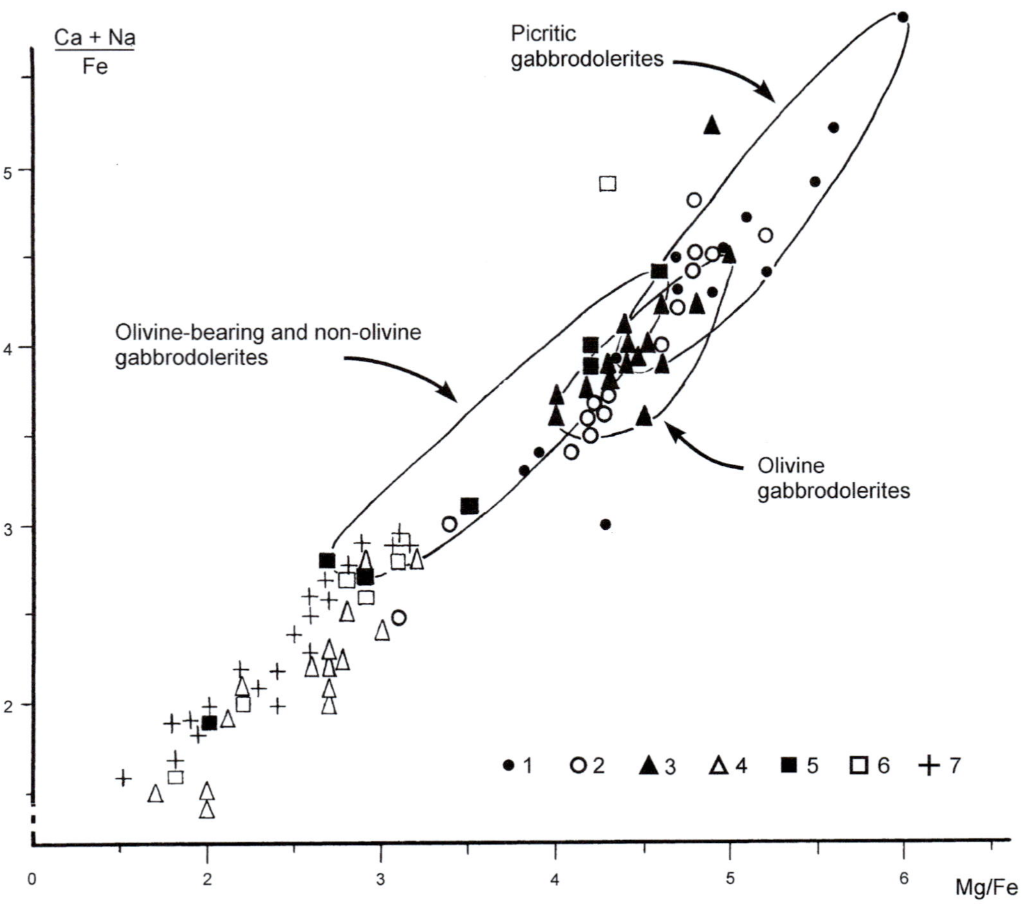

Fig. 4.8 Diagram of (Ca + Na)/Fe versus Mg/Fe for clinopyroxenes from gabbrodolerites of the Upper Talnakh (UTI) and Mt. Pegmatitovaya (PGI) Intrusions. *1* and *2* green and brownish pyroxenes from picrite gabbrodolerites of UTI, respectively; *3* and *4* the same minerals from olivine gabbrodolerites of UTI; *5*, *6* the same minerals from olivine-bearing and olivine-free gabbrodolerites of UTI; *7* brownish pyroxenes of PGI

ordered arrangement of analyses on plots as a trail with a minor dispersion (Ryabov 1974, 1978a). The compositional fields of Cpx from different rocks of the intrusion regularly and with an overlapping shift along the trail of dots have a general tendency to decrease in calcium and magnesium contents of the mineral and increase in iron content during crystallization differentiation (Fig. 4.8). The composition of brownish Cpx from mafic rock horizons of the Noril'sk Intrusions shows a similarity with the mineral of mafic rock intrusions.

A comparison of the compositions of Ol and Cpx coexisting within a single rock (Fig. 4.9) shows that the analyses of iron content in the cores of Ol phenocrysts and central parts of Cpx crystals form a trail with minor dispersion. A gentle slope of the trail suggests a higher iron content of Ol compared to associated Cpx. A second field (contoured by a dashed line, Fig. 4.9) corresponds to analyses of iron content of small Ol grains from the matrix and frontal rims of Cpx crystals. It is worth noting that in this field one can observe Ol and Cpx with similar iron content, and in general the iron content of Ol is closer to that of Cpx, than in the former case.

Of particular interest is the distribution of Ni in zoned clinopyroxene crystals in the Upper Talnakh Intrusion (Fig. 4.10). Variations in Ni content from the center to periphery of crystals, shown by the arrows on the diagram, undergo inversion at the border of the mafic and ultramafic parts of the intrusion. Crystallization of clinopyroxene in barren mafic rock was accompanied by a decrease in Ni content. Although, in the ore-bearing horizons of picrite and taxite gabbrodolerite, Ni content increased as the mineral crystallized.

4.1.2.3 Orthopyroxene

Rhombic pyroxene in traps is distributed nonuniformly (Sobolev 1986; Godlevsky 1959). It is absent from many rocks and increased content to 1–3% (less often 5–7%) is found in olivine-rich gabbrodolerite or adjacent rocks. In effusive traps, small amounts of orthopyroxene (Opx)

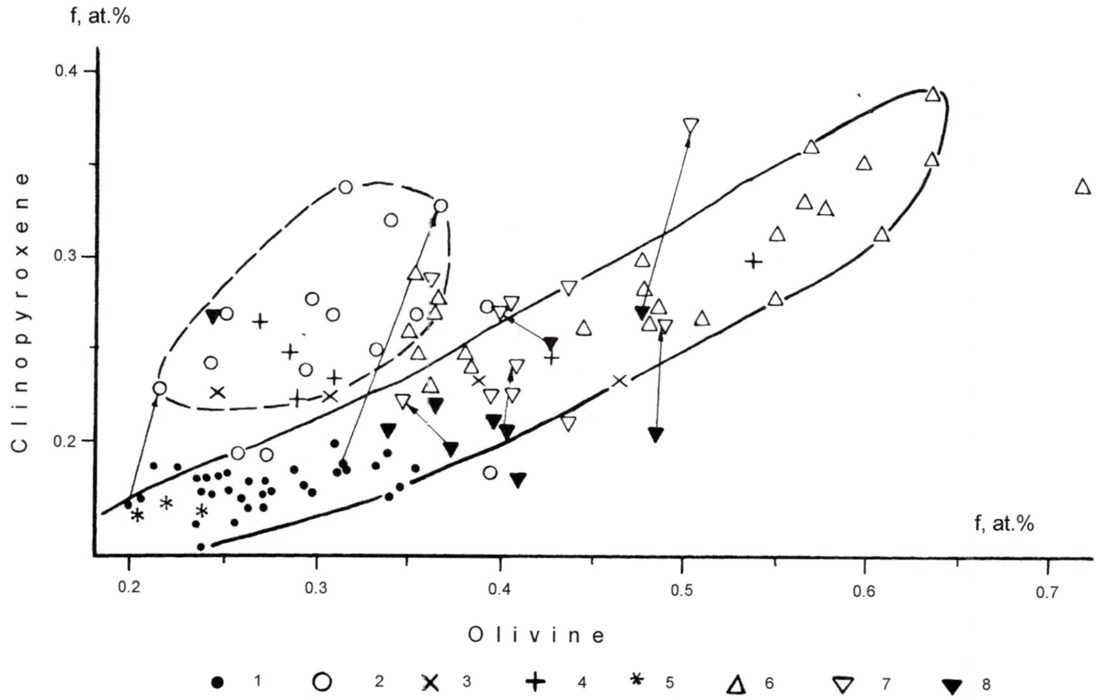

Fig. 4.9 Diagram of iron content in associated clinopyroxene versus olivine from rocks of different intrusions and picrite basalts of the northwestern part of the Siberian Platform. Noril'sky-type intrusions: *1* large Ol crystal in comparison with Cpx (core), *2* small Ol crystal in comparison with Cpx (rim), *3* Magnitny Stream Intrusion, *4* Pyasinsko–Vologchansky Intrusions, *5* picrite basalt, *6–8* Khungtukun Intrusion *6* large Ol crystal in comparison with Cpx (rim), *7* small Ol crystal in comparison with Cpx (rim), *8* small Ol crystal in comparison with Cpx (core). Pairs of minerals present in single sample are connected by *arrows*

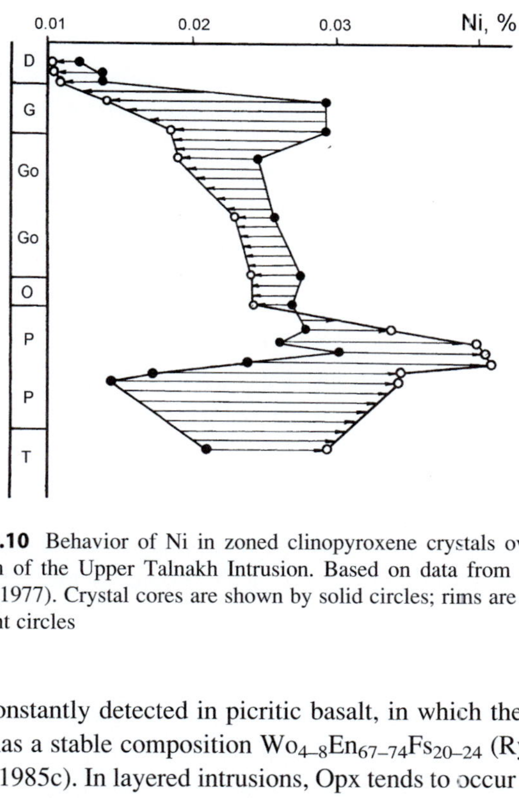

Fig. 4.10 Behavior of Ni in zoned clinopyroxene crystals over the section of the Upper Talnakh Intrusion. Based on data from Distler et al. (1977). Crystal cores are shown by solid circles; rims are shown by light circles

are constantly detected in picritic basalt, in which the mineral has a stable composition $Wo_{4-8}En_{67-74}Fs_{20-24}$ (Ryabov et al. 1985c). In layered intrusions, Opx tends to occur in the picritic gabbrodolerite horizon or concentrated in overlapping and underlying rocks: olivine, troctolite, and taxite gabbronorite. In the Noril'sk Region, Opx is widespread in gabbronorite of the Zubovsky Intrusion (Godlevsky 1959).

The forms of Opx in rocks are diverse. In picritic basalt, along with independent grains, one can observe rims of Opx near Cpx. In the poorly differentiated Khungtukun and Maymechinsky Intrusions, Opx forms in olivine gabbrodolerite, but it is most abundant in mineralized rocks (Ryabov et al. 1985a; Ryabov 1989b). The specific features of the occurrence of Opx in the Noril'sk Intrusions have been described in many petrographic works (Zolotukhin et al. 1975; Stepanov 1975; Zolotukhin and Shchedrin 1977). In quartz norites from the top of the Zubovsky Intrusion, Opx forms idiomorphic bronzite crystals, and lower in the section, in olivine gabbrodolerite, Opx has varying iron content Fs_{21-32} (Godlevsky 1959). In picrite and taxite gabbrodolerite, Opx forms rims around olivine and less often independent poikilocrysts.

Data on the chemical composition of Opx from traps are dispersed in different sources (Genkin et al. 1970; Ivanov et al. 1971a, b; Dodin et al. 1972; Zolotukhin et al. 1975; Ryabov and Zolotukhin 1977; Zolotukhin and Shchedrin 1977; Stepanov 1977; Genkin et al. 1981; Zolotukhin 1984a; Ryabov et al. 1985a; Tuganova 1988; Ryabov 1989b). In spite of the long list of works, each of them contains only a few analyses of the mineral.

Judging from microprobe analyses, the iron content of Opx, observed as independent crystals and overgrowing rims around Cpx, is Fs_{20-24} and is Fs_{10-12} in associated Cpx (Ryabov et al. 1985a). In the Noril'sk Intrusions, the iron content of Opx increases from the base upward through the intrusion from picrite (Fs_{22}) to olivine-bearing gabbrodolerite (Fs_{36}) (Stepanov 1977). The total iron content in picritic gabbrodolerite varies from 20 to 23 at% in the Upper Talnakh Intrusion, 18–24 at% in the Manturovsky Intrusion (Zolotukhin and Shchedrin 1977), and amounts to 25 at% in the Lower Noril'sky Intrusion (Genkin et al. 1981). These authors report that in olivine and taxite gabbrodolerite, the iron content in Opx is in the range of 27–29 and 31–32 at%, respectively, in the Upper Talnakh Intrusion and 28 and 27 at% in taxites and troctolites of the Manturovsky Intrusion.

Ferruginous Opx Fs_{54} was detected by its refraction index in ferrohypersthene ferrogabbro from the Alamdzhakh (Masaitis 1958) and Anakitsky (Reverdatto 1963) Intrusions. Microprobe analyses of high-iron Opx (up to Fs_{82}) in mineralized rocks from Khungtukun were performed earlier (Ryabov et al. 1985a; Ryabov 1989b).

Some idea on the composition of Opx and associated minerals can be gained from the diagrams presented earlier. In the picrite horizon of the Upper Talnakh Intrusion, the ratios of pyroxene end members in Opx and Cpx vary insignificantly (see Fig. 3.22). From the center to periphery of zoned Opx crystals, one can observe a slight decrease in the En component content and an increase in the Wo and Fs components. The titanium and chromium content of Opx are lower than in associated Cpx. On the Hess trapezium diagram (see Fig. 4.7), Opx forms a wide region of compositions, extending as a narrow trail from magnesium to iron-rich varieties. The former are connected with olivine-rich rocks mainly in the Noril'sk Intrusions (see Fig. 4.7, field B); the latter occur sporadically in gabbrodolerite of poorly differentiated mafic rock intrusions, and high-iron varieties were found in ferrogabbro and among reaction formations in the a-Fe ores from the Khungtukun Intrusion.

The investigation into the olivine–hypersthene equilibrium, performed by Stepanov (1977), showed that in picritic gabbrodolerite, the iron content in Opx is higher than in olivine (Fa_{21}–Fs_{22}), in taxite gabbrodolerite, they are equal (Fa_{29}–Fs_{29}), whereas olivine gabbrodolerite are characterized by inverse relations (Fa_{34}–Fs_{32} and Fa_{40}–Fs_{36}). The boundary of ultramafic and mafic (trap) associations of minerals is a narrow compositional field of equilibrated olivine–orthopyroxene with equal iron content (Fa_{29}–Fs_{29}) (Stepanov 1975).

Elevated Opx concentrations in the Zubovsky Intrusions were related by Godlevsky (1959) to the supersaturation of melt with alumina, which resulted from the assimilation of argillaceous sediments. Later, Stepanov (1977) arrived at the conclusion that orthopyroxene–olivine equilibrium is temperature and pressure-independent but rather depends on the alteration of chemical potentials of SiO_2 and FeO, owing to which Opx acts as a buffer additive to olivine, supporting a certain level of the chemical potential of SiO_2.

Taking into consideration the limited assimilation capability of tholeiite-basalt melts in hypabyssal chambers, the appearance of Opx in rocks might have resulted not from the supersaturation of melt with alumina, but from the change in chemical potential of SiO_2 and FeO, as reported by Stepanov (1977). Most likely, the reason for the latter is the high-temperature oxidation of early olivine (autometamorphic or with participation of fluids), as assumed for Hawaiian picrites (Muir et al. 1957) and supported experimentally (Haggerty and Barker 1967). The formation of Opx in the Noril'sk picrites was related by Sobolev (1986) to the resorption of olivine, after interaction with a residual melt.

4.1.3 Plagioclase

Plagioclase (Pl) along with pyroxene is the major rock-forming mineral of traps. Content varies from 0 to 100%, with composition including the entire isomorphous series from albite to anorthite. In addition, in alkaline silicic rocks, K–Na feldspars are not uncommon, which at times occupy a large volume (Ryabov and Zolotukhin 1977).

Typically the content of Pl in traps ranges 40–60%. Lower content occurs in melanocratic rocks, in which the proportion of olivine and pyroxene increases. High Pl concentrations are observed in mafic pegmatoids, leucogabbros, gabbropegmatites, taxite gabbrodolerite, etc. In addition, the occurrence of monomineral labradorite is known in the composition of differentiated picrite flow of the Khrebtovy Stream (Kahaelakh plateau) (Ryabov et al. 1985c) and peculiar chromite–plagioclase rocks, in the UZ of the Upper Talnakh Intrusion (Ryabov 1984a, b).

Effusive traps feature sheets of plagioclase-phyric and glomeroporphyritic basalt. The porphyritic appearance of Pl is not rare in the traps of the Upper Talnakh, Noril'sk-I, Lower Noril'sk, and Lower Talnakh Intrusions; Morongovsky and Zubovsky Complexes; in the poorly differentiated Khungtukun, Maymechinsky, and Dzhaltul Intrusions; and some other nondifferentiated intrusions.

The nature of plagioclasite bodies in gabbrodoleries and horizons of plagioclase-rich leucogabbro in the Noril'sk Intrusions was debated for a long time (Likhachev 1965; Tarasov 1976). Oleinikov (1979) attached much significance to the presence of leucogabbro or anorthite gabbro, which is evidence of an anorthite preference in melt differentiation in the prechamber period of their evolution.

Many geoscientists paid attention and tried to explain the relationship between the richest sulfide ores and plagioclase-rich rocks—taxite gabbrodolerite (Godlevsky 1959; Rogover 1959; Korovyakov et al. 1963; Zolotukhin 1964; Smirnov

1966; Ivanov et al. 1971a, b; Tarasov 1976). Sobolev (1986) called the Noril'sk taxite gabbrodolerite "sulfide pegmatites." Taxite gabbrodolerite and gabbro are the most important criteria of mineralization in layered trap intrusions (Zolotukhin 1978b). The problem regarding the genesis of feldspathic rocks (mafic pegmatites) is one of the key problems in elucidating the nature of the Noril'sk ore-bearing intrusions.

Regular transitions from taxitized picrites to magmatogenic breccia, taxite gabbrodolerite, and leucogabbro and gradual changes in the material composition of these rocks are evidence of a single mechanism for their formation; the interaction of melt and rocks with volatile elements (Zolotukhin 1964; Zolotukhin et al. 1975; Natorkhin et al. 1977).

The features of Pl occurrence in traps are discussed in many works (including Ryabov and Zolotukhin 1977). Study of Pl composition showed that its maficity, as a rule, correlated with the magnesium composition of rocks. It was also mentioned that, when analyzing some particular rocks and intrusion sections, changes in mineral–rock composition is significantly complicated by the presence of many generations of Pl and their zoning.

Zoning of Pl in traps is normally direct. In nondifferentiated traps: basalt, dolerite, and gabbrodolerite zoning is less pronounced than in layered bodies. It is most clearly manifest in the Noril'sk Intrusions. Multizoned tabular Pl vary widely in composition, whereas zoning is less developed in small prisms and is absent in laths. In picritic gabbrodolerite from the Upper Talnakh Intrusion (according to the measurements from the Fedorov table) from the center to periphery of zoned Pl crystals, the An component content varied as follows: (98–80%), (68–62%), (56–50%), (43–42%) (Zolotukhin et al. 1975). In olivine gabbrodolerite, from the same source, An component content in zoning varies: (95–85%), (82–70%), (68–55%), (50–42%). Whereas, in olivine-free varieties, An component content in zoning varies: (82–70%), (60–48%), (42–35%), (25–20%). In taxite gabbrodolerite, Pl composition from core to periphery, An component varies from 92% down to 54–25%.

In spite of the multizonality of Pl, the intrusion reveals a general tendency for deoxidation of mineral from the base to the top owing to an increase in alkaline components in the mineral composition. This was established by the chemical analysis of monofractions of Pl, representing an admixture of all zones and by recalculation of analyses to normative composition. Thorough analyses of Pl composition using the Fedorov table or microprobe suggest strong differentiation of melt within the intrusion, within single horizons and even within single crystals.

According to optical studies, Pl composition in picrite–troctolite gabbrodolerite from intrusive traps of the Siberian Platform varies from An_{96} to An_{70} (Zolotukhin 1978a). The Pl composition conforms to An_{84-94} in picritic basalt and to An_{70-80} in picritic gabbrodolerites from the Lower Fokinsky Intrusion; the Pl composition of phenocrysts from picrite-troctolite intrusions of the Morongovsky Complex and intergrowths in olivine and prisms in the groundmass have the composition An_{90-80} and An_{60-70}, respectively. In troctolite gabbrodolerite from the Lower Talnakh Intrusion, phenocrysts are An_{94}, whereas prisms are An_{60-70}, and in the same varieties of rocks from the Lower Noril'sk Intrusion, the composition of prismatic crystals is An_{60-85}. In olivine-poor and olivine-free gabbrodolerite, Pl composition deoxidizes from labradorite to andesine–oligoclase and even albite. To compare, in the Skaergaard Intrusion, the whole range of variations in Pl composition is An_{77}–An_{34} (Wager and Brown 1968).

Different research on layered intrusions (Upper Talnakh, Khungtukun, Morongovsky Complex, and Lower Talnakh) provides evidence of stable composition for Pl phenocrysts independent of their Upper Talnakh Intrusion position in the body and composition of the host rock. These are anorthite–-bytownite An_{98-82} in the Upper Talnakh Intrusion and bytownite An_{87-78} in the Khungtukun Intrusion (Lyul'ko et al. 1972; Ryabov et al. 1985a). This suggests that high-mafic Pl was earlier a liquidus phase. In the Khungtukun Intrusion, this refers to the whole magmatic body and to mafic layer in the Upper Talnakh Intrusion. In the ultramafic layer of Upper Talnakh Intrusion, Pl composition was governed by the composition of liquid.

In many petrographic works, the results of studies of Pl zonality and variations in composition in intrusions (e.g., in the Upper Talnakh Intrusion) show that the range of compositional variations in one crystal is virtually identical to the whole range of variation in mineral composition in the whole intrusion. Microprobe analyses of plagioclase in traps support the tendency of variations in their composition established by optical methods. The mineral compositions determined by different methods were similar, ruling out the high-mafic anorthite–bytownite plagioclase that, according to data from X-ray analysis, shifted to the bytownite–labradorite field. Nevertheless, the available data provide evidence of the active deoxidation of Pl, which suggests rapidly increasing alkalinity of the melt during its crystallization, both within one horizon of rocks and from the base upward through the section.

In addition to the maficity of Pl, the indicators of melt differentiation are trace elements in the mineral. It is known that the most abundant trace elements in Pl are Fe, Mg, Ba, Sr, and B. Some features of the geochemistry of the last three elements were discussed in Anastasenko (1973), Nesterenko and Almukhamedov (1973), Feoktistov and Elisar'eva (1976).

The rapid Pl crystallization in basalts and rocks outpoured to the surface from hypabyssal intrusions is the reason why crystals trap allogeneic chemical elements, for example, Fe. In addition, we plotted a diagram of FeO versus An component (Fig. 4.11), which shows that the content of

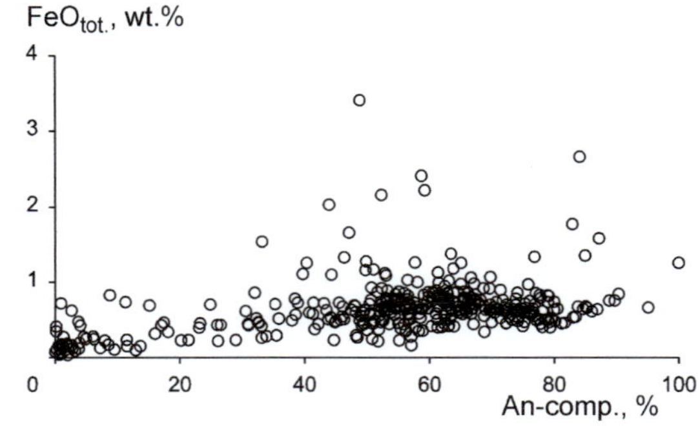

Fig. 4.11 Diagram FeO versus an component for feldspars from traps of the Noril'sk Region

FeO in plagioclase varies from 0 to 1.0wt%; higher values are less typical. With increasing anorthite in plagioclase, FeO also increases.

Boron, strontium, and barium are isomorphously incorporated into the feldspar lattice; minor amounts of strontium and barium also enter pyroxene lattices.

Our analyses of boron in Pl from effusive and intrusive traps show that the initial boron content in different melts differs considerably. For example, in Pl from glomeroporphyritic basalts of two different formations boron content was 51 and 10 ppm. In differentiated streams of picritic lava, boron concentration in metadiorite was higher than in basalt. In layered intrusions, the boron content regularly increases in minerals and in rocks from the base upward through the intrusion (Fig. 4.12). The boron content of Pl from mafic pegmatoids is typically higher than that in gabbrodolerite in layered rocks.

Study of Ba and Sr in Pl showed the highest concentrations (1,700 and 2,000 ppm, respectively) in pegmatoids from the UZ of the Pyasinsko–Vologchansky Intrusion. In Pl from olivine gabbrodolerite in the middle part of the intrusion the concentrations of these elements are 108 and 341 ppm, and in near-bottom taxites, 160 and 290 ppm. A similar tendency was observed in Pl from the Lower Talnakh Intrusion.

Isomorphism of Ba and Sr in Pl is related to the mineral maficity, as a result of which the increase in the alkalinity of mineral and rocks from bottom upward the section of the Upper Talnakh Intrusion is accompanied by increasing content of these elements and growth of Ba/Sr values (see Fig. 4.12). Considerable variations in the content of B (7–65 ppm), Ba (34–2,100 ppm), and Sr (97–2,000 ppm) in plagioclase and changes in their concentrations in layered intrusions reflect the process of crystallization differentiation of the melt.

4.1.4 Spinel Group Minerals

In the Siberian Platform traps, spinel group minerals include magnetite, spinel, and chromite.

The minerals of magnetite series are represented by titanomagnetite, magnomagnetite, magnoferrite, and magnetite. Titanomagnetite is the most abundant variety both in effusive rocks and intrusive rocks. It is found in minor amounts in virtually all rock type, and its crystal forms in basalt, dolerite, and gabbrodolerite are diverse. Normally, it forms small isometric subidiomorphic and xenomorphic grains, skeletal crystals, poikilocrysts with intergrowths of plagioclase, crystals with and without exsolution structures, graphic intergrowths with pigeonite, balls in glass, and finally sites of interstitial sideronite dissemination that coalesce with silicates.

Magnetite deposits, ore manifestations, and minor accumulations are widely known in connection with traps. The elevated concentrations of magnetite (up to massive compact ore) are known from basalt, eruption pipes, in the intrusions of the Noril'sky, Kuz'movsky, and Angarsky Complexes (Urvantsev 1959; Pavlov 1961; Khryanina

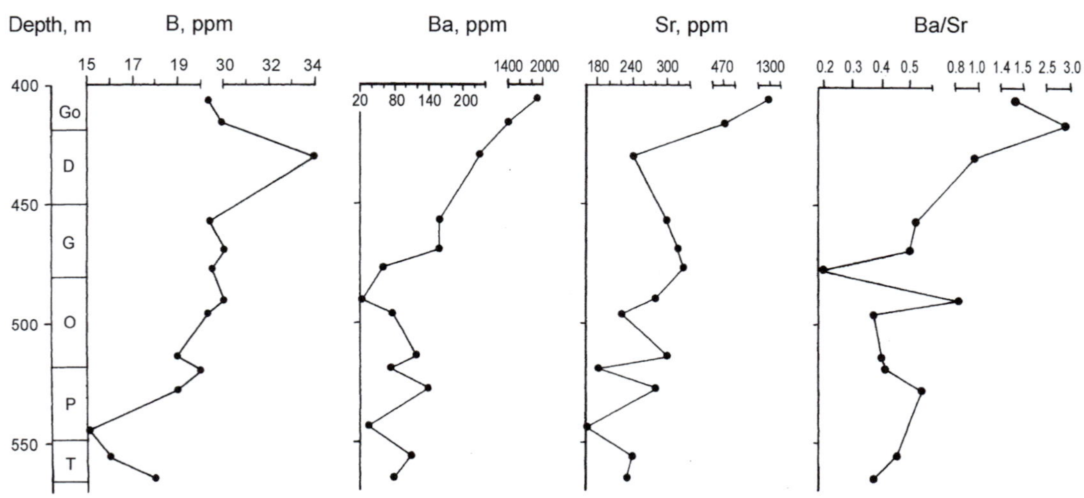

Fig. 4.12 Behavior of boron, barium, and strontium in the section of the Upper Talnakh Intrusion, drillhole KZ-274

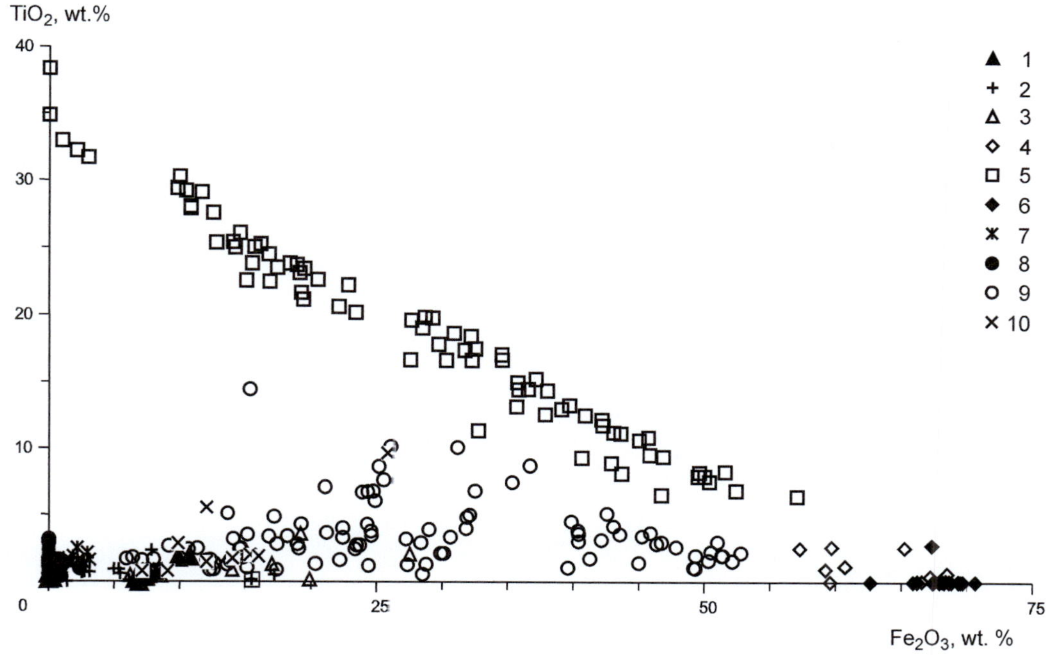

Fig. 4.13 Diagram of TiO_2 versus Fe_2O_3 (calculated data) for spinel from magmatic rocks of the Noril'sk Region. *1* hercynite; *2* pleonaste; *3* spinel; *4* magnetite; *5* titanomagnetite, ulvospinel; *6* magnomagnetite; *7* magnesialuminochromite; *8* magnesiochromite; *9* chromite; *10* chromepicotite

1962; Godlevsky 1967; Kavardin 1976; Ryabov 1976; Yudina et al. 1977; Ryabov and Zolotukhin 1977).

The diversity of forms of magnetite, its variable composition, and the wide spectrum of associated minerals suggest crystallization of magnetite in a wide range of – conditions, starting during the magmatic stage and ending in low-temperature hydrothermal conditions (Ryabov and Zolotukhin 1977). Here, we shall touch upon only magmatic accessory magnetite.

Most minerals of magnetite series in effusive and intrusive traps of the Noril'sk Region are titanomagnetite.

The TiO_2–Fe_2O_3 diagram (Fig. 4.13) shows the composition of spinel group minerals. Titanomagnetites form a band of compactly arranged dots with wide variations in TiO_2. The maximum oxide content (32–38wt%) is present in ulvospinel, followed by a swarm titanomagnetite dots containing a gradual decrease in TiO_2 from 30 to 7wt% that are replaced by magnetites with an admixture of TiO_2 up to 4wt%. Elevated TiO_2 content of 10wt% (and rarely higher) is found in chromite and chromopicotite.

In the derivatives of a tholeiite-basalt melt, the highest titanium magnetites are found in effusives and in pegmatoid accumulations of layered intrusions. In titanomagnetite from gabbrodolerite of different magmatic complexes, the TiO_2 content is highly variable. The mineral contains 9.57–15.01wt% TiO_2 in the rocks of the Daldakansky Complex, 14.42–16.66wt% in gabbrodolerites of the Mountain Pegmatitovaya Intrusion (Morongovsky Complex), 17.10–20.20wt% in the Ogonersky Complex, and 2.58–22.57wt% in the Kruglogorsky type (Noril'sk Complex). Titanomagnetite with a high ulvospinel component 70–80% (TiO_2; 19–27wt%) was found in gabbrodolerites of the Khungtukun and Maymechinsky Intrusions (Ryabov 1989b).

In derivatives of trachybasalt melts, the content of TiO_2 in titanomagnetite varies from 15.27 to 34.87wt% with a predominance of high titanium content. In some cases, the mineral composition is similar to the stoichiometric composition of ulvospinel Fe_2TiO_4, which must contain 35.7% TiO_2 and 64.3% FeO. Virtually pure spinels were established by microprobe analysis in trachydolerites of the Ergalakhsky Complex. On the diagram (Fig. 4.13), they are isolated as a group of dots in the region of the highest TiO_2 concentration. Titanomagnetites containing more than 70% of ulvospinel component are found in some effusives and intrusions of the normal series of different complexes.

Titanomagnetites contain minor amounts of trace elements. The amount of Al_2O_3 is typically lower than 2.0wt%, seldom up to 3.30wt%, and MgO is no more than 1.5wt%, occasionally up to 2.63wt%. Some insight into the distribution of MnO and V_2O_5 in titanomagnetites can be gained from Fig. 4.14. The diagram shows the results of microprobe analysis of titanomagnetites and chrome spinels, which are listed in the Rock indications provided in Volume 2. One can see from the diagram that the oxide content in minerals varies widely. The MnO content in titanomagnetite is normally lower than 1.7wt% but might amount to 3.8wt%.

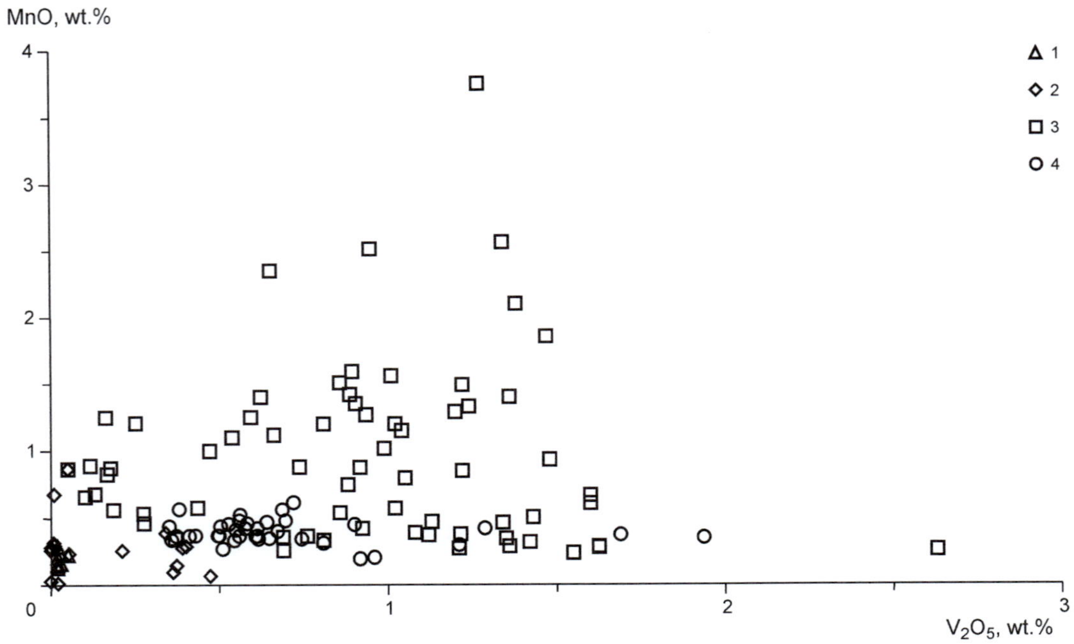

Fig. 4.14 Diagram MnO versus V_2O_5 for spinel from magmatic rocks of the Noril'sk Region. *1* spinel, *2* magnetite, *3* titanomagnetite, *4* chromite

The V_2O_5 content varies from 0 to 1.7 wt%, in one case reaching 2.7wt%.

Magnomagnetites are a particular variety of magnetite that form ore lavas in sheets of trachybasalts of the Yuryakhsky Unit (Ryabov 1997). The mineral features an elevated MgO content (3.58–5.82wt%). A more detailed characterization of magnomagnetites is given in Chap. 3.7.3. Magnetites poor in and free from trace elements in traps occur in late and postmagmatic formations. Specific features of their occurrence and composition are discussed in Kavardin (1976), Ryabov and Zolotukhin (1977), and Yudina and Sandomirskaya (1980).

In addition to magnomagnetites of ore lava, individual minerals were found in fragments of magmatogenic breccia of the Upper Talnakh Intrusion (see Rock indication 130) and in picritic gabbrodolerite of the Lower Noril'sk Intrusion (see Rock indication 155). The former contained 12.63wt% and the latter 2.56wt% MgO. Uncommon composition was detected in the oxide-ore phase in picritic gabbrodolerite of the Pyasinsko–Vologochansky Intrusion (see Rock indication 169). It contained 29.27wt% TiO_2, 1.61wt% Al_2O_3, 12.69wt% Fe_2O_3, 31.97wt% FeO, 0.32wt % MnO, 15.48wt% MgO, 1.10wt% Cr_2O_3, 0.12wt% NiO, and 0.69wt% V_2O_5.

The minerals of spinel series in traps are represented by a number of varieties: true spinel ($MgAl_2O_4$) through compositionally intermediate pleonaste ($(Mg,Fe)Al_2O_4$) to hercynite ($FeAl_2O_4$). Pleonaste includes spinel, in which $Mg:Fe^{2+}$ ranges from 3 to 1; this value in spinel is >3 and in hercynite is <1 (Deer et al. 1966). The stoichiometric composition of spinel contains 28.2% MgO and 71.8% Al_2O_3, and the composition of pure hercynite must contain 41.3% FeO and 58.7% Al_2O_3. Spinel minerals commonly occur in compositionally different pegmatoid formations of layered intrusions: in leucogabbros, taxite gabbrodolerite and magmatogenic breccia. Their occurrence was described in the Noril'sk, Zub, and Khungtukun Intrusions (Genkin et al. 1976; Ryabov and Pavlov 1984; Ryabov 1989b; Sluzhenikin et al. 1994).

In pegmatoids of the Noril'sk Intrusions, spinels are green, gray-green, honey yellow, and brownish. The spinel content varies widely from single grains to 10–15% rock content. Nevertheless, green spinel in magmatic rocks are very scarce, and until recently, the only observations of green spinel in thin sections and insoluble residue of the chemical decomposition of olivine were reported by Likhachev (1965) and Stepanov (1975, 1977). It is worth noting that the above-mentioned authors do not doubt the magmatogenic origin of green spinel in gabbrodolerite.

The features of spinel occurrence are described in Chap. 3.6.1 and illustrated in the thin section photographs. It is worth emphasizing once more that the forms of its occurrence are similar to those of chrome spinel, which it may coexist in one rock or even within one thin section (see Rock indication 107).

Red and violet hercynite and pink-lilac pleonaste were described in taxite gabbrodolerite, in glasses and melilite segregations of the Khungtukun Intrusion (Ryabov et al. 1985a; Ryabov 1989b). They were also found in the pegmatoid horizon of the UZ in the Dzhaltul massif.

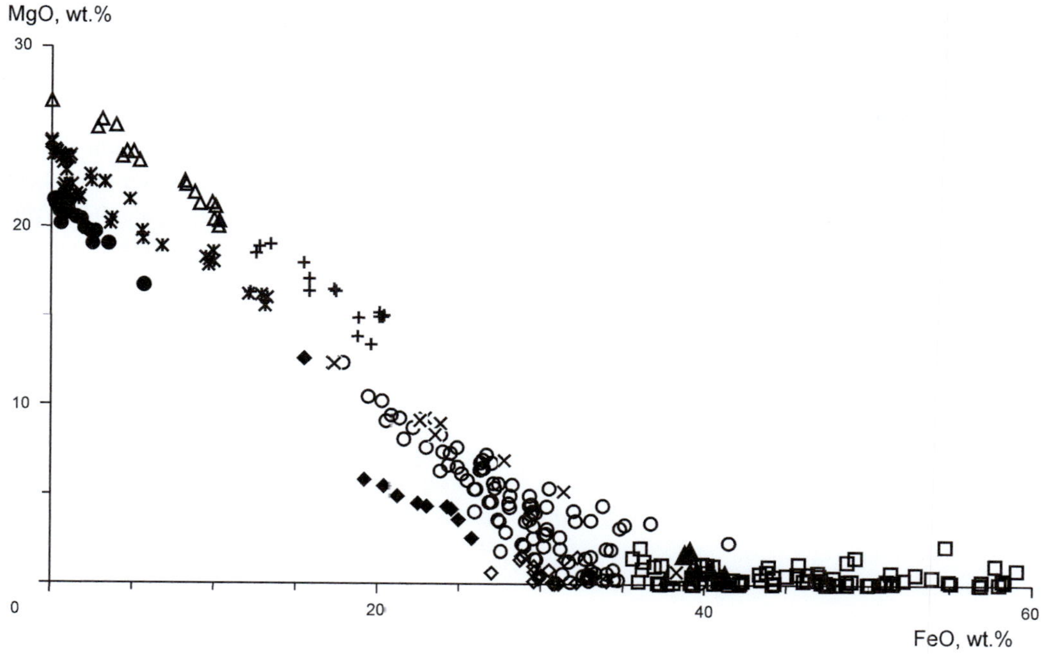

Fig. 4.15 Diagram MgO versus FeO (calculated data) for spinel from magmatic rocks of the Noril'sk Region. See Fig. 4.13 for legend

Their forms are identical to those of spinel from the Noril'sk Intrusion. Hercynite and pleonaste were found as inclusions in olivine, pyroxene, plagioclase, and in native iron. Thus, the magmatogenic origin of these spinels is beyond doubt.

The composition of spinel series minerals in traps varies widely and includes virtually the entire series of varieties. In the Noril'sk Region, spinels form in pegmatoids of the UZ, in fragments of magmatogenic breccia, and chromium-free endoskarns related to the Noril'sk Intrusion (see Rock indication 111). Pleonaste crystals are present in the magmatogenic breccia of the Kharaelakhsky branch and in pegmatoids from the UZ of the Dzhaltul–Khungtukunsky-type intrusions, and hercynite is found in pegmatoids of the UZ of the Noril'sky-type and Dzhaltul–Khungtukunsky-type intrusions. The MgO–FeO diagram (Fig. 4.15) shows the variety of spinel compositions from the Noril'sk Region. According to the ratio of MgO and FeO, the spinel composition dots are arranged in three bands. The first band is a series of points of spinel–pleonaste–hercynite composition, the second of magnesioalumochromite and chromepicotite–chromite, and the third of magnesiochromite–magnomagnetite. The content MgO in titanomagnetites and magnetites is normally no more than 1–1.5wt%.

The minerals of spinel series have a high-Al_2O_3 content from 45.67 to 68.82wt%. The MgO content ranges from 20.05–26.96wt% in spinel, 13.44–19.05wt% in pleonaste, and 0.31–1.89wt% in hercynite (in single cases up to 11.77wt%). The FeO content varieties from 0–10.24wt%, 12.2–20.35wt%, 37.87–41.27wt% (in single cases 24.01wt%), respectively. The concentration of TiO_2 in the minerals varies from 0 to 3.62wt%, and the MnO, NiO, and V_2O_5 content is more than a few fractions of a percent.

In the halo of layered rocks of the Upper Talnakh Intrusion, there are metamorphic rocks with green spinel, forsterite, phlogopite, and fassaite, as described by Yudina (1965), Turovtsev, Batuev, Zolotukhin, Ryabov, and Zotova, but they are not discussed in the present work.

Chromite series minerals were found in picritic basalt, picrite dolerites of the Mikchandinsky neck, and in picritic gabbrodolerite of high-magnesium intrusions: Lower Fokinsky, Magnitny Stream, Ergalachny Stream, and Upper Pyasinsky Intrusions. The forms of occurrences of chromite in picrites from the above-mentioned areas are identical, which, along with other mineralogic–petrographic features, allows one to classify these rocks as a single genetic type. Chromite in these rocks forms a uniformly dispersed dissemination of small octahedral crystals with smoothed peaks that are observed in plagioclase, clinopyroxene, and (less often) in olivine. Observation of chromite is a good criterion for distinguishing picrites of picritic basalt melts from picrites as derivatives of tholeiite-basalt magmas. In picrite and troctolite gabbrodolerite from intrusions of the Morongovsky Complex, Zubovsky Intrusion, the middle course of the Yenisey River, and many others, chromite is extremely rare.

In the layered intrusions of the Noril'sky-type, chromite mineralization was found long ago, and some data on the forms of its occurrence are reported in Ivanov et al. (1971a, b), Zolotukhin et al. (1975), and Ryabov and Zolotukhin

(1977). The first detailed studies were undertaken by Genkin et al. (1979, 1981). They showed that in the Noril'sky-type intrusions, chromite is concentrated in picritic gabbrodolerite. It was established that, according to reflectivity, two varieties can be distinguished, dark and light, and their distribution can be defined as dispersed dissemination and accumulation. Later, we established another horizon in the Noril'sk Intrusions with chromite mineralization in the contact zones (Ryabov et al. 1982; Ryabov 1984a, b). Study of chromite mineralization in the chamber roof of Noril'sk Intrusions showed their great similarity with chromite from in the near-basal parts of massifs. In the UZ, as in lower picrites, we have found light and dark chromites, dispersed dissemination and accumulations. Dark chromite forms grains with smoothed rounded shapes, and light chromite makes up independent idiomorphic crystals with sharp peaks and overgrowth rims around the former. The composition of dark chromite is magnesium–aluminum-rich, and the light chromite is iron–titanium-rich and depleted in chromium.

Our data, and data from other researchers, show that in the near-basal parts of the Noril'sk Intrusion chromite is found in the picritic gabbrodolerite horizon (Ivanov et al. 1971a, b; Genkin et al. 1979, 1981; Kudelina et al. 1983). In all other rocks, including those adjacent to picrites of olivine and lower taxite gabbrodolerite, chromite is absent or is rare to absent. Opinions differ on the estimation of chromite distribution within the picrite horizon and differentiation mechanisms. Some believe that the highest content and the largest accumulations occur in the lower parts of the horizon (Genkin et al. 1979). Others think that the highest chromite concentrations occur in the upper part of a picrite layer (Ivanov et al. 1971a, b; Kudelina et al. 1983). Our research (Ryabov and Yakobi 1981) supports the latter observations and provides evidence that the accumulation of chromite up to 20–30 vol% always occurs in the upper part of a picrite horizon, in the roof of the intrusion, or close to it. From the top down through the picrites, chromite content drops sharply, and the mineral is absent from the base of the layer and underlying rocks.

As mentioned above, the features of the occurrence of chromite in UZ are very similar to those described in the horizon of lower picrites (Genkin et al. 1979). Accumulations of the mineral to 30–40% and even to 60% were found in gabbrodolerite of various compositions. The sites with dense chromite dissemination have sharp borders with ore-free parts and form round and lenticular segregations and layers. Chromite grains are arranged in clinopyroxene, plagioclase, biotite, and amphibole. Chromite forms overgrowths on olivine and phenocrysts of early plagioclase. Similar to near-basal picrites, the border of chromite accumulation might pass through a single mineral (clinopyroxene, plagioclase, or biotite), as a result of which one part of the mineral features abundant in chromite intergrowths and the other is completely chromite-free. Similar observations for chromite accumulations were earlier described in lower picrites (Zolotukhin et al. 1975; Genkin et al. 1979) and are used by geoscientists as a proof of the liquation nature of ore segregations (Fenogenov 1987a, b). For the lower picrites, Genkin et al. (1979) report that some drop-shaped sulfide globule impregnations localized within chromite accumulations are composed of pyrrhotite, chalcopyrite, and pentlandite often intergrown with prehnite. Moreover, sulfides cement chromite grains, and late chrome-bearing magnetite in picrites is localized among interstitial sulfides. Thus, joint occurrence of compositionally different ore accumulations, the globular shapes of their occurrence and relative position suggest the existence of immiscible ore liquids (Genkin et al. 1979; Fenogenov 1987a, b; Ryabov 1989b).

The Chemical Composition of Chrome Spinels

Investigations show that the composition of spinels in traps varies widely, especially in the Noril'sk Intrusion. Some insight into the mineral composition of chromite series can be gained from the diagrams plotted for the Siberian Platform traps and the comparison of mineral composition from different formations.

The Cr_2O_3–Al_2O_3 diagram (Fig. 4.16) shows that spinels have wide compositional variations and differing tendencies in their behavior. Spinels of the composition magnesiochromite–magnesioalumochromite–chromopleonaste–chromepicotite (chrome hercynite) form a trail of dots whose position indicate a strong relationship between Cr and Al and suggests isomorphism in the mineral structure. Magnesiochromite contains 56.4–69.63wt% Cr_2O_3 and 1.72–14.73wt% Al_2O_3. In magnesioalumochromite, the concentrations of these oxides are 30.84–50.72wt% and 20.54–34.97wt%, respectively. Chrome-poor Al-spinels in traps are scarce. On the diagram, this variety of mineral is represented by single analyses of chromepicotite (9.91wt% Cr_2O_3) from pegmatite of the UZ of the Upper Talnakh Intrusion (see Rock indication 107) and chromehercynite (2.0wt% Cr_2O_3) from pegmatoids of the UZ of the Khungtukun Intrusion (Ryabov 1989b). Magnesiochromites and magnesioalumochromites, as was mentioned above, form in magnesium calcic and magnesium apotaxite endoskarns of the UZ of the Upper Talnakh Intrusion.

The groundmass of spinel from lower and upper picrites and upper taxite gabbrodolerite contains chromites and chromepicotites, which have variable Cr_2O_3 content (11.05–46.13wt%), Al_2O_3 content (0.51–27.34wt%), and a direct parabolic relationship between oxides (see Fig. 4.16). The Cr_2O_3 content grows rapidly to 30wt% with increasing Al_2O_3 to 10wt%, and, with further increase in the chrome

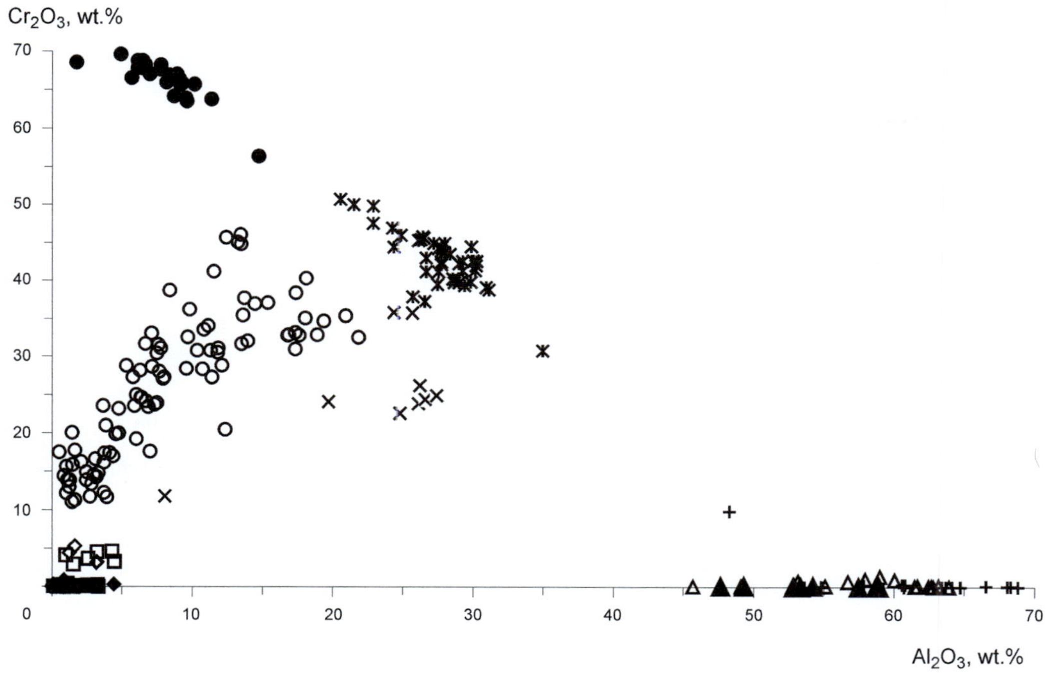

Fig. 4.16 Diagram Cr_2O_3 versus Al_2O_3 for spinel from magmatic rocks of the Noril'sk Region. See Fig. 4.13 for legend

oxide content from 30 to 40wt%, Al_2O_3 increases from 10 to 25wt%. The highest Cr_2O_3 content of 46.13wt% was found in chromite from picritic basalt.

In titanomagnetites from bubbly oxide-ore formations in leucogabbros of UZ of the Upper Talnakh Intrusion, the Cr_2O_3 content varies from 2.97 to 4.66wt%, and Al_2O_3 ranges from 0.99 to 4.45wt%. In magnetites of olivine–biotite gabbrodolerite from layered intrusions, the amount of these oxides ranges within 3.12–5.33 and 1.22–3.18wt% (see Fig. 4.16).

On the ternary diagram Al–Cr–(Fe^{3+} + 2Ti) (Fig. 4.17), the crystallization trend of chromite from picritic basalt shows a gradual decrease in the Cr and Al content in the mineral and an increase in Fe^{3+} and Ti. When comparing the compositions of chromite from picritic basalt of the Ayansky sheet with similar rocks in the Gudchikhinsky Suite in the Noril'sk Region (Table 4.5), one can see a similarity, whereas the minerals from volcanic ultramafic rocks of the Maymecha–Kotuysky Province are richer in Mg, Cr, and Fe^{3+} but poorer in Al, Mn, and Fe^{2+}.

Chrome spinel from the Noril'sk Intrusion show different tendencies of compositional variations on the ternary diagram: chromite from segregations in picritic gabbrodolerite have a rather stable composition (see Fig. 4.17), the field with specks and disseminated chromite form a trail, which indicates a crystallization trend (dashed line) similar to picritic basalt. Our data (solid dots) follow the trend based on the data from (Genkin et al. 1979). A specific compositional feature of spinel from the Noril'sk Intrusion is the common changes in their composition toward enrichment in Al and the existence of magnesium–iron-bearing Al-spinels.

The composition of chromite varies even within one chromite segregation in picritic gabbrodolerite. Microprobe analyses of 28 chromite grains from one sample were carried out by Genkin et al. (1979). The oxide content of chromite varied as follows: 2.1–9.7wt% TiO_2, 5.4–21.0wt% Al_2O_3, 33.7–41.1wt% Cr_2O_3, 4.9–12.2wt% Fe_2O_3, 24.1–35.5wt% FeO, and 3.2–8.4wt% MgO. More variations were observed in the composition of spinel in the upper pegmatoids. A clear example is the analyses of spinel from bubbly spinel formations shown in the microphotograph in Rock indication 107. Microprobe was used to determine the composition of spinel rings among the bubbles localized in a 1-cm-long area. It was found that the composition of spinel from different bubbles varies in all mineral-forming oxides. In this thin section, three stable compositions of spinel were observed. Oxide concentrations were distributed as follows: Al_2O_3 24.80–26.59wt%, 1.01–4.13wt%, and 58.90–60.77wt%; Cr_2O_3 22.64–24.46wt%, 12.21–17.44wt%, and 0.29–0.33wt%; $FeO_{tot.}$ 35.52–37.96wt%, 64.66–75.84wt%, and 20.68–22.10wt%; and MgO 8.35–9.27wt%, 0.10–1.81wt%, and 16.44–16.54wt% (see Rock indication 107). It is also worth noting that the composition of mineral-forming oxides, F and Cl, in micas from different bubbles varied considerably. These data provide evidence that non-equilibrium conditions of mineral formation existed with the composition of spinel from upper pegmatoids faithfully reproducing the changes in the composition of mineral from the lower picritic gabbrodolerite.

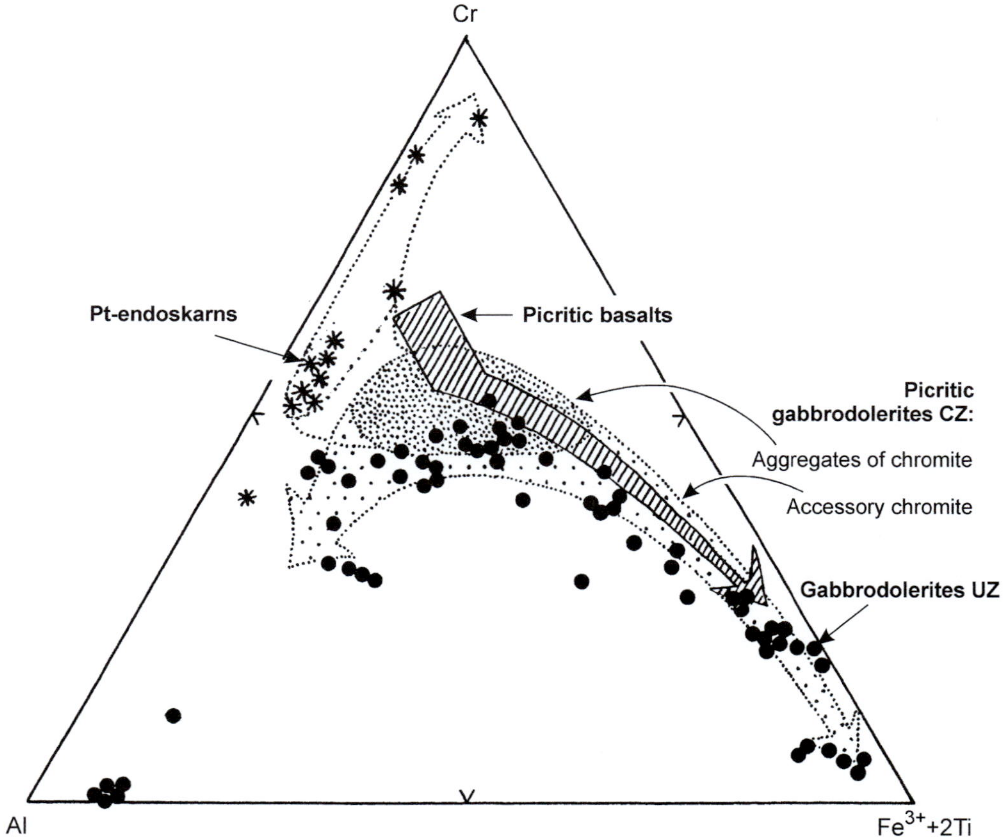

Fig. 4.17 Specific features of the evolution of spinel composition from picrite basalts and rocks from the Noril'sky-type intrusions

Table 4.5 Major mineral-forming oxide ranges in Cr spinel from traps of the Siberian and African Platforms

	1	2	3	4	5	6	7
TiO_2	1.42–6.23	3.23–6.34	1.33–2.07	1.9–12.5	2.00–9.70	0.90–11.97	0.47–10.39
Al_2O_3	3.04–14.46	9.35–11.06	5.25–7.47	5.1–20.0	5.4–21.0	5.77–26.72	5.66–27.90
Cr_2O_3	22.46–46.12	34.32–42.17	42.69–53.90	18.9–38.6	33.70–44.30	11.28–40.33	18.17–49.43
Fe_2O_3	6.73–36.16	–	13.57–29.04	10.5–27.5	4.90–13.10	8.58–28.25	2.40–28.44
FeO	21.85–29.29	33.62–40.44	9.18–18.25	18.4–38.4	21.40–35.50	18.55–43.17	21.92–36.97
MnO	0.17–0.97	0.58–0.59	0.34–0.59	0.2	0.17–0.20	0.25–0.81	0.24–0.61
MgO	4.37–8.09	6.80–7.88	9.42–15.49	1.9–12.2	3.10–9.50	1.52–12.38	2.77–9.21
NiO	–	–	–	0.1–0.17	0.05–0.24	–	0.08–0.30

Notes: 1—picritic basalt of the Noril'sk Region (Ryabov et al. 1985a, b, c), 2—picritic basalt of the Ayan River (Vasil'ev 1988), 3—volcanic ultramafic rock of the Maymecha–Kotuysky Province (Bagdasarov et al. 1979), 4—disseminated chromite from picritic gabbrodolerite of the Noril'sk Intrusions (Genkin et al. 1979), 5—chromite accumulations of the Noril'sk Intrusions, 6—gabbrodolerite UZ of Noril'sk Intrusions (Ryabov 1984a, b), 7—Karoo dolerites (Ealles et al. 1980)

Analyzing the composition of chromite in the rock groundmass (as inclusions in olivine and plagioclase) (Table 4.6) revealed considerable variations in the composition of mineral-forming oxides. Comparison of chromite from inclusions in olivine and plagioclase shows that in the latter, it has higher average values of Al, Cr, Mg, and low Fe^{2+}. Chromite inclusions in olivine have minimum Ti content and maximum Fe^{3+} content.

An increase in the iron and titanium content in chrome spinel and decrease in chromium and alumina content during crystallization differentiation were established in different layered intrusions. A similar trend in oxide content in chrome spinel from Noril'sk gabbroic rocks was observed for chrome spinel from the Novosibirk pre-Ob' region (Glotov 1985) and dolerites from Karru (Ealles et al. 1980). We detected a continuous range of composition between chromite and titanomagnetite and determined their crystallization stages (Ryabov 1992b).

Comparison of Noril'sk chromite with minerals from rocks and intrusions of the region has been already carried

Table 4.6 Variation ranges and average mineral-forming oxide content in the upper zone of Noril'sk Intrusions chromites

Oxides	Inclusions in the bulk rock	Inclusions in olivine	Inclusions in plagioclase
TiO_2	3.20–7.16 / 5.18	0.90–2.93 / 1.69	0.92–7.68 / 2.97
Al_2O_3	6.29–7.96 / 7.12	7.90–16.78 / 12.95	5.77–23.08 / 14.54
Cr_2O_3	27.35–28.24 / 27.80	27.15–36.62 / 33.42	27.31–38.50 / 34.27
Fe_2O_3	16.01–20.80 / 18.40	10.95–28.25 / 19.10	10.56–27.65 / 15.75
FeO	33.30–43.17 / 38.24	23.50–28.20 / 25.61	19.77–32.15 / 24.87
MnO	0.72–0.81 / 0.76	0.54–0.72 / 0.64	0.45–0.68 / 0.60
MgO	1.52–3.57 / 2.54	4.28–9.66 / 6.26	4.41–11.26 / 7.48

Note: Numerator shows variation range, denominator shows the arithmetic mean

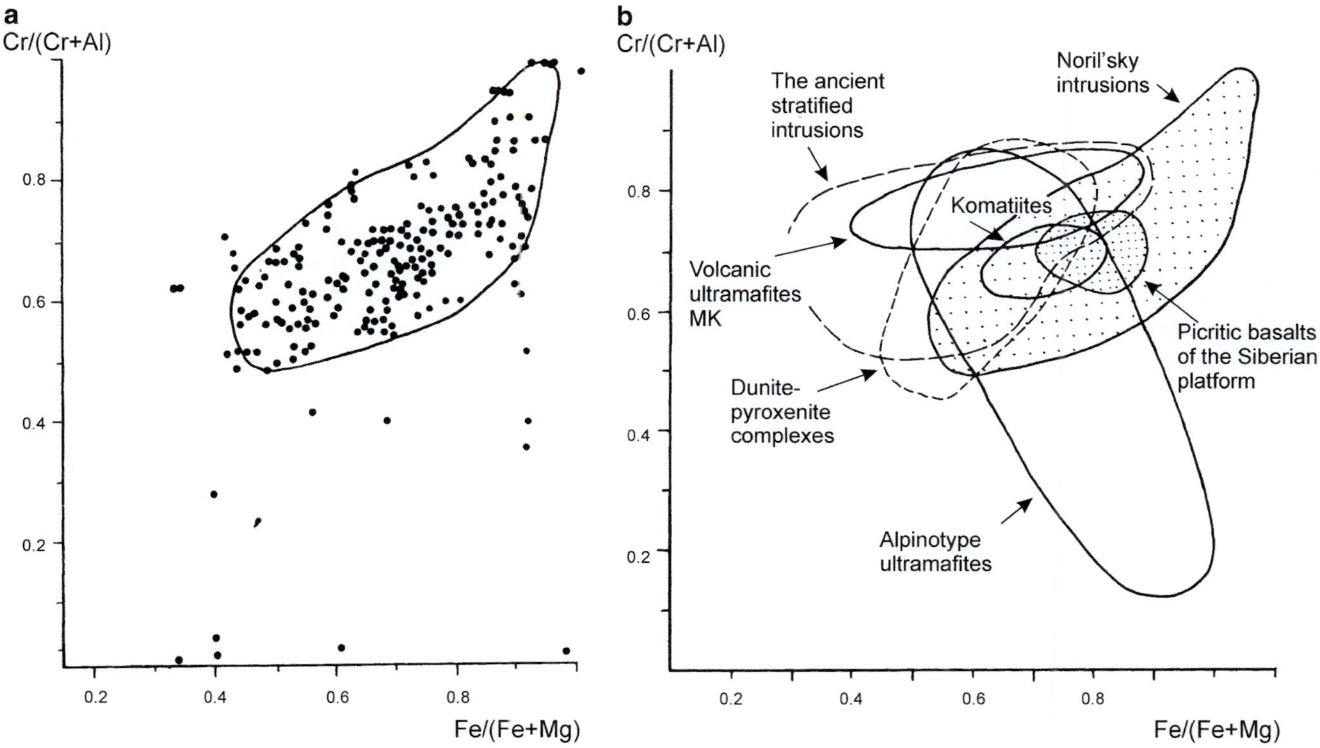

Fig. 4.18 Diagram of Cr/(Cr + Al) versus Fe/(Fe + Mg) (at%): (**a**) for chrome spinel from traps; (**b**) comparison of compositions of chrome spinel from traps and rocks from different intrusions (After Bagdasarov et al. 1979). *MK* Maymecha–Kotuysky Province

out (Genkin et al. 1979; Kudelina et al. 1983). Chromite from ultramafic rocks of the Urals and alpine-type ultramafic rocks from the intrusions of the Bushveld, Stillwater, and Great Dike have a stable composition and differ from trap chromite, containing lower TiO^{2+} and Fe^{3+} content. In addition, pegmatites from the eastern Bushveld contain a variety of spinels whose TiO_2 content ranges from 0.82 to 20.09wt% and Cr_2O_3 from 0.42 to 42.7wt% (Cameron and Glover 1973).

One of the specific features of chromite from the Noril'sk Intrusions is the existence of high-titanium (TiO_2 to 17wt%) varieties. This makes Noril'sk chromites similar to the chromites from lunar rocks, kimberlites, alkaline olivine basalts from Iceland, alkaline–ultramafic rocks from the Maymecha–Kotuysky Province (Genkin et al. 1979; Kudelina et al. 1983; Ryabov 1984a, b). The similarity in composition and variation tendencies of chromite indicated by the authors above show that features of the Noril'sk Intrusions and derivatives of alkaline magmatism determined the specific nature of the mineral in Noril'sky-type ore-bearing intrusions. On the whole, the traps of the Noril'sk Region manifest a complete crystallization trend of spinel (see Fig. 4.13), which spans the high-chromium and high-titanium varieties as a single genetic series.

The C/(C + 1)–Fe/(Fe + Mg) diagram (Fig. 4.18) shows the analyses of chrome spinel from the Noril'sk Intrusions. Compositional fields of the mineral from other sources are shown for comparison. One can see from comparison that in spite of the contact and intersection of the compositional fields of chrome spinel from different sources, each has peculiar features. In the coordinate system, the Noril'sk chrome spinels are similar to komatiites, although varieties with equivalent chromium content have higher iron content than those from ancient layered intrusions, dunite-- clinopyroxenite complexes, volcanogenic complexes, and volcanogenic ultramafic rocks in the Maymecha–Kotuysky Province. Chrome spinel from the rocks of trap formations and from alpine-type ultramafic rocks have opposite composition evolution tendencies.

Finishing the characterization of spinel group minerals, we want to emphasize the wide variations in their mineral-forming components. Magnetite series minerals are found in all varieties of effusive and intrusive traps. They form titanomagnetites with a varying TiO_2 content (up to stoichiometrically pure ulvospinel). Chromite is a typomorphic accessory mineral of derivatives of picritic basalt melts, and it does not commonly occur in olivine-rich cumulates of tholeiite-basalt melts. Spinel series minerals, specifically from the spinel–hercynite series, and accumulations of chrome spinel form in mafic pegmatoids of layered intrusions, which suggests crystallization from fluidized melts.

4.1.5 Ilmenite

Ilmenite forms in virtually all effusive and intrusive rocks as tabular, thick, plate-like crystals, latticed clusters, and irregularly shaped crystals. Ilmenite forms in association with titanomagnetite, wustite, native iron, sulfides, and various silicates. The wide variety of associated minerals suggests crystallization of ilmenite in a wide range of physicochemical conditions.

Ilmenite composition in igneous rocks from the Noril'sk Region varies. The mineral contains 40.16–54.74wt% TiO_2, 40–56.32wt% FeO, 0.20–5.14wt% MnO, and 0–1.1wt% V_2O_5. The compositional peculiarities of ilmenite in the Eragalakh Complex are given in Table 3.2, and an overview of rocks in the region is presented in Fig. 4.19.

The theoretical composition of ilmenite includes 47.34% FeO and 52.66% TiO_2. Higher concentrations of TiO_2 in Noril'sk ilmenites (Fig. 4.19a) seem to be related to the admixture of rutile, whereas the excess iron is most likely due to an isostructural admixture of Fe_2O_3. A MnO content exceeding 3wt% was observed in ilmenite from tholeiite basalts of the Syverminsky Suite, glomeroporphyritic basalts from the Gudchikhinsky Suite, fragments of the Kharaelakh magmatogenic breccia, pegmatoids of the UZ, and syenites from the flanks of the Upper Talnakh Intrusion.

Elevated MgO content was observed in vanadium-poor ilmenite (Fig. 4.19b) in trachydolerite of the Avamsky Complex, and in vanadium-rich ilmenite in picritic basalt of the Tuklonsky Suite of Sunduk Mountain and Mikchadin lava flows. The two latter rocks and picritic gabbrodolerite from the Zubovsky featured the highest V_2O_5 content.

4.1.6 Minerals of Mica Group

These minerals are represented by biotite and phlogopite. They are typically brownish, with colorless and green varieties less common. Small amounts of mica are found in layered intrusions and the haloes of altered rocks rimming intrusions. Micas are scarce in isotropic intrusions and are rare to absent in effusive rocks.

Micas typically form scarce small flakes and small aggregations often in association with magnetite or sulfides. Elevated mica content (to 1–3%) is observed throughout the Zelenaya Griva Intrusion. In other regions, accumulations of biotite were observed in Noril'sky-type intrusions where they occur in mineralized gabbrodolerite and pegmatoids. Highest mica concentrations (to 5–7%) were found in upper taxite gabbrodolerite. A remarkable feature of mica in picrite–dolerite of the CZ is its occurrence in caps over drop-shaped sulfide impregnations often in association with other minerals containing volatiles.

A series of diagrams in Fig. 4.20 present the compositional features of micas in traps. Some symbols show biotites and phlogopites from gabbrodolerite of layered ore-bearing and ore-free intrusions, from pegmatoids of UZ and from trachydolerites. The composition of micas from metasomatites in the halo of altered rocks rimming the Noril'sk Intrusion is provided for comparison. The Mg:Fe = 2:1 ratio (Deer et al. 1966) is taken as the conventional separation of phlogopites and biotites.

The Fe–Mg diagram (Fig. 4.20a) shows wide compositional variations of the mineral, a single geochemical trend, and a gradual transition from biotites to phlogopites. A strong correlation between Fe and Mg suggests isomorphism in the mineral structure. The total iron content of micas ranges from 2.3 to 86.7 at% (Fig. 4.20b). Biotites and phlogopites of varying iron content are found in all groups of rocks and intrusions designated on the diagram. The highest magnesium micas form in pegmatoids of the UZ and metasomatites, as well as in picritic gabbrodolerite. High-iron biotites were found in metadiorite and in olivine in gabbrodolerite intrusions of the Noril'sk and Kruglogorsky-type and Ogoner Complex. The content of all mineral-forming oxides in micas vary widely (Fig. 4.20c–f). The SiO_2 content ranges from 33.35 to 41.64wt%, and Al_2O_3 ranges from 9.5 to 14.47wt%. Wide variations in Si and Al are observed both in biotite and phlogopite. On the (K + Na)/

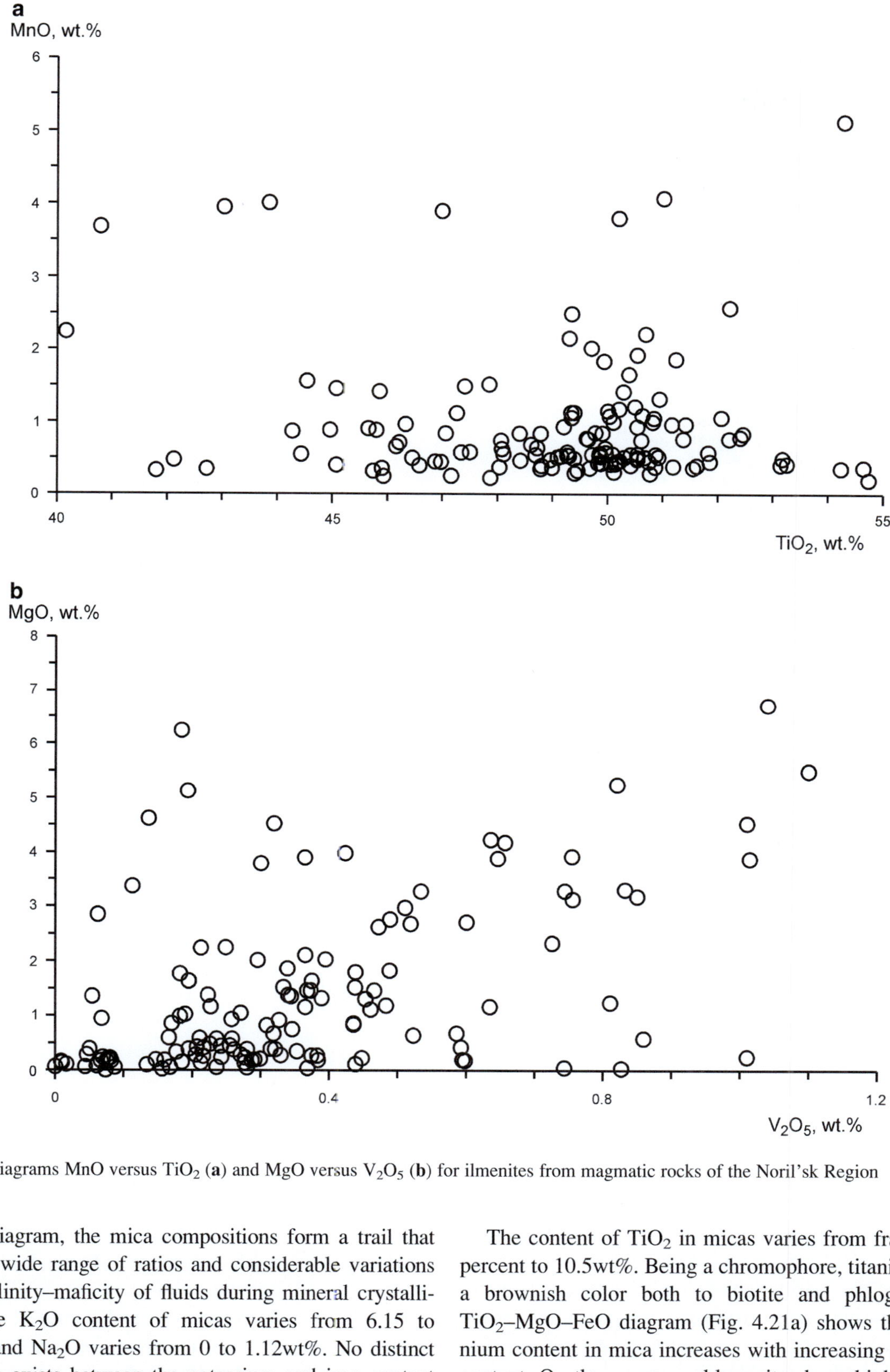

Fig. 4.19 Diagrams MnO versus TiO_2 (**a**) and MgO versus V_2O_5 (**b**) for ilmenites from magmatic rocks of the Noril'sk Region

Al–Si/Al diagram, the mica compositions form a trail that suggests a wide range of ratios and considerable variations in the alkalinity–maficity of fluids during mineral crystallization. The K_2O content of micas varies from 6.15 to 9.40wt%, and Na_2O varies from 0 to 1.12wt%. No distinct relationship exists between the potassium and iron content of mica (Fig. 4.20e). The halogen content in mica increases with increasing Si/Al. This is more pronounced in phlogopite than in biotite.

The content of TiO_2 in micas varies from fractions of a percent to 10.5wt%. Being a chromophore, titanium imparts a brownish color both to biotite and phlogopite. The TiO_2–MgO–FeO diagram (Fig. 4.21a) shows that the titanium content in mica increases with increasing magnesium content. On the average, phlogopites have higher titanium content, and the highest concentrations of TiO_2 were found in micas from pegmatoids of the UZ. The distribution of TiO_2 versus the halogen content is given in the TiO_2–Cl–F

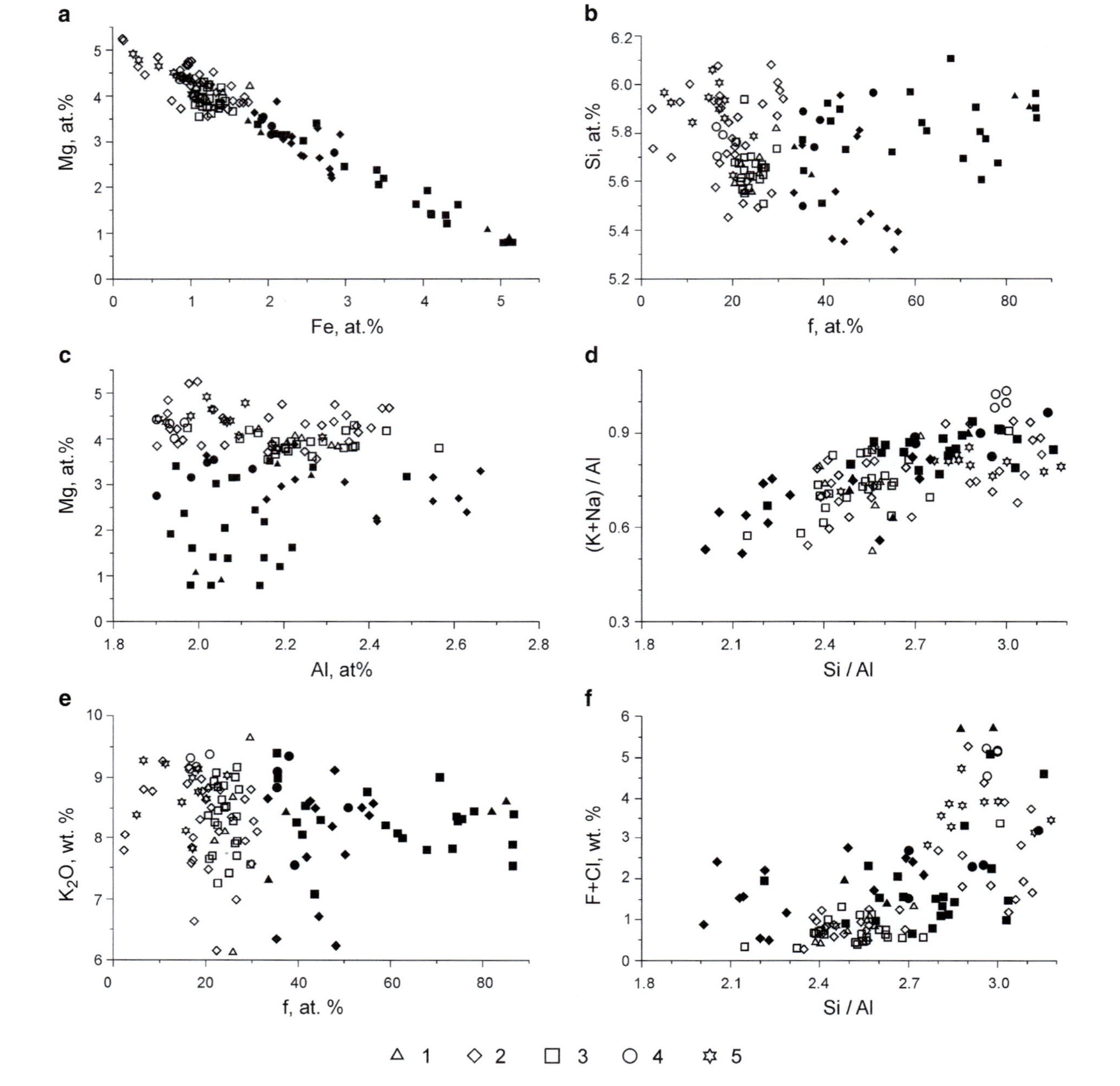

Fig. 4.20 Composition diagram of micas from rocks of the Noril'sk Region. *1* layered series of rocks of the Noril'sky-type intrusions; *2* pegmatoids of UZ, the same region; *3* ore-free trap intrusions; *4* intrusion of trachydolerites; *5* metasomatites of the aureole of Noril'sky-type intrusions. Biotite is shown as *solid shapes*; phlogopite is shown as *hollow shapes*. Explanations are given in the text

diagram (Fig. 4.21b). An increase in the Cl fraction in fluids has a stronger influence on the content of TiO_2 in mica than the F content.

An idea of the distribution of Cl and F in micas from the Noril'sk Intrusions and surrounding metasomatites can be gained from Fig. 4.22a. For comparison, the composition of micas from the Bushveld, Stillwater, and Skaergaard Intrusions is shown in the inset. Biotite from the gabbrodolerite layered series of rocks from the CZ of the Upper Talnakh Intrusion, including picritic gabbrodolerite and micas from the metasomatites, have similar Cl and F content and similar variations in their ratios. Three main variations of Cl and F content are observed: (1) a drastic increase in Cl content with constant low F content, (2) simultaneous increase in Cl and F content with a predominance of F, and (3) increased F content with low Cl content. The first and the second trends are found in the Upper Talnakh, Bushveld, and Stillwater Intrusions, and the third in the Upper Talnakh and Skaergaard Intrusions. The concentration and the ratios of Cl and F in micas from

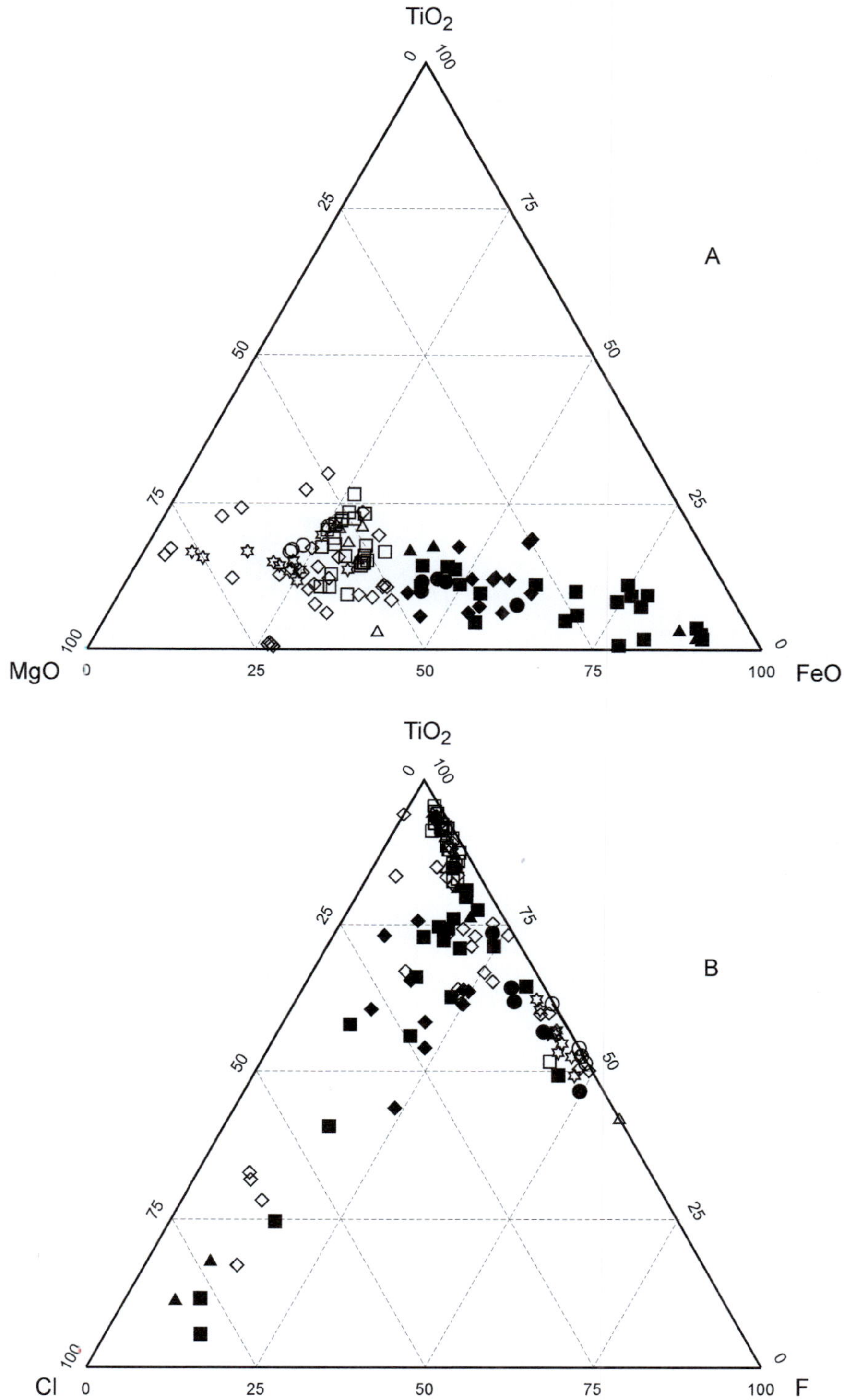

Fig. 4.21 Diagrams of TiO_2–MgO–FeO and TiO_2–Cl–F (wt%) for mica from rocks of the Noril'sk Region (see Fig. 4.20 for legend)

pegmatoids of the UZ vary widely and form two branches on the diagram. One indicates a drastic increase in Cl content and the other in F content. Remarkably, micas from the upper and lower contacts of the intrusion are separated according to their composition. It is worth noting that the lower contact accompanies rich and massive ores and micas form relics in them. Metasomatites of the upper contact are typically ore-free and contain a sterile pyrrhotite mineralization.

The ternary Fe/(Fe + Mg)–Cl–F diagram (Fig. 4.22b) shows wide variations in iron content of mica from the UZ pegmatoids and a tendency to increase Cl content with

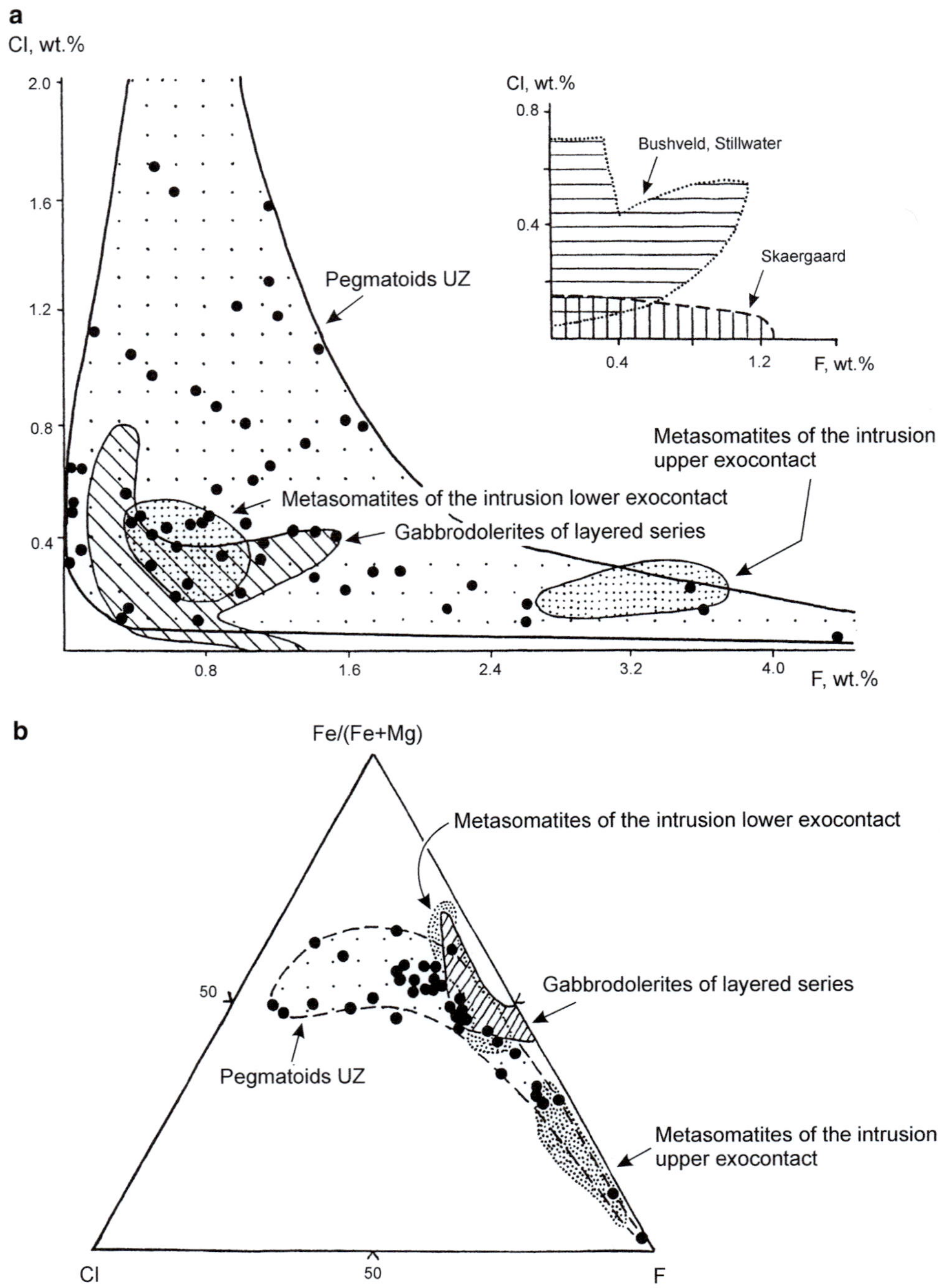

Fig. 4.22 Diagrams of Cl–F (**a**) and Fe/(Fe + Mg)–Cl–F (**b**) for mica from gabbrodolerites of the Upper Talnakh Intrusion (UTI) and surrounding metasomites. For comparison the compositions of biotites from rocks of the Bushveld, Stillwater, and Skaergaard Intrusions are adopted from the data of Boudreau et al. (1986)

increasing iron content. Micas in metasomatites also differ: in the lower contact, they feature higher iron content than in the upper contact.

As micas are found in Pt–Cu–Ni sulfide ore and Pt low-sulfide ore, it is important to define the specific distribution of halogens in the ores. The average F and Cl content in mica from sulfide-type ores are 0.70 and 0.13wt% in pyrite gabbrodolerite, 0.92 and 0.23wt% in taxite gabbrodolerite, 0.71 and 0.37wt% in metasomatites of the lower contact, and 3.22 and 0.20wt% in metasomatites of the upper contact. Concentration of F and Cl in micas of Pt low-sulfide ores reaches 5.23 and 5.21wt%, and the average content (excluding extreme values) is 1.64 and 0.42wt% in the upper taxite gabbrodolerite and 1.17 and 0.93wt% in leucogabbro. Judging from the distribution of halogens in the hydroxyl group of micas in the following series of rocks: picritic gabbrodolerite, lower taxite gabbrodolerite, upper taxite gabbrodolerite, and upper leucogabbro, during the formation of Pt ores the fraction of F + Cl increased (0.83–1 to 15–2wt%, and 0.6–2.1wt%, respectively), which was followed by an increase in F content and a decrease in Cl content. This is also reflected in the successive increase in F/Cl values in the micas of the above-mentioned series of rocks as 5.3, 4.0, 3.9, and 1.25wt%.

4.2 Distribution Features of Trap Ore Genesis and Associated Problems of Differentiation

Problems associated with the geochemistry of traps of the Siberian Platform have been discussed in many published works. They provide a considerable body of information on the microcomponent composition of some rocks, ores, and intrusions in general. The most complete summary of the geochemistry is available in Eliseev (1959) Nesterenko and Almukhamedov (1973), Arkhipova (1975), and Oleinikov (1979). In recent years, precise analytical data on the Siberian traps have been published by many authors (Lightfoot et al. 1990, 1994; Naldrett et al. 1992; Czamanske et al. 1994; Distler and Kunilov 1994).

4.2.1 Chromium

The behavior of chromium during the evolution of a basaltic melt is of great theoretical interest; chromium is a sensitive indicator of the depth of magma formation. The problems of chromium geochemistry are discussed in Dodin (1967), Feoktistov (1978), Nesterenko and Almukhamedov (1973), Kopylova and Oleinikov (1973), Arkhipova (1975), Kravchenko (1977), and Oleinikov (1979).

These authors research has revealed a number of important features, especially the relationship between their chromium and magnesium content. Picritic gabbrodolerite from the Noril'sk Intrusion are considered high chromium, and this was a reliable criterion for sorting intrusions into potentially ore-bearing and ore-free, using the weighted average Cr content and ratios with other elements. In addition, geoscientists understand the deviations from the major geochemical dependence Cr–MgO, for example, in ore-free low-chromium high-magnesium intrusions of the Morongovsky Complex and Lower Talnakhsky-type intrusions, but the reasons for their origin have not been established. It was also established earlier that the greatest variations in chromium content are typical of the layered Noril'sk Intrusions, with maximum concentrations found in picrites and minimum concentrations found in silicic rocks. The ratios of Cr content to other components are an indicator of differentiation (Arkhipova 1975).

Nesterenko and Almukhamedov (1973) focused upon the existence of two concentration maximums of chromium accumulation in the middle and near-basal parts of the Noril'sk Intrusions. They believe that under intrachamber conditions chromium is redistributed due to high-temperature formation of chromium oxy compounds and gravitational precipitation of early phases. As a result, a lower contact zone of accumulation is formed in picritic gabbrodolerite in the near-basal parts of intrusions, whereas the appearance of a second indistinct concentration maximum is related to the precipitation of minerals less enriched in chromium. All Noril'sk geoscientists accept this point of view, and the evidence is provided by Goldshmidt (1954) of the high-temperature crystallization of chromium oxides, with spinel minerals the first to precipitate.

Our results, obtained from studies of the geochemistry of chromium in a number of locations, differ from traditional concepts (Ryabov and Yakobi 1981). The typical associated variation in chromium and magnesium content of rock is preserved only in effusive traps and nondifferentiated intrusions that rapidly lost the fluid phase (Fig. 4.23a). In addition, it is in volanogenic rocks that we found unambiguous evidence for the redistribution of Cr with participation of fluids (Ryabov et al. 1977). These are dendrite-like megacrysts of chromous pyroxene (Cr_2O_3 to 0.68%). The megacrysts resulted from the resorption of chromite and olivine in picrite under the influence of fluids at $T = 1{,}200°C$. Of similar nature are the plumose megacrysts of pyroxene from the Lower Fokinsky Intrusion in which $Cr_2O_3 = 1.06\%$.

In layered trap intrusions, the relationship between Cr distribution and MgO content becomes more complicated (Fig. 4.23b). The diagram shows that high-Cr_2O_3 content occurs in rocks with variable magnesium content. To reveal the reasons of this phenomenon, we studied the behavior of Cr in the intrusion with varying degrees of differentiation (Ryabov and Yakobi 1981). It was established that the behavior of Cr depends on a number of factors, most importantly the composition of initial melt, its temperature, and fluid regime. These factors are responsible for the absolute content of Cr, forms of its occurrence, and regions of concentration.

In intrusions with different degrees of differentiation, at different intersections of a single body and in different parts of the magmatic column, the evident tendencies of Cr behavior are preserved, though the scales of accumulation widely vary. Earlier this was shown in the example of weakly differentiated sills of Great Lake (Tasmania), Kaerkan, and some layered intrusions of the Noril'sk Region (Ryabov and Yakobi 1981). Among weakly

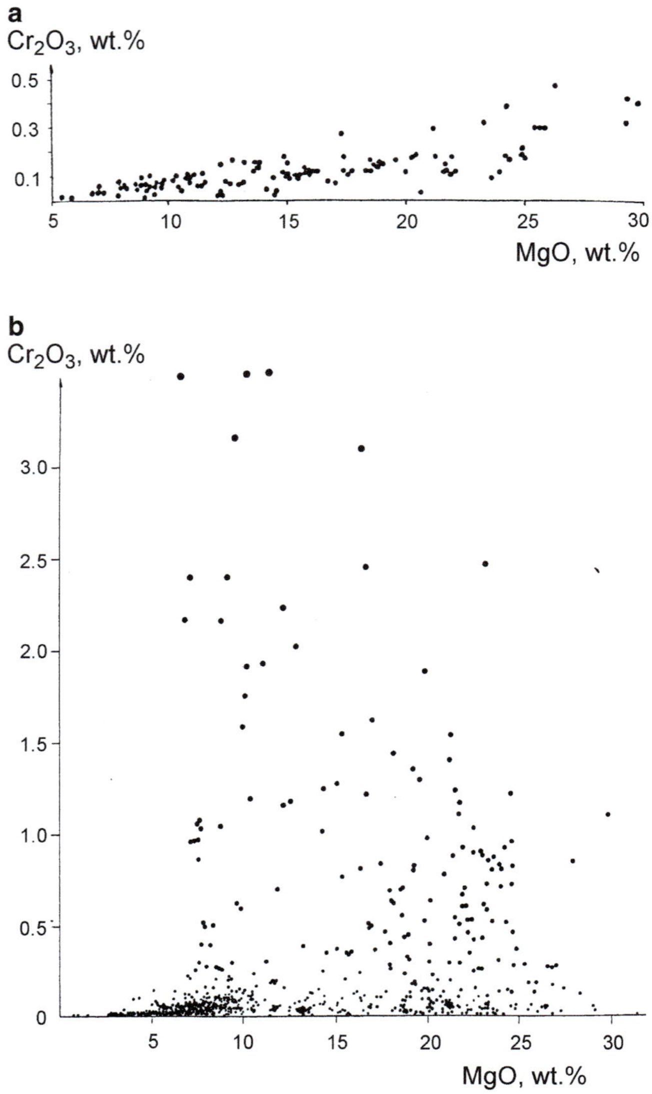

Fig. 4.23 Diagram of Cr_2O_3 versus MgO contents in ore-free traps (**a**) and gabbrodolerites of the Noril'sky-type intrusions (**b**)

differentiated high-magnesium intrusions, two types of massifs are distinguished by Cr_2O_3 content in picrite–troctolite gabbro. The first type is the Cr-enriched intrusions: Lower Fokinsky, Ergalachny Stream, Magnitny Stream, Manturovsky, and Upper Pyasinsky. Picrites of these bodies have $Cr_2O_3 = 0.27–0.47$wt% (Fig. 4.23a).

Weakly differentiated low-Cr intrusions of the second type include the massifs of Upper Talnakh, Lower Noril'sk, Zelenaya Griva, and Morongvsky group of intrusions (Mountain Pikritovaya, Mountain Putanaya, Mountain Morongo, and Mountain Ruinnaya Intrusions), in which the Cr_2O_3 content of troctolite gabbrodolerite is no more than 0.1wt%. Similar to this type of intrusion are Zubovsky, Pyasinsko–Vologchansky, Gorstroy, and Tulaek–Taas River Intrusions, whose picrite–troctolite displays typically low Cr_2O_3 content but increases in Cr_2O_3 content with increasing MgO content. A particular feature of these high-magnesium intrusions, in addition to an extremely low chromium content of picrite–troctolite gabbrodolerite, is the elevated chromium content of olivine-poor rocks (Fig. 4.24). The main concentrator of chromium in these intrusions is pyroxene (Ryabov and Yakobi 1981).

The behavior of chromium was best displayed in parts of the Noril'sk Intrusion. Here, one can observe up to five concentration maximums or states with stable chromium content, each characterizing a particular genetic origin. Two are associated with the upper and lower contact gabbrodolerite and reflect the Cr_2O_3 content in the initial melt. Between contact facies in the intrusion, there are three more concentration maximums: (I) in pegmatoids of the UZ, (II) in pyroxene-rich olivine gabbrodolerite, and (III) in picritic gabbrodolerites (Figs. 4.25, 4.26 and 4.27). The existence of the class II and III maximums was reported in Nesterenko and Almukhamedov (1973), and the highest (I)

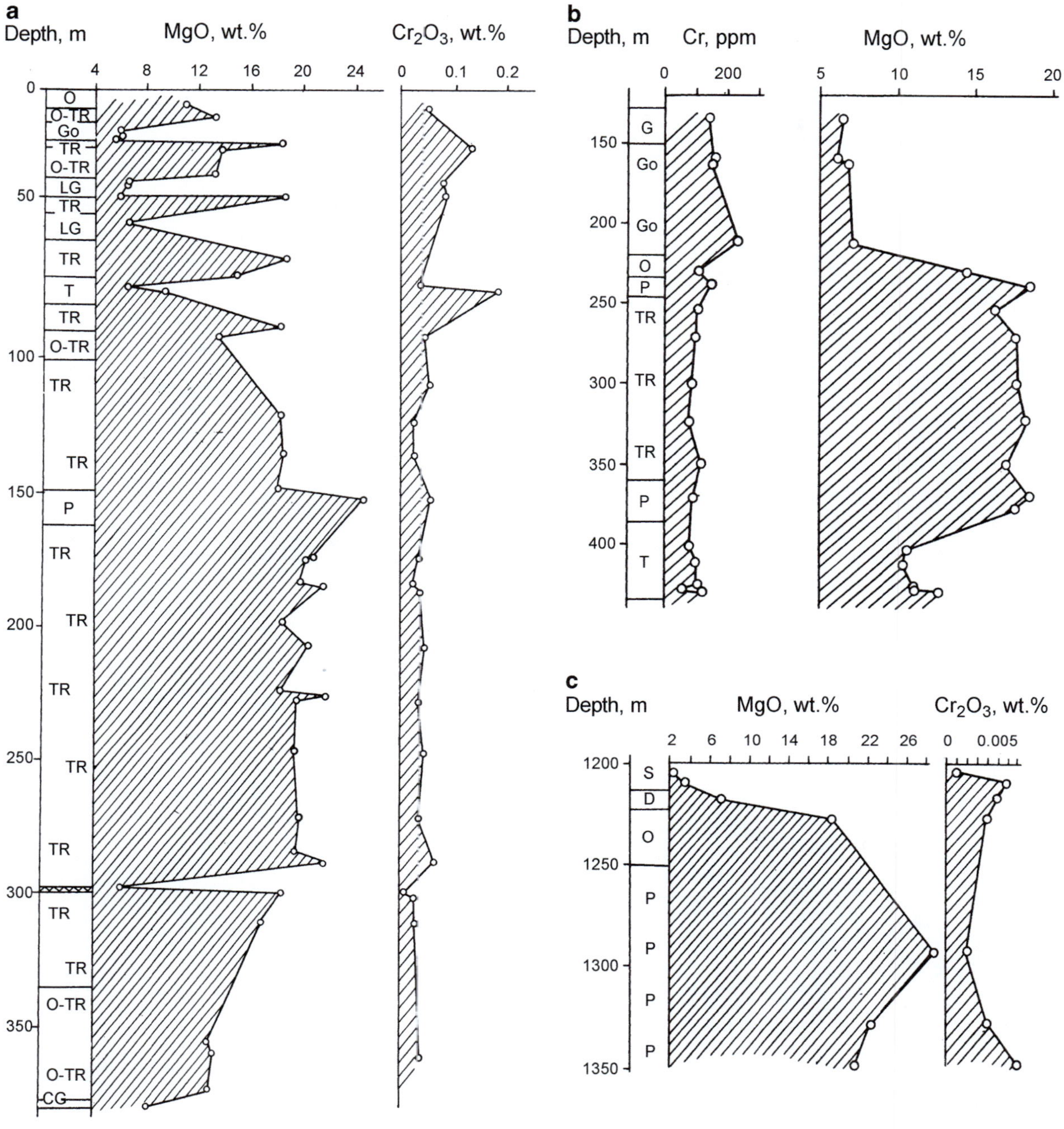

Fig. 4.24 Variations in the contents of chromium and magnesium in the sections of intrusions: (**a**) Mt. Putanaya, drillhole E-36, (**b**) Pyasinsko–Vologochansky, drillhole AS-6, (**c**) Lower Noril'sk, drillhole NP-4 (Ryabov and Yakobi 1981)

maximum was detected and characterized in Ryabov et al. (1982) and Ryabov (1984a, b).

4.2.1.1 Lower Concentration Maximum (III) in Picritic Gabbrodolerite

In nondifferentiated and weakly differentiated magmatic bodies in the picrite–troctolite gabbrodolerite horizons, the distribution of Cr is virtually uniform throughout the section, but in the Noril'sk Intrusions, these rocks in some parts are depleted and in others are enriched in chromium: an order of magnitude higher or lower than abundance ratio.

It was suggested in Genkin et al. (1979) that chromite (chromium) accumulates in the base of the picrite horizon. Further, this fact was repeatedly reported in the genetic constructions of other authors. Our studies showed that in a picrite horizon, chromium (chromite) is concentrated not in the base but in the roof of a layer (see Figs. 4.25, 4.26 and 4.27). The same conclusion was drawn by Shatkov (1973). The taxite gabbrodolerite underlying the picrite horizons have very low chromium content. Study of the behavior of Cr in the picrite horizons suggests that the shift to maximum concentration in the roof is controlled by the horizon of

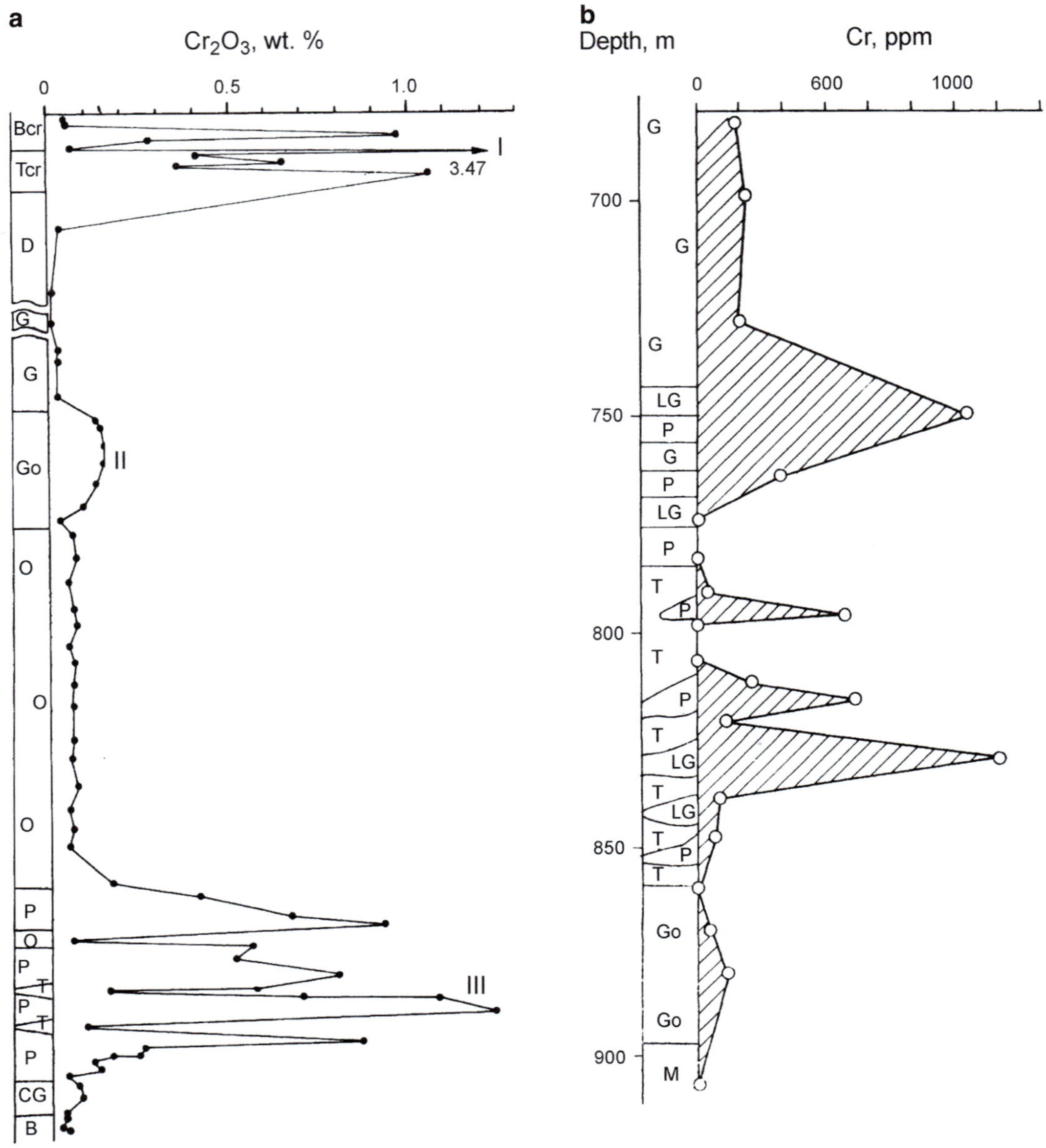

Fig. 4.25 Behavior of chromium in the sections of the east Noril'sk Branch of the Noril'sk-I Intrusion (**a**) and Kharaelakhsky Branch of the Upper Talnakh Intrusion (**b**). (**a**) Drillhole PE-35 (Ryabov 1984a, b), (**b**) drillhole KZ-585

underlying mafic pegmatoids, or to be more exact, by related fluids. As a result, in the close contact with taxite gabbrodolerite, picrites are depleted in chromium which was accumulated at a distance from pegmatoids. Alternation of picrites with other rocks in some cases complicates the distribution of chromium and the configuration of the oxide variation diagram. Calculations showed that picritic gabbrodolerite with $Cr_2O_3 < 0.1$wt% in the Upper Talnakh Intrusion amount to 58%, and in the Manturovsky, Imangdinsky, Mountain Chernaya, and Noril'sk-II Intrusions, to 32% (Fig. 4.27). This conclusion is important as it eliminates the search characteristic of elevated chromium content in the Noril'sk picrites. Moreover, detection of chromium-depleted picrite–troctolite gabbrodolerite in the Noril'sk ore-bearing intrusions is very important in connection with the wide occurrence of similar rocks in the Lower Talnakh and Lower Noril'sk Intrusions, which suggests a genetic relationship between the Upper Talnakh–Lower Talnakh and Noril'sk-I–Lower Noril'sk Intrusions (Ryabov and Yakobi 1981). This also allows one to use the low-chromium ore-free intrusions of the Lower Talnakh type as a criterion for finding ore-bearing bodies, which concentrated oregenic components that are now absent (Ryabov 1981).

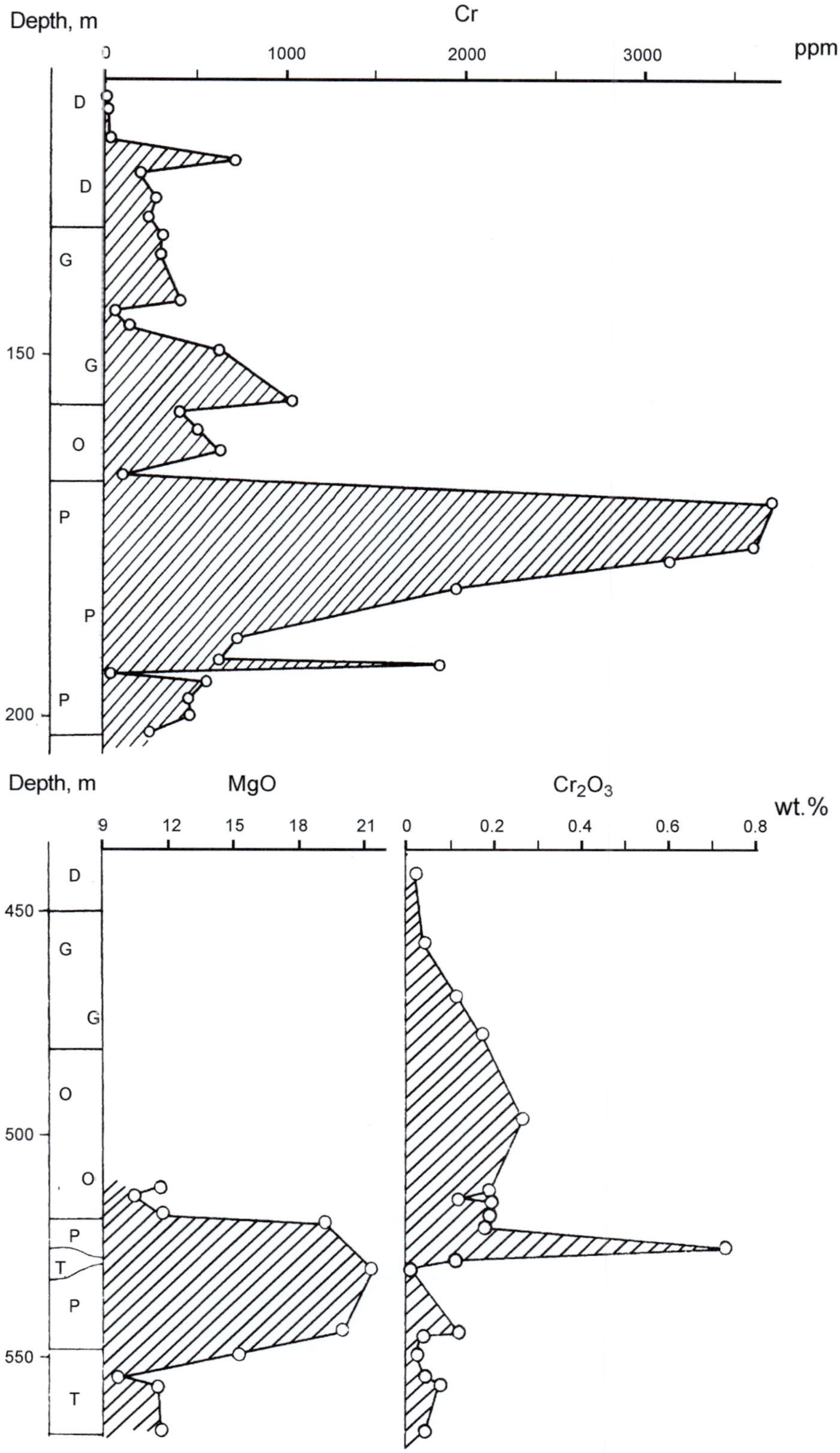

Fig. 4.26 Behavior of chromium in the sections of the Upper Talnakh Intrusion. (**a**) Drillhole KZ-184 (Ryabov and Yakobi 1981), (**b**) drillhole KZ-169

4.2.1.2 Concentration Maximum (II) in Olivine Gabbrodolerite

Elevated Cr_2O_3 content is persistently found in olivine gabbro not only in the Noril'sk Intrusions but in most weakly differentiated sills. This is related to the enrichment of rocks in clinopyroxene, which is an early silicate phase that trapped chromium as an isomorphous admixture that was at minor concentration in the melt. The average Cr_2O_3 content in green pyroxene in olivine and taxite gabbrodolerite is 0.47% and 0.45% and is 0.61% in picrite

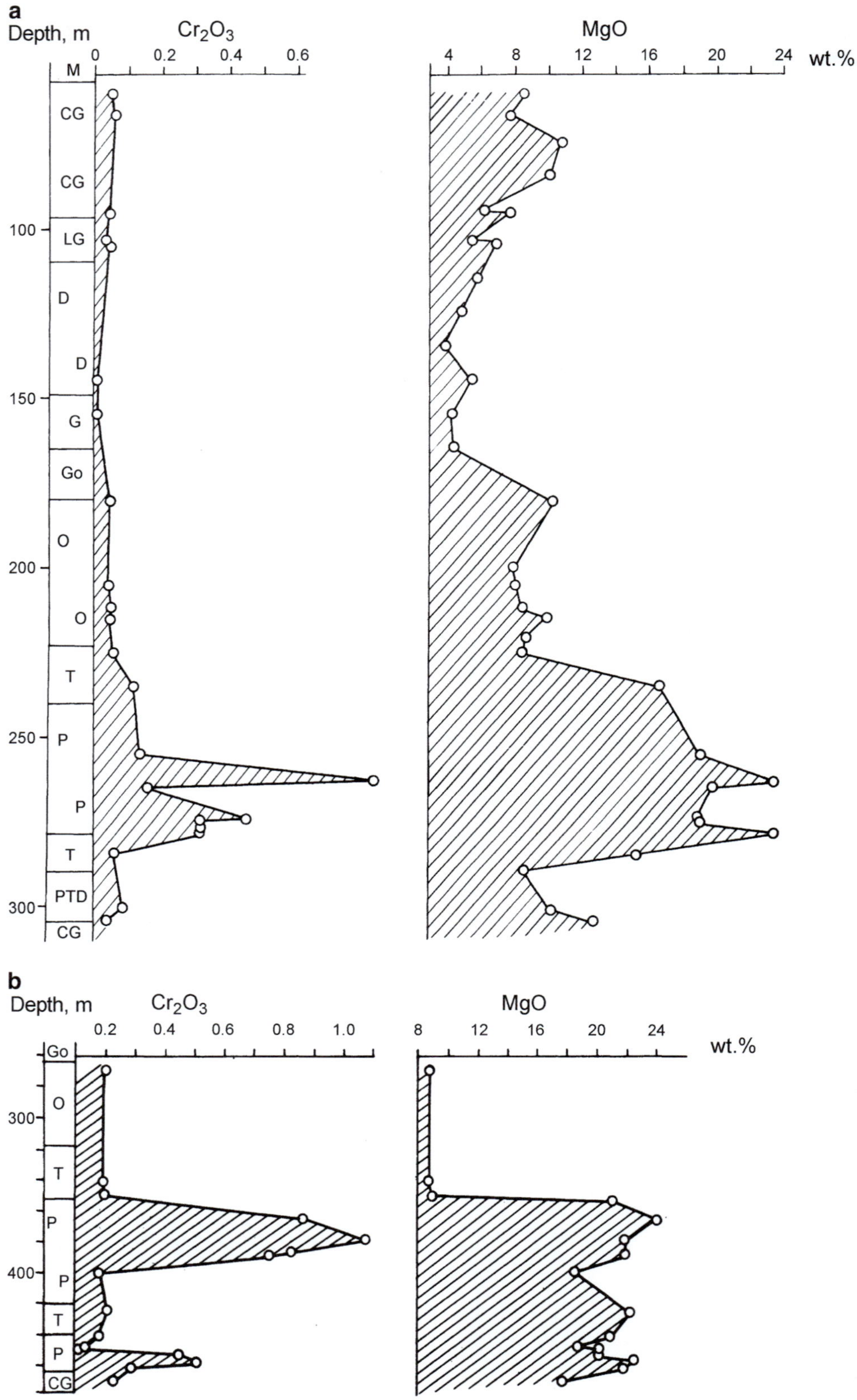

Fig. 4.27 Behavior of Cr_2O_3 and MgO in the sections of the Manturovsky Intrusion, drillhole IM-39 (**a**) and Drillhole IM-40 (**b**). Based on data from Zolotukhin and Shchedrin (1977)

gabbro (Godlevsky et al. 1971). In weakly differentiated mafic intrusions, clinopyroxene is the main concentrator of Cr_2O_3, and Cr_2O_3 content may be very high.

4.2.1.3 Highest Concentration Maximum (I) in the Horizon of the UZ

Detailed geochemical studies showed the presence of this maximum in massifs with a varying degree of differentiation. However, the most pronounced occurrence was described for the Noril'sk Intrusions (Ryabov and Yakobi 1981; Ryabov et al. 1982; Ryabov 1983, 1984a, b). In different magmatic bodies, and in their different parts (leading, middle, or near base), the maximum and mean Cr_2O_3 content changes. In the leading part of the Upper Talnakh Intrusion, the geochemical maximum is almost indefinable, and, with depth, in the middle parts of the body, the absolute content of Cr gradually increases and is clearly pronounced in the upper taxites. The content of Cr_2O_3 in the rocks reaches 10%.

The geochemical maximum of chromium content in the rocks of the UZ is nonuniform, forming a saw shape on the plot. Picritic gabbrodolerite localized in the chromite-bearing horizon in some cases have a rich "rash" of chromite ($Cr_2O_3 = 1.21\%$) (Rock indication 103) and in others are virtually chromium-free ($Cr_2O_3 = 0.029\%$, Rock indication 104). The thickness of the chromite-bearing horizon might vary from few to tens of centimeters and to several meters.

In the Kharaelakhsky branch of the Upper Talnakh Intrusion (see Fig. 4.25), one can observe dramatic chromium depletion in relict taxitized picrite in magmatogenic breccia and preservation of low concentrations in some fragments. Extraction of chromium from rocks, its redistribution, and accumulation in the roof of the picrite horizon can be undoubtedly related solely to the participation of fluids.

4.2.2 Platinum Metals

The elevated levels of abundance ratios of platinum-group elements are known to be associated with ultramafic rocks. Study of the behavior of platinoids in mafic–ultramafic complexes allowed geoscientists to discover the specific order of fractionation of elements according to the composition of a magmatic melt and methods of differentiation. The concentration and ratio of platinum-group elements are believed to reflect their concentrations in the regions of magma formation. Transferred by melt into present-day magma chambers, they are distributed in different minerals and, in the presence of sulfur and other components, are redistributed and concentrated in sulfides or other phases. The problems of the geochemistry of platinoids are discussed in detail by Cousins (1973), Page et al. (1976), Glasunov et al. (1977), Naldrett and Cabri (1976), and Crocket and MacRae (1986).

It was empirically established that in ultramafic rocks, Pt typically predominates over Pd, and in the rocks of the gabbroic–basaltic series, Pd typically predominates over Pt. The behavior of these elements is to a considerable degree controlled by the content of siderophile elements (Marakushev, I979). The greater mobility of Pd compared to Pt is due to a greater sulfur affinity of the former and with the appearance of sulfur in the system Pd is one of the first to migrate. The predominance of Pd over Pt in some mafic bodies is the result of lower temperature and shallower melting of magma compared to the ultramafic rocks in which Pt > Pd. This suggests deeper, high-temperature magmas, where melting also involved heat-resistant minerals (Naldrett and Cabri 1976).

The problems of the geochemistry of platinoids in traps are discussed in Ginsburg and Rogover (1960), Genkin (1968), Smirnov (1966), Goryainov and Ushakov (1970), Genkin et al. (1973), Dodin et al. (1973), Distler et al. (1976), and Konkina and Rasin (1976). Most consider the behavior of platinum metals in the Noril'sk Deposits. Distribution of platinoids in ore-free effusive and intrusive traps of varying alkalinity–maficity is described in Dodin et al. (1973) and Distler et al. (1976).

In studying the behavior of platinoids in traps, we had three tasks: to reveal the specific features of the distribution of elements in high-magnesium rocks of the platform, to establish the character of their variations in layered intrusions, and to study their behavior in the UZ of these intrusions. The results of analyses of noble and nonferrous metals, chromium, and sulfur in the lumps of ores are listed in the Rock indications (see Volume 2) and in the Appendix. It is worth noting that, in general, the data obtained in the UIGGM of the Siberian Branch of the RAS agree well with data in the literature.

4.2.2.1 Platinoids in High-Magnesium Traps

The goal-oriented study of platinoids in this type of sampling is conducted for the first time and therefore is of great interest. To solve this problem, we have performed geochemical research of single-type petrographic varieties of picritic basalt from various volcanic troughs (Lamsky, Imangdinsky, Noril'sky, Vologchansky, and Kharaelakhsky), from picritic gabbrodolerite intrusions of the Fokinsky Complex (Lower Fokinsky, Magnitny Stream, Upper Pyasinsky, Mikchandinsky neck), from the same rocks of the Noril'sky-type intrusions (Noril'sk-I, Upper Talnakh, Mountain Chernaya), picrite–troctolite gabbrodolerite intrusions of the Lower Talnakhsky type (Lower Talnakh, Lower Noril'sk) accompanying the ore-bearing rocks, as well as from intrusions related to small deposits and ore occurrences (intrusions of Magnitny Stream, Imangdinsky, Zubovsky,

Table 4.7 Distribution of ore-forming metals in Mg-rich igneous rocks of the Siberian Platform

Rock type	Pt/Pd	Rh/Ru	Pt/Rh	Au/(Pt + Pd)	Au/Ag	Cu/Ni	Ni/Co
Taxitic gabbrodolerite UZ, Upper Talnakh Intrusion	0.30	1.64	12.0	0.17	1.17	1.56	19.7
Taxitic gabbrodolerite, Noril'sk-I Intrusion	0.37	0.21	12.8	0.04	0.64	0.78	10.3
Picritic basalt lava	3.33	0.46	5.7	1.16	0.33	0.12	36.2
Picritic gabbrodolerite and dolerite of the Fokinsky Complex	1.98	0.35	24.8	0.43	–	0.15	12.9
Picritic gabbrodolerite of the Noril'sky-type	0.43	5.83	11.8	0.04	–	2.13	18.9
Picritic gabbrodolerite of the Lower Talnakhsky type	1.93	0.69	3.8	0.22	0.14	0.68	5.7
Picrite–troctolite gabbrodolerite, Imangdinsky, Manturovsky, Mountain Chernaya	0.33	4.00	5.9	0.07	–	1.06	18.7
Picrite–troctolite gabbrodolerite of the Zubovsky type	0.65	1.75	6.4	1.64	–	0.06	7.8
Picrite–troctolite gabbrodolerite of the Zubovsky type, ore-bearing	0.66	3.67	11.1	0.14	–	2.38	25.4
Troctolite gabbrodolerite of the Morongovsky Complex	0.97	0.60	6.4	0.24	–	0.11	–
Limburgite, Bokovoe Lake	3.64	–	6.7	7.19	–	0.10	11.6
Nonfeldspar picrite, dyke, Murukakit River	1.73	–	–	0.21	–	0.13	16.8
Meymechite, Maymecha River	0.24[a]	–	0.8[a]	1.27	–	0.03	93.3
Dunite, Gulincky Intrusion	–	0.04	–	29.9	–	0.01	24.4
Pyroxenite ore-bearing, in the same place	3.43	–	–	90.32	–	0.07	31.6
Olivinite, Bor-Uryakh	–	–	–	8.12	–	0.09	11.1

[a]Average from two samples (Distler 1982), other samples from the authors

Table 4.8 Distribution of ore-forming metals in mafites of the Siberian Platform

Name of the rocks	Pt/Pd	Pt/Rh	Au/(Pt + Pd)	Au/Ag	Cu/Ni	Ni/Co
Basalts of the "normal series"	0.71	17.5	–	–	0.92	1.47
Subalkaline basalts	0.37	20.0	–	–	0.49	1.15
Near-conduit basalt facies of the Mikchandinsky paleovolcano	0.31	3.1	24.4	–	–	–
Nondifferentiated dolerites of the "normal series"[a]	0.70	4.1	–	–	1.08	1.97
Nondifferentiated dolerite, subalkaline	0.64	3.8	–	–	1.01	0.83
Olivine gabbrodolerite, Khungtukun Intrusion	0.21	–	0.3	0.25	1.38	1.59
Olivine gabbrodolerite, Manturovsky Intrusion	0.24	8.5	–	–	1.50	12.80
Olivine gabbrodolerite, Upper Talnakh Intrusion[b]	0.51	2.0	0.1	–	–	–
Olivine gabbrodolerite, nondifferentiated intrusions[b]	0.57	0.6	0.2	–	–	–

[a]Data from V.V. Distler et al. (1976)
[b]Data from D.A. Dodin et al. (1973)

Pyasinsko–Vologochansky, Mountain Morongo, Mountain Putanaya, Mountain Picritovaya, Mountain Ruinnaya). To compare, we have analyzed some sample of dunite, meimechite, and alkaline–ultramafic rocks from Maymecha–-Kotuysky (Yu. R. Vasiliev's samples) and Kamensky (our collection) Provinces.

The studies showed that in ore-free derivatives of picritoid melts, Pt predominate over Pd (Table 4.7). A similar situation is observed in limburgites and ore pyroxenites of the Kamensky and Maymecha–Kotuysky Provinces. In sulfide-bearing picritic gabbrodolerite from the Noril'sk Intrusions, Pt/Pd < 1. In picrite and troctolite gabbrodolerite, which are cumulative products of tholeiite-basalt melts, Pd predominates over Pt, and with the appearance of sulfides in gabbrodolerite, the element ratio decreases further. This is well seen from the comparison of the Morongovsky Complex and Zubovsky type. Most likely, the ratio Pt/Pd $= 1.93$ in the Upper Talnakh-type intrusions reflects a geochemical memory and their relationship with picritoid melts. In olivinites from the Talnakh magmatogenic breccia, Pt/Pd changes from 0.23 to 3.0, and the content of all other components varies as well. Moreover, most of other samples had extremely low platinoid content—lower than the test sensitivity, which suggests a depletion of rocks and redistribution of elements as a result of fluid interaction.

The highest Rh/Ru values were found in layered intrusions with sulfide mineralization, whereas the lowest were found in lava picrites, picrites from the Fokinsky Complex, and dunites from the Gulinsky massif. The ratio Au/(Pt + Pd) > 1 in the rocks of normal series were established only in picrite lavas and picrite–troctolite gabbrodolerite of the Zubovsky-type intrusions. In the alkaline–ultramafic rocks of the Kamensky and Maymecha–Kotuysky Provinces, Au predominates over total platinoids and only in a picrate dyke did it occur in a lower amount than platinoids.

Mafic rocks, both effusive and intrusive, have Pt/Pd < 1 (Table 4.8). Additional information on the distribution of platinoids in the rocks of this series is available in Dodin et al. (1973) and Distler et al. (1976).

4.2.2.2 The Behavior of Platinoids in the Layered Intrusions

In the CZ and LZ of intrusions, the absolute content of platinum-group elements varies widely. The maximum concentrations typically formed in sulfide-enriched picrite and taxite gabbrodolerite and massive ores. Within individual horizons, the quantities of metals also vary.

Within the intrusions, the mafic and ultramafic parts of the layered series differ distinctly in platinum and nonferrous metals, Au, Ag, and S content (Fig. 4.28a, b). Analysis of the behavior of each in the intrusion shows they have a nonuniform distribution. In the mafic rock part, the Pt maximum is observed in metadiorite, with a decrease in olivine-free gabbrodolerite, an increase in olivine gabbrodolerite, and a decrease at the boundary with picrite gabbro. Sulfide-bearing picritic gabbrodolerite features a dramatic accumulation of Pt but drops sharply in leucocratic segregations. The Pd maximum is found in metadiorite and at the boundary of olivine-free and olivine gabbrodolerite. The content of Rh in the mafic rock part remains nearly the same and the behavior of Ru is close to that described for Pd; the only difference is that the upper maximum appears to be shifted to the lower part of metadiorite and upper part of olivine-free gabbrodolerite, whereas in olivine gabbrodolerite it increases toward the bottom of the layer. This tendency is also preserved in underlying rocks. The Pt/Pd value has two maximums, which occurs in metadiorite and olivine gabbrodolerite.

In a separate picrite horizon, the maximum accumulation of Pt, Pd, Rh, as well as Au and Cr is found in the middle part of the layer, and the S maximum is shifted to the base (Fig. 4.28b). The Pt/Rh value throughout the layer varies weakly and is no higher than 0.38. It is worth noting that the maximum concentrations of Pt, Pd, Rh, and Cr are found in the same picrite horizon found in different parts of the body.

Figure 4.28c shows the Kharaelakhsky Branch of the Upper Talnakh Intrusion. The saw-toothed shape of the element content plot in the breccia horizon is due to the rocks multicomponent composition. At the same time, the values of Pt/Pd and (Pt + Pd) appear to be stable for the separate rocks from fragments or from the cement of breccia (Fig. 4.29). In these rocks, we want to emphasize once again that the fragments of magmatogenic breccia are highly depleted in platinum and nonferrous metals and that this part of the Upper Talnakh Intrusion is the richest in sulfide and massive ores (Gulin and Sukhov 1974), which suggests the complete extraction of oregenic elements from the rocks with sulfur-bearing fluids.

4.2.2.3 Platinoids in the UZ of the Noril'sky Intrusions

The coexistence of the geochemical maxima of sulfur, nonferrous, and noble metals in the upper taxite gabbrodolerite of the Noril'sk Intrusion was known to geoscientists long ago. Data on this are reported in the initial works devoted to the Noril'sk-I Intrusion, including Kotul'sky (1948), Rogover (1959), Korovyakov et al. (1963), and Smirnov (1966) and in other later articles. Poor occurrences of sulfides in the rocks from the UZ of the Noril'sk-I Intrusions were only of mineralogic interest rather than practical significance. In addition, the above authors wrote that where sulfide mineralization was poor the rocks of the UZ had elevated platinum-group element content and their ratio to nonferrous metals is several times higher than in impregnation ores of the lower ore-bearing zone (Ginsburg and Rogover 1960; Smirnov 1966). However, the main orientation to produce studies on the copper–nickel sulfide ores was the reason that studies of lean ores were fragmentary. With the discovery of the Talnakh Deposit (1960) with rich massive sulfide deposits at the base of the intrusion, interest in poor and lean sulfide ores of the UZ was disengaged completely. By analogy with the Noril'sk-I Intrusion, the upper horizons of the Talnakh Intrusion were considered unpromising for mineralization. This partly resulted from the zoned appearance of the Upper Talnakh Intrusion, where early drilling (in the 1960s) encountered the top part of the intrusion that featured a thick cap of alkaline–silicic rocks with accessory sulfide mineralization in the roof. Only with drilling of the middle and then basal parts of intrusion in the UZ (that features the upper taxite gabbrodolerite) was a wide set of petrographic types of rocks that host sulfide and chromite ore mineralization established.

As mentioned above, the sulfide-poor mineralization in the UZ of the Upper Talnakh Intrusion was similar to that of the Noril'sk-I Intrusion and was therefore not of interest for ore geologists captivated by the abundant information on diverse ores of the lower horizons. As a result, the UZ of the Noril'sk Intrusions was poorly studied up to the beginning of the 1980s, and data of detailed studies on the ore mineralization of these intrusions are absent from the geologic literature. In particular, the monograph "Copper-nickel sulfide ores from Noril'sk ore deposits" (Genkin et al. 1981), which summarized substantial body of work from the IGEM, lacks any information on the mineralogy and geochemistry of oregenic elements in the UZ.

Our study of the geochemistry of chromium in traps allowed us to reveal the existence of the upper concentration maximum of chromium accumulation (Ryabov and Yakobi 1981). The chromium content of rocks of the UZ of different intrusions was found to vary and, depending on the position in the magmatic column, initial melt composition, degree of

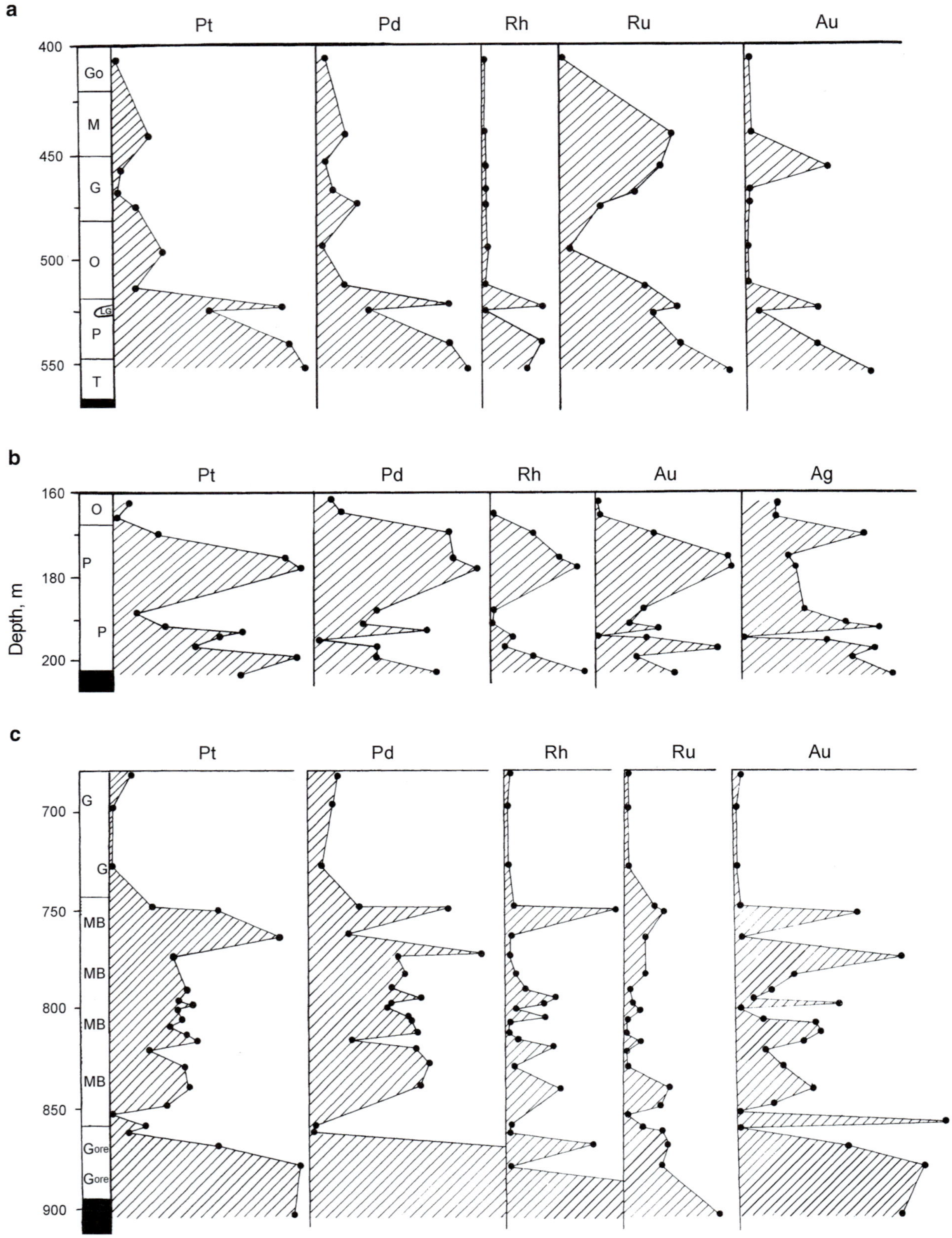

Fig. 4.28 Behavior of noble metals in the sections of the Upper Talnakh Intrusion. (**a**) Over the intrusion section, drillhole KZ-274; (**b**) over the picrite horizon section of the intrusion, drillhole KZ-184; (**c**) over the section of the Kharaelakhsky Branch of intrusion, drillhole KZ-585; massive sulfide ore is shown in column as solid

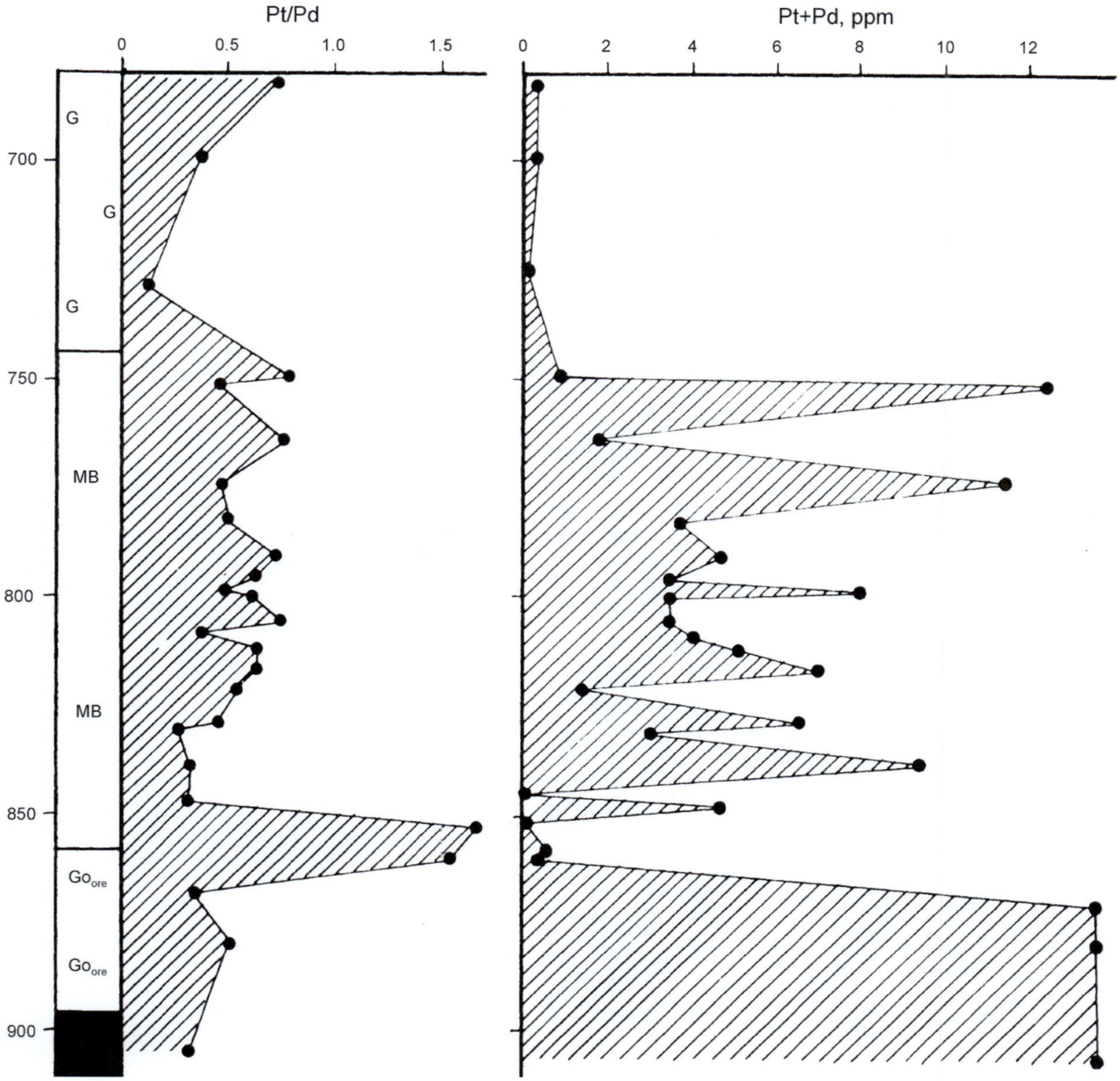

Fig. 4.29 Diagram of variations in Pt/Pd and Pt + Pd in the section of the Kharaelakhsky Branch of the Upper Talnakh Intrusion, drillhole KZ-585

differentiation, etc., may be different. In the Upper Talnakh Intrusion, the maximum chrome content increases when passing from the top to middle parts of the intrusion. At the top of the intrusion (Fig. 4.26a), the maximum chromium content is observed as a weak peak among virtually chromium-free rocks, but further down, the content increases and becomes comparable with or even exceeds the chromium content of picritic gabbrodolerite of the LZ.

Taking into account the empirically established relationship between the chromium and platinum contents of rocks known in the platinum-bearing belt of the Urals, Bushveld, and ultramafic rocks from different regions of the world, we analyzed the rocks from the UZ for platinoids and nonferrous metals. As a result, we found a platinum–chromite type of mineralization, which is new for the Noril'sk deposits (Ryabov et al. 1982; Ryabov 1983, 1984a, b, 1985, 1994).

The content of platinum metals in the UZ of the Noril'sk Intrusions vary widely and range up to 60 ppm (Sluzhenikin et al. 1994; Dodin et al. 1994). Insight into the distribution of elements in the UZ of the Upper Talnakh Intrusion can be gained from Fig. 4.30. The diagram shows that the horizon of upper pegmatoids accumulates quite a number of elements that manifests as a differentiated distribution in the rocks.

The diagrams connecting Cr and S with Pt + Pd + Rh are, in essence, classificational for the separation of low-sulfide Pt ores (Ryabov 1992b; 1994). Mineralogical–geochemical studies of the upper pegmatoids of trap intrusions

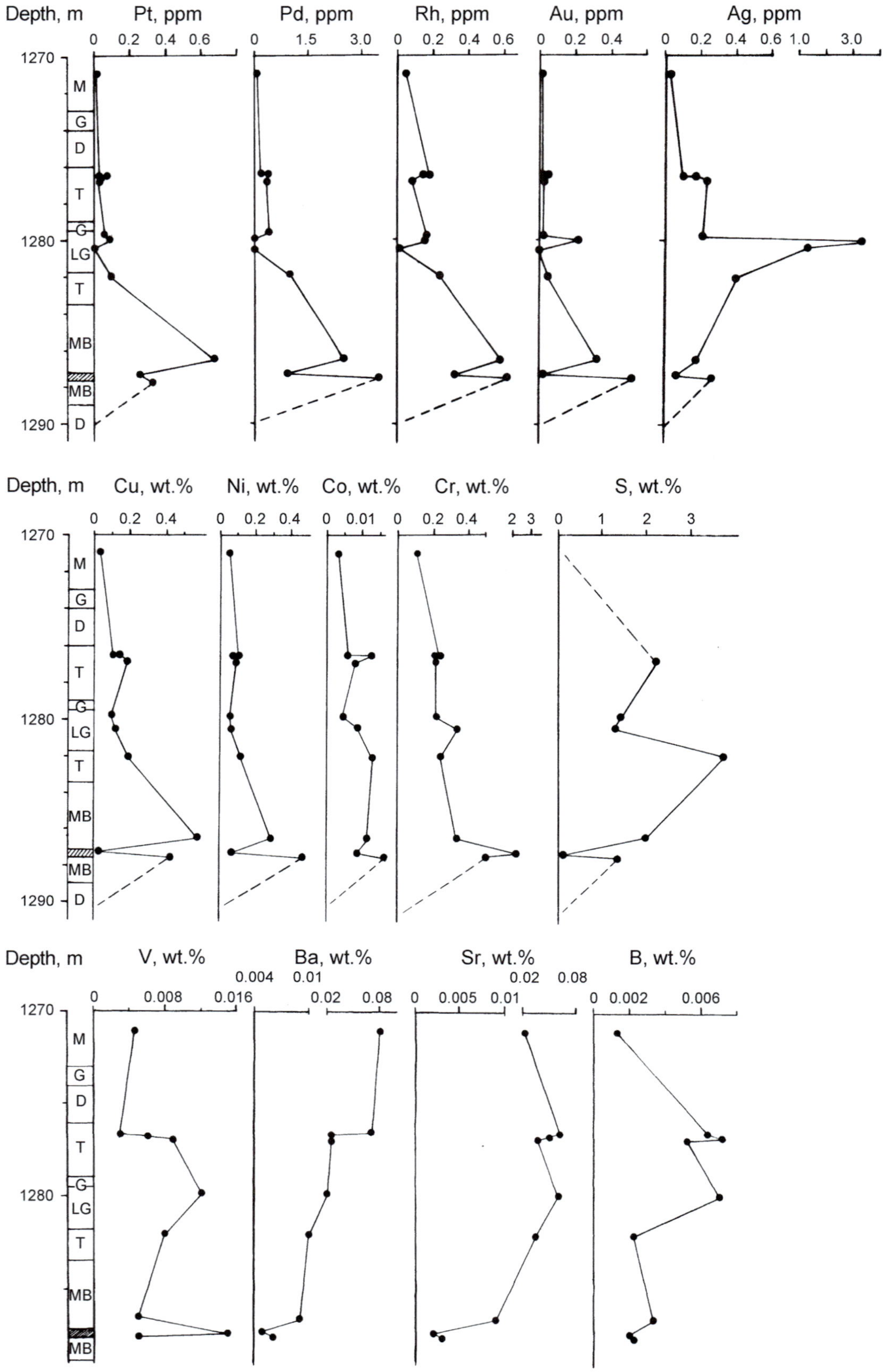

Fig. 4.30 Distribution of some chemical elements in the section of the upper zone of the Upper Talnakh Intrusion, drillhole KZ-1441

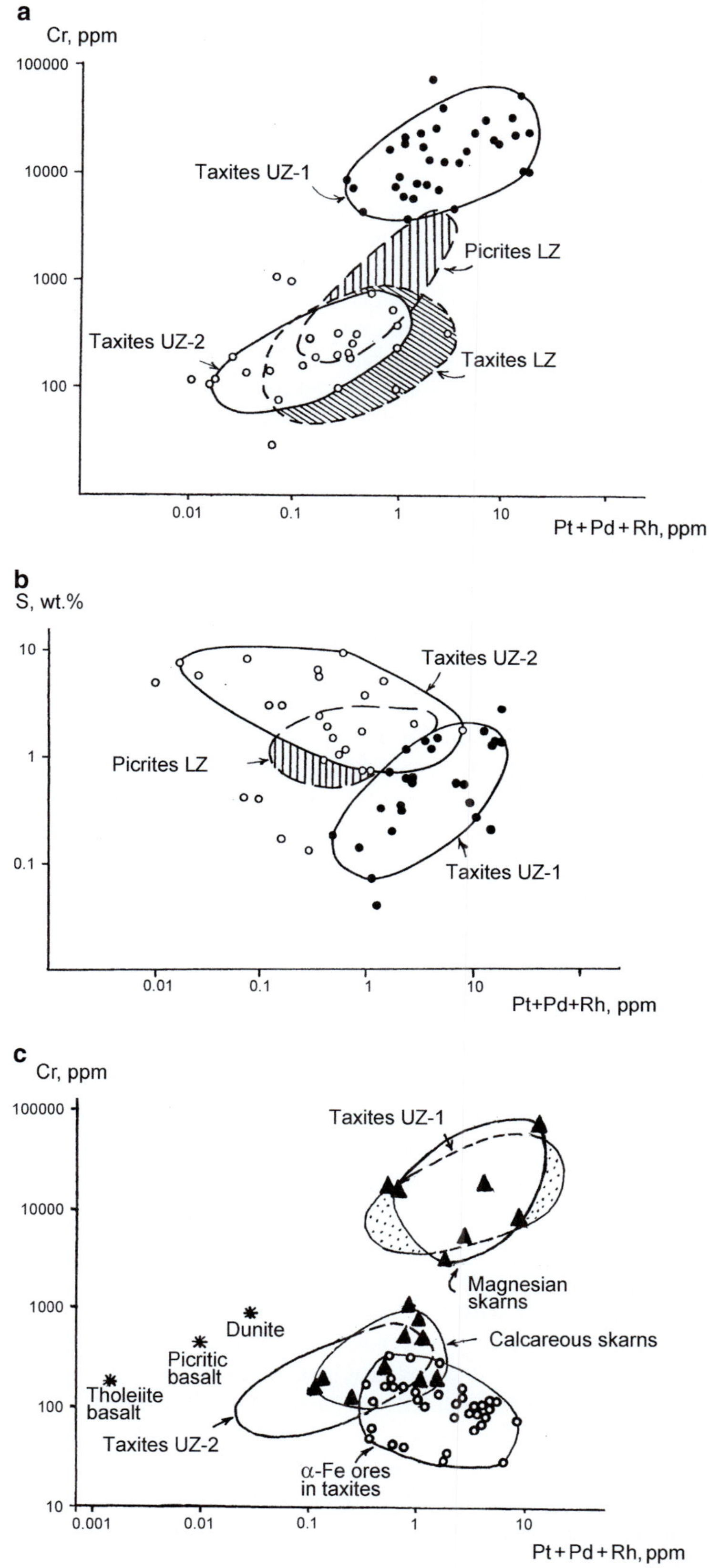

Fig. 4.31 Diagram of PGE content versus content of chromium and sulfur in different types of Pt low-sulfide ores in the rocks from the UZ of trap intrusions of the Siberian Platform. Explanations are provided in the text

Table 4.9 Concentration of PGE and Au (ppm) in trap rock-forming minerals

Mineral	Rock	Intrusion	Pt	Pd	Rh	Ru	Au
Olivine	Picritic gabbrodolerite	Noril'sk-I	0.30	0.64	0.05	–	0.025
	Troctolitic gabbrodolerite	Upper Talnakh	0.09	0.64	0.07	0.10	0.023
	Troctolitic gabbrodolerite	Mountain Putanaya	0.73	0.05	0.013	–	0.063
	Taxitic gabbrodolerite	Mountain Chernaya	0.095	0.26	–	–	0.033
Clinopyroxene (green)	Olivine–biotite gabbrodolerite	Noril'sk-I	0.083	0.083	0.005	–	0.027
	Taxitic gabbrodolerite	Noril'sk-I	0.057	0.06	0.012	–	0.086
	Melanocratic basalt	Mikchanda river	0.036	0.05	0.009	–	0.064
Clinopyroxene	Olivine gabbrodolerite	Arylakhsky	0.070	0.074	–	–	0.009
	Olivine gabbrodolerite	Arylakhsky	0.035	0.009	–	–	0.009
	Olivine gabbrodolerite	Upper Ergalakh	0.018	0.056	–	–	0.018
	Taxitic gabbrodolerite	Noril'sk-I	0.63	0.080	–	–	0.021
Plagioclase	Troctolitic gabbrodolerite	Noril'sk-I	0.03	0.52	0.007	–	0.018
	Troctolitic gabbrodolerite	Mountain Putanaya	0.03	0.04	0.22	–	
	Olivine–biotite gabbrodolerite	Noril'sk-I	0.05	0.04	0.013	–	0.63
	Taxitic gabbrodolerite	Noril'sk-I	0.05	0.07		–	0.018

Dodin et al. (1973)

allowed us to distinguish three main types of Pt low-sulfide ores: platinum–chromite, platinum–chromite-free, and platinum–iron metal or platinum–graphite. The transformation of Pt low-sulfide ores at the final stages of pegmatite phase resulted in Pt-skarn low-sulfide ores, among which the most notable are Pt–chromite in magnesium skarns and Pt–chromite-free in calcic skarns (Ryabov et al. 1996; Ryabov 1999a, b). All these five types of ores are shown on the diagram associating Cr and S with Pt + Pd + Rh (Fig. 4.31). To compare, we show the composition fields of Pt sulfide ores in picritic gabbrodolerite of the CZ, taxite gabbrodolerite of the UZ, and the composition of some ore-free rocks of the platform.

Chromium content of Pt–chromite ores differs substantially from Pt–chromite-free ores (Fig. 4.31a). Their sulfur content overlaps, but an increase in S content in the Pt–chromite type of ores is accompanied by an increase in Pt + Pd + Rh, and increasing S in Pt–chromite-free type leads to a decrease in the concentration of platinum metals (Fig. 4.31b). The composition of Pt ores in skarns and native iron in comparison with the ores mentioned above are shown in Fig. 4.31c. The position of the composition fields of Pt ores in skarns (solid triangles) suggests there is a typical succession of chromium and platinoid ratios and in ore-bearing pegmatoids replaced by them. The composition field of Pt–Fe-metal ores by Cr content partly overlaps with the Pt–chromium-free ores but differs substantially in the sulfur content from all types of rocks as this type of ore is virtually sulfide-free (Ryabov and Anoshin 1999). The diagrams, considering the specific features of ore composition, are basically classification plots, and they suggest that the ore formation of Pt low-sulfide ores depended on the fluid regime of an ore–magmatic system.

Forms of Platinoid Occurrence

Platinum-group metals in sulfide rocks are assumed to be isomorphously incorporated into the crystal lattices of rock-forming and ore minerals with a further transition at the magmatic or postmagmatic stage into independent ore phases. In the presence of sulfur, platinoids are preferable concentrated in sulfides (Naldrett and Cabri 1976). In the Noril'sk Intrusion, it was established that Pt accumulates in magnetite and olivine and to a lesser degree in pyroxene and plagioclase. The main Pd concentrator is olivine (Table 4.9) (Dodin et al. 1973; Distler et al. 1976).

In rocks of the UZ, we analyzed the noble metal content in potential concentrators magnetic chrome spinel and sulfides. Spinel contained only Ag (7.4–8.2) × 10^{-5}, and in sulfides, the Pt content varied (0.7–11) × 10^{-5}, Pd (4.6–130) × 10^{-5}, Rh (0.89–50) × 10^{-6}, and Ru (5.5–16) × 10^{-6}. The main concentrator of platinum in ores of the UZ is sperrylite, and the main concentrators of palladium are mayakite, vysotskite, stibiopalldinite, and arsenoantimonides (Ryabov 1992b, 1994; Sluzhenikin et al. 1994).

The problem of the origin of ore concentrations of noble metals in ultramafic and mafic–ultramafic rocks has been debated by geoscientists for many years (Page et al. 1982; Naldrett and Cabri 1976; Glasunov et al. 1977; Crocket and MacRae 1986). In the Noril'sk deposits, most researchers identify the accumulation of noble metals with the mechanisms of formation of Cu–Ni ores and consider them a product of liquation (Kotul'sky 1946; Godlevsky 1959, 1968; Smirnov 1966; Genkin et al. 1981). Studying the zonal distribution of noble metals in the Upper Talnakh Intrusion, Konkina and Rasin (1976) come to the conclusion that it reflects the liquation of ore-bearing magma with further crystallization differentiation of sulfide liquid and epigenetic transformation of the initial zonality. The authors indicate the increase in the concentration of noble metals from bottom up through the ore deposits is related to the participation of volatiles in the redistribution of components. The problems concerning the behavior of different metals in sulfide liquids in respect to the Noril'sk ores are discussed in the works of Genkin, Distler, Razin, Ershov, Popova, Kavardin, Mitenkov, and Karpenkov.

Most likely, the initial Pt/Pd value in the ultramafic rocks from the Noril'sk Intrusion exceeded unit, that is, it was close to the values in the products of picritoid melts (see Table 4.7). The redistribution of platinoids with an increase in palladium content and a stable ratio Pt/Pd < 1 seems to reflect the stabilization of the conditions of ore formation and, probably, is due to the high fS_2 value and association of ultramafic rocks with mafic rocks, the latter feature a similar ratio of platinoids.

Opinions differ on the origin of the platinum-bearing rocks at Stillwater. A very popular recent model was suggested by T. Irvine and further developed by Campbell, Barns, and Naldrett (Irvine et al. 1982, 1983; Keith et al. 1982). According to their concept, an incompletely solidified magmatic body was intruded by an additional melt that featured liquation separation of sulfide liquid that took place during the ascent of the melt. The turbulent regime of this intrusion promoted extraction of platinoids from the large volume of silicate magma.

Moreover, when interpreting the formation of ores with platinoids of Stillwater, many geologists attach importance to processes involving the participation of fluids (Boudreau 1988). The reason for this is the frequent occurrence of volatile-bearing phases, including those with Cl (phlogopite, pargasite, apatite, and graphite), the pegmatoid appearance of the rocks accompanying mineralization, the presence of fluid microinclusions, and the redeposition of sulfides along tectonic zones. In connection with the importance of fluids, it is interesting to note the detection of high fluid saturation of platinum minerals in the granulite complex (Tugovik et al. 1982). Fluids were found to contain H_2O, CO_2, H_2, N_2, and CH_4. Platinum minerals are assumed to form with the participation of S and As-bearing fluid. Some authors believe that platinoids redistributed under the influence of fluid flows as element-organic compounds (Slobodskoy 1981; Rudashevsky 1983). Taking into account the position of the platinum-bearing horizon in the UZ of the Noril'sk Intrusion, the bubbly and globular fabric of the rocks, and the abundance of volatile-bearing minerals, it is reasonable to suggest that fluids played an important part in the formation of the above-described ores.

4.2.2.4 Distribution of PGE in Traps

In constructing genetic models, one of the main parameters indicating the source of metals, size, and tendencies and factors of PGE redistribution during melt differentiation and formation of magmatic deposits is the content of metals in primary magmas. This information is provided by the determination of PGE in the rocks of lava sheets and isotropic intrusions of varying composition, comparison of effusive and intrusive comagmates, and the composition of ore-free and rocks and intrusions mineralized to varying degrees.

The absolute content of PGE in lava amounts to tens of mg/t; the concentration of metals irregularly decreases in the series of derivatives of picritic basalt, tholeiite-basalt, and trachybasalt melts. Moreover, the first differ from the second and the third featuring a predominance of Pt over Pd. Comparison of comagmates showed that in ore-free petrochemically similar rocks from different facies Pt/Pd > 1, and in the derivatives of tholeiite basalt and in trachybasalt rocks Pt/Pd < 1 (Fig. 4.32, see Tables 4.7 and 4.8). In Noril'sky-type layered mafic intrusions differentiated to picrite and troctolite gabbrodolerite, all have Pt/Pd < 1. Mafic–ultramafic intrusions have the same Pt/Pd ratio, despite the fact that their picritic gabbrodolerite are probably the derivatives of picritic basalt melt intruded simultaneously with mafic magma. There is reason to suggest that the change in the Pt/Pd ratio in picritic gabbrodolerite is related to the presence of sulfide mineralization.

It is known that the behavior of PGE is, to a considerable degree, controlled by the content of oxophilic–chalcophile elements and depends on the sulfur regime in an ore–magmatic system. Thus, the enrichment of rocks in sulfides is accompanied by an increase in PGE content and in the significance of palladium over platinum. As a result, in the mineralized picritic gabbrodolerite of the Noril'sk Intrusion and in rocks from other intrusions, there is a tendency for a decrease in Pt/Pd value with increasing sulfide mass. This can be observed when comparing the positions of composition fields of picritic basalt and picritic gabbrodolerite of the Noril'sk Intrusion (Fig. 4.32).

4.2.3 Nickel

4.2.3.1 Some Problems of Ore Formation

The behavior of nickel is known to be the most fundamental problem of the genesis of the Noril'sk deposits. Closely connected with this is the problem of the source of Ni in sulfide ores, the solution of which is significant for the development of strategies and tactics in the search

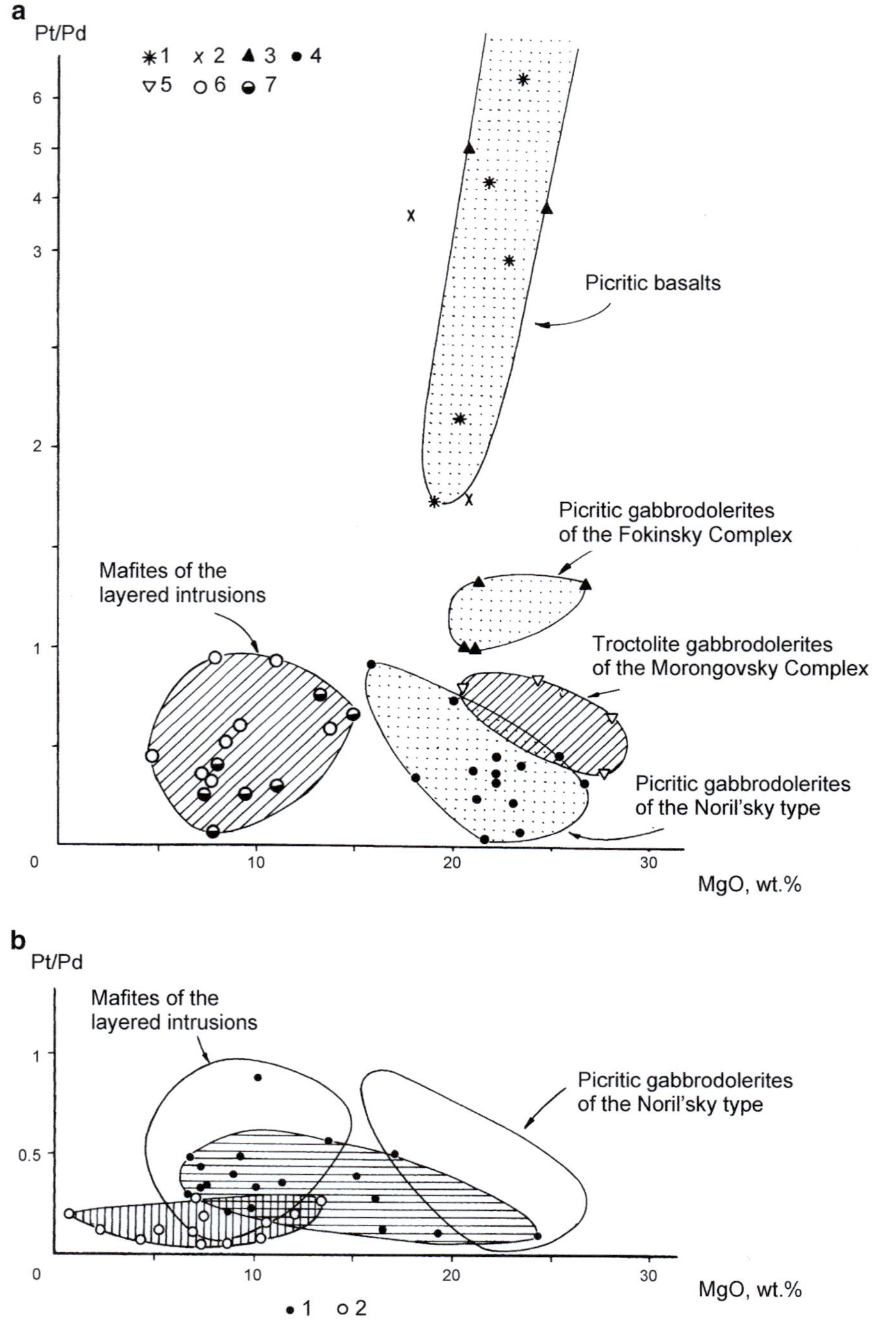

Fig. 4.32 Diagrams of Pt/Pd versus MgO in traps of the Siberian Platform. (**a**): *1* picrite basalts of lava flows; *2* effusive limburgites and augitites from the Kamensky and Maymecha–Kotuysky Provinces; *3* picrite dolerites and gabbrodolerites of Pikritovy Stream, Magnitny Stream, Lower Fokinsky, Upper Pyasinskaya Intrusions and Mikchandinsky neck; *4* mineralized picrite gabbrodolerites from the Upper Talnakh and Noril'sk-I Intrusions; *5* troctolite gabbrodolerites of Mt. Morongo, Mt. Putanaya, and Mt. Ruinnaya Intrusions and apophyses studied from drillholes; *6* olivine-poor gabbrodolerites of the Morongovsky Complex Intrusions, as well as Khungtukun and Maymechinsky Intrusions; *7* the same rocks with sulfide (more than 10%) mineralization. Derivatives of tholeiite-basalt melt are shaded; picrite-basalt field is shown as speckled pattern. (**b**) *1* magmatogenic breccia and *2* taxitic gabbrodolerites of the Noril'sky-type intrusions

for new deposits. The geochemistry of Ni in traps was described previously by Eliseev (1959) Korovyakov (1960), Tomanovskaya and Pol'kin (1970), Dodin et al. (1971), Kopylova and Oleinikov (1973), Nesterenko and Almukhamedov (1973), Distler et al. (1977), and Oleinikov (1979).

The distribution of Ni and its accumulation in some phases of ore-free traps depends mainly on the composition and temperature of melt and the fluid regime of the magmatic system and agrees well with theoretical assumptions (Marakushev and Bezmen 1972). Elevated Ni concentrations occur in high-temperature crystal phases and increases when the MgO content in the melt exceeds 4%, which is supported by its accumulations in mafic minerals (Vogt 1923). This phase in Skaergaard, Karoo Dolerites, and traps of the Siberian Platform is typically represented by olivine (Korovyakov 1948, 1960; Godlevsky 1959; Wager and Brown 1968; Kravchenko 1977).

Opinions differ on the origin of the Noril'sk deposits and the source of Ni. Some consider them to be magmatic, others suggest metasomatic, and still others, of hydrothermal origin. The first opinions on the origin of the Noril'sk deposits were presented by Kotul'sky (1946), Godlevsky (1959), and Urvantsev (1959), who concluded that the origin of sulfides was the result of melt liquation. According to these authors, the Ni-bearing sulfide magma separated from silicate magma in situ, in a shallow chamber (Korovyakov 1948, 1960), or intermediate chamber (Godlevsky 1967; Oleinikov 1979) or in a deep chamber (Kotul'sky 1946). The mechanism of formation of the Noril'sk ores by liquation was supported by experimental evidence (Olshansky 1951; Nekrasov and Gorbachev 1978), by physical–chemical substantiation (Godlevsky 1959; Genkin et al. 1981), and was further developed by numerous authors who supported the concept. The similarity in the content and ratios of nonferrous metals in picritic basalt and picritic gabbrodolerite with poor ore led Smirnov (1966) to consider that ultramafic magma was the original source for the ores. In the opinion of Eliseev'a (1959), lean ores are syngenetic and industrial ores epigenetic. A similar conclusion was reached by Kavardin and Goryainov, Sukhov, and other researchers.

The metasomatic hypothesis of formation of Noril'sk ores was developed by Zolotukhin (1964, 1965, 1971), Gulin et al. (1968), Goryainov (1971), and Tarasov (1976). This concept was also discussed in detail by Zolotukhin (1965). It is assumed that as a result of remelting some part of the rock in the picrite horizon and their transformation into taxite–gabbrodolerite (under the effect of gas emanation), silicate Ni was leached from olivine and further precipitated as sulfides. The most intense leaching was caused by postmagmatic melts. The author believes that the effect of this mechanism is consistent with the acid–mafic interaction in solutions reported by Korzhinsky.

A hydrothermal hypothesis of formation of sulfide Cu–Ni ores was suggested by Eliseev (1959). His work is the most thorough study of the geochemistry of Ni in sulfide deposits of what was known as the USSR. The main argument in favor of the hydrothermal formation of ores is the association of sulfides with various posterior minerals.

The source of nickel and sulfur in sulfide ores has been debated throughout the history of studies of the Noril'sk deposits. The main objection to liquation is the low solubility of sulfur in silicate melts, including basalt, known from experimental works (Almukhamedov and Medvedev 1982). Indeed, the volume of sulfide mass in the deposits is evidence that such large quantities cannot be dissolved in silicate melts. Godlevsky, Urvantsev, and others report the origin of discrete immiscible liquids in deep chambers and segregation of sulfide drops proceeded during movement to present-day chambers and during magma differentiation (Godlevsky 1971; Distler et al. 1977; Godlevsky and Likhachev 1979). Another mechanism suggests the supply of sulfur as early as the magmatic stage and the formation of ores as a result of magmatogenic sulfurization of melt in deep chambers (Almukhamedov and Medvedev 1982).

A very popular concept for the source of sulfur is the assimilation of sulfate-bearing rocks, resulting from the long-term residence of magma in an intermediate chamber (Godlevsky and Grinenko 1963; Goryainov and Yalovoy 1972; Kovalenker et al. 1974; Grinenko 1987). This mechanism also explains the shift of isotope ratio S^{32}/S^{34} toward an increase in the fraction of heavy isotope typical of sedimentary sulfates when compared against the chondritic ratio. However, new problems arise subsequently, such as the change in the sulfur isotope ratios in sulfides occurring throughout the intrusion (Kovalenker et al. 1974), the mechanism of reduction of sulfate, and the problem of CaO release during assimilation of sulfates. Rocks of ore-bearing intrusions are known to be practically isochemical with ore-free bodies and have low calcium content.

All these problems are, on the one hand, easily solved using metasomatic concepts and can be explained by the evolution of sulfur-bearing solutions but, on the other, still do not answer the question of the source of nickel (Genkin et al. 1971; Gulin and Sukhov 1973) nor explain the lack of appreciable metasomatic crystallization in picrite and taxite gabbrodolerite (Godlevsky 1968). Being inclined to the ore-forming role of postmagmatic solutions related to taxites and having estimated the balance of nickel in the Noril'sk-I Deposit, Tarasov (1976) thought that nickel content is low in the intrusion and 90% of nickel was, most likely, supplied from a deep source.

4.2.3.2 Distribution of Nickel in Rocks and Minerals

The main Ni concentrator in trap minerals is olivine, and the Ni distribution in this mineral was described in Chap. 4.1. Variable and, in places, high Ni content was detected in biotite (to 0.6wt%), in spinel (to 0.45wt%), and in pyroxene (to 0.092wt%). Data on Ni content in minerals is available in Genkin et al. (1971, 1981), Ivanov et al. (1971a, b), Dodin et al. (1972), Distler et al. (1977), and Ryabov and Zolotukhin (1977).

The behavior of Ni in olivine and the distribution of Ni in rocks in trap intrusions can be observed on the variation diagrams presented in this work. Comments on these diagrams were given when characterizing olivines and now let us consider the behavior of Ni in rocks. From the top down through the Pegmatitovaya Mountain Intrusion (see Fig. 3.39), one can observe a simultaneous increase in the Ni_{Ol} and Ni_{rock} content, suggesting the leading role of silicate nickel in rocks, which is also supported by the low S content in gabbrodolerite. The sulfide-free section of the Putanaya Mountain Intrusion (see Fig. 3.36) also manifests a simultaneous variation of Ni in the mineral and in the rock. In the Noril'sk-I (see Fig. 3.20), the content of Ni features some jumps but gradually increases from the base upward. It is seen here that the gradual growth of Ni_{rock} takes place only in mafic rocks, whereas in picrites Ni content dramatically increases, which corresponds with the dramatic increase in S content.

In the Noril'sk Intrusions (see Fig. 3.18), Ni_{rock} consists of silicate and sulfide Ni, the latter being predominant in the mineralized rocks at the base. In picrite and taxite gabbrodolerite, there is no direct relationship between Ni_{Ol}, Ni_{rock}, and S; the composition of the rocks is significantly affected by zonation (Distler et al. 1977). Nevertheless, in drillholes, one can observe a higher Ni content in ultramafic horizons than in the mafic horizons, higher concentrations of S in the former, and considerable nonuniformity of the Ni-bearing ore horizon.Nickel is a sensitive indicator of the composition of initial melts, their differentiation processes, magmatic system fluid regime, and ore formation. In ore-free traps, the major Ni-bearing phase is olivine, the composition and content of which are responsible for the Ni content of rocks. In sulfide-bearing gabbrodolerite, the situation is complicated by the occurrence of sulfide nickel in addition to silicate nickel, and the redistribution of the metal between the mineral phases and fluid.

The highest concentrations of silicate NiO are found in the derivatives of picritic basalt melts, that is, in picritic basalt and gabbrodolerite from the intrusions of the Fokinsky Complex (Fig. 4.33). In the layered intrusions of the Morongovsky Complex, Zubovsky type, and others that are derivatives of tholeiite-basalt melts, NiO in olivine increases with the increasing magnesium content. In picritic gabbrodolerite of the Noril'sk Intrusions, the NiO content of olivine varies widely in a rather narrow range with changing MgO and, judging from the orientation of the composition fields on the diagram, with decreasing NiO in olivine the MgO content tends to increase. The composition field of varying composition of troctolite gabbrodolerite from the Talnakh-type intrusions displays considerable variations in MgO content with a steady low NiO content in olivine. The diagram shows that the rocks with similar magnesium content may have different NiO content in olivine.

When comparing the total nickel content with magnesium content, the distribution of composition fields of ore-free derivatives of picrite- and tholeiite-basalt melts (Fig. 4.33) is preserved: highest nickel content is found in picritic basalt and gabbrodolerite from the Fokinsky Complex, and in layered mafic rock intrusions, NiO content increases with increasing olivine. In picritic gabbrodolerite of the Noril'sk Intrusions, the NiO content is mainly controlled by sulfide content, whereas troctolite gabbrodolerite of the Lower Talnakh type are depleted in Ni in spite of the high amount of olivine and high magnesium content. Subalkaline traps (the figure shows only the composition field) have a low NiO content and a poor tendency for increasing Ni with increasing MgO content.

The diagram (Fig. 4.34) shows the relationship between the NiO content in olivine and NiO in the rock (Ryabov 1992b). The field covered with specks shows the main geochemical trend of NiO in ore-free traps. Judging by its elongate shape, NiO content in the rock has a direct relationship with the NiO content of olivine. Denser specks on the diagram designate the fields of other groups of rocks. The mafic rock part of the Noril'sk Intrusions is localized within the major geochemical trend of ore-free traps. Two shaded fields deviating from the overall tendency correspond to picritic gabbrodolerite from the Noril'sk Intrusions, which are the leading part of a magmatic column, and to troctolite gabbrodolerite intrusions of the Upper Talnakh type, which occur in the rear part of a magmatic column. The arrangement of fields shows that, when compared to picritic basalt, NiO content decreases in picritic gabbrodolerite of the Upper Talnakh and Noril'sk-I Intrusions and becomes lower still in troctolite gabbrodolerite of the Lower Talnakh and Lower Noril'sk Intrusions. In addition, the total NiO content of ore-bearing intrusions is higher than the norm established for ore-free rocks and is related to the presence of Ni sulfides. The low NiO content in olivine of troctolite gabbrodolerite of the Lower Talnakh type, which corresponds to the level of olivine-poor rocks from ore-free intrusions, is due to the removal of Ni by fluids with further accumulation in the rocks of the leading parts of the magmatic column, that is, ore-bearing intrusions (Ryabov 1992b).

4.2.3.3 The Behavior of Nickel in Ore Formation

The available analytical data and geologic material provide evidence for the sequence of ore formation.

On the basis of evidence of comagmatism of picritic basalt and picritic gabbrodolerite in the Noril'sk Intrusions, Ni content in picritoid melts and olivines from different facies were initially the same. Under the effects of sulfur-bearing fluids on olivine, silicate Ni and some Fe were extracted with and further transformed into a sulfide form. The high temperature of the sulfurization was the reason for the preservation of olivine as a mineral species, and the removal of Fe and Ni led to an increase in the magnesium content (to create forsterite). It is natural that in the zones most saturated with fluids, maximum extraction of silicate Ni and Fe from olivine is anticipated. In this way, the known Ni-rich ores in picrite horizons (Godlevsky 1959; Rogover 1959; Smirnov 1966) and the decrease in Ni content at some distance from these rocks are symptomatic. Noteworthy is also the relationship between Ni-depleted olivines from olivinites of the magmatogenic breccia of the

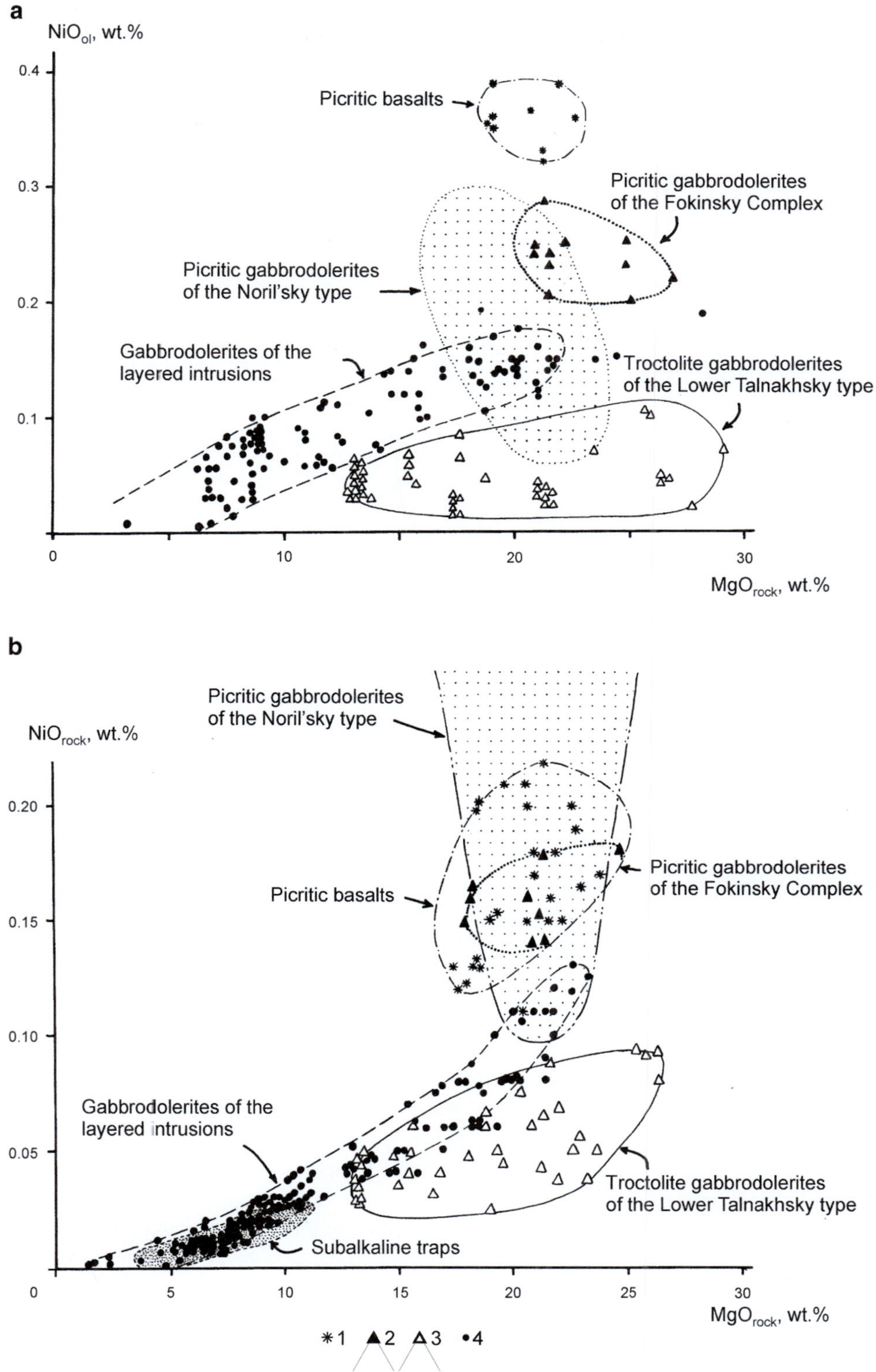

Fig. 4.33 Diagrams of NiO content in olivines (**a**) and in rocks (**b**) versus magnesium content in traps. *1* basalts of the Gudchikhinsky Suite; *2* gabbrodolerites of the intrusions: Lower Fokinsky, Magnitny Stream, and Pikritovy Stream; *3* gabbrodolerites of the Lower Talnakh and Lower Noril'sk Intrusions; *4* gabbrodolerites of the Mt. Putanaya, Mt. Pegmatitovaya, and Zubovsky Intrusions. Explanations are given in the text

Kharaelakhsky Branch of the Upper Talnakh Intrusion and the weighted average content of sulfides that is highest of all Noril'sk bodies (Table 4.10). The Kharaelakhsky branch is characterized by a rich to solid sulfide mass bordering on poor impregnated ores, which suggests outflow of sulfide liquid from the region where it was formed.

The reality of sulfurization process was supported by experiments (Kullerud and Yoder 1965; Almukhamedov

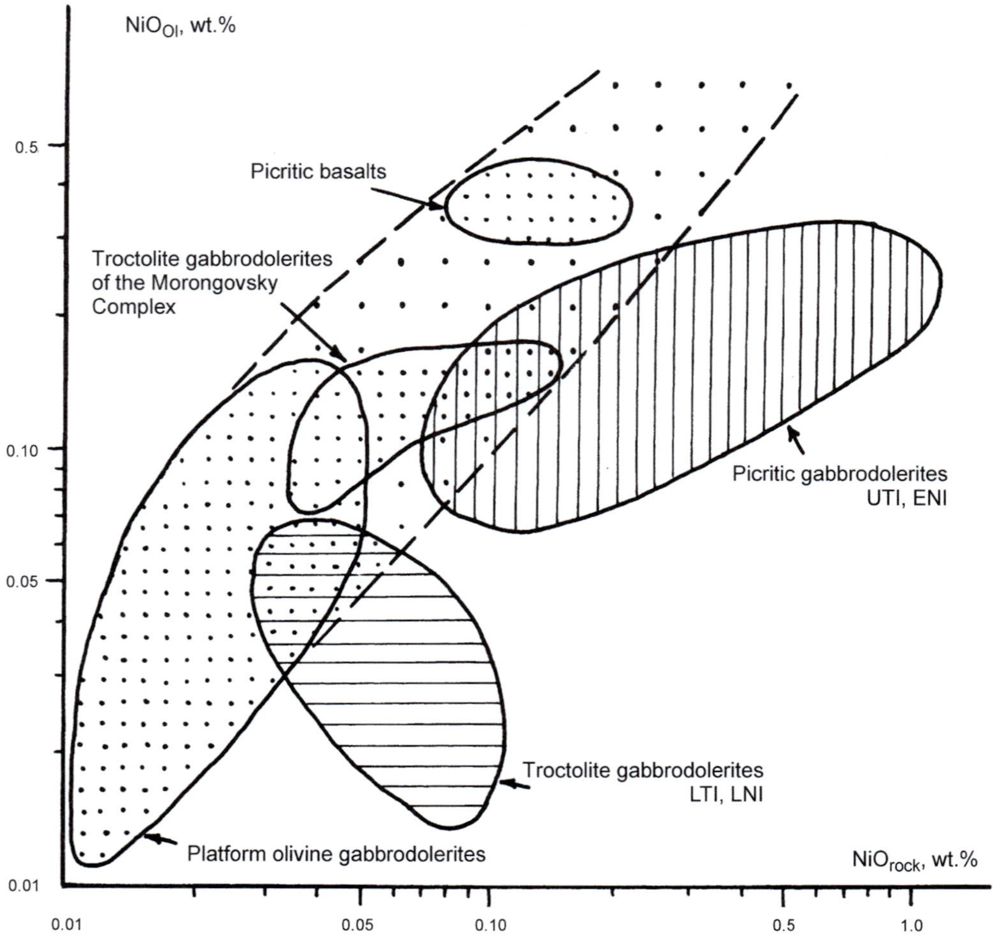

Fig. 4.34 Cumulative diagram of nickel content in olivine versus the total nickel content in ore-free and sulfide-bearing traps of the Siberian Platform. The general geochemical trend for ore-free rocks is shown by lighter speckles, whereas the fields of some types of rocks are distinguished by denser speckles. Shaded are fields of compositions of ore-bearing Upper Talnakh (UTI) and Noril'sk-I Intrusions, Eastern Noril'sk Branch (ENI), and accompanying nickel-depleted Lower Talnakh (LTI) and Lower Noril'sk (LNI) Intrusions

and Medvedev 1982), which demonstrated the extraction of Ni and Fe from solid ultramafic rocks from melt. It is worth noting that the amount of Fe in the liquid decreased proportionally to excess S. It is safe to assume that the system with excess sulfur will provide crystallization of mafic minerals with a lower Fe content than a melt without sulfur.

Experimental study of Ni distribution between olivine and sulfide liquid (Arutyunyan, I986) showed that in sulfur-undersaturated melts, the high affinity of Ni with S results in the concentration of metal in sulfides and depletion in olivine. With growing sulfur content in melt, the depletion of olivine in Ni continues to a limit. In sulfur-saturated liquids, if no sulfides precipitate, that is, if sulfides remain in the system or some proportion outflow, the Ni content of olivine increases with increasing sulfide suspension in melt. The change in from depletion of olivine in nickel to enrichment takes place after the fractionation of 26% olivine in the presence of 0.5% sulfides. Additionally, an increase in the concentration of suspended sulfide to 0.7% results in an increase in the Ni content of olivine.

Evidence for the variation in occurrence of Ni from silicate to sulfide comes from the above-mentioned data on olivines and other data from Ryabov (1992b, c). These show the wide variation in Ni content in minerals from one horizon of rocks (and even from individual samples), the increase in the Ni_{Ol} content near the pegmatoid horizon, and the variable concentration of Ni_{Ol} in sulfide-rich and sulfide-poor areas of rock.

It is worth noting that any of the suggested hypotheses for ore formation are compelled to admit the supply of S and Ni to the deposit. For Ni, this may be a juvenile source or redistribution of metal from within the magmatic column. Based on the supposed mechanism of ore formation, Ni was supplied from ultramafic material within the magmatic column, and with the participation of fluids was redistributed from the rear to leading parts of the column.

Table 4.10 Average sulfide concentration in differentiated gabbrodolerite (Gulin and Sukhov 1973)

Intrusion, intrusive branch	Number of holes	Sulfides concentration (wt%)
Noril'sk-I	350	10.7
Upper Talnakh, Southwestern	97	12.4
Upper Talnakh, Northeastern	40	13.8
Upper Talnakh, Kharaelakhsky	91	26.8

Note: Sulfide content calculated for typical ore material (vol%):
close set ores—85
commercial ores—15
ore-bearing rocks—0.5

This may explain the joint occurrence of the Noril'sk Intrusion and the highly magnesium Lower Talnakh and Lower Noril'sk Intrusions that are depleted in oregenic components.

The influence of fluid on melts or solid phases, with selective extraction of Ni and Fe, probably cannot be called metasomatic transformation. Here, we speak about a unidirectional diffusion differentiation of components controlled by the degree of affinity of chemical elements to S (Marakushev and Bezmen 1972). Thus, sulfurization reflects the specific character of the kinetics of formation and the composition of compounds, whereas liquation reflects the mechanism of separation of generated ore substance from silicate matter. Therefore, it makes no sense to oppose these concepts (Almukhamedov and Medvedev 1982; Zolotukhin 1988).

An increase in pS_2 in silicate melts, on one hand, accelerates and intensifies the extraction of Ni from olivine and melt and, on the other, stimulates the liquation ore–silicate immiscibility of the liquid. Liquation may take place only when the content of S exceeds its solubility limit in the melt. This is confirmed by the observations of Skinner and Peck (1969) for basalt from Kilauea in Hawaii, in which sulfide balls form only in interstitial melts when crystallization in residual melts resulted in supersaturation with sulfur.

In the Noril'sk Intrusions, a gradual increase in fS_2 took place along with a successive sulfurization processes with liquation separation of liquids, coalescence of a fine emulsion into large drops, and the migration of ore liquid and further crystallization. Separation of immiscible sulfide liquid under a generated liquation situation, which could occur under abyssal conditions, proceeded with the movement of melt into the present-day chamber and crystallized within it. Liquation in a deep chamber seems to be least likely. This evidenced by the absence of sulfides in picritic basalts, which are comagmatic with Noril'sk picritic gabbrodolerites (Zolotukhin 1988) and by the decrease in ore saturation and Ni content in the near-basal part of intrusions. The comparatively irregular distribution of ore dissemination in the picrite horizon as well as the drop-shaped and interstitial forms of its occurrence suggest, most likely, coalescence of a finely dispersed emulsion into larger accumulations rather than a separation of ore liquid into drops. The physical evidence of the isotopy of S in the Noril'sk deposits does not contradict the concepts of sulfurization with further liquation of melts.

4.2.4 Titanium

In each magmatic body—in basalt flows, dykes, or sills—the behavior of titanium has specific features, scales of accumulation and is dependent of a number of factors including the composition of the initial melt and crystallization conditions. Some trap intrusions clearly exhibit an increase in TiO_2 content from the base upward through the bodies with an accumulation of titanium oxide in rocks most enriched in alkali elements. This is well observed in the above-given variation diagrams of oxides for different intrusions, which also suggest that titanium is a sensitive indicator of melt differentiation. This was discussed in many works (Nesterenko and Almukhamedov (1973) and Arkhipova (1975)).

The main concentrators of titanium in traps are titanomagmatite, ilmenite, and chrome spinel, and in silicates are biotite and clinopyroxene, as well as scarce rutile and perovskite.

The titanium content of rocks depends on the composition of the initial melt and differentiation features and is controlled by the amount of titanium-bearing mineral phases. The behavior of titanium is most distinctly displayed in layered intrusions. This can be observed on the oxide variation diagrams plotted for intrusions of different complexes.

Distribution of titanium in rocks of differing composition and magmatic bodies can be observed from the TiO_2–MgO diagrams (Fig. 4.35). Part of Fig. 4.35a shows the composition of rocks from the weakly differentiated mafic Khungtukun and Maymechinsky Intrusions. In the layered series of rocks from these intrusions, titanium content slightly increases as magnesium content decreases. In the groundmass of these rocks, TiO_2 is 0.7–1.5wt%. Elevated concentrations of TiO_2 2.0–3.1wt% are found in mafic pegmatoids, whose composition stand apart from the swarm of points for rocks from layered intrusions. Figure 4.35b shows the composition of isotropic sills of trachydolerite from the Egralakhsky Complex and high-magnesium intrusions of the Morongovsky Complex. Figure 4.35c illustrates the composition of pegmatoids from the sills of trachydolerite and rocks from the clearly stratified

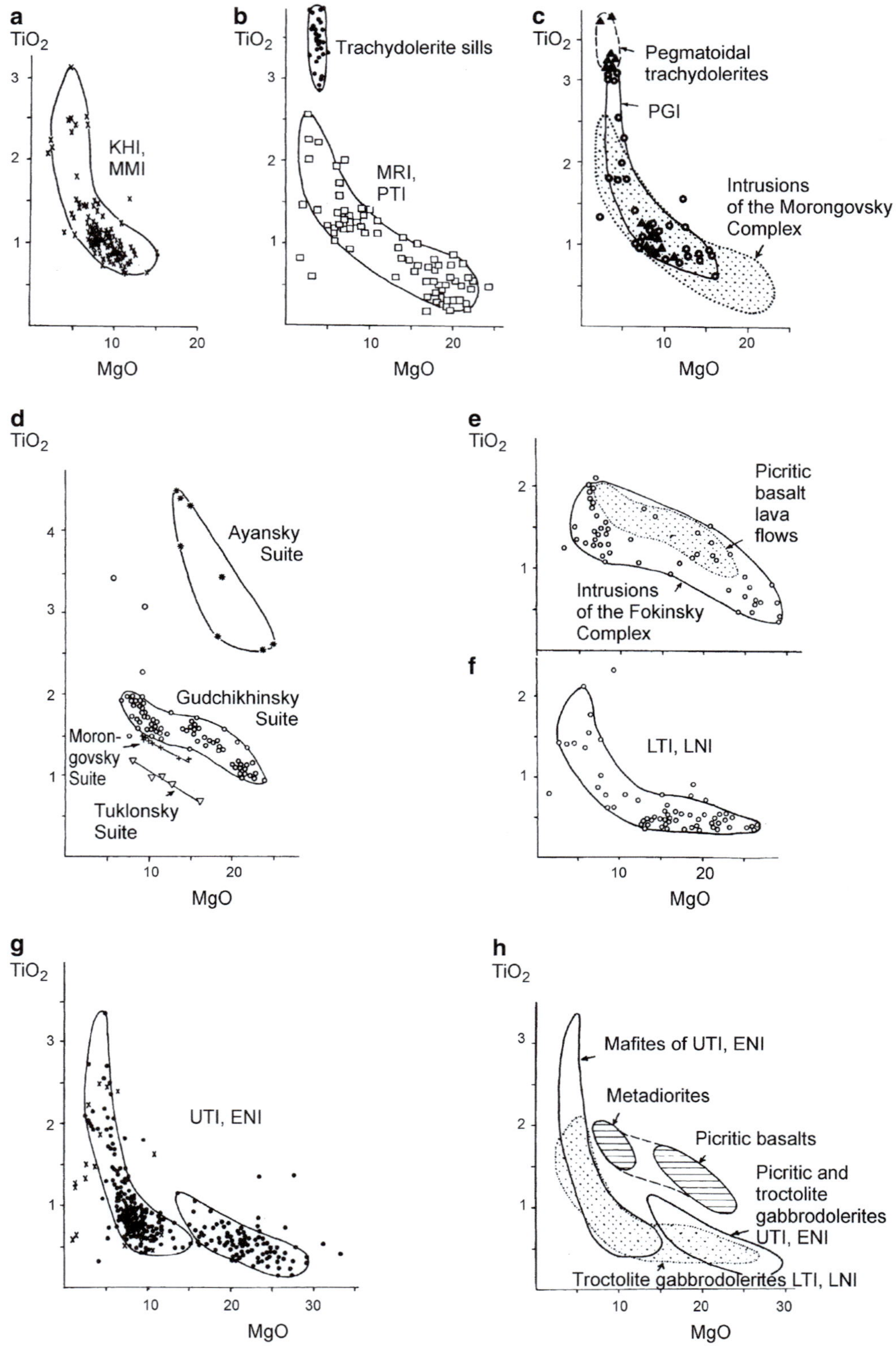

Fig. 4.35 Diagram TiO_2 versus MgO (wt%) in differentiated traps. (**a–c**) mafic intrusions, (**d**) lavas of picrite basalts and ankaramites, (**e**) Fokinsky Complex Intrusions compared to lava picrites, (**f**) Lower Talnakhsky-type intrusions, (**g**) Noril'sky-type intrusions (*left field*—mafic rocks, *right field*—picrites), (**h**) comparative diagram for different layered formations. Explanations are given in the text. Intrusions: *KHI* Khungtukun, *MMI* Maymechinsky, *MRI* Mt. Morongo, *PTI* Mt. Putanaya, *PGI* Mt. Pegmatitovaya, *LTI* Lower Talnakh, *LNI* Lower Noril'sk, *UTI* Upper Talnakh, *ENI* Noril'sk-I, Eastern Noril'sk Branch

intrusions of the Pegmatitovaya Mountain that is the leading part the Morongovsky group of intrusions. The points on Fig. 4.35c present the field of rocks whose analyses points were shown on Fig. 4.35b. Analysis of Fig. 4.35b, c shows that trachydolerites and their pegmatoids have similar high titanium content. In gabbrodolerite from the Mountain Morongo and Mountain Putanaya Intrusions, the titanium content decreases with increasing MgO content, and there are two groups of rocks with stable MgO and TiO_2 content. The content of these oxides is 19.0 and 0.5wt% in troctolite gabbrodolerite and 7.0 and 1.3wt% in olivine gabbrodolerite. Some analyses of gabbro show up to 2.5wt % TiO_2. In the rocks from the Pegmatitovaya Mountain Intrusion, TiO_2 increased to 3.3wt%, and the analyses fell into the field of pegmatoid trachydolerites. The arrangement of composition fields of rocks from the leading (Mountain Pegmatitovaya Intrusion) and near-basal (Mountain Morongo and Mountain Putanaya Intrusions) parts of the extended magmatic body suggests an increase in the titanium content of rocks in the roof of the magmatic column and the probability of the formation of tholeiite-basalt rocks from melt, which featured a TiO_2 and MgO content similar to trachydolerites.

Figure 4.35d–f shows the composition of some high-magnesium effusive rocks and intrusions. The titanium content of effusive rocks successively increases from olivine-phyric (picrite) basalts of the Tuklonsky Suite, through the ankaramites from Arylakh of the Morongovsky Suite, picritic basalt of the Gudchikhinsky Suite, and picritic basalt of the Ayansky Suite.

The titanium content of rocks in the Fokinsky Complex Intrusions gradually decreases with increasing MgO (see Fig. 4.35e). Comparison of gabbrodolerite from intrusions of this complex with the rocks from differentiated flows of picritic basalt shows great similarity, which might be additional evidence supporting their comagmatism. In troctolite gabbrodolerite from the Lower Talnakh and Lower Noril'sk Intrusions (see Fig. 4.35f), the TiO_2 content (about 0.5–0.6wt%) remains in rocks with varying MgO content 12–27wt%. In the mafic part of these intrusions, as the MgO content decreases, there is an increase in TiO_2 content from 2.1 to 2.3wt%. Figure 4.35e, f shows that the rocks from the Fokinsky Complex and Lower Talnakh Intrusion have a similar TiO_2 content (0.3–0.4 to 2.0–2.1wt%) and MgO content (3–4 to 27–29wt%).

In the Noril'sky-type intrusions (Fig. 4.35g) in mafic rocks one can observe that, when MgO content ranges 3–8wt%, the TiO_2 content dramatically increases from 0.4 to 3.4wt%. In the ultramafic horizon, MgO ranges 14–28wt %, whereas TiO_2 ranges from 0.2 to 1.2wt%. The composition of pegmatoids (shown by crosses on the diagram) is similar to that of the gabbrodolerite among which they occur. The composition fields of rocks of the Noril'sky and Lower Talnakhsky-type intrusions are very similar (see Fig. 4.35h) and, by analogy with the Morongovsky group of intrusions, show an accumulation of titanium in the leading part of magmatic column. Differentiated lava flows of picritic basalt, when compared to the Noril'sky- and Lower Talnakhsky-type intrusions, feature higher titanium content in rocks of equivalent magnesium content.

When comparing picrites from magmatic formations as supposed comagmates, one can see a decrease in titanium content in the series from picritic basalt to picritic gabbrodolerite of the Fokinsky Complex Intrusion and a further decrease to picrite gabbrodolerite of the Upper Talnakh Intrusion and, finally, the lowest titanium content is found in picrites of the Lower Talnakh and Lower Noril'sk Intrusions (see Fig. 4.35e).

In general, the intrusions and effusives presented on the diagrams show initially different titanium content in their parent magma and then increase in titanium concentration with increasing rock alkalinity. The highest TiO_2 content is found both in independent melts (trachydolerites) and in their differentiation products. The accumulation of TiO_2 is the result of successive crystallization differentiation of melt (Pegmatitovaya Mountain Intrusions) and segregations of fluid-saturated pegmatite melt (layered sill of trachydolerites in drill hole F-221).

4.2.5 Sulfur

The intrusions in the Noril'sk Region form ore deposits following enrichment in sulfur relative to nickel (Godlevsky 1967). This conclusion is obvious considering that sulfur can extract nonferrous metals and concentrated them as sulfides in a wide range of conditions during magmatogenic, metasomatic, and hydrothermal processes (Kullerud and Yoder 1965; Almukhamedov and Medvedev 1982). Therefore, one of the main issues related to the origin of Pt–Cu–Ni sulfide deposits is the behavior of sulfur in the ore-forming process: the source of sulfur, its migration path, and the form, place, and conditions of sulfide formation. Connected to this, sulfur isotopes are the most important characteristic of the geochemical history of ore–magmatic processes.

The problems of the geochemistry of sulfur in basalt melts, in gabbrodolerite of trap intrusions with varying sulfide ore saturation, including the ores from the Noril'sk deposits, have been discussed by many authors. On the Siberian Platform, the major body of information on the isotopic composition of sulfur concerns the deposits of the Noril'sk and Talnakh ore bodies, for which abundant analytical data have been accumulated. The number of isotopic analyses of sulfur only for the Talnakh ore body is more than 800.

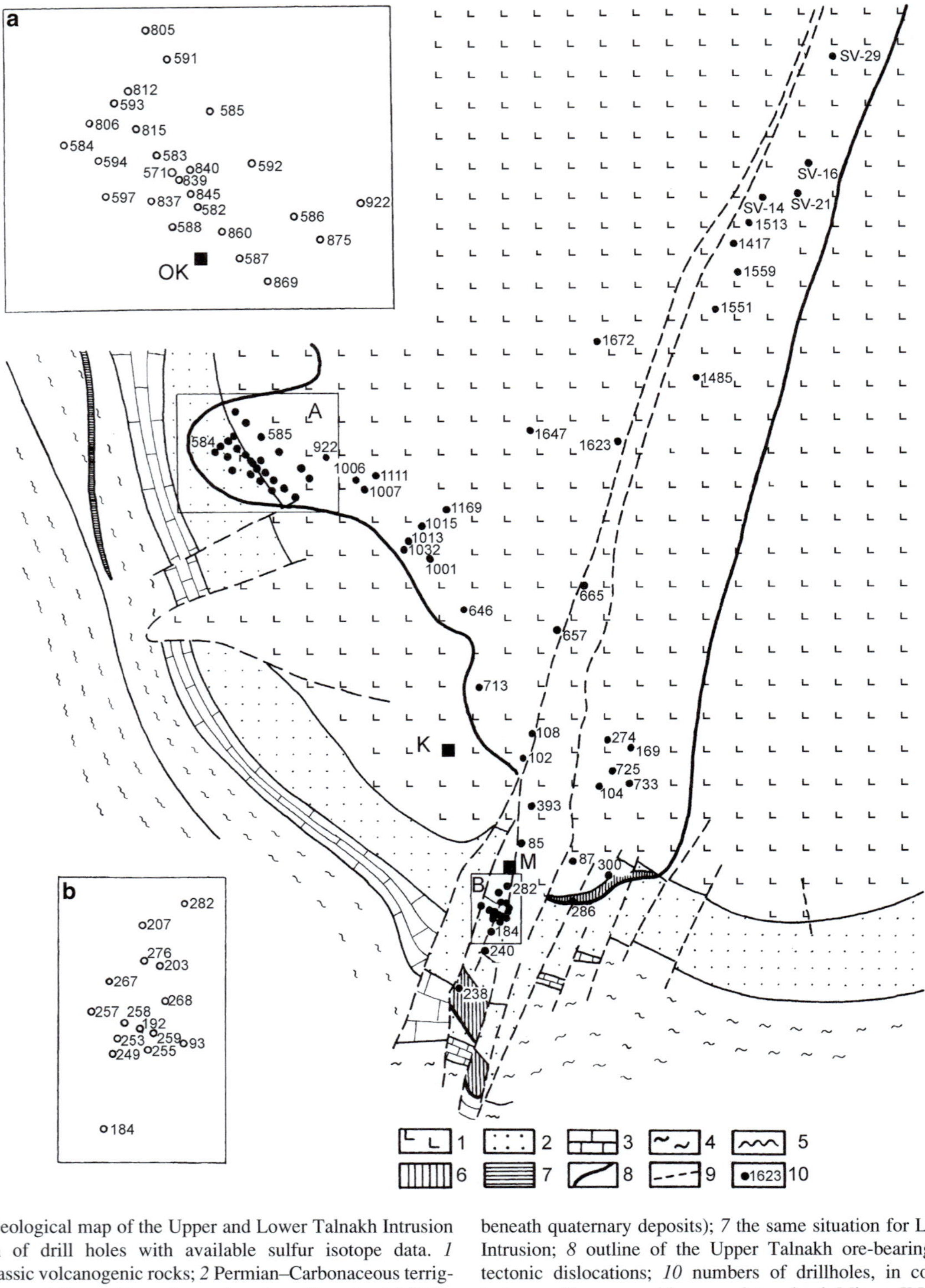

Fig. 4.36 Geological map of the Upper and Lower Talnakh Intrusion and location of drill holes with available sulfur isotope data. *1* Permian–Triassic volcanogenic rocks; *2* Permian–Carbonaceous terrigenous sediments; *3* Upper Devonian carbonate sediments; *4* Middle Devonian carbonate-, clay-, and sulfate-bearing rocks; *5* similar Lower Devonian rocks, *6* Upper Talnakh ore-bearing intrusion (outcrop beneath quaternary deposits); *7* the same situation for Lower Talnakh Intrusion; *8* outline of the Upper Talnakh ore-bearing intrusion; *9* tectonic dislocations; *10* numbers of drillholes, in cores of which isotope sulfur composition was determined. Mines: *K* Komsomolsky, *M* Mayak, *OK* Oktyabr'sky

In the plan of the ore body, they are distributed so that the best studied sections are the Oktyabr'sky Deposit (A) and the southern part of the Talnakhsky Deposit (B) (Fig. 4.36). The schematic geologic map of the Talnakh ore body illustrates the numbers of drill holes for which isotopic sulfur composition of sulfides and sulfates are available in the literature. The sampling points are arranged so that in the lower part of the scheme, including that in inset B, are concentrated representing the southern part of the Talnakh Deposit; data published in the work by Grinenko (1966). In inset A, representing the western part of the Oktyabr'sky Deposit, one can observe data of Gorbachev and Grinenko (1973). Between sites A and B, data from Kovalenker et al. (1974) and Avgustinchik (1981) are arranged. The drill holes tested by the authors occur in different parts of the ore body and are traced in northeastern direction.

Information on the geochemistry of sulfur and its isotopes in Traps is available in works by Avgustinchik, Almukhamedov, Godlevsky, Gorbachev, Goryainov, Grinenko, Grinenko, Zotov, Kovalenker, Kuz'min, Likhachev, Medvedev, Staritsyna, and Tuganova.

The low solubility of sulfur in basalt melts was the reason for proposing sulfate from sedimentary rocks as an additional sulfur source (Vinogradov and Grinenko 1964). It is assumed that in the Noril'sk deposits about 2/3 of sulfur was sourced, together with deep magma material, from subcrustal chambers and 1/3 sourced from sedimentary gypsum–anhydride rocks (Godlevsky 1967). Geoscientists explain that sulfur from sedimentary sulfates is responsible for the heavier (compared to meteoritic) isotope sulfur composition of sulfides from the Noril'sk deposits. When passing from ore occurrences to small, medium, and large deposits on the Siberian Platform, sulfides demonstrate a tendency to be associated with an increase in sulfur content and composition fraction of δ^{34}S (Fig. 1.13).

In the literature, the sulfur isotope composition of sulfides in different ores varies widely. The highest δ^{34}S values in the Noril'sk Region were found in magmatogenic sulfide veins and impregnations in basalt from the Loganchi Mountain (Lake Khantayskoe), which were 21.5 and 23.5‰, respectively (Ryabov 1997). In addition, it is assumed that the Noril'sk deposits featured intermediate magma chambers where homogenization of sulfur isotopes took place; as a result the sulfides of impregnated and massive ores of the Noril'sk deposits gained identical, stable isotope ratios. Therefore, some geoscientists think that the isotopic composition of sulfur in sulfides will not reveal genetic differences and are independent of the depth of occurrence of magmatic bodies, their occurrence in different horizons of intrusion or in different parts of the same horizons, and of the form of ore impregnation. Finally, different types of impregnated and massive ores have the same isotope composition (Grinenko 1966, 1985, 1987; Gorbachev and Grinenko 1973). The variations in sulfur isotope composition are related to the different scales of magma contamination of sedimentary sulfates and incomplete homogenization of crustal sulfur in an intermediate chamber (Grinenko 1967). Moreover, the similarity of sulfur isotope composition of sulfides from rocks and ores of the Noril'sk deposits is assumed to be evidence of a single source for ore and similar mechanism for magmatogenic formation (Grinenko 1987).

The sulfur isotope compositions from the Oktyabr'sky, Talnakhsky, and Noril'sk-I Deposits have the following values (δ^{34}S): impregnated ores 12–8.2–8.2‰; massive ores 11.5–8.8–8.1‰ (Gorbachev and Grinenko 1973). On the basis of studies, these authors come to the conclusion that regular variations in δ^{34}S in sulfides along the intrusion are absent, which is evidence of perfect homogenization of sulfide and sulfate sulfur in the magma chamber. They believe that in the intermediate chamber δ^{34}S was present as in ore of the Talnakhsky Deposits. In the Oktyabr'sky Deposits, additionally up to 50% sedimentary sulfur with δ^{34}S 15.2‰ was accumulated. During the successive formation of the intrusive branches of the Upper Talnakh Intrusion, δ^{34}S increased to 8.4‰ in the Southwestern branch, 9.0‰ in the Northeastern, and 11.7‰ in the Northwestern branch. The above-mentioned authors attribute this to the fact that later portions of magmatic melt were enriched in heavy isotopes at the expense of more complete reduction of sulfate during the evolution of the chamber. In their opinion, if the formation had taken place as a continuous process, the enrichment of sulfides in δ^{34}S would have occurred gradually in the direction of magma movement from leading to basal parts, which was not observed in the Upper Talnakh Intrusion.

The behavior of δ^{34}S in the Upper Talnakh Intrusion is shown on the variation diagrams (Fig. 4.37). In the ore-bearing horizon of the Talnakh Deposit (left plot), one can observe a rather stable sulfur composition in pyrrhotite except for the upper part of the picrite horizon and olivine gabbrodolerite. Chalcopyrite differs with a higher isotope composition of sulfur, compared to coexisting pyrrhotite. The difference in the value of δ^{34}S between the minerals is insignificant, but it is unidirectional and increases dramatically in the horizon of mafic pegmatoids. Data from Gorbachev and Grinenko (1973) evidence that in the Oktyabr'sky Deposit, the saw-shaped parts of the variation diagram are confined to the horizon of mafic pegmatoids. Here, one can also observe a regular increase of heavier sulfur isotopes from the top downward through the massive ore deposit, which is related to the variation in the composition of ores from chalcopyrite with pyrrhotite through chalcopyrite–pyrrhotite to cubanite–pyrrhotite.

Features of the isotopic composition of major types of ores are shown in Fig. 4.38, which is based on our data and data from the literature. The fraction of δ^{34}S in sulfides varies from 4 to 12–13‰, its variation proceeding regularly in coexisting chalcopyrite and pyrrhotite. Mafic pegmatoids reveal the widest variations in isotope composition, whereas in picritic gabbrodolerite, the variability of composition is much less pronounced. Massive ores of the Southwestern branch of the Talnakh Deposit differ from similar ores of the Oktyabr'sky Deposit featuring a lighter isotope sulfur composition of sulfides. The data for these two deposits

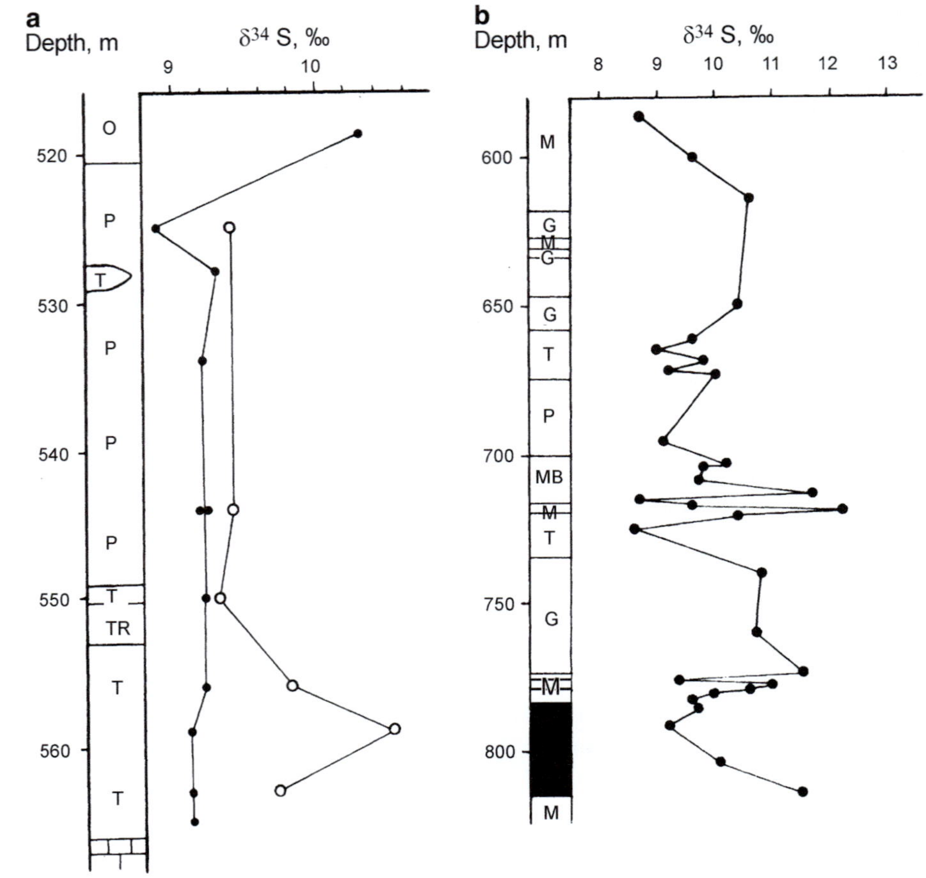

Fig. 4.37 Variation diagrams demonstrating behavior of sulfide sulfur isotopes in the sections of the Talnakhsky Deposit ((**a**) for drillhole KZ-274) and Oktyabr'sky Deposit ((**b**) for drillhole KZ-583). Based on data from Gorbachev and Grinenko (1973). (**a**) Pyrrhotite is shown by *solid circles* and chalcopyrite is shown by *light circles*; (**b**) composition of sulfide mixture is shown as *solid circles*

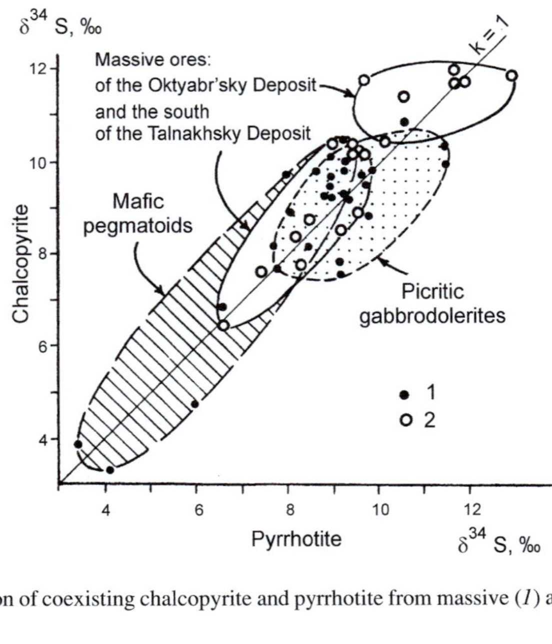

Fig. 4.38 Diagram of isotope composition of coexisting chalcopyrite and pyrrhotite from massive (*1*) and disseminated (*2*) ores of the Talnakh ore junction

Table 4.11 S-isotope data ($\delta^{34}S$ ‰) in Talnakhsky ore junction sulfide and sulfate minerals

Sample	Rock	Pyrrhotite	Chalcopyrite	Anhydrite
	Upper Talnakh Intrusion, CZ and LZ			
KZ-274/518.9	Olivine gabbrodolerite	10.3	–	–
KZ-274/525.0	Picritic gabbrodolerite	8.9	9.4	–
KZ-274/528.0	Picritic gabbrodolerite	9.3	–	–
KZ-274/534.0	Picritic gabbrodolerite	9.2	–	–
KZ-274/544.0	Picritic gabbrodolerite	9.2	9.4	–
KZ-274/550.0	Taxitic gabbrodolerite	9.2	9.3	–
KZ-274/556.0	Taxitic gabbrodolerite	9.2	9.8	–
KZ-274/559.0	Taxitic gabbrodolerite	9.1	10.5	–
KZ-274/563.0	Taxitic gabbrodolerite	9.1	9.7	–
KZ-274/565.0	Taxitic gabbrodolerite	9.1	–	–
KZ-184/182.0	Picritic gabbrodolerite	10.3	–	–
K-42	Picritic gabbrodolerite	8.6	9.8	–
SV-16/1848.5	Olivine-bearing gabbrodolerite	10.5	10.9	–
SV-16/1850.0	Picritic gabbrodolerite	9.7	9.5	–
SV-16/1852.0	Picritic gabbrodolerite	9.6	9.7	–
SV-16/1854.2	Picritic gabbrodolerite	9.8	9.8	–
SV-16/1860.0	Taxitic gabbrodolerite	12.8	–	–
SV-16/1862.0	Taxitic gabbrodolerite	9.3	9.2	–
SV-16/1866.1	Taxitic gabbrodolerite	9.3	10.1	–
	Upper Zone of Upper Talnakh Intrusion			
SV-21/1670.5	Leucogabbro	10.1	–	–
KZ-1647/1757.0	Leucogabbro	10.7	–	–
KZ-1441/1280.0	Leucogabbro	9.7	–	–
KZ-1441/1282.0	Leucogabbro	11.1	–	–
KZ-1441/1287.5	Picritic gabbrodolerite	7.8	–	–
KZ-1623/1514.5	Taxitic gabbrodolerite	6.3	–	–
KZ-1447/1413.0	Taxitic gabbrodolerite	9.5	–	–
KZ-1672/1910.0	Taxitic gabbrodolerite	10.0	–	–
KZ-1485/1198.5	Taxitic gabbrodolerite	9.7	–	–
KZ-1778/1190.6	Taxitic gabbrodolerite	13.4	–	–
SV-29/1971.0	Olivine-bearing gabbrodolerite	13.7	–	–
SV-29/1974.0	Olivine-bearing gabbrodolerite	14.3	–	–
KZ-1169/1472.0	Olivine gabbrodolerite	8.1	–	–
KZ-1765/1052.7	Non-olivine gabbrodolerite	11.3	–	–
KZ-1417/1403.9	Pegmatoid prismatic-granular gabbrodolerite	12.8	–	–
KZ-1485/1196.3	Pegmatoid prismatic-granular gabbrodolerite	8.3	–	–
KZ-1559/1302.0	Subalkaline gabbrodolerite	14.9	–	–
KZ-1701/647.3	Pyrrhotite–graphite cement of microdolerite breccia	14.0	–	–
	Lower Talnakh Intrusion			
SV-16/1750.0	Picritic gabbrodolerite	11.6	–	–
SV-16/1764.0	Picritic gabbrodolerite	–	11.6	–
	Upper exocontact zone of Upper Talnakh Intrusion			
KZ-1417/1392.5	Biotite–feldspar rock	15.7	–	–
KZ-1551/1139.4	Apodolerite albitite	13.2	–	–
KZ-1441/1271.0	Potassium feldspar–albite rock	10.2	–	–
KZ-1513/1466.2	Albite rock with graphite	9.5	–	–
KZ-1417/1391.8	Albite rock with graphite	14.1	–	–
SV-14/1550.0	Albite rock with graphite	16.8	–	–
	Sulfide–sulfate association of Oktyabr'sky Deposit			
KZ-588/733.4	Stratified drop-shaped phenocryst in picritic gabbrodolerite	13.1	–	21.2
KZ-585/674.3	Stratified drop-shaped phenocryst in picritic gabbrodolerite	12.7	–	19.2
KZ-837/660.7	Stratified drop-shaped phenocryst in picritic gabbrodolerite	–	10.3	18.1

(continued)

Table 4.11 (continued)

Sample	Rock	Pyrrhotite	Chalcopyrite	Anhydrite
KZ-806/615.4	Stratified drop-shaped phenocryst in picritic gabbrodolerite	–	–	18.5
KZ-593/664.9	Pyrrhotite–anhydrite macrocrystalline rock	12.4	–	19.0
KZ-593/681.0	Chalcopyrite–anhydrite emulsive accretion	–	9.7	–
KZ-657/716.5	Magnetite ore with anhydrite and pyrite	(pyrite −1.08)	–	21.0
	Sedimentary sheet of gypsum from area Maymecha River			
KH-2/28	Porphyritic phenocryst	–	–	29.8
KH-2/28	Matrix rock grains			30.2

Analyses are performed in IGG SB RAS, analytics A.P. Pertseva

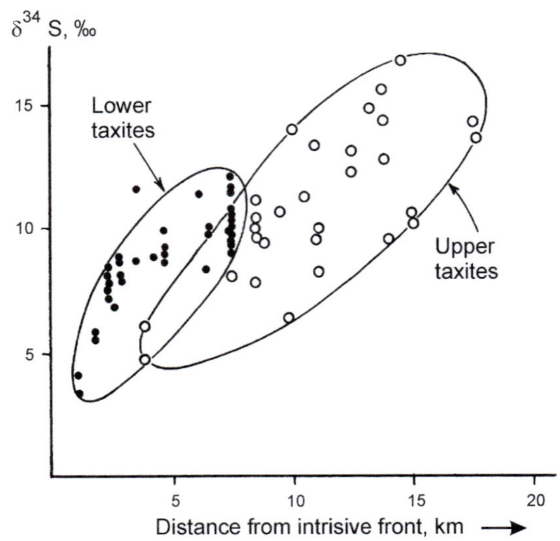

Fig. 4.39 Variations in sulfur isotope composition of pyrrhotite in the mafic pegmatoid horizons of the Upper Talnakh ore-bearing intrusion from the leading to basal parts of the massif

suggest a direct relationship between the amount of sulfides and the content of δ^{34}S in them as well as δ^{34}S enrichment of the deep parts of the magma chamber (Oktyabr'sky Deposit) compared to its leading parts (Talnakh Deposit) (Gorbachev and Grinenko 1973; Zotov 1979). However, these conclusions appear not to be completely correct considering that the Talnakh Deposit is localized in a sulfate-free series, and the Oktyabr'sky Deposit lies in a sulfate-bearing deposit, and its gabbrodolerite, metasomatites, and ores are found in gypsum and ahnydrite (Ryabov and Zolotukhin 1977).

To rule out the possible influence of sulfate-bearing rocks on the sulfur isotope composition of sulfides and impregnated ores in gabbrodolerite, we carried out sampling of the horizons of taxite gabbrodolerite of the Upper Talnakh Intrusion occurring in terrigenous deposits. We picked out monofractions of pyrrhotite from the upper and lower taxite gabbrodolerite down the dip of the body every 400–600 m over a distance of more than 15 km: from drill hole KZ-184 in the south to drill hole SV-29 in the north (see Table 4.11 and Fig. 4.36). Contrary to the concept of homogenization of sulfur isotope composition in intermediate chamber and absence of compositional differences in sulfides of one horizon, we established there was fractionation of sulfur isotopes in pegmatoid horizons from the base up through the extended magmatic body. The upper and lower taxites reveal a similar tendency to lighter sulfur isotopes from near-basal to leading parts of magmatic column. Upper pegmatoids have a lighter isotope composition than lower ones (Fig. 4.39).

In the Lower Talnakh Intrusion, the range of variations in δ^{34}S is 3.1–7.9, less often 8.7–11.8‰ (Avgustinchik 1981). According to the published data, up the dip of the Lower Talnakh Intrusion, there is a tendency toward lighter sulfur isotopes. When studying a magmatic column that formed part of the Noril'sk intrusions, Likhachev (1988) reports the sulfur isotope composition of high-magnesium members with the following δ^{34}S values: picritic basalt 8‰, ore-free picritic gabbrodolerite of Lower Fokinsky Complex 9–11‰, ore-bearing picritic gabbrodolerite of the Noril'sky-type 7–10‰, and ore-free picritic gabbrodolerite of the Lower Talnakhsky type in the central parts 4–8‰ and edge facies 9–13‰.

These data suggest fractionation of sulfur isotopes during ore formation. This issue will be discussed in more detail in Chap. 5.

References

Almukhamedov AI, Medvedev AY (1982) Geochemistry of sulfur in evolution processes in mafic magmas (ed Chernyshov LV). Nauka, Moscow, 148 p (in Russian)

Anastasenko GF (1973) Distribution of boron in rocks and minerals in the north-west of the Siberian Platform. Geokhimiya 10:1481–1489 (in Russian)

Arkhipova AI (1975) Geochemical characteristic of intrusive traps of the Noril'sk plateau. Trudy NIIGA (ed Kavardin GI). Nedra, Leningrad, 135 p (in Russian)

Ashikhmina NA, Bogatikov OA, Dikov YP, Frikh-Har DI, Cimbalnikova A, Zemcik T, Klechka M, Kos M, Mikesh Y, Palivcova M, Zilek I (1987) Natural glasses as indicators of geochemical processes. Nauka, Moscow, 157 p (in Russian)

Avgustinchik IA (1981) On composition of sulfide mineralization of the Lower Talnakh intrusion. In: Likhachev AP (ed) Genesis and localization conditions of copper-nickel mineralization, vol 162. Trydy TSNIGRI, Moscow (in Russian)

Bagdasarov EA, Landa EA, Markovsky BA (1979) Chemical composition and crystallization conditions of chrome-spinel from volcanic ultramafics and other rocks of mafic-ultramafic series. Zapiski VMO 108(5):524–535 (in Russian)

Boudreau AE (1988) Investigation of the Stillwater Complex. IV. The role of volatiles in the petrogenesis of the J-M reef, Minneapolis adit section. Can Min 26(1):193–208

Boudreau AE, Mathez EA, McCollum IS (1986) Halogen geochemistry of the Stillwater and Bushveld Complexes: evidence for transport of platinum group elements by Cl-rich fluids. J Petrol 27(4):967–986

Cameron EN, Glover ED (1973) Unusual titanium-chromium spinels from the eastern Bushveld Complex. Am Min 58(3–4):72–188

Cousins CA (1973) Notes on the geochemistry of the platinum group elements. Trans Geol Soc S Afr 76(1):77–81

Crocket JH, MacRae WE (1986) Platinum – group element distribution in komatiitic and tholeitic volcanic rocks from Munro Township, Ontario. Econ Geol 81(5):1242–1251

Czamanske GK, Wooden JL, Zientek ML, Fedorenko VA, Zenko TE, Kent J, King B-SW, Knight RJ, Siems DG (1994) Geochemical and isotopic constraints on the petrogenesis of the Noril'sk–Talnakh ore-forming system. In: Lightfoot PC, Naldrett AJ (eds) Proceedings of the Sudbury-Noril'sk symposium. Ontario Geol Surv Spec Pub 5:313–341

Deer WA, Howie RA, Zussman J (1966) Rock-forming minerals [Russian translation of 1962 English edition: An Introduction to the Rock-Forming Minerals, Longman, London]. Mir, Moscow, vol 3, 317 p

Distler VV (1982) Nickel, copper and platinum-group metals. In: Fedorchuk VP (ed-in-chief) Geochemistry of ore-forming processes. Nauka, Moscow, pp 6–38 (in Russian)

Distler VV, Kunilov VE (eds) (1994) Geology and ore deposits of the Noril'sk Region. VII internat platinum symposium guidbook. Moscovskiy Contact Moscow, 67 p (in Russian)

Distler VV, Nikol'skaya NN, Eshkova ZA (1976) Platinum group elements in traps of the Noril'sk region. In: Geochemistry of ore-forming elements of mafic and ultramafic rocks. Nauka, Moscow, pp 110–128 (in Russian)

Distler VV, Malevskiy AY, Laputina IP (1977) Distribution of platinoids between pyrrhotite and pentlandite during crystallization of sulfide melt. Geochimiya 11:1646–1659 (in Russian)

Dobretsov NL, Kochkin YN, Krivenko AP, Kutolin VA (1971) Rock-forming pyroxenes. Nauka, Moscow, 454 p (in Russian)

Dodin DA (1967) Petrology of traps of the East Khraelakh (Noril'sk region). Unpublished Ph.D. thesis, VSEGEI, Leningrad, 31 p (in Russian)

Dodin DA, Batuev BN (1971) Geology and petrology of Talnakh differentiated intrusions and their metamorphic aureole. In: Urvantsev NN (ed) Petrology and Metallogeny of Talnakh and Noril'sk differentiated intrusions. Nedra, Leningrad, pp 31–100 (in Russian)

Dodin DA, Shatkov VA (1971) On composition of clinopyroxenes of the Talnakh nickeliferous intrusion (Noril'sk region). Doklady Akad Nauk USSR 200(2):435–438 (in Russian)

Dodin DA, Batuev BN, Mitenkov GA, Izoitko VM (1971) Atlas of rocks and ores of Noril'sk copper-nickel deposits. Nedra, Leningrad, 560 p (in Russian)

Dodin DA, Shatkov VA, Batuev BN, Izoitko VM (1972) New data on minerals of the Talnakh ore junction. In: Copper-nickel ores of the Talnakh ore junction. Nedra, Leningrad, pp 35–51 (in Russian)

Dodin DA, Solomina LN, Shatkov VA (1973) Some aspects of geochemistry of platinum minerals in connection with exploration of copper–nickel ores. In: The North-Siberian nickeliferous region and its industrial meaning. Nedra, Leningrad, pp 108–115 (in Russian)

Dodin DA, Chernyshov NM, Polferov DV, Tarnovetsky LL (1994). Platinum metal deposits of the world. Vol 1: Low-sulfide platinum metal deposits in rhythmically-layered complexes. Geoinformmark, Moscow, 279 p (in Russian)

Ealles HV, Reynolds IV, Gouws DA (1980) The spinel-group minerals of the Central Karoo tholeitic province. Trans Geol Soc S Afr 83(2):243–253

Eliseev EN (1959) Geochemistry of the main sulfide copper-nickel provinces of USSR. In: Lazarenko EK (ed) Problems of geochemistry, issue 1. Lviv Univer Pub, Lviv, pp 5–184 (in Russian)

Fenogenov AN (1987a) Petrological role of immiscibility in generation of igneous rocks. Izv vysshikh uchebnykh zavedeniy: Geologiya i Razvedka 2:25–39 (in Russian)

Fenogenov AN (1987b) Petrological role of immiscibility in generation of igneous rocks. Izv vysshikh uchebnykh zavedeniy: Geologiya i Razvedka 4:31–40 (in Russian)

Feoktistov GD (1978) Petrology and conditions for generation of trap sills (ed Zolotukhin VV). Nauka, Novosibirsk, 168 p (in Russian)

Feoktistov GD (1984) Computer calculation program for crystallization of silicate melt. Application of mathematical methods in geology and geophysics. Earth's Crust Institute SB AS USSR Irkutsk, pp 160–178 (in Russian)

Feoktistov GD, Elisar'eva TI (1976) Distribution of barium and strontium in minerals from intrusive traps at the south of the Siberian Platform. In: Problems of mineralogy and geochemistry of igneous rocks of Eastern Siberia. Irkutsk, pp 103–111 (in Russian)

Feoktistov GD, Ushchapovskaya ZF, Vasil'ev VK, Lakhno TA, Elizar'eva TI (1975) Mineralogy of traps in the South of the Siberian platform (ed-in-chief Feoktistov GD). Nauka, Novosibirsk, 87 p (in Russian)

Frikh-Khar DI (1977) Crystallization of magmatic glass and some aspects of petrogenesis. Nauka, Moscow, 119 p (in Russian)

Genkin AD (1968) Platinum metal minerals and their associations in Noril'sk copper–nickel ores. Nauka, Moscow, 106 p (in Russian)

Genkin AD, Dudykina AS, Telesheva LR (1970) Some data on the composition of rock-forming pyroxenes and olivines from the Noril'sk-I gabbrodolerites intrusion. In: Minerals of basites in connection with aspects of petrogenesis. Nauka, Moscow, pp 40–56 (in Russian)

Genkin AD, Gladyshev GD, Distler VV, Filimonova AA, Khitrov VG (1971) Regularities of distribution and occurrence forms of platinum group elements in Noril'sk ore deposits. In: Abstracts. IGEM, Moscow, pp 17–19 (in Russian)

Genkin AD, Distler VV, Laputina IP, Filimonova AA (1973) Toward geochemistry in copper-nickel ores. Geochimiya No 9:1336–1346 (in Russian)

Genkin AD, Staritsyna GN, Goryainov IN (1976) Peculiarities of copper-nickel mineralization at the northeast of the Tunguskaya syneclise (based on sulfur isotopic composition of ores and rocks). Geokhimiya 11:1662–1672 (in Russian)

Genkin AD, Distler VV, Laputina IP (1979) Chromite mineralization of differentiated trap intrusions. In: Conditions of formation of magamatic ore deposits. Nauka, Moscow, pp 105–127 (in Russian)

Genkin AD, Distler VV, Gladyshev GD, Filimonova AA, Evstigneeva TL, Kovalenker VA, Laputina IP, Smirnov VA, Grokhovskaya TL (1981) Copper-nicel sulfide ores from Noril'sk ore deposits (ed Shadlun TL). Nauka, Moscow, 234 p (in Russian)

Ginsburg VL, Rogover GB (1960) Regularities of distribution of non-ferrous metals in major minerals and silicates. Sov Geol 3:48–60 (in Russian)

Glasunov OM, Lavrov SM, Terekhovich SL (1977) Platinum in ultramafic rocks. In: Geochemistry of endogenous processes. Institute of Geochemistry, Irkutsk, pp 55–59

Glotov AI (1985) Nickeliferrous picrite-dolerite formation of the Novosibirsk Ob' region. Unpublished Ph.D. thesis, Institute of Geology and Geophysics SB AS USSR, Novosibirsk, 17 p (in Russian)

Godlevsky MN (1959) Traps and ore-bearing intrusions of the Noril'sk region. Gosgeoltekhizdat, Moscow, 68 p (in Russian)

Godlevsky MN (1967) Sources of mineralization associated with traps of the Siberian platform. In: Urvantsev NN (ed) Petrology of traps of Siberian platform. Nedra, Leningrad, pp 173–189 (in Russian)

Godlevsky MN (1968) Magmatic ore deposits. In: Genesis of endogenous ore deposits. Nedra, Moscow, pp 8–73 (in Russian)

Godlevsky MN (1971) The proportions between sulfide and silicate parts in a coarse of evolution of mafic magma. In: Abstracts of I Internat Geochem Congress, vol 1. Moscow, pp 65–77 (in Russian)

Godlevsky MN, Grinenko LN (1963) Some data on sulfur isotope compositions from sulfides of the Noril'sk deposits. Geokhimiya 6:491–499 (in Russian)

Godlevsky MN, Likhachev AP (1979) Conditions for initiation of crystallization of ore magmas forming copper-nickel deposits. In: Kuznetsov VA (ed) The main parameters of processes of endogenous mineralization, vol 1. Nauka, Novosibirsk, pp 109–118 (in Russian)

Godlevsky MN, Polushkina AP, Stepanov VK (1971) Monoclinic pyroxenes of the Talnakh differentiated intrusion. Zapiski VMO 100, Second Series, issue 5, pp 545–557 (in Russian)

Goldshmidt VM (1954) Geochemistry (ed A Muir). Univ Press, London-Oxford, 730 p

Gorbachev NS (1989) Fluid-magmatic interaction in sulfide-silicate systems (ed Zyryanov VN). Nauka, Moscow, 126 p (in Russian)

Gorbachev NS, Grinenko LN (1973) Sulfur isotope compositions from sulfides and sulfates of the Oktyabrsky nickel-copper deposits (Noril'sk region) and some aspects of their genesis. Geokhimiya 8:1127–1136 (in Russian)

Goryainov IN (1969) On genesis of layered intrusions by example of the Talnakh massif. In: Zaridze GM (ed) Magmatism, metamorphism and metasomatism. Central Committee Communist Party of Georgia Pub, Tbilisi, pp 47–61 (in Russian)

Goryainov IN (1971) About genesis of the Talnakh ore deposit. In: Urvantsev NN (ed) Petrology and mineralization of Talnakh and Noril'sk intrusions. Nedra, Leningrad, pp 182–196 (in Russian)

Goryainov IN, Yalovoy AA (1972) Sulfur isotopes fractionation in sulfides of the Talnakh ore deposit. In: Urvantsev NN (ed) Copper-nickel ores of the Talnakh ore junction. NIIGA, Leningrad, pp 58–62 (in Russian)

Goryainov IN, Ushakov VI (1970) Regularities of platinum and palladium distribution in high-grade ores of Talnakh ore deposit. Uchenye Zapiski NIIGA, issue 17, pp 67–83 (in Russian)

Grinenko LN (1966) Sulfur isotope composition from sulfides of the Talnakh copper–nickel deposit and some aspects of their genesis. Geologiya rudnykh mestorozhdeniy 4:15–31 (in Russian)

Grinenko LN (1967) Sulfur isotope composition from sulfides of some copper–nickel deposits and ore occurrences of the Siberian Platform. In: Petrology of traps of the Siberian Platform. Nedra, Leningrad, pp 221–229 (in Russian)

Grinenko LN (1985) Sources of sulfur of the nickeliferous and barren gabbrodolerites intrusions in the northwest part of the Siberian platform. Geologiya rudnykh mestorozhdeniy 1:3–15 (in Russian)

Grinenko LN (1987) Genetic model of formation of copper-nickel deposits based on isotope-geochemical data. In: Sotnikov VI (ed) Construction of models of ore-forming systems. Nauka, Novosibirsk, pp 119–128 (in Russian)

Gulin SA, Sukhov LG (1973) Some remarks on liquation-magmatic hypothesis concerning formation of the Norlsk-type copper-nickel deposits. Sovetskaya Geologiya 2:24–35 (in Russian)

Gulin SA, Sukhov LG (1974) Criteria for prediction of copper-nickel deposits based on their metasomatic origin conceptions. In: Lind EN (ed) The state and trends of investigations into metallogeny of traps. Krasnoyarsk TGU, Krasnoyarsk, pp 67–69 (in Russian)

Gulin SA, Pogrebitskiy YuE, Sukhov LG (1968) Zonal-metasomatic nature of Noril'sk copper–nickel deposits and criteria for exploration apparent from it. In: Prediction and methods of exploration for deposits of nickel, tin, and diamonds in Soviet Arctic zone. NIIGA, Leningrad, pp 29–32 (in Russian)

Haggerty SE, Barker I (1967) The alteration of olivine in basaltic and associated lavas. Contrib Mineral Petrol 16(3):233–257

Irvine TN, Keith DW, Todd SG (1982) Formation crystallization and liquid mixing in a stratified magma body. Annual Rept Dir Geophys Lab 1981–1982. Carnegie Inst, Washington DC, pp 286–294

Irvine TN, Keith DW, Todd SG (1983) The J-M platinum–palladium reef of the Stillwater Complex, Montana. II. Origin by double-diffusive convective magma mixing and implications for the Bushveld Complex. Econ Geol 78(7):1287–1334

Ivanov MK, Ivanova TK, Shatkov VA (1971a) Some facial peculiarities of formation of intrusive and effusive picrites. In: Urvantsev NN (ed) Geology and mineral resources of the Noril'sk region, NIIGA, Leningrad, pp 83–84 (in Russian)

Ivanov MK, Ivanova TK, Tarasov AV, Shatkov VA (1971b) Peculiarities of petrology and mineralization of differentiated intrusions of the Noril'sk ore junction (Noril'sk-I, Noril'sk-II and Mt. Chernaya deposits). In: Urvantsev NN (ed) Petrology and metallogeny of the Talnakh and Noril'sk differentiated intrusions. Nedra, Leningrad, pp 197–304 (in Russian)

Kavardin GI (1976) Metallogeny of the northwestern part of the Siberian Platform (ed Urvantsev NN). Nedra, Leningrad, 158 p (in Russian)

Keith DW, Todd SG, Irvine TN (1982) Setting and compositions of the J-M platinum–palladium reef and other sulfide zones in the banded series of the Stillwater Complex. Annual Rept Dir Geophys Lab 19811982. Carnegie Inst, Washington, DC, pp 281–286

Khryanina LP (1962) Trap magmatism of the Bakhta River basin and the Lower Podkamennaya Tungusska River and association of magnetite mineralization with it. Trudy IGEM AS SSSR, Moscow, issue 71, pp 3–89 (in Russian)

Konkina OM, Rasin LV (1976) Zonation in distributions of palladium, platinum and gold in the Talnakh intrusion. Trudy TSNIGRI 122:98–107 (in Russian)

Kopylova AG, Oleinikov BV (1973) Distribution of nickel, cobalt, chrome, vanadium, and copper in rocks and minerals of multiple-aged trap intrusions. In: Geology and geochemistry of basites of the east part of The Siberian Platform. Nauka, Moscow, pp 164–191 (in Russian)

Korovyakov IA (1948) On picritic effusive traps of the northwestern part of the Siberian Platform. Doklady Akad Nauk USSR LXII (1):129–133 (in Russian)

Korovyakov IA (1960) Behavior of nickel in Siberian traps. In: Mineral'noe Syr'e, issue 1. VIMS, Moscow, pp 169–183 (in Russian)

Korovyakov IA, Nelyubin AE, Raikova ZA, Khortova LK (1963) Origin of the Noril'sk trap intrusions hosting sulfide copper-nickel ores. In: Trudy VIMS (ed Gon'shakova VI), New Series, issue 9, Gosgeoltekhizdat, Moscow, 102 p (in Russian)

Kotul'sky VK (1946) Toward the problem of origin of magmatic copper-nickel deposits. Doklady Akad Nauk SSSR 51(5):381–383 (in Russian)

Kotul'sky VK (1948) The state-of-art problem of genesis of copper-nickel sulfide deposits. Sov Geol 29:11–24 (in Russian)

Kovalenker VA, Gladyshev GD, Nosik LP (1974) Isotope composition of sulfur from sulfides from deposits of the Talnakh ore junctionwith special reference to their selenium content. Izv Akad Nauk SSSR Geol Ser 2:80–91 (in Russian)

Kravchenko SM (1977) Fractionation of minor elements during differentiation of mafic magmas. Nauka, Moscow, 218 p (in Russian)

Krivenko AP, Polyakov GV, Lavrentyev YG (1974) Zoning of augites and some aspects of petrogenesis of sienite-gabbro formation. Doklady Akad Nauk SSSR 216(4):895–898 (in Russian)

Kudelina MB, Kudryavtseva VP, Garanin VK (1983) Peculiarities of the chemical and phase compositions of spinellides and ilmenite in the Talnakh intrusion (Noril'sk region). Geologiya rudnykh mestorozhdeniy 25(5):24–37 (in Russian)

Kullerud GA, Yoder HS (1965) Sulfide-silicate relation and their bearing on ore formation under magmatic, postmagmatic and metamorphic condition. In: Proceedings of symposium: problems of postmagmatic ore deposition, vol 2, Prague, pp 327–331

Lebedev AP (1955) Trap formation of the central part of the Tunguska basin. AS USSR, Moscow, 197 p (in Russian)

Lebedev AP (1957) On differentiation types in traps of the Siberian platform. Izv Akad Nauk SSSR Geol Ser 2:55–74 (in Russian)

Lightfoot PC, Naldrett AJ, Gorbachev NS, Doherty W, Fedorenko VA (1990) Geochemistry of the Siberian trap of the Noril'sk area USSR, with implications for the relative contributions of crust and mantle to flood basalt magmatism. Contrib Mineral Petrol 104:631–644

Lightfoot PC, Naldrett AJ, Gorbachev NS, Fedorenko VA, Hawkesworth CJ, Hert J, Doherty W (1994) Chemostratigraphy of Siberian trap lavas, Noril'sk district Russia: implications and source of flood basalt magmas and their associated Ni-Cu mineralization. In: Lightfoot PC, Naldrett AJ (eds) Ontario Geological Survey special volume, Proceedings of the Sudbury - Noril'sk symposium, Ministry of Northern Development and Mines, Toronto, vol 5, pp 283–312

Likhachev AP (1965) Role of leucocratic gabbro in the generation of Noril'sk differentiated intrusions. Izv Akad Nauk SSSR Geol Ser 10:75–89 (in Russian)

Likhachev AP (1977) On crystallization conditions of trap magmas of the northwestern Siberian platform. Zapiski VMO 5:594–606 (in Russian)

Likhachev AP (1988) Genetic models of sulfide-nikeliferous formations and their relation to other endogenous formations. In: Obolensky AA, Sotnikov VI, Sharapov VN (eds) Ore formation and genetic models of endogenous ore formations. Nauka, Novosibirsk, pp 158–165 (in Russian)

Likhachev AP (1997) Trap magmatism and Pt–Cu–Ni ore formation in the Noril'sk region. Otechestv Geol 10:8–19 (in Russian)

Lurje ML, Ledneva VP, Selivanovskaya TV, Semenov LS, Tuganova EV, Ryabchenko AA, Komarova MZ, Staritsyna GN, Tomanovskaya YuI (1976) Structures of traps of the Siberian platform (ed ML Lurje). Trudy VSEGEI, New Series, vol 235. Nedra, Leningrad, 171 p (in Russian)

Lyul'ko VA, Nesterovsky VS, Goverdovskaya TG (1972) Magmatogenic breccias of nickeliferous trap intrusions. In: Urvantsev NN (ed) Copper-nickel ores of the Talnakh ore junction. Nedra, Leningrad, pp 123–127 (in Russian)

Marakushev AA (ed) (1986) Petrography: Part III. Moscow Univ Pub, Moscow, 287 p (in Russian)

Marakushev AA, Bezmen NI (1972) Thermodynamics of sulfides and oxides in connection with rock formation problems. Nauka, Moscow, 229 p (in Russian)

Masaitis VL (1958) Petrology of the Alamdzhakh trap intrusion (the Vilyuy River basin) (ed Lurje ML). Trudy VSEGEI Leningrad, New Series, No 22, 133 p (in Russian)

Mathison Charter I (1987) Pyroxene oikocrysts in troctolitic cumulates evidence for supercooled crystallisation and postcumulus modification. Contrib Mineral Petrol 97(2):228–236

Muir JD, Tilley CE, Scoom JH (1957) Contributions to the petrology of Hawaiian basalt. 1. The Picrite basalts of Kilauea. Am J Sci 255(4):241–253

Naldrett AJ, Cabri LJ (1976) Ultramafic and related mafic rocks: Their classification and genesis with special reference to the concentration of nickel sulfides and platinum-group elements. Econ Geol 71:1131–1158

Naldrett AJ, Lightfoot PC, Fedorenko VA, Goebachev NS, Doherty W (1992) Geology and geochemistry of intrusions and flood basalts of the Noril'sk region USSR with implications for the origin of the Ni–Cu ores. Econ Geol 87(4):975–1004

Natorkhin IA, Arkhipova AI, Batuev BN (1977) Petrology of Talnakh intrusions (ed Egorov LS). Nedra, Leningrad, 236 p (in Russian)

Nekrasov IY, Gorbachev NS (1978) Physical-chemical conditions of formation of the Noril'sky-type differentiated intrusions and copper-nickel ores. Ocherki fisiko-khimicheskoy petrologii 7:92–123 (in Russian)

Nesterenko GV, Almukhamedov AI (1973) Geochemistry of differentiated traps (Siberian platform). Nauka, Moscow, 198 p (in Russian)

Oleinikov BV (1979) Geochemistry and ore genesis of platform basites (ed Bazhenov AI). Nauka, Novosibirsk, 264 p (in Russian)

Oleinikov BV, Pankov VYu (1985) Melted inclusions and minerals-prisoners in products of basalt melt protocrystallization. In: Petrological–geochemical features of the deep evolution of a substance from kimberlitic and mafic magmatic systems. Yakutsk Affiliated Branch AS USSR Pub, Yakutsk, pp 130–164 (in Russian)

Oleinikov BV, Okrugin AV, Tomshin MD, Levashov VK, Varganov AS, Kopylova AG, Pankov VYu (1985) Native metal formation in platform basites. Yakutsk Affiliated Branch AS USSR Pub, Yakutsk, 188 p (in Russian)

Olshansky YI (1951) The Fe–FeS–FeO–SiO_2 system. Izv Akad Nauk SSSR, Geol Ser 6:128–176

Page NS, Rowe JJ, Haffty J (1976) Platinum metals in the Stillwater Complex, Montana. Econ Geol 71(7):1352–1363

Page NJ, Von Guenewaldt G, Haffty J, Aruscavage PJ (1982) Comparison of platinum, palladium and rhodium distributions in some layered intrusions with special reference to the late differentiates (upper zone) of the Bushveld Complex, South Africa. Econ Geol 77(6):1405–1418

Pavlov NV (1961) Magnomagnetite deposits of the Tungusskaya syneclise of the Siberian Platform. IGEM AS SSSR, Moscow, 224 p (in Russian)

Poroshin EE, Ledneva VP (1977) Clinopyroxenes of effusive traps of the East Siberia. Geologiya i Geofisika 7:133–137 (in Russian)

Reverdatto VV (1963) Petrology of Anakitsky differentiated trap massif and its structure. Geologiya i geofisika 10:79–92 (in Russin)

Rogover GB (1959) The Noril'sk-I deposit. Gosgeoltekhizdat, 168 p (in Russian)

Rudashevsky NS (1983) New peculiarities of differentiation of platinum-group elements in the Earth's crust. Doklady Akad Nauk SSSR 268(1):201–206 (in Russian)

Ryabov VV (1974) Plagioclases and clinopyroxenes of layered trap intrusions as indicators of differentiation of magmatic melt. Doklady Akad Nauk SSSR 219(1):187–200 (in Russian)

Ryabov VV (1976) Magnetites from matasomatites of the Talnakh ore junction. In: Transactions on genetic and experimental mineralogy, vol 9. Nauka, Novosibirsk, pp 113–125 (in Russian)

Ryabov VV (1978a) Peculiarities of principal rock-forming minerals as exploration criteria. In: Sobolev VS (ed) Criteria for exploration of the Noril'sky-type sulfide ores. Nauka, Novosibirsk, pp 66–85 (in Russian)

Ryabov VV (1978b) About some peculiarities in the behavior of chrome and titanium in magmatic clinopyroxenites from different formations. In: Proc on petrology and mineralogy of ultramafic and mafic rocks. Nauka, Novosibirsk, pp 119–133 (in Russian)

Ryabov VV (1981) High-magnesian trap intrusions depleted in chrome as an exploration criterion for the Norl'sk type ore-hosting intrusions. In: Petrology of lithosphere and mineralization (Abstracts). Leningrad, pp 324–325 (in Russian)

Ryabov VV (1983) Compositional peculiarities of endocontact zones of Noril'sk intrusions. In: Trap magmatism of the Siberian platform in

connection with tectonics and exploration of mineral resources. Krasnoyarsk, pp 186–188 (in Russian)
Ryabov VV (1984b) Peculiarities of mineralogy of magnesian basites in the Noril'sk region. In: Sobolev VS (ed-in-chief) Magnesian basites of the western Siberian platform and aspects of their nickel content. Nauka, Novosibirsk, pp 150–158 (in Russian)
Ryabov VV (1984a) On composition of upper contact zones of Noril'sk intrusions hosting chromite-rich mineralization. In: Polyakov GV (ed) Petrochemistry: criteria for mineralzation of magmatic complexes. IGG SB AS USSR, Novosibirsk, pp 124–142 (in Russian)
Ryabov VV (1985) Peculiarities of native mineral formation at the roof of chambers of trap intrusion. In: Oleinikov BV (ed) Formation of native elements under endogenous conditions (abstracts). Yakutsk, pp 18–21 (in Russian)
Ryabov VV. (1988b) Immiscible liquid in traps of Siberian platform. In: 2nd International Conference on Natural Glasses, Prague, 21–25 September 1987, pp 401–405
Ryabov VV (1988a) Immiscible differentiation in traps. In: Goncharenko AI (ed) Actual problems of Siberian geology, TGU, Tomsk, vol 1. pp 132–134 (in Russian)
Ryabov VV (1989a) Chemical composition of immiscible liquids in natural glasses from traps. Geologiya i Geofisika 11:69–78 (in Russian)
Ryabov VV (1989b) Immiscibility in natural glasses (by example of traps) (ed Zolotukhin VV). Nauka, Novosibirsk, 223 p (in Russian)
Ryabov VV (1992a) Olivines of Siberian traps as indicators of petrogenesis and ore formation (ed Zolotukhin VV). Nauka, Novosibirsk, 116 p (in Russian)
Ryabov VV (1992b) Nontraditional types of platinum metal ores in the Noril'sky-type sulfide deposits. Russ Geol Geophys 33(11):79–88 (in Russian)
Ryabov VV (1992c) Types of low-sulfide of platinum-group metal ores in contact zones of trap intrusions. In: Distler VV, Evstigneeva TL, Kamshilina EM (eds) Geology and genesis of platinum metal deposits. IGEM Publication, Moscow, pp 40–41 (in Russian)
Ryabov VV (1994) Platinum of Siberian traps. UIGGM SB RAS, Novosibirsk, 32 p (in Russian)
Ryabov VV (1997) Ore melts among basalts on the Putorana plateau (Siberian Platform). Russ Geol Geophys 38(12):1932–1949
Ryabov VV (1999a) Fluid regime of trap magmatism and ore formation (petrological aspect). Russ Geol Geophys 40(10):1437–1452
Ryabov VV (1999b) Ore geology of the Noril'sk region on the threshold of millennium. In: Simonov ON (ed) Abstracts of the regional symposium: trends in geological-exploration works on nickel, copper and platinoids in the Taimyr Autonomous Area in 2000–2005. Taimyrkomprirodresursy, Noril'sk, pp 25–26 (in Russian)
Ryabov VV, Anoshin GN (1999) Platinum–iron mineralization in intrusive traps of the Siberian Platform. Russ Geol Geophys 40(2):163–176
Ryabov VV, Pavlov AL (1984) Physicochemical principles of the formation of magnesial scarn paragenesis in chambers of stratified intrusives of Noril'sky-type. Sov Geol Geophys 25(3):51–56
Ryabov VV, Yakobi NYa (1981) Behavior of chrome during differentiation of trap magma. In: Sobolev VS (ed) Problems of genetic petrology. Nauka, Novosibirsk, pp 85–111 (in Russian)
Ryabov VV, Zolotukhin VV (1977) Minerals of differentiated traps (ed Sobolev VS). Nauka, Novosibirsk, 392 p (in Russian)
Ryabov VV, Bakumenko IT, Fominykh IM (1977) Dendritic metacrysts in traps of the Noril'sk region and some aspects of their formation. In: Sobolev VS (ed) Proc on petrology and mineralogy. Nauka, Novosibirsk, pp 47–71 (in Russian)
Ryabov VV, Tsimbalist VG, Yakobi NA (1982) Concentration of chrome and platinoids at the roof of the Noril'sky-type layered intrusions. Doklady Akad Nauk SSSR 266(2):350–353 (in Russian)
Ryabov VV, Pavlov AL, Lopatin GG (1985a) Native iron of Siberian traps (ed Zolotukhin VV). Nauka, Novosibirsk, 169 p (in Russian)
Ryabov VV, Konenko VF, Krasov NF (1985b) Immiscibility phenomena in glasses from ores of native iron of Khungtukun intrusion. Doklady Akad Nauk SSSR 285(4):982–987 (in Russian)
Ryabov VV, Konenko VF, Khmel'nikova OS (1985c) Rock-forming minerals of picritic basalts of the Noril'sk region. Sov Geol Geophys 26(4):77–84
Ryabov VV, Shevko AY, Simonov ON, Anoshin GN (1996) Composition of Pt-bearing Cr-rich skarns in Talnakh (Norrilsk region). Russ Geol Geophys 37(7):57–72 (in Russian)
Shatkov VA (I973) Rock-forming minerals and some regularities in formation of ore-hosting intrusions of the Noril'sk region. Unpublished Ph.D. thesis. IGG SB AS SSSR Novosibirsk, 24 p (in Russian)
Skinner BJ, Peck DL (I969) An immiscible sulfide melt from Hawaii. Econ Geol Monogr 4(3):10–32
Slobodskoy RM (1981) Hetero-organic compounds in magmatogenic and ore-forming processes. Nauka, Novosibirsk, 134 p (in Russian)
Sluzhenikin SF, Distler VV, Duzhikov OA, Kravtsov VF, Kunilov VE, Laputina IP, Turovtsev DM (1994) Low-sulfide minertalization in the Noril'sk differentiated intrusions. Geologiya rudnykh mestorozhdeniy 36(3):195–217 (in Russian)
Smirnov MF (1966) Structure of the Noril'sk nickeliferous intrusions and sulfide ores. Nedra, Moscow, 58 p (in Russian)
Sobolev VS (1949) Introduction into mineralogy of silicates. Lviv University Publication, Lviv, 341 p (in Russian)
Sobolev VS (1986) Petrology of traps: Izbrannye trudy. Nauka, Novosibirsk, 209 p (in Russian)
Staritsyna GN, Tolmachevsky YN, Kravtsova LI (1972) Intrusive traps of the northeastern side of the Tungusskaya sineclise (ed Egorov LS). Nedra, Leningrad, 210 p (in Russian)
Stepanov VK (1975) Rock-forming minerals of the Talnakh intrusion and analysis of their paragenesises with development of criteria of mineralization. Unpublished PhD thesis, Moscow, 25 p (in Russian)
Stepanov VK (1977) Olivines and hypersthenes of the Talnakh intrusion. In: Geology, petrology and geochemistry of copper-nickel deposits. Trudy TSNIGRI, vol 127. TSNIGRI, Moscow, pp 17–34 (in Russian)
Tarasov AV (1976) On formation mechanisms of the Noril'sk intrusions and related sulfide bodies. In: Pospelov GL (ed) Replacement and intrusion during magmatism and ore formation. Nauka, Novosibirsk, pp 123–276, in Russian
Tomanovskaya YuI, Pol'kin YaI (1970) Regularities of distribution of microelements in vokcanic rocks of trap formation at the northwest of the Siberian Platform. Trudy NIIGA (ed Urvantsev NN), vol 162, issue 2. Nedra, Leningrad, pp 96–115 (in Russian)
Tuganova EV (1988) Petrological peculiarities of formation of Noril'sky-type nickeliferous intrusions. In: Mitrofanov FP, Gorbunov GI (eds) Presence of nickel in mafic-ultramafic complexes. Apatites, pp 36–40 (in Russian)
Tugovik GI, Zaikin ID, Kirasirova VI (1982) Fluid composition of inclusions in platinoids from granulitic complex and some aspects of their genesis. Doklady Akad Nauk SSSR 266(3):704–706 (in Russian)
Urvantsev NN (1959) The Yenisei ore field. Collected papers on Arctic region geology. Trudy NIIGA (ed Markov FG), vol 102, issue 10. Nedra, Leningrad, pp 28–48 (in Russian)
Vasil'ev YR (1970) Chromous clinopyroxene from Lower Fokinsky trap intrusion (north-west of the Siberian Platform). Geologuya I Geofisika 7:130–134 (in Russian)
Vasil'ev YuR (1988) Feldspar picrites at the Ayan River (north of the Siberian platform). Geologiya i geofizika, 4:68–75 (in Russian)
Vasil'ev YR, Zolotukhin VV (1975) Petrology of ultrafamites at the north of the Siberian Platform. Nauka, Novosibirsk, 271 p

Vedernikova GA (1961) Differentiated trap massif of Padunskie rapids of the Angara River. Geologiya i Geophisika 1:66–76 (in Russian)

Vilensky AM (1967) Petrology of intrusive traps at the north of the Siberian platform (ed Lebedev AP). Nauka, Moscow, 270 p (in Russian)

Velinsky VV, Bannikov OL (1986) Olivines of alpinotype ultrarbasites. In: Proceedings of IGG SB USSR, IGG SBRAS, Novosibirsk, issue 641, 103 p (in Russian)

Vilensky AM, Kavardin GI, Kravtsova LI, Staritsina GN (1964) Petrology of trap intrusions on the right-bank of the Yenisey River lower reach (ed Kazakov AN). Nauka, Moscow, 237 p (in Russian)

Vilensky AM, Shkodzinsky VS, Zedgenizov AN (1970) On the acidity-alkalinity effect on the composition of magmatic clinopyroxenes. In: Geology and petrology of intrusive traps of the Siberian Platform. Nauka, Moscow, pp 120–143 (in Russian)

Vinogradov AP, Grinenko LN (1964) About effect of country rocks on the sulfur isotopic composition in ore sulfides. Geokhimiya 6:491–499 (in Russian)

Vladimirov BM (1962) Petrography of Padunsky and Margudolsly trap intrusions. Trudy ESF SB AS USSR, issue 10, Irkutsk, 151 p (in Russian)

Vogt JHL (1923) Nickel in igneous rocks. Econ Geol 18(4):307–353

Wager LR, Brown GM (1968) Layered igneous rocks. Oliver and Boyd, Edinburgh/London, 588 p

Yudina VV (1965) Some metamorphism and metasomatism phenomena associated with the Talnakh gabbrodolerite intrusion (Noril'sk region). In: The proportion between magmatism and metamorphism in genesis of ultrafamites. Nauka, Moscow, pp 112–175 (in Russian)

Yudina VV, Sandomirskaya SM (1980) On composition of ore minerals of sulfide-magnetite occurrences in the Noril'sk region. Zapiski VMO 2:187–194 (in Russian)

Yudina VV, Lyul'ko VA, Nemenenok TI (1977) The "Magnetitovaya" diatreme in the Noril'sk area. Sov Geol Geophys 18(6):73–81

Zolotukhin VV (1964) Main regularities of prototectonics and aspects of origin of ore-bearing trap intrusions (ed Sobolev VS). Nauka, Moscow, 192 p (in Russian)

Zolotukhin VV (1965) On peculiarities in distribution of nickel in the Noril'sk-I intrusion. Doklady Akad Nauk SSSR 162(6):1390–1393 (in Russian)

Zolotukhin VV (1971) On genesis of so called "liquation" of copper-nickel ores in the light of new data (about infiltration-autometasomatic hypothesis). Geologiya i Geofisika 9:12–22 (in Russian)

Zolotukhin VV (1978a) Differentiated ore-bearing trap intrusions of the Noril'sk and adjacent regions. In: Zolotukhin VV, Vilensky AM (eds) Petrology and perspectives of metallogeny of traps at the north of the Siberian platform. Nauka, Novosibirsk, pp 121–160 (in Russian)

Zolotukhin VV (1978b) Taxitic rocks and gabbro as an important criterion of metallogeny of differentiated trap intrusions. In: Sobolev VS (ed) Criteria for Exploration of the Noril'sk-type Sulfide Ores. Nauka, Novosibirsk, pp 52–64 (in Russian)

Zolotukhin VV (1984b) Magnesian basites and sulfide mineralization. In: Sobolev VS (ed-in-chief) Magnesian basites of the western Siberian platform and aspects of their nickel content. Nauka, Novosibirsk, pp 176–200 (in Russian)

Zolotukhin VV (1984a) Petrology of magnesian basites of the Tunguska (Middle Yenisei) province. In: Sobolev VS (ed-in-chief) Magnesian basites of the western Siberian platform and aspects of their nickel content. Nauka, Novosibirsk, pp 48–149 (in Russian)

Zolotukhin VV (1988) A generalized model of sulfide copper-nickel ore formation as a process of sulfurization. In: Obolensky AA, Sotnikov VI, Sharapov VN (eds) Mineralization and genetic models of endogenous ore formations. Nauka, Novosibirsk, pp 172–181 (in Russian)

Zolotukhin VV, Podgornykh NM (1998) On probable overheat of Permo-Triassic trap melts of the Siberian platform. Doklady Akad Nauk SSSR 363(2):226–229 (in Russian)

Zolotukhin VV, Shchedrin NF (1977) Differentiated intrusions of the Imangdinsky ore junction (ed Sobolev VS). Nauka, Novosibirsk, 135 p (in Russian)

Zolotukhin VV, Vasil'ev YR (1967) Peculiarities of formation mechanism of ore-bearing trap intrusions at the northwest of Siberian platform (ed Sobolev VS). Nauka, Moscow, 231 p (in Russian)

Zolotukhin VV, Ryabov VV, Vasil'ev YR, Shatkov VA (1975) Petrology of the Talnakh ore-bearing differentiated intrusion (ed Sobolev VS). Nauka, Novosibirsk, 432 p (in Russian)

Zolotukhin VV, Vilensky AM, Duzhikov OA (1986) Basalts of the Siberian platform (eds Sobolev VS, Sobolev NV). Nauka, Novosibirsk, 255 p (in Russian)

Zolotukhin VV, Vasil'ev YuR, Duzhikov OA (1989) Diversity of traps and initial magmas. Nauka, Novosibirsk, 247 p (in Russian)

Zotov IA (1979) Genesis of trap intrusions and metamorphic formations of the Talnakh (ed Korzhinsky DS). Nauka, Moscow, 155 p (in Russian)

5 Basalt Magma Differentiation as a Source of Variety in Traps

Abstract

The origin of layered intrusions and related ore formation is one of the main problems of the magmatic and ore geology of traps of the Siberian Platform. This chapter gives the critical review of available genetic conceptions. The model of the fluid–magmatic differentiation of basalt melt and ore formation in traps is proposed as an alternative to these conceptions. The leading role in these processes is assigned to three mutually related factors: tectonics, magmatism, and fluid regime of ore–magmatic systems. For validation of the model, a cause-and-effect relationship between these factors is established, the role of each is identified, and a system of natural evidences is formulated. The tectonics of the Noril'sk Region is dominated by faults that govern the placement of layered intrusions, ore zones, ore accumulations, and deposits. The long-continued tectonic activity of deep faults resulted in formation of thick zones of disintegrated rocks. Formation fluids from sedimentary rocks of the Platform cover were attracted into these zones due to decompression with fluidized zones formed along the faults. The formation fluids served as a source of S, F, Cl, CH_4, H_2, H_2O, and other volatile components, whereas the basaltic magma was a source of ore-forming metals. Formation fluids activated by a melt intruded through the fault zones extracted metals from a basaltic magma, and so became ore-forming fluids. The fluid–magmatic interaction of a melt with volatiles facilitated prechamber differentiation of the melt into coexisting immiscible silicate–silicate and ore–silicate melts of variable composition, and crystallization of the melt completed the process of petrogenesis and ore formation.

It is a well-established fact that the variety of rocks of the trap formation on the Siberian Platform is governed by the composition of an initial magmatic melt and the physical–-chemical conditions of its differentiation and crystallization (Sobolev 1986; Masaitis 1964; Vilensky 1967; Zolotukhin and Vasil'ev 1976; Lurje et al. 1976; Likhachev 1978; Oleinikov 1979). Data on initial magmatic melts was obtained from studies of lava flows and undifferentiated intrusions, as well as from contact quenched crusts. The composition of initial magma of undifferentiated bodies is typically inferred from the composition of contact facies and calculations of average weighted compositions (Korovyakov et al. 1963). The chemical composition of rocks in undifferentiated trap bodies is practically identical to the calculated average composition throughout a body. However, in undifferentiated intrusions, significant discrepancy between these data is observed, which is most pronounced in the Noril'sk Intrusions.

Judging from the literature and available data (see Appendix), the magmatic melt of most traps of the Siberian Platform was compositionally close to the "average trap" (according to Kutolin 1969) and an insignificant portion of the melt corresponds to feldspar picrites, subalkaline mafic rocks, and other types of rocks that are rare, which can be probably regarded as the differentiation products of tholeiitic basalt magma.

V.V. Ryabov et al., *Trap Magmatism and Ore Formation in the Siberian Noril'sk Region: Volume 1. Trap Petrology; Volume 2. Atlas of Magmatic Rocks*, Modern Approaches in Solid Earth Sciences 3, DOI 10.1007/978-94-007-5022-7_5,

5.1 Fractional Crystallization of Mafic Melts

Field observations in combination with known experimental and thermodynamic data give us an insight into the mechanism of formation of both lava flows and intrusions (Zolotukhin and Vasil'ev 1967; Sharapov and Golubev 1976; Sharapov and Cherepanov 1986; Feoktistov 1978).

Gravitational crystallization differentiation, which is believed by some to proceed in upper-level chambers and by others in deep-seated or intermediate chambers, or completed in situ (Godlevsky 1959; Masaitis 1958; Zolotukhin 1964; Oleinikov 1979), is the most theoretically and experimentally justified mechanism of formation of mafic intrusions.

The main geological arguments in favor of gravitational crystallization differentiation are four interrelated properties (Sobolev 1986; Godlevsky 1959; Zolotukhin 1964; Vilensky 1967; Dodin and Batuev 1971; Nesterenko and Almukhamedov 1973; Arkhipova 1975; Oleinikov 1979):

1. Asymmetric intrusion structure with a regular distribution of olivine-free through olivine-bearing to olivine and picritic gabbrodolerites downward through the intrusion
2. Directional variation in chemical composition and behavior of impurity elements in the magmatic body
3. Increase in olivine downward through the intrusion
4. Invisible layering that manifests as regular compositional changes in mineral associations, including increased Fe content in mafic minerals and alkalinity of plagioclase during crystallization, the changes being compatible with the succession of rock formation from the base to the roof of the intrusion

Advances in the understanding of crystallization differentiation are known to be incommensurable with the current level of knowledge of the petrology of layered trap intrusions that invoke a differentiation mechanism. However, as paradoxically as it may seem, a scrupulously and thoroughly studied intrusion that could be the undoubted standard of differentiation has not been described in the literature. Most petrologic data reflect only general trends in variation in rock composition through layered intrusions, although these data turn out to be compatible with cases of crystallization differentiation. Moreover, studies of the composition of pyroxene in the Noril'sk Intrusions (considered a classic example of fractional differentiation) have allowed some geologists to infer that this mineral has no way of indicating the melt composition evolution during crystallization. Presently, no comprehensive data on the possible range and depth of chamber differentiation, invisible layering, crystallization trends of mafic minerals, or geochemical indicators of the process are available. The fact that geologists prefer to use the average content of petrogenic oxides and trace elements (calculated for large groups of rocks, individual horizons, and intrusions) has aligned different rocks due to common peculiarities in their differentiation. Experience has shown that only in-depth study of rocks directly associated with the intrusion can be informative for identification of differentiation characteristics.

The trend of the crystallization differentiation process is evident at the micro-level in the succession of mineral crystallization. It is also seen in variations of mineral compositions from core to rim in zoned grains, as well as in minerals formed in the rock matrix (usually in compliance with the Bowen principle). In this way, plagioclase composition becomes progressively more alkaline over the course of crystallization, Fe content in pyroxene increases, and the residual melt takes on a tholeiite glass composition with increased silica content or crystallizes to form micropegmatite or granophyre. The Mt. Pegmatitovaya Intrusion is a classic example of intrachamber gravitational crystallization differentiation. Investigations of the intrusion have shown regular variations in elemental composition throughout the body, and variation diagrams also demonstrate changes in petrochemical characteristics. The crystallization series of layered rocks from mafic intrusions was mapped onto diagrams showing the fractionation coefficient (C.f.) versus SiO_2 as variation curves and shows an increase in C.f. with insignificant SiO_2 variation (Fig. 5.1). Such a crystallization scheme has been established for the Skaergaard Intrusion (Wager and Brown 1968), and it is peculiar to trap sills (Zolotukhin 1964; Feoktistov 1978). AFM diagrams indicate that differentiation of the melt proceeds with increasing magnesium content and decreasing alkali-element and iron content. Branching crystallization trends on the diagrams is associated with formation of granophyres and ferrogabbro at the final stages of intrusion.

The invisible layering is conformed by unified crystallization trends for clinopyroxene and olivine throughout the intrusion. In the Mt. Pegmatitovaya Intrusion the full range of crystallization differentiation of mafic intrusions in the platform has been traced for the first time using the widest series of progressively varying clinopyroxene and olivine compositions. The Fe content of clinopyroxene in the Mt. Pegmatitovaya Intrusion covers a range from Fs14 to Fs56 ($f = 23$–69 at. %), and olivine ranges from Fa35 to Fa70 with a regular decrease in NiOOl from 970 to 40 g/t and an increase in CaOOl content from 600 to 2,800 g/t (Figs. 3.39 and 3.40).

The crystallization trend for clinopyroxene from mafic intrusions of the Siberian Platform on the Hess (1941) ternary diagram trapezoid (Fig. 4.7) occupies a position close to the Skaergaard differentiation trend according to Wager and Brown (1968). During the crystallization, a consistent increase in Fe content in clinopyroxene with that in olivine is observed, and the alkalinity of plagioclase increases, ranging from bytownite to oligoclase (Ryabov 1984b, 1989c; Ryabov et al. 1985c).

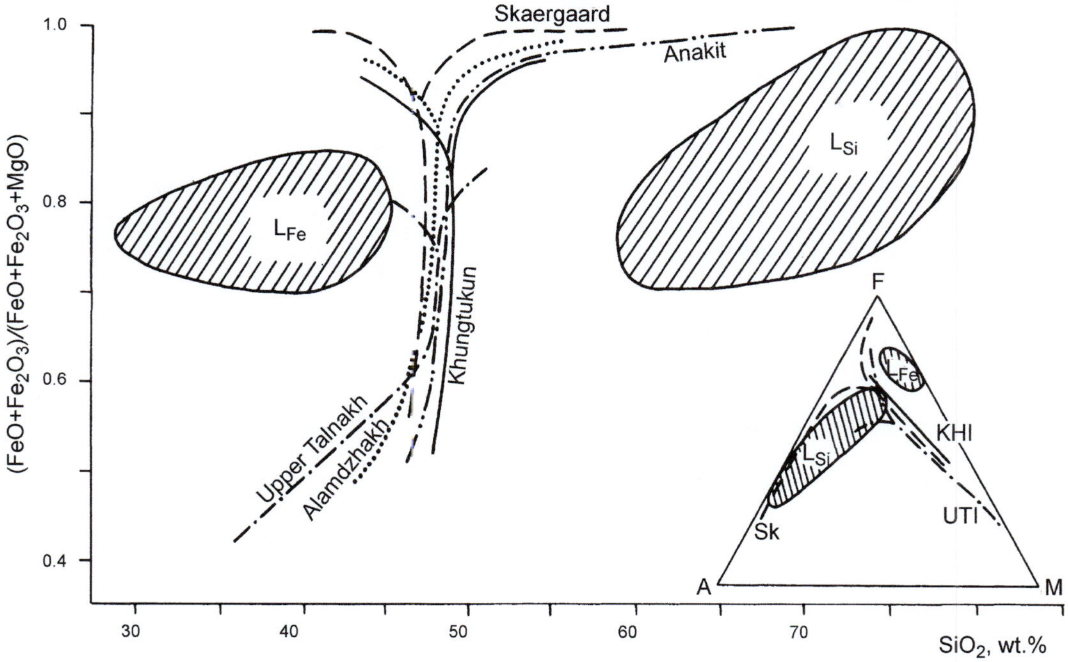

Fig. 5.1 Fractionation coefficient (C.f.) diagrams SiO_2 versus AFM (wt%) for selected layered intrusions and immiscible glasses of the Khungtukun Intrusion (Ryabov 1989b). Glasses: L_{Fe} mafic, L_{Si} felsic. Intrusions: *SK* Skaergaard, *T* Upper Talnakh, *X* Khungtukun

Crystallization differentiation manifests itself as a regular distribution of minor and foreign chemical components throughout a body. In zoned clinopyroxene crystals, an increase in Ti content and decrease in Cr content is observed with increasing Fe content. The same trend is traced upward through the intrusion. Differentiation is also manifest in a change in (Ca + Na)/Al – Si/Al and (Ca + Na)/Fe – Mg/Fe ratios in zoned clinopyroxene crystals. Composition of the intrusion indicates an increase in alkalinity and silica acidity of the melt during crystallization (Ryabov 1974).

Variation in plagioclase composition and Ba and Sr content in plagioclase compared to whole rock composition upward through the intrusion (Figs. 3.15, 4.12, and others) serves as an indicator of differentiation of a melt under chamber conditions.

5.2 Nature of High-Magnesian Traps

If the composition of contact gabbrodolerites in slightly differentiated mafic intrusions shows a strong similarity to the calculated average weighted composition of these bodies, then it is obvious that melts with similar magnesium content will fall short of generating olivine-rich rocks when these data are compared in the Noril'sk Intrusions (Table 5.1). Zolotukhin (1964) shows that to generate olivine, a parent magma should have the composition of olivine or troctolitic gabbrodolerite. Only if the MgO content in the melt is at least 10% (Stepanov 1975) will a horizon of picritic gabbrodolerite be formed during fractionation.

When calculations of the average weighted composition for Upper Talnakh Intrusions (Table 5.1) are compared to data from the literature, significant fluctuations in the "initial" magma compositions are observed in different parts of the same body, which is inexplicable in the context of crystallization differentiation and it seems to be a doubtful indicator of initial melt composition. Wide variations in the average weighted composition are conditioned by both significant variations in the thickness of some rock horizons in different parts of the intrusion (Dodin and Batuev 1971; Likhachev 1998) and the absence of a correlation between them (Natorkhin et al. 1977; Stepanov 1981). As mentioned above, some geologists explain the incompatibility between average weighted composition within an intrusion and in contact facies as evidence of mafic–ultramafic layering of melts in a deep-seated chamber (Marakushev et al. 1982). Others explain it by an influx of a "low-melting" magma that formed the contact gabbrodolerite as compared to an exfoliating batch of magnesian liquid (Zolotukhin and Laguta 1985), which, in essence, also supports the introduction of early differentiated melts into the chamber. Taking this into consideration, it is not unreasonable to assume the simultaneous influx of mafic and ultramafic melts with widely varying volume. Thus, wide variations in the thickness of picritic gabbrodolerites in layered intrusions can be ascribed to the composition of mafic and ultramafic melts (see Appendix, Table 4.9).

Table 5.1 The average weighted composition of the Upper Talnakh differentiated intrusion

Source, drill hole	SiO_2	TiO_2	Al_2O_3	Fe_2O_3	FeO	MnO	MgO	CaO	Na_2O	K_2O	P_2O_5	LOI
Urvantsev (1970)	40.74	0.81	14.01	4.53	9.07	0.17	13.23	9.36	1.70	0.68	0.23	3.63
Komarova and Lyul'ko (1967)	42.04	0.78	13.65	4.96	8.33	0.15	12.93	9.38	1.74	0.82	0.21	4.14
Zolotukhin et al. (1975)	43.40	0.96	14.47	3.73	9.24	0.18	11.39	9.16	1.74	0.88	0.14	3.58
Dodin and Batuev (1971)	43.95	0.95	14.03	4.23	8.38	0.18	10.40	9.50	1.83	1.05	0.17	3.60
Vilensky (1978)	43.83	0.84	14.18	4.54	8.25	0.14	11.48	9.56	1.80	0.60	0.14	2.86
Southwestern branch of the Upper Talnakh Intrusion, KZ-184 ($n = 27$)	43.90	0.84	12.22	4.30	8.57	0.21	13.67	8.35	1.70	0.97	0.09	4.76
Northeastern branch of the Upper Talnakh Intrusion, KZ-274 ($n = 24$)	43.16	1.00	14.48	3.34	9.50	0.16	11.49	9.84	1.77	0.75	0.22	2.93
Kharaelakhsky branch of the Upper Talnakh Intrusion, KZ -585 ($n = 46$)	42.05	0.68	13.22	4.32	6.78	0.19	15.69	9.58	1.68	0.63	0.12	3.74
Kharaelakhsky branch board of the Upper Talnakh Intrusion, T-56 ($n = 12$)	46.81	0.97	16.16	3.58	9.50	0.20	7.21	10.32	2.36	0.91	0.13	2.60
North of the Upper Talnakh Intrusion, SV-16 ($n = 19$)	44.60	1.00	14.20	4.45	8.02	0.19	10.71	9.48	1.99	0.76	0.12	3.98
Contact gabbrodolerites ($n = 52$), Ryabov (1989b)	46.79	1.18	14.83	3.69	8.46	0.19	7.92	10.30	2.14	1.14	0.16	2.80

Comagmatism of picritic basalt and layered intrusions of the Noril'sk type is thought to be well established, beginning with early works by Korovyakov (1948), Kotul'sky (1946, 1948), and Godlevsky (1959) to the present. Additionally, the principal question of whether picritic lavas are comagmatic to layered intrusions as a whole or only to picritic gabbrodolerites is yet to be answered.

Most geologists are inclined to believe that sheets of picritic basalts are comagmatic with the Noril'sk Intrusions (Dodin 1962, 1967; Duzhikov 1971; Marakushev et al. 1982; Zolotukhin et al. 1986; Duzhikov et al. 1988). They consider that this concept is supported by the similarity of mafic–ultramafic nature of differentiation inherent to both rock types, the closeness of their mineralogical–petrochemical compositions, and similarity of formation mechanism. Some geologists explain the layering of differentiated picritic sheets and the Noril'sk Intrusions by a classic mechanism of crystallization differentiation with gravitational accumulation of olivine in the near-basal parts of bodies (Zolotukhin et al. 1986), whereas others associated this with liquid immiscibility in a deep-seated chamber (Marakushev et al. 1982).

Ivanov et al. (1971a) consider that picritic basalts are comagmatic with only picritic gabbrodolerites of the Noril'sk Intrusions, and the formation of the latter proceeded as successive two-phase influxes of ultramafic and mafic magmas (Ivanov et al. 1971b).

Our investigations have shown that picritic basalts are comagmatic to picritic gabbrodolerites rather than to intrusions as a whole. In flows of picritic basalt, layering into metadiorite and basalt occurred as a result of fluid–magmatic differentiation as proposed by Marakushev et al. (1982), not in a deep chamber, but under intersheet conditions as shown by olivine porphyries of upper mandelstones overlying the metadiorite. In situ fluid–magmatic differentiation in basalt flows is supported by the sharp contact between compositionally contrasting metadiorites and picrites. Olivine-ophitic and/or ankaramite basalts, which are present in a number of cases, are secondary reaction layers formed at the boundary between a fluid-saturated melt of metadiorite and picrite. Observations along the section through a single flow and inside the picritic horizons show that gravitational accumulation of olivine does not take place. Picritic lavas are enriched in volatiles. Defining rocks as metadiorites fits the ideas related by Loewinson-Lessing (1933). In essence, these rocks are the products of metamagmatism according to Korzhinsky (1973), which is formed as a consequence of the interaction of melt with fluid. The mineral associations and structure of metadiorites from basalt flows bear witness to the relatively low-temperature layering in a chamber rather than in deep-seated chambers. High-temperature phases are absent in metadiorites, and principal minerals include andesine, chromic needle-shaped clinopyroxene (Fs11), ilmenite, and large volumes of palagonite.

The unevenness of metadiorite thickness and the sporadic nature of their occurrence in basalt flows are also an important point. Metadiorites show a strong similarity to "blowholes," typically occurring as lenses, or as loaf-like or finger-like forms. The fact that the petrographic compositions of intrusion contact facies are inconsistent is an argument against comagmatism of picritic lavas and the Noril'sk Intrusions. In the first case, these are vitreous olivine porphyries (mandelstones), and in the second case, these are olivine-free contact dolerites or gabbrodolerites.

Our studies show that the layering in sheet has a strong similarity to the immiscibility features of picrite-like and leucocratic melts described in picritic basalts of the Pechenga area, which are interpreted to be the result of pre-crystallization immiscibility (Smolkin et al. 1987).

As opposed to differentiated basalt flows in the Noril'sk Intrusions, the section composed of mafic rocks demonstrates a strong similarity to common traps of a normal series layered into olivine-free, olivine-bearing, and olivine gabbrodolerites. Metadiorites are restricted, as a rule, to the leading part of a body and represent products of metamagmatic alteration of basaltic melt. Along dip, these rocks change to prismatically grained gabbrodolerites. Geologists regard these as in situ rocks rather than rocks from depth. Palagonite, oligoclase–albite, and titanium clinopyroxene are ubiquitous in the rocks, and green amphibole developed after pyroxene is frequently observed (Dodin and Batuev 1971; Ivanov et al. 1971b; Zolotukhin et al. 1975).

The Lower Fokinsky Intrusion, which shows a strong structural and compositional similarity to lavas, as well as a number of undifferentiated picrite bodies are comagmatic to picritic basalts. Fluid–magmatic differentiation in the Lower Fokinsky Intrusion manifests itself as gabbro-pegmatites and in the horizon of olivine-free gabbrodolerites as elements with a pegmatitic fabric. These rocks can be regarded as analogs of metadiorites in other intrusions. It can be assumed that the fluid–magmatic mechanism of differentiation causes an additional increase in the magnesium content of a melt in the ultramafic portion of layered liquids that stimulates olivine crystallization. In this way, one can explain enrichment in olivine and increased MgO content of picritic basalts in the sheets that feature metadiorites, that is, layering takes place in the melt.

As pointed out above, most geologists explain the pseudostratified structure of the Noril'sk Intrusions as a result of gravitational crystallization differentiation mechanism (Godlevsky 1959; Zolotukhin 1964; Dodin 1967; Stepanov 1975).

It has long been thought that the theory of crystallization differentiation would have lost its appeal and universality if the gravitational effect had been excluded from it. One of the most important arguments for the Noril'sk Intrusion, supporting the fact that their layering is due to the above mechanism, is the increase in the olivine and MgO content in rocks from olivine-free through olivine-bearing to picritic gabbrodolerite downward through the intrusion. Indeed, the gradual transition in olivine and MgO content is observed between mafic members of the intrusion, but the olivine content sharply increases from 10–12% to 40–60% between olivine gabbrodolerite and picritic gabbrodolerite. This has long been observed by geologists and is the basis of the idea of repetitive influxes of melts with different compositions.

In order to ascertain when olivine crystals will form in a melt and the time picritic horizons take to form, geologists commonly make calculations using the Stokes formula using constants for crystal size and melt viscosity. For the Noril'sk-I Intrusion, in particular, the crystal size was taken to be 0.5 cm (as observed in picritic gabbrodolerite), and viscosity was adopted from the Volarovich diagram (Zolotukhin and Vasil'ev 1967). However, Goryainov (1975) questioned the feasibility of using the Stokes formula with reference to the Noril'sk Intrusions since other, equally valid, values for grain size and viscosity were not taken into account in the calculations. It can be argued, from observations of contact facies of the Lower Talnakh Intrusion and mandelstones from picritic basalts, that the original size of olivine crystals in a heterogenic melt that flowed onto the surface or moved into a hypabyssal chamber was less than 0.1 cm. Gravitational settling of such small crystals could not be implemented, and they were suspended in the melt. Successive growth of crystals (to 0.5–0.7 cm) was accompanied by the increase in viscosity of the melt; therefore, large crystals could not be gravitationally settled down because of a magma viscosity. So the likelihood of formation of a picritic horizon due to gravitational settling of olivine protocrystals is, at least, doubtful. The possibility of such complete separation of olivine by gravitation (manifested in the Noril'sk Intrusions by a sharp transition from olivine gabbrodolerite to picritic gabbrodolerite) is also improbable. The consistent optical orientation of olivine crystals in the picritic horizon points to a "directional" introduction olivine (Zolotukhin 1964; Likhachev 1996), though it does not necessarily imply gravitational accumulation of olivine in situ. In the opinion of Smirnov (1966), the gravitational accumulation of olivine in the Noril'sk Intrusions can be suggested only for the mafic part of an intrusion in the rock series from olivine-free through olivine-bearing to olivine gabbrodolerite.

Experimental investigations (Fig. 1.39) have shown that during sulfurization of a tholeiitic basalt melt, its magnesium content increases (Almukhamedov and Medvedev 1982), the high-magnesium basaltic melt is layered, and liquids of tholeiitic basalt and peridotite compositions are formed (Gorbachev 1989). Additionally, an increase in Cl, F, and S in fluid induces fluid–magmatic interactions and enhances the extractive ability of fluids with respect to elements with mafic properties (alkali metals, MgO, CaO, FeO), which results in increased mineralization and redistribution of these elements and, in particular, in an increase in the MgO content of fluidized melt.

One of the concepts of the Bowen (1922) theory of crystallization differentiation is the reaction series based on the succession of mineral segregations from a magmatic melt. The series is usually applied as an explanation of trap petrogenesis. As was noted by different researchers, the order of mineral segregation according to the Bowen reaction series is not universal, and it is not infrequently violated. The reason for these deviations depends on the physical–chemical peculiarities of the melt. Examples of deviations from the reaction series with decreasing basicity in traps are ophitic gabbros and other gabbrodolerites, in which plagioclases crystallize before mafic minerals, and magnetite (that is commonly an early crystalline phase) is not an infrequent product of the final phases of a magmatic process (Loewinson-Lessing 1935). One noteworthy observation is the association of olivine and quartz in gabbrodolerites of the Khungtukun Intrusion (Ryabov 1989b) and a number of other intrusions, the mantling of clinopyroxene by olivine in rocks of Mikchandinsky flow sequence, the mantling of clinopyroxene by olivine in Maymechinsky Intrusion or orthopyroxene in picrites of the Noril'sk Region, the increase in magnesium content of olivine in the matrix as compared to its more iron-rich phenocrysts in the Khungtukun Intrusion (Ryabov 1989c), and the inverse zoning in biotites (Sobolev 1986; Godlevsky 1959). These examples provide additional reasons for doubt of the universality of the crystallization differentiation mechanism.

In spite of Bowen's belief that the reaction series in petrogenesis proposed by him is unique, many of its positions have been complimented, corrected, refined, as well as criticized. The universality of the theory has been called into question by geologists more than once. Grigor'ev (1935) drew attention to the fact that the members of a discontinuous reaction series may produce discontinuous series themselves, for example, clinoenstatite–diopside, and this, in turn, may cause the melting of one or more compounds to be incongruent and the whole reaction series to be violated. The incongruent melting becomes melting without decomposition when some components are added, for example, FeO (Ginsberg 1951). Some examples of deviations from the Bowen reaction series (Bowen 1922) in traps have been mentioned above, and these may be found in a number of petrographic studies on the magmatism of the platform.

The discontinuity of both mafic and ultramafic portions of the Noril'sk Intrusions has long been discussed by geologists. To explain this, two-phase and even multiphase influxes of melts of various compositions have been assumed, giving reason why two or multiple rock horizons were formed (Rogover 1959; Smirnov 1966; Ivanov et al. 1971b). However, from this standpoint, the order of rocks encountered in cross sections of layered intrusions seems to be unintelligible. Another model of development of the Noril'sk Intrusions was based on the mixing two initial magmas in a recent chamber with consequent gravitational separation of crystalline phases and liquid sulfides in the picritic gabbrodolerite horizon (Tuganova 1988). However, as we have seen, known petrographic and mineralogical (ontogenetic) data concerning crystallization features of chromites, pyroxenes, and olivines are underestimated, and the physical–chemical nature of the process remains unclear; it is difficult to find a method of first mixing tholeiitic basalt magma with ultramafic magmas in an intermediate chamber and then layering this mixed magma into the same mafic (tholeiitic basalt) and ultramafic liquids. This is much better understood as layering into mafic and ultramafic liquids at depth due to immiscibility (Marakushev et al. 1982).

Our investigations confirm the assumption that two immiscible melts of both basaltic and picrite-like compositions were simultaneously introduced into recent magmatic chambers of the Noril'sk Intrusions. Marakushev et al. (1982) also proposed the same. The feasibility of layering magma into mafic and ultramafic liquids has been experimentally demonstrated by Bezmen (1984), and occurrences of joint influxes of aphyric and picritic basalts in the Mikchandinsky paleovolcano serve as geologic evidence in support of simultaneous introduction of melts of various compositions. Some of the evidence for the simultaneous introduction of immiscible mafic and ultramafic liquids into a recent chamber in the Noril'sk Intrusions is presented below.

1. No quenched, sharp contact at the boundary between olivine and picritic gabbrodolerite (Rogover 1959; Smirnov 1966; Ivanov et al. 1971b; Ryabov and Yakobi 1981) as observed in petrographic and petrochemical studies of the Noril'sk Intrusions
2. Layering of olivine and picritic gabbrodolerites (Rogover 1959; Smirnov 1966; Ivanov et al. 1971b; Ryabov and Yakobi 1981; Dodin, and Batuev 1971; Zolotukhin et al. 1975; Natorkhin et al. 1977; Marakushev et al. 1982)
3. The absence of a correlation between olivine-rich and olivine-poor rocks (Smirnov 1966; Natorkhin et al. 1977; Stepanov 1981) that is manifest as an incompatibility between the thickness of mafic and ultramafic rocks in different parts of a body
4. Incompatibility between contact gabbrodolerite compositions, undifferentiated flanks of Noril'sk Intrusions, and the assumed average weighted composition of a body (Marakushev et al. 1982; Ryabov 1989b)
5. Abrupt variations in amount, composition, and minerals at the boundary between olivine and picritic gabbrodolerite (Ivanov et al. 1971b; Stepanov 1975)
6. The dual nature and distribution of typomorphic chemical elements in Noril'sk Intrusions, where mafic parts in the middle of the intrusion have features similar to those in common traps and lower mafic rocks have features similar to ultramafic rocks (Godlevsky 1967)
7. Compositional, mineralogical, and petrochemical similarity (Smirnov 1966) of picritic basalts to picritic gabbrodolerite with poor sulfide dissemination
8. Existence of independent apophyses of gabbrodolerite in the Noril'sk-I and Upper Talnakh Intrusions (Rogover 1959; Smirnov 1966; Ivanov et al. 1971b; Zolotukhin et al. 1975)

Nevertheless, Ivanov et al. (1971b) explain some petrological and geochemical differences between picritic basalt and picritic gabbrodolerite in the Noril'sk Intrusions by the fact that the rocks belong to different facies. Effusive picrites have lost their connection with their parent magma chamber, whereas comagmatic intrusive rocks retained a connection and experienced alteration under the action of fluids. In particular, rocks from Iceland that show increased sulfur content have been observed in intrusive facies and rocks with low sulfur content observed in comagmatic effusive facies (Gerasimovsky and Kuznetsova 1974). The authors attribute this to loss of volatiles (decontamination) during lava inflow.

It is possible that during simultaneous introduction of two immiscible liquids into a hypabyssal chamber of the Noril'sk Intrusions, the relatively less viscous and more mobile mafic melt preceded the ultramafic melt and then facilitated its advance by providing a conduit within the mafic melt path, lubricating the ultramafic flow and allowing it to maintain a high temperature. The stream of homogeneous then heterogeneous picrite-like melt flowed down to the basal part of a magmatic chamber during subhorizontal motion due to higher density. Once in the chamber, the altered PT conditions initiated pre-crystallization differentiation in each stream. Crystallization from the mafic liquid began, as well as in the ultramafic liquid, governed by fractional differentiation mechanisms for the entire body, although this was complicated by the presence of fluids. During this process, each batch of a melt had its own evolutionary peculiarities: a mafic layer began to crystallize, forming olivine, which was precipitated in the near-basal part of the layer; in a picritic layer, olivine continued to grow, as well as disseminated chromite crystallized. But as a whole, the course of crystallization in situ followed Godlevsky's (1971) image-bearing expression, and equalization of conditions in the magmatic system took place and resulted in an alignment of compositions of both rocks and minerals under action of powerful crystallization differentiation. Since pyroxene and most plagioclases are minerals typically formed in chamber crystallization (according to petrographic data), then it is reasonable that each of these minerals follows a unified crystallization trend throughout the central zone of a body including both mafic and ultramafic horizons. Despite the equalizing role of crystallization differentiation, both horizons have retained, to a certain extent, their identity. Many geologists have drawn attention to this, and it has been the subject of discussions for many years.

Gravitational crystallization differentiation of a mafic part of the Noril'sk Intrusions resulted in regular distribution of mineral phases through a body, compositional variation of rocks and minerals, and crystallization trends (more magnesian rocks represented by picritic gabbrodolerite). The differentiation of mafic rocks of the Noril'sk Intrusions did not reach the range of intrusions like those at Mt. Pegmatitovaya or Alamdzhakh, having minerals with high Fe content and featuring granophyre or ferrogabbro rock types. Here fluids played a role in the creation of features peculiar to the rocks (Masaitis 1958; Stepanov 1975).

The location of olivine-rich rocks, including picritic gabbrodolerite, at the base of mafic bodies is attributed to the specific weight of liquids. In a number of cases where no "sagging" of an ultramafic stream to the base of a magmatic body is observed, olivine-rich gabbrodolerite occupy a position "suspended" relative to the base of the intrusion. This, in particular, has been described by Sukhanova (1968) when she first characterized the Kharaelakhsky branch of the Upper Talnakh Intrusion (Oktyabrsky deposit) pointing to a "reversal" of layering in cross section.

Based on petrologic investigations, three genetic types of olivine-rich rocks can be recognized (Ryabov 1989a).

I. *Picritic basalts and gabbrodolerites* as crystallization products of picrite-basalt melts. In barren varieties, minor elements content (in particular, chromium, nickel, and platinum group elements) is close to Clarke of ultramafic rocks, as well as the typomorphic association of mafic minerals with low Fe content such as olivine, ortho- and clinopyroxene, calcium plagioclase (bytownite–anorthite composition), and chromite are characteristic (Stepanov 1975; Ryabov et al. 1985a). In the intrusive facies, rocks of this type frequently occur in the Lower Fokinsky, Magnitny Stream, Upper Pyasinsky, Upper Talnakh, Eastern Noril'sk, and other intrusions. In many places, in environments rich in fluids, these rocks undergo strong metasomatic alteration, which overprints the original composition and is substantially conditioned by volatile composition and the PT state of a system. Degeneration of the rocks is distinct in the picritic horizon in the Noril'sk Intrusions, for example, as a redistribution of Cr and Ni and variation in mineral composition. The degeneration of rocks increased in independent high-magnesian intrusions such as Lower Talnakh and Lower Noril'sk, where picritic–troctolitic gabbrodolerites represent a compositionally stable rock subtype depleted in oregenic components (Ryabov 1981; Ryabov and Yakobi 1981).

II. *Picritic–troctolitic gabbrodolerites* are cumulative products of fractional crystallization of tholeiitic basalt melts. Variations in the chemical composition of rocks, minerals, and microelements through layered mafic intrusions show a unified trend induced by crystallization differentiation with gravitational separation of early mineral phases. As is evident from work by Nesterenko and Almukhamedov (1973), Arkhipova (1975), Kravchenko (1977), and Oleinikov (1979), Clarks of chemical elements of this type of picrite-basalt troctolite and the nature of their distribution are of the same order and have variation trends typical to mafic rocks. This rock type is represented in the Mt. Pegmatitovaya, Mt. Moromgo, Mt. Putanaya, and Mt. Picritovaya Intrusions and in bodies of the Middle Yenisey River (Zolotukhin 1984a). The typomorphic association of minerals in these rocks consists of olivine, clinopyroxene that both are more ferrous and less nickeliferrous than those in type I, and plagioclase of labradorite composition; the absence of chromite is specific.

III. *Secondary picritic–troctolitic gabbrodolerites* as products of fluid–magmatic differentiation or meta-magmatic processes in recent magmatic chambers (Ryabov 1989b). Fluids played a dominant role in rock formation, increasing partial pressure, which at the magmatic stage resulted in the selective extraction of iron and nickel and the generation of sulfides, oxides, carbides, or native ore phases, and additionally in the crystallization of olivine with low Fe content from a relatively high magnesium liquid. The main role of fluids was the redistribution of elements that led to an accumulation of magnesium. Experimental techniques and thermodynamic support for the feasibility of these processes under environmental conditions have been considered by a number of authors: Almukhamedov and Medvedev (1982), Naldrett et al. (1984), and Ryabov and Pavlov (1984). Upper picritic and troctolitic gabbrodolerites in the Upper Zone of the Khungtukun mafic intrusion with no olivine-rich rocks at the base of the body provide strong evidence for the existence of this rock type. Troctolitic gabbrodolerites of the Mt. Ozernaya Intrusion have been described by Zolotukhin and Vasil'ev (1967) they are known in the Mt. Pegmatitovaya Intrusion (Fig. 3.37), rocks unusually enriched in olivine have been found in the upper zones of a number of trap intrusions in the Tungussky syneclise (Lebedev 1955), and lastly there is every reason to believe that upper picrites of the Noril'sk Intrusions have been formed by a similar fluid–magmatic mechanism. Most obviously, it is impossible to use crystallization differentiation as an explanation as to why these rocks, especially mafic ones, have been formed in the UZ of the intrusion. The geochemical character of a melt strongly affects the compositional peculiarities of rocks in the UZ; in this way, in the Khungtukun Intrusion, picrites are formed with nickel-free chrysolite in association with α-Fe, similar to the Upper Talnakh Intrusion where chrysolite–forsterite crystals with variable nickel content, chromites, and platinum minerals are formed. Thin interlayers of picrite–troctolite are known to have formed due to fluid interaction along sub-bedded proto-cracks in a number of type III mafic intrusions. The final example is chamber restites, which are essentially olivine rocks from the Kharaelakhsky magmatogenic breccia with a high magnesium content largely conditioned by the outflow of sialic component under the action of volatiles (Ryabov and Pavlov 1984). It is important to note that in the same intrusion, one or more genetic types of high-magnesium rocks can occur.

The vast volume of olivine-rich rocks in the Lower Talnakh and Noril'sk Intrusions, which we consider as root portions of extended intrusions of the Noril'sk type (Ryabov and Yakobi 1981), can be explained by the action of leading mafic melts in magmatic columns and lagging high-magnesium liquids (Zolotukhin and Laguta 1985). The same may be assumed for the Morongovsky Intrusions, with the Mt. Pegmatitovaya Intrusion acting as the leading magmatic column and the intrusions of Mt. Putanaya and Mt. Picritovaya the rear portion of the column. Aspects of horizontal zonation of extended intrusions are also described by Ivanova (1975), Sukhareva and Kuznetsova (1983), and Stepanov (1981).

Summarizing the above, it may be assumed that fluid–magmatic differentiation has been a leading factor in forming high-magnesium traps.

5.3 Liquid Immiscibility in Glasses from Traps

The discovery of silicate–silicate and ore–silicate immiscibility in glasses has been an important finding in magmatic petrology. Immiscibility as a characteristic phenomenon of glasses from traps of the Siberian Platform has been published in detail by Ryabov (1989c). Only some aspects that show peculiarities of immiscible differentiation are described here with their possible consequences. Some insight into the occurrence and compositions of coexisting immiscible glasses is shown in the thin section photograph of Rock indication 226 (see Volume 2).

Immiscibility in glasses of the Siberian Traps are observed in inclusions in olivine that is itself enclosed in porphyritic plagioclase, as well as in clinopyroxene, ore matrix, and basalt matrix (Pankov 1984; Ryabov et al. 1985a; Tikhonenkov et al. 1985; Ryabov 1989b). Electron microprobe analyses of immiscible glasses, plotted on Fig. 5.2, indicate a resemblance to liquid immiscibility data experimentally obtained by Roedder (1951), although the data occupies a somewhat larger area being shifted toward the $R_2O + Al_2O_3$ corner. Particular analyses of traps, plotted on this diagram for comparison, fall into the field of immiscible glasses, and data from compositionally contrasting rocks coexisting in the same magmatic body show similar trends of variation in glass composition. The arrangement of data within the zone of immiscibility allows their composition to be interpreted as potentially inclined toward layering or to consider them as products of layering (Ryabov 1989b). Separation coefficients of chemical elements between mafic and sialic components calculated for glasses turn out to be similar, and the variation in Fe content in coexisting immiscible liquids is illustrated as a near-linear plot (Fig. 5.3).

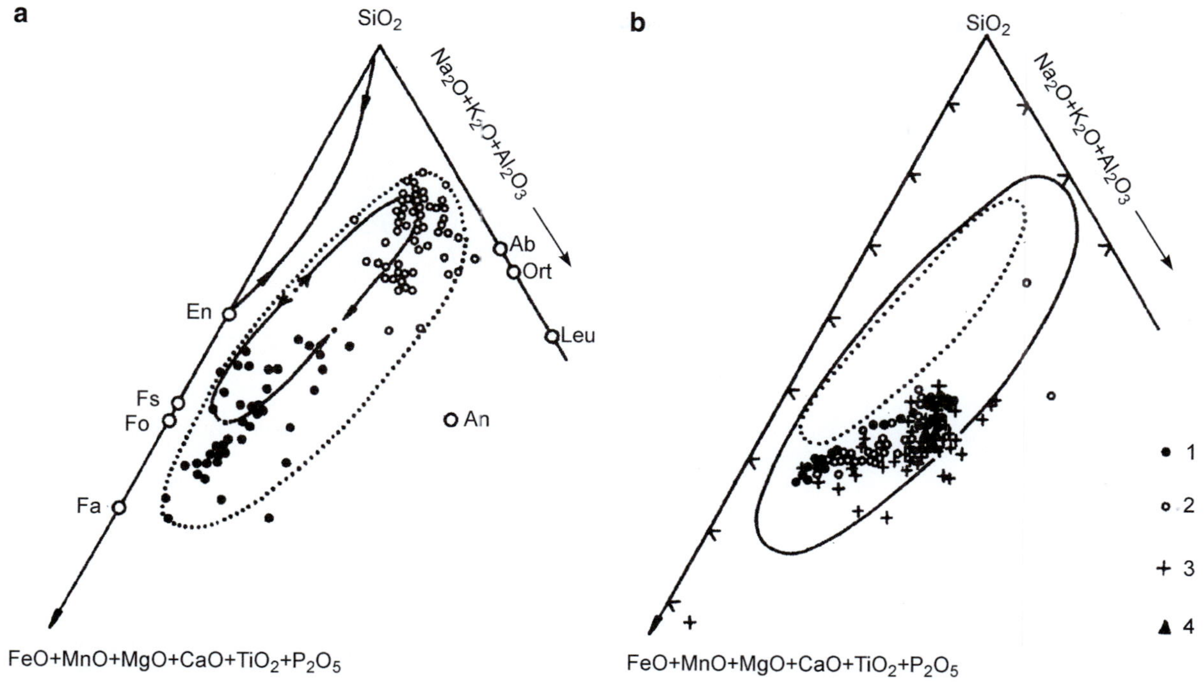

Fig. 5.2 Pseudo-ternary diagram for immiscible liquids and rocks of trap intrusions. (**a**) Glasses in traps: 1 = felsic, 2 = mafic. Stable immiscibility fields after Greig (1928), Roedder (1951), and Weiblen and Morrey (1975, 1980) are outlined by a *solid line*, and the field of immiscible glasses in traps is outlined by a *dashed line*. (**b**) Rock composition from the Upper Talnakh (1) and Lower Talnakh (2) Intrusions, (3) UZ of these intrusions, and (4) sills of titanaugite dolerite are shown on a background showing the low-temperature immiscibility field (*dashed line*). (**c**) Coexisting contrasting rocks in traps. *Dashed line* outlines the field of immiscible glass compositions from traps. Coexisting rocks are connected by lines

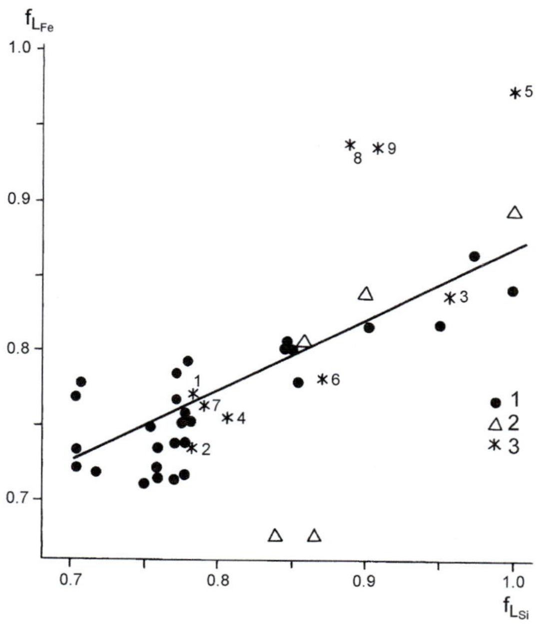

Fig. 5.3 Variations in total Fe content $f = FeO/(FeO + MgO)$, wt% in coexisting mafic (L_{Fe}), and sialic (L_{Si}) liquids (Ryabov 1989b). 1 = globules featuring an emulsion of L_{Fe} glass in L_{Si} glass in traps (GG), 2 = globules of glass with crystallites (GC), 3 = glasses from different formations

The comparison between immiscible glasses from traps and analogous substances from effusive terrestrial rocks of different types and lunar basalts has revealed their strong similarity in the composition and tendency of its variation (Ryabov 1988a, b, 1989b).

According to Marakushev's classification, the most widely developed layering of silicate liquids in traps is sialic–mafic in nature (Fig. 5.4a). During layering, one liquid is enriched in SiO_2, Al_2O_3, Na_2O, and K_2O and the other in FeO, MgO, CaO, MnO, TiO_2, and P_2O_5. Crystallization of layered liquids proceeds with increasing an alkalinity; silicic acid is seen on plots showing an analysis of glass composition with crystallites (GC) and interstitial glasses (IG), which occur in the matrix of crystallized aggregates. Pairs of coexisting contrasting rocks plotted on the diagram (Fig. 5.4b), compositions of which are given by Ryabov (1989b), show trends for the compositional variations similar to those of immiscible glasses.

Chemical compositions of different types of glasses, rock-forming, immiscible without crystallites (outlined), partly crystallized globules, interstitial, and homogenous, have been plotted on the Zavaritsky diagram (Fig. 5.5). For comparison, the diagram for principal types of magmatic rocks

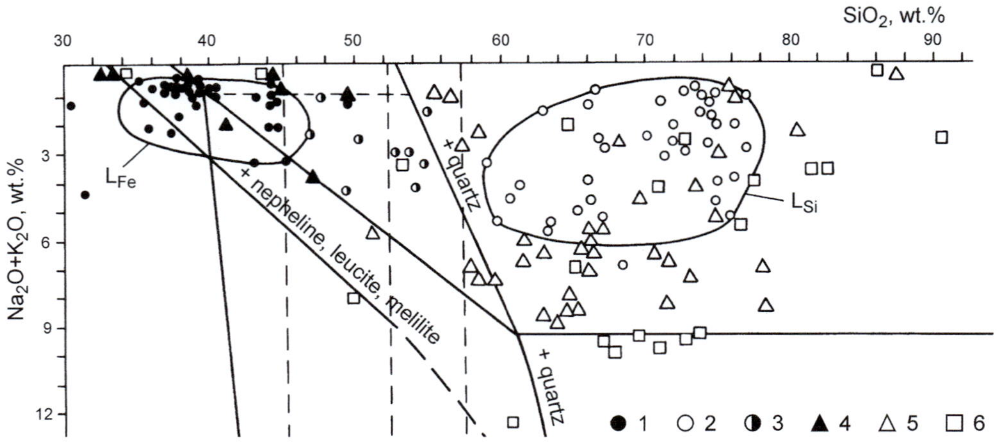

Fig. 5.4 (Na_2O + K_2O) versus SiO_2 diagram for immiscible glasses. A and for coexisting rocks contrasting in composition B (Ryabov 1990). (**a**) Glasses: mafic (1) and sialic (2) GG, 3 = homogenous HG, 4, 5 = mafic (4) and sialic (5) GC, interstitial IG. (**b**) Pairs of rocks: 1 = spinel–melilitic rock – taxitic gabbrodolerite, Khungtukun Intrusion; 2 – ferrogabbro–granophyre, Alamdzhansky Intrusion (Masaitis 1958); 3 = ferrogabbro–granophyre, Anakitsky Intrusion (Reverdatto 1963); 4 = mineralized and barren variolites from UTI bottom; 5 = picritic and olivine-free gabbrodiorite, Morongovsky Intrusion; 6 = picritic gabbrodolerite–leucogabbro, UZ UTI; 7 = chromitite–leucogabbro; 8 = picritic basalt–labradorite, Khrebtovy River; 9–11 = melanocratic and leucocratic parts of magmatogenic breccias from UTI; 12 = olivine-bearing granodiorite–granophyre, Middle Dudinsky Intrusion; 13 = picritic gabbrodolerite – dolerite–pegmatite, Khungtukun Intrusion; rocks after Daly (1933): a = basalt, b = andesite, c = dacite, d = liparite are given for comparison

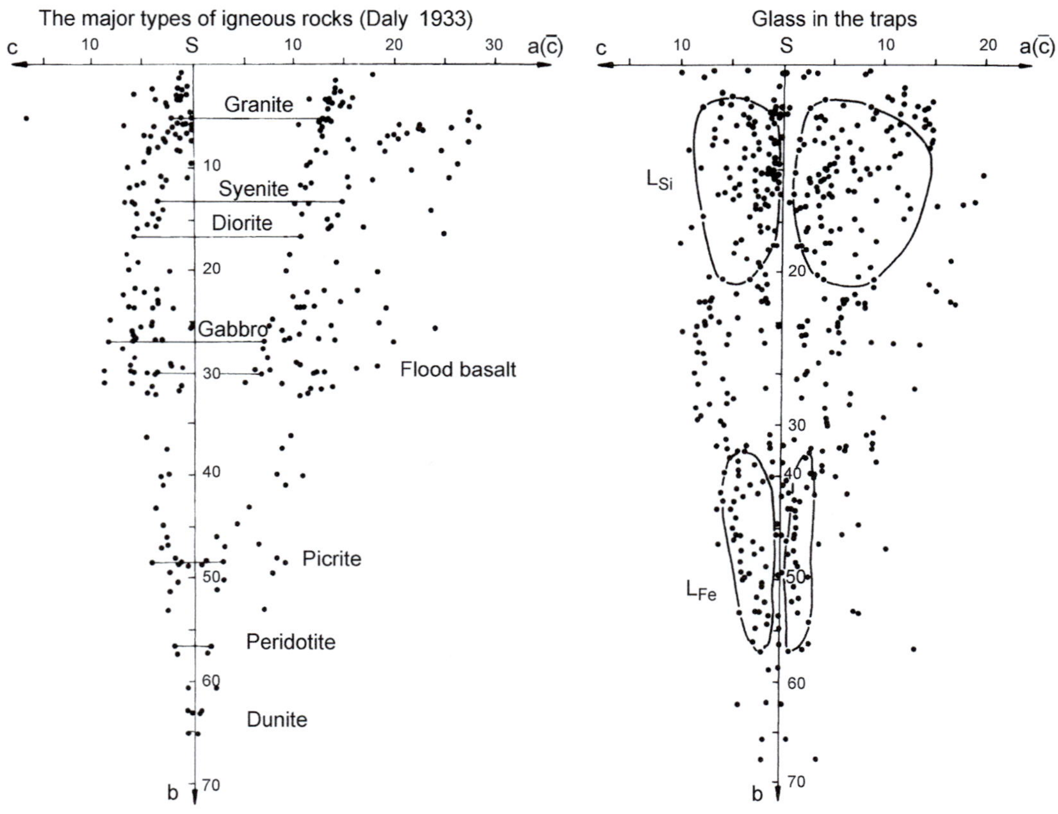

Fig. 5.5 Comparative petrochemical diagram by Zavaritsky (1947) for main types of igneous rocks by Daly (1933) (*left*) and glasses from traps of the Siberian Platform. The fields show figurative points of mafic (L_{Fe}) and sialic (L_{Si}) immiscible glasses (Ryabov 1989a, b, c)

according to Daly (1933) is given at the left. Glass composition points form a tail area completely analogous to that of crystalline rocks that allows the assumption that different types of rocks may be formed as a result of immiscible differentiation of an initial mafic melt. The variety in liquid compositions can be significantly increased if gravitational separation of early crystalline phases that is evident from partly crystallized glasses is considered.

Ore globules found in immiscible and rock-forming glasses in the Khungtukun and Maymechinsky Intrusions are of great ore-genetic importance. It is strong evidence for the existence of ore–silicate immiscibility in traps at the magmatic stage, since only phases occurring in glasses (as opposed to crystallized rocks) provide objective information on phenomena that took place in melts during the pre-crystallization period, whereas the known occurrence of ore phases in crystallized rocks do not rule out different interpretations for their formation. Despite the importance of ore phases in glass, at the present, no more than ten publications are available in the geological literature. Sulfide globules in the interstitial glass from Hawaii Island described by Skinner and Peck (1969) can be considered as one example. Sulfur content in the melt was 0.038 wt%, and sulfide liquid appeared only during significant crystallization of a basalt melt with relative oversaturation of sulfur at a temperature no less than 1,065°C. When studying mineralization associated with the Khungtukun Intrusion (Ryabov 1989b) and considering known experimental investigations on the solubility of different chemical elements in salt and metallic melts (Vol 1962), ideas about immiscibility in ore liquids have been demonstrated. Immiscible ore liquids have been established to form emulsions, emulsions inside other emulsions, and layered drop-like disseminations involving α-Fe, native copper, pyrrhotite, chalcopyrite, wüstite, and magnetite.

Immiscible liquids appear in a primarily homogenous melt at a temperature higher than the liquidus or after stages of crystallization (Marakushev 1981) with metastable immiscibility in the residual melt (Philpotts 1982; Manankov and Sharapov 1985). Later, immiscibility can appear in low-temperature salt melts including sulfide liquids (Zharikov et al. 1986; Naumov et al. 1986; Godlevsky 1959). A review of literature sources and data on traps indicates that the temperature range at the onset of liquation is practically unlimited and is governed by the physical–chemical state of a system (Ryabov 1989b).

A number of factors can stimulate the onset of melt separation into immiscible liquids, most important of which are: falling temperature, increasing concentration of immiscibility agents, saturation of a melt with fluids or, inversely, a sharp reduction in solubility and dissociation of volatiles, or change in the fugacity of oxygen or sulfur. Immiscible differentiation can proceed in dry ferriferous melts and liquids enriched in titanium and chromium, but layering may proceed more actively and more frequently when volatiles are involved into the process (fluid–magmatic differentiation). The presence of volatiles in a melt promotes its depolymerization, appearance of chemical and physical inhomogeneities, decrease in crystallization temperature of a melt and its viscosity and enhances activity and migration ability of chemical elements (Ryabov 1989b).

Comparing the diversity of immiscibility products in traps, it can be seen that this mechanism takes place and acts within a wide range of physical–chemical conditions at the magmatic and postmagmatic stages and has a multistage character. Immiscibility differentiation precedes crystallization differentiation, accompanies it, and follows it. The existence of stable pre-crystallization immiscibility in interstitial residual liquids has been proved by authors (Pankov 1984; Tikhonenkov et al. 1985; Ryabov et al. 1985a).

Temperature conditions of immiscible separation can be judged from thermometric experiments and thermodynamic calculations and may also be gained by a comparison of natural paragenesis of crystallized globules, which have a preserved emulsion pattern of immiscible glasses, with data from known fusibility diagrams. Homogenization of anomalous inclusions preserved in pyroxenes was performed by N.F. Krasov (Krasov et al. 1985) at the Institute of Geology and Geophysics SB AS SSSR. In this experiment, a liquid was separated into ore and silicate melts, which coexisted at a temperature range from 1,200 to 960°C (solidification temperature of the inclusion) (Ryabov et al. 1985a). Thermodynamic calculations for ore phases allowed us to determine that cohenite was formed at a temperature above 900–750°C, and paragenesis of α-Fe+ graphite was formed at a temperature lower than 900°C. Comparison between eutectoid intergrowths of minerals occurring in ores of the Khungtukun Intrusion with known eutectics of these minerals in appropriate artificial systems allowed us to assume the following crystallization: α-Fe + FeO + Fe_2SiO_4 at 1,177°C, FeO + Fe_2SiO_4 at 1,170°C, α-Fe + FeO at 1,140–906°C, FeO + Fe_2SiO_4 + $FeAl_2O_4$ at 1,148°C, α-Fe + Fe_3C at 1,147°C, FeO + FeS at 950°C, and α-Fe + FeO + Fe_2SiO_4 + FeS at 920°C (Ryabov 1989b). In other words, all eutectoid intergrowths were formed within a temperature range of 1,177–906°C, that is, under realistic conditions of a magmatic process in recent magmatic chambers. Lastly, the quaternary paragenesis LSi + LFe + Fe_2SiO_4 + SiO_2 occurring in the Khungtukun Intrusion that, according to the experimental data, was formed at a temperature 1,163°C (Freestone and Powell 1983) is a convenient reference.

Fluid inclusion studies of immiscibility phenomena in inclusions in olivine from the Ust-Khan'insky Intrusion have shown that layering of liquids proceeded at a temperature somewhat above 1,140–1,170°C (Pankov 1984). In the matrix of the Putorana plateau basalts, immiscibility occurred at temperatures lower than 1,200–1,190°C (Tikhonenkov et al. 1985).

The lower temperature bound of magmatogenic immiscibility of a sulfide liquid appears to be confined by the crystallization temperature of sulfides, estimated as being within 600–450°C, according to Godlevsky (1959), Eliseev (1959), and Smirnov (1966).

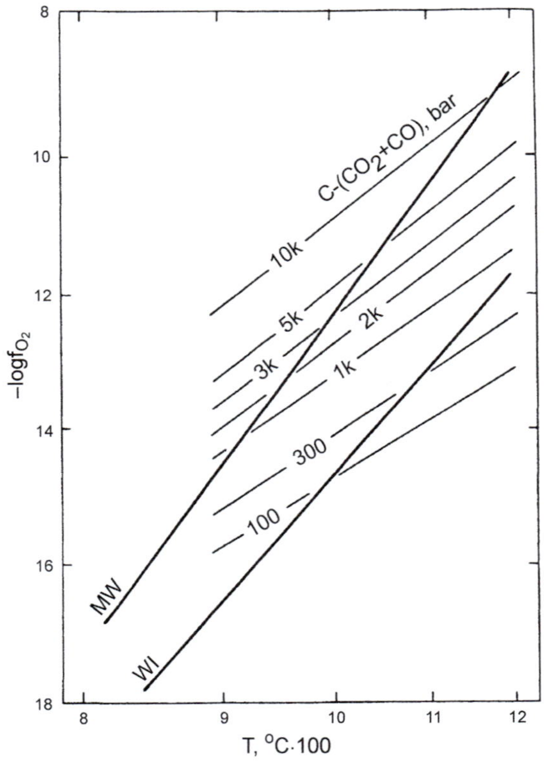

Fig. 5.6 Plot of log fO_2 versus T in the system C – (CO_2 + CO) showing the stability field for metallic iron (Sato and Valenza 1980)

In connection with crystallization of ore phases, it should be noted that high fusion temperatures of a number of native minerals and carbides, and occasionally very highly reducing conditions during their generation in experiments, provide our current understanding of why they are regarded as being products of deep-seated hearths up to now (Oleinikov et al. 1985), although α-Fe and Fe_3C are known from zeolites and serpentines (Chukhrov 1960). Figure 5.6 shows that with increasing temperature from 1,000 to 1,200°C, α-Fe may be formed under hypabyssal conditions, with low oxygen fugacity and a highly reducing fluid composition. This is an experimental support for the fact that the minerals can be formed under chamber conditions.

5.4 Fluid Regime of Trap Magmatism

Crystallization of any magmatic melt is accompanied by separation of dissolved volatiles, and the gravitational effect promotes migration of volatiles upward through a magmatic body from the root parts of a magmatic column to its leading portion. Accordingly, it is reasonable that a zone in the cross section of a sub-sheeted intrusion, mostly enriched in volatiles, should be located under a screen of upper quenched crusts, that is, UZ. As crystallization of a melt proceeds, volatile components are concentrated in the UZ and form fluid-saturated areas or entire horizons, depending on their volume. In effusive traps, they form "gas bubbles" as rounded, lens-like, or finger-shaped accumulations characterized by a well-crystallized form, enriched in oxide-ore phases and abundant amygdales. The sizes of these fluid-saturated areas in effusives traps vary from several centimeters to large lenses and whole horizons several meters thick at the roof of flows. Different flows have varying thicknesses of amygdaloidal zones and amygdale saturation. In some places, the amygdale content reaches 40–60%, and the rock takes a blistered, holystone-like pattern. Despite the opinion that trap magmas are dry, such large quantities of volatiles are evidence that they can be accumulated at the roof of basalt flows forming zones saturated with fluids and form an independent bubble phase. This also takes place in intrusions, but in view of other physical–chemical conditions, fluidized melts crystallize as schlieren-like bodies of mafic pegmatoid varying in size and thickness and continuous caps of pegmatoid at the roof of intrusions. Sharp contacts with country rocks are distinctive features of pegmatoid bodies.

The involvement of volatiles during the development of pegmatoids is undoubted. A good consensus exists among geologists that slightly differentiated traps are depleted in volatiles, and schlieren in the traps are either absent or they scattered as bodies of pegmatoid. At the same time, it is considered that magma of the Noril'sk Intrusions was enriched in volatiles, which manifested as pegmatoids and the thick aureole of metasomatite (Zolotukhin et al. 1975).

Two main types of pegmatoids have been recognized in the Noril'sk Intrusions, some of which are located in the upper part of a layered rock series as coarse-grained prismatically granular gabbrodolerite, and others are restricted to the UZ and LZ horizons providing a variety of evenly and unevenly granular ataxitic pegmatoid. The composition of the former corresponds to rocks of the intrusions, whereas the latter have an anomalous composition. When pegmatoids of the first type formed, the only local volatile resource was elements dissolved in the adjoining melt, whereas pegmatoids of the second type, taxites, and related rocks were formed at intense inflows of fluids from deep-seated parts of a magmatic column.

Merging upper and lower taxites traced the flanks of the Noril'sk-I, Noril'sk-II, and Upper Talnakh Intrusions, combining evenly and unevenly grained rocks (taxites and leucogabbros) in the same horizon, with a gradual transition between them—evidence of the genetic affinity between these pegmatoid rocks (Zolotukhin 1964; Zolotukhin and Vasil'ev 1967; Ryabov 1969; Zolotukhin et al. 1975). Pegmatoids in contact zones are formed in opening cracks created by a penetrating magmatic wedge at the boundary

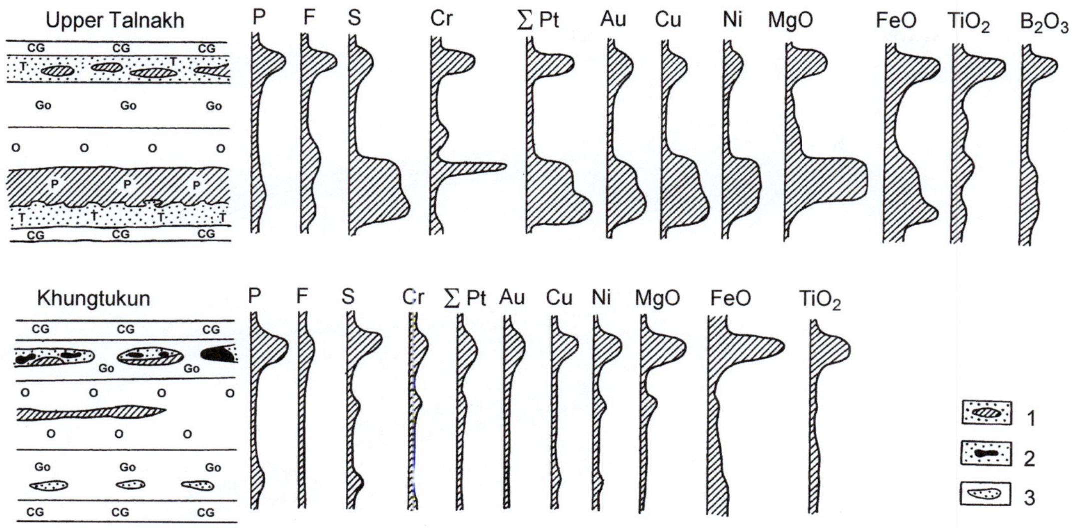

Fig. 5.7 Behavior of selected elements in intrusions with differing differentiation degree (Ryabov 1989b). 1 = contact gabbrodolerite, 2 = olivine-bearing gabbrodolerite, 3 = olivine gabbrodolerite, 4 = picritic and troctolitic gabbrodolerite, 5 = taxitic gabbrodolerite, 6 = melanocratic inclusions in pegmatoid, 7 = ore nodules, and 8 = pegmatoid schlieren

with contact gabbrodolerites (Zolotukhin 1964). It should be noted that if upper contact gabbrodolerites served as a screen in the UZ beneath which fluids accumulated and migrated, then the horizon of picritic gabbrodolerites or a stream of unconsolidated, heterogeneous picrite-like melt played the role of screen in the LZ.

As there is a lack of volatiles dissolved in silicate melt, significant alteration and active large-scale fluid–magmatic differentiation with subsequent petrogenesis and ore formation may be possible only in certain parts, zones, or horizons of magmatic bodies where volatiles and oregenic elements separated from a crystallizing melt and/or these volatiles were represented by transmagmatic gas streams. Additional mobilization of volatiles can be implemented by a magmatic melt while it moved through sedimentary cover rocks.

Geochemical investigations of trap intrusions have shown that the crystallization of a magmatic melt in the UZ of different bodies resulted in an increase in Fe content in the residual melt, as well as an accumulation of a number of chemical elements including F, Cl, H_2S, CO_2, H_2, H_2O, CH_4, B_2O_3, P_2O_5, Cr_2O_3, TiO_2, nonferrous and noble metals, and others (Fig. 5.7). The accumulation of particular elements within an intrusion varies, and Fig. 5.7 shows that two concentration maxima exist for all listed elements in intrusions cross sections.

Considering that the Noril'sk Intrusions are enriched in volatiles, understanding the conditions of petrogenesis and ore formation in these zones elucidates the source and composition of fluids, fluid–magmatic interaction, and role of fluids in processes of redistribution of chemical elements, as well as petrogenesis and ore formation.

5.4.1 Paths of Migration, Sources, and Composition of Fluids in Ore-Bearing Intrusions

Melts that have formed layered ore-bearing intrusions of the Noril'sk type on the Siberian Platform are mostly enriched in volatiles. Migration paths of volatiles can be obtained from the principal scheme of the structure of a magmatic column referenced to the Talnakhsky ore junction (Fig. 5.8). The figure also shows a generalized cross section of sedimentary rocks for the Siberian Platform through which all basalt melts intrude. Both Upper Talnakh (UT) ore-bearing and barren Lower Talnakh (LT) Intrusions are shown in the figure. The distinctive feature of the Noril'sk ore junction is the stable association of ore-bearing and high-magnesium barren intrusions that are composed of picrite-like and troctolitic gabbrodolerite depleted in oregenic metals (Ryabov 1992b).

Crystallization of the melt in a hypabyssal chamber was accompanied by separation of volatiles, which in the gently dipping part of a magmatic body migrated upward (byanalogy with basalt sheets), and following accumulation gave birth to coarse-crystalline rocks such as prismatically granular gabbrodolerites and gabbrodiorites. Volatiles of near-basal parts of a magmatic column moved forward and up, forming fluid conductors in the consolidating body, and elevated the pressure at the column head. As volatiles moved upward through the magmatic column, their volume increased. Fluid channels were limited by screening horizons, formed by contact gabbrodolerite near the top and by an olivine mixture in the picritic gabbrodolerite horizon at the base; the olivine

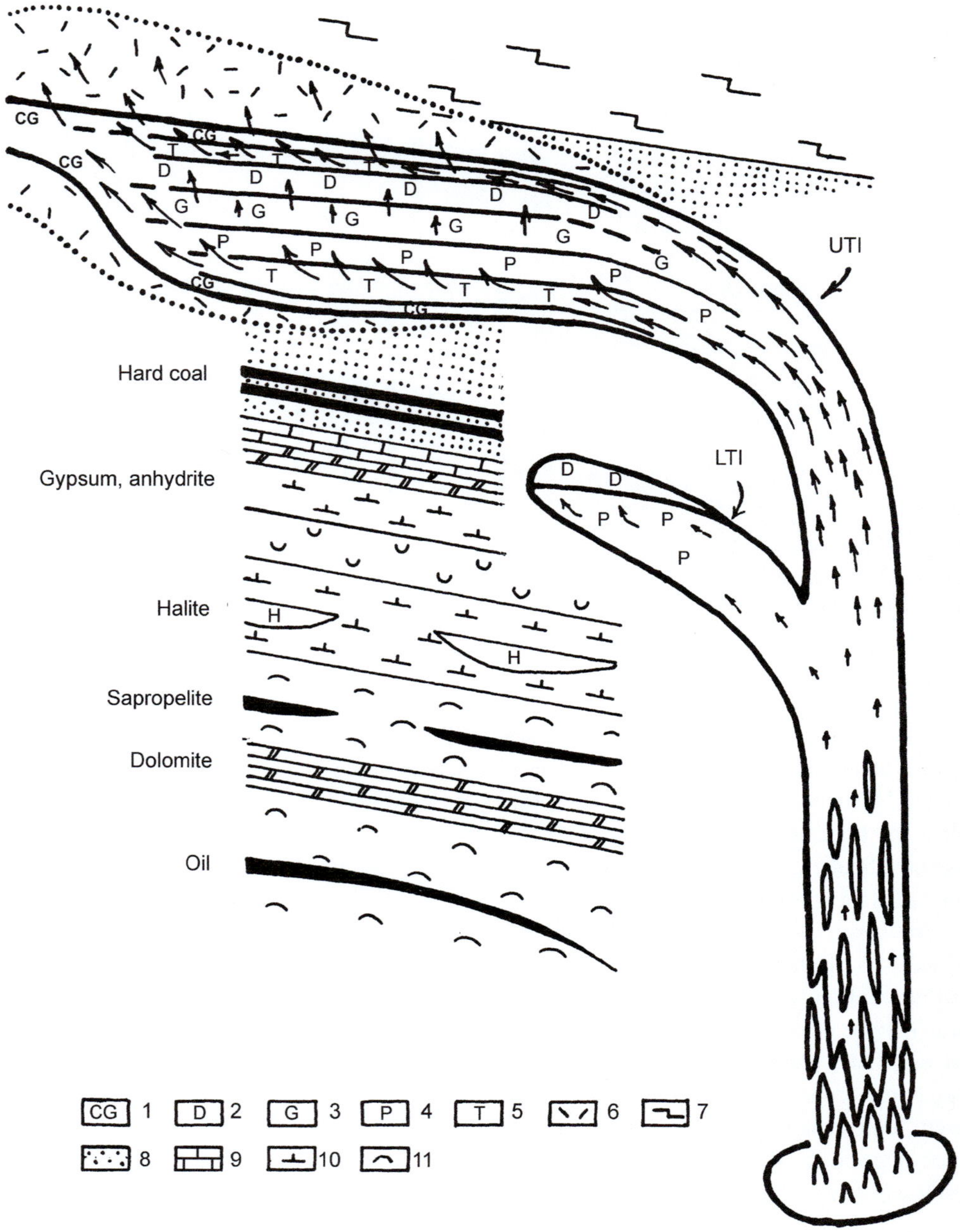

Fig. 5.8 Magmatic column model based on the Talnakhsky ore junction. 1 = contact facies, 2–3 = mafic layer: gabbrodiorites (2) and olivine-free, olivine-bearing, and olivine gabbrodolerites (3), 4 ultramafic layer, 5 = pegmatoids, 6 = metasomatites (intrusion aureole), 7 = effusive traps, 8–11 = country rocks: sandstones (8), limestones (9), marls and argillites (10), clay and bituminous limestones and dolomites (11). *UTI* Upper Talnakh Intrusion, *LTI* Lower Talnakh Intrusion

mixture in the picritic gabbrodolerite horizon was already becoming viscous. In the leading part of the intrusion, the lower fluid channel cut through the whole intrusion rising from its base to the roof and terminates as a confined "explosion" in the form of magmatogenic, metasomatic, and ore breccias.

Taking into account different sources of volatiles, two main types of pegmatoids can be recognized here (Ryabov 1999a). The first type was formed at the expense of fluids dissolved in a melt, and as the melt cooled and crystallized in a hypabyssal chamber, the pegmatoid migrated into the upper part. These pegmatoids are located in the upper part of a layered series of rocks and are represented by prismatically granular gabbrodolerite and gabbropegmatite with or without quartz and/or micrographophyre. They are known from a number of trap intrusions. The second type of pegmatoids are leucocratic and olivine gabbro, taxitic gabbrodolerite, and magmatogenic breccia, which commonly occur in layered ore-bearing intrusions (Zolotukhin 1997; Distler and Kunilov 1994; Ryabov 1992b, 1999a). This type of pegmatoid was formed after volatiles separated from a melt at the rear of a magmatic column and mobilized from nearby sedimentary rocks. These are specific transmagmatic fluids, whose nature has been long discussed in the geological literature (Korzhinsky et al. 1987; Marakushev and Perchyuk 1971; Zolotukhin 1997). The appearance of these fluids in intrusion chambers retarded formation of the layered series of rocks, and the pegmatoids associated with them cut gabbrodolerites without evidence of quenching. The potential ability of transmagmatic fluids that had "washed" a magmatic melt and pegmatoids formed with the involvement of these fluids is obviously significantly higher than those in chambers.

Erratic bodies of leucogabbro in the intrusion are an unusual type of pegmatoid, and different researchers have attempted to shed light on their genesis. Likhachev (1965) considers the bodies to be an indicator of earlier formation of leucogabbro as compared to the layered series of rocks. Oleinikov (1979) regards them as xenoliths and considers that they may be evidence of the anorthositic trend for crystallization in an intermediate hearth, and Tarasov (1976) consider them products of the magmatic replacement, gabbro creation, or metasomatic leucocratization. The nature of these leucocratic pegmatoid bodies is believed to resemble gas bubbles in lava sheets; that is, they are areas of a fluidized melt not immiscible with a matrix melt depleted in volatiles.

Involvement of transmagmatic fluids in petrogenesis is manifest in the wide variety of evenly grained and ataxitic pegmatoids and unusual magmatogenic breccias. Pegmatoids of the UZ are typically composed of leucogabbro and taxitic gabbrodolerite with rare poor sulfide dissemination, picritic gabbrodolerite and magmatogenic breccia (Ryabov 1984a; Sluzhenikin et al. 1994). Accumulations of chrome spinel in pegmatoids of the UZ form rounded and irregular, subhorizontally elongated lenses in rocks free from chromite and resemble gas bubbles. Pegmatoids of the LZ are commonly represented by taxitic gabbrodolerite enriched in sulfides; in essence these pegmatoids are ore pegmatites. As a rule, chrome spinel is absent in these rocks. Downward through the intrusion in the near-basal part, a common sequence of rocks can be observed: picritic gabbrodolerite, taxitized picrite, magmatogenic breccia, melanocratic taxitic gabbrodolerite, taxitic gabbrodolerite, leucogabbro, and contact gabbrodolerite.

Magmatogenic breccia manifests most distinctly in the Kharaelakhsky branch of the Upper Talnakh Intrusion where a fluid channel cuts through a layered series of rocks that rise from the base to the roof of the magmatic body. The effect of volatiles on rocks that have variable composition is traced by the character of alteration of fragments of gabbrodolerite in the central parts, where the initial substrate is preserved in some places (Ryabov 1992b). Fragments of picritic gabbrodolerite with droplet-like sulfides are banded by granulated olivine or are completely transformed into olivinites. Olivine-poor gabbrodolerite is replaced by clinopyroxenite. Both types of fragments are cemented by leucogabbro. Green spinel and diopside enriched in Al_2O_3 occur in variable quantities. Mineralogical–geochemical studies leave no doubt that olivinites bear no relation to cumulates or to depth restites brought up from a magmatic hearth. The olivine crystals were formed as a result of recrystallization of early generations of the mineral under the action of volatiles (Ryabov 1992b).

The sources of volatiles and their volume in magmatogenic and ore-forming systems have always posed a problem in concepts of fluid–magmatic differentiation and ore formation. Current understanding has established that one of the principal exceptions against the petrogenetic and ore-forming role of sulfides is the low solubility of volatiles in melts that cannot explain the vast scope of ore accumulations. However, the problem is not insurmountable assuming that fluidized melts are liquids and that they are extremely saturated with volatiles and contain them as a phase in equilibrium with the melt. It can also be assumed that fluids dissolved in a melt contained gases in excess of their solubility, which were mobilized by the melt from country rocks (Fig. 5.8). These volatiles played an ore-generating role and were agents of liquation, extraction of metals, and their transfer and accumulation.

Parent basalt magmas and fluidized magmas that had actively interacted with country rocks contain variable volatile content. Following extrusion through the sedimentary cover, fluid-saturated magma selectively extracted, mobilized, and activated components of country rocks including volatiles, which elevated the pressure of volatiles in a magmatic system and expanded the phase composition of fluids. Taking into account the composition of country rocks, it can be assumed with confidence that magma rising from deep-seated hearths might well be enriched in hydrogen and hydrocarbons, sulfur, carbon monoxide and dioxide, chlorine, water, and other volatiles.

Composition of Fluids in Traps: Current understanding indicates the composition of fluids including fS_2 and fO_2 plays a defining role in ore mineralization (Naldrett 1969). Judging from the available data, a magmatic melt that formed traps contained a wide variety of volatiles of varying composition and concentration (Ryabov 1989b). The wide variation in fO_2 and fS_2 is indirect evidence for the wide variety of mineral associations that occur in traps. Undoubtedly, formation of α-Fe ores containing cohenite proceeded under highly reducing conditions (Ryabov et al. 1985c), and for crystallization of sulfide ores, the oxidation potential was bound to be low with the melt featuring a reducing nature (Genkin et al. 1981). An increase in the ability of fluids to oxidize can be traced from near-basal parts of a magmatic body to its leading parts (Ryabov 1984a, b, 1986).

Examining the gas chromatography data and theoretical arguments, it is certain that the mafic mass of volatiles primarily had a highly reducing composition (Marakushev and Perchyuk 1974; Nikol'sky 1978) that was gradually oxidized in the course of migration within magmatic columns. Thermodynamic calculations made by Pavlov (Ryabov et al. 1985c) have shown that gases containing CH_4, CO, H_2, and H_2S were the most probable agents during the formation of α-Fe ores (Ryabov et al. 1985c). A similar set of volatiles has been determined analytically by Shchyukolyukov in mineralized rocks from Mt. Ozernaya (Oleinikov et al. 1985). Judging from the calculations, the thermal dissociation of hydrocarbons at 630–700°C played a crucial role in reduction reactions of iron to a native state, and the rate of reaction sharply increases with increasing the temperature. Transmagmatic gases could be a source of reducing fluids and host mafic intrusions themselves. In rocks underlying the Khungtukun Intrusion, the extended horizon of sapropelitic coals is known. Their high gas saturation (up to 91% of volatiles) and instability to thermal effects allow the assumption that degassing of sapropelites at the contact with a magmatic melt could result in incorporation of volatiles into the magma. This assumption is supported by results of isotope analyses of carbon from sapropelites, as well as cohenite and graphite from α-Fe ores, corresponding to $d^{13}C_{PDB}$ values of −2.36, −2.12, −2.96, −2.36, and −2.42. The close resemblance of isotopic ratios is evidence of the possible involvement of hydrocarbons from sapropelitic coals in the mineralization process (Ryabov et al. 1985c).

In recent years, a number of different researchers have found there is evidence that hydrocarbons and hydrogen are present in products of trap magmatism of the Siberian Platform (Mukhina 1985; Oleinikov et al. 1985; Vorontsov et al. 1986; Kucher et al. 1986; Martikhaeva et al. 1986; Neruchev 1987; Starosel'tsev and Starosel'tsev 1987). The overall pollution of traps by unbound carbon had been established earlier by Lebedev (1957). Additionally, the frequent observations of anthraxolite in effusives and crystalline graphite, including a cubic octahedron habit at the contact of mafic dykes with tuffaceous rocks at the roof of the Noril'sk-I and Anakitsky Intrusions, and in ores of the Upper Talnakh and Khungtukun Intrusions, as well as traces of bitumen and oil in mines of the Upper Talnakh Intrusion and in effusives of the Noril'sk and Putorana Plateaus serve as an indirect evidence for the presence of carbon-bearing fluids in traps (Ivanov et al. 1971a, b; Zolotukhin and Ryabov 1977; Oleinikov et al. 1985; Ryabov et al. 1985c).

Study of the isotopic composition of carbon from high-temperature (500–750°C) fluid inclusions from the Upper Talnakh Intrusion has shown an average dC13 deficiency of 20.6% that allows one to assume that the source of the carbon was juvenile as opposed to the Khungtukun Intrusion (Kucher et al. 1986).

Many researchers of ore deposits on the Siberian Platform drew attention to the fact that halogens and halogen compounds may be of great importance in the formation of sulfide and magnetite ores (Pavlov 1961; Godlevsky 1967; Zolotukhin 1974; Kuz'min 1978; Pavlov 1979). It has long been known that fluorine and chlorine act as active agents of immiscibility (Grigor'ev 1935; Fisher 1950), strong extracting agents, and carriers of nonferrous and noble metals (Wicks and Block 1963).

Pavlov (1992) places emphasis on the fact that iron-bearing silicate melts enriched in halogenides of alkali metals gain the capacity to generate salt halogenide melts highly concentrated with oregenic components and enriched in volatiles, which after crystallization of a parent intrusion kept their ability to separation with subsequent migration to zones of ore deposition at the postmagmatic stage.

Zolotukhin (1971, 1974) suggested that probable carriers of oregenic metals for Noril'sk ores could be concentrated solutions of the (K, Na)2SxnMeS type and soluble compounds of the (Na, K)Me(F, Cl)3 type, where Me is represented by Cu, Ni, Fe, and others.

Investigations undertaken by Wicks and Block (1963) indicate the high volatility of halogenides within the temperature range. The melting temperature of halogenides of Ni, Co, Cu, and Cr is lower than melting temperatures of basalts at 200–1,000°C, and sublimation and decomposition of Pt halogenides take place at 200–600°C.

The involvement of chlorine in the formation of Noril'sk sulfide ores has been repeatedly emphasized by Genkin (1968). This is supported by the discovery of chlorinated djerfisherite (Genkin et al. 1969) and occurrence of apatite in sulfide ores described by Ryabov and Zolotukhin (1977). Droplet-like disseminations are often accompanied by a cap of reaction products including serpentine, chlorite, anhydrite, biotite, and carbonates (Ryabov and Zolotukhin 1977). The fact that reaction caps are formed with involvement of volatiles leaves no doubt in the minds of geologists. These compounds contain 0.25–1.2% Cl^-, up to 0.15 and 0.07% F^-. and B_2O_3, respectively (Nekrasov and Gorbachev 1978). Micro-inclusions of fluids with chlorides of Fe, Ni, Cu, and Al in sulfides of Talnakhsky ores, as well as the increased Cl- ion content from inclusions in a series of rocks from a barren trap to mineralized picritic and taxitic gabbrodolerites, are evidence of fluid composition (Goryainov and Andreeva 1972; Goryainov et al. 1973; Aplonov and Moskalyuk 1978). The discovery of the ore phase $Pd_4Bi_5Cl_3$ (Karpenkov et al. 1981) is evidence of the involvement of chlorine in Pt mineralization. Distinct evidence of the involvement of halogens in ore-forming processes is offered by the halite-magnetite ore type in the Angaro–Ilimsky magnetite deposits.

Currently, the active role of chlorine-bearing fluids is used by geologists to explain the genesis of platinum metal deposits of the Bushweld and Stillwater Complexes (Boudreau and McCallum 1985; Boudreau et al. 1986; Mathez et al. 1985). Biotite and apatite from mineralized rocks enriched in chlorine have afforded ground for these ideas.

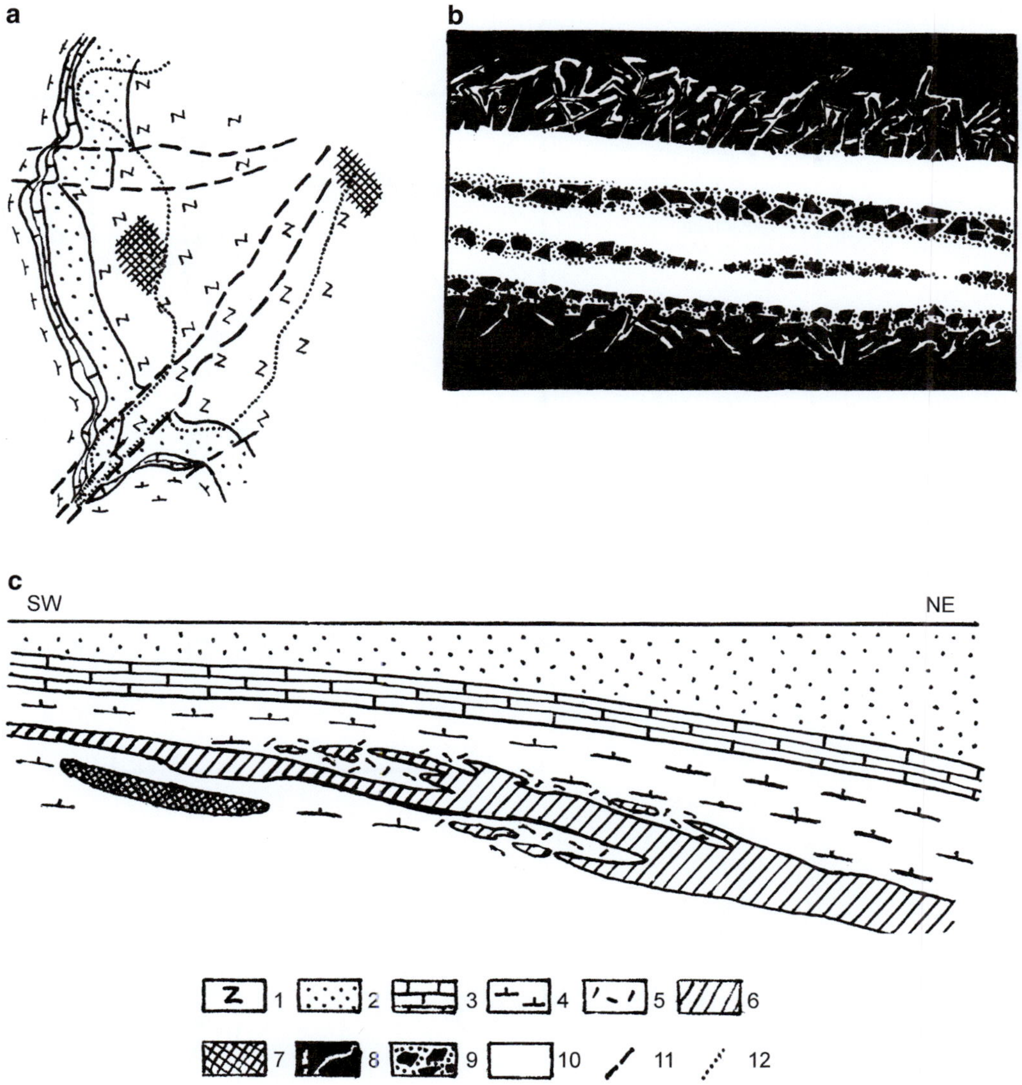

Fig. 5.9 Location of halite accumulation within the boundaries of the Upper Talnakh Intrusion (**a**), structure of the western salt-bearing accumulation (**b**), and its location in cross section (**c**). 1 = basalts, 2 = terrigenous rocks, 3 = carbonate deposits, 4 = anhydrite–marl deposits, 5 = metasomatites, 6 = intrusion gabbrodolerites, 7 = ore-bearing strata, 8 = halite veinlets in marls, 9 = breccia-like salt, 10 = homogenous salt, 11 = tectonic offsets, 12 = intrusion boundary

Apatite, biotite, and amphibole in traps and vesuvianite, apophyllite, and fluorite in addition to those in metasomatically altered rocks are carriers of fluorine and chlorine (Ryabov and Zolotukhin 1977; Ryabov 1989b). Taking into account peculiarities in the distribution of these minerals in sequences of ore-bearing intrusions, the upper and lower zones of halogenide accumulation are recognized, which coincide with horizons of ore-bearing rocks and frame the last. The content of halogen-bearing minerals in the UZ sometimes reaches several percent.

It should also be noted that deposits of rock salts not specific for the northwestern part of the Siberian Platform were discovered by the authors and later by Noril'sk geologists in the regions near the Mikchanda and Imangda Rivers (Ryabov and Sokolova 1970). The location of halite deposits within the Upper Talnakh Intrusion aureole (Fig. 5.9) are undoubtedly of interest with respect to the possible involvement of Na and Cl in the processes of petrogenesis and as a source for thick aureoles of albitites around the intrusion, as well as chlorine as an agent for ore formation.

From results of studies of noble gas isotopes (Fig. 5.10), fluids related to the Upper Talnakh ore-bearing intrusion had insignificant mantle helium and a major proportion of atmospheric argon (Neruchev and Prasolov 1995). This is an argument in favor of a core source for the mafic mass of fluids, that is, for mobilization of volatiles from sedimentary rock strata.

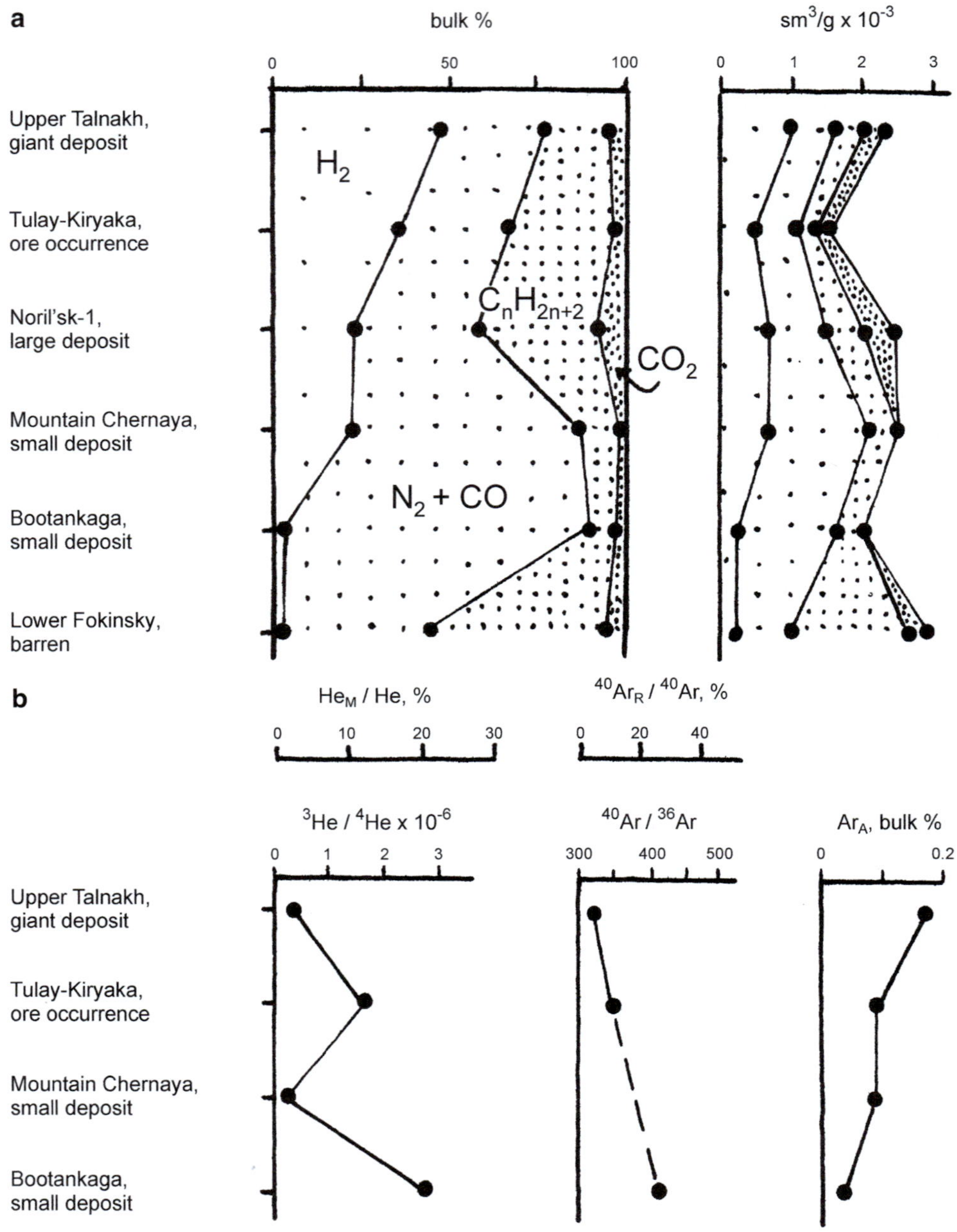

Fig. 5.10 Composition of gas micro-inclusions in minerals and closed pores in gabbrodolerites from trap intrusions with varying ore saturation (**a**) and isotope ratio of noble gases in the intrusions (**b**) (Neruchev and Prasolov 1995). He_M mantle helium, Ar_R radiogenic argon, Ar_A atmospheric argon

The composition of a gas and its reactivity when leaking through a melt change with the evolution of physical–chemical parameters of the magmatic system (Gorbachev 1989; Nikol'sky 1987). During the process of migration, volatiles selectively extract oregenic elements from a melt forming areas of melt (and later rocks) depleted in these elements in the rear of a magmatic column and accumulating ores at the head of the column.

5.4.2 Fluid–Magma Differentiation in Traps

The presence of volatiles affects the evolution of a melt to a considerable degree by promoting fluid–magma differentiation (Marakushev et al. 1982; Zharikov 1976), metamagmatism (Korzhinsky 1973), stimulating ferritization and sulfurization (Pavlov 1979, 1983; Almukhamedov and Medvedev 1982), and reducing crystallization temperature (Zavaritsky and Sobolev 1961). In addition to catalytic involvement in various reactions, fluids, when interacting with a melt, play a significant role in the extraction and transfer of different chemical components and determination of their form. Consider some consequences of fluid–magmatic differentiation in traps outlined below.

Limited solubility of volatiles in mafic melts causes separation of fluid-saturated melts from "dry" melts following the pattern of immiscible liquids (Lebedev 1957; Ryabov 1989b). In this way, gas bubbles, that is, areas with plentiful amygdales in basalts, and formation of metadiorites in sheets of picritic basalts can be

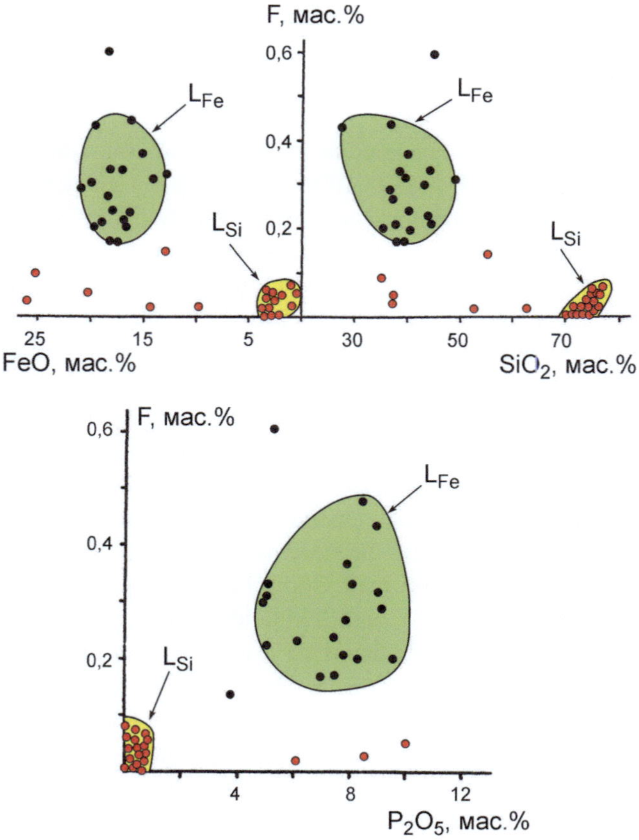

Fig. 5.11 Partitioning of fluorine and phosphorus between mafic (L_{Fe}) and sialic (L_{Si}) liquids in coexisting immiscible glasses (Ryabov 1989c)

considered fluid-saturated liquids separated from "dry" melts. This is similar for schlieren of gabbropegmatoids in poikilo-ophitic gabbrodolerites in a number of trap intrusions (Lebedev 1957; Masaitis 1958; Ryabov 1989b).

Considering the tendency to liquid immiscibility in ferrous melts (Roedder 1951, 1979), especially with high concentrations of fluorine and phosphorus (Grigor'ev 1935; Fisher 1950), the wide distribution of fluid–magma immiscible differentiation in trap intrusions enriched in volatiles may be assumed in the UZ of trap intrusions (Ryabov 1989b). Fluorine and phosphorus are active agents of liquid immiscibility, and an increase in their concentrations stimulates melt layering into immiscible liquids. Observation of "glass-in-glass" and ore globules in glass serve as conclusive evidence of the existence of silicate–silicate and ore–silicate (silicate–salt) liquid immiscibility (Rock Indication 226).

Coexisting immiscible glasses have a pyroxene-normative or ferrogabbroic mafic composition or a quartz–feldspar or granophyric sialic composition (Fig. 5.11). The chemical compositions of homogenous and partly crystallized liquation glasses (see Fig. 5.5) cover the whole variety of igneous rocks according to Daly (1933), highlighting the importance of the fluid–magmatic immiscible differentiation mechanism. With reference to traps, mafic–ultramafic and ferrogabbroic separation of a melt has been supported experimentally.

Basalt and trachybasalt melts are potentially inclined to liquid immiscibility. This manifests, in particular, as "pea-like" textures (poikilo-ophitic textures) that are widely developed in these rocks. The "peas" and the matrix represented by clinopyroxene and tholeiitic glass (or palagonite) are products of immiscible separation of a basalt melt into coexisting pyroxene-normative and quartz–feldspar liquids (Ryabov 1989b). The "pea-like" texture is characteristic for melts enriched in volatiles and fluids in the melts are agents of immiscibility. Increasing volatile content upward the Mt. Pegmatitovaya Intrusion reflects an increase in apatite and P_2O_5 and in "pea" diameter (Fig. 3.37).

In immiscible glasses, the accumulation of volatiles is distinctly observed in silicate liquids after they have been separated, forming a melt oversaturated with fluids. This can be seen in Fig. 5.11 where it is obvious that P_2O_5 and F are concentrated in mafic liquids—LFe (up to 9.58 and 0.6 wt%, respectively). Due to the high content of volatiles, the melt is obviously oversaturated with F that leads to the formation of ore liquid as well as separation of an individual gas phase that is in equilibrium with the fluidized melt. It may be assumed that this is the process by which free fluids occur and are then additionally enriched in volatile components; following interacting with country rocks, these then extract oregenic metals from a melt and transfer them to places of accumulation.

An important consequence of fluid–magma differentiation is that the volatile phase forms as a result of separation of silicate magma, is then concentrated in ultramafic liquid, and then forms a fluidized melt potentially inclined to the extraction and transfer of various chemical elements. This oversaturation of ultramafic liquid is supported by the wide occurrences of intersheet fluid–magma differentiation widely developed in picritic lavas and the confinement of droplet-like sulfide disseminations to picritic gabbrodolerite in the Noril'sk Intrusion.

An example of the petrogenetic role of immiscible differentiation in mafic rocks of the Siberian Platform is the association of rocks contrasting in composition in the UZ such as granophyres and ferrogabbros in the Alamdzhakh, Anakitsky, Middel Dudinsky, and other intrusions. Crystallization of mafic melt is considered to proceed according to the Skaergaard scheme that trap intrusions follow (Fig. 5.1), and at the final stage of crystallization, it sharply changes direction to either increase or decrease SiO_2, accompanied by an insignificant increase in fractionation coefficient. The division of the variation line in two opposing directions at the final stages of crystallization is caused by silicate–silicate separation of a mafic melt into two liquids approximate in composition to ferrogabbro and granophyre. In the

AFM diagram, this layering is manifest by branches, one branch points to accumulation of iron and second to that of alkalis.

Experimental investigations by MacBirney and Nakamura (1974) give a basis for this interpretation of the crystallization trend. These authors, with reference to the Skaergaard Intrusion, obtained two immiscible liquids compositionally approximate to granophyre and ferrogabbro under close to natural conditions. More support for the immiscible nature of these rocks in traps is that their composition is close to mafic and sialic glasses, as shown in Fig. 5.1.

The direct determination of homogenization temperatures in inclusions is estimated at 1,250–850°C, while silicate-–silicate immiscibility takes place (Pankov 1984; Ryabov et al. 1985a, c; Tikhonenkov et al. 1985). These data are completely comparable with known data on homogenization temperatures from rocks from other regions (Philpotts 1979, 1982; Krasov and Clocciatti 1979; Babansky et al. 1985).

The resemblance of the component composition of rocks from the UZ of the Upper Talnakh and Khungtukun Intrusions demonstrates the similarity of processes that took place (Ryabov et al. 1985c; Ryabov 1988a, 1989b). Fluid–magma differentiation that proceeded with active involvement of volatiles in the UZ of the Upper Talnakh Intrusion was similar to that in the UZ of the Khungtukun Intrusion, but significant differences in geochemical features of melts were responsible for other rock and mineralization features in each. In the UZ of the Khungtukun Intrusion, native iron and nickel–iron with isomorphic platinoids, copper, sulfides, and carbides are common, and the Upper Talnakh Intrusion features platinoids, native copper, sulfides, and chromites. It is important to note that in the Khungtukun Intrusion, nickel-rich phases such as kamacite and accumulations of platinoids up to 11 g/t (Oleinikov et al. 1997, 1999; Ryabov and Anoshin 1999) are revealed despite the low Clarke of nickel, which also supports mineralization mechanisms in the UZ of trap intrusions.

Fluid–magma differentiation in the UZ of the Noril'sk Intrusions is supported by the wide distribution of globular texture formed with the involvement of volatile-bearing minerals (see passports 96 and 107), as well as accumulations of chrome spinel up to 40–60 vol% and unusual chromite plagioclasites (Ryabov 1984a, b, 1989b). As we have seen, formation of platinum–chromite mineralization in the UZ of the Noril'sk Intrusions can be explained by any other mechanism except fluid–magma differentiation with fluid transfer of oregenic elements.

Picritic gabbrodolerite in the UZ of the Khungtukun and Upper Talnakh Intrusions is symptomatic. Picrites formed in association with α-Fe nodules in the Khungtukun Intrusion indicates they may be products of melt crystallization following the involvement of fluids, which depleted a silicate melt in iron and increased the magnesium content up to 21.03 wt% (Ryabov 1989b). A similar mechanism is proposed to explain upper picrites of the Noril'sk Intrusions, although an independent stream of picritoid melt immiscible with a mafic picritoid melt is not inconceivable as was seen in Mikchanda.

The increase in magnesium content in rocks from the UZ of intrusions, as well as the relationship between sulfides ore and mafic pegmatoids and wide variations in mafic pegmatoid composition (from feldspar leucogabbro-taxites to essentially olivine and pyroxene magmatogenic breccia fragments) can be explained as a result of the interaction with a magmatic melt (homogenous or heterogeneous). It has been established experimentally that during the sulfurization process, iron content in a melt decreases in proportion to excess sulfur, resulting in increased magnesium content in the residual liquid (Fig. 1.11). From this, it can be seen that mafic minerals with higher magnesium content will crystallize in systems with excess sulfur as compared to sulfur-free melt. Gorbachev (1989) also showed that an increase in S, F, and Cl in fluids leads to an increase in MgO content; that is, fluids serve as carriers of MgO and may cause an increase in the MgO content of the melt. Additionally, fluids promote the formation of feldspar-rich rocks that can be seen in a diopside–anorthitite–water diagram (Zavaritsky and Sobolev 1961), which shows that the ratio between diopside and anorthite in the anhydrous system is 60:40 and at $pH_2O = 5{,}000$ bar changes to 28:72. The principal scheme for the change in composition of a melt during formation of taxites is given by Korzhinsky (1973).

Thermodynamic calculations have shown that in chambers of the Noril'sk Intrusions, the redistribution formed spinel–forsterite paragenesis of magnesian scarns depleted in iron, as well as feldspar and/or clinopyroxene rock enriched in magnetite that may be the result of sulfurization and ferritization under local PT conditions (Ryabov and Pavlov 1984). Calculations allow us to explain not only the ore saturation of taxitic horizons but also the wide occurrence of olivines with low iron content (up to forsterites) in magmatogenic breccia of the Kharaelakh Intrusion without undergoing melt conditions.

Similar to the UZ of intrusions, formation of near-basal parts of intrusions was also controlled by a fluid regime. The greatest ore saturation of these rocks and directional variations in composition throughout the rocks can be explained by considering the high fluid saturation of the horizon of lower taxites. An increase in the extent of rock alteration is noted from the horizon of pegmatoids through picrites, taxitic picrites to taxites with relics of granulated olivine, which can be accounted for by the action of fluids. A similar depletion in chrome and silicate nickel is noted for the same transition. Inversely, away from taxites, chromium content increases through the sequence from the picritic horizon to gabbrodolerite of the upper part (a layer enriched in chromite) and biotitized rocks that underlie and overlie the taxitic horizon and form together with ores.

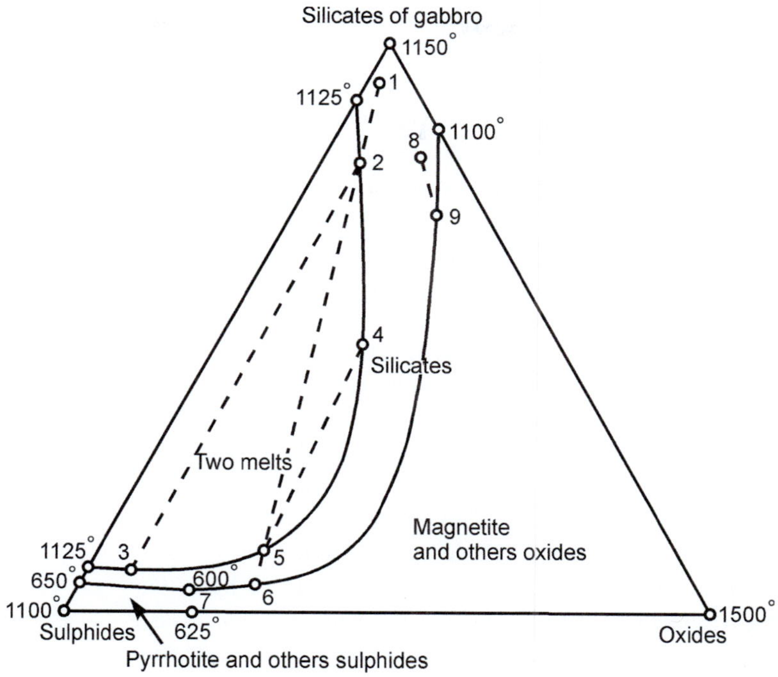

Fig. 5.12 Approximate fusion diagram of the pyrrhotite–magnetite–silicate system in gabbro as a ternary system independent of pressure (Smith 1968). Subsidiary *dashed lines* illustrate liquid composition variation during crystallization from points 1 and 8. *Arrows* on monovariant lines indicates decreasing temperature

5.5 Ore Formation in Traps

In order to explain the genesis of a deposit, it is necessary to establish the source for the ore, as well as to deduce the possible physical–chemical conditions for the transfer and deposition of ore components. Ryabov and Pavlov (1984), Ryabov et al. (1985c) attempted to solve this problem, to find an explanation for the genesis of α-Fe ores. This has determined that the source of iron was the mafic magma itself; immiscible separation of a melt in the UZ of the intrusion took place under reducing conditions with ore liquid segregated from the melt. The feasibility of this process is supported by geologic data, thermodynamic calculations, and is in compliance with known experimental data (Ryabov 1989b).

Pt–Cu–Ni sulfide ores and Pt low-sulfide deposits are confined to the Noril'sk layered intrusions. Many geologists have considered that the intimate relation between platinoids and sulfides in these deposits reveals the source of the ore accumulations.

5.5.1 Sulfide Deposits

The origin of sulfide ores in the Noril'sk deposits has been unresolved for many years and is widely discussed in geological literature. Among the proposed genetic concepts, two main groups can be distinguished: magmatic and post-magmatic (metasomatic and hydrothermal).

Immiscibility is the most popular magmatic explanation among geologists, which is presented in early works on the Noril'sk deposits by Kotul'sky (1948), Godlevsky (1959, 1967, 1968), and Urvantsev (1959). Based on this hypothesis, a primarily homogenous nickel-bearing magma was separated into immiscible sulfide and silicate liquids due to changes in physical–chemical conditions (temperature, partial crystallization, assimilation of country rocks, etc.). Differences in the specific weight of liquids led to gravitational accumulation of sulfides at near-basal parts of intrusions as disseminated and massive ores. Some geologists believe that sulfide–silicate layering proceeded in recent chambers, whereas others believe layering proceeded in intermediate or deep-seated hearths (Korovyakov et al. 1963; Likhachev 1978; Oleinikov 1979; Genkin et al. 1981).

Experimental investigations by Olshansky (1951) in the Fe–FeS–FeO–SiO_2 system give some support for the concept. Later, this system was documented in detail, and data on many other systems involving different ore components were used by researchers. The principal scheme of melt evolution in the process of crystallization with immiscible separation is shown on the model diagram: silicates gabbro–sulfides–oxides (Fig. 5.12), constructed with diagrams of two- and three-component systems (Smith 1968). The diagram gives some idea of the succession of melt crystallization with different primarily saturation with sulfides and variable Fe 3+/Fe 2+ ratio.

The geological evidence for the immiscible nature of Noril'sk ores is considered in a number of works by supporters of this concept. Droplet-like sulfide disseminations in gabbrodolerite, their layering, the stable ore–silicate paragenesis of minerals, and the zonal texture of ore are the main arguments for immiscibility (Genkin et al. 1981). Godlevsky (1961) conclusively states that syngenetic disseminated ores support the notion of immiscible magmatic separation, opposing all explanations based on hydrothermal processes. No one has argued against until now.

The metasomatic (infiltration autometasomatic) process has been developed over many years by Zolotukhin (1964, 1971) and other geologists (Gulin et al. 1968; Goryainov 1969; Gulin and Sukhov 1973; Tarasov 1976). In this hypothesis, the formation of sulfide deposits occurs due to the active interaction of deep-seated fluids containing already crystallized but very hot rocks along permeable zones. Metasomatic ore formation proceeds in compliance with acid–alkaline interaction in postmagmatic solutions according to Korzhinsky (Zolotukhin 1965).

The feasibility of the metasomatic origin of fluids has been supported experimentally by Kullerud and Yoder (1965) as a result of the reaction of sulfur with solid ultramafic rocks. Sulfide formation (sulfurization), regardless of its state in solution (as a gas state or in a condensed state), due to interaction with iron in silicates or with the melt has been shown to occur experimentally. This process may proceed in a wide range of PT conditions of magmatic and postmagmatic processes (Petrenko et al. 1974; Nekrasov and Gorbachev 1978; Almukhamedov and Medvedev 1982; Arutyunyan 1986).

Korzhinsky (1973) relates the formation of mineralized taxitic gabbrodolerites of the Noril'sk deposits with metamagmatic sulfurization with the active involvement of fluids. The author supposes that the same process was responsible for the formation of immiscible disseminated ores (Petrov et al. 1981). Natorkhin et al. (1977) follow the same concepts.

The concept of magmatogenic sulfurization as applied to the Noril'sk deposits is developed by Almukhamedov and Medvedev (1982). The authors consider that at depth melts are undersaturated with sulfur and their saturation increases in intermediate hearths at the expense of sulfur from sedimentary rocks. Saturation of a silicate melt with sulfur leads to the formation of sulfide liquid, which experiences gravitational separation in intrusion chambers. Magmatic sulfurization, according to the authors, is akin to ore–silicate immiscibility. On the assumption of low sulfur solubility in basalt melts, they assign the formation of an insignificant amount of sulfide liquid to immiscibility. In addition, the authors refer to experiments by Haughton et al. (1974), who "have shown that saturation of basaltic melt with sulfur through a gas phase is accompanied by formation of immiscible sulfide liquid." The figures in Almukhamedov and Medvedev (1982) also show that sulfide liquid in silicate melt forms thin emulsions; that is, experiments on magmatic sulfurization were completed by the appearance of an immiscibility pattern.

Characterizing magmatic sulfurization, Zolotukhin (1988) considers that "if sulfur is introduced into a melt by deep fluids or is derived from country rocks, then it will interact with ore elements of a silicate melt forming sulfides, but the process will no longer be immiscibility, but sulfurization." However, this process is not ore–silicate immiscibility (Zolotukhin and Malyuk 1985) even during interaction of sulfur with a melt.

As we have seen, it is beyond reason to contrast magmatic sulfurization to sulfide–silicate immiscibility. These concepts reflect a unified phenomenon: sulfurization, i.e., chemical reaction leading to sulfide formation, and immiscible separation as a result of interrelation between two liquids being formed: sulfide and silicate ones. The same phenomena occur in ferrous–silicate systems that form oxide ores and ore liquids (Pavlov 1979).

Chromium, titanium, iron, and other metals in excess concentrations can act as agents of immiscibility; under other favorable conditions, they can stimulate separation of fluids (Ryabov 1989b). The strongest agents of immiscibility are fluids of different compositions.

Liquid immiscibility also occurs when a melt is saturated with volatiles. This can occur after partial crystallization of a melt primarily undersaturated with sulfur, for example, tholeiites from the Alae crater (Skinner and Peck 1969) that featured additional inflow of sulfur-bearing volatiles from deep-seated parts of a crystallizing magmatic column. Enrichment of a melt with volatiles leads to separation of fluid-saturated melts from dry melts and to different phenomena inside liquids enriched in volatiles: fluid–magmatic or silicate–salt differentiation, immiscibility, magmatic ferritization, and sulfurization (sulfidization) described by Loewinson-Lessing (1935), Korzhinsky (1973), Zolotukhin (1964, 1988), Pavlov (1979), Almukhamedov and Medvedev (1982), and Marakushev et al. (1982). These phenomena reflect, in essence, a unified many-sided process of interaction between fluid and melt, and using one or other term is most likely connected with a desire to emphasize physical, chemical, genetic, and other aspects of a process.

Currently, in addition to fluid–magmatic differentiation, the concept of fluid–magmatic interaction is widely used. This interaction takes into consideration the dynamic nature of the interaction between fluids and a melt in an ore–magmatic system. As a result of fluid–magmatic interaction, immiscible separation may occur or a redistribution of chemical elements, which can also lead to separation of liquids. This concept may be applied, in particular, to the formation of the Talnakhsky ore junction and Lower Talnakh Intrusion depleted in oregenic elements and the Upper Talnakh ore-bearing intrusion.

Many questions on the formation of mafic pegmatoids and ore bodies may be precluded under the assumption that a layered melt being intruded into a recent chamber was enriched in volatiles. The volume of volatiles during consolidation increased, and the migration of volatiles proceeded along thermal fluid conductors (such as both the UZ and LZ). The large proportion of sulfur in the fluid leads to magmatic sulfurization that could take place anytime in the magmatic process: early in deep-seated hearths, in channels during melt migration, or in a hypabyssal chamber. Judging from disseminated ore textures, one may think the main mass of sulfides was formed as a result of sulfurization and sulfide–silicate immiscibility. The presence of sulfide ores in picritic gabbrodolerite of the Noril'sk Intrusions and the absence of sulfide segregations in comagmatic effusive picritic basalt, as well as the increase in ore saturation in near-basal parts, serve as an argument for the appearance of sulfides only in intrusive chambers or feeder conduits and an absence in deep-seated or intermediate hearths (Zolotukhin and Malyuk 1985). Inclusions of sulfide globules in plagioclase, pyroxene, and regenerated olivine of the Upper Talnakh Intrusion (Vortsepnev 1978), which crystallized under hypabyssal conditions, the association of nickel-rich ores with horizons of picritic gabbrodolerite (Smirnov 1966; Ivanov et al. 1971a, b), and nonequilibrium associations of ore and silicate phases bear witness to the chamber nature of ores. Droplet-like ores are formed as a result of adhesion, that is, agglomeration of a fine-grained emulsion into globules and drops instead of breaking up a stream of sulfide liquid into disseminations relatively uniformly distributed in the picrite horizon. Formation of disseminated ores in a recent chamber and their magmatic nature offer an explanation for all ore features, such as vertical zonality in mineralized rock composition in a deposit (Godlevsky 1971; Smirnov 1966; Natorkhin et al. 1977; Genkin et al. 1981).

The gravitational effect and low viscosity of a sulfide melt (Olshansky 1948; Arutyunyan 1977; Traskin et al. 1988) may lead to the formation of large ore accumulations and the migration of mobile ore magmas. Fluid inclusion studies of micro-inclusions (glass + ore phase + gas) from remelted minerals located near the contact of massive sulfide ores of the Upper Talnakh Intrusion indicated the onset of glass melting at 820–840°C and complete homogenization at 920–950°C (Zotov et al. 1977). In the opinion of Zotov et al. (1977), this supports the notion that massive ores were formed from sulfide melts.

It is reasonable to assume that as the intrusion consolidates and syngenetic magmatogenic ores are formed, fluids do not disappear completely and continue to be involved into postmagmatic processes, specifically the formation of metasomatic ore and silicate minerals and rocks. The active involvement of fluids at this stage manifests in the redistribution of chemical elements, appearance of the metasomatic zonality and metasomatic ores, which are not considered in this book.

The view that sulfur from sedimentary sulfates may be involved in ore-forming processes satisfactorily explains the origin of sulfur for the Noril'sk ores (Godlevsky 1967). It was noted that 2/3 of sulfur is removed from subcrustal hearths and 1/3 is assimilated from sedimentary gypsum–anhydrite rocks. The huge volume of sulfide in the Noril'sk deposits, as well as the increase in heavy sulfur isotope content in sulfides, may be explained by this premise (Godlevsky 1967; Grinenko 1967, 1987). Additionally, the assimilation of calcium sulfate should significantly increase the CaO content of the melt in ore-bearing intrusions compared to barren intrusions. This would also be expressed by appearance of certain high-calcic mineral phases. However, known petrochemistry data in the literature show there are no significant differences in the CaO contents in similar rock types, nor in ore-bearing and barren targets. This means that if sulfur was taken from sedimentary rocks, then calcium remained inactive. Besides, if sedimentary sulfates are used as a source of sulfur, there is a requirement for a mechanism to reduce of sulfur to sulfide. Geologists believe the reducing action of coal beds from rocks of the Permian–Carboniferous Tungussky series may provide a mechanism. However, it should be noted that the main mass of sulfide ores of the Noril'sk deposits are found in coal-free Devonian rocks, and, therefore, the role of coals as a reducing agent is questionable.

When the gas phase in rocks of the Noril'sk Region and Kola Peninsula was studied, it was established that hydrogen and nitrogen predominated in ore-bearing intrusions and methyl hydride in barren intrusions (Neruchev 1987, 1988; Neruchev and Prasolov 1995). Based on this data, the authors inferred that H_2 played an important role in ore-bearing intrusions such as the Upper Talnakh Intrusion and CH_4 in barren intrusions of the Lower Fokinsky type (see Fig. 5.10). It may be assumed that hydrocarbons primarily predominated in fluids of parent magmas of all the intrusions mentioned above and that proportions of gases obtained by mass-spectrometric analysis reflect the situation in magmatic chambers. In barren intrusions, the composition of a gas phase has been preserved without significant changes, and in ore-bearing intrusions, the gas phase experienced some transformation. After leaking through sulfate-bearing strata, fluids changed their composition according the exchange reaction and reaction of sulfide formation:

$$CaSO_4 + CH_4 = CaCO_3 + H_2S + H_2O$$

$$H_2S + Me = MeS + H_2$$

These reactions explain the specific hydrogen character of ore-bearing intrusions, the source of sulfur of sulfide deposits, and the origination depth of sulfide liquid, which is governed by the appearance of sulfates in country rocks. This also provides a satisfactory explanation to why the loss of calcium sulfate did not increase the CaO content that would be expected in the case of assimilation of anhydrite by magma; rather, it formed carbonate in situ in the sedimentary rocks. Lastly, water forms as a consequence of the reactions and acts as a solvent of halite deposits, an active catalyst for different reactions, and an important component of many acids (without which chlorine is inactive even with respect to iron) (Pavlov 1975; Ryabov 1997, 1999a).

Pt low-sulfide deposits in trap intrusions are restricted to pegmatoids of the UZ. Three ore types are recognized: Pt–chromite, Pt–chromite-free, and Pt–graphite (Fig. 4.31) (Ryabov 1992a, c, 1994). One characteristic feature of the ores is the high PGE/S ratio (g/t:wt%) varying from 3 to 805, averaging 50. The Pt + Pd concentration of 12–15 g/t is observed in rocks containing from 1.5 to 2.3 wt% Cr and from 0.19 to 0.25 wt% S. With increasing sulfide (to more than 7%) content, PGE content decreases; massive sulfide ores in the UZ contain no valuable components. Despite low sulfur concentrations in Pt low-sulfide deposits, their origin in the Noril'sk deposits, by analogy with foreign ones, is considered in the context of the sulfide model, although the high activity of H_2, HCl, HF, and H_2O has been noted (Sluzhenikin et al. 1994; Dodin et al. 1994).

It would be unlikely that such low sulfur content in fluids could produce such vast and efficient extraction in large volumes of melt and create deposits. Most likely, sulfur in addition to other volatiles (CH_4, N_2 + CO, and others) was simply part of the fluid, and sulfides served as collectors of noble metals. Halogens played the main extracting and transporting role.

The important role of halogens in ore formation has long been discussed in the literature with respect to Pt low-sulfide ores of the Stillwater and Bushveld Complexes (Boudreau et al. 1986). In the Siberian Platform, the active involvement of halogens had been established in the study of the Angaro–Ilimsky iron-ore deposits (Pavlov 1975) and magnetite lava in the Khantayskoe Lake region (Ryabov 1997). The involvement of halogens in ore formation in the Noril'sk deposits manifests in occurrences of F- and Cl-bearing minerals: apatite, amphibole, Cl-djerfisherite, apophyllite, iowaite, F-graphite, and a phase of platinum chloride. Chlorides consistently present in gas–liquid inclusions preserved in ore and nonmetalliferous minerals (Goryainov and Andreeva 1972). The highest concentrations of F (10 wt %) have been determined in graphite and Cl (5–7 wt%) in apatite and biotite (Fig. 4.19). The range in composition of biotite from Pt low-sulfide ores of the UZ covers the whole variety of mica compositions found in the intrusion and the aureole of altered rocks surrounding the intrusion. The distribution of halogens in biotite is evidence that when passing from Pt sulfide to Pt low-sulfide ores, the proportion of halogens in the fluid composition increases as the role of Cl increases and the role of F and S decreases. It should be noted that there is evidence of an increased role for As, Sb, Se, and Te judging from the mineral composition of Pt low-sulfide ores (Ryabov 1992a; Sluzhenikin et al. 1994).

The Pt graphite (Ryabov 1992c) or Pt–Fe metal type of ores associated with native iron and restricted to pegmatoids of the UZ of intrusions of the Dzhaltul–Khungtukunsky type are of a great genetic interest. Mafic minerals in these ores include: α-Fe (ferrite), cohenite, kamacite, graphite, hercynite, high-iron silicates, plagioclase, apatite, and sporadic troilite (Oleinikov et al. 1985; Ryabov 1989b).

Noble metals are concentrated in kamacite and reach up to 17.0 wt% Pt, 42.0 wt% Pd, 6.3 wt% Rh, 0.49 wt% Os, 12.2 wt% Au, and 2.32 wt% Ag (Kopylova and Oleinikov 1997). According to our data, the Pt + Pd + Rh content in bulk ore samples from the Khungtukun Intrusion reaches 9.20 g/t, and that from the Dzhaltul–Mt. Ozernaya Intrusion is 11.5 g/t. There is a direct correlation between the concentration of noble and nonferrous metals. Sulfur content in α-Fe ores varies from 0 to 0.3 wt%. In most of α-Fe samples, the sulfides content does not exceed several parts of a percent; more frequently, sulfides are absent. Noble metals are concentrated in ferrite as an isomorphic impurity (Kopylova and Oleinikov 1997; Ryabov and Anoshin 1999).

Based on eutectoid intergrowths of minerals, data on thermodynamic calculations, and experiments on homogenization of melted inclusions (gas + ore phase + silicate phases) in clinopyroxenes, it has been established that formation of platiniferrous ores of native iron proceeded as a result of immiscibility in a mafic melt in a hypabyssal chamber with the involvement of reducing fluids at 1,220–900°C (Nikol'sky 1987).

The composition of volatiles in α-Fe ores of the Mt. Ozernaya Intrusion is H_2 (75%), CH_4 (8%), N_2 + CO (15%), and CO_2 (2%) (Shchukolyukov et al. 1981). The carbon isotope composition ($\delta^{13}C_{PDB}$) in minerals from α-Fe ores of the Khungtukun Intrusion corresponds to ($\delta^{13}C_{PDB}$) (−1.35 to −2.42) in graphite and (−2.12 to −2.96) in cohenite and is very close to the isotope composition of carbon from sapropelites (−2.37), beds of which are located in sedimentary dolomites underlying the intrusion (Ryabov et al. 1985c). These data confirm the possibility of mobilization of volatiles from sedimentary rocks by a melt and for their fixation in different mineral forms. It can be also assumed that dominance of hydrogen in gas composition in α-Fe ores is conditioned by the linking hydrogen in crystalline phases of Fe_3C and graphite. These data support the mobilization of volatiles from enclosing rocks by a melt.

This type of ore is not only an alternative to the sulfide model but also points to the overestimation of the role of reducing fluids, as well as the role of H_3P and NH_3 in extraction, transfer, and accumulation of PGE in Pt-ore deposits. Geochemical investigations (Oleinikov et al. 1920) allow the assertion that fluid of the H-C-facies actively transfer not only Ni, Co, and Cu but also Ge, Sn, and Au and PGE, which are concentrated in a metallic phase.

5.6 The Role of Mineralogical–Geochemical Investigations in Solving the Problems of Petrogenesis and Ore Formation

Many researchers repeatedly point out that the main rock-forming minerals from mafic rocks can be very informative as indicators of the differentiation of a magmatic melt (Sobolev 1986; Wager and Brown 1968). These were widely covered in a number of our works (Ryabov 1974, 1978a, b, 1984b; Ryabov and Zolotukhin 1977, 1978), and, therefore, only two mineral forms (olivine and chromite) will be considered in detail in order to show how detailed ontogenetic investigations are important for understanding the problems of petrogenesis and ore formation.

5.6.1 The Origin and Development of Olivine in Traps

The data presented in this work indicates that the morphology of crystals, volume of mineral in rock, its distribution, inclusions in minerals, alteration character, zonality, interrelations with other phases, timing of mineral generation and associations, and the mineral composition and impurities are very important for the solution of genetic problems. Underestimation of these characteristics leads to inaccurate inferences, and their use in combination with another data represents a reliable ground for genetic reconstructions.

The origin and development of olivine is related to the initial composition of a melt, interchamber differentiation, and ore formation. Occurrences of 20–25% olivine in mandelstones of picritic basalts and vitreous quenched crusts in the Lower Fokinsky Intrusion are evidence of the existence of independent picritic melts. The increase in olivine content downward through the Mt. Pegmatitovaya Intrusion, the decrease in magnesium content from the roof to the base of the body, the increase in NiO_{Ol}, NiO_{rock}, and MgO in the rock from the roof to the base of the body, as well the regular change in the clinopyroxene composition in this direction, provide evidence for the existence of gravitational crystallization differentiation and invisible layering.

The absence of olivine in the contact facies of the Noril'sk Intrusions indicates that a melt intruded into a recent chamber does not have a composition of olivine or troctolitic gabbrodolerite as many geologists believe. Moreover, the distribution of olivine through the body and the composition of the mineral and phases associated with it allow one to assume with certainty that emplacement of immiscible melts of tholeiitic basalt and picritoid compositions has taken place.

Attention has been given to the fact that petrographic investigations revealed abundant intergrowths, that is, solid solution breakdown structures in olivine. Those were most frequently observed in picritic gabbrodolerite of ore-bearing intrusions and were relatively rare in barren rocks.

Geologists have long been aware of regularly oriented intergrowths in olivine from a variety of magmatic rocks. These intergrowths were first described by Judd (1885) in olivines from peridotites of Rhum Island (Inner Hebrides). From that time, the number of descriptions has increased and their geography extended. It has been established that these intergrowths may be composed of magnetite, ilmenite, chrome spinel, pyroxene, and ore–silicate phases. Reviews on this are available in a number of works including Agafonov et al. (1974), who also make some judgments about the nature of these intergrowths. In addition, it should be noted that similar intergrowths have been described in olivines from lunar samples (Putnis 1979).

At the present time, the concept that breakdown structures in olivine are a result of increased fO_2 in a melt, that is, high-temperature oxidation, is the most justified geologically, theoretically, and experimentally (Haggerty and Barker 1967; Champness 1970; Goode 1974; Nitsan 1974; Putnis 1979; Johnston and Stout 1984). Investigations undertaken by these researchers have shown that the morphology and composition of high-temperature oxidation products are foremost related to the kinetics of a process. Depending on the temperature and fO_2, forsterite, enstatite, SiO_2 phase, magnetite, and hematite may be oxidation products. Kolterman in experiments on olivine oxidation at Pgeneral = 1 atm obtained forsterite, amorphous SiO_2, and hematite at T = 820°, and forsterite, enstatite, cristobalite, and magnetite at 1,080°C (Champness 1970). As a consequence of oxidation of ferriferous olivine, the fayalite component breaks down to form enstatite, and the released silica reacts with forsterite to also form enstatite (Goode 1974). Haggerty and Barker (1967) traced the whole succession of olivine oxidation experimentally. Depending on the fO_2 regime, oxidation can proceed fast and completely, or incompletely with delays at one or more stages. Haggerty and Barker (1967) also traced experimentally the transformation of olivine within a wide temperature range. It has been revealed that small olivine grains change quickly into Fo + Fe_2O_3 at 600–1,000°C and larger grains survive within the range 830–1,000°C for short time, before crystallizing into an association of metastable phases En + Fe_3O_4 + Fo + Fe_2O_3, which over time changes to the stable association Fo + Fe_2O_3.

Thermodynamic calculations by Nitsan (1974) of olivine oxidation reactions at 1 atm total pressure indicated that stability areas for olivines of particular composition were established under varying T and fO_2 conditions and showed wide possibilities for the formation of magnesian olivine with increasing fO_2 in a wide temperature range. A similar plot of log oxygen volatility versus temperature is shown in Fig. 1.6. In addition, we emphasize that the Fe content in olivine decreases at constant temperature with not only increasing fO_2 but also with strongly decreasing fO_2. This case seems to take place in the UZ of the Khungtukun Intrusion with the involvement of reducing fluids (Ryabov 1989b).

In picrites from the Hawaiian Islands, reaction rims of orthopyroxene and opacite surrounding olivine have been described (Muir et al. 1957), which were formed as a consequence of the reaction $3Fe_2SiO_4 + 3Mg_2SiO_4 + O_2 = 2 Fe_3O_4 + 6MgSiO_3$. Orthopyroxene Fe–Mg ferrite symplektites developed on lherzolite xenolith rims in lavas from Kauai Island in the same region were formed by a similar mechanism (Johnston and Stout 1984). Walker and Poldervaart (1949) inferred that in dolerites from Karoo, olivine may be transformed into orthopyroxene or pigeonite at the magmatic stage. Some geoscientists consider reaction rims of orthopyroxene surrounding olivine in coronas (or symplektitic structures) as oxidation products (Ol + H_2O).

Sobolev (1986) noted that the wide alteration of olivine is specific for traps of the Siberian Platform. He emphasizes that as a result of this alteration, first-generation olivine is absent in normal traps and is partly resorbed in picritic Noril'sk Intrusions, and early olivine in hypersthene is completely dissolved in the reaction with residual melt forming orthopyroxene.

The tendency of olivine for alteration is due to the structure of the mineral that consists of isolated SiO_2 tetrahedra connected by Fe and Mg. The structure turns to be very unstable to variation in physical-chemical conditions under which olivine is easily oxidized and resorbed.

During the crystallization of a magmatic melt, fO_2 steadily increases and this may lead to ferritization and formation of magnetite ores (Pavlov 1979). Oxidation processes proceed most actively with the involvement of fluids, which increase in volume as the melt crystallizes and inflow of fluids from hypeal parts of the magmatic column takes place.

It can be assumed that high-temperature oxidation of olivines and the formation of more magnesian varieties are very common in traps. Taking this into account, reaction rims of orthopyroxene surrounding olivine or complete replacement of olivine by orthopyroxene in picritic basalts and gabbrodolerite in a number of intrusions can be satisfactory explained. Accordingly, a new explanation can be proposed that the wide development of orthopyroxene in gabbrodolerites of the Mt. Zub Intrusion took place due to oxidation, as opposed to assimilation of argillaceous deposits (Godlevsky 1959). The progress of high-temperature oxidation can be seen, in general terms, in picritic lavas where a fast reduction in temperature at high fO_2 leads to complete or partial opacitization of olivine or replacement of the olivine by bowlingite; crystallization of the central parts of a flow under stabilized conditions manifests as olivine mantled with orthopyroxene, and in ankaramite varieties, olivine is replaced by clinopyroxene. Ankaramites of Arylakh appear to be an example of high-temperature oxidation of olivine. Lastly, the presence of magnesian olivine as intratelluric olivine in rocks approaching "medium" traps (as detailed by Oleinikov 1979) cannot be unambiguously attributed to high-temperature oxidation.

Experimental investigations have shown that an increase in fS_2 in a melt and in a system with olivine, or with elevated fO_2, causes an increase in the magnesium content of the residual liquid and olivine and leads to the formation of iron sulfide instead of iron oxide (Almukhamedov and Medvedev 1982; Naldrett et al. 1984; Arutyunyan 1986). The intensity of the process is associated with the pressure of fluids under favorable conditions.

In the Upper Talnakh Intrusion, the thermodynamics of this process has been considered earlier (Ryabov and Pavlov 1984), where it was shown that the appearance of high-magnesian olivines (before forsterite) and formation of the magnesian skarn, paragenesis of minerals may take place under the action of sulfur-bearing fluids on picritic gabbrodolerite.

The iron content of olivine, as well as the behavior of isomorphic impurities in olivine (silicate nickel in particular), are strongly affected by the fluid regime. Silicate nickel in olivine as the source for sulfide ores of the Noril'sk deposits has been considered by Korovyakov (1960) and Zolotukhin (1965, 1988). However, criticisms of these concepts are also known (Genkin et al. 1970; Gulin and Sukhov 1974; Tarasov 1976).

Experimental investigations on the partitioning of Ni between olivine and sulfide liquid have shown that the behavior of Ni depends on the fO_2 and fS_2 regimes (Naldrett et al. 1984; Arutyunyan 1986). In melts undersaturated with sulfur, the high affinity between Ni and S leads to concentration of the metal in sulfides and to depletion of Ni in olivine (Arutyunyan 1986). With increasing S content in a melt, the depletion of nickel in olivine continues to a certain limit. In liquids saturated in sulfur, the Ni content in olivine increases as the sulfide suspension content in the melt increases. If no sulfides have settled or insignificant sulfide settling takes place, then the Ni content in olivine increases as sulfide suspension in the melt increases. The change from depletion of nickel in olivine to enrichment takes place after fractionation of 26% olivine in the presence of 0.5% sulfides. With an increase in suspended sulfide content phase up to 0.7%, the Ni content starts to increase. Examples of this in nature have been described by Ryabov (1992b).

When considering the formation of the Noril'sk ores, the following should be noted. Picritic gabbrodolerite, by analogy with picritic basalt, contained olivine enriched in silicate nickel. Olivine is unstable in an environment rich in fluids. As a result of fluid–magmatic interaction under the action of sulfur-bearing gases, magmatogenic sulfurization with extraction of silicate nickel and partial extraction of iron from olivine took place, forming a Ni-bearing sulfide liquid and olivine with higher magnesium content. This mechanism of ore formation is in compliance with the fact that the nickel content of picritic basalts and sulfide-poor ores of picritic gabbrodolerite are similar (Smirnov 1966). Also, nickel ores are restricted to horizons of picritic gabbrodolerite (Rogover 1959; Zolotukhin 1965), and high ore concentrations are associated with high-magnesium olivine depleted in nickel, and similar rocks, for example, in the Kharaelakhsky branch of the Upper Talnakh Intrusion.

One strong objection against sulfide–silicate immiscibility in an intrusion chamber is the inconsistency between volumes of ores and silicate mass and between sulfide and silicate nickel in ores and in a prospective source, that is, olivines in picritic gabbrodolerite (Zolotukhin 1971; Gulin and Sukhov 1973; Tarasov 1976). This objection could be ignored on the basis of the dynamic sulfurization model, which includes immiscibility arising when fluid flow increases and oregenic elements are redistributed throughout the magmatic column. In this case, an increase in nickel concentration in ore in the leading parts of the magmatic column is accompanied by depletion in nickel in the rear portion of the column. We can see this phenomenon when ore content of the Upper Talnakh or Noril'sk-I Intrusions is compared to the Lower Talnakh Intrusion. The importance of the redistribution of oregenic components within the magmatic column has been noted by Ryabov (1981) with detailed petrologic investigations of the behavior of Cr by Ryabov and Yakobi (1981).

Based on the study of Ni/Mg partition coefficients in mafic-ultramafic melts with different sulfide saturation, Naldrett et al. (1984) concluded that melts saturated with sulfides are faster depleted in Ni relative MgO than undersaturated ones, and the equilibrium of magma with sulfides leads to significant decrease in the ratio of the Ni content to the content of forsterite component. Observing massifs with different ore saturation, the authors have shown that Ni content in olivine may be used as an indicator of the sulfide ore grade quality. Krivenko et al. (1986) recommend predicting ore-bearing intrusions using criteria based on an estimation of sulfur saturation of the melt based on experiments by Naldrett and other data from both barren and ore-bearing layered gabbroid massifs. To assess the potential nickel content of intrusions, the authors propose using $(Ni_{Ol} * fOl) - Ni_{rock}$. The authors consider that "the more volume of sulfide mass is present in the system, the poorer is the coexisting olivine in nickel" (p. 35), whereas increased Ni concentration in rock and low Ni content in olivine is evidence of the presence of scattered sulfides coexisting with early olivine.

In this way, it should be noted that the depletion of one or more oregenic component has long been known as a search criterion for Noril'sk targets. However, the mechanism of the process is more complex than a simple redistribution of silicate and sulfide Ni occurrence forms in olivine and sulfides. In spite of the resemblance of trends in nickel behavior in both barren and mineralized traps to those obtained in experiments, on which the criteria are based (Naldrett et al. 1984; Krivenko et al. 1986), use of data from experiments in the form proposed by the authors may not be reasonable. First, the basis for the criterion suggested by the authors is equilibrium crystallization, but pronounced nonequilibrium crystallization has been observed. Second, different trends for Ni behavior in liquids undersaturated and oversaturated in sulfur (Fig. 5.13) have been obtained in experiments (Arutyunyan 1986). And third, in order to judge opinions on nickel content, it is first necessary to know the geochemistry of olivines in particular types of rocks, intrusions, and formations (which have their own peculiarities), and this should be closely related to the development of the mineral and petrology of the intrusion as a whole.

The study of silicate nickel behavior in olivines of traps (Fig. 4.2) shows a normal geochemical trend in barren rocks and a deflection in this trend for both ore-bearing and concomitant minerals and rocks depleted in nickel. Extraction of Ni from olivine during magmatic sulfurization is supported by data in Fig. 4.2: a region of points showing a gradual decrease in nickel content with an increase in magnesium content in the series from taxitic picrites to taxites, troctolites, and magmatogenic breccia. The data representing olivine from mafic pegmatoids intercepts the normal geochemical trend of NiO versus FeO in olivines of traps, and olivines of picritic–troctolitic gabbrodolerite depleted in nickel in Lower Talnakh and Lower Noril'sk Intrusions fall into the field of pegmatoids, showing their genetic affinity and the involvement of fluids. Thus, the results of study on olivine development in traps, and the

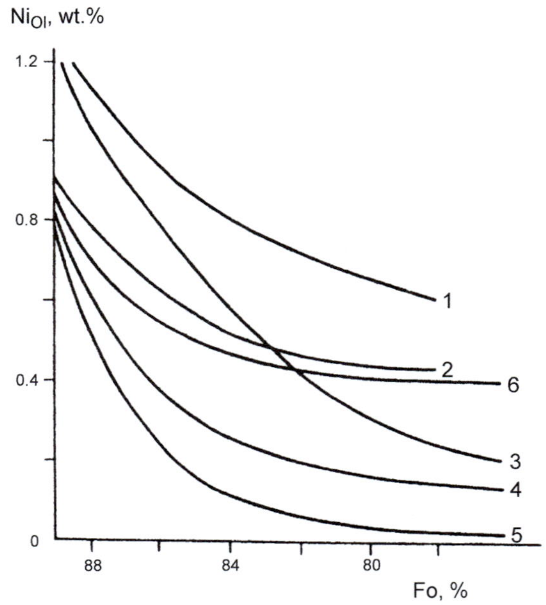

Fig. 5.13 Ni concentration versus the magnesium content in olivines formed during equilibrium crystallization of a melt (Arutyunyan et al. 1988). 1 = silicate melt, 2 = sulfide–silicate system, 3 = fractionation in silicate system, 4–6 = sulfide–silicate system with complete settling of sulfide phase (4), 50% settling (5), and without sulfide settling (6)

study of silicate nickel geochemistry, are very important in determining the solution to genetic problems. They also represent mineralogical criterion for estimating potentially nickeliferous bodies and to reveal sulfide nickel ores.

5.6.2 Features of Chromite Crystallization in Traps

Study of Cr geochemistry and peculiarities in the occurrence of Cr-bearing mineral phases is of great petrologic and oregenic importance. The main Cr concentrator in derivatives of tholeiitic basalt melts is clinopyroxene, and those in picritoid melts are chrome spinels and clinopyroxene (to a lesser extent) (Ryabov and Zolotukhin 1977).

It has been established experimentally that Cr solubility increases from 200 to 1,460 g/t in a tholeiitic melt, at Pgeneral = 1 atm and $fO_2 = 10^{-9}$ atm, as temperature increases from 1,180 to 1,300°C (Colette and Pierre 1982). Additionally, the increase in Cr solubility from 800 to 1,460 g/t takes place at a constant temperature of 1,300°C, with increasing fO_2.

The low Clarke of Cr in tholeiitic basalt melts is responsible for the absence of chrome spinel in mafic rocks, and low concentrations of Cr isomorphically enter lattices in early crystalline phases. Clinopyroxenes, which can accumulate significant concentrations of Cr_2O_3 (up to 1.44 wt%) in the cores of crystals, are most favorable for this. If the chemical composition of a melt, T and fO_2 are optimal for the crystallization of large amounts of pyroxene, then the main mass of Cr may be concentrated in its structure (Ryabov and Yakobi 1981). This is the case in the Muskox Intrusion (Irvine and Smith 1969).

Some authors propose that accumulations of chrome spinel in traps were formed by leaching pyroxenes (Zolotukhin and Vasil'ev 1975). Shatkov (1973) considers that the transition from Cr-rich cores to titanium-rich rims in zoned crystals is associated with the onset of chromite crystallization in the Noril'sk Intrusions. Godlevsky et al. (1971) believe that in the course of magmatic system evolution, spinel becomes unstable and decomposes and Cr passes into the melt and is preserved before the onset of green clinopyroxene crystallization. Our investigations have shown that Cr geochemistry has its own specific character in crystallization products of tholeiitic basalt and picritoid melts and it is significantly complicated with involvement of fluids. The high Clarke of Cr in picitoid melts is associated with higher solubility in magnesian liquids and is represented by accessory chromite in crystallization products of these melts. Chrome spinels are absent in picritic and troctolitic gabbrodolerite that are cumulates of tholeiitic basalt melts characterized by the low Clarke of chromium despite high magnesium content. Picritic lavas and gabbrodolerite of the Fokinsky Complex represent a primary picrite-basalt melts example, and picritic–troctolitic gabbrodolerites of the Morongovsky Complex and intrusions of the Middle Yenisey River may serve as an example of derivatives of tholeiite-basalt magmas.

The timing, placement, and crystallization conditions of accessory chromite in picrites have been discussed previously; Oleinikov (1979) indicates formation at the earliest phase in an intermediate chamber, whereas (Genkin et al. 1979) believe they formed in a recent chamber. Stepanov (1975) has inferred that spinel is stable in all systems including the anorthite–forsterite rocks, then spinel should be one of the earliest phases crystallized; the appearance of spinel is governed by the initial composition of a melt. In this way, spinel should be found in the Noril'sk Intrusions, but not in typical trap intrusions. Chromite, in Stepanov's opinion, is related to hysteron-magmatic minerals.

Judging from calculated data, formation of picritic gabbrodolerites in different intrusions containing accessory chromite proceeded at decreasing temperatures from 1,350 to 1,100°C and fO_2 from 10^{-4} to 10^{-7} atm (Zolotukhin 1964; Zolotukhin and Vasil'ev 1967). Other data indicates that formation of dispersed chromites took place at 1,400–1,150°C and fO_2 from 10^{-6} to 10^{-7} atm (Genkin et al. 1979).

Our data on the conditions under which dispersed chromite formed fit well with concepts expressed by Genkin et al. (1981) who regard the mineral as an early phase. The

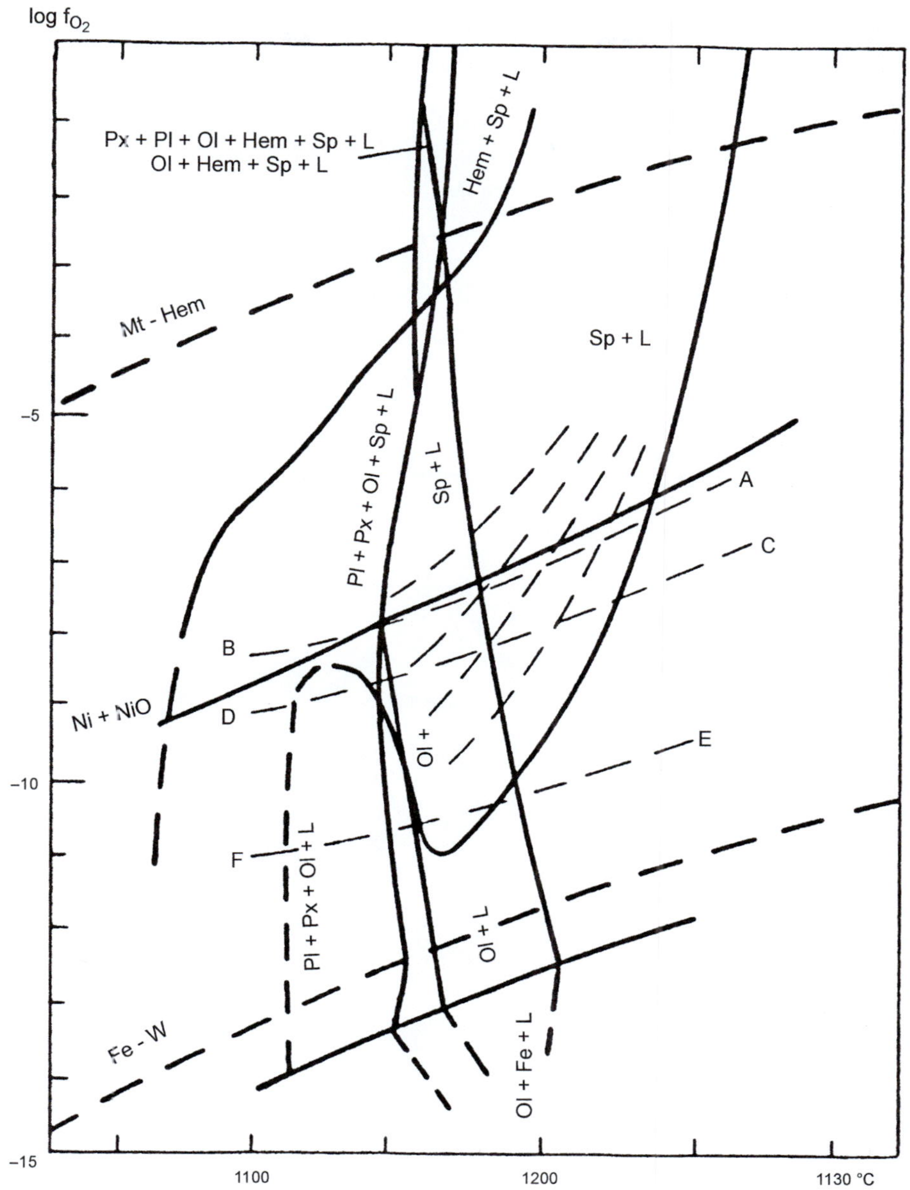

Fig. 5.14 Log fO_2 versus T for spinels and coexisting phases (Hill and Roedder 1974). *L* melt, *Ol* olivine, *Pl* plagioclase, *Px* pyroxene, *Sp* spinel, *Hem* hematite, *W* wüstite, *Mt* magnetite, *Fe* metallic iron

crystallization path of spinel and coexisting phases can be traced on Fig. 5.14, based on a diagram by Hill and Roedder (1974). The picritic melt of composition A will evolve along the curve A with a progressive decrease in and increase in fO_2. Here, chromite is the earliest spinel phase, and during crystallization, the Cr, Mg, and Al content all decrease with an increase in Ti and Fe content. We can see this in picritic lava in the Noril'sk Region and in dispersed chromite of picritic gabbrodolerite (Genkin et al. 1979; Ryabov 1984a). As can be seen in Fig. 5.14, a change in the magmatic system state along the A line leads to a successive paragenesis of (Sp + L) → (Sp + Ol + L) → (Pl + Px + Sp + Ol + L).

The formation of chromite accumulations in the horizon of picritic gabbrodolerites proceeded with the involvement of fluids, which stimulated immiscible ore–silicate fractionation of liquids. This is supported by globular ore accumulations, sharp contacts with barren matrix, and globules containing volatile minerals and sulfides (Genkin et al. 1981). We share this judgment concerning the immiscible mechanism for the formation of chromite accumulations, but we cannot agree that these phenomena took place at the base of a picritic layer. Chromite accumulations formed at the top of the picritic horizon and this is an additional argument in favor of their fluid–magmatic nature. On the contrary, picritic gabbrodolerite depleted in chromium are located at the base of the layer. This also supports a fluid mechanism for redistribution of chromium and indicates the source of the mineral and where extraction of chromium took place under chamber conditions.

Fig. 5.15 Sketches of thin sections from rocks of the Noril'sk Intrusions. Explanations are given in the text. 1 = olivine crystals in chromite rash, 2 = layered chalcopyrite (*at the top*)–pyrrhotite disseminations (*at the bottom*), 3 = minerals containing volatiles, 4 = plagioclase, 5 = pyroxene, 6 = crystallites in palagonite, 7 = poor (*rash*) and rich (*shaded*) small chrome spinel crystals with bubbles filled with volatile components, 8 = rims surrounding gas bubbles: chrome spinel crystals (*shaded*) and *green* Mg–Al spinel (*unshaded*), augmented fragment of picture "D", 9 = *black* chrome spinel crystals (*shaded*) and transparent cherry-red magnesiochromites (*unshaded*) in Pt endoskarns, 10 = zoned garnets: uvarovite (*with specks*) and grossular (*without specks*), 11 = granoblastic chrome-diopside aggregate, 12 = magnesian–calcareous skarn

Kudelina et al. (1983) paid attention to the fact that maximum chromium concentrations were found at the top of the picritic horizon. Strange as it may seem, these authors propose a gravitational crystallization differentiation mechanism, which settled out early refractory phases (olivine and chromite) at the bottom of a magmatic chamber that serves as the top of the picritic horizon. They explain the absence of chromite at the bottom of picrites and in taxites by the fact that this part of intrusion represented a near-contact zone, which had already formed by the time chromite had settled, and chromite could not reach the base of the intrusive chamber. In the context of this mechanism, the extreme depletion in chromium (lower than Clarke) in lower parts of the horizon, which they regarded as cumulates, remains to be explained, as well as the chromite form, existence of two chemical types of the chromite, wide variations in its composition, and the possibility for gravitational settling of the finest chromite grains and high degree of chromite separation at the boundary between olivine gabbrodolerite and picritic gabbrodolerite.

The crystallization timing of spinels during petrogenesis can be determined with accuracy in petrographic investigations. At the top of the horizon of picritic gabbrodolerite, a rich "rash" of Cr-Sp appeared after the crystallization of olivine and the formation of droplet-like sulfide disseminations (Fig. 5.15a). The "rash" was formed before crystallization of clinopyroxene and plagioclase as evidenced by the distribution of Cr-Sp in a rock irrespective of phase boundaries of these silicates and even within individual crystals (Fig. 5.15b, c). In leucogabbro of the UZ, the "rash" of Cr-Sp grains was formed in interstices after crystallization of large, first-generation, tabular mafic plagioclase crystals (Fig. 5.15d), and the blebby texture of the rock with biotite and amphiboles in bubbles, as well as concentration of spinel around palagonite bubbles (Fig. 5.15c, e), is evidence of the association of ore mineralization with volatiles. Formation of magnesiochromites, chrome-diopsides, and uvarovites containing Cr_2O_3 up to 68.6, 5.2, and 12.5 wt%, respectively, proceeded by the replacement of Cr-Sp accumulations in the upper taxites under action of high-temperature fluids (Ryabov et al. 1996c).

The leading role of fluids in the redistribution of chromium and a number of other oregenic elements is clearly manifest in the platinum–chromite mineralization in the UZ of the Noril'sk Intrusions. The wide distribution of ore accumulations in spherical and globular forms, and minerals containing volatiles that are clearly seen on microphotographs of rocks from the UZ (see Volume 2), allows one to assume with certainty that fluid–magmatic differentiation and pneumatolysis took place.

Contrary to long-held concepts on the inert nature of Cr during petrogenesis, many publications on the redistribution of Cr within the bounds of a magmatic body and variation in its form are available in the geological literature. In connection with this, the reactive nature of chrome-clinopyroxene relative top chromite has been noted by Irvine (1967) in the Muskox Intrusion; the resorption of chromite from dunites and recrystallization of rocks to form olivinites and chrome-diopside bodies has been studied in the Inagli Massif (Glagolev et al. 1974), and a similar situation is observed in the Kugda and Odikhincha Massifs in the Maymecha--Kotuysky Province (Vasil'ev and Zolotukhin 1975). Rocks of the Mikchandinsky Sheet may also serve as a good example of the redistribution of Cr (Ryabov et al. 1977). Here the resorption of chromite and olivine in picritic basalts, as well as formation of chromous augite, can be traced completely.

Taking into account the mobility of chromium in a fluid phase, for example, as a chloride (Wicks and Block 1963), the mechanism for the formation of chromite-bearing horizons in the UZ at the roof of the Noril'sk Intrusions can be understood. An accumulation of fluids and chromium has led to immiscible differentiation of a fluidized melt in the UZ. At the same time, the resemblance in form and composition of chromites from the UZ and picritic horizon assists in the identification of their formation mechanism, that is, both can be considered as products of a fluid–magmatic process. High concentrations of chromite in olivine-free gabbrodolerites and low concentrations in olivine gabbrodolerites, as well as unusual chromite anorthosites, can be explained by this mechanism. Therefore, there is no need to seek compromise with experimental data on chromium solubility in dry basalt melts. Now, it is obvious that the variety of traps, the formation of layered intrusions, and ore formation cannot be explained by fusibility curves in dry and low-component systems. Reliable geological data bear witness to the greater role of fluids in petrogenesis and ore formation, the existence of fluidized melts with their own evolutionary paths, to different mechanisms of differentiation and large-scale redistribution of oregenic elements, which with involvement of fluids lead to the formation of deposits.

5.7 Model of Fluid–Magmatic Differentiation of Melt and Ore Formation

It may be assumed on the basis of investigations using known natural and experimental data (Ryabov 1999a) that the action of volatiles produced in deep-seated magmatic hearths continued at varying rates in feeder conduits and magmatic chambers and was accompanied by the extraction of different chemical elements from both the melt and country rocks and transferred into zones of accumulation. Differentiation of the melt occurred due to its immiscible differentiation into coexisting liquids, which were stable under existing PT conditions. A melt during its evolution exhibits a tendency to equilibrium, that is, it proceeds under the same laws for all systems. A change in parameters of a magmatic system, such as a decrease in temperature and pressure, supernormal enrichment in metals (e.g., Fe, Mg, Cr, or Ti), and/or in volatile components, dissociation of volatiles, or change in phase composition and partial pressure, leads to differentiation of the melt into immiscible liquids that are stable under local conditions. Subsequent evolutionary changes in physical–chemical conditions later destabilized the system, which responded through immiscible differentiation tailored to new conditions with a new composition for coexisting liquids. This continued sequentially, permanently, and repeatedly until crystallization begins and continues during crystallization of silicate, ore, and salt melts (Ryabov 1989b).

The evolution of parameters in a magmatic system reflects the dynamics of fluid–magmatic interaction in deep-seated chambers, during melt movement in feeder conduits, and in hypabyssal magmatic chambers. The main propositions of this process and references of naturally alternating events are listed below.

1. Accumulation of volatiles in magmatic melts leads to the formation of fluidized melts that are apt to experience immiscible silicate–silicate differentiation when PT conditions are changed.
2. As a result of the differentiation of a fluidized melt, immiscible liquids are formed, one that is relatively dry and one fluid-saturated under new conditions. It contains an independent gas phase that exceeds its solubility in the melt and is in equilibrium with the latter.
3. Filtration of the concomitant free gas phase from deep-seated chambers to hypabyssal chambers is accompanied

by additional enrichment in volatiles from country rocks and, depending on conditions, by evolution and change in the reaction activity.

4. The presence of a free phase in fluidized melts is governed by the potential ore presence.
5. The fluid regime of a magmatic chamber (concentration and composition of fO_2, conditions, and other parameters) that is responsible for the phase composition of silicate and ore liquids (sulfide, oxide, or native metal) plays a dominant role in the extraction and transfer of metals and their accumulations in ore-forming elements.
6. The fluid regime of a magmatic system plays a crucial role in the extraction and redistribution of oregenic and petrogenic chemical elements and the origination of melts and rocks enriched and depleted in these elements, as well as for the phase composition of ore liquids (sulfide, oxide, or native metal).

Volatiles accompanying a melt had a multicomponent composition (H_2, CH_4, N_2, H_3P, H_2S, F, Cl, CO, CO_2, H_2O, and others) with variable concentration. Differences between the thermodynamic and chemical properties of gases were responsible for their differential activity during a magmatic process; this activity being complicated by interaction between gases. The evolution of a fluid regime is reflected in successive changes in the activities of various volatiles, and each stage of evolution is characterized by the dominant role of a particular volatile that is linked to mineral forms. Finally, this volatile gave way to another volatile and stimulated further activity.

One of the important factors of fluid–magmatic interaction is the close genetic relationship between rocks that are anomalously depleted or rich in ore components. During migration through a magmatic melt, the fluids extract and remove oregenic metals, but leave traces of these metals in the rocks leaving them depleted with respect to Clarke. This is a possible way to trace fluid flow migration and to define potential sources of ore accumulation.

Ore liquids are formed as a result of basalt magma immiscibility that proceeds as volatile activity increases during their migration through the melt. The ore emulsion coalesces in a moving melt, forming droplets that merge into lenses and streams, and finally experiences gravitational separation that forms an ore melt. One can judge the consistency of this kind of ore magma from tuffites of the Angaro–Ilimsky iron-ore deposits described in Kalugin et al. (1994) and from blebby palagonite–magnetite lava of the Magnetitovy Stream (Ryabov 1997). In the last case, the eruption of ore magma took place as a fluidized ore–silicate emulsion, where volatiles acted as an independent phase facilitating migration of a heavy ore liquid upward through the magmatic channel. In this way, the association of minerals bearing these volatiles in the form of "caps," that is, gas bubbles, above droplet-like sulfide disseminations (Fig. 5.15a) can be explained.

For the Noril'sk deposits, the saturation of a melt in sulfur and existence of an independent gas sulfur-bearing phase is obvious. The high partial pressure of sulfur suppressed the activity of other volatiles, and the high degree of sulfur affinity with iron, nonferrous, and noble metals enabled their extraction from the melt and led to the formation of a sulfide liquid, which migrated as an emulsion together with the fluidized melt into hypabyssal magmatic chambers. The formation of a sulfide liquid decreased fS_2 and increased the activity of other volatiles including halogens. This resulted, in particular, in the extraction of chromium and noble metals and the redistribution of noble metals as volatile halide compounds within the magmatic body. All this is reflected in the formation of Cr-Sp accumulations at the roof of the picritic horizon and Pt–chromite and chromite-free low-sulfide ores in upper pegmatoids.

The dominant role of hydrocarbons, hydrogen, nitrogen, and phosphorus practically without involvement of sulfur was manifest in the Pt–Fe (Pt–graphite) type of ores in schlieren ore pegmatites in Dzhaltul–Khungtukunsky-type intrusions, that is, nodules of native platiniferous α-Fe. The establishment of this ore type is supported by the involvement of reducing fluids in the development of Pt low-sulfide ores that is assumed to have occurred in the Bushveld, Stillwater, and other intrusions (Volborth and Housley 1984).

Summarizing the above, it can be inferred that fluids have impacted traps of the Siberian Platform taking part in formation of a wide variety of layered intrusions, and it is hard to overstate the importance of volatiles in petrogenesis. Volatiles can be argued to play the dominant role in magmatogenic ore formation and that all magmatogenic deposits on the platform were formed owing to fluidized melts.

5.8 Geodynamic Control Model of Traps Magmatism and Ore Formation of the Siberian Platform

The problems of trap magmatism and ore formation have attracted geological interest over the decades because of large scale of flood basalt magmatism and the unique giant Noril'sk-type Ni–Cu–PGE deposits. A large volume of geological and analytical information on traps of the Siberian Platform and related ore deposits has been gathered, but in spite of this, there is no consensus among geologists regarding the main genetic problems. Many differing concepts have been suggested for these problems. As we can see, the basic reasons for this plethora of hypotheses are not only the complexity of the problems but also the specialized nature of aspects of the problems. There is also insufficient knowledge of the targets under investigation and also underestimation and neglect of earlier studies that have asserted geological facts. As a result, we have presentation of genetic concepts without satisfactory recourse to real evidence that would confirm the validity of their positions. The large volume of precise analytical data obtained recently in Russian and western laboratories, as well as interpretation of these data, unfortunately have not shed new light on these genetic problems and have actually increased the number of competing hypotheses. The volume of information continues to increase, the competing hypotheses may be increased ad infinitum, and the large amount of high-precision analyses provides no answer to genetic problems. Analytical data and empirical observations are not solely numerical data and facts, and in order to interpret these data correctly, ideas are needed to integrate them into unified logical succession. Therefore, an ore-petrological model should be developed that would integrate all known data into a unified evolutionary chain, reveal cause-and-effect relationships, and provide a physical–chemical basis for the ore–magmatic process. Many parts of the model have already been investigated in detail, the sulfide system, for example, and plenty of empirical formulae have been established. So, it is necessary to find and validate places in the general evolutionary chain for these parts.

Recent concepts on the mechanism of formation of flood basalts and the Noril'sk deposits have been stated practically by all leading Russian and foreign researchers (Letnikov et al. 1998; Distler et al. 1999; Aplonov 2001; Dodin 2002, 2005; Dodin and Duzhikov 2005; Likhachev 2006; Marakushev et al. 2002, 2007; Naldrett 2003). The novelty of models recently proposed by many researchers is in the attraction of giant mantle plumes rising from the lower mantle or from the lower mantle–core interface to explain all ore–magmatic features of the Siberian Platform. Unfortunately, followers of the plume concept and its numerous defenders discuss (in most cases) general aspects and provide no conclusive natural evidence in support of this process.

Some view of formation of the giant Noril'sk deposits under action of a superplume as well as the petrological–geodynamic model of these deposits are considered in studies by Dodin (2002) and Dodin and Duzhikov (2005). These authors consider the following principal criteria in favor of the involvement of superplume: the triple junction of rifts, the size and volume of igneous rocks, the accumulation of ferro-titanous basalts with komatiitic tendency for crystallization, certain isotopic characteristics of the mantle, high heat flow, pre-rift tectonic regime with accumulation of alkaline mafic rocks at the early stage of magmatism, fluid regime, and Mohorovicic discontinuity. These authors assign empirical regularities to the criteria, but they give no arguments in support of their cause-and-effect relationship with respect to the genesis of the deposits. This does not rule out any different explanations, as some criteria are generally uncertain.

Taking into consideration the above, we have undertaken a series of geological and metallogenic studies, which enabled us to develop a "working" genetic ore-petrological model. Our investigations were grounded on the available data in literature on tectonics, trap magmatism, metallogeny, and ore formations of the region, as well as on our own data. A territory of the northwestern part of the Siberian Platform was investigated, bounded by the Platform border in the north and by the latitude of the Nizhnyaya Tunguska River to the south. In our model, we made an attempt to find and justify the cause-and-effect relationship between various geological phenomena. At present, we continued to itemize links in the evolutionary chain of the ore–magmatic system using selected targets and regions of the Siberian Platform.

5.8.1 Geological Structure of Northwest of the Siberian Platform

The schematic geological map (Fig. 5.16) gives a view of the geological structure of the region. The Tungussky syneclise and the Noril'sko–Kharaelakhsky depression separated by the Khantaysko–Rybninsky uplift are the major structural elements of the region. The depression is traced along the entire northern border of the Siberian Platform and extends east from the Noril'sk Province into the Kamensky Province and farther into the Maymecha–Kotuysky Basin. To the north, the depression is separated by the Kystyktakh–Ayan–Ambardakhsky uplift from the Tungussky syneclise. The western side of the syneclise is also complicated by a basin and uplift that frame the central part of the syneclise. Marginal troughs at the Platform periphery are composed of a system of depressions or volcanic basins.

A network of regional faults, as well as local faults of varying size, is observed in the trap field of the Tungussky syneclise. The Imangdino–Letninsky Fault (more than 1,000 km long), Noril'sko–Kharaelakhsky Fault (350 km long), Yenisey–Khatangsky Fault (north Kharaelakh) (more than 500 km long), and other faults within a system of branched faults are the largest and most extensive. Zones of fractured rocks, up to 10–12 km wide, are located along the faults and in intersection regions. Deep faults are considered as the main magma- and ore-controlling structures of the region.

The Noril'sko–Kharaelakhsky depression that includes the Noril'sky and Talnakhsky ore junctions with giant PGE deposits are most extensively studied. The basin had a long history and being located at the turn of large geological structures was three times exposed to rifting and collision (Strunin et al. 1994). During these processes, the depression was dissected by a system of deep faults,

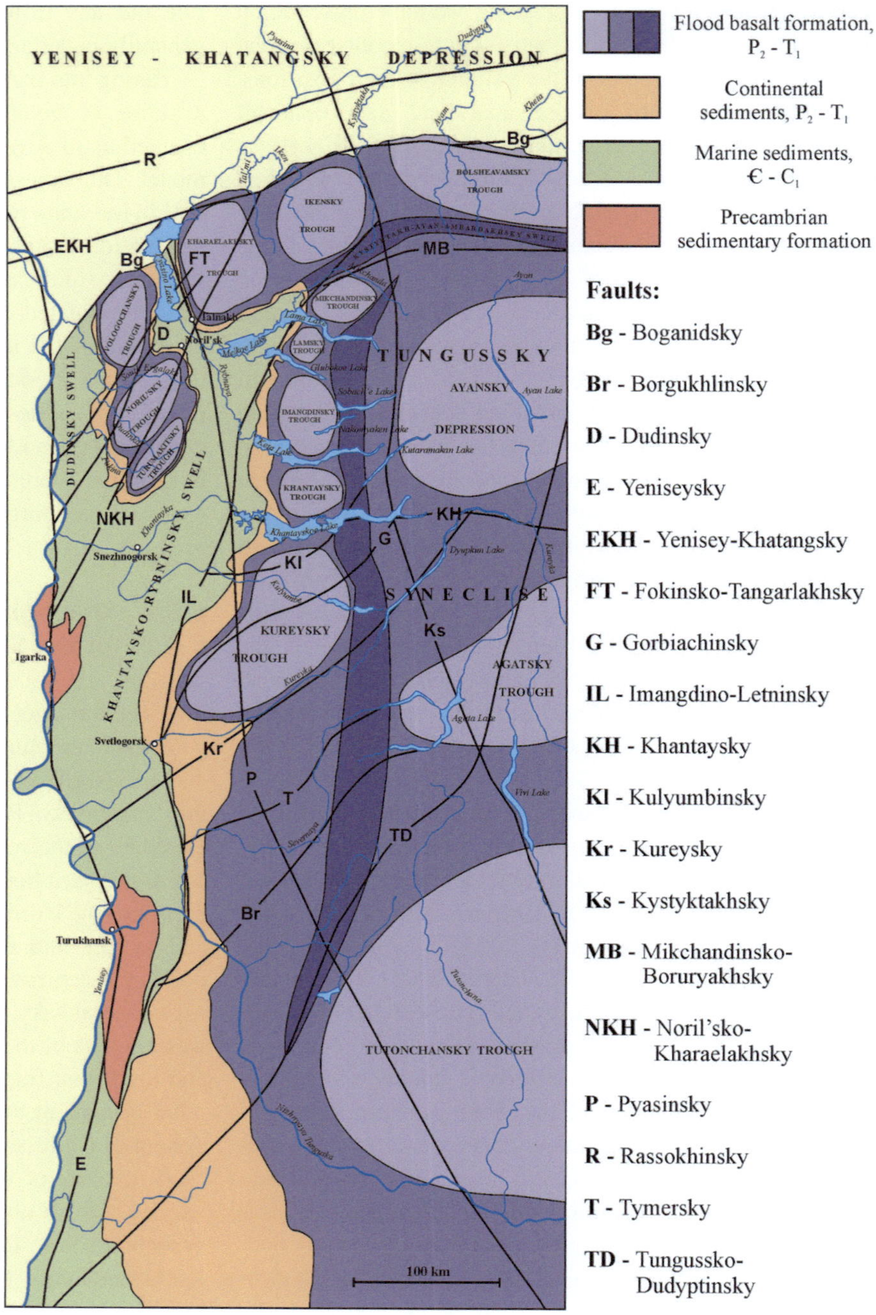

Fig. 5.16 Schematic structural–tectonic map of the Yeniseysky nickeliferous province of the northwest of the Siberian Platform

including the major Noril'sko–Kharaelakhsky, north Kharaelakhsky, Boganidsky, Pyasinsky, Fokinsko–Tangaralakhsky, Dydinsky, and Abagalakhsky Faults, which are accompanied by smaller tectonic offsets.

The Noril'sko–Kharaelakhsky Fault intersects the Noril'sk and Kharaelakh troughs along the depression axis, and the fault is complicated by near-fault disjunctive and plicative offsets and accompanied by fissure zones. The location of the Noril'sko–Kharaelakhsky depression under extreme conditions of unloading in the direction from rift structures and faults in the regions of high permeability of Earth's crust is one of the main reasons for its unique mineralization. The geological data give evidence of the extremely unstable environment within the Noril'sko–Kharaelakhsky depression for long periods of time. Many geologists consider the peculiar structural–tectonic location of the Noril'sko–Kharaelakhsky ore region close to the triple junction of rifts (Yeniseysky, Yenisey–Khatangsky, and Noril'sko–Kharaelakhsky) to be one of the main factors responsible for formation of giant deposits (Strunin et al. 1994; Dodin et al. 2002; Dodin and Duzhikov 2005). Different publications give detailed characteristics of the deep-seated structure of the depression and its development features based on the wealth of geological and geophysical data. Unfortunately, no consideration has been given in the studies as to the cause-and-effect relationship between intrusions of the Noril'sk-type and giant sulfide deposits with rift structures and what elements of geodynamics in this region (factors and processes) were responsible for formation of these unique natural targets.

5.8.2 "Abnormal Formations" in Flood Basalts

Abnormal formations have been important arguments for creation of a genetic model of basalt melt differentiation and ore formation. The location of these formations within the flood basalts and their internal structure, form, and composition enable the geodynamic environment in the region to be assessed during their formation. The features most important for creation of the genetic model are discussed below.

The local occurrence of highly magnesian and subalkaline lavas in the flood basalt field, the increase in thickness near faults, as well as the variable composition of these lavas in time and space indicate that faults formed conduits feeding magma as well as evidence for the depths of magmatic hearths and their repeated excitation.

Komatiite-like rocks form sheets in near-vent facies of paleovolcanoes. The interlayering of picritic and tholeiitic basalts has been strong evidence for prechamber differentiation of basalt melts and the possibility of eruption and/or emplacement of immiscible liquids. This fact is important as a possible explanation for the nature of layered mafic–ultramafic intrusions.

Numerous dykes of the northern Platform confirm the possibility for formation of lava strata as a result of wide-area eruption. The diversity in composition of the dykes and their multiphase structure with a multiple phase composition reflect the wide development of the prechamber fluid–magmatic differentiation of trachybasalt melt with involvement of volatiles and its heteromorphous crystallization due to overcooling.

Magnetite lava is indicative of the formation of ore liquid during fluid–magmatic differentiation of trachybasalt melt in magma exit channels. Ore spherules formed due to liquid immiscibility and migrated along the channels together with a fluidized melt and then, as PT conditions change, coalesced to form magnetite liquid.

Volcanic structures such as tuff volcanoes, diatremes, and calderas are discharge areas of endogenous fluids. The composition and geochemical features of fluids vary widely. This reflects in the diversity of ore occurrence (sulfide, magnetite, copper, bitumen, B-bearing, sulfate, carbonate, rare earth, spar-bearing, etc.) that form in paleovolcanoes and fissured zones.

5.8.3 Tectonic Factors in Metallogenic Zoning

Tectonic factors govern the placement of not only "abnormal formations" in the volcanogenic unit but also the distribution of layered intrusions and Cu–Ni–PGE sulfide deposits, PGE low-sulfide and PGE–Fe metal occurrences and deposits. Near-fault metasomatites and a series of mineralogical–geochemical anomalies of varying composition that form small localities and zones, as well as vast fields, are associated with faults.

The magma-feeding and ore-controlling role of faults leaves no doubt in the mind of most of geologists (Maslov 1963). All known layered intrusions on the Siberian Platform, as well as highly magnesian intrusions and intrusions with various ore saturation, are located in fault zones. Tectonic factors control the morphology and internal structure of intrusive bodies; generally swelled parts of intrusions with a pronounced differentiated structure are located in a fault zone. Further away from the fault, the differentiation degree and igneous body thickness decreases, and the igneous body changes to a thin non-differentiated sill. Magma-feeding faults control features of prototectonics (streaky, linear, and trachitoidal textures) in intrusion rocks (Vilensky et al. 1964), with most pronounced traces of flowing melt observed near exit channels.

The results of regional geological investigations revealed that ore junctions and metallogenic zones with various geochemical features are confined to fault zones (Fig. 5.17). Several hundreds of different occurrences and deposits that have been united into the Yeniseysky nickeliferous or ore province (Urvantsev 1959; Kavardin et al. 1967; Kavardin and Golubkov 1968; Kavardin 1976; Sukhov et al. 1981; Ryabov 2007) are located along the northwestern border of the Siberian Platform. The province occupies the northwestern part of the Platform and is confined by the Yenisey–Khatangsky rift trough to the north, by the latitude of the Nizhnyaya Tunguska River in the south, by the West Siberian plate in the west, and by the line of lakes Lama, Khantayskoe, Agata, and the Tutonchana River in the east. As a result of metallogenic zonation, the province was divided into five regions: the Noril'sko–Kharaelakhsky ore region and Lamsko–Khantaysky, Kureysko–Severorechensky, Nizhnetungusssky, and Igarsky ore-bearing regions. Within these regions, 14 ore zones and 34 ore junctions and fields have been distinguished (Fig. 5.17).

Analysis of geological data enabled the following characteristic features for fault zones of the region to be distinguished:

Thick intrusions (more than 100–200 m)
Extensive intrusions feature swelled sections (up to 500–600 m)
Intrusions are often layered
Highly magnesian intrusions

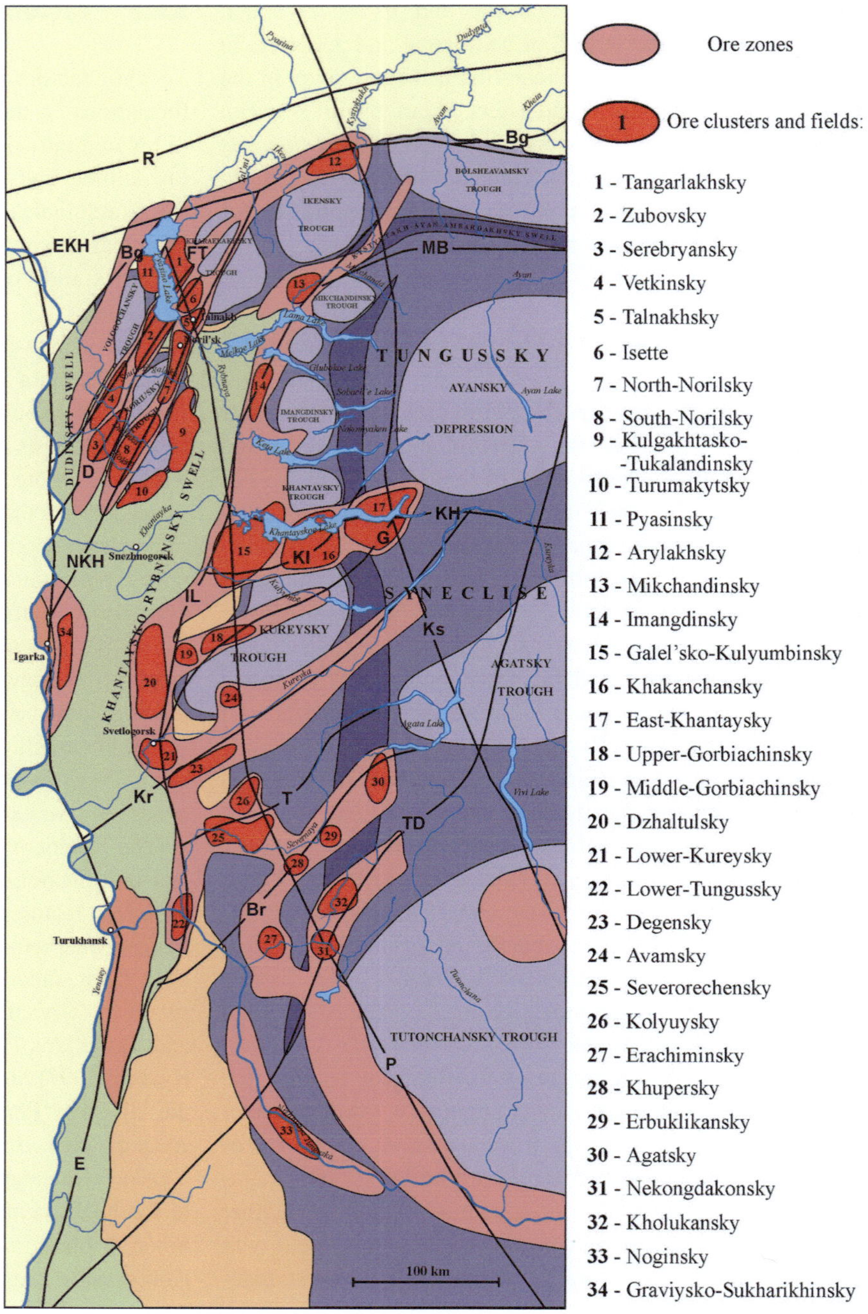

Fig. 5.17 Schematic metallogenic map of the northwest of the Siberian Platform (From Urvantsev 1959; Kavardin et al. 1967; Kavardin 1976). Fault names are given in Fig. 16

Metasomatites and hydrothermalites
Mineralized cracked rock zones
Metallogenic zones
Ore deposits
Geochemical anomalies
Hydrochemical anomalies
Mineral springs
Outgassing
Geothermal anomalies
Karst phenomena
Taliks and plentiful vegetation

Data on these features are scattered in numerous works (Godlevsky 1959; Korovyakov et al. 1963; Vilensky et al. 1964; Zolotukhin and Vasil'ev 1967; Kavardin et al. 1968; Dodin and Batuev 1971; Ivanov et al. 1971a, b; Kravtsov et al. 1971; Natorkhin et al. 1977; Turovtsev 2002). Most geologists are in no doubt concerning the association of these features with faults. The sole exception is metasomatites where there are many different ideas.

The prevailing view today is that thick caps of metasomatite at the roof of differentiated intrusions, which sometimes reach hundreds of meters thick, were formed by the action of volatiles separated from the magmatic melt as it crystallized. From this, we can see that the peculiarity of layered ore-bearing intrusions of the Noril'sky type is the increased volatile content as opposed to most trap sills, which commonly lack metasomatites.

In addition, it is known that thick caps of metasomatite are present in aureoles of both ore-bearing and barren intrusions (Yudina 1965; Turovtsev 2002). Generally, metasomatites occur independently of intrusions (Golubkov 1970). Intrusions have contact quenched zones that serve as screens for volatiles and evidence that the volatiles dissolved in the melt did not leave the magmatic chamber. Moreover, metasomatites, including those in the aureole of the Upper Talnakh Intrusion, are superimposed on contact gabbrodolerites at the contact. It should also be noted that the solubility of volatiles in basaltic melts was rather low. Therefore, there was little likelihood that thick metasomatites were formed around intrusions due to the interaction of rocks and volatiles that separated from the melt when it crystallized. The fact that intrusions and metasomatites were incompatible in thickness is demonstrated in the Tomulakh region north of the Oktyabrsky deposit. Here the thickness of the Upper Talnakh Intrusion is 2–35 m, whereas the thickness of the metasomatites aureole around the intrusion is 330–450 m (Turovtsev 2002). Many examples of thick zones and fields of metasomatites that are not associated with intrusions are also known on the Siberian Platform. These zones and fields are located, as a rule, in fault zones. For example, fields of metasomatites occur in the Pyasino Lake, Avam, Tiptur-Orokta, Severnaya, and Gorbichian River regions. Drilling in the metasomatites revealed that their thickness reaches several hundred meters and there was found no intrusion beneath them. There are different ideas concerning the origin of these metasomatites. Most geologists accept a traditional point of view suggesting that metasomatites were formed under the action of a hypothetic deep-lying igneous body that was not subject to erosion and was not tapped by drilling. Other geologists consider, quite reasonably by our opinion, that metasomatites relate to intrusions paragenetically rather than genetically and they recognize a separate type of near-fault metasomatites, which were formed due to fluid activity (Golubkov 1970). Thus, the fluid source explanation is topical not only of ores but also metasomatites.

5.8.4 Ore Formations

The characteristics of principal types of ore formations associated with platform magmatism have been the subject of numerous studies (Kavardin 1976; Genkin et al. 1981; Ryabov 1992a; Ryabov and Anoshin 1999; Sluzhenikin et al. 1994; Duzhikov 1988), and these have been augmented by new data. All deposits and ore occurrences in the northwest of the Siberian Platform are situated in fault zones (see Fig. 5.17) and form the following ore formations:

- Cu–Ni–PGE sulfide of the Noril'sk type
- PGE low-sulfide of the Noril'sk type
- PGE–Fe metal of the Dzhaltul–Khungtukunsky type
- Iron ore
- Pyrrhotite
- B-bearing
- Rare earth
- Spar-bearing
- Graphite
- Copper-zeolitic
- Copper sandstone
- Copper-porphyry

Peculiarities in the petrology of layered intrusions of the Noril'sk-type and related PGE–Cu–Ni sulfide deposits (Dodin and Batuev 1971; Ivanov et al. 1971b; Natorkhin et al. 1977; Genkin et al. 1981; Distler et al. 1988) and PGE low-sulfide ores and deposits (Ryabov 1984a, 1992a; Sluzhenikin et al. 1994), as well as their association with fault zones, have been described in the geological literature many times. The PGE–Fe metal deposit has been established in trap intrusions of the Dzhaltul–Khungtukunsky type in Maymecha–Kotuysky and Kureysko–Letninsky Provinces near deep faults (Ryabov and Anoshin 1999; Oleinikov and Kopylova 2000; Ryabov and Lapkovsky 2010).

The iron-ore formation includes magnesioferrite, magnetite, and sulfide–magnetite deposits and occurrences. These extend as a band more than 150 km along the Imangdino–Letninsky Fault. Other than deposits associated with diatremes of the Vetka and Tiptur–Orokta Rivers (Angaro–Ilimsky type), these deposits are commonly represented by steeply dipping magnetite veins within skarn rocks (Kavardin 1976). The veins lie at the contact with dolerites or gabbrodolerites and far away from the gabbrodolerites within metasomatites (Fig. 5.18). Some cases are known when ore veins form in the absence of intrusions and thick veins are observed in association with thin dolerite dikes (Pavlov 1961). The metasomatites and magnetite corrode dolerites, and the thickness of ore deposits is independent of the intrusion thickness. It is thought that intrusion of dolerites and magnetite veins coincide only spatially because these used the same tectonically depressed zones during their formation.

The pyrrhotite mineralization is confined to cracked rocks in fault zones and subsidence calderas. Examples of this type of mineralization include ore occurrences in the Kureysko–Gorbichiansky area and the south Noril'sk ore junction (Ryabov and Lapkovsky 2009). The mineralized rocks in fault zones extend several hundred meters to several kilometers (Fig. 5.19) and can form massive,

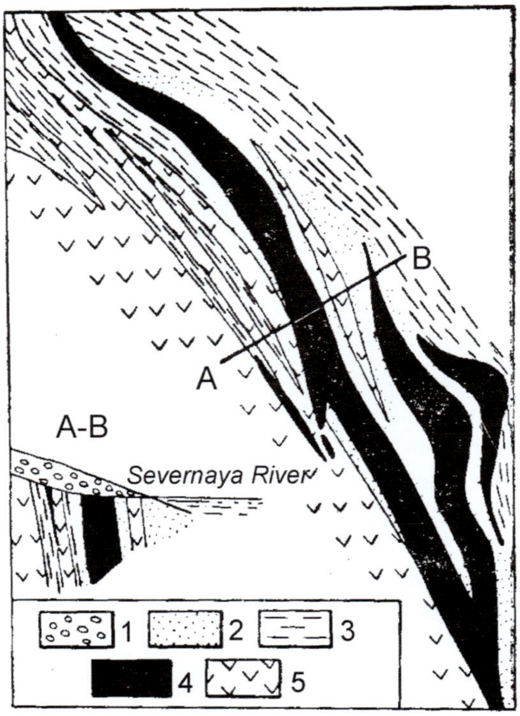

Fig. 5.18 Magnetite occurrence along the Severnaya River (From Kavardin 1976). 1 = Quaternary deposits, 2 = skarn limestone, 3 = limestone, 4 = magnetite ore, 5 = amphibolized gabbrodolerite

impregnated, and stockwork ores. The ore mineralization and concomitant metasomatites extend to a depth of several tens of meters. The massive ores often form veins filling cracks. Disseminated ores and metasomatites substitute sedimentary, volcanogenic, and intrusive rocks of various compositions. Deep-seated intrusions, such as near-fault metasomatites, that were never tapped by drilling are supposed to be sources of solutions for such ore occurrences.

It is believed that magnetite and pyrrhotite ore formations, as well as metasomatites, were formed due to pneumatolytic–hydrothermal activation in long-living fault zones. It is not infrequently observed that intrusions, ore mineralization, and fluids coincide spatially as they use the same tectonically depressed zones for intrusion of a melt and migration of volatiles (Ryabov 2007; Ryabov and Lapkovsky 2009).

Graphite formation is represented by occurrences and deposits of graphite (Kurgan'kov and Serdyuk 2002). It is traditionally considered that graphite forms under hydrothermal action of dolerite sills on coal. The immense Tunguska coal-bearing basin was thus a favorable environment for the formation of graphite deposits since dolerite sills often intruded along coal beds (Mazor and Pronina 1983). In spite of this, graphite occurrences are relatively scarce. The thermal effect of dolerite on coal is restricted, as a rule, by the formation of coke and/or anthracite (Fig. 5.20). This demonstrates that the sole thermal effect of dolerite on a coal bed would not be sufficient to form graphite.

Fig. 5.19 Outcrop near the Kan'onny River (junction of the Tiptur–Orokta Rivers)

Fig. 5.20 20 Open pit Kureysky graphite mine

The Kureysky and Noginsky deposits are the largest graphite deposits on the Platform and are confined to the junction of the Imangdino–Letninsky Fault with the Kureysky and Nizhnetungussky Faults. In these deposits, graphite beds up to 16 m thick are situated at the roof of trap sills within slightly metamorphosed rocks and locally within metasomatites, which are imposed on both country rocks and dolerite sills. In the Noginsky deposit, the graphite bed changes strike into the hard coal bed. Drilling the Ozernoe graphite deposit in the Dzhaltul region revealed a 20-m-thick graphite bed within metasomatites, 95 m from the nearest dolerite sill in the sequence (Turovtsev 2002). This deposit cannot be explained by thermo-metamorphic action of the intrusion on the coal bed. Coal beds in the western side of the Tungussky syneclise are observed in different situations: sometimes far from dolerite sills, in slightly metamorphosed sedimentary rocks, surrounded by metasomatites or in contact with thin sills that were not able to create a temperature effect required for graphitization.

Taking this into consideration, it has been assumed that the graphitization of hard coal proceeded under action of gaseous hydrocarbons migrating through fault zones (Kontorovich et al. 1997). Numerous bitumen occurrences as veinlets and amygdales in basalts and sedimentary rocks support the migration of hydrocarbons through fissured zones (Ryabov et al. 2006). The graphite deposits are restricted to zones of large faults where migration of hydrocarbons is most probable. A thermal field produced by thick sills appears to be of importance in forming largest graphite deposits.

Deposits and occurrences of copper in the region belong to the copper-zeolite and copper-porphyry formations and the formation of cuprous sandstones. We note only some of the characteristic features of these deposits and occurrences without assessing the validity of formation theories. The Arylakhsky copper deposit is largest example of a copper-zeolitic formation (Duzhikov 1988). In addition to this deposit, a series of ore occurrences is known along Nerakachi, Uokhir', and Ketere Rivers, which are located in a similar geological setting. Copper nuggets vary from submicroscopic to several centimeters across. The weight of individual nuggets may be as much as 24 kg (Rad'ko 2007; Savushkin 2008). Geologists investigating the Arylakhsky deposit have found that many features of the deposit are common to those of the Lake Superior deposits in the USA (White 1968), as well as to similar deposits of the Kola Peninsula and Koryksko–Kamchatsky regions (Duzhikov 1988). This enables us to identify their origin and to distinguish stratiform deposits of the copper-zeolitic formation on the Siberian Platform.

Disseminated native copper is also found in tuffs and in compact and amygdaloidal basalts, located along some fault zones. The main ore bodies of this type are traced as a band along the Abagalakhsky Fault and also as a field within the Bolsheavamsky trough (see Fig. 2.34) (Ryabov et al. 2005). The fact that native copper is restricted to subsidence calderas of paleovolcanoes (of the Aralykhsky deposit type and others) and to cracked zones in volcanic rocks suggests that the migration of endogenous fluids along tectonically depressed zones and unloading of volcanic structures may be the primary ore-forming factor.

The Graviysky deposit and the Sukharikhinsky ore occurrence are examples of the formation of cuprous sandstones (Rzhevskiy et al. 1980; Gablina et al. 1986; Lurje 1988; Shklyarik and Oleshkevich 1988; Gablina 2002). These are situated on the Igarsky promontory at the junction of the Noril'sko–Kharaelakhsky Faults with faults of the Yeniseysky rift system. Mineralization here extends through a sequence of sedimentary rocks including the Riphean age Izluchinsky Suite and the Vendian age Sukharikinsky Suite. The principal minerals are chalcosine, bornite, covellite, jurleite, cuprite, fahl ores, and native copper. The characteristic mineralogical feature of the rocks is the presence of Ge-bearing copper sulfate–renierite in association with tennantite and fahl ore (Nikiforova and Kulagov 1988). Accumulations of anthraxolite are noted in the ores. The origin of the ores and the formation of cuprous sandstones in the Noril'sk Region are considered from the viewpoint of a traditional "sedimentary" model (Gablina et al. 1986; Lurje 1988; Gablina 2002). Alongside with this preconceived hypothesis, there exists another model of ore formation that suggests a possibility for formation of deposits due to action of ancient smokers (Ryabov 2007).

Extensive data are available on a subject of present-day black and light smokers, hot vent outflows, and cold fluid seeps on the ocean floor. These produce various reduced compounds including sulfides and ore-forming metals through faults (Fouquet et al. 1997; Borodaev et al. 2004; Lein et al. 2004). It is proposed that smokers acted on the Fennoscandian Shield during different geological epochs (Kulikova and Kulikov 2002). Available evidence indicates that smokers are confined to rifts and faults through which subsurface fluids are discharged. Discharge of hot fluids from smokers produces a thermal field that favorably affects the appearance and rapid growth of rift structures. Smokers also discharge volatile compounds including CH_4, H_2S, H_2, and metals, which are concentrated in adjacent areas, forming ore accumulations. Taking this fact into account, one can assume that the Graviysky deposit and Sukharikhinsky occurrence were formed by ancient smokers. They are confined to a reef situated in a rift structure and the intersection of the Yeniseysky and Noril'sko–Kharaelakhsky Faults (Shklyarik and Oleshkevich 1988). Copper mineralization appears to be associated with long-lived fluid systems based on the type of smokers.

Copper-porphyry formation on the Siberian Platform is associated with the Bolgokhtokh granite stock. The copper and molybdenum mineralization here is restricted to metasomatites. Comparative analysis of the geological setting and geochemical features of the Bolgokhtokh ore occurrence with known copper-porphyry deposits of America, Aleutian Islands, Bulgaria, Iran, and the Philippines suggests a genetic similarity (Strunin et al. 1988, 1994).

It seems probable that structural–tectonic factors played an important role in formation of copper–molybdenum mineralization (see Sect. 3.10). The stock is located in the intersection zone of the large Boganidsky and Kraevoy Faults, the latter of which trends in a near north–south direction along the edge of the basalt field of the Siberian Platform. Both faults are characterized by an increased fluid saturation. Accumulations of liquid and hard bitumen in igneous and sedimentary rocks are associated with the Boganidsky Fault zone. In the Kraevoy Fault zone, a series of paleovolcanoes, including diatremes of the Vetka River, contain magnetite mineralization and pyrrhotite ores. The Middle Fokinsky volcano-tectonic structure contains Cu–Ni sulfide mineralization in gabbrodolerite intrusions that are located in the Kraevoy Fault zone about 100 km from the granite stock.

B-bearing and rare-earth formations are mainly restricted to paleovolcanoes of the central and fractured types, as well as to zones of fractured rock in fault zones. These formations are known in the Severorechensky, Kureysko–Gorbichiansky, and Avamsky areas (Osipova 1966; Vasil'ev et al. 1992; Turovtsev 2002) and consist of carbonatites, metasomatites, and hydrothermalites with accumulations of apatite and datolite.

Spar-bearing formations are represented by small and large ore occurrences and large ore fields of optic calcite, which form in different horizons of volcanogenic units in the basalt field of the Siberian Platform. Iceland spar is observed in amygdales, miarolitic cavities, globular lavas, and fracture zones of basalt and tuff. The largest reserves have accumulated in the spar-bearing province situated along the Nizhnyaya Tunguska River Fault, an area of more than 100 km^2 (Kurgan'kov and Serdyuk 2002). Here, optic calcite forms bunches and veinlets in tuff breccias, lava breccias, and fractured basalts. Crystals are typically 10–70 cm but can grow bigger, and some crystals grow to 90 kg (rarely up to 180 kg, and one crystal was found that was 310 kg). The Iceland spar forms scalenohedral and rhombohedral crystals and varies from light yellow to light brown. The material is of very high quality: described as "unique" or "first grade." In association with Iceland spar, bitumen forms veinlets and inclusions, and natural paraffin (evinkite $C_{23}H_{48}$) forms in cracks in basalts (Skropyshev 1953; Kotel'nikova et al. 2004). It can be assumed that the occurrences and deposits of optic calcite are associated with oxidation of hydrocarbon fluids in fractured rock zones controlled by faults.

5.8.5 Source of Metals in Ore Deposits

One of the problems of ore geology is the source of ore-forming components: metals and volatiles. This problem is especially acute with respect to large deposits and

Table 5.2 Composition of metals in basaltic melts

Magma	Cu, ppm	Ni, ppm	Co, ppm	(Pt + Pd + Rh), ppb	Fe_2O_3, wt%
Tholeiite-basaltic	130	110	40	100	13
Picrite-basaltic	90	700	90	180	16
Trachybasaltic	30	20	30	60	14

emphasizes the continuing importance of the above-mentioned ore formations related to traps.

The geological structure of iron-ore deposits including the Angaro–Ilimsky type, the compositional features of ores, and the mechanism for their formation have been described previously (Pavlov 1961; Pavlov 1975; Kalugin et al. 1994). The most popular idea is that these deposits were formed with active involvement of halite deposits and salt brines from enclosing rocks. Iron was extracted from a basaltic melt and was transported by chloride, whereas dolomite served as a source of magnesium for the formation of magnesioferrite ore accumulations. This is supported by the fact that natural areas of magnesioferrite deposits and saliferous sources coincide on geological maps (Pavlov 1961), as well as by experimental data and thermodynamic calculations (Kalugin et al. 1994). It has been reckoned that the large Korshunovsky magnetite deposit might be formed if only 5% iron was transported by chloride from a trap intrusion, which is an area no more than 3 × 4 km in extent and 100 m thick (1.2 km^3 volume) (Godlevsky 1967).

The origin of the Noril'sk giant Pt–Cu–Ni deposits has long been debated in the literature. Without going analyzing current points of view, we note that the problem is a long way off being solved. In respect to the ore formation of the Noril'sk deposits, it is believed that the source of metal was basaltic magma. From available data, the metal content is variable in compositionally variable basalts, but they are of the same order of magnitude (Godlevsky 1967; Dodin et al. 1994; Neruchev and Prasolov 1995; Fedorenko et al. 1996; Ryabov et al. 2001; Almukhamedov et al. 2004). Average metal content from numerous sources are given in Table 5.2.

Calculations have shown that a trap intrusion of 1 km^3 volume (10 km long, 1 km wide, and 100 m thick) with average concentrations of Pt and Pd 0.0005 g/t in the melt contains about 1.5 t of Pt and 1.5 t Pd (Neruchev and Prasolov 1995). For comparison, the authors note that the volume of the Noril'sk-I Intrusions is similar in volume to lavas erupted during eruption of the Tolbachik volcano on the Kamchatka Peninsula in 1975–1976. During the eruption of this volcano, Pt, Pd, and Au were detected in high-temperature sublimates (1,000°C) and in fumaroles (600–700°C). Based on determination of the total water coming during eruption of the volcano and the average concentration of Pt in condensate water, it has been established that hundreds of tons of Pt were carried to the surface. The relationship between lava composition and accompanied gases is notable. Reduced gases such as H_2, H_2S, CO, and CH_4, as well as C_2H_6 and C_3H_8, played a significant role in exhalations of the Northern breach of the Tolbachik volcano, and oxidized gases such as H_2O, CO_2, and SO_2, as well as HCl and HF, predominated in exhalations of the Southern breach (Menyailov et al. 1984). Lavas, bombs, and ash from the Northern breach have mainly a magnesian, moderately alkaline composition: 9.84–10.21 wt% MgO, 13.43–13.48 wt% Al_2O_3, 2.38–2.44 wt% Na_2O, and 0.97–1.02 wt% K_2O. Those of the southern breach have mainly a subalkaline-aluminiferous composition: 4.58–4.98 MgO, 16.62–16.92 wt% Al_2O_3, 3.52–3.62 wt% Na_2O, and 2.13–2.18 wt% K_2O (Flerov et al. 1984).

Other researchers show that the volume of flood basalt on the Siberian Platform ranges from 700 to 1.5–2 million km^3. Concentrations of nonferrous and noble metals in barren traps are given in a number of works (Godlevsky 1967; Arkhipova 1975; Dodin et al. 1994; Distler and Kunilov 1994; Neruchev and Prasolov 1995; Almukhamedov et al. 2004). Taking into account Clarks of nonferrous and noble metals in traps and the volume of lava, potential resources of metal in the erupted lavas can be roughly evaluated.

Reserves of Ni and Pt + Pd + Rh within the above-mentioned volume may be about 5–8 and 0.5–7 million tons, respectively. Naturally, these values are approximate; nevertheless, they allow one to judge the potential resources of a mantle tholeiite-basaltic melt and to provide an answer to the question of the source of ore-forming metals.

It was mentioned previously that intrusions of the Noril'sk type become ore deposits during enrichment in sulfur rather than enrichment in Ni (Godlevsky 1967). Taking into account the wide range of ore-forming metals in basaltic melts, it is thought that all the deposits associated with trap magmatism were formed due to the involvement of fluids. Thus, the main problem of ore formation in traps is a source of fluids and a mechanism of interaction between fluids and melts.

5.8.6 Source of Fluids

If the source of metals was a basaltic magma, then the formation of the deposits is dependent on mechanisms to

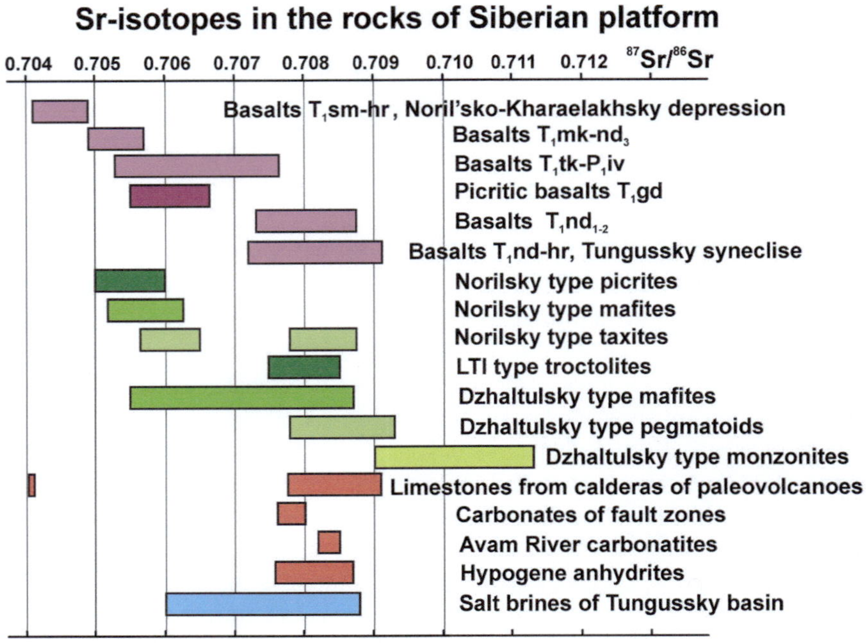

Fig. 5.21 Isotopic strontium composition in formations of the northwestern Siberian Platform. Volcanic rock suites: *Sm* Samoedsky, *Kh* Kharaelakhsky, *Mr* Morongovsky, *Nd* Nadezhdinsky, *Tk* Tuklonsky, *Gd* Gudchikhinsky, *Iv* Ivakinsky. *LTI* Lower Talnakh Intrusion (From data of Almukhamedov et al. 2004; Lightfoot et al. 1993; Sharma et al. 1992; Distler and Kunilov 1994; Koroleva and Oleinikov 1998; Vasil'ev et al. 1992; Shvartsev 2000; and our data)

extract metals from melts, as well as their transportation and accumulation. This has been successfully resolved for cases where volatiles were involved. With reference to deposits of the Siberian Platform including the deposits of the Noril'sk type, four alternative sources of volatiles are proposed: (1) basaltic melt, (2) transmagmatic (intratelluric) flows, (3) plumes, and (4) sedimentary rocks of the platform cover.

The following points should be noted regarding trap melts as potential sources of volatiles. The vast majority of dolerites from sills on the Siberian Platform have poikilo-ophitic texture that shows crystallization of the rocks was fast and that basalts melts were poor in volatiles. In addition, intrusions have contact quench zones that show volatiles did not leave the igneous chamber and metasomatites accompanying the intrusions were superimposed on igneous rocks at contacts. Thick quench zones of sills and thin contact hornfels zones that frame most trap sills show that that the basaltic melts were depleted in volatiles. Metasomatites at sill contact zones are rarely, if ever, found, and if metasomatites did form in intrusion aureoles, their thickness is not correlated with the thickness of the igneous body. Experimental data demonstrate that the saturation point of volatiles, including sulfur in basaltic melts, is sufficiently low so that even small ore deposits could be formed with involvement of volatiles, but there exist no basaltic melts oversaturated with volatiles. Thus, it is unlikely that a basaltic melt could be a source of volatiles for the formation of large deposits and thick metasomatite halos. Lenses and, more rarely, discontinuous horizons of pegmatoids can be considered as the only occurrence of volatiles in trap sills.

The nature of transmagmatic and intratelluric fluids flows, described in the literature (Korzhinsky 1973; Korzhinsky et al. 1987; Kuznetsov 1990) and used by geologists in genetic concepts with reference to metasomatites and ore formation (particularly for the Noril'sk deposits), is not clear. If it is thought that fluids appeared under mantle conditions such as trap melts, then rocks and ores being formed with involvement of these fluids would most probably retain isotopic depth markers. However, evidence for this fact cannot be found in the literature.

A large volume of information on the isotopic composition of a whole spectrum of chemical elements has been accumulated on both ore-bearing and barren traps in the northeast of the Siberian Platform. Significant variations in isotopic ratios for the range of chemical elements have been established in effusive and intrusive traps, as well as in layered ore-bearing and barren intrusions (Shchukolyukov et al. 1981; Sharma et al. 1992; Lightfoot et al. 1993; Wooden et al. 1993; Czamanske et al. 1994; Distler and Kunilov 1994; Landa et al. 1994; Neruchev and Prasolov 1995; Hawkesworth et al. 1995; Koroleva and Oleinikov 1998). Most geologists attribute this impurity to a crustal component in mantle melts. One indicator of rock formation depth is the 3He/4He ratio, and investigations have shown that the portion of mantle helium in rocks and ores of the Upper Talnakh and Dzhaltul Intrusions does not exceed 3–5% (Shchukolyukov et al. 1981; Neruchev and Prasolov 1995). The isotopic composition of strontium is a sensitive indicator of crustal material in various igneous rocks. Some variations in 87Sr/86Sr ratios are shown in Fig. 5.21

constructed on the basis of hundreds of analyses from different publications and our data. The 87Sr/86Sr ratios in hypogene carbonates, anhydrites, and brines are provided for comparison in the figure. An elevated heavy strontium isotope level in igneous rocks as compared with mantle levels (87Sr/86Sr 0.704 ‰) suggests that formation fluids from sedimentary rocks were involved in ore and magmatic processes. Variations in 87Sr/86Sr ratios in flood basalts appear to be due to variations in fluid saturation of intruding magma in feeder channels, which varies progressively during the eruption of lava flows in a single volcanic cycle.

Analysis of isotopic composition data of chemical elements reveals particular regularities. Undifferentiated barren traps retain mantle isotopic markers, whereas isotopic ratios in layered intrusions and ore-bearing traps shift toward crustal values. This can be considered the result of interaction of a basaltic melt with formation fluids during magma flow through fluid-saturated faults.

Plume models consider that fluids were sourced from the lower mantle–core region. Unfortunately, there is no conclusive natural evidence for this source of fluids in Siberian Traps, as well as for intratelluric flows. All known isotopic composition data from traps indicate that basaltic melts were generated in the upper mantle, but most of these and ore-bearing and layered intrusions have crustal markers and favor the involvement of formation fluids from sedimentary rocks.

The crustal source of volatiles is an alternative to hypotheses of mantle and plume fluid flows. Many geologists accept the involvement of sedimentary rocks following interaction with core–mantle material. However, there is no consensus regarding the mechanism of the interaction or the form of involvement of sedimentary rocks in ore–magmatic processes since they are different from core–mantle interaction. The understanding that assimilation and contamination of sedimentary rocks took place is prevalent, and we think that a melt interacted with formation or interstitial fluids of sedimentary rocks.

Sedimentary rocks of the platform consist of terrigenous, lagoon-terrigenous, and marine deposits including limestones, dolomites, graptolithic shales, sapropelites phosphorites, argillites, siltstones, sandstones, coals, anhydrites, gypsum, and halites. These rocks serve as inexhaustible source of volatiles: CH_4, H_2, H_2O, CO, CO_2, H_2S, SO_2, F, Cl, P_2O_5, and B_2O_3. Mineralogical–geochemical studies of undifferentiated traps and related ores provide evidence that one or more of these volatiles were derived from sedimentary rocks, and data on the isotopic composition for a series of chemical elements are indicative of the their crustal isotopic ratios.

5.8.7 Formation Fluids

Geologists explain the unusual composition of rocks that form in trap intrusions (hybrid metasomatic, granite-like, syenite-like, monzonites, and others) by incorporating sedimentary rocks of the platform cover. This also explains a number of features of ore formations and the crustal values of isotopic ratios for a range of elements in barren and ore-bearing traps. Many geologists consider that crustal isotopic values occur due to interaction of basaltic magma with surrounding sedimentary rocks, which are assimilated and contaminated by the magma. The country rock is completely absorbed and melted by the magma without retention of any rock structure, and a portion of the products of metasomatism is transferred with volatiles into the magma. Hence, the chemical composition of the rocks would be expected to be significantly changed. The composition of basaltic magma will significantly change when a basaltic magma assimilates or is contaminated by sandstones, argillites, limestones, anhydrites, or other rocks. The content of silica, aluminum, lime, and other elements sharply increases. Additionally, a relatively uniform, stable composition of flood basalts and dolerite sills on the platform can be considered evidence that there was no significant assimilation of sedimentary rocks. Unfortunately, the capability of basaltic magmas to assimilate material is insignificant. For example, in the case of assimilation of sulfates by the Noril'sk Intrusions, the melt would be expected to be enriched in lime. However, this was not observed, and the petrochemical composition of barren traps and those with disseminated ores is not different (Godlevsky 1967; Ryabov 1999a; Likhachev 2006). Many examples have been described in the literature (Zolotukhin 1964; Zolotukhin and Vasil'ev 1967; Tarasov 1976) where eruptive breccia and xenoliths of sandstones, siltstones, argillites, anhydrites, and limestones were not resorbed by basaltic magma; at best only rock fragments were slightly sub-melted or hornfelsed on rims, so there was no large-scale assimilation of sedimentary rocks. Therefore, there must be other mechanisms to release volatiles from sedimentary rocks. Moving basaltic magma does not assimilate sedimentary rocks, but reacts with formation fluids of particular compositions (e.g., containing sulfate ions), which accumulated within unconsolidated fluidized fault zones. It is these formation fluids or interstitial solutions of sedimentary rocks that were carriers of crustal isotopic markers. These markers were fixed in lavas, as well as in differentiated and ore-bearing traps during the process of fluid–magmatic interaction.

It is possible that sulfur from sulfates in gypsum–anhydrite rocks is the source of sulfur assimilated by a basaltic melt. One argument in favor of this concept is the increased proportion of the heavy sulfur isotope in sulfide ores of the Noril'sk deposits (Godlevsky 1967; Grinenko and Grinenko 1974). Inexhaustible sulfate resources in the Siberian Platform cover sequence solve the question of the source of the large-scale sulfur volume necessary for huge sulfide deposits. However, there is an associated need for a mechanism to reduce sulfates to sulfides. It was assumed that reduction in the Noril'sk-I and Talnakhsky deposits could have been caused by Permo-Carboniferous coal measures (Godlevsky 1967; Naldrett 2003). However, this reduction by coal is not acceptable for the Oktyabrsky deposit since its ores are found in marine carbonates and shales, and sulfate-bearing deposits of Devonian age lie stratigraphically lower than the Permo-Carboniferous coal measures. In this case, sulfate reduction can be solved if the wide involvement of hydrocarbons in petrogenesis and ore formation in traps were accounted for (Ryabov et al. 2006). It seems likely that hydrocarbons were reductants and that the process proceeded in accordance with the Engler–Gefer reaction: $CaSO_4 + CH_4 \rightarrow CaCO_3 + H_2S$.

5.8.8 Sources of Hydrocarbons

Many geologists consider that hydrocarbons were involved in the ore-forming processes. In the most ore-genetic models, the sources of hydrocarbons are deep fluids from the upper and lower mantle (Korzhinsky et al. 1987; Kuznetsov 1990; Dobretsov et al. 2001). No doubt these deep fluids exist, but it is more likely that the main source of hydrocarbons in ore–magmatic processes were coal-bearing rocks and carbonate deposits. The involvement of coal beds as a reducing agent is proposed reasonably only in a few publications (Godlevsky 1967; Farfel 1988; Naldrett 2003). It seems certain that intrusions erupted through coal-bearing sedimentary rocks had a thermo-metamorphic effect on coal-bearing sedimentary rocks. This was accompanied by depletion of the coals and loss of hydrocarbons. Hydrocarbon gases formed as a result of heating of coal migrated according to decompression principles into integrated rock fault zones and were involved in ore–magmatic processes.

One can assume that carbonate rocks were affected by breaking, crushing, destruction, dispersion, and dissociation in fault zones and by thermal activity during emplacement of trap intrusions. Heating of carbonate rocks affected by a basaltic magma led first to the dissociation of dolomite at 800°C and then calcite at 940°C (Deer et al. 1966). Dissociation products were MgO, CaO, and CO_2. The thermal dissociation of calcite proceeds at pressures below 100 atm and will not decompose at 170 atm and higher. Calcite will melt at 1,290°C (Smirnov and Rasumova 2002). It may be expected that most CO_2 will be released when carbonate rocks are subject to mechanical dispersion in tectonic zones rather than to thermal decomposition.

Synthesis of hydrocarbons from oxidized forms of - carbon may take place due to various catalytic reactions (Storch et al. 1951). The following two reactions involving hydrogen and water play the most important role:

$$nCO + (2n+1)H_2 \rightarrow C_nH_{2n+2} + nH_2O \text{ (Fisher -- Tropsch synthesis)}$$

$$(3n+1)CO + (n+1)H_2O \rightarrow C_nH_{2n+2} + (2n+1)CO_2 \text{(Kolbel -- Engelhardt synthesis)}$$

The reactions are characterized by a high catalytic activity and significant heat generation.

Generation of hydrocarbons from extremely oxidized carbon compounds within a wide temperature range has been proved and reproduced experimentally (Cherskiy et al. 1986). Synthesis of hydrocarbons follows:

$$CO_3^{2-} + 7H_2O + 8e^- \rightarrow CH_4 + 10\ OH^-$$

In the given reaction, CO_3 represents a carbon donor (carbonates, CO_2, carbonate ions dissolved in formation waters), H_2O is a hydrogen donor, and e^- are free electrons generated during polarization of the matrix-fluid system. The sources of free electrons are reductants (compounds of carbon, iron, and sulfur), as well as extremely oxidized compounds such as SiO_2 and CH_4, a wide series of liquid and gaseous hydrocarbons. Cherskiy et al. (1986) indicate that synthesis of hydrocarbons is stimulated during mechanical activation of rocks under tectonic–seismic factors, circulation of formation fluids, an increase in temperature up to 80–90°C, and generation of thermogradient fields. Hydrocarbons are formed at temperatures lower than 250–300°C.

In natural environments including the Siberian Platform, decomposition of carbonate could take place in fault zones during periods of tectonomagmatic activation, which was accompanied by pressure drop and temperature increase. Owing to disjunctive dislocations, destruction and dissociation of carbonates and local pressure decrease took place, resulting in the formation of zones of disintegrated rocks along faults. Dispersed carbonate rocks served as a favorable material for catalytic reactions.

5.8.9 Role of Tectonic Factors

Long-lived regional faults and associated lower-order compensating faults were occasionally activated over a long period of time. This caused breaking, grinding, and dispersion of sedimentary rocks, hindered growth of new minerals in fissures, and facilitated migration of formation fluids within tectonic zones. Disintegration of rocks in fault zones encouraged migration of formation fluids into these zones due to decompression. Ground, dispersed, and dissociated material from sulfate and carbonate rocks was favorable for chemical transformations.

It is known that sulfate in an ionized state can be relatively easily reduced to sulfide with the involvement of hydrocarbons at low pressures and temperatures. Dissociation products of limestone and dolomite serve as a source of various forms of carbon, which are transformed into hydrocarbons by the Fisher–Tropsch and Kolbel–Engelhardt catalytic reactions (Storch et al. 1951) or in accordance with reactions documented by Cherskiy–Mel'nikov–Tsarev (Cherskiy et al. 1986). Taking this into account, it can be assumed that dissociated material of sedimentary rocks in fault zones was mobilized by fluids. This material was the source of sulfur, hydrocarbons, magnesium, strontium, and other chemical elements when fluids interacted with a melt. This was reflected in the composition of igneous rocks and element isotopic composition of the rocks and ores.

The high permeability of rocks in tectonic zones, very deep faults, and their periodic activation enabled deep heat conduction and increased circulation of formation fluids and groundwaters in the zones. As a consequence, geothermal anomalies, mineral springs, gas emissions, and so on are restricted to these fault zones. This is supported by geothermal investigations, performed in 60 drillholes in permafrost in a region near the Talnakhsky ore junction in the Noril'sko–Kharaelakhsky Fault zone (Iogin 1971). Results from these studies show that the temperature in drillholes increases near tectonic deformations. The largest geothermal gradient (4.0°C per 100 m) was fixed in localized zones of differentiated deeply lying intrusions. The average temperature gradient was 2.40°C per 100 m, in hole ST-2 it was as high as 43.5°C at 1,740 m. In mine workings of the Talnakhsky ore junction, the temperature in deep horizons was close to 50 °C.

5.8.10 Role of Fluids

Fluids accumulated and migrated through tectonically depressed zones of faults that melts used during intrusion. This predetermined the genetic links between fluids and melt. Formation fluids mobilized and activated by a melt in fault zones extracted metals from a basaltic magma, and so became ore-forming fluids that transported metals and formed ore accumulations and ore deposits, as well as formed metasomatic aureoles at intrusion contacts, along cracked zones, and around ore accumulations.

Extraction of metals proceeded most extensively when fluids interacted with liquid melt rather than with crystallized rock. Migration of compositionally variable fluids through a melt corrected the redox potential of the system. Fluids represented an individual phase that was in equilibrium with silicate melt and therefore the volume of volatiles was not limited by their solubility in the melt. Fluids concomitant with the basaltic melt had a multicomponent composition which was responsible for the geochemical features of ores, and the fluid volume controlled deposit size.

Volatile components stimulated fluid–magmatic differentiation of a basaltic melt, which began and proceeded most intensively in feeder channels. The fluid–magmatic interaction of the melt with volatiles of variable composition caused silicate–silicate and ore–silicate layering of the melt, its sulfurization, ferritization, metallization, and immiscible segregation of sulfide, magnetite, and iron metal liquids. Additionally, volatiles provided transport of ore liquid and facilitated accumulation of ores. Layered liquids finally crystallized in magmatic chambers with local redistribution of chemical elements forming differentiated ore-bearing intrusions (Ryabov 1999a).

5.8.11 Main Geodynamic Factors

The performed investigations enable determination of the main geodynamic factors of ore-forming systems in traps of the Siberian Platform (Ryabov 2007).

The long duration of tectonic activity in the region caused the formation of deep faults with thick zones of cracked rock. Multiple periods of fault rejuvenation impeded healing of these cracks, and formation fluids were attracted into tectonic zones of disintegrated rocks due to decompression with fluidized zones formed along faults. Basaltic magma favored these tectonic zones for intrusion, and magma mobilized and activated fluids concentrated here. Interaction between magma and fluids caused fluid–magmatic differentiation into coexisting immiscible silicate–silicate and ore–silicate liquids. Fluids activated by the melt played the ore-forming role in this system. Basaltic magma was the

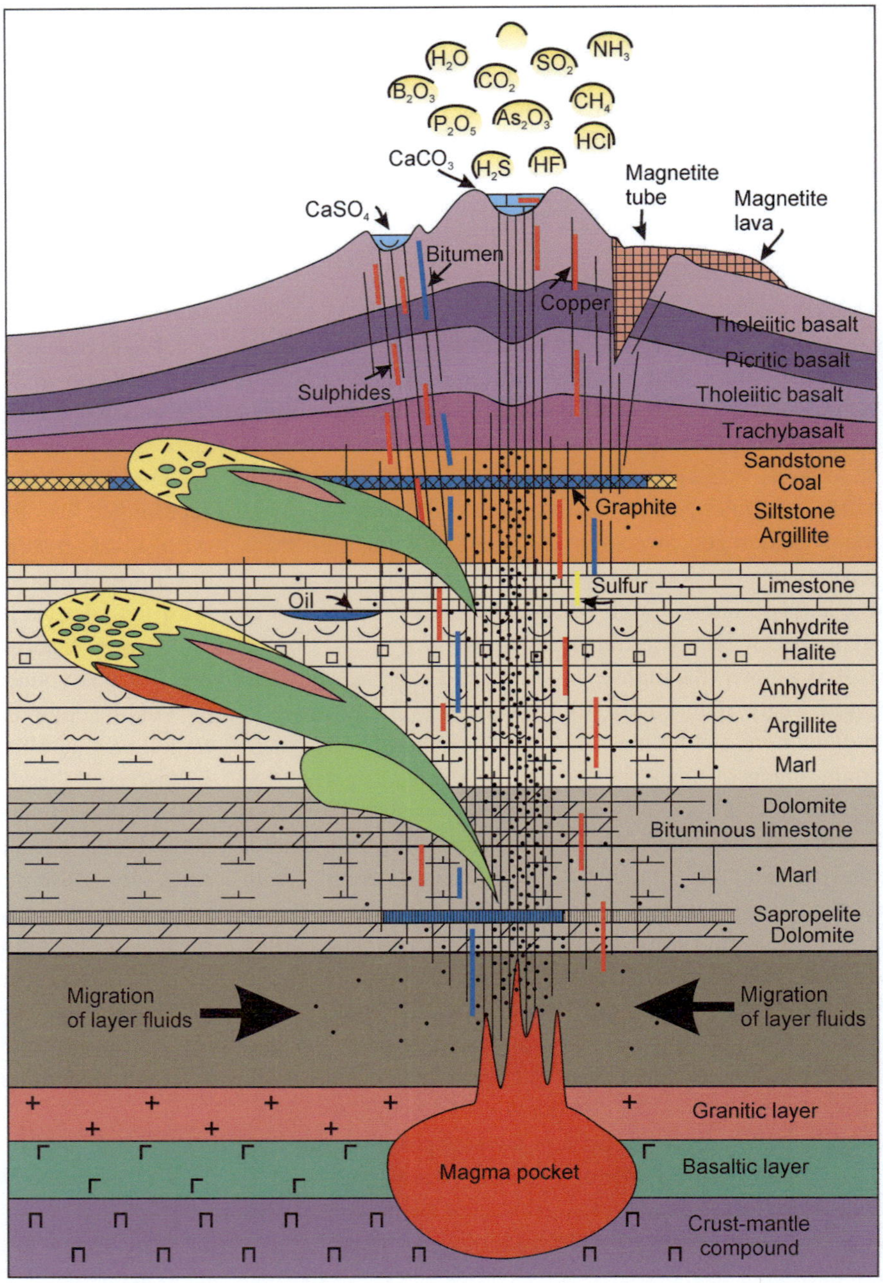

Fig. 5.22 Model for ore deposit formation associated with traps of the Siberian Platform (Explanations are given in the text)

source of ore-forming metals, and the fluid regime controlled extraction of metals from the melt, their transport and accumulation. The composition and volume of fluids were responsible for the geochemical features and size of the deposits in accordance with the physicochemical state parameters of the ore–magmatic system.

Based on this, a working ore–magmatic model for the formation of deposits associated with traps of the Siberian Platform has been proposed (Ryabov 2004, 2007). The general view of the model is given in Fig. 5.22. Here, a cross section of rocks in the Platform cover sequence is shown in which all potential carriers of hydrocarbons, sulfur, halogens, and other volatile components are present. At bottom of the figure, the structure of the deeper parts of the Siberian Platform and a magmatic chamber are shown. Vertical lines denote rock fractures, which increase in density at fault junctures. The direction of formation fluid migration through disintegrated zones is shown as a concentration of points. Intrusions are shown for the Talnakhsky ore junction as xenoliths located at different hypsometric levels (green), for the Upper Talnakh, Kharaelakh, and Lower Talnakh Intrusions (downward through the levels). In the leading parts of the intrusions, metasomatite aureoles (yellow) are shown, horizons of pegmatoids with Pt low-sulfide mineralization (pink) are shown in the upper parts of igneous bodies, and Cu–Ni–PGE sulfide ores (red) are shown at the base of the bodies. Degassing leads to formation of hydrothermal mineralization (bitumen, sulfides, copper, sulfur, and others) along cracked zones and free liberation of volatile components into the atmosphere. Within a fault zone, graphitization of coal strata is common, and accumulations of liquid and hard bitumen are formed along cracks.

In summary, the following key points can be noted in the genetic model for ore formation:

- The source of ore-forming metals for deposits is a basaltic magma.
- The source of volatile components is sedimentary rocks of the Siberian Platform cover.
- The principal volatile components of ore-forming fluids are H_2S, CH_4, H_2O, HCl, HF, CO_2, CO, P_2O_5, B_2O_3, SO_2, and NH_3.
- The involvement of volatile components is responsible for the fluid–magmatic differentiation of a basaltic melt.
- Fluids extract ore-forming metals from a melt and transport and accumulate them.
- The formation of ore deposits is governed by a cause-and-effect relationship with phenomena such as tectonics, fluid regime, and magmatism.

References

Agafonov LV, Chepurov AI, Lavrentyev YG, Pokachalova OS (1974) Regularly oriented inclusions in olivines from ultrabasites of Kamchatka. Geologia i Geophisika, No 6:49–60 (in Russian)

Almukhamedov AI, Medvedev AY (1982) Geochemistry of sulfur in evolution processes in mafic magmas (ed Chernyshov LV). Nauka, Moscow, 148 p (in Russian)

Almukhamedov AI, Medvedev AY, Zolotukhin VV (2004) Spatial and temporal evolution of composition of Permo-Triassic basalts in the Siberian Platform. Petrologiya 12(4):339–353 (in Russian)

Aplonov VS (2001) Thermobarogeochemical model of the Talnakh Ni-PGE deposit. VNIIOkeangeologiya, St Petersburg, 234 p (in Russian)

Aplonov VS, Moskalyuk AA (1978) Evolution of chemical behavior of solutions forming Noril'sk copper-nickel ores. In: Theory and practice of fluid inclusion study. Nauka, Moscow, pp 119–122 (in Russian)

Arkhipova AI (1975) Geochemical characteristic of intrusive traps of the Noril'sk plateau. Trudy NIIGA (ed Kavardin GI). Nedra, Leningrad, 135 p (in Russian)

Arutyunyan LA (1977) About one peculiarity of behavior of sulfide concretions in sulfide-silicate melts. Doklady Akad Nauk USSR 236(4):957–960 (in Russian)

Arutyunyan LA (1986) Geochemical regularities of nickel behavior in endogenous systems. Unpublished Doctor Sc thesis, GEOKHI RAS, Moscow, 51 p (in Russian)

Arutyunyan LA, Sargsyan GO, Smolkin VF (1988) On nickeliferous olivines crystallizing in silicate and sulfide-silicate systems. In: Nickeliferous mafic-ultramafic complexes of Ukraina, Siberia and Far East. Apatity, pp 78–81 (in Russian)

Babansky AD, Solovova IP, Ryabchikov ID (1985) Microliquation heterogeneity in melted inclusions and in the mesostasis of andesite. In: Thermobarometry and geochemistry of ore-forming fluids: abstracts of VII All-Union conference, Lviv, pp 123–124 (in Russian)

Bezmen NI (1984) Experimental study of liquid differentiation of mafic-ultramafic melts. In: Abstracts of international geol conference, vol 4, Moscow, pp 264–265

Borodaev YuS, Mosgova NN, Gablina IF, Bogdanov YuA, Starostin NI, Fardust F (2004) Zonal pipes of black smokers from the Rainbow hydrothermal field (Mid-Atlantic Ridge, latitude 36°14′ N). Vestnik Moskovskogo Univer, Ser 4. Geology 3:35–48 (in Russian)

Boudreau AE, McCallum SI (1985) Composition of apatite and biotite from the Stillwater Complex, Montana: evidence for hydrothermal transport of PGE and REE in Cl-bearing solutions (Abstract) Lunar and Planet Sci, vol 16: Pap 16ht Conf 11–15 Mar 1985. Pt1, Houston, TX, pp 85–86

Boudreau AE, Mathez EA, McCollum IS (1986) Halogen geochemistry of the Stillwater and Bushveld Complexes: evidence for transport of platinum group elements by Cl-rich fluids. J Petrol 27 (4):967–986

Bowen NL (1922) The reaction principle in petrogenesis. J Geol 30:177–198

Champness PE (1970) Nucleation and growth of iron oxides in olivines, $(Mg, Fe)_2SiO_4$. Mineral Mag 37(291):790–800

Cherskiy NV, Melnikov VP, Tsarev VP (1986) The process of generation of hydrocarbons from extremely oxidized compounds of carbon and water. Doklady Akad Nauk SSSR, Nauki o Zemle Section 288:201–209 (in Russian)

Chukhrov FV (ed-in-chief) (1960) Minerals: handbook, vol I. Nauka, Moscow, 617 p (in Russian)

Colette M, Pierre M (1982) An experimental determination of chrome solubility in mafic silicate liquids. Application: distribution of chrome in liquids of chromiferous spinels. C R Acad Sci, Ser 3, 295:371–373 (in French)

Czamanske GK, Wooden JL, Zientek ML, Fedorenko VA, Zenko TE, Kent J, King B-SW, Knight RJ, Siems DG (1994) Geochemical and isotopic constraints on the petrogenesis of the Noril'sk–Talnakh ore-forming system. In: Lightfoot PC, Naldrett AJ (eds) Proceedings of the Sudbury-Noril'sk symposium. Ontario Geol Surv Spec Pub 5:313–341

Daly R (1933) Igneous rocks and depths of the Earth. McGraw-Hill, New York London, 533 p

Deer WA, Howie RA, Zussman J (1966) Rock-forming minerals, vol 3 [Russian translation of 1962 English edition: an Introduction to the rock-forming minerals. Longman, London]. Mir, Moscow, 317 p

Distler VV, Kunilov VE (eds) (1994) Geology and ore deposits of the Noril'sk Region. VII internat platinum symposium guidbook. Moscovskiy Contact Moscow, 67 p (in Russian)

Distler VV, Grokhovskaya TL, Evstigneeva TL, Sluzhenikin SF, Filimonova AA, Duzhikov OA, Laputina IP (1988) Petrology of magmatic sulfide ore formation. Nauka, Moscow, 232 p (in Russian)

Distler VV, Sluzhenikin SF, Cabri LJ, Krivolutskaya NA, Turovtzev DM, Golovanova TA, Mokhov AV, Knauf VV, Oleshkevich OI (1999) Platinum ores of the Noril'sk layered intrusions: interrelation of magmatic and fluid concentration of noble metals. Geologiya rudnykh mestorozhdeniy 41(3):241–265 (in Russian)

Dobretsov NL, Kirdyashkin AG, Kirdyashkin AA (2001) Depth geodynamics, SB RAS, Affiliated Branch GEO, Novosibirsk, 408 p (in Russian)

Dodin DA (1962) Oceanites and chrysotile-asbestos occurrences in the basin of the Upper Sukhoy Balek River (northwestern margin of the Siberian Platform). Inform Bull NIIGA. NIIGA, Leningrad, issue 29, pp 39–54 (in Russian)

Dodin DA (1967) Petrology of traps of the East Khraelakh (Noril'sk region). Unpublished Ph.D. thesis, VSEGEI, Leningrad, 31 p (in Russian)

Dodin DA (2002) Metallogeny of Taimyr-Norilsk region. Nauka, St Petersburg, 822 p (in Russian)

Dodin DA (2005) Sustainable development of the Arctic Region (problems and perspectives). Nauka, St Petersburg, 283 p (in Russian)

Dodin DA, Batuev BN (1971) Geology and petrology of Talnakh differentiated intrusions and their metamorphic aureole. In: Urvantsev NN (ed) Petrology and metallogeny of Talnakh and Noril'sk differentiated intrusions. Nedra, Leningrad, pp 31–100 (in Russian)

Dodin DA, Duzhikov OA (2005) Large and unique PGE-bearing deposits of volcanogenic pluton-related complexes: geodynamics, metallogeny, models of formation. In: Platinum of Russia, OOO Geoinformmark Pub, vol VI. Moscow, pp 150–165 (in Russian)

Dodin DA, Chernyshov NM, Polferov DV, Tarnovetsky LL (1994) Platinum metal deposits of the world, vol 1: Low-sulfide platinummetal deposits in rhythmically-layered complexes. Geoinformmark, Moscow, 279 p (in Russian)

Dodin DA, Dodina TS, Duzhikov OA, Neruchev SS (2002) Noril'sk-Talnakh giant ore deposit. In: Russian Arctic region: geological history, minerageny, geoecology. Nauka, St. Petersburg, pp 664–682 (in Russian)

Duzhikov OA (1971) On picritic basalts of the Noril'sk region. Doklady Akad Nauk 197(6):1406–1410

Duzhikov OA (1988) Copper-zeolite ore formation. In: Geology and metalogeny of sulfide deposits of Noril'sk Region USSR. Nauka, Moscow, pp 173–179 (in Russian) [refer to English translation of 1988 Russian edition by Duzikov OA, Distler VV, Strunin VM, Mkrtychyan AK, Sherman ML, Sluzhenikin SF, Lurje AM. Soc Econ Geol Spec Pub 1:242 p]

Duzhikov OA, Distler VV, Sherman MK, Sluzhenikin SF (1988) Sulfide copper-nickel ore formation. In: Geology and metallogeny of sulfide deposits of Noril'sk region USSR. Nauka Moscow, pp 77–172 (in Russian) (refer to English translation of 1988 Russian edition by Duzikov OA, Distler VV, Strunin VM, Mkrtychyan AK, Sherman ML, Sluzhenikin SF, Lurje AM. Soc Econ Geol Spec Pub 1:242 p)

Eliseev EN (1959) Geochemistry of the main sulfide copper-nickel provinces of USSR. In: Problems of geochemistry, issue 1. Lviv Univer Pub, Lviv, pp 5–184 (in Russian)

Farfel LS (1988) Prediction of ore deposits. Nedra, Moscow, 150 p (in Russian)

Fedorenko VA, Lightfoot PC, Naldrett AJ, Czamanske GK, Hawkesworth CJ, Wooden JL, Ebel DS (1996) Petrogenesis of the flood-basalt sequence at Noril'sk North Central Siberia. Int Geol Rev 38:99–135

Feoktistov GD (1978) Petrology and conditions for generation of trap sills (ed Zolotukhin VV). Nauka, Novosibirsk, 168 p (in Russian)

Fisher R (1950) Entmichungen in Schmelzen aus Schwermetallxyden, Silikaten und Phospaten, Ihre geochemische end lagerstatten-kundliche Bedeutung. Neues Jahrb, fur Min 81(3):315–364

Flerov GB, Andreev VN, Budnikov VA, Tsyurupa AI (1984) Petrology of igneous products. Great Tolbachinskiy fissure eruption (1975–1976, Kamchatka Peninsula). Nauka, Moscow, pp 223–284 (in Russian)

Fouquet YY, Charlou J-L, Ondreas H, Radford-Kroery J, Donval JP, Dulville E, Apprioural R, Cambon P, Pelle H, Landure JY, Normand A, Ponsevera E, German C, Person L, Barriga F, Costal I, Relvas J, Ribero A (1997) Discovery and first submersible investigations on the Rainbow hydrothermal field on the MAR (36°14′ N). EOS Am Geophys Union Trans 78(46), 832 p

Freestone IC, Powell R (1983) The low temperature field of liquid immiscibility in system K_2O–Al_2O_3–FeO–SiO_2 with special reference to the join fayalite-leucite-silica. Contrib Mineral Petrol 81(2–3):291–299

Gablina IF (2002) Paragentic association of cuprous sandstones and shales with coal-bearing deposits: degassing of the Earth: geodynamics, geofluids, oil and gas. In: Proceedings of the international conference in memory of academician P.N. Kropotkin, GEOS, Moscow, 20–24 May 2002, pp 107–109 (in Russian)

Gablina IF, Rzhevskiy VF, Vasilovskaya LV (1986) Epigenetic ore-controlling zonation of the Graviisky cuprum deposit. In: Genesis of rare element and lead-zinc stratiform deposits. Nauka, Moscow, pp 147–168 (in Russian)

Genkin AD (1968) Platinum metal minerals and their associations in Noril'sk copper–nickel ores. Nauka, Moscow, 106 p (in Russian)

Genkin AD, Troneva NV, Zhuravlev NN (1969) The first finding of potassium, iron, and copper sulfide—djierfisherite in ores. Geologiya rudnykh mestorozhdeniy No 5:57–65 (in Russian)

Genkin AD, Dudykina AS, Telesheva LR (1970) Some data on the composition of rock-forming pyroxenes and olivines from the Noril'sk-I gabbrodolerites intrusion. In: Minerals of basites in connection with aspects of petrogenesis. Nauka, Moscow, pp 40–56 (in Russian)

Genkin AD, Distler VV, Laputina IP (1979) Chromite mineralization of differentiated trap intrusions. In: Conditions of formation of magamatic ore deposits. Nauka, Moscow, pp 105–127 (in Russian)

Genkin AD, Distler VV, Gladyshev GD, Filimonova AA, Evstigneeva TL, Kovalenker VA, Laputina IP, Smirnov VA, Grokhovskaya TL (1981) Copper-nicel sulfide ores from Noril'sk ore deposits (ed Shadlun TL). Nauka, Moscow, 234 p (in Russian)

Gerasimovsky VI, Kuznetsova SY (1974) On sulfur content in effusive rocks of Iceland. Geokimiya No 8:1239–1242 (in Russian)

Ginsberg AS (1951) Experimental petrography. Leningrad Univer Publ., Leningrad, 270 p (in Russian)

Glagolev AA, Korchagin AM, Kharchenko AG (1974) Arbarastakh and Inagli alkaline-ultramafic massifs. Nauka, Moscow, 175 p (in Russian)

Godlevsky MN (1959) Traps and ore-bearing intrusions of the Noril'sk region. Gosgeoltekhizdat, Moscow, 68 p (in Russian)

Godlevsky MN (1961) Review of hypotheses concerning origin of sulfide copper-nickel deposits. Trudy, VSEGEI, Leningrad, New Series, issue 45, pp 71–82 (in Russian)

Godlevsky MN (1967) Sources of mineralization associated with traps of the Siberian platform. In: Urvantsev NN (ed) Petrology of traps of Siberian platform. Nedra, Leningrad, pp 173–189 (in Russian)

Godlevsky MN (1968) Magmatic ore deposits. In: Genesis of endogenous ore deposits. Nedra, Moscow, pp 8–73 (in Russian)

Godlevsky MN (1971) The proportions between sulfide and silicate parts in a coarse of evolution of mafic magma. In: Abstracts of the international geochemical congress, vol 1, Moscow, pp 65–77 (in Russian)

Godlevsky MN, Polushkina AP, Stepanov VK (1971) Monoclinic pyroxenes of the Talnakh differentiated intrusion. Zapiski VMO 100. Second Series, issue 5, pp 545–557 (in Russian)

Golubkov VS (1970) Mesozoic group of formations of near-fault metasomatism in the Yenisei ore province. Geology and mineral resources of north-western part of the Siberian Platform. Nedra, Leningrad, pp 23–49 (in Russian)

Goode ADT (1974) Oxidation of natural olivines. Nature 248 (5448):500–501

Gorbachev NS (1989) Fluid-magmatic interaction in sulfide-silicate systems (ed Zyryanov VN). Nauka, Moscow, 126 p (in Russian)

Goryainov IN (1969) On genesis of layered intrusions by example of the Talnakh massif. In: Zaridze GM (ed) Magmatism, metamorphism and metasomatism. Central Committee Communist Party of Georgia Pub, Tbilisi, pp 47–61 (in Russian)

Goryainov IN (1975) Viscosity of basalt melts in connection with the problem of gravitational fractionation. In: Copper-nickel ores of the northwestern part of the Siberian Platform. NIIGA, Leningrad, pp 108–117 (in Russian)

Goryainov IN, Andreeva TV (1972) Fe, Ni, Cu and Al chlorides in copper–nickel ores of Talnakh sulfide ore deposits. Doclady Akademii Nauk 204(6):1456–1459 (in Russian)

Goryainov IN, Aplonov VS, Moskalyuk AA (1973) Composition of gas–liquid inclusions in rocks of the Talnakh intrusion. In: North-

Siberian nickeliferous region and its perspectives for mining. NIIGA, Leningrad, pp 97–103 (in Russian)

Greig JW (1928) On the evidence which has been presented for liquid silicate immiscibility in the laboratory and in the rocks of Agate Point. Am J Sci 215:375–402

Grigor'ev DP (1935) Immiscibility of silicate systems compositionally close to natural rocks. Zapiski VMO 64(1):250–267 (in Russian)

Grinenko LN (1967) Sulfur isotope composition from sulfides of some copper-nickel deposits and ore occurrences of the Siberian platform. In: Petrology of traps of the Siberian platform. Nedra, Leningrad, pp 221–229 (in Russian)

Grinenko LN (1987) Genetic model of formation of copper-nickel deposits based on isotope-geochemical data. In: Sotnikov VI (ed) Construction of models of ore-forming systems. Nauka, Novosibirsk, pp 119–128 (in Russian)

Grinenko VA, Grinenko LN (1974) Geochemistry of sulfur isotopes (ed Vinogradov AP). Nauka, Moscow, 274 p (in Russian)

Gulin SA, Pogrebitskiy YuE, Sukhov LG (I968) *In* Zonal-metasomatic nature of Noril'sk copper-nickel deposits and criteria for exploration apparent from it. Prediction and methods of exploration for deposits of nickel, tin, and diamonds in Soviet Arctic zone. NIIGA Leningrad, pp 29–32 (in Russian)

Gulin SA, Sukhov LG (1973) Some remarks on liquation-magmatic hypothesis concerning formation of the Norlsk-type copper-nickel deposits. Sovetskaya Geologiya 2:24–35 (in Russian)

Gulin SA, Sukhov LG (1974) Criteria for prediction of copper-nickel deposits based on their metasomatic origin conceptions. In: Lind EN (ed) The state and trends of investigations into metallogeny of traps. Krasnoyarsk TGU, Krasnoyarsk, pp 67–69 (in Russian)

Haggerty SE, Barker I (1967) The alteration of olivine in basaltic and associated lavas. Contrib Mineral Petrol 16(3):233–257

Haughton DR, Roeder PL, Skinner BJ (1974) Solubility of sulfur in mafic magmas. Econ Geol 69(4):451–467

Hawkesworth CJ, Lightfoot PC, Fedorenko VA, Blake S, Naldrett AJ, Doherty W, Gorbachev NS (1995) Magma differentiation and mineralization in the Siberian continental flood basalts. Lithos 34: 61–88

Hess HH (1941) Pyroxenes of common mafic magmas. Am Mineral 26:515–535

Hill RET, Roedder PL (1974) The crystallization of spinel from basaltic liquid as a function of oxygen fugacity. J Geol 82(6):167–180

Iogin CM (1971) Geothermal gradient of the Talnakh ore junction. In: Geology and mineral recourses of the Noril'sk region (transactions of conference). NIIGA, Leningrad, pp 345–346 (in Russian)

Irvine TN (1967) Chromian spinel as a petrogenetic indicator. Part 2. Petrologic applications. Can J Earth Sci 4:71–103

Irvine TN, Smith CH (1969) Primary oxide minerals in the layered series of the Muskox intrusion. In Magmatic ore deposits. The Econ Geol Pub Company, Lancaster, pp 67–86

Ivanov MK, Ivanova TK, Shatkov VA (1971a) Some facial peculiarities of formation of intrusive and effusive picrites. In: Urvantsev NN (ed) Geology and mineral resources of the Noril'sk region. NIIGA, Leningrad, pp 83–84 (in Russian)

Ivanov MK, Ivanova TK, Tarasov AV, Shatkov VA (1971b) Peculiarities of petrology and mineralization of differentiated intrusions of the Noril'sk ore junction (Noril'sk-I, Noril'sk-II and Mt. Chernaya deposits). In: Urvantsev NN (ed) Petrology and metallogeny of the Talnakh and Noril'sk differentiated intrusions. Nedra, Leningrad, pp 197–304 (in Russian)

Ivanova TK (1975) Application of structural-facial analysis in volcanic reconstructions in the Noril'sk region, In: Urvantsev NN (ed) Copper-nickel ores at the northwest of the Siberian Platform. NIIGA, Leningrad, pp 52–75 (in Russian)

Johnston A Dana, Stout James H (1984) Development of orthpyroxene -Fe/Mg ferrite symlecites by continuous olivine oxidation. Contrib Miner and Petrol 88(2):196–202

Judd JW (1885) On the Tertiary and older Peridotites of Scotland. Qurt Journ Geol Soc London 41:354–418

Kalugin IA, Tret'yakov GA, Von-der-Flaas GS (1994) Origin of iron ores in traps. OIGGM SB RAS, Novosibirsk, 46 p (in Russian)

Karpenkov AM, Rudashevsky NS, Shumskaya NI (1981) Natural palladium and bismuth chloride: the phase of composition Pd4Bi5Cl3. Zapiski VMO 1:86–91 (in Russian)

Kavardin GI (1976) Metallogeny of the northwestern part of the Siberian Platform (ed Urvantsev NN). Nedra, Leningrad, 158 p (in Russian)

Kavardin GI, Golubkov VS (1968) Metallogenic zonation of the northwestern Siberian Platform. Doklady Akad Nauk SSSR 178(2): 412–415 (in Russian)

Kavardin GI, Golubkov VS, Ivanova AM, Staritsyna GN (1967) Metallogenic zonation of the Yenysei nickel-bearing province. Uchenye Zapiski NIIGA, Regional'naya Geologiya 11:43–137 (in Russian)

Kavardin GI, Staritsyna GN, Golubkov VS, Goryainov IN, Kravtsova LI (1968) Traps of the Yenisei ore province (ed Urvantsev NN). Nedra, Leningrad, 208 p (in Russian)

Komarova MZ, Lyul'ko TP (1967) On subdivision of trap intrusions of the Noril'sk region. In: Petrology of traps of the Siberian Platform. Nedra, Leningrad, pp 43–54 (in Russian)

Kontorovich AE, Pavlov AL, Khomenko AV, Tret'yakov GA (1997) Physico-chemical conditions of graphitization of carbon-bearing rocks (by example of the Siberian Platform). Geokhimiya 6:563–570 (in Russian)

Kopylova AG, Oleinikov BV (1997) Noble metals in native iron of continental basites. In: Geological structure and mineral resources of Sakha Republic (Yakutiya), vol 3. IG Yakutsk Affiliated Branch SB RAS, Yakutsk, pp 56–58 (in Russian)

Koroleva OV, Oleinikov BV (1998) Geochemistry and genesis of monzonitoids of the Dzaltul'sky trap intrusive. Geologiya i Geofisika 39(2):178–189 (in Russian)

Korovyakov IA (1948) On picritic effusive traps of the northwestern part of the Siberian Platform. Doklady Akad Nauk USSR LXII (1):129–133 (in Russian)

Korovyakov IA (1960) Behavior of nickel in Siberian traps. In: Mineral'noe Syr'e, issue 1. VIMS, Moscow, pp 169–183 (in Russian)

Korovyakov IA, Nelyubin AE, Raikova ZA, Khortova LK (1963) Origin of the Noril'sk trap intrusions hosting sulfide copper-nickel ores. In: Trudy VIMS (ed Gon'shakova VI), New Series, issue 9. Gosgeoltekhizdat, Moscow, 102 p (in Russian)

Korzhinsky DS (1973) Metamagmatic processes. Izv Akad Nauk SSSR Geol Ser 12:3–6 (in Russian)

Korzhinsky DS, Zotov IA, Pertsev NN (1987) Transmagmatic fluids, metamagmatism, and ore formation. In: Regularities of metamagmatism and metamorphism. Nauka, Moscow, pp 5–28 (in Russian)

Kotel'nikova EN, Filatov SK, Chukanov NV (2004) Evenkite: symmetry, chemical composition, identification, and behavior on heating. Zapiski VMO CXXXIII(3):80–93 (in Russian)

Kotul'sky VK (1946) Toward the problem of origin of magmatic copper-nickel deposits. Doklady Akad Nauk SSSR 51(5):381–383 (in Russian)

Kotul'sky VK (1948) The state-of-art problem of genesis of copper-nickel sulfide deposits. Sov Geol 29:11–24 (in Russian)

Krasov NF, Clocciatti R (1979) Immiscibility of silicate melt and its possible petrogenetic role (based on data from study of melted inclusions). Doklady Akad Nauk SSSR 248(1):201–204 (in Russian)

Krasov NF, Ryabov VV, Konenko VF (1985) Immiscibility of natural melts from the data of fluid inclusion study. Study of fluid inclusions and geochemistry of ore-forming fluids. Abstracts

of VII All-Union conference, Lviv AS USSR, pp 111–112 (in Russian)
Kravchenko SM (1977) Fractionation of minor elements during differentiation of mafic magmas. Nauka, Moscow, 218 p (in Russian)
Kravtsov VF, Sedykh YuN, Gor YuG (1971) Geologo-structural features of the Talnakh ore junction. In: Petrology and metallogeny of the Talnakh and Noril'sk differentiated intrusions. Nedra, Leningrad, pp 8–31 (in Russian)
Krivenko AP, Polyakov GV, Glotov AI, Izokh AE, Balykin PA (1986) Criteria for the nickel content of differentiated ultramafic intrusions and estimate of perspectives for particular massifs in the southern Krasnoyarsk region and in Western Mongolia. In: Geochemistry in a local metallogenic analysis. Abstracts of the All-Union symposium, vol 2. IGG SB AS USSR, Novosibirsk, pp 34–35 (in Russian)
Kucher MI, Fridman AI, Vlasov VV (1986) Study of carbon isotope composition in high-temperature fluid inclusions in intrusive rocks of the Talnakh ore junction. In: Abstracts of the 11th All-Union symposium on isotope geochemistry, Moscow, pp 212–214 (in Russian)
Kudelina MB, Kudryavtseva VP, Garanin VK (1983) Peculiarities of the chemical and phase compositions of spinellides and ilmenite in the Talnakh intrusion (Noril'sk region). Geologiya rudnykh mestorozhdeniy 25(5):24–37 (in Russian)
Kulikova VV, Kulikov VA (2002). Smokers in Precambrian—? Degassing of the Earth: geofluids, oil and gas. In: Proceedings of the international conference in memory of academician P.N. Kropotkin, GEOS, Moscow, 20–24 May 2002, pp 172–175 (in Russian)
Kullerud GA, Yoder HS (1965) Sulfide-silicate relation and their bearing on ore formation under magmatic, postmagmatic and metamorphic condition. In: Proceedings of symposium: problems of postmagmatic ore deposition, vol 2, Prague, pp 327–331
Kurgan'kov PP, Serdyuk SS (2002) Mineral raw material base for graphite and Iceland spar of the North of the Central Siberia and perspective of its development. In: Russian Arctic Region: geological history, minerageny, geoecology. Nauka, St. Petersburg, pp 614–619 (in Russian)
Kutolin VA (1969) A statistic study of chemical peculiarities of basalts from different formations. Nauka, Novosibirsk, 140 p (in Russian)
Kuz'min EE (1978) Hydrogeochemical criteria for exploration of copper–nickel deposits. In: Sobolev VS (ed) Criteria for exploration of the Noril'sk-type sulfide ores. Nauka, Novosibirsk, pp 116–121 (in Russian)
Kuznetsov YuA (1990) Problems of origin and formation analysis of igneous formations. Selected transactions, vol . III. Nauka, SB RAS, Novosibirsk, 292 p (in Russian)
Landa EA, Neruchev SS, Prasolov EM (1994) Isotope-geochemical criteria of typification, identification of ore material sources, and predictive estimate of PGE-metal deposits. In: Platinum in Russia. Problems of development of mineral raw material base of PGE metals. AO Geoinformmark, Moscow, pp 196–206 (in Russian)
Lebedev AP (1955) Trap formation of the central part of the Tunguska basin. AS USSR, Moscow, 197 p (in Russian)
Lebedev AP (1957) On differentiation types in traps of the Siberian platform. Izv Akad Nauk SSSR Geol Ser 2:55–74 (in Russian)
Lein AY, Maslennikov VV, Maslennikova SP, Ul'yanova NV, Zaikov VV, Spiro B (2004) Isotopes of sulfur and carbon in near-hydrothermal ecosystems of black smokers of the Ural paleovolcano. Geochimiya 7:770–784 (in Russian)
Letnikov FA, Gantimurova TP, Sizykh NV (1998) The evaluation of the fluid regime of continental volcanism in some Eurasia regions (Perm-Miocene). Geokhimiya 5:448–455 (in Russian)
Lightfoot PC, Hawkesworth CJ, Hergt J, Naldrett AJ, Gorbachev NS, Fedorenko VA, Doherty W (1993) Remobilisation of continental lithosphere by mantle plumes: major trace element and Sr- Nd- and Pb-isotope evidence for picritic and tholeiitic lavas of the Noril'sk district Siberian Trap, Russia. Contrib Mineral Petrol 114:171–188
Likhachev AP (1965) Role of leucocratic gabbro in the generation of Noril'sk differentiated intrusions. Izv Akad Nauk SSSR Geol Ser 10:75–89 (in Russian)
Likhachev AP (1978) On conditions of generation of ore-bearing and barren mafic-ultramafic magmas. Doklady Akad Nauk SSSR 238 (2):447–450 (in Russian)
Likhachev AP (1996) Toward dynamics of emplacement of Talnakh ore-bearing intrusions and related platinum–copper–nickel deposits. Otechestv Geol 8:20–26 (in Russian)
Likhachev AP (1998) The Talnakh intrusion and its platinum–copper–nickel ores. Rudy i metally 1:36–46 (in Russian)
Likhachev AP (2006) Platinum-copper-nickel and platinum deposits. Eslan, Moscow, 496 p (in Russian)
Loewinson-Lessing FYu (1933) Petrography. ONTI, Moscow, 462 p (in Russian)
Loewinson-Lessing, FYu (1935) Four coryphaei of petrography: Washington, Dupark, Sederholm, and Vogt. Izv Akad Nauk SSSR, Series VII, No. 3 (in Russian)
Lurje (1988) Formation of cuprous sandstones and shales. In: Geology and metallogeny of sulfide deposits of Norill'sk region USSR. Nauka, Moscow, pp 225–238 (in Russian) [refer to English translation of 1988 Russian Edition by Duzhikov OA, Distler VV, Strunin VM, Mkrtchyan AK, Sherman ML, Sluzhenikin SF, Lurje AM. Soc Econ Geol Spec Pub, No 1, 242 p]
Lurje ML, Ledneva VP, Selivanovskaya TV, Semenov LS, Tuganova EV, Ryabchenko AA, Komarova MZ, Staritsyna GN, Tomanovskaya YuI (1976) Structures of traps of the Siberian platform (ed ML Lurje). Trudy VSEGEI, New Series, vol 235. Nedra, Leningrad, 171 p (in Russian)
MacBirney AR, Nakamura Y (1974) Immiscibility in late-stage magmas of the Skaergard intrusion. Year book 73 for 1973–1974. Carnegie Institution, Washington, DC, pp 348–352
Manankov AV, Sharapov VN (1985) Kinetics of phase transitions in mafic melts and magmas, vol 602. Trudy IGG SB AS USSR, Novosibirsk, 175 p (in Russian)
Marakushev AA (ed) (1981) Petrography: Part II. Moscow Univ Pub, Moscow, 328 p (in Russian)
Marakushev AA, Perchyuk LL (1971) Origin and evolution of transmagmatic and metamorphic fluids. In: Abstracts of international geochemical congress, vol 2, Moscow, pp 513–514 (in Russian)
Marakushev AA, Perchyuk LL (1974) Thermodynamic model of the Earth's fluid regime. Ocherki Fisiko-khimicheskoy petrologii, issue 4. Nauka, Moscow, pp 102–130 (in Russian)
Marakushev AA, Fenogenov AN, Emel'yanenko PF, Duzhikov OA, Skripnichenko VA (1982) Genesis of the Noril'sk-type layered intrusions. Vestnik MGU, Geol Ser 1:3–19 (in Russian)
Marakushev AA, Paneyakh NA, Zotov IA (2002) Specific features of formation of copper–nickel sulfide deposits in trap provinces (by example of Pechenga and Noril'sk). Doklady Akad Nauk 382 (5):668–673 (in Russian)
Marakushev AA, Paneyakh NA, Zotov IA (2007) Petrologic model of formation of sulfide deposits. Doklady Akad Nauk 416(2):232–235 (in Russian)
Martikhaeva DKh, Leont'eva VG, Mitrofanova AYu, Vorontsov AE (1986) An experience of chromate-mass-spectrometric study of compositions of carbonaceous matters from basaltic diatremes of the Siberian Platform. In: Second All-Union conference on carbon geochemistry, Moscow, 29 Sept–1 Oct 1986, pp 290–291 (in Russian)
Masaitis VL (1958) Petrology of the Alamdzhakh trap intrusion (the Vilyuy River basin) (ed Lurje ML). Trudy VSEGEI Leningrad, New Series, No 22, 133 p (in Russian)
Masaitis VL (1964) Magmatic trap subprovinces of the Siberian Platform. In: Transactions for geology of the Eastern Siberia. VSEGEI, Leningrad, pp 137–158 (in Russian)
Maslov (1963) Tectonics of the Igarka-Noril'sk region and ore-controlling structures. In: Tectonics of Siberia, vol 2. Nedra, Novosibirsk, pp 336–350 (in Russian)

Mathez EA, Boudreau AE, McCallum IS (1985) Apatite and biotite from the Stillwater Complex, Montana. Can Mineral 23, 308 p

Mazor YuR, Pronina NV (1983) Generation of graphites under conditions of magmatogene metamorphism (by example of the Tunguska basin). In: Coal fields and conditions of their formation. Nauka, Moscow, pp 142–145 (in Russian)

Menyailov IA, Nikitina LP, Shapar' VN (1984) Geochemical features of volcanic gases: great Tolbachinsky fissure eruption (1975–1976, Kamchatka Peninsula). Nauka, Moscow, pp 285–308 (in Russian)

Muir JD, Tilley CE, Scoom JH (1957) Contributions to the petrology of Hawaiian basalt. 1. The Picrite basalts of Kilauea. Am J Sci 255 (4):241–253

Mukhina AM (1985) Investigation of composition of gases from tuffs in the south-eastern part of the Tungusskaya syneclise. In: Geology and mineral resources of the Eastern Siberia. Institute of Earth's Crust Pub, Irkutsk, pp 62–63 (in Russian)

Naldrett AJ (1969) A portion of the system Fe–S–O and its application to sulfide ore magmas. J Petrol 10:171–201

Naldrett AJ (2003) Magmatic Sulfide Deposits of Nickel-Copper and Platinum-metal ores. St Petersburg University, St Petersburg, 487 p

Naldrett AJ, Duke JM, Lightfoot PC, Thompson JFH (1984) Quantitative modeling of the segregation of magmatic sulfides: an exploration guide. CIM Bull 77(864):46–56

Natorkhin IA, Arkhipova AI, Batuev BN (1977) Petrology of Talnakh intrusions (ed Egorov LS). Nedra, Leningrad, 236 p (in Russian)

Naumov VB, Kovalenko VI, Sobolev AV, Tikhonenkov PI, Samoilov SS (1986) Immiscibility of silicate and salt melts based on evidence derived from studies of inclusions in high-temperature fluorine. Doklady Akad Nauk SSSR 2:453–456 (in Russian)

Nekrasov IY, Gorbachev NS (1978) Physical-chemical conditions of formation of the Noril'sk-type differentiated intrusions and copper-nickel ores. Ocherki fisiko-khimicheskoy petrologii 7:92–123 (in Russian)

Neruchev SS (1987) Gas phase of some differentiated intrusions of the Noril'sk region. In: Conference of young researches, PGO Sevmorgeologiya, Leningrad, April 1986, pp 15–21 (in Russian)

Neruchev SS (1988) The gas phase composition in rocks of some nickeliferous intrusions. Geochemistry and mineralogy of the Noril'sk region. Nauka, Leningrad, 59 p (in Russian)

Neruchev SS, Prasolov EM (1995) Fluid-geochemical model of platinoid deposits associated with trap magmatism In: Orlov VP (ed) Platinum in Russia, Geoinformmark, Moscow vol II, book 1, pp 94–106 (in Russian)

Nesterenko GV, Almukhamedov AI (1973) Geochemistry of differentiated traps (Siberian platform). Nauka, Moscow, 198 p (in Russian)

Nikiforova TS, Kulagov EL (1988) Rare minerals from stratiform cuprum ores of the Igarsky region. In: Geochemistry and mineralogy of ore formations in the Noril'sk region. Nauka, Leningrad, pp 98–101 (in Russian)

Nikol'sky NS (1978) Thermodynamics of mineral equilibria of basites (ed Govorov IN). Nauka, Moscow, 177 p (in Russian)

Nikol'sky NS (1987) Fluid regime of endogenous mineral formation. Nauka Moscow, 198 p (in Russian)

Nitsan U (1974) Stability field with respect to oxidation and reduction. J Geophys Res 79(5):706–711

Oleinikov BV (1979) Geochemistry and ore genesis of platform basites (ed Bazhenov AI). Nauka, Novosibirsk, 264 p (in Russian)

Oleinikov BV, Kopylova AG (2000) The new type of noble metal mineralization in trap intrusives of the northern Siberian Platform. OtechestvGeol 5:21–23 (in Russian)

Oleinikov BV, Okrugin AV, Tomshin MD, Levashov VK, Varganov AS, Kopylova AG, Pankov VYu (1985) Native metal formation in platform basites. Yakutsk Affiliated Branch AS USSR Pub, Yakutsk, 188 p (in Russian)

Oleinikov BV, Kopylova AG, Barashkov YuP (1997) Noble metals in early magmatic native iron of continental basites and sulfide globules of kimberlites. In: The principal problems in study of magmatogenic ore deposits. Abstracts of international symposium in honor of the centenary of Academician AG Betekhtin, IGEM RAS, Moscow, 8–10 Apr 1997, pp 293–294

Oleinikov BV, Kopylova AG, Korobeinikov AF, Kolpakova NA (1999) Platinum and palladium in the metallic phase of earth's basites. Geokhimiya 364(1):107–109 (in Russian)

Olshansky YI (1948) About significant flowability of sulfide melts and probable geological significance of this phenomenon. Doklady Akad Nauk USSR 63(2):187–190 (in Russian)

Olshansky YI (1951) The Fe–FeS–FeO–SiO_2 system. Izv Akad Nauk SSSR, Geol Ser 6:128–176

Osipova GN (1966) Analcime-scapolite vent facies of diatremes at the Severnaya River (North-west of the Siberian Platform). In: Geology and petrology of intrusive traps of the Siberian Platform. Nauka, Moscow, pp 135–155 (in Russian)

Pankov VYu (1984) Genetic types of inclusions in protocrysts of gabbro-dolerites from the Siberian Platform. In: Geochemistry and mineralogy of basites and ultrabasites of the Sibarian Platform. Yakutsk, pp 43–53 (in Russian)

Pavlov NV (1961) Magnomagnetite deposits of the Tungusskaya syneclise of the Siberian Platform. IGEM AS SSSR, Moscow, 224 p (in Russian)

Pavlov DI (1975) Magnetite ore formation with exogenous chloride liquids involved. Nedra, Moscow, 246 p (in Russian)

Pavlov AL (1979) Thermodynamics of sulfidization processes in silicate melts and problems of sulfide formation. Nauka, Novosibirsk, 80 p (in Russian)

Pavlov AL (1983) Genesis of magmatic magnetite deposits (ed Sharapov VN). Nauka, Novosibirsk, 220 p (in Russian)

Pavlov AL (1992) Physico-chemical modeling of magmatogene fluid ore-forming systems. UIGGM SB RAS, Novosibirsk, 115 p (in Russian)

Petrenko GV, Arutyunyan LA, Zhangurov AA, Mityunin YK, Predovsky AA (1974) About possibility of nickel removal from olivinites under hydrothermal conditions. Geokhimiya 8:1185–1192 (in Russian)

Petrov VP, Bogatikov OA, Petrov RP (eds) (1981) Petrographic dictionary. Nedra, Moscow, 495 p (in Russian)

Philpotts AR (1979) Silicate liquid immiscibility in tholeiitic basalts. J Petrol 20(1):99–118

Philpotts AR (1982) Compositions of immiscible liquids in volcanic rocks. Contr Miner Petrol 80(3):201–218

Putnis A (1979) Electron petrography of high-temperature oxidation in olivine from the Layered Intrusion. Mineral Mag 43(326):293–296

Rad'ko VA (2007) About indications of a large cuprum deposits at the north of the Putorana plateau. Geology and mineral deposits of the Krasnoyarsk Region, Krasnoyarsk, issue 8, pp 126–130 (in Russian)

Reverdatto VV (1963) Petrology of Anakitsky differentiated trap massif and its structure. Geologiya i geofisika 10:79–92 (in Russian)

Roedder E (1951) Low-temperature liquid immiscibility in the system K_2O–FeO–A_2O_3–SiO_2. Am Mineral 36:282–286

Roedder E (1979) Silicate liquid immiscibility in magmas. In: Yoder HS Jr (ed) The evolution of Igneous magmas. Princeton Univ Press, Princeton, pp 15–57

Rogover GB (1959) The Noril'sk-I deposit. Gosgeoltekhizdat, Moscow, 168 p (in Russian)

Ryabov VV (1969) About origin of taxitic gabbrodolerites, leucocratic gabbro, and magmatic breccia in the Kharaelakhsky branch of the Talnakh intrusion. Geologiya i Geofisika 2:51–58 (in Russian)

Ryabov VV (1974) Plagioclases and clinopyroxenes of layered trap intrusions as indicators of differentiation of magmatic melt. Doklady Akad Nauk SSSR 219(1):187–200 (in Russian)

Ryabov VV (1978a) About some peculiarities in the behavior of chrome and titanium in magmatic clinopyroxenites from different

formations. In: Proceedings on petrology and mineralogy of ultramafic and mafic rocks. Nauka, Novosibirsk, pp 119–133 (in Russian)
Ryabov VV (1978b) Peculiarities of principal rock-forming minerals as exploration criteria. In: Sobolev VS (ed) Criteria for exploration of the Noril'sk-type sulfide ores. Nauka, Novosibirsk, pp 66–85 (in Russian)
Ryabov VV (1981) High-magnesian trap intrusions depleted in chrome as an exploration criterion for the Norl'sk type ore-hosting intrusions. In: Petrology of lithosphere and mineralization (Abstracts). Leningrad, pp 324–325 (in Russian)
Ryabov VV (1984a) On composition of upper contact zones of Noril'sk intrusions hosting chromite-rich mineralization. In: Polyakov GV (ed) Petrochemistry: criteria for mineralzation of magmatic complexes. IGG SB AS USSR, Novosibirsk, pp 124–142 (in Russian)
Ryabov VV (1984b) Peculiarities of mineralogy of magnesian basites in the Noril'sk region. In: Sobolev VS (ed-in-chief) Magnesian basites of the western Siberian platform and aspects of their nickel content. Nauka, Novosibirsk, pp 150–158 (in Russian)
Ryabov VV (1986) Formation peculiarities of mafic pegmatoid trap intrusions. In: Origin and evolution of magmatic formations in the Earth's history. Proceedings of VII All-Union petrographic symposium, vol 2. IGG SB AS USSR, Novosibirsk, pp 57–58 (in Russian)
Ryabov VV (1988a) Immiscible differentiation in traps. In: Goncharenko AI (ed) Actual problems of Siberian geology, vol 1, TGU, Tomsk, pp 132–134 (in Russian)
Ryabov VV. (1988b) Immiscible liquid in traps of Siberian platform. In: 2nd international conference on natural glasses, Prague, 21–25 Sept 1987, pp 401–405
Ryabov VV (1989a) Genetic types of high-magnesian traps. In: Oleinikov BV (ed) Mafic magmatism of the Siberian platform and its metallogeny. Yakutsk, pp 78–79 (in Russian)
Ryabov VV (1989b) Immiscibility in natural glasses (by example of traps) (ed Zolotukhin VV). Nauka, Novosibirsk, 223 p (in Russian)
Ryabov VV (1989c) Chemical composition of immiscible liquids in natural glasses from traps. Geologiya i Geofisika 11:69–78 (in Russian)
Ryabov VV (1990) Differentiation of magnesian melts with reference to traps of the Siberian Platform, Unpublished DSc thesis, Institute of Geology and Geophysics SB USSR AS, Novosibirsk, 31 p (in Russian)
Ryabov VV (1992a) Nontraditional types of platinum metal ores in the Noril'sk-type sulfide deposits. Russ Geol Geophys 33(11): 79–88 (in Russian)
Ryabov VV (1992b) Olivines of Siberian traps as indicators of petrogenesis and ore formation (ed Zolotukhin VV). Nauka, Novosibirsk, 116 p (in Russian)
Ryabov VV (1992c) Types of low-sulfide of platinum-group metal ores in contact zones of trap intrusions. In: Distler VV, Evstigneeva TL, Kamshilina EM (eds) Geology and genesis of platinum metal deposits. IGEM Publication, Moscow, pp 40–41 (in Russian)
Ryabov VV (1994) Platinum of Siberian traps. UIGGM SB RAS, Novosibirsk, 32 p (in Russian)
Ryabov VV (1997) Ore melts among basalts on the Putorana plateau (Siberian Platform). Russ Geol Geophys 38(12):1932–1949
Ryabov VV (1999a) Fluid regime of trap magmatism and ore formation (petrological aspect). Russ Geol Geophys 40(10):1437–1452
Ryabov VV (1999b) Ore geology of the Noril'sk region on the threshold of millennium. In: Simonov ON (ed) Abstracts of the regional symposium: trends in geological-exploration works on nickel, copper and platinoids in the Taimyr Autonomous Area in 2000–2005. Taimyrkomprirodresursy, Noril'sk, pp 25–26 (in Russian)
Ryabov VV (2004) Noril'sk Cu–Ni–PGE giant deposits: problems of relationships between tectonics, magmatism and ore formation. In: Petrology of magmas and metamorphic complexes. Proceedings of petrographic scientific conference, Tomsk, issue 4, pp 160–165 (in Russian)
Ryabov VV (2007) Geodynamic control on trap magmatism and oreformation (Siberian Platform). In: Large igneous provinces of Asia, mantle plumes and metallogeny. International symposium, Novosibirsk, pp 190–192
Ryabov VV, Anoshin GN (1999) Platinum–iron mineralization in intrusive traps of the Siberian Platform. Russ Geol Geophys 40 (2):163–176
Ryabov VV, Lapkovsky AA (2009) Volcano-tectonic structures in Siberian traps as areas of accumulation and discharge of formation fluids during the process of tectono-magmatic activation of the region. In: Volcanism and geodynamics: transactions of IV All-Russian symposium on volacanology and paleovocanology, vol 2, pp 490–493 (in Russian)
Ryabov VV, Lapkovsky AA (2010) Native iron (–platinum) ores from the Siberian Platform trap intrusions. Aust J Earth Sci 57: 707–736
Ryabov VV, Pavlov AL (1984) Physicochemical principles of the formation of magnesial scarn paragenesis in chambers of stratified intrusives of Noril'sk type. Sov Geol Geophys 25(3):51–56
Ryabov VV, Sokolova GA (1970) Findings of halite at the exocontact of the Talnakh ore-hosting intrusion and its role in postmagmatic processes. In: Problems of regional geology and petrography in Siberia and methods of geochemical and geophysical investigations. Nauka, Novosibirsk, pp 79–82 (in Russian)
Ryabov VV, Yakobi NYa (1981) Behavior of chrome during differentiation of trap magma. In: Sobolev VS (ed) Problems of genetic petrology. Nauka, Novosibirsk, pp 85–111 (in Russian)
Ryabov VV, Zolotukhin VV (1977) Minerals of differentiated traps (ed Sobolev VS). Nauka, Novosibirsk, 392 p (in Russian)
Ryabov VV, Zolotukhin VV (1978) Composition of rock-forming minerals of traps as an indicator of important petrologic processes. In: Abstracts of XI congress of internat mineralogical association, vol 1, Novosibirsk, pp 51–53 (in Russian)
Ryabov VV, Bakumenko IT, Fominykh IM (1977) Dendritic metacrysts in traps of the Noril'sk region and some aspects of their formation. In: Sobolev VS (ed) Proceedings on petrology and mineralogy. Nauka, Novosibirsk, pp 47–71 (in Russian)
Ryabov VV, Konenko VF, Krasov NF (1985a) Immiscibility phenomena in glasses from ores of native iron of Khungtukun Obolensky AA (ed) intrusion. Doklady Akad Nauk SSSR 285(4):982–987 (in Russian)
Ryabov VV, Konenko VF, Khmel'nikova OS (1985b) Rock-forming minerals of picritic basalts of the Noril'sk region. Sov Geol Geophys 26(4):77–84
Ryabov VV, Pavlov AL, Lopatin GG (1985c) Native iron of Siberian traps (ed Zolotukhin VV). Nauka, Novosibirsk, 169 p (in Russian)
Ryabov VV, Shevko AYa, Anoshin GN, Simonov ON, Krakovetsky YuK (1996a) Dyumtaley as a new classical example of a layered mafic intrusion with sulfide and magnetite mineralization. In: Abstracts of Ist international symposium "Large and unique deposits of rare and noble metals (problems of genesis and implementation)", St Petersburg, 8–11 Oct 1996, pp 77–79 (in Russian)
Ryabov VV, Shevko AYa, Lyul'ko VA, Dushatkin AB (1996b) Compositional diversification of rocks in small intrusions as a consequence of magmatic differentiation of trachybasalt melt. In: Magmatism and geodynamics of Siberia. Abstracts of scientific conference in honor of 75 anniversary of MP Kortusov, Tomsk, pp 80–81 (in Russian)
Ryabov VV, Shevko AYa, Simonov ON, and Anoshin GN (1996c) Composition of Pt-bearing Cr-rich skarns in Talnakh) (Norrilsk region). Russ Geol Geophys 37(7): 57–72 (in Russian)
Ryabov VV, Gora MP, Grib DE, Shevko AYa (2001) Multiphase limburgite dikes with phenomena of fluid-magmatic interaction.

In: Abstracts of IV Russian conference on experimental mineralogy, Chernogolovka, p 65 (in Russian)
Ryabov VV, Shevko AYa, and Zateeva SN (2005) "Abnormal formations" in Traps of the Siberian platform as indicators of geodynamic conditions of flood basalt formation. Litosfera 4: 165–177 (in Russian)
Ryabov VV, Ponomarchuk VA, Talibova AG (2006) Bitumen, graphite and carbon-bearing matter in rocks and ores of the northwestern Siberian Platform: occurrence and carbon isotope composition. In: Topical problems of ore formation and metallogeny. Obolensky AA (ed) Abstracts of international meeting devoted to 100th anniversary of academician. GEO, Novosibirsk, pp 189–190 (in Russian)
Rzhevskiy VF, Miroshnikov AE, Dushatkin AB, Shklyarik GK (1980) Cuprous upper Cambrian deposits of the Igarsky region. In: Processes of sedimentary and volcanogenic-sedimentary accumulation of non-ferrous metals (Siberia and Far East). Nauka, Novosibirsk, pp 81–84 (in Russian)
Sato M, Valenza M (1980) Oxygen fugacities of the layered series of Skaergaard intrusion, East Greenland. Am J Sci 230A:134–158
Savushkin MP (2008) From the history of native cuprum occurrences in the west-northern Siberian platform. In: Geology and mineral deposits of the Krasnoyarsk Region. Krasnoyarsk, issue 9, pp 265–266 (in Russian)
Sharapov VN, Cherepanov AN (1986) Dynamics of magma differentiation (eds Polyakov GV, Kirdyashkin AG). Nauka, Novosibirsk, 186 p (in Russian)
Sharapov VN, Golubev VS (1976) Dynamics of interaction between magma and rocks. Nauka, Novosibirsk, 237 p (in Russian)
Sharma M, Basu AR, Nesterenko GV (1992) Temporal Sr-, Nd-, Pb-isotopic variations in the Siberian flood basalts: implications for the plume-source characteristics. Earth Planet Sci Lett 113:365–381
Shatkov VA (1973) Rock-forming minerals and some regularities in formation of ore-hosting intrusions of the Noril'sk region. Unpublished Ph.D. thesis. IGG SB AS SSSR Novosibirsk, 24 p (in Russian)
Shchukolyukov YA, Verkhovsky AB, Drubetsky EK, Oleinikov BV, Okrugin AV, Bibakova EV, Makarov VA, Kirnozova TI, Prasolov EM, Meshik AP, Subbotin EP, Zuev BK (1981) Exploration of isotopic indications of mantle origin of native element substance in trap rocks. Geokhimiya 10:1442–1452 (in Russian)
Shklyarik GK, Oleshkevich OI (1988) Mineralizing rift complex of rocks in the Graviisky stratiform cuprum deposit (Igarsky region). In: Geochemistry and mineralogy of ore formations of the Noril'sk region. Leningrad, Nauka, pp 87–90 (in Russian)
Shvartsev SL (2000) The chemical composition and strontium isotopes in brines of the Tunguska basin in connection with a problem of their formation. Geokhimiya 11:1170–1184 (in Russian)
Skinner BJ, Peck DL (1969) An immiscible sulfide melt from Hawaii. Econ Geol Monogr 4(3):10–32
Skropyshev AV (1953) About paraffin from polymetallic vein. Doklady Akad Nauk SSSR 88(4):717–719 (in Russian)
Sluzhenikin SF, Distler VV, Duzhikov OA, Kravtsov VF, Kunilov VE, Laputina IP, Turovtsev DM (1994) Low-sulfide minertalization in the Noril'sk differentiated intrusions. Geologiya rudnykh mestorozhdeniy 36(3):195–217 (in Russian)
Smirnov MF (1966) Structure of the Noril'sk nickeliferous intrusions and sulfide ores. Nedra, Moscow, 58 p (in Russian)
Smirnov VA, Rasumova NP (2002) Towards the interaction between magma and carbonate rocks. Ural Geol J 4(28):35–43 (in Russian)
Smith FG (1968) Physical geochemistry. Nedra, Moscow, 476 p (Russian translation) [English version is published: Addition-Wesley, Reading, 476 p]
Smolkin VF, Borisov AE, Marakushev AA (1987) Indications of differentiation and immiscibility of melts in Pechenga picritic basalts. Doklady Akad Nauk SSSR 294(3):669–673 (in Russian)
Sobolev VS (1986) Petrology of traps: Izbrannye trudy. Nauka, Novosibirsk, 209 p (in Russian)
Starosel'tsev KV, Starosel'tsev VS (1987) Fossil anthraxolie-zeolite fumarole in Triassic volcanites of the north Tungusskaya sineclise. Geologiya i Geofisika 3:113–117 (in Russian)
Stepanov VK (1975) Rock-forming minerals of the Talnakh intrusion and analysis of their paragenesises with development of criteria of mineralization. Unpublished Ph.D. thesis, Moscow, 25 p (in Russian)
Stepanov VK (1981) Dynamic model of emplacement, crystallization and mineralization of the Noril'sk ore-hosting intrusions. In: Genesis and localization conditions of copper-nickel mineralization. Trudy TSNIGRI, TSNIGRI, Moscow, vol 162. pp 13–19 (in Russian)
Storch HH, Golumbic N, Anderson R (1951) The Fischer-Tropsh and related synthesis. Willey, New York, 554 p
Strunin BM, Distler VV, Mkrtychyan AK (1988) Copper-porphyry ore formation. In: Geology and metalogeny of sulfide deposits of Noril'sk region USSR. Nauka, Moscow, pp 205–224 [refer to English translation of 1988 Russian edition by Duzikov OA, Distler VV, Strunin VM, Mkrtychyan AK, Sherman ML, Sluzhenikin SF, Lurje AM. Society of Econ Geologists Special Publication 1, 242 p]
Strunin BM, Duzhikov OA, Barmina OA, Komarov VV (1994) Geological map of the Noril'sk region of 1:200,000 scale. In: Sherman ML, Mkrtychyan AK, Malich NS (eds) Explanatory note. AO Geoinformmark, Moscow, 118 p
Sukhanova EN (1968) About internal structure, chemical behavior and metalogeny of the Northwestern (Kharaelakhsky) branch of the Talnakh intrusion. In: Sorokov SD (ed) Geology and mineral deposits of the Noril'sk Mining region. NGMK, Noril'sk, pp 112–114 (in Russian)
Sukhareva MS, Kuznetsova NP (1983) Toward a question of relation between differentiated intrusions of the Talnakh ore junction (by example of northern flanks). In: Abstracts of trap magmatism of the Siberian Platform with respect to its tectonics and exploration of mineral deposits. Krasnoyarskgeologiya Pub, Krasnoyarsk, pp 89–92 (in Russian)
Sukhov LG, Dudenko LN, Natorkhin IA (1981) Quantitative methods for prediction of endogenous ore deposits. Nedra, Leningrad, 139 p (in Russian)
Tarasov AV (1976) On formation mechanisms of the Noril'sk intrusions and related sulfide bodies. In: Pospelov GL (ed) Replacement and intrusion during magmatism and ore formation. Nauka, Novosibirsk, pp 123–276 (in Russian)
Tikhonenkov PI, Nesternko GV, Naumov VB (1985) Specific character of immiscibility of flood basalts of the Siberian platform. In: Abstracts of VII All-Union conference: thermometry and geochemistry of fluid ore formation. Lviv, pp 140–141 (in Russian)
Traskin VY, Arutyunyan LA, Savchenko AN, Neradovsky YN (1988) About wetting ability and moveability of sulfide melts. Geokhimiya 3:418–423 (in Russian)
Tuganova EV (1988) Petrological peculiarities of formation of Noril'sk-type nickeliferous intrusions. In: Mitrofanov FP, Gorbunov GI (eds) Presence of nickel in mafic-ultramafic complexes. Apatites, pp 36–40 (in Russian)
Turovtsev DM (2002) Contact metamorphism of the Noril'sk intrusions. Nauchniy Mir, Moscow, 318 p
Urvantsev NN (1959) The Yenisei ore field. Collected papers on Arctic region geology. Trudy NIIGA (ed Markov FG), vol 102, issue 10. Nedra, Leningrad, pp 28–48 (in Russian)
Urvantsev NN (1970) Geologic-tectonic peculiarities of generation of Noril'sk copper-nickel ores. Uchenye Zapiski NIIGA, Regional'naya Geologiya, issue 18. pp 39–55 (in Russian)
Vasil'ev YR, Zolotukhin VV (1975) Petrology of ultrabasites at the north of the Siberian Platform. Nauka, Novosibirsk, 271 p
Vasil'ev YR, Laguta ON, Pavlov VG, Shevko AY, Ponomarchuk VA (1992) Isotope composition of carbon, oxygen, and strontium in carbonate occurrences of the Avam River basin (north-west of the Siberian Platform). Doklad Akad Nauk SSSR 325(5):1034–1038 (in Russian)

Vilensky AM (1967) Petrology of intrusive traps at the north of the Siberian platform (ed Lebedev AP). Nauka, Moscow, 270 p (in Russian)

Vilensky AM (1978) Peculiarities of interchamber differentiation of trap intrusions and some general features of magmatism of the Noril'sk region. In: Sobolev VS (ed) Criteria for Exploration of the Noril'sk-type sulfide ores. Nauka, Novosibirsk, pp 30–44 (in Russian)

Vilensky AM, Kavardin GI, Kravtsova LI, Staritsina GN (1964) Petrology of trap intrusions on the right-bank of the Yenisey River lower reach (ed Kazakov AN). Nauka, Moscow, 237 p (in Russian)

Vol AE (1962) Structure and properties of double metallic systems, vol 2. Fizmatgiz, Moscow, 981 p

Volborth A, Housley RM (1984) Preliminary description of complex graphite, sulfide, arsenide, and platinum-group element mineralization in a pegmatoid pyroxenite of the Stillwater Complex, Montana, USA. TMPM Tscher Miner Peter Mitt 33:213–230

Vorontsov AE, Razvozzhaeva EA, Syngaevskiy ED, Khlebnikova AA (1986) Geochemical investigation of carbonaceous matters from diatremes of the Siberian platform. Geokhimiya 2:226–235 (in Russian)

Vortsepnev VV (1978) Conditions of formation of the Talnakh ore-bearing intrusion on the base of data on inclusions. Geologiya rudnykh mestorozhdeniy 2:64–70 (in Russian)

Wager LR, Brown GM (1968) Layered igneous rocks. Oliver and Boyd, Edinburgh/London, 588 p

Walker F, Poldervaart A (1949) Karoo dolerites of the Union of South Africa. Geol Soc Am Bull 60(4):591–706

Weiblen PW, Morey GB (1975) The Duluth complex – a petrologic and tectonic summary. 36th Ann Duluth Minning Symp. Dept of Conferences and Continuing Education Univ of Minnesota, pp 72–95

Weiblen PW, Morey GB (1980) A summary of the stratigraphy, petrology and structure of the Duluth complex. Amer J Science 280-A: 88–133

White WS (1968) The native copper deposits of the northern Michigan. In: Ridge JD (ed) Ore deposits of the United States, 1933–1967. American Institute of Mining Engineers, New York, V.1, pp 303–325

Wicks CE, Block FE (1963) Thermodynamic properties of 65 elements—their oxides, halides, carbides, nitrides. USA Bureau of Mines, Bull 605. USA Government Printing Office, Washington, DC, 240 p

Wooden JL, Czamanske GK, Fedorenko VA, Arndt NT, Chauvel C, Bouse RM, King B-SW, Knight RJ, Siems DF (1993) Isotopic and trace element constraints on mantle and crustal contributions to characterization of Siberian continental flood basalts Noril'sk area Siberia. Geochim Cosmochim Acta 57:3677–3704

Yudina VV (1965) Some metamorphism and metasomatism phenomena associated with the Talnakh gabbrodolerite intrusion (Noril'sk region). In: The proportion between magmatism and metamorphism in genesis of ultrabasites. Nauka, Moscow, pp 112–175 (in Russian)

Zavaritsky AN (1947) On pegmatites as transitional rocks between igneous rocks and veins. Zap VMO 76:36–50 (in Russian)

Zavaritsky AN, Sobolev VS (1961) Physical-chemical grounds of petrography of igneous rocks. Gosgeolteckhizdat, Moscow, 383 p (in Russian)

Zharikov VA (1976) Grounds of physical-chemical petrology. Moscow Univ Pub, Moscow, 420 p (in Russian)

Zharikov VA, Gorbachev NS, Ishbulatov RA (1986) Fluid-magmatic differentiation of mafic magamas. Sov Geol Grophys 27(7):25–29

Zolotukhin VV (1964) Main regularities of prototectonics and aspects of origin of ore-bearing trap intrusions (ed Sobolev VS). Nauka, Moscow, 192 p (in Russian)

Zolotukhin VV (1965) On peculiarities in distribution of nickel in the Norilsk-I intrusion. Doklady Akad Nauk SSSR 162(6):1390–1393 (in Russian)

Zolotukhin VV (1971) On genesis of so called "liquation" of copper-nickel ores in the light of new data (about infiltration-autometasomatic hypothesis). Geologiya i Geofisika 9:12–22 (in Russian)

Zolotukhin VV (1974) Complex fluoride compounds as probable carriers of metals as related to problems of origin of ore associations. Sov Geol Geophys 15(12):9–17

Zolotukhin VV (1984) Petrology of magnesian basites of the Tunguska (Middle Yenisei) province. In: Sobolev VS (ed-in chief) Magnesian basites of the western Siberian platform and aspects of their nickel content. Nauka, Novosibirsk, pp 48–149 (in Russian)

Zolotukhin VV (1988) A generalized model of sulfide copper-nickel ore formation as a process of sulfurization. In: Obolensky AA, Sotnikov VI, Sharapov VN (eds) Mineralization and genetic models of endogenous ore formations. Nauka, Novosibirsk, pp 172–181 (in Russian)

Zolotukhin VV (1997) Mafic pegmatoids of the Noril'sk ore-bearing intrusions and the problem of genesis of the Platinum-copper-nickel mineralization of the Noril'sk type (ed Polyakov GV). SB RAN NITS OIGGM, Novosibirsk, 88 p (in Russian)

Zolotukhin VV, Laguta ON (1985) On fractionation of magnesian mafic melts and diversification of traps in the Siberian platform. Doklady AN USSR 280(4):967–972 (in Russian)

Zolotukhin VV, Malyuk BI (1985) On genesis of copper-nickel sulfide deposits. Sov Geol Geophys 26(10):54–63 (in Russian)

Zolotukhin VV, Ryabov VV (1977) Problems of genesis of sulfide copper-nickel ores and principal parameters of the process. In: Kuznetsov VA, Berzina AP (eds) Principal parameters of processes of endogenous ore formation, vol 1. Nauka, Novosibirsk, pp 50–53 (in Russian)

Zolotukhin VV, Vasil'ev YR (1967) Peculiarities of formation mechanism of ore-bearing trap intrusions at the northwest of Siberian platform (ed Sobolev VS). Nauka, Moscow, 231 p (in Russian)

Zolotukhin VV, Vasil'ev YR (1976) Problems of platform magmatism: the depths of the magma chambers in the upper mantle and their part in creating the variety of igneous rocks. Sov Geol Geophys 16(2):1–6

Zolotukhin VV, Ryabov VV, Vasil'ev YR, Shatkov VA (1975) Petrology of the Talnakh ore-bearing differentiated intrusion (ed Sobolev VS). Nauka, Novosibirsk, 432 p (in Russian)

Zolotukhin VV, Vilensky AM, Duzhikov OA (1986) Basalts of the Siberian platform (eds Sobolev VS, Sobolev NV). Nauka, Novosibirsk, 255 p (in Russian)

Zotov AI, Vortsepnev VV, Mel'nikov FP (1977) Magmatic nature of massive sulfide ores of the Talnakh). In: Kuznetsov VA, Berzina AP (eds) The main parameters of natural processes of endogenous ore formation, vol 1. Nauka, Novosibirsk, pp 53–54 (in Russian)

Conclusions

Analysis of the data in this book allows the formulation of principal conclusions:

1. Petrologic investigations including microprobe analyses of mineral phases in combination with data on the internal structure of magmatic bodies bear important genetic information and indicate the initial composition of a melt, mechanisms for its differentiation, and crystallization features. The morphology of crystals, their composition, zonality, and structural interrelations with other phases reflect also the evolution of a magmatic system in the course of ore formation. It can be assumed that progress in the field of magmatic and ore geology will be defined in the future at a deeper level, studying rocks to determine the geochemical and genetic petrography of targets, as opposed to the descriptive method prevalent up to now. This new direction in petrography must be based on microprobe geochemical study of rocks coexisting in nature and varying in composition as phases and intergrowths. This will allow an understanding of the multiplicity of natural processes involved. On the basis of petrographic investigations and from available data, the importance of pre-crystallization differentiation of a melt may already be assumed to be high. For picritic basalt lavas and layered intrusions in particular, it can be said that the nucleation of olivine crystals took place in feeder conduits rather than in deep-seated hearths as the temperature of a magma decreased in transit from abyssal depths to hypabyssal depths and continued under chamber conditions. In lava sheets and traps, hypogeal protocrysts are virtually absent, and porphyritic accumulations were formed either in transit to or within chambers. Gravitational separation of crystalline phases has been poorly developed in hypabyssal intrusions, and the layered structure of magmatic bodies originated in a liquid state under pre-chamber conditions.
2. Tholeiitic basalt magma is the parent magma for the whole variety of traps in the northwest of the Siberian Platform, forming daughter high-magnesium and subalkaline melts. Differentiation of these melts in deep-seated hearths and feeder conduits, as well as their crystallization under chamber conditions, is responsible for trap variety. All derivative melts bear the genetic memory of the parent tholeiitic basalt magma at the mineralogical–geochemical level. In the Noril'sk Region, crystallization products of tholeiitic basalt melts are most frequent, whereas picrite basalt and subalkaline basalt melts are less abundant. Compositionally different melts flowing onto the surface were intruded through volcanic–terrigenous strata individually or collectively as immiscible liquids.
3. Basalt melts, especially magnesian–ferrous melts with elevated Cr, Ti, and P content, were inclined to feature liquid layering. Agents for fluid–magmatic differentiation in traps are F, Cl, S, H_2, CH_4, CO_2, P_2O_5, H_2O, and B_2O_3. A change in the physical–chemical state of a magmatic system is responsible for the presence of immiscible liquids. This can include fluid saturation of a melt or reduction of both the temperature and pressure, which takes place when the magma passes from the abyssal level to the hypabyssal one. This is accompanied by a change in solubility and volatile activity, dissociation of volatiles, and involvement of the volatiles in the redistribution of rock-forming and oregenic elements, the development of fluidized melts, and their separation. This resulted in the appearance of immiscible silicate–silicate and ore–silicate liquids. The former have a sialic–mafic-layered nature that manifests in both tholeiitic and subalkaline basalt melts, whereas the composition of ore liquids such as sulfide, oxide, or metallic is governed by the C-H-S-O system and geochemical features of the melt. Thus, magmatic (immiscible, fluid–magmatic) differentiation was implemented within a wide range of PT conditions, taking place in deep-seated hearths, in hypabyssal chambers, and in lava sheets. Pre-crystallization liquid differentiation was responsible for the composition of subsequent crystallization products.
4. Crystallization differentiation is a driving mechanism for chamber evolution of a melt. It levels the liquid layering that took place earlier, partially retaining some features, and manifests as peculiarities in rock structures and mineral composition. As the temperature decreases, a wide variety of rocks are formed in the sequence of layered intrusions with regular and step-less variations in rock-forming mineral composition and in the petrogenic oxide and impurity content. Invisible layering may be traced by

V.V. Ryabov et al., *Trap Magmatism and Ore Formation in the Siberian Noril'sk Region: Volume 1. Trap Petrology; Volume 2. Atlas of Magmatic Rocks*, Modern Approaches in Solid Earth Sciences 3, DOI 10.1007/978-94-007-5022-7,

changes in mineral composition during crystallization that is not infrequently complicated by volatiles. The composition of major rock-forming minerals in traps includes olivine from forsterite to fayalite and plagioclase from anorthite to albite, and the range of pyroxene composition is controlled by wide variations in the Fe content and the mineral-forming oxide contents in diopside, pigeonite, and orthopyroxene.

5. The fluid regime of a magmatic system is a governing factor in the formation of layered intrusions and mineralization associated with these intrusions. Interaction between volatiles and magmatic melts began in deep-seated hearths and continued in feeder conduits and in magmatic chambers. Different batches of magmatic melt in lavas and intrusions contained variable volatile content, and the most fluidized magmas were inclined to layering and were potentially ore bearing. The involvement of volatiles in fluid–magmatic differentiation of a melt manifests as selective extraction of different chemical elements from melts and country rocks and transportation this elements to accumulation zones. In this respect, ferritization, sulfurization, or metallization of the melt took place depending on the fluid composition, and evolution of the fluid regime led to changes in various volatile component activities. The rocks depleted or enriched in some or other rock components are regular members of a unified fluid–magmatic system. One efficient mechanism for formation of fluidized silicate melts is immiscible differentiation, which leads to separation of a liquid enriched in volatiles. Volatiles that are in excess of their solubility in basalt melts play an ore-forming role.
6. Sulfide Pt–Cu–Ni deposits of different sizes are associated with trap intrusions. The limited solubility of sulfur in a basaltic melt affords the assumption that an independent sulfur-bearing phase can exist and this can explain the origin of large sulfide deposits of the Noril'sky type. A complex of volatile components is supplemented during migration of a basalt magma through sedimentary rocks. Volatiles are involved in the extraction of petrogenic and ore-forming components and their transport with transmagmatic fluids. The ore-forming role of halogens manifests in the formation of pegmatoid-related Pt-low-sulfide ores, which are known in the Noril'sk and other layered intrusions. Detection and proof of existence of Pt–Fe-metal (Pt–graphite) types of ores in traps provide evidence in favor of the accumulation of noble metals with a dominant role played by hydrocarbons, hydrogen, nitrogen, and phosphorous, practically without involvement of sulfur. Observed sulfide-free platinum ore enhances the prospectivity of trap intrusions and increases the possibility of revealing nontraditional types of platinum ores and deposits.

Appendix: Analytical Tables

Gudchikhinsky Suite

Talnahsky Differentiated Picritic Basalt Flow Sequence

Table A.1 Chemical composition of rocks, drill hole KZ-491

	Depth (m)									
Component	9	10B	12Aa	13	14	15A[b]	15B[a]	16	17	18
SiO_2	41.14	51.50	60.00	42.12	35.30	42.34	47.90	40.70	42.22	44.88
TiO_2	1.45	1.65	0.91	1.55	1.55	1.47	1.95	1.65	1.85	1.85
Al_2O_3	9.20	9.00	7.70	10.25	10.30	8.79	11.25	10.79	11.65	12.67
Fe_2O_3	4.80	4.67	3.58	3.24	5.23	5.14	4.95	3.76	3.78	3.37
FeO	6.65	7.21	6.23	9.03	5.18	5.62	6.66	8.12	6.58	7.49
MnO	0.17	0.07	0.06	0.12	0.18	0.24	0.10	0.14	0.16	0.16
MgO	10.42	8.76	8.07	11.40	7.93	7.72	7.92	9.67	8.19	9.23
CaO	11.30	5.62	4.80	8.40	19.00	12.41	7.03	9.28	9.20	6.20
Na_2O	1.13	0.79	0.73	0.70	1.59	0.83	1.21	2.19	3.39	3.15
K_2O	0.03	0.35	0.61	0.02	0.09	0.23	0.31	0.04	0.14	0.18
P_2O_5	0.20	0.28	0.15	0.18	0.08	0.08	0.30	0.22	0.14	0.21
CuO	0.03	0.03	0.06	0.02	0.03	–	0.03	0.05	0.01	0.02
NiO	0.08	0.10	0.08	0.07	0.09	–	0.10	0.05	0.04	0.04
Cr_2O_3	0.05	0.09	0.02	0.06	0.07	0.34	0.17	0.05	0.05	0.05
LOI	12.35	8.88	6.40	12.50	13.10	14.04	9.12	12.69	11.65	10.19
Total	99.00	99.00	99.40	99.66	99.72	99.25	99.00	99.40	99.05	99.69
C.f.	52.4	57.6	54.9	51.8	56.8	58.2	59.4	55.1	55.8	54.1
S	0.26	–	0.19	0.10	0.02	0.32	0.30	0.04	0.18	0.15
CO_2	5.57	–	–	–	5.52	–	–	5.14	3.90	–
Rock	MA	MA	MA	MA	PB	MA	MA	MA	MD	MD
	Depth (m)									
Component	19	20	21	21.5	22	23	24	**25B**	28	33
SiO_2	45.30	45.92	46.00	44.48	46.00	42.00	43.20	40.48	40.10	42.26
TiO_2	1.85	1.95	1.70	2.00	1.65	1.32	1.40	1.00	0.95	0.95
Al_2O_3	11.90	12.70	11.00	12.21	12.34	9.81	7.78	5.52	6.62	6.58
Fe_2O_3	3.82	2.99	5.69	5.40	5.61	7.16	4.78	6.74	6.35	2.60
FeO	7.62	6.02	6.02	6.37	7.07	7.28	8.32	7.84	8.19	10.34
MnO	0.13	0.11	0.12	0.15	0.11	0.12	0.17	0.15	0.17	0.19
MgO	9.05	8.00	8.03	9.27	10.47	14.77	20.65	21.66	23.89	21.22
CaO	7.10	9.00	10.50	10.34	7.26	8.20	5.97	6.80	5.94	6.97
Na_2O	3.51	1.83	1.79	1.89	1.60	0.98	0.92	1.10	0.93	0.77
K_2O	0.29	0.21	0.22	0.15	0.19	0.12	0.22	0.17	0.08	0.19
P_2O_5	0.22	0.16	0.21	0.18	0.19	0.15	0.30	0.14	0.12	0.16
CuO	0.01	0.04	0.04	0.03	0.04	0.03	0.01	0.03	0.03	–
NiO	0.04	0.04	0.04	0.05	0.04	0.08	0.21	0.16	0.17	0.18

(continued)

V.V. Ryabov et al., *Trap Magmatism and Ore Formation in the Siberian Noril'sk Region: Volume 1. Trap Petrology; Volume 2. Atlas of Magmatic Rocks*, Modern Approaches in Solid Earth Sciences 3, DOI 10.1007/978-94-007-5022-7,

Table A.1 (continued)

	Depth (m)									
Component	19	20	21	21.5	22	23	24	**25B**	28	33
Cr_2O_3	0.06	0.05	0.05	0.06	0.04	0.11	0.18	0.12	0.12	0.18
LOI	9.70	7.36	7.87	6.64	6.22	6.62	5.04	7.66	6.19	6.91
Total	100.60	96.38	99.28	99.22	98.83	98.75	99.15	99.57	99.85	99.50
C.f.	55.8	53.0	59.3	55.9	54.8	49.4	38.8	40.2	37.8	37.9
S	0.07	0.11	0.16	0.03	0.04	0.07	0.05	0.08	0.03	–
CO_2	–	2.22	2.52	1.68	–	–	–	–	–	–
Rock	MD	OB	OB	AK	OB	AK	PB	PB	PB	PB
	Depth (m)									
Component	37	39	40	41	43	49	51	52	53	54
SiO_2	43.36	34.56	49.80	44.76	45.84	45.26	42.28	43.20	43.80	42.20
TiO_2	1.00	1.55	1.95	1.75	1.75	1.55	1.42	1.46	1.50	1.55
Al_2O_3	6.45	9.81	10.30	11.04	10.06	9.82	9.32	9.75	10.06	10.00
Fe_2O_3	2.13	6.25	7.41	8.72	7.14	8.06	6.77	6.29	6.99	7.36
FeO	10.77	5.25	3.85	3.29	4.62	4.69	5.67	5.05	6.59	5.89
MnO	0.18	0.20	0.15	0.18	0.14	0.17	0.16	0.18	0.13	0.15
MgO	21.94	8.63	7.17	9.42	12.56	9.69	11.48	12.10	15.41	14.38
CaO	6.29	16.70	8.92	11.80	9.32	9.80	12.70	11.57	6.79	8.04
Na_2O	0.73	1.77	1.73	1.86	1.64	1.48	1.59	1.58	1.18	1.33
K_2O	0.19	0.20	0.12	0.08	0.16	0.06	0.11	0.14	0.05	0.06
P_2O_5	0.14	0.12	0.23	0.16	0.19	0.18	0.18	0.19	0.16	0.16
CuO	–	0.03	0.06	0.03	0.04	0.03	0.04	0.05	0.03	0.03
NiO	0.18	0.08	0.05	0.05	0.07	0.07	0.10	0.11	0.10	0.10
Cr_2O_3	0.18	0.07	0.03	0.05	0.07	0.06	0.07	0.09	0.09	0.09
LOI	6.65	14.60	7.05	5.64	4.96	7.71	7.68	7.76	6.20	7.46
Total	100.19	99.82	98.82	98.83	98.56	98.63	99.57	99.52	99.08	98.80
C.f.	37.0	57.1	61.1	56.0	48.4	56.8	52.0	48.4	46.8	48.0
S	0.02	–	0.49	0.08	0.06	0.12	0.09	0.04	0.03	0.05
CO_2	–	7.68	–	2.68	0.27	2.98	3.33	0.41	–	–
Rock	PB	MA	MA	OB	AK	MA	OB	OB	AK	AK
	Depth (m)									
Component	55	56[b]	56A[a]	57	58	59	61	63	64	65
SiO_2	49.14	44.84	44.24	41.40	42.95	44.00	43.04	42.64	44.50	44.24
TiO_2	2.25	1.55	1.40	1.55	1.57	1.50	1.32	1.42	1.65	1.40
Al_2O_3	13.37	10.92	10.18	10.43	11.50	10.10	8.95	9.44	9.53	10.18
Fe_2O_3	6.32	6.02	5.75	7.09	5.63	6.23	6.31	5.75	9.24	5.75
FeO	5.19	4.28	6.74	6.32	6.38	6.73	6.81	7.16	3.70	6.74
MnO	0.16	0.14	0.19	0.15	0.14	0.16	0.18	0.21	0.17	0.19
MgO	9.04	11.01	15.92	15.54	15.08	16.10	18.53	17.61	15.16	15.92
CaO	7.07	11.09	8.17	8.24	8.38	7.90	6.84	7.07	7.26	8.17
Na_2O	2.25	1.72	1.35	1.31	1.42	1.33	1.08	1.14	1.00	1.35
K_2O	0.16	0.09	0.58	0.05	0.12	0.38	0.32	0.55	0.57	0.58
P_2O_5	0.27	0.16	0.17	0.18	0.20	0.18	0.12	0.15	0.26	0.17
CuO	0.04	0.04	0.03	0.08	0.03	–	0.03	0.03	0.01	0.03
NiO	0.04	0.08	0.10	0.08	0.08	0.17	0.13	0.12	0.13	0.10
Cr_2O_3	0.01	0.10	0.11	0.11	0.10	0.12	0.12	0.11	0.15	0.11
LOI	3.76	6.70	4.73	6.68	5.16	4.86	5.65	5.77	4.46	4.73
Total	99.07	98.74	99.66	99.21	98.74	99.76	99.43	99.17	97.79	99.66
C.f.	56.0	48.3	44.0	46.3	44.3	44.6	41.5	42.3	46.0	44.0
S	0.07	0.14	0.03	0.04	0.04	0.03	0.04	0.04	0.06	0.03
CO_2	–	2.00	–	–	–	–	–	–	–	–
Rock	MD	AK	AK	AK	AK	AK	PB	PB	AK	AK

Note: Detailed analytical data on samples indicated in bold are shown in Volume 2
[a]With amygdalae eliminated
[b]With amygdalae

Table A.2 Chemical composition of minerals, drill hole KZ-491

	Depth (m)												
Component	**25.0**					28.0	29.0	31.0		33.0	35.0		37.0
Chemical composition of olivine													
SiO_2	37.35	39.35	38.75	38.21	38.52	39.36	39.14	38.36	39.05	39.19	38.05	38.48	39.06
FeO	18.26	20.12	17.91	20.64	19.69	19.52	18.82	23.36	19.22	21.73	22.50	21.05	19.27
CaO	0.28	0.24	0.27	0.29	0.26	0.08	0.03	0.05	0.25	0.24	0.29	0.26	0.25
MgO	42.42	40.06	42.79	40.69	40.47	39.98	40.47	37.78	40.76	37.70	39.07	39.87	40.65
NiO	–	–	–	–	–	0.36	0.37	0.32	0.39	0.33	0.35	0.36	0.39
Total	98.31	99.77	99.72	99.83	98.94	99.30	98.83	99.87	99.67	99.19	100.26	100.02	99.62
Fo	81	78	81	78	79	79	79	74	79	76	76	77	79
Fa	19	22	19	22	21	22	21	26	21	24	24	23	21
Rock	PB	PB	PB	PB	PB	PB	PB	PB	PB	PB	PB	PB	PB

	Depth (m)											
Component	10.0		23.0	24.0	**25.0**		28.0		29.0	31.0	33.0	35.0
Chemical composition of spinel group minerals												
SiO_2	–	–	–	–	0.22	0.10	–	–	–	–	–	–
TiO_2	2.00	1.95	3.64	1.50	2.43	1.67	1.49	1.42	1.91	2.84	2.05	6.23
Al_2O_3	13.12	11.32	4.36	11.79	11.53	12.94	12.81	11.84	11.79	12.03	11.33	6.19
FeO	32.44	29.73	62.04	21.32	25.07	28.81	31.26	31.10	34.89	30.86	40.37	45.12
MnO	0.68	0.62	0.49	0.62	0.97	0.93	0.61	0.63	0.22	0.17	0.62	0.14
MgO	6.65	7.25	2.27	5.42	5.78	8.09	6.26	6.17	5.23	7.64	4.81	4.78
CaO	–	–	–	–	0.08	0.09	–	–	–	–	–	–
Cr_2O_3	43.53	44.46	18.20	45.52	41.20	45.72	45.65	46.12	43.93	42.99	38.25	30.62
Total	98.42	95.33	91.00	86.17	87.28	98.35	98.08	97.28	97.97	96.53	97.43	93.08
Rock	MA	MA	AK	PB	PB	PB	PB	PB	PB	PB	PB	PB

Mikchandinsky Differentiated Basalt Flow

Table A.3 Chemical composition of rocks

	Numbers of samples with index MCH									
Component	200	201/1	201/2	60/	202	203	204/1	204/2	205	206
SiO_2	51.95	49.85	51.52	44.76	47.92	49.70	51.70	50.59	50.69	50.80
TiO_2	0.987	0.443	0.43	0.47	0.557	0.641	0.428	0.562	0.722	0.385
Al_2O_3	14.84	11.59	8.10	8.29	10.31	10.05	8.02	9.61	10.28	7.40
Fe_2O_3	9.80	9.63	7.46	11.94	12.2	10.31	7.33	8.36	8.77	6.99
MnO	0.16	0.161	0.17	0.2	0.206	0.174	0.184	0.178	0.177	0.179
MgO	7.34	11.53	14.00	20.35	15.21	14.24	14.33	13.59	12.49	14.55
CaO	10.48	12.63	16.4	8.27	9.08	10.86	15.70	13.56	13.01	13.01
Na_2O	2.03	1.35	0.99	0.89	1.40	1.04	0.78	1.05	1.13	0.36
K_2O	1.05	0.38	0.28	0.37	0.56	0.56	0.42	0.50	0.72	0.37
P_2O_5	0.117	0.051	0.031	0.11	0.072	0.064	0.03	0.053	0.081	<0.003
Ba	0.032	0.008	<0.003	–	0.016	0.015	0.019	0.013	0.019	<0.003
LOI	1.30	2.26	0.56	4.56	1.90	2.02	0.84	1.78	1.70	1.46
Total	100.09	99.88	99.94	100.21	99.43	99.67	99.78	99.85	99.79	95.50
C.f.	57.2	45.5	34.8	37.0	44.5	42.0	33.8	38.1	41.3	32.5
Rock	AB	BO	DB-I	PB	SB	SB	DB-II	SB	SB	DB-III
Cu	40	32	34	30	800	110	140	50	24	16
Ni	20	92	59	41	210	200	160	190	59	69
Cr	280	1,100	2,300	85	1,000	1,400	2,300	1,700	1,400	2,400
Sr	200	150	96	–	130	130	160	120	130	150

(continued)

Table A.3 (continued)

Component	Numbers of samples with index MCH								
Component	200	201/1	201/2	60/	202	203	204/1	204/2	205	206
Ba	360	440	180	–	210	210	400	170	260	250

Component	Numbers of samples with index MCH								
Component	207	208/1	208/2	209	210	211	212	213/1	213/2
SiO_2	48.72	47.98	51.39	51.26	47.54	50.43	49.64	48.02	51.51
TiO_2	0.525	0.515	0.647	0.471	0.547	0.665	0.658	0.525	0.54
Al_2O_3	10.15	10.84	9.05	9.15	10.36	11.51	11.25	11.34	11.21
Fe_2O_3	9.49	10.74	8.71	7.8	10.77	10.15	10.71	10.57	7.65
MnO	0.186	0.165	0.191	0.17	0.181	0.163	0.173	0.181	0.158
MgO	13.43	13.7	12.92	13.35	15.54	12.12	13.07	13.41	11.95
CaO	12.3	10.71	13.82	15.42	8.92	11.01	9.75	10.67	13.87
Na_2O	1.12	0.92	0.96	0.95	0.81	1.62	1.71	1.06	1.12
K_2O	0.48	0.42	0.79	0.28	0.51	0.73	0.73	0.44	0.76
P_2O_5	0.053	0.049	0.078	0.038	0.061	0.083	0.09	0.058	0.054
Ba	0.012	0.01	0.017	<0.003	0.014	0.018	0.021	0.012	0.016
LOI	3.34	3.50	1.52	1.10	4.44	1.26	1.90	3.66	1.12
Total	99.81	99.55	100.09	99.99	99.69	99.76	99.70	99.95	99.96
C.f.	41.4	43.9	40.3	36.9	40.9	45.6	45.0	44.1	39.0
Rock	SB	SB	SB	DB-IV	SB	DB-V	SB	SB	DB-VI
Cu	34	38	28	32	28	48	32	32	21
Ni	88	72	59	95	170	78	92	85	56
Cr	1,300	960	1,600	2,200	1,000	940	840	940	1,400
Sr	120	120	130	110	110	140	120	120	200
Ba	210	180	260	180	170	190	200	180	160

Component	Numbers of samples with index MCH							
Component	214	215	216	217/1	217/2	217/3	218	219/1
SiO_2	48.37	51.76	48.52	48.12	51.13	49.36	48.39	48.13
TiO_2	0.6	0.417	0.531	0.579	0.436	0.614	0.65	0.606
Al_2O_3	11.64	8.7	11.37	11.99	10.29	12.01	12.11	11.96
Fe_2O_3	9.81	7.23	9.09	10.54	7.83	11.15	11.01	10.5
MnO	0.196	0.168	0.175	0.169	0.166	0.192	0.186	0.177
MgO	11.52	13.57	11.68	12.94	12.92	11.6	12.46	12.87
CaO	12	15.02	12.7	9.7	14.67	10.68	9.72	9.8
Na_2O	1.28	1.26	1.22	1.38	0.89	1.30	1.45	1.27
K_2O	0.58	0.52	0.51	0.58	0.34	0.59	0.63	0.59
P_2O_5	0.071	0.031	0.06	0.063	0.036	0.067	0.077	0.071
Ba	0.012	0.008	0.009	0.015	0.006	0.021	0.015	0.012
LOI	3.94	1.24	3.76	3.83	1.28	2.73	3.08	3.84
Total	100.02	99.92	99.63	99.91	99.99	100.31	99.78	99.83
C.f.	46.0	34.8	43.8	44.9	37.7	49.0	46.9	44.9
Rock	SB	DB-VII	SB	SB	DB-VIII	SB	SB	DB-IX
Cu	36	28	33	35	58	92	40	120
Ni	75	50	79	82	57	63	88	94
Cr	740	2,100	850	610	1,400	590	500	630
Sr	150	160	160	130	130	170	150	140
Ba	26	250	250	210	220	260	220	250

Component	Numbers of samples with index MCH							
Component	219/2	220A	221	222	223	225	224	227
SiO_2	52.17	49.13	48.97	50.52	50.97	50.44	50.61	51.27
TiO_2	0.531	0.634	0.703	0.753	0.798	0.832	1.00	0.994
Al_2O_3	9.82	11.12	12.05	13.56	15.00	15.04	15.09	15.13
Fe_2O_3	7.68	9.74	10.67	8.85	9.53	9.71	10.38	10.55
MnO	0.173	0.17	0.176	0.166	0.17	0.163	0.161	0.174

(continued)

Table A.3 (continued)

Component	Numbers of samples with index MCH							
	219/2	220A	221	222	223	225	224	227
MgO	12.43	13.03	12.08	8.72	7.68	8.06	7.30	7.31
CaO	14.92	10.64	9.26	11.49	10.13	10.13	10.46	10.21
Na_2O	1.08	1.30	1.54	1.78	2.30	2.22	2.41	2.08
K_2O	0.53	0.64	0.85	0.79	0.87	0.83	0.89	1.04
P_2O_5	0.059	0.073	0.087	0.084	0.091	0.095	0.111	0.116
Ba	0.007	0.014	0.02	0.019	0.024	0.022	0.024	0.028
LOI	0.64	3.13	3.54	3.02	2.42	2.6	1.80	1.52
Total	100.04	99.62	99.95	99.75	99.98	100.14	100.24	100.42
C.f.	38.2	42.8	46.9	50.4	55.4	54.6	58.7	59.1
Rock	SB	SB	SB	B	B	B	B	B
Cu	33	69	73	53	41	54	137	46
Ni	57	88	85	57	31	38	19	28
Cr	1,400	760	500	450	240	260	230	260
Sr	150	140	180	180	190	200	210	220
Ba	230	270	290	290	270	270	310	330

Top to base in a section

Basic Section of Khonnamakitsky Suite and Yuryakhsky Unit

(The western limb of the Tungussky syneclise, the right bank of the Khantayskoe Lake)

Table A.4 Chemical composition of rocks

Component	Khonnamakitsky Suite								
	Numbers of samples with index KHZH								
	22	104	105	107	108M	108ZH	109	110	111
SiO_2	48.4	47.55	47.83	48.37	47.91	47.67	47.69	48.44	48.33
TiO_2	1.26	1.363	1.368	1.441	1.375	1.409	1.481	1.47	1.443
Al_2O_3	14.3	14.1	15.2	15.03	14.96	15.54	15.35	15.63	15.12
Fe_2O_3	13.02	12.56	12.66	13.9	10.35	11.57	13.82	12.83	13.38
MgO	4.99	5.95	6.62	7.23	6.9	5.94	6.28	6.57	7.4
CaO	10.86	10.58	10.69	10.36	13.11	11.87	10.91	10.67	10.76
Na_2O	1.99	3.98	1.88	1.99	2.64	3.08	2.31	1.99	2.16
K_2O	0.28	0.84	0.59	0.52	0.13	0.12	0.31	0.37	0.37
P_2O_5	0.14	0.112	0.124	0.123	0.037	0.102	0.137	0.13	0.131
Ba	–	0.02	0.021	0.026	0.016	0.014	0.02	0.023	0.019
LOI	3.9	3.14	2.9	1.1	2.62	2.62	2.12	1.8	0.9
Total	99.14	100.20	99.88	100.09	100.05	99.94	100.43	99.92	100.01
C.f.	72.3	67.9	65.7	65.8	60.0	66.1	68.8	66.1	64.4
Cu	0.015	0.0071	0.022	0.016	0.0036	0.005	0.014	0.017	0.017
Ni	0.01	0.009	0.011	0.011	0.011	0.011	0.013	0.013	0.011
Co	–	0.004	0.0045	0.005	0.0025	0.0039	0.0049	0.0043	0.0042
Cr	0.023	0.016	0.02	0.021	0.027	0.021	0.034	0.029	0.02
Li[a]	–	–	5	–	11	12	–	–	7
Be[a]	–	–	0.55	–	0.65	0.75	–	–	0.35
B[a]	–	–	3.7	–	9.6	9.5	–	–	6.5
F[a]	–	–	550	–	260	500	–	–	380
Sc[a]	–	–	40	–	44	50	–	–	40
V[a]	–	–	560	–	230	400	–	–	350
Cr[a]	–	–	250	–	200	280	–	–	240
Co[a]	–	–	34	–	18	27	–	–	37

(continued)

Table A.4 (continued)

	Khonnamakitsky Suite								
	Numbers of samples with index KHZH								
Component	22	104	105	107	108M	108ZH	109	110	111
Ni[a]	–	–	110	–	110	130	–	–	140
Cu[a]	–	–	200	–	34	57	–	–	180
Zn[a]	–	–	110	–	43	340	–	–	79
Rb[a]	–	–	14	–	3	2	–	–	8
Sr[a]	–	–	230	–	320	200	–	–	190
Zr[a]	–	–	89	–	88	93	–	–	95
Nb[a]	–	–	4	–	6	8	–	–	4
Ag[a]	–	–		–			–	–	
Sn[a]	–	–	2.2	–	1.4	1.8	–	–	1.9
Ba[a]	–	–	140	–	100	120	–	–	130
Pb[a]	–	–	3.1	–	3	6.2	–	–	2.8
Rock	B	B	B	B	B	Gl	B	B	B

	Khonnamakitsky Suite								
	Numbers of samples with index KHZH								
Component	113	114	115	116	117	118	119	120	121
SiO_2	48.13	46.31	47.44	48.94	48.83	49.75	48.26	48.28	48.51
TiO_2	1.424	1.41	1.464	1.305	1.131	1.361	1.337	1.371	1.48
Al_2O_3	14.3	15.47	15.45	15.39	15.99	14.05	14.95	14.64	14.84
Fe_2O_3	12.4	13.32	12.99	12.93	10.69	12.84	13.11	12.49	13.11
MgO	7.33	6.88	6.46	7.14	6.48	7.01	7.81	6.71	7.22
CaO	9.75	8.1	11.07	11.09	10.25	5.82	5.51	10.78	10.54
Na_2O	2.85	3.25	2.4	2.35	3.22	3.5	3.47	2.39	2.13
K_2O	0.32	0.28	0.21	0.39	1.31	2.32	1.38	0.29	0.36
P_2O_5	0.129	0.115	0.13	0.112	0.073	0.129	0.121	0.128	0.133
Ba	0.022	0.016	0.013	0.017	0.024	0.295	0.092	0.022	0.019
LOI	3.26	4.78	2.2	0.26	1.99	2.7	3.8	2.9	1.4
Total	99.92	99.93	99.83	99.92	99.99	99.78	99.84	100.00	99.74
C.f.	62.8	65.9	66.8	64.4	62.3	64.7	62.7	65.1	64.5
Cu	0.0071	0.02	0.015	0.013	0.0041	0.0094	0.014	0.016	0.016
Ni	0.029	0.01	0.011	0.012	0.014	0.01	0.007	0.006	0.012
Co	0.0045	0.0063	0.0057	0.0045	0.0043	0.0043	0.0052	0.0045	0.0043
Cr	0.017	0.016	0.02	0.033	0.039	0.019	0.013	0.018	0.021
Li	–	–	–	–	14	40	52	–	–
Be	–	–	–	–	0.4	0.75	1.1	–	–
B	–	–	–	–	11	35	4.4	–	–
F	–	–	–	–	300	440	360	–	–
Sc	–	–	–	–	38	62	27	–	–
V	–	–	–	–	170	380	370	–	–
Cr	–	–	–	–	500	160	170	–	–
Co	–	–	–	–	34	36	50	–	–
Ni	–	–	–	–	180	140	100	–	–
Cu	–	–	–	–	30	130	150	–	–
Zn	–	–	–	–	50	89	130	–	–
Rb	–	–	–	–	30	78	48	–	–
Sr	–	–	–	–	280	640	620	–	–
Zr	–	–	–	–	62	92	88	–	–
Nb	–	–	–	–	3	2	1	–	–
Ag	–	–	–	–		0.09		–	–
Sn	–	–	–	–	1.4	2.6	1.8	–	–
Ba	–	–	–	–	220	2,130	540	–	–

(continued)

Table A.4 (continued)

	Khonnamakitsky Suite								
	Numbers of samples with index KHZH								
Component	113	114	115	116	117	118	119	120	121
Pb	–	–	–	–	2.4	4.9	3	–	–
Rock	B	B	B	B	B	B	B	B	B
	Khonnamakitsky Suite								
	Numbers of samples with index KHZH								
Component	93	94	122	70	71	74A	74B	77	78
SiO_2	48.74	46.05	47.92	48.71	52.26	48.31	44.35	48.84	45.9
TiO_2	1.354	1.258	1.395	1.228	1.049	1.107	1.073	1.181	1.119
Al_2O_3	15.02	14.05	15.28	15.22	14.99	15.23	15.45	15.21	15.18
Fe_2O_3	13.21	12.12	12.8	11.63	10.05	12.59	12.65	12.29	9.58
MgO	7.22	6.47	6.71	7	6.26	7.04	7.6	7.41	5.65
CaO	10.8	9.96	8.86	11.37	5.77	11.88	10.85	11.54	14.78
Na_2O	1.9	1.94	3.53	1.85	2.84	1.64	1.51	1.94	1.9
K_2O	0.55	0.54	0.53	0.43	0.34	0.18	0.13	0.3	0.24
P_2O_5	0.121	0.14	0.123	0.131	0.108	0.108	0.106	0.116	0.114
Ba	0.02	0.033	0.031	0.018	0.017	0.012	0.007	0.018	0.011
LOI	0.96	7.26	2.72	2.42	6.06	2.16	6.27	1	5.39
Total	99.90	99.82	99.90	100.01	99.74	100.26	100.00	99.85	99.86
C.f.	64.7	65.2	65.6	62.4	61.6	64.1	62.5	62.4	62.9
Cu	0.015	0.012	0.01	0.013	0.013	0.013	0.0079	0.012	0.013
Ni	0.012	0.01	0.013	0.011	0.007	0.013	0.013	0.013	0.01
Co	0.0043	0.005	0.0043	0.0047	0.0033	0.0044	0.0042	0.0048	0.0042
Cr	0.02	0.018	0.017	0.015	0.016	0.019	0.019	0.02	0.017
Li	10	–	–	4	–	4	–	4	–
Be	0.9	–	–	0.5	–	0.95	–	0.75	–
B	5.1	–	–	5.4	–	7.6	–	5.9	–
F	330	–	–	230	–	240	–	320	–
Sc	22	–	–	52	–	45	–	45	–
V	300	–	–	400	–	350	–	290	–
Cr	170	–	–	170	–	180	–	200	–
Co	30	–	–	44	–	40	–	31	–
Ni	140	–	–	180	–	150	–	130	–
Cu	160	–	–	140	–	140	–	130	–
Zn	120	–	–	87	–	150	–	94	–
Rb	12	–	–	7	–	2	–	4	–
Sr	200	–	–	230	–	190	–	200	–
Zr	89	–	–	85	–	77	–	89	–
Nb	6	–	–	21	–	3	–	4	–
Ag		–	–	0.08	–		–		–
Sn	2.1	–	–	2.2	–	2.4	–	3	–
Ba	160	–	–	200	–	90	–	130	–
Pb	4.1	–	–	7.1	–	3.6	–	5.1	–
Rock	B	B	B	B	Tf	B	Gl	B	B
	Khonnamakitsky Suite					Yuryakhsky Unit			
	Numbers of samples with index KHZH								
Component	81	85	89	124	125	40	42	127	**128**
SiO_2	48.24	48.04	48.31	48.56	41.29	49.25	49.94	48.34	49.65
TiO_2	1.307	1.297	1.226	1.314	1.09	2.181	2.234	2.199	2.237
Al_2O_3	15.33	15.15	14.92	15.14	12.83	13.83	13.45	13.71	13.36
Fe_2O_3	12.36	12.25	12.12	12.87	10.85	14.89	15.39	14.74	15.08
MgO	7.09	7.17	7.71	6.77	7.08	3.95	4.08	3.5	3.7
CaO	11.32	11.31	10.79	11.37	13.4	8.48	8.06	8.42	8.84

(continued)

Table A.4 (continued)

	Khonnamakitsky Suite					Yuryakhsky Unit			
	Numbers of samples with index KHZH								
Component	81	85	89	124	125	40	42	127	**128**
Na_2O	1.8	1.8	1.72	2.03	1.6	2.16	2.72	3.01	2.86
K_2O	0.22	0.26	0.47	0.35	0.6	1.8	1.91	1.92	1.54
P_2O_5	0.133	0.13	0.122	0.124	0.114	0.528	0.53	0.534	0.536
Ba	0.017	0.016	0.017	0.022	0.011	0.082	0.085	0.062	0.082
LOI	2.18	2.42	2.32	1.44	11.06	2.72	1.44	3.58	2.2
Total	100.00	99.84	99.73	99.99	99.93	99.87	99.84	100.02	100.09
C.f.	63.5	63.1	61.1	65.5	60.5	79.0	79.0	80.8	80.3
Cu	0.016	0.0078	0.019	0.015	0.012	0.0025	0.0031	0.0021	0.0028
Ni	0.011	0.011	0.01	0.01	0.01	0.001	0.005	0.0008	0.001
Co	0.0035	0.0039	0.0042	0.0045	0.0042	0.0035	0.0035	0.0038	0.0043
Cr	0.016	0.016	0.017	0.016	0.016	0.003	0.022	0.0052	0.003
Li	4	4	6	–	–	8	12	16	–
Be	0.5	0.5	0.6	–	–	2.4	2.4	3.6	–
B	4.3	4.7	4.4	–	–	5.2	13	8.4	–
F	310	280	350	–	–	650	870	660	–
Sc	35	45	31	–	–	40	50	20	–
V	250	700	250	–	–	280	300	170	–
Cr	230	160	170	–	–	37	220	50	–
Co	40	42	34	–	–	22	14	19	–
Ni	160	160	120	–	–	28	55	21	–
Cu	170	110	190	–	–	34	55	25	–
Zn	96	110	83	–	–	120	240	150	–
Rb	2	3	10	–	–	26	44	36	–
Sr	210	230	220	–	–	390	390	270	–
Zr	98	90	87	–	–	263	272	272	–
Nb	7	7	7	–	–	7	25	21	–
Ag				–	–				–
Sn	2.2	2.4	2.1	–	–	4.1	4.9	4.2	–
Ba	170	160	170	–	–	690	600	440	–
Pb	3.7	4.4	4.1	–	–	1.2	12	10	–
Rock	B	B	B	B	Tf	TB	TB	TB	TB

	Yuryakhsky Unit								
	Numbers of samples with index KHZH								
Component	129/A	129	131	132	**133**	133/A	134	135	138-1
SiO_2	49.81	46.17	47.88	48.36	49.58	49.58	49.93	46.66	49.65
TiO_2	2.207	2.309	2.362	2.292	2.077	2.165	2.114	2.257	2.283
Al_2O_3	13.62	14	14.48	13.63	13.67	13.57	13.24	14.47	13.45
Fe_2O_3	14.8	16.48	15.07	14.36	14.23	14.52	14.56	16.08	15.46
MgO	3.26	3.94	3.61	3.1	3.35	3.15	3.72	3.39	3.89
CaO	8.68	8.85	8.3	9.39	9.27	8.96	8.44	9.14	6.78
Na_2O	2.81	2.22	2.64	2.7	2.7	2.87	2.37	2.54	2.88
K_2O	1.46	1.07	2.01	1.67	1.4	1.29	1.64	1.89	2.11
P_2O_5	0.534	0.541	0.594	0.583	0.513	0.542	0.52	0.532	0.573
Ba	0.091	0.089	0.085	0.078	0.085	0.091	0.062	0.077	0.081
LOI	2.68	4.46	3.02	3.82	2.74	2.98	3.39	2.54	2.68
Total	99.95	100.13	100.05	99.98	99.62	99.72	99.99	99.58	99.84
C.f.	81.9	80.7	80.7	82.2	80.9	82.2	79.6	82.6	79.9
Cu	0.0025	0.0025	0.0029	0.007	0.0029	0.0029	0.0027	0.0025	0.0032
Ni	0.0031	0.0031	0.0031	0.0036	0.0028	0.0047	0.0039	0.02	0.0024

(continued)

Table A.4 (continued)

	Yuryakhsky Unit								
	Numbers of samples with index KHZH								
Component	129/A	129	131	132	**133**	133/A	134	135	138-1
Co	0.0033	0.003	0.0031	0.0038	0.0034	0.0035	0.0033	0.0032	0.0028
Cr	0.0069	0.011	0.0066	0.0066	0.0059	0.0066	0.0066	0.006	0.0062
Li	–	–	–	–	9	–	–	–	15
Be	–	–	–	–	2.1	–	–	–	3.7
B	–	–	–	–	9.3	–	–	–	12
F	–	–	–	–	680	–	–	–	1,200
Sc	–	–	–	–	28	–	–	–	26
V	–	–	–	–	360	–	–	–	330
Cr	–	–	–	–	50	–	–	–	42
Co	–	–	–	–	30	–	–	–	20
Ni	–	–	–	–	32	–	–	–	29
Cu	–	–	–	–	40	–	–	–	38
Zn	–	–	–	–	180	–	–	–	110
Rb	–	–	–	–	32	–	–	–	40
Sr	–	–	–	–	420	–	–	–	230
Zr	–	–	–	–	267	–	–	–	278
Nb	–	–	–	–	24	–	–	–	21
Ag	–	–	–	–		–	–	–	
Sn	–	–	–	–	5.4	–	–	–	5.3
Ba	–	–	–	–	610	–	–	–	420
Pb	–	–	–	–	9.6	–	–	–	6.9
Rock	TB	TB	TB	TB	TB	TB	TB	TB	TB

	Yuryakhsky Unit								
	Numbers of samples with index KHZH								
Component	138-2	139	141	144	145/1	159B	160A	160B	147
SiO_2	44.55	46.19	49.12	49.57	47.6	49.02	49.54	49.8	50.12
TiO_2	2.313	2.322	2.231	2.154	2.179	2.192	2.14	2.238	2.122
Al_2O_3	13.87	13.32	13.96	13.79	14.02	13.67	13.63	13.4	13.28
Fe_2O_3	18.85	15.44	14.56	15.02	14.59	14.51	14.79	14.31	13.12
MgO	4.39	4.45	3.39	3.86	3.96	3.61	3.4	3.32	3.74
CaO	4.5	7.85	8.97	8.63	8.53	8.9	8.47	8	7
Na_2O	1.63	2.86	3.26	3.12	2.98	2.51	2.81	2.95	3.67
K_2O	2.9	2.09	1.72	1.89	1.77	1.31	1.87	2.13	2.72
P_2O_5	0.867	0.561	0.556	0.527	0.524	0.542	0.498	0.582	0.515
Ba	0.076	0.078	0.074	0.07	0.06	0.081	0.07	0.088	0.107
LOI	5.45	4.92	2.08	1.32	3.7	3.22	2.48	3.2	3.48
Total	99.40	100.08	99.92	99.95	99.91	99.57	99.70	100.02	99.87
C.f.	81.1	77.6	81.1	79.6	78.7	80.1	81.3	81.2	77.8
Cu	0.0043	0.0031	0.0027	0.0036	0.0025	0.0027	0.0027	0.0027	0.0024
Ni	0.0008	0.0044	0.002	0.0047	0.0022	0.0028	0.0028	0.0008	0.0035
Co	0.0023	0.0032	0.003	0.003	0.0031	0.0029	0.0032	0.0028	0.003
Cr	0.0052	0.0059	0.0052	0.0069	0.0045	0.0066	0.0066	0.0055	0.009
Li	–	–	–	–	11	–	–	8	10
Be	–	–	–	–	1.85	–	–	1.3	1.6
B	–	–	–	–	9.2	–	–	6.4	10
F	–	–	–	–	680	–	–	960	1,500
Sc	–	–	–	–	33	–	–	30	30
V	–	–	–	–	300	–	–	320	220
Cr	–	–	–	–	35	–	–	34	57

(continued)

Table A.4 (continued)

	Yuryakhsky Unit								
	Numbers of samples with index KHZH								
Component	138-2	139	141	144	145/1	159B	160A	160B	147
Co	–	–	–	–	24	–	–	22	19
Ni	–	–	–	–	20	–	–	21	24
Cu	–	–	–	–	58	–	–	45	30
Zn	–	–	–	–	210	–	–	190	77
Rb	–	–	–	–	34	–	–	48	48
Sr	–	–	–	–	310	–	–	390	230
Zr	–	–	–	–	268	–	–	298	265
Nb	–	–	–	–	22	–	–	23	22
Ag	–	–	–	–		–	–		0.03
Sn	–	–	–	–	5.3	–	–	3.9	3.9
Ba	–	–	–	–	590	–	–	590	740
Pb	–	–	–	–	11	–	–	10	5.3
Rock	Plg	TB	TB	TB	TB	TB	TB	TB	TB
	Yuryakhsky Unit								
	Numbers of samples with index KHZH								
Component	148	149	173	152	153	174	155	158	156
SiO_2	49.3	50.58	41.79	50.43	49.43	46.29	51.44	50.27	50.16
TiO_2	2.197	2.328	1.267	2.196	2.164	1.088	2.253	2.204	2.285
Al_2O_3	13.71	13.73	12.91	12.93	12.7	14.53	14.09	13.77	13.55
Fe_2O_3	14.73	14.57	15.26	13.75	14.43	14.42	11.44	13.4	11.48
MgO	3.59	3.82	9.67	4.45	4.52	8.86	3.75	4.14	4.89
CaO	8.67	6.49	9.18	7.56	7.66	3.94	8.25	8.17	8.35
Na_2O	3.36	5.14	0.6	5.77	5.56	3.19	5.68	5.07	4.96
K_2O	1.48	1.04	1.21	0.63	0.24	0.41	0.61	0.18	1.05
P_2O_5	0.551	0.564	0.097	0.519	0.53	0.032	0.544	0.541	0.543
Ba	0.087	0.05	0.003	0.062	0.013	0.005	0.118	0.003	0.133
LOI	2.58	1.96	7.85	1.88	2.72	6.76	1.82	2.02	2.54
Total	100.26	100.27	99.84	100.18	99.97	99.53	100.00	99.77	99.94
C.f.	80.4	79.2	61.2	75.5	76.1	61.9	75.3	76.4	70.1
Cu	0.0024	0.005	0.0021	0.0035	0.0094	0.02	0.0059	0.0038	0.0025
Ni	0.0047	0.0024	0.02	0.0024	0.0024	0.021	0.0027	0.0041	0.0036
Co	0.0027	0.0033	0.0025	0.0027	0.0042	0.0042	0.0028	0.0026	0.0026
Cr	0.019	0.008	0.019	0.0052	0.0093	0.018	0.0052	0.011	0.0069
Li	–	–	–	8	–	–	–	19	–
Be	–	–	–	3.1	–	–	–	1.7	–
B	–	–	–	10	–	–	–	49	–
F	–	–	–	960	–	–	–	900	–
Sc	–	–	–	27	–	–	–	46	–
V	–	–	–	320	–	–	–	330	–
Cr	–	–	–	40	–	–	–	90	–
Co	–	–	–	23	–	–	–	20	–
Ni	–	–	–	18	–	–	–	31	–
Cu	–	–	–	34	–	–	–	61	–
Zn	–	–	–	58	–	–	–	80	–
Rb	–	–	–	11	–	–	–	5	–
Sr	–	–	–	360	–	–	–	290	–
Zr	–	–	–	271	–	–	–	275	–
Nb	–	–	–	26	–	–	–	22	–
Ag	–	–	–		–	–	–		–
Sn	–	–	–	3.2	–	–	–	3.6	–

(continued)

Table A.4 (continued)

	Yuryakhsky Unit								
	Numbers of samples with index KHZH								
Component	148	149	173	152	153	174	155	158	156
Ba	–	–	–	440	–	–	–	120	–
Pb	–	–	–	4.1	–	–	–	4.3	–
Rock	TB	TB	Plg	TB	TB	Plg	TB	TB	TB
	Yuryakhsky Unit								
	Numbers of samples with index KHZH								
Component	162B	165	169	170	**180**	**182**	184	183	**181**
SiO_2	50.07	50.28	49.37	49.41	48.84	49.44	50.7	49.71	49.39
TiO_2	2.447	2.269	2.212	2.209	2.247	2.186	2.3	2.228	2.264
Al_2O_3	14.38	14	13.42	13.49	13.21	13.7	13.42	13.81	13.42
Fe_2O_3	14.45	13.88	15.18	15.21	15.35	14.73	14.71	14.29	13.01
MgO	3.29	4.51	3.67	3.66	3.75	3.82	3.33	3.68	4.06
CaO	4.62	6.31	8.82	8.61	7.04	8.22	7.84	7.83	7.58
Na_2O	4.45	3.72	2.98	2.81	4.13	2.95	3.2	3.96	4.47
K_2O	2.03	2.15	1.97	1.94	2.29	1.95	2.64	2.14	2.27
P_2O_5	0.59	0.54	0.524	0.516	0.546	0.539	0.554	0.543	0.559
Ba	0.127	0.072	0.078	0.072	0.108	0.081	0.181	0.166	0.145
LOI	3.22	2.5	1.84	2.23	2.18	2.3	1.38	1.4	2.76
Total	99.67	100.23	100.06	100.16	99.69	99.92	100.26	99.76	99.93
C.f.	81.5	75.5	80.5	80.6	80.4	79.4	81.5	79.5	76.2
Cu	0.0042	0.0027	0.0027	0.0027	0.0031	0.0024	0.0029	0.0029	0.0031
Ni	0.0008	0.0038	0.0027	0.0044	0.0016	0.0016	0.0058	0.0027	0.0028
Co	0.004	0.0032	0.0032	0.0033	0.0032	0.003	0.0038	0.0029	0.0032
Cr	0.0062	0.015	0.0062	0.017	0.0066	0.008	0.016	0.0073	0.01
Li	8	–	–	14	–	12	–	–	7
Be	1.8	–	–	1.8	–	1.85	–	–	2
B	7.6	–	–	12	–	9.8	–	–	11
F	1,500	–	–	1,000	–	900	–	–	960
Sc	34	–	–	37	–	40	–	–	32
V	280	–	–	350	–	240	–	–	360
Cr	30	–	–	160	–	64	–	–	100
Co	18	–	–	20	–	24	–	–	32
Ni	14	–	–	49	–	27	–	–	32
Cu	43	–	–	53	–	34	–	–	36
Zn	75	–	–	210	–	100	–	–	60
Rb	34	–	–	40	–	36	–	–	44
Sr	330	–	–	390	–	400	–	–	370
Zr	305	–	–	261	–	268	–	–	285
Nb	30	–	–	20	–	21	–	–	23
Ag		–	–		–		–	–	0.09
Sn	4.4	–	–	4	–	5.2	–	–	3.7
Ba	720	–	–	770	–	650	–	–	940
Pb	6.6	–	–	8.1	–	7.9	–	–	6.5
Rock	TB	TB	TB	TB	TB	TB	TB	TB	TB

Notes: Concentration of components marked ([a]) undertaken by the Institute of Geochemistry at the Siberian Branch of the Russian Academy of Science using Zr. Nb—X-ray fluorescence analysis; Be, B, F, Sc, V, Cr, Co, Ni, Cu, Zn, Sr, Ag, Ba, Pb—atomic emission spectrometry; Li, Rb—flame photometry. Trace element concentration is given in ppm. All the others—in weight percent

Top to base in a section

Ergalakhsky Intrusive Complex

Layered Intrusion of Trachydolerite, drill hole F-221

Table A.5 Chemical composition of rocks, drill hole F-221

	Depth (m)										
Component	**447.0**	461.0	475.0	480.0	480.0	485.0	**493.0**	503.0	**512.5**	**525.0**	546.8
SiO_2	46.13	46.54	47.20	45.06	47.36	47.82	49.96	47.99	48.32	47.45	47.43
TiO_2	3.000	2.800	3.171	2.628	3.041	2.894	2.933	3.271	3.753	3.188	2.836
Al_2O_3	14.75	14.33	14.72	12.46	13.83	14.47	12.80	14.27	12.42	14.53	14.27
Fe_2O_3	13.80	13.13	13.47	15.59	14.26	13.05	14.30	12.91	15.23	13.70	14.06
MnO	0.210	0.157	0.197	0.271	0.206	0.160	0.214	0.177	0.218	0.212	0.210
MgO	4.59	5.56	4.39	3.24	4.92	4.34	2.31	3.56	3.30	4.39	4.77
CaO	7.87	7.95	7.89	7.86	8.15	7.74	6.26	8.81	7.65	8.55	8.08
Na_2O	4.08	3.39	3.93	3.31	3.26	3.40	4.66	4.08	4.07	3.86	3.95
K_2O	1.21	1.32	1.70	2.27	1.54	1.96	2.58	1.59	2.02	1.44	1.51
P_2O_5	0.557	0.510	0.573	0.997	0.540	0.538	1.030	0.581	0.774	0.549	0.568
Ba	0.072	0.071	0.097	0.108	0.084	0.089	0.128	0.087	0.116	0.081	0.083
LOI	3.84	3.44	2.26	5.77	2.32	2.98	2.12	1.78	1.50	1.06	1.46
Total	100.11	99.20	99.60	99.56	99.51	99.44	99.30	99.11	99.37	99.01	99.23
C.f.	75.0	70.3	75.4	82.8	74.4	75.0	86.1	78.4	82.2	75.7	74.7
Rock	CT	TD	TD	TDP	TD	TD	TDP	TDP	TDP	TDo	TDo
	Depth (m)										
Component	555.4	**563.5**	575.7	583.1	**590.0**	601.5	620.0	**640.0**	650.0	665.0	676.6
SiO_2	46.87	48.41	47.12	47.18	47.81	47.44	47.52	47.60	48.55	47.74	48.36
TiO_2	3.095	2.743	3.094	2.736	2.766	2.666	2.563	2.499	2.714	2.949	2.928
Al_2O_3	14.65	14.58	15.79	14.96	16.25	15.15	15.21	15.42	14.72	14.25	14.63
Fe_2O_3	13.57	14.13	12.42	13.72	12.06	12.92	13.30	12.42	13.41	13.57	13.68
MnO	0.186	0.194	0.187	0.209	0.170	0.181	0.197	0.184	0.196	0.200	0.188
MgO	4.42	5.39	4.56	5.75	4.35	5.20	5.84	5.74	4.99	5.13	4.75
CaO	8.19	7.75	8.57	7.97	8.55	8.11	8.43	8.22	8.19	8.49	8.34
Na_2O	3.66	3.08	3.98	3.49	3.89	3.20	3.69	3.28	3.23	3.25	3.26
K_2O	1.44	1.60	1.48	1.58	1.40	1.44	1.34	1.29	1.41	1.49	1.51
P_2O_5	0.583	0.550	0.543	0.515	0.461	0.488	0.457	0.451	0.485	0.510	0.533
Ba	0.079	0.085	0.083	0.086	0.074	0.076	0.070	0.070	0.073	0.081	0.079
LOI	2.35	1.30	1.94	1.48	1.78	2.60	1.24	2.30	0.84	1.82	1.08
Total	99.09	99.81	99.77	99.68	99.56	99.47	99.86	99.47	98.81	99.48	99.34
C.f.	75.4	72.4	73.1	70.5	73.5	71.3	69.5	68.4	72.9	72.6	74.2
Rock	TDo	OTD	TDo	OTD	TDo	OTD	OTD	OTD	TDo	TDo	CT

Table A.6 Chemical composition of olivine, drill hole F-221

	Depth (m)							
	525.0			**563.5**			583.1	
Component	Core	Rim	gm	Core	Core–rim	Rim	Core	Rim
SiO_2	35.06	35.08	33.76	34.85	34.84	33.64	35.16	34.75
TiO_2	0.032	0.030	0.045	0.042	0.034	0.048	0.038	0.065
Cr_2O_3	0.000	0.000	0.000	0.011	0.000	0.000	0.005	0.000
FeO	39.67	40.36	46.42	40.03	41.56	45.63	35.76	41.08
MnO	0.566	0.563	0.753	0.639	0.621	0.703	0.487	0.611
MgO	24.54	24.12	18.99	23.69	23.33	19.37	26.93	23.39
CaO	0.400	0.380	0.426	0.346	0.362	0.365	0.420	0.368
NiO	0.000	0.000	0.000	0.013	0.002	0.005	0.012	0.021

(continued)

Table A.6 (continued)

	Depth (m)							
	525.0			**563.5**			583.1	
Component	Core	Rim	gm	Core	Core–rim	Rim	Core	Rim
Total	100.27	100.53	100.39	99.62	100.75	99.76	98.81	100.29
Fo	52	52	42	51	50	43	57	50
Fa	48	48	58	49	50	57	43	50
Rock	TDo	TDo	TDo	OTD	OTD	OTD	OTD	OTD

	Depth (m)								
	620.0			**640.0**				665.0	
Component	Rim	gm	gm	Core	Rim	gm	gm	gm	gm
SiO_2	34.60	33.40	33.91	33.34	33.91	34.24	33.70	33.42	33.01
TiO_2	0.034	0.040	0.075	0.025	0.032	0.048	0.040	0.025	0.033
Cr_2O_3	0.000	0.007	0.000	0.000	0.000	0.003	0.001	0.000	0.000
FeO	40.20	44.56	45.87	43.33	44.08	45.07	45.46	47.57	48.45
MnO	0.572	0.690	0.738	0.668	0.676	0.710	0.715	0.802	0.776
MgO	24.28	20.03	19.03	20.81	20.69	19.90	19.68	17.72	16.89
CaO	0.356	0.366	0.275	0.399	0.296	0.299	0.234	0.277	0.274
NiO	0.013	0.000	0.003	0.000	0.009	0.000	0.000	0.000	0.000
Total	100.06	99.09	99.90	98.57	99.69	100.27	99.83	99.81	99.43
Fo	52	44	43	46	46	44	44	40	38
Fa	48	56	57	54	54	56	56	60	62
Rock	OTD	OTD	OTD	OTD	OTD	OTD	OTD	TDo	TDo

Note: *gm* grain in groundmass

Table A.7 Chemical composition of pyroxene, drill hole F-221

	Depth (m)									
Component	**447.0**		475.0			**493.0**				
SiO_2	47.51	47.87	49.54	49.86	49.46	50.02	49.75	49.56	49.97	49.55
TiO_2	2.80	2.63	1.83	1.87	2.02	1.16	1.00	0.626	0.920	0.834
Al_2O_3	4.06	3.80	2.77	2.59	2.83	1.15	0.83	0.283	0.745	0.563
Cr_2O_3	0.131	0.041	0.068	0.015	0.011	0.023	0.011	0.013	0.000	0.003
FeO	11.88	12.52	10.79	11.06	11.14	17.82	19.73	23.63	19.89	21.54
MnO	0.220	0.276	0.241	0.231	0.245	0.459	0.477	0.562	0.486	0.513
MgO	12.12	12.00	13.21	13.11	13.07	9.00	7.59	5.78	7.45	6.70
CaO	19.77	19.42	20.15	20.29	20.11	20.20	19.92	18.75	19.93	19.85
Na_2O	0.448	0.454	0.410	0.371	0.405	0.355	0.393	0.473	0.391	0.398
K_2O	0.009	0.004	0.006	0.009	0.000	0.001	0.002	0.032	0.001	0.008
Total	98.95	99.02	99.02	99.41	99.29	100.19	99.70	99.71	99.78	99.96
f	35.5	36.9	31.4	32.1	32.4	52.6	59.3	69.6	60.0	64.3
Wo	43	42	43	43	43	43	43	41	43	43
En	37	36	39	38	38	27	23	18	22	20
Fs	20	22	18	19	19	30	34	41	35	37
Rock	CT	CT	TDo	TDo	TDo	TDP	TDP	TDP	TDP	TDP

	Depth (m)									
Component	**493.0**	503.0		**512.5**		**525.0**			546.8	
SiO_2	48.68	49.97	49.65	49.88	49.54	49.00	50.51	47.45	49.86	48.54
TiO_2	0.943	1.59	1.46	1.61	1.24	1.90	1.08	1.36	1.88	2.50
Al_2O_3	0.629	2.19	1.72	2.17	1.47	2.40	1.14	1.19	2.53	3.27
Cr_2O_3	0.000	0.011	0.011	0.014	0.004	0.015	0.021	0.000	0.016	0.014
FeO	22.12	12.50	15.60	13.38	17.08	13.33	16.21	22.51	11.85	12.29
MnO	0.507	0.242	0.373	0.308	0.438	0.319	0.374	0.468	0.245	0.255
MgO	6.90	12.37	10.66	12.07	9.93	11.64	10.50	5.84	12.01	11.91

(continued)

Table A.7 (continued)

	Depth (m)									
Component	**493.0**	503.0		**512.5**		**525.0**			546.8	
CaO	19.65	20.89	19.53	19.90	19.34	20.18	19.77	19.87	20.83	19.76
Na_2O	0.459	0.379	0.426	0.383	0.366	0.411	0.335	0.488	0.396	0.442
K_2O	0.006	0.022	0.002	0.005	0.005	0.012	0.000	0.026	0.000	0.001
Cl	0.006	–	–	–	–	–	–	0.001	–	–
F	0.151	–	–	–	–	–	–	0.205	–	–
Total	100.05	100.16	99.43	99.72	99.41	99.21	99.94	99.41	99.62	98.98
f	64.3	36.2	45.1	38.3	49.1	39.1	46.4	68.4	35.6	36.7
Wo	42	43	42	42	41	43	42	43	44	43
En	20	36	32	35	30	34	31	18	36	36
Fs	38	21	26	23	29	23	27	39	20	21
Rock	TDP	TDP	TDP	TDP	TDP	TDo	TDo	TDo	TDo	TDo

	Depth (m)									
Component	**563.5**		575.7		583.1			**590.0**		601.5
SiO_2	49.12	49.59	49.26	49.52	49.70	48.99	49.35	49.05	49.16	49.28
TiO_2	1.95	1.84	1.89	1.85	1.78	2.09	1.72	2.11	1.75	1.80
Al_2O_3	2.73	2.59	2.68	2.58	2.60	3.09	2.29	3.02	2.44	2.57
Cr_2O_3	0.026	0.003	0.004	0.036	0.001	0.030	0.003	0.013	0.000	0.131
FeO	10.87	12.32	11.47	11.79	11.58	11.58	12.83	12.05	13.43	10.21
MnO	0.241	0.270	0.248	0.278	0.240	0.264	0.281	0.288	0.302	0.213
MgO	12.67	12.52	12.39	12.29	12.75	12.68	11.84	12.42	11.91	13.27
CaO	20.49	19.62	20.33	20.25	20.05	19.99	20.09	19.84	19.55	20.50
Na_2O	0.415	0.463	0.421	0.440	0.409	0.429	0.414	0.433	0.444	0.394
K_2O	0.013	0.009	0.003	0.005	0.000	0.001	0.013	0.000	0.013	0.009
Total	98.53	99.23	98.70	99.04	99.11	99.14	98.83	99.22	99.00	98.38
f	32.5	35.6	34.2	35.0	33.8	33.9	37.8	35.2	38.8	30.2
Wo	44	42	43	43	43	43	43	42	42	44
En	38	37	37	37	38	37	35	37	35	39
Fs	18	21	20	20	19	20	22	21	23	17
Rock	OTD	OTD	TDo	TDo	OTD	OTD	OTD	TDo	TDo	OTD

	Depth (m)									
Component	601.5		620.0		**640.0**		665.0		676.6	
SiO_2	49.19	49.61	50.65	49.50	49.24	49.35	50.21	49.95	45.36	48.55
TiO_2	2.05	1.96	1.20	1.99	1.94	1.98	1.54	1.48	3.88	2.07
Al_2O_3	3.19	2.86	1.46	2.93	2.91	2.84	2.10	2.11	5.08	2.64
Cr_2O_3	0.222	0.045	0.018	0.035	0.067	0.025	0.014	0.024	0.131	0.013
FeO	10.34	10.56	11.33	11.32	10.76	11.45	11.93	12.69	12.86	14.44
MnO	0.233	0.203	0.264	0.249	0.215	0.251	0.296	0.299	0.215	0.306
MgO	13.25	12.93	13.49	12.97	13.26	12.86	12.64	12.52	10.69	10.70
CaO	19.92	20.47	20.06	20.00	20.27	20.02	20.43	20.10	20.03	20.16
Na_2O	0.419	0.401	0.350	0.465	0.405	0.420	0.408	0.531	0.551	0.502
K_2O	0.000	0.000	0.007	0.000	0.001	0.000	0.011	0.008	0.008	0.042
Total	98.81	99.04	98.83	99.46	99.07	99.20	99.58	99.71	98.81	99.42
f	30.5	31.4	32.0	32.9	31.3	33.3	34.6	36.3	40.3	43.1
Wo	43	44	42	43	43	43	43	42	44	43
En	39	38	39	38	39	38	37	37	33	32
Fs	18	18	19	19	18	19	20	21	23	25
Rock	OTD	OTD	OTD	OTD	OTD	OTD	TDo	TDo	CT	CT

Table A.8 Chemical composition of feldspars, drill hole F-221

	Depth (m)									
Component	**447.0**			475.0			**512.5**		**525.0**	
SiO_2	52.75	53.50	53.46	54.36	55.41	60.11	56.97	59.95	52.94	53.78
TiO_2	–	–	–	–	–	–	0.116	0.091	–	–
Al_2O_3	29.40	29.10	28.90	28.10	27.65	24.56	25.71	23.97	29.42	28.96
Cr_2O_3	0.000	0.000	0.000	0.000	0.000	0.000	0.003	0.006	0.012	0.000
FeO	0.650	0.740	0.666	0.680	0.680	0.447	0.531	0.458	0.648	0.657
MnO	0.020	0.011	0.000	0.018	0.003	0.000	0.031	0.017	0.022	0.012
MgO	0.159	0.098	0.114	0.059	0.061	0.026	0.088	0.050	0.063	0.073
CaO	12.21	11.27	11.40	10.72	10.05	6.06	8.73	6.50	11.87	11.44
BaO	0.024	0.058	0.033	0.060	0.054	0.177	–	–	0.040	0.051
SrO	0.000	0.000	0.005	0.000	0.000	0.000	–	–	0.019	0.000
Na_2O	4.31	4.92	4.65	4.92	5.27	7.22	5.56	6.58	4.38	4.65
K_2O	0.332	0.392	0.443	0.506	0.598	0.837	0.808	1.25	0.322	0.436
Total	99.86	100.09	99.67	99.42	99.78	99.44	98.55	98.87	99.74	100.06
Ab	38	43	41	44	47	65	51	60	39	41
Ort	2	2	3	3	4	5	5	7	2	3
An	60	55	56	53	49	30	44	33	59	56
Rock	CT	CT	CT	TDo	TDo	TDo	TDP	TDP	TDo	TDo
	Depth (m)									
Component	**590.0**				**640.0**			676.6		
SiO_2	53.86	53.91	52.52	62.90	52.72	53.29	53.50	52.91	53.10	56.66
Al_2O_3	28.31	27.71	29.13	22.07	29.40	28.74	28.88	28.93	28.64	26.65
Cr_2O_3	0.000	0.000	0.000	0.000	0.002	0.000	0.000	0.000	0.000	0.000
FeO	0.771	1.090	0.743	0.426	0.699	0.727	0.757	0.635	0.624	0.731
MnO	0.012	0.005	0.005	0.000	0.012	0.008	0.013	0.001	0.000	0.016
MgO	0.067	0.220	0.039	0.060	0.068	0.101	0.066	0.147	0.125	0.077
CaO	10.83	10.69	12.13	3.39	12.06	11.67	11.28	11.63	11.45	8.77
BaO	0.039	0.049	0.070	0.551	0.034	0.041	0.035	0.062	0.051	0.084
SrO	0.000	0.000	0.008	0.000	0.010	0.005	0.000	0.026	0.001	0.000
Na_2O	4.92	4.89	4.20	6.64	4.31	4.50	4.84	4.52	4.73	6.02
K_2O	0.506	0.516	0.319	3.78	0.379	0.390	0.415	0.378	0.409	0.687
Total	99.32	99.08	99.16	99.82	99.69	99.47	99.79	99.24	99.13	99.70
Ab	44	44	38	60	39	40	43	40	42	53
Ort	3	3	2	23	2	2	2	2	2	4
An	53	53	60	17	59	58	55	58	56	43
Rock	TDo	TDo	TDo	TDo	OTD	OTD	OTD	CT	CT	CT

Table A.9 Chemical composition of spinel group minerals, drill hole F-221

	Depth (m)										
Component	**447.0**	**493.0**	503.0	**512.5**		546.8	583.1	**590.0**	**640.0**	665.0	676.6
TiO_2	16.68	25.28	24.39	27.58	28.07	27.26	24.81	34.87	19.04	20.89	20.62
Al_2O_3	0.518	0.357	1.46	0.815	0.841	0.866	1.77	0.614	2.09	1.64	1.00
Cr_2O_3	0.170	0.000	0.010	0.005	0.000	0.000	0.010	0.000	0.050	0.066	0.109
FeO	76.14	66.67	69.21	66.29	64.61	61.51	67.53	59.04	71.39	71.62	71.26
MnO	0.367	0.464	0.384	0.565	0.682	1.14	1.18	0.537	1.15	1.23	0.645
MgO	0.502	0.280	0.018	0.138	0.206	0.931	0.827	0.860	1.15	0.081	0.044
NiO	0.014	0.000	0.000	0.011	0.010	0.010	0.006	0.023	0.012	0.017	0.007
V_2O_5	0.760	0.276	0.473	0.186	0.131	0.414	0.897	0.275	1.04	0.680	0.722
Total	95.15	93.33	95.95	95.59	94.55	92.13	97.03	96.22	95.92	96.22	94.41
Fe_2O_3[a]	34.80	16.24	18.33	12.55	10.82	10.89	18.04	0.00	28.80	25.39	25.31
FeO[a]	44.83	52.06	52.72	55.00	54.87	51.71	51.30	59.04	45.47	48.77	48.49
Rock	CT	TDP	TDP	TDP	TDP	TDo	OTD	TDo	OTD	TDo	CT

[a]Hereinafter calculated data

Table A.10 Chemical composition of ilmenite, drill hole F-221

	Depth (m)								
Component	**447.0**	475.0		**493.0**	503.0	**525.0**		546.8	**563.5**
TiO_2	48.78	50.09	49.39	50.14	50.14	50.42	50.86	50.22	49.68
Al_2O_3	0.020	0.062	0.018	0.010	0.059	0.060	0.075	0.060	0.095
Cr_2O_3	0.000	0.000	0.001	0.001	0.004	0.000	0.003	0.000	0.010
FeO	47.60	48.05	47.30	47.65	46.93	45.97	45.62	45.83	46.16
MnO	0.838	0.307	0.303	0.423	0.398	0.399	0.378	0.391	0.359
MgO	0.481	0.714	0.574	0.063	0.991	1.36	1.64	1.41	2.11
NiO	0.000	0.000	0.012	0.001	0.013	0.002	0.002	0.000	0.009
V_2O_5	0.226	0.063	0.137	0.044	0.121	0.344	0.375	0.377	0.365
Total	97.95	99.29	97.74	98.33	98.66	98.56	98.95	98.29	98.79
Rock	CT	TD	TD	TDP	TDP	TDo	TDo	TDo	OTD

	Depth (m)									
Component	575.7	583.1	**590.0**	601.5		620.0		**640.0**	665.0	676.6
TiO_2	49.75	49.86	49.16	49.93	49.27	49.53	49.26	49.46	49.85	48.20
Al_2O_3	0.100	0.106	0.063	0.100	0.103	0.109	0.111	0.075	0.113	0.022
Cr_2O_3	0.000	0.000	0.000	0.000	0.000	0.011	0.000	0.000	0.007	0.000
FeO	46.60	45.93	47.46	45.84	45.79	45.72	45.90	45.68	45.39	48.25
MnO	0.440	0.362	0.526	0.349	0.339	0.337	0.352	0.331	0.352	0.315
MgO	1.25	2.10	0.589	2.20	2.32	2.35	2.18	2.24	2.29	0.356
NiO	0.015	0.027	0.034	0.006	0.008	0.000	0.042	0.013	0.000	0.004
V_2O_5	0.263	0.236	0.211	0.345	0.338	0.281	0.162	0.213	0.274	0.219
Total	98.42	98.62	98.04	98.77	98.17	98.34	98.01	98.01	98.28	97.37
Rock	TDo	OTD	TDo	OTD	OTD	OTD	OTD	OTD	TDo	CT

Table A.11 Chemical composition of apatite, drill hole F-221

	Depth (m)									
Component	**447.0**	475.0		**493.0**			503.0			**512.5**
SiO_2	1.42	0.232	0.233	0.146	0.244	0.284	0.173	0.195	0.215	0.131
FeO	0.724	0.243	0.361	0.302	0.646	0.281	0.302	0.360	0.288	0.448
MnO	0.029	0.031	0.028	0.024	0.044	0.039	0.002	0.007	0.021	0.007
MgO	–	0.124	–	0.254	0.238	–	0.242	0.239	–	0.258
CaO	53.66	55.09	54.90	55.86	55.60	55.15	57.21	56.57	55.97	56.68
SO_3	0.000	–	0.028	–	–	0.013	–	–	0.000	–
Cl	0.066	0.509	0.122	0.225	0.404	0.036	1.450	0.301	0.055	0.262
F	3.36	3.64	3.59	4.13	3.71	3.87	3.88	3.60	3.42	3.88
P_2O_5	39.90	40.57	41.25	41.21	40.88	41.13	40.88	40.89	41.24	40.91
CeO_2	0.438	0.438	0.057	0.088	0.119	0.182	0.064	0.075	0.110	0.088
Total	99.60	100.88	100.57	102.24	101.89	100.99	104.20	102.24	101.32	102.66
Rock	CT	TD	TD	TDP	TDP	TDP	TDP	TDP	TDP	TDP

	Depth (m)									
Component	**512.5**		**525.0**				546.8		563.5	
SiO_2	0.159	0.242	0.321	0.373	0.263	0.545	0.257	0.273	0.267	0.281
FeO	0.416	0.703	0.420	0.393	0.540	0.681	0.469	0.607	0.324	0.385
MnO	0.009	0.052	0.025	0.033	0.031	0.032	0.027	0.049	0.000	0.049
MgO	0.267	–	0.216	–	–	–	0.210	–	0.247	–
CaO	57.13	55.65	55.23	56.02	56.62	54.68	54.89	54.54	55.26	54.76
SO_3	–	0.016	–	0.001	0.000	0.000	–	0.006	–	0.028
Cl	1.47	0.048	0.357	0.054	0.050	0.039	0.000	0.042	0.265	0.052
F	3.98	3.46	4.24	3.74	3.62	3.79	3.86	3.50	3.90	3.47
P_2O_5	40.25	41.32	40.84	41.45	41.00	40.66	40.92	41.13	40.71	40.75
CeO_2	0.050	0.150	0.151	0.153	0.126	0.196	0.123	0.085	0.111	0.152

(continued)

Table A.11 (continued)

	Depth (m)									
Component	**512.5**		**525.0**				546.8		563.5	
Total	103.73	101.64	101.80	102.22	102.25	100.62	100.76	100.23	101.08	99.93
Rock	TDP	TDP	TDo	TDo	TDo	TDo	TDo	TDo	OTD	OTD

	Depth (m)										
Component	575.7	583.1	**590.0**		601.5	620.0	**640.0**	665.0		676.6	
SiO_2	0.297	0.410	1.80	0.246	0.552	0.327	0.328	0.378	0.301	3.67	1.14
FeO	0.395	0.373	0.284	0.403	0.433	0.312	0.391	0.342	0.389	1.15	0.695
MnO	0.027	0.041	0.023	0.034	0.023	0.037	0.035	0.035	0.032	0.008	0.037
CaO	55.89	55.34	54.43	56.11	55.01	54.78	55.06	55.01	55.86	52.95	54.29
SO_3	0.056	0.000	0.013	0.000	0.049	0.024	0.006	0.030	0.023	0.025	0.020
Cl	0.061	0.081	0.094	0.062	0.083	0.074	0.087	0.447	0.127	0.067	0.065
F	3.51	3.70	4.37	3.98	3.64	3.94	3.67	2.45	4.09	3.52	3.23
P_2O_5	40.92	40.58	39.44	40.85	40.18	40.89	40.71	40.43	41.02	39.25	40.14
CeO_2	0.115	0.243	0.243	0.201	0.221	0.151	0.139	0.329	0.177	0.284	0.434
Total	101.27	100.77	100.7	101.89	100.19	100.54	100.43	99.45	102.02	100.92	100.05
Rock	TDo	OTD	TDo	TDo	OTD	OTD	OTD	TDo	TDo	CT	CT

Noril'sky Intrusive Complex

Noril'sky-Type Intrusions

Southwestern Intrusive Branch of the Upper Talnakh Intrusion

Table A.12 Chemical composition of rocks, drill hole KZ-184

	Depth (m)								
Component	**116.0**	**123.0**	125.0	128.5	131.0	133.5	136.0	142.5	144.0
SiO_2	45.30	48.00	48.10	48.10	47.60	49.00	47.90	47.60	45.10
TiO_2	2.55	0.85	0.96	1.15	1.21	1.08	1.06	0.89	0.79
Al_2O_3	12.00	15.20	14.60	14.20	15.40	14.40	15.20	15.90	16.90
Fe_2O_3	7.01	3.98	2.90	2.47	3.30	3.46	3.70	4.66	4.51
FeO	11.42	5.06	7.11	8.22	7.83	7.51	6.03	5.53	5.03
MnO	0.33	0.41	0.24	0.21	0.26	0.28	0.18	0.21	0.18
MgO	4.84	7.50	6.84	6.84	6.50	6.50	6.67	7.17	8.34
CaO	8.35	10.67	10.90	10.67	8.81	9.05	10.67	8.81	8.12
Na_2O	2.65	2.35	2.48	2.57	3.13	3.75	3.13	3.00	2.35
K_2O	1.47	2.04	1.58	1.53	1.79	0.71	1.21	2.00	1.47
P_2O_5	0.18	0.10	0.10	0.12	0.14	0.13	0.10	0.09	0.09
LOI	3.14	3.33	3.30	3.03	3.08	3.44	3.21	3.39	5.97
H_2O^-	0.24	0.15	0.09	0.05	0.17	0.08	0.12	0.09	0.53
Total	99.48	99.64	99.20	99.16	99.22	99.39	99.18	99.34	99.38
C.f.	79.2	54.7	59.4	61.0	63.1	62.8	59.3	58.7	53.4
S	0.20	0.15	0.08	0.11	0.11	0.11	0.09	0.15	0.14
CuO	0.020	0.007	0.006	0.017	0.015	0.010	0.011	0.011	0.009
NiO	0.001	0.014	0.010	0.008	0.008	0.008	0.009	0.012	0.021
Cr_2O_3	0.010	0.140	0.033	0.043	0.038	0.050	0.050	0.052	0.017
Rock	MD	MD	MD	MD	MD	G	G	G	G
	Depth (m)								
Component	146.0	149.0	156.0	157.5	160.0	**163.0**	**166.5**	170.0	176.0
SiO_2	48.60	45.10	46.70	45.00	45.00	46.30	45.20	41.00	39.10
TiO_2	0.92	0.76	0.70	0.74	0.76	0.75	1.05	0.62	0.60

(continued)

Table A.12 (continued)

Component	Depth (m) 146.0	149.0	156.0	157.5	160.0	**163.0**	**166.5**	170.0	176.0
Al_2O_3	14.10	17.10	14.10	16.40	16.60	18.10	11.60	8.90	8.30
Fe_2O_3	3.86	3.71	2.79	3.50	3.83	2.38	0.94	5.39	7.12
FeO	6.54	3.95	5.14	6.21	6.18	7.04	14.36	8.83	7.90
MnO	0.24	0.11	0.14	0.14	0.15	0.17	0.25	0.21	0.19
MgO	7.84	6.33	9.34	10.34	9.17	9.17	15.68	21.55	22.21
CaO	10.20	15.08	13.69	10.20	12.06	10.21	7.89	6.03	5.56
Na_2O	2.35	1.80	1.60	1.64	1.80	1.79	1.79	0.85	0.75
K_2O	2.00	1.21	1.21	1.00	1.16	0.75	0.37	0.31	0.33
P_2O_5	0.09	0.08	0.07	0.08	0.10	0.10	0.12	0.07	0.08
LOI	3.26	3.98	3.41	3.60	2.85	3.56	1.80	6.40	7.31
H_2O^-	0.12	0.07	0.18	0.45	0.16	0.26	0.08	0.12	0.36
Total	100.12	99.28	99.07	99.30	99.82	100.58	101.13	100.28	99.81
C.f.	57.0	54.8	45.9	48.4	52.2	50.7	49.4	39.8	40.3
S	0.08	0.14	0.08	0.09	0.11	0.12	0.09	0.90	1.30
CuO	0.013	0.005	0.008	0.010	0.014	–	–	–	–
NiO	0.010	0.012	0.019	0.041	0.041	–	–	–	–
Cr_2O_3	0.025	0.140	0.210	0.080	0.100	0.094	0.015	0.543	0.529
Rock	G	G	G	O	O	O	O	P	P

Component	Depth (m) 178.0	189.0	**192.0**	192.6	**194.0**	195.0	197.5	199.5	202.3
SiO_2	36.90	40.70	44.40	36.70	38.90	41.10	38.70	40.10	39.00
TiO_2	0.60	0.56	0.64	0.43	0.62	0.66	0.52	0.58	0.60
Al_2O_3	7.70	7.70	11.00	6.50	7.90	8.70	6.60	7.80	7.10
Fe_2O_3	7.14	5.51	2.71	2.44	4.02	4.84	7.73	7.09	5.19
FeO	8.69	8.54	9.26	14.72	15.37	11.85	10.41	11.34	11.34
MnO	0.21	0.21	0.21	0.22	0.24	0.22	0.25	0.24	0.25
MgO	22.21	23.35	16.35	26.82	20.08	21.01	23.18	21.18	22.18
CaO	4.87	5.57	8.32	3.94	5.63	5.10	4.41	5.34	4.87
Na_2O	0.40	0.70	1.07	0.60	0.83	0.85	0.50	0.70	0.55
K_2O	0.35	0.31	1.00	0.20	0.42	0.58	0.37	0.40	0.40
P_2O_5	0.06	0.08	0.06	0.05	0.08	0.08	0.06	0.07	0.08
LOI	9.46	7.17	4.68	7.52	5.69	5.71	7.65	5.25	7.27
H_2O^-	0.70	0.16	0.21	0.15	0.26	0.16	0.29	0.19	0.42
Total	99.29	100.56	99.91	100.29	100.04	100.86	100.67	100.28	99.25
C.f.	41.6	37.6	42.3	39.0	49.1	44.3	43.9	46.5	42.7
S	1.48	0.49	0.82	2.32	2.69	1.72	2.08	1.63	2.32
CuO	–	–	–	–	0.190	–	–	–	–
NiO	–	–	–	–	0.170	–	–	–	–
Cr_2O_3	0.460	0.108	0.094	0.273	0.140	0.084	0.067	0.070	0.037
Rock	P	P	P	P	P	P	P	P	P

Table A.13 Chemical composition of olivine, drill hole KZ-184

Component	Depth (m) 170.0		176.0		189.0		**192.0**	
SiO_2	38.28	39.18	38.84	39.90	39.14	36.85	38.48	38.82
TiO_2	0.08	0.08	0.06	0.06	0.07	0.011	0.07	0.08
FeO	19.90	20.92	19.37	18.99	21.84	22.51	23.40	24.75
MgO	41.23	40.06	41.15	41.56	39.59	36.73	37.82	36.56
CaO	0.13	0.11	0.10	0.10	0.18	–	0.22	0.24
NiO	0.23	0.24	0.25	0.21	0.13	–	0.07	0.08

(continued)

Table A.13 (continued)

Component	Depth (m) 170.0		176.0		189.0		**192.0**	
Cr_2O_3	0.08	0.08	0.06	0.06	0.07	0.03	0.07	0.08
Total	99.93	100.67	99.83	100.88	101.02	96.13	100.13	100.61
Fo	79	77	79	80	76	74	74	72
Fa	21	23	21	20	24	26	26	28
Rock	P	P	P	P	P	P	P	P

Component	Depth (m) 192.6	**194.0**		195.0	197.5	199.5	
SiO_2	38.22	38.10	38.42	38.40	39.00	38.31	38.14
TiO_2	0.06	0.09	0.08	0.07	–	–	–
FeO	21.94	24.44	24.90	23.93	22.17	24.06	24.78
MgO	39.09	37.19	37.32	37.77	39.19	36.98	36.55
CaO	0.21	0.16	0.14	0.17	0.05	0.13	0.12
NiO	0.12	0.07	0.08	0.06	0.000	0.029	0.007
Cr_2O_3	0.06	0.09	0.08	0.07	–	–	–
Total	99.70	100.14	101.02	100.47	100.41	99.51	99.60
Fo	76	73	73	74	76	73	72
Fa	24	27	27	26	24	27	28
Rock	P	P	P	P	P	P	P

Table A.14 Chemical composition of clinopyroxene, drill hole KZ-184

	Depth (m) **116.0**		**123.0**		131.0		142.5		144.0	149.0	156.0	157.5
Component	Core	Rim	Core	Rim	Core	Rim	Core	Core	Rim	Core	Core	Core
SiO_2	51.35	50.91	51.32	51.69	52.26	51.90	52.18	52.53	50.29	51.97	51.83	51.73
TiO_2	0.79	0.61	0.73	0.56	0.43	0.57	0.46	0.39	1.19	0.50	0.43	0.48
Al_2O_3	2.22	1.58	2.05	2.44	2.38	2.30	2.59	1.97	3.06	3.36	2.83	2.93
FeO	12.95	14.39	12.16	8.45	6.98	9.77	6.95	6.86	9.66	6.25	5.96	6.43
MgO	14.58	14.77	15.03	16.57	16.55	16.05	16.52	17.51	14.78	16.06	16.50	16.53
CaO	18.57	17.28	18.78	20.19	20.68	19.51	21.05	20.35	20.58	20.99	21.36	20.73
Na_2O	0.32	0.30	0.34	0.29	0.24	0.30	0.28	0.27	0.48	0.34	0.30	0.33
Cr_2O_3	0.00	0.03	0.02	0.07	0.34	0.01	0.46	0.33	0.01	0.85	0.52	0.93
Total	100.78	99.87	100.43	100.26	99.86	100.41	100.49	100.21	100.05	100.32	99.73	100.09
Fe_2O_3[a]	2.34	3.14	2.85	3.21	1.35	2.44	2.06	2.48	3.77	0.84	1.96	1.89
FeO[a]	10.85	11.57	9.59	5.56	5.77	7.57	5.09	4.63	6.27	5.50	4.19	4.73
f	33.3	35.3	31.2	22.2	19.1	25.5	19.1	18.0	26.8	17.9	16.9	17.9
Wo	38	35	38	41	42	40	43	40	42	44	44	43
En	41	42	43	46	47	45	46	49	42	46	47	47
Fs	21	23	19	13	11	15	11	11	16	10	9	10
Rock	MD	MD	MD	MD	MD	MD	G	G	G	G	G	O

	Depth (m) 157.5	160.0		**163.0**	**166.5**		170.0	176.0	182.0		189.0	
Component	Rim	Core	Rim	Core	Core	Rim	Core	Core	Core	Rim	Core	Rim
SiO_2	51.16	52.03	50.86	52.27	51.36	52.06	51.92	51.77	51.75	51.89	53.14	51.63
TiO_2	1.25	0.50	1.22	0.39	0.56	0.61	0.67	0.53	0.58	1.15	0.43	1.54
Al_2O_3	2.31	2.47	2.45	2.15	3.25	1.54	2.55	2.90	2.78	2.10	2.19	2.12
FeO	10.12	6.95	10.55	6.10	6.57	9.40	6.73	5.98	6.20	7.48	5.54	7.12
MgO	15.82	16.65	16.18	17.05	16.13	15.13	16.61	16.60	16.49	17.25	17.26	16.76
CaO	19.01	20.44	18.32	20.63	20.42	20.24	19.86	20.38	20.38	19.13	20.88	19.20
Na_2O	0.35	0.33	0.32	0.32	0.31	0.32	0.33	0.31	0.34	0.37	0.24	0.32
Cr_2O_3	0.05	0.70	0.04	0.71	1.03	0.15	0.90	1.07	1.01	0.63	0.94	0.70

(continued)

Table A.14 (continued)

	Depth (m)											
	157.5	160.0		**163.0**	**166.5**		170.0	176.0	182.0		189.0	
Component	Rim	Core	Rim	Core	Core	Rim	Core	Core	Core	Rim	Core	Rim
Total	100.07	100.07	99.94	99.62	99.63	99.45	99.57	99.54	99.53	100.00	100.62	99.39
Fe_2O_3[a]	2.22	1.87	2.71	1.73	1.12	1.10	0.70	0.82	0.98	1.53	0.27	0.00
FeO[a]	8.12	5.27	8.12	4.54	5.57	8.41	6.10	5.24	5.32	6.11	5.30	7.12
f	26.4	19.0	26.8	16.7	18.6	25.9	18.5	16.8	17.4	19.6	15.3	19.3
Wo	39	42	37	42	42	42	41	42	42	39	42	40
En	45	47	46	48	47	43	48	48	48	49	49	48
Fs	16	11	17	10	11	15	11	10	10	12	9	12
Rock	O	O	O	O	O	O	P	P	P	P	P	P

	Depth (m)											
	192.0		192.6		**194.0**		195.0			197.3	199.5	
Component	Core	Rim	Core	Rim	Core	Rim	Core	Rim	Rim	Core	Core	Rim
SiO_2	52.56	52.59	51.98	51.30	52.85	52.55	51.96	52.95	52.45	52.53	51.87	52.17
TiO_2	0.56	1.11	0.49	1.02	0.56	1.13	0.58	1.47	1.48	1.03	0.55	1.17
Al_2O_3	2.60	2.17	2.05	1.69	1.92	1.89	3.10	1.67	1.66	2.00	2.88	1.80
FeO	5.26	6.18	4.86	5.78	6.00	7.20	6.95	7.03	7.25	6.75	6.32	7.29
MgO	16.57	16.55	16.42	15.74	17.34	16.87	16.45	17.04	17.07	17.01	16.50	16.99
CaO	21.01	20.75	21.59	21.27	20.30	19.44	20.32	19.67	19.66	19.69	20.74	19.68
Na_2O	0.27	0.27	0.26	0.28	0.24	0.30	0.28	0.32	0.31	0.30	0.28	0.32
Cr_2O_3	0.85	0.32	1.25	0.27	0.77	0.48	1.18	0.40	0.39	0.58	1.01	0.43
Total	99.68	99.94	98.9	97.35	99.98	99.86	100.82	100.55	100.27	99.89	100.15	99.85
Fe_2O_3[a]	0.00	0.00	0.45	0.14	0.37	0.00	1.06	0.00	0.28	0.00	1.21	0.93
FeO[a]	5.26	6.18	4.46	5.65	5.67	7.20	6.00	7.03	6.99	6.75	5.24	6.45
f	15.1	17.3	14.2	17.1	16.3	19.3	19.2	18.8	19.2	18.2	17.7	19.4
Wo	44	43	45	45	41	40	42	40	40	40	43	40
En	48	47	47	46	49	48	47	49	48	49	47	48
Fs	8	10	8	9	10	12	11	11	12	11	10	12
Rock	P	P	P	P	P	P	P	P	P	P	P	P

[a]Hereinafter calculated data

Table A.15 Chemical composition of orthopyroxene, drill hole KZ-184

	Depth (m)		
Component	170.0	197.3	202.5
SiO_2	55.00	56.14	55.25
TiO_2	0.57	0.40	0.55
Al_2O_3	1.24	0.42	0.76
FeO	13.60	14.58	15.32
MgO	29.39	29.22	28.61
CaO	1.92	1.20	1.51
Na_2O	0.04	0.05	0.02
Cr_2O_3	0.33	0.04	0.03
Total	102.09	102.05	102.05
Fe_2O_3[a]	2.62	1.05	1.99
FeO[a]	11.24	13.63	13.53
f	20.6	21.9	23.1
Wo	4	2	3
En	76	76	75
Fs	20	22	22
Rock	P	P	P

[a]Hereinafter calculated data

Table A.16 Chemical composition of spinel group minerals, drill hole KZ-184

	Depth (m)											
Component	**163.0**		**166.5**			170.0		176.0			182.0	
TiO_2	6.85	0.44	2.62	2.51	2.95	4.96	2.70	3.74	4.97	2.20	0.83	2.53
Al_2O_3	3.66	0.50	3.20	3.18	1.28	8.10	6.39	3.18	10.89	3.33	22.47	13.61
FeO	59.74	90.10	86.00	83.01	84.63	49.81	54.44	63.32	45.03	65.96	31.96	45.58
MnO	0.62	0.08	0.44	0.14	0.12	0.69	0.75	0.86	0.58	0.84	0.39	0.66
MgO	5.36	0.99	1.49	1.36	0.91	6.00	3.23	3.04	8.37	2.58	10.06	5.77
Cr_2O_3	23.56	0.11	3.23	3.12	3.49	29.71	30.93	26.08	30.07	24.50	34.25	32.20
Total	99.79	92.22	96.98	93.32	93.38	99.27	98.44	100.22	99.91	99.41	99.96	100.35
$Fe_2O_3{}^a$	32.51	67.24	59.72	57.30	58.17	24.10	27.40	35.80	22.01	39.66	12.43	20.13
FeO^a	30.48	29.60	32.26	31.45	32.29	28.12	29.79	31.11	25.22	30.28	20.78	27.47
f	86.2	98.1	97.0	97.2	98.1	82.3	90.4	92.1	75.1	93.5	64.1	81.6
Rock	O	O	O	O	O	P	P	P	P	P	P	P

	Depth (m)										
Component	182.0	189.0			**192.0**		192.6	**194.0**	195.0	197.5	
TiO_2	2.50	1.07	1.05	–	2.86	5.59	3.15	5.15	2.98	6.45	0.22
Al_2O_3	1.18	24.23	23.38	–	26.23	19.70	17.84	13.52	17.63	8.18	0.07
FeO	74.19	27.38	26.95	95.26	36.63	42.25	36.67	44.28	36.95	39.20	91.19
MnO	0.70	0.34	0.13	0.24	0.42	0.15	0.57	0.61	0.54	0.70	0.06
MgO	1.33	10.27	9.65	0.25	6.92	5.18	5.93	4.10	5.46	4.60	0.98
Cr_2O_3	19.75	37.54	36.79	0.16	26.29	24.17	35.90	31.71	37.80	40.03	0.07
Total	99.65	100.83	97.95	95.91	99.35	97.04	100.06	99.37	101.36	99.16	92.59
$Fe_2O_3{}^a$	46.38	6.88	6.32	71.03	9.88	12.06	9.34	13.65	8.58	8.76	68.55
FeO^a	32.45	21.19	21.26	31.34	27.74	31.40	28.26	32.00	29.23	31.32	29.51
f	96.9	59.9	61.0	99.5	74.8	82.1	77.6	85.8	79.2	82.7	98.1
Rock	P	P	P	P	P	P	P	P	P	P	P

[a]Hereinafter calculated data

Table A.17 Chemical composition of mica group minerals, drill hole KZ-184

	Depth (m)							
	116.0	157.5	160.0					
Component	Core	Core	Core	Rim		**163.0**	**166.5**	
SiO_2	35.54	37.54	38.38	39.56	40.51	37.63	35.22	38.16
TiO_2	1.24	6.85	2.17	0.78	5.33	0.95	0.72	5.33
Al_2O_3	10.10	12.56	11.76	12.42	11.73	11.75	10.39	12.65
FeO	34.52	13.99	12.46	11.40	9.66	13.66	36.45	11.36
MgO	4.27	14.80	19.38	20.57	18.93	18.26	3.59	18.25
K_2O	8.44	8.70	8.95	9.27	7.18	9.65	8.61	6.12
NiO	0.004	0.106	0.019	0.027	0.052	0.018	0.000	0.035
F	0.658	1.25	1.21	1.27	1.27	1.32	0.484	0.692
Cl	5.05	0.246	0.002	0.01	0.331	0.002	5.21	0.281
Total	99.82	96.04	94.33	95.31	94.99	93.24	100.67	92.88
H_2O^a	1.99	3.35	3.40	3.44	3.42	3.25	2.02	3.61
f	81.9	34.7	26.5	23.7	22.3	29.6	85.1	25.9
Rock	MD	G	O	O	O	O	O	O

	Depth (m)						
				197.5		202.5	
Component	176.0	**192.0**	192.6	Core	Rim	Core	Rim
SiO_2	37.46	38.90	39.23	38.58	39.26	36.56	38.26
TiO_2	6.85	7.31	5.22	7.13	7.35	6.43	5.85
Al_2O_3	11.83	12.85	13.49	12.97	13.10	12.68	13.28
FeO	8.94	8.97	8.47	8.13	8.37	14.28	8.27
MgO	18.79	17.99	19.24	18.20	17.89	14.73	18.90

(continued)

Table A.17 (continued)

	Depth (m)						
				197.5		202.5	
Component	176.0	**192.0**	192.6	Core	Rim	Core	Rim
K_2O	6.04	7.94	7.98	7.34	5.72	8.75	8.62
NiO	0.125	0.047	0.095	0.035	0.032	0.071	0.015
F	0.563	0.912	0.595	1.10	0.924	1.14	0.847
Cl	0.165	0.166	0.194	0.07	0.066	0.212	0.12
Total	90.76	95.09	94.51	93.56	92.71	94.85	94.16
H_2O[a]	3.66	3.64	3.78	3.53	3.63	3.35	3.64
f	21.1	21.9	19.8	20.0	20.8	35.2	19.7
Rock	P	P	P	P	P	P	P

[a]Hereinafter calculated data

Northeastern Intrusive Branch of the Upper Talnakh Intrusion

Table A.18 Chemical composition of rocks, drill hole KZ-274

	Depth (m)											
Component	406.0	416.0	430.0	441.0	457.0	469.0	477.5	481.0	490.0	496.0	501.0	514.0
SiO_2	44.72	47.20	47.87	48.67	49.01	48.07	48.20	45.36	48.33	42.93	44.48	45.10
TiO_2	3.34	2.70	1.61	1.80	1.02	0.82	0.75	0.62	0.70	1.14	0.62	0.89
Al_2O_3	13.63	14.03	13.22	14.10	15.20	17.27	16.47	15.77	15.47	14.17	16.68	15.64
Fe_2O_3	5.35	2.35	5.93	5.88	2.55	2.58	3.06	3.40	1.61	5.75	2.75	2.16
FeO	8.22	8.86	8.31	8.37	7.22	5.28	4.71	5.54	5.64	7.55	6.97	10.02
MnO	0.23	0.22	0.22	0.24	0.22	0.12	0.12	0.14	0.18	0.19	0.10	0.18
MgO	4.96	5.33	6.11	5.69	8.02	7.65	8.48	8.90	9.45	13.61	10.37	11.55
CaO	8.88	9.17	10.00	10.11	12.15	12.63	13.69	15.06	13.47	11.43	12.49	9.49
Na_2O	3.47	3.42	2.39	2.39	2.11	2.17	1.93	1.62	1.84	1.46	1.73	1.58
K_2O	3.06	2.52	0.88	0.72	0.43	0.72	0.43	0.33	0.43	0.22	0.50	0.43
P_2O_5	1.35	1.06	0.18	0.21	0.12	0.10	0.08	0.16	0.08	0.06	0.21	0.11
LOI	2.48	3.03	2.46	1.64	1.62	2.35	1.82	2.22	2.26	1.37	2.68	2.49
H_2O^-	–	–	–	–	–	–	–	–	–	–	–	–
Total	99.69	99.89	99.18	99.82	99.67	99.76	99.74	99.12	99.46	99.88	99.58	99.64
C.f.	73.2	67.8	70.0	71.5	54.9	50.7	47.8	50.1	43.4	49.4	48.4	51.3
S	0.12	0.048	0.328	0.076	0.136	0.1	0.096	0.108	0.212	0.048	0.16	–
CuO	0.021	0.004	0.02	0.023	0.013	0.012	0.011	–	0.088	0.01	–	0.05
NiO	0.022	0.01	0.022	0.022	0.022	0.019	0.027	–	0.06	0.047	–	–
Cr_2O_3	0.015	0.017	0.007	0.011	0.035	0.053	0.093	–	0.114	0.132	–	0.19
Rock	Go	Go	MD	MD	G	G	G	G	O	O	O	O
	Depth (m)											
Component	514.4	514.5	518.9	519.9	525.0	528.0	530.0	544.0	549.4	554.0	556.0	566.0
SiO_2	45.20	45.56	45.14	36.00	36.93	44.80	32.06	35.44	40.50	39.44	41.00	33.90
TiO_2	0.75	0.72	0.84	0.74	0.42	0.97	0.42	0.32	0.54	0.91	0.86	0.52
Al_2O_3	18.66	17.98	15.70	16.37	7.19	15.60	7.04	10.73	12.23	18.10	13.25	13.15
Fe_2O_3	0.77	2.11	2.62	0.90	8.08	3.52	b.a.	0.57	0.59	1.85	6.02	9.75
FeO	8.62	7.67	9.88	13.51	8.97	7.77	19.47	16.23	14.22	12.27	11.50	11.11
MnO	0.07	0.08	0.20	0.18	0.16	0.09	0.16	0.16	0.15	0.15	0.20	0.16
MgO	10.50	11.14	11.67	19.22	24.28	8.83	21.45	20.10	15.26	9.80	11.63	11.74
CaO	11.04	10.30	9.35	6.80	4.21	11.85	4.16	5.20	7.70	9.30	8.73	8.94
Na_2O	0.50	1.48	1.50	1.46	0.54	1.83	1.18	1.64	2.00	1.98	1.35	0.97
K_2O	2.70	0.31	0.37	0.40	0.25	0.58	0.35	0.44	0.61	0.46	0.37	0.54
P_2O_5	0.15	0.21	0.15	0.12	0.15	0.20	0.09	0.11	0.13	0.16	0.10	0.08
LOI	0.78	2.32	1.52	3.70	8.88	3.22	6.13	3.80	2.05	1.95	3.83	5.74

(continued)

Table A.18 (continued)

	Depth (m)											
Component	514.4	514.5	518.9	519.9	525.0	528.0	530.0	544.0	549.4	554.0	556.0	566.0
H_2O^-	0.14	–	1.22	0.23	–	–	0.21	0.23	0.12	0.18	2.8	3.08
Total	99.88	99.88	100.16	99.63	100.06	99.26	92.72	94.97	96.10	96.55	101.64	99.68
C.f.	47.2	46.8	51.7	42.9	41.3	56.1	47.6	45.5	49.3	59.0	60.1	64.0
S	0.14	0.16	–	0.73	1.692	1.04	4.65	2.35	2.27	1.97	–	–
CuO	0.03	0.023	0.15	0.09	–	0.081	1.89	0.85	1.08	0.76	1.1	3.68
NiO	0.01	0.039	0.11	0.14	–	0.043	1.08	0.56	0.46	0.44	0.74	1.14
Cr_2O_3	0.12	0.097	0.19	0.18	–	0.055	b.a.	0.04	0.02	0.04	0.04	0.04
Rock	O	O	O	P	P	GP	P	P	T	T	T	T

Table A.19 Trace element contents of rocks, drill hole KZ-274

	Depth (m)									
Component	406.0	416.0	427.0	430.0	437.5	441.0	445.0	452.0	457.0	469.0
S	1,200	–	–	–	–	760	–	–	1,360	1,000
Cu	170	–	186	–	142	235	161	155	107	126
Ni	170	–	71	–	107	74	97	79	88	114
Co	42	–	50	–	54	41	56	54	41	36
Pb	–	–	17	–	20	b.a.	17	41	b.a.	b.a.
Zn	–	–	82	–	88	125	91	105	138	71
Cr	102.6	–	50	–	48	55	116	90	144	387
V	100	95	366	250	302	–	318	302	210	160
Ba	1,900	1,400	285	230	173	–	201	208	160	160
Sr	1,300	470	312	240	229	–	208	260	300	320
B	23	29	–	34	–	–	–	–	24	30
Pt	0.0001	–	0.01	–	0.01	0.035	0.000	0.042	0.0125	0.005
Pd	0.011	–	0.011	–	0.016	0.02	0.013	0.032	0.0132	0.015
Rh	0.0001	–	–	–	–	0.0004	–	–	0.0001	0.002
Ru	0.0001	–	–	–	–	0.034	–	–	0.03	0.022
Au	0.0038	–	0.02	–	0.007	0.0084	0.005	0.008	0.11	0.0045
Ag	–	–	0.036	–	0.033	–	0.12	0.13	–	–
Rock	Go	Go	MD	MD	MD	MD	MD	G	G	G
	Depth (m)									
Component	477.5	481.0	490.0	496.0	501.0	506.0	511.0	514.5	514.5	518.9
S	960	–	–	480	–	–	–	–	–	–
Cu	535	70	–	78	110	168	138	15,242	183	–
Ni	276	220	–	298	250	303	261	5,571	304	–
Co	39	44	–	49	46	54	48	54	47	–
Pb	сл	53	–	сл	241	8	50	8	сл	–
Zn	55	187	–	99	468	67	164	66	81	–
Cr	587	850	–	900	560	628	896	882	675	–
V	140	151	120	130	159	159	159	119	110	100
Ba	62	125	230	78	146	118	153	114	120	74
Sr	330	192	280	220	229	187	276	187	300	180
B	25	–	30	23	–	–	–	–	19	20
Pt	0.024	0.013	–	0.04	0.01	0.01	0.01	–	0.0225	–
Pd	0.045	0.025	–	0.011	0.012	0.054	0.024	–	0.024	–
Rh	0.0001	–	–	0.005	–	–	–	–	0.002	–
Ru	0.015	–	–	0.01	–	–	–	–	0.025	–
Au	0.0057	0.015	–	0.0036	0.013	0.01	0.005	0.01	0.007	–

(continued)

Table A.19 (continued)

Component	Depth (m)									
	477.5	481.0	490.0	496.0	501.0	506.0	511.0	514.5	514.5	518.9
Ag	–	0.37	–	–	0.2	0.12	0.23	0.24	–	–
Rock	G	G	O	O	O	O	O	O	O	O

Component	Depth (m)										
	525.0	527.0	528.0	533.0	544.0	544.0	550.0	556.0	556.0	563.0	566.0
S	16,300	–	9,700	–	–	–	–	–	–	–	–
Cu	4,872	7,500	648	4,375	8,387	9,454	2,068	13,790	31,784	7,903	–
Ni	3,682	2,620	348	3,470	5,760	6,024	10,208	6,071	9,126	5,071	–
Co	145	70	49	162	231	265	327	183	273	183	–
Pb	b.a.	17	b.a.	25	197	b.a.	17	29	29	25	–
Zn	104	39	71	104	50	112	66	88	116	91	–
Cr	2,498	385	378	1,255	585	410	285	305	253	325	–
V	–	40	180	114	118	–	151	199	95	246	100
Ba	–	215	140	56	56	–	69	136	110	136	84
Sr	–	291	280	99	73	–	135	156	240	166	230
B	–	–	19	–	–	–	–	–	16	–	18
Pt	0.6	0.7	0.037	0.41	0.4	0.725	1.92	–	1.38	0.58	–
Pd	0.88	0.5	0.05	0.74	0.59	1	2.04	–	2.26	0.83	–
Rh	0.1	0.072	0.009	0.1	0.1	0.1	0.29	–	0.074	0.08	–
Ru	0.036	–	0.027	–	–	0.037	–	–	0.064	–	–
Au	0.1	0.062	0.0135	0.019	0.035	0.086	0.06	0.076	0.396	0.046	–
Ag	–	2	–	1.1	2.5	–	6.5	4.2	–	2.2	–
Rock	P	P	GP	P	P	P	T	T	T	T	T

Table A.20 Chemical composition of pyroxene, drill hole KZ-274 (partial analysis)

Component	Depth (m)											
	474.0	474.0	474.0	474.0	544.0	544.0	544.0	544.0	544.0	544.0	550.0	565.0
Mineral	Cpx	Cpx	Cpx	Cpx	Opx	Opx	Opx	Opx	Opx	Opx	Cpx	Opx
CaO	19.66	20.05	19.36	17.82	1.52	1.57	1.56	1.60	1.44	1.41	20.15	1.78
MgO	16.30	16.60	16.88	16.77	28.27	28.80	29.30	29.05	28.33	28.57	15.50	22.75
FeO	5.87	6.12	6.36	7.09	13.48	13.51	13.53	13.60	13.62	13.59	7.86	21.01
NiO	0.01	0.024	0.035	0.018	0.068	0.042	0.054	0.029	0.067	0.049	0.047	0.065
f	16.9	17.1	17.5	19.2	21.1	20.8	20.5	20.7	21.2	21.0	22.3	34.0
Wo	42	42	41	38	3	3	3	3	3	3	42	3
En	48	48	49	50	77	77	77	77	77	77	45	64
Fs	10	10	10	12	20	20	20	20	20	20	13	33
Rock	G	G	G	G	P	P	P	P	P	P	T	T

Table A.21 Chemical composition of olivine, drill hole KZ-274

Component	Depth (m)									
	469.0	469A	477.0		481.0	486.0	496.0			
SiO_2	38.04	37.44	38.50	36.80	38.37	37.81	38.91	38.57	39.18	39.23
TiO_2	–	–	–	–	–	–	0.000	0.086	0.000	0.000
FeO	23.28	26.36	21.55	30.03	29.93	25.21	23.81	24.99	24.55	24.41
MnO	–	–	–	–	–	–	0.392	0.421	0.436	0.428
MgO	38.59	36.04	39.06	32.80	36.87	36.56	36.73	36.10	36.89	37.36
CaO	0.16	0.21	0.09	0.15	0.17	0.21	0.283	0.126	0.161	0.268
Na_2O	–	–	–	–	–	–	0.000	0.000	0.087	0.000
K_2O	–	–	–	–	–	–	0.072	0.008	0.039	0.000
NiO	0.078	0.102	0.134	0.217	0.093	0.125	0.000	0.374	0.032	0.037
Cr_2O_3	–	–	–	–	–	–	0.000	0.037	0.043	0.160

(continued)

Table A.21 (continued)

	Depth (m)									
Component	469.0	469A	477.0		481.0	486.0	496.0			
Total	100.15	100.15	99.33	100.00	105.43	99.92	100.20	100.71	101.42	101.89
Fo	75	71	76	66	69	72	73	72	73	73
Fa	25	29	24	34	31	28	27	28	27	27
Rock	G	G	G	G	G	O	O	O	O	O

	Depth (m)									
Component	496.0	506.0	509.0	511.0	514.5				518.9	528.0
SiO_2	38.95	37.70	37.54	37.57	37.21	37.37	37.61	36.86	37.60	37.03
TiO_2	0.085	–	–	–	–	0.000	0.011	0.000	–	–
FeO	24.55	23.34	25.15	26.19	28.22	25.56	23.94	24.96	26.04	28.85
MnO	0.446	–	–	–	–	0.444	0.252	0.356	–	–
MgO	37.60	39.04	37.33	36.02	34.29	35.22	36.16	34.21	36.08	34.04
CaO	0.300	0.23	0.18	0.15	0.16	0.130	0.301	0.115	0.18	0.16
Na_2O	0.000	–	–	–	–	0.330	0.049	0.181	–	–
K_2O	0.000	–	–	–	–	0.019	0.000	0.029	–	–
NiO	0.077	0.098	0.138	0.104	0.106	0.301	0.015	0.094	0.111	0.032
Cr_2O_3	0.000	–	–	–	–	0.175	0.170	0.147	–	–
Total	102.01	100.41	100.34	100.03	99.99	99.55	98.51	96.95	100.01	100.11
Fo	73	75	73	71	68	71	73	71	71	68
Fa	27	25	27	29	32	29	27	29	29	32
Rock	O	O	O	O	O	O	O	O	O	GP

	Depth (m)									
Component	533.0	537.0	544.0							
SiO_2	38.40	38.10	37.79	38.72	38.51	38.40	38.69	38.94	38.69	38.97
TiO_2	–	–	–	–	–	–	–	–	–	–
FeO	22.03	23.09	22.42	21.93	22.07	22.68	22.21	21.84	21.61	21.17
MnO	–	–	–	–	–	–	–	–	–	–
MgO	39.16	38.53	40.18	38.50	38.99	38.47	38.36	38.07	39.01	38.83
CaO	0.1	0.16	0.14	0.14	0.000	0.000	0.000	0.13	0.08	0.15
Na_2O	–	–	–	–	–	–	–	–	–	–
K_2O	–	–	–	–	–	–	–	–	–	–
NiO	0.215	0.185	0.207	0.188	0.187	0.193	0.185	0.177	0.208	0.177
Cr_2O_3	–	–	–	–	–	–	–	–	–	–
Total	99.91	100.07	100.74	99.48	99.76	99.74	99.45	99.16	99.60	99.30
Fo	76	75	76	76	76	75	75	76	76	77
Fa	24	25	24	24	24	25	25	24	24	23
Rock	P	P	P	P	P	P	P	P	P	P

	Depth (m)										
Component	559.0					563.0	565.0				
SiO_2	37.69	36.78	36.99	36.94	37.14	36.74	36.30	31.79	37.10	36.98	36.75
TiO_2	–	–	–	–	–	–	–	–	–	–	–
FeO	31.27	32.03	31.49	31.45	31.73	30.15	33.18	31.23	29.82	29.54	31.00
MnO	–	–	–	–	–	–	–	–	–	–	–
MgO	29.20	30.54	30.54	30.67	30.01	32.91	30.09	31.42	32.45	33.14	31.82
CaO	0.000	0.000	0.14	0.17	0.000	0.16	0.16	0.15	0.1	0.12	0.16
Na_2O	–	–	–	–	–	–	–	–	–	–	–
K_2O	–	–	–	–	–	–	–	–	–	–	–
NiO	0.206	0.203	0.201	0.208	0.207	0.167	0.205	0.172	0.206	0.172	0.151
Cr_2O_3	–	–	–	–	–	–	–	–	–	–	–
Total	98.37	99.55	99.36	99.44	99.09	100.13	99.94	94.76	99.68	99.95	99.88
Fo	62	63	63	63	63	66	62	64	66	67	65

(continued)

Table A.21 (continued)

Component	Depth (m) 559.0					563.0	565.0				
Fa	38	37	37	37	37	34	38	36	34	33	35
Rock	T	T	T	T	T	T	T	T	T	T	T

Component	Depth (m) 565.0										
SiO_2	36.10	35.75	36.30	35.90	36.65	38.17	37.66	38.51	38.24	38.96	37.96
TiO_2	0.103	0.000	0.040	0.022	0.047	0.004	0.077	0.000	0.038	0.000	0.130
FeO	31.14	31.08	30.65	31.50	27.71	28.03	28.95	25.19	24.92	25.10	24.74
MnO	0.572	0.655	0.773	0.728	0.448	0.440	0.576	0.266	0.327	0.439	0.456
MgO	29.87	30.03	30.21	30.09	32.36	34.12	33.34	37.01	35.55	37.04	36.14
CaO	0.208	0.092	0.198	0.163	0.168	0.063	0.113	0.207	0.113	0.136	0.050
Na_2O	0.000	0.000	0.000	0.000	0.000	0.000	0.000	0.000	0.000	0.012	0.000
K_2O	0.024	0.057	0.051	0.022	0.000	0.000	0.000	0.000	0.000	0.000	0.037
NiO	0.406	0.112	0.246	0.044	0.373	0.138	0.155	0.205	0.109	0.335	0.435
Cr_2O_3	0.067	0.226	0.105	0.000	0.000	0.123	0.091	0.101	0.000	0.056	0.009
Total	98.49	98.00	98.57	98.47	97.76	101.09	100.96	101.49	99.30	102.08	99.96
Fo	63	63	64	63	68	68	67	72	72	72	72
Fa	37	37	36	37	32	32	33	28	28	28	28
Rock	T	T	T	T	T	T	T	T	T	T	T

Table A.22 Chemical composition of mica group minerals, drill hole KZ-274

Component	Depth (m) 525.0		527.0		544.0		556.0 Core	Rim	
SiO_2	37.17	37.88	38.12	39.28	37.41	37.85	37.32	38.98	37.30
TiO_2	7.78	5.26	7.98	2.80	4.06	8.01	6.25	6.03	6.90
Al_2O_3	12.39	13.14	12.70	14.19	11.50	12.48	12.75	12.59	12.69
FeO	8.73	8.57	10.64	8.75	9.13	8.13	15.14	14.15	8.26
MgO	17.02	18.43	16.39	20.36	21.58	17.28	14.23	15.70	17.95
K_2O	8.45	8.68	8.76	8.07	4.65	9.00	8.43	7.30	8.84
NiO	0.164	0.07	0.191	0.045	0.11	0.095	0.217	0.254	0.083
F	0.706	0.527	0.839	0.867	0.806	0.64	1.54	1.06	0.537
Cl	0.153	0.169	0.24	0.322	0.121	0.116	0.413	0.33	0.167
Total	92.56	92.73	95.86	94.68	89.37	93.60	96.29	96.39	92.73
H_2O	3.61	3.71	3.62	3.61	3.51	3.70	3.14	3.48	3.70
f	22.3	20.7	26.7	19.4	19.2	20.9	37.4	33.6	20.5
Rock	P	P	P	P	P	P	T	T	T

Kharaelakhsky Intrusive Branch of the Upper Talnakh Intrusion

Table A.23 Chemical composition of rocks, drill hole KZ-585

Component	Depth (m) 681.7	691.0	694.3	696.2	699.2	701.8	705.3	710.4	714.2
SiO_2	48.77	47.47	48.28	48.99	46.76	48.01	48.67	48.27	48.20
TiO_2	0.72	0.72	0.68	0.68	0.68	0.60	0.76	0.49	0.76
Al_2O_3	15.36	16.10	15.33	15.94	16.78	15.80	15.76	14.32	16.06
Fe_2O_3	5.13	4.97	3.88	3.45	3.39	4.78	3.47	4.76	3.63
FeO	6.54	6.36	8.71	8.44	6.66	7.77	8.06	7.23	8.38
MnO	0.16	0.13	0.17	0.20	0.29	0.20	0.20	0.16	0.28
MgO	9.80	8.06	8.14	7.95	7.59	8.04	7.77	7.77	7.80
CaO	7.94	10.52	10.51	10.72	8.09	8.88	10.23	11.15	8.87

(continued)

Table A.23 (continued)

Component	Depth (m) 681.7	691.0	694.3	696.2	699.2	701.8	705.3	710.4	714.2
Na_2O	2.43	2.64	2.63	2.45	3.88	3.52	2.44	2.90	3.32
K_2O	1.50	1.08	1.14	0.72	0.84	0.60	1.21	0.70	1.32
P_2O_5	0.12	0.11	0.11	0.07	0.09	0.13	0.14	0.07	0.14
LOI	1.16	1.80	1.23	0.63	4.94	2.55	0.86	1.33	1.33
Total	99.63	99.96	100.81	100.24	99.99	100.88	99.57	99.15	100.09
H_2O^-	0.38	0.69	0.37	0.74	0.93	0.51	0.88	0.39	0.46
C.f.	54.4	58.4	60.7	59.9	57.0	61.0	59.7	60.7	60.6
Rock	G	G	G	G	G	G	G	G	G

Component	Depth (m) 724.7	727.7	730.4	738.7	746.4	749.5	752.0	764.0	766.7	768.0	772.5
SiO_2	47.86	46.30	47.47	47.52	19.12	43.10	46.34	39.23	39.87	44.96	47.86
TiO_2	0.62	0.44	0.80	0.68	0.50	0.30	0.99	0.98	0.92	1.05	0.76
Al_2O_3	16.08	16.40	15.77	15.13	4.73	24.03	16.11	15.04	11.54	12.34	16.25
Fe_2O_3	4.77	6.88	5.16	5.96	4.46	3.25	6.20	6.63	7.36	2.65	5.47
FeO	7.27	4.98	6.15	5.25	2.96	3.44	4.91	5.80	5.24	4.11	5.89
MnO	0.20	0.22	0.02	0.21	0.16	0.04	0.22	0.42	0.16	0.15	0.19
MgO	8.42	9.81	7.03	7.53	42.45	4.06	7.35	13.20	10.35	8.53	8.40
CaO	8.74	9.37	10.49	10.48	4.80	14.83	11.87	12.66	17.46	22.07	10.69
Na_2O	2.45	2.71	3.24	2.97	0.11	1.71	2.92	0.54	0.76	0.88	2.44
K_2O	1.20	1.50	1.32	1.32	0.04	1.83	0.66	0.18	0.04	–	1.32
P_2O_5	0.12	0.13	0.13	0.13	0.05	0.07	0.12	0.12	0.05	0.06	0.13
LOI	1.82	1.98	2.16	2.02	14.64	3.64	2.55	5.98	3.50	2.50	1.33
Total	99.55	100.72	99.74	99.20	94.02	100.30	100.24	100.78	97.25	99.30	100.73
H_2O^-	0.80	0.56	0.38	0.25	2.18	1.37	0.71	0.83	0.21	0.69	0.52
C.f.	58.9	54.7	61.7	59.8	14.9	62.2	60.2	48.5	54.9	44.2	57.5
Rock	G	G	G	G	MB	MB	G	MB	MB	MB	MB

Component	Depth (m) 774.0	776.0	788.3	798.0	808.2	812.6	818.0	822.8	828.5	**831.0**
SiO_2	46.38	46.81	40.87	39.82	36.30	36.70	40.29	39.10	46.94	34.40
TiO_2	1.05	0.48	0.60	0.25	0.52	0.40	0.44	0.31	1.05	0.41
Al_2O_3	16.53	16.72	10.77	9.25	8.98	9.99	21.64	9.07	17.28	7.15
Fe_2O_3	4.96	4.10	5.22	6.04	2.38	8.25	7.22	4.54	4.25	3.94
FeO	6.13	4.61	6.13	7.24	7.77	5.98	3.60	10.74	6.96	6.66
MnO	0.16	0.17	0.21	0.34	0.15	0.27	0.09	0.24	0.27	0.14
MgO	8.00	8.75	10.77	24.47	31.17	24.01	2.99	27.62	4.71	33.19
CaO	10.10	11.52	20.26	7.40	4.45	6.60	15.47	5.93	10.81	3.06
Na_2O	2.45	2.37	0.34	0.47	0.12	0.27	1.22	0.47	3.38	0.18
K_2O	1.50	0.96	0.06	0.30	0.13	0.24	0.90	0.12	0.66	0.16
Li_2O	–	–	–	–	0.02	–	–	–	–	–
P_2O_5	0.16	0.06	0.03	0.05	0.20	0.07	0.06	0.06	0.03	0.14
LOI	2.46	2.97	3.40	4.08	7.46	6.26	4.71	2.61	2.52	9.46
Total	99.88	99.52	98.66	99.71	99.65	99.04	98.63	100.81	98.86	98.89
H_2O^-	–	0.49	–	0.83	0.24	0.84	0.39	0.87	0.44	0.28
H_2O^+	–	–	–	–	5.83	–	–	–	–	8.46
S	–	–	3.33	0.92	0.43	2.40	3.56	–	–	0.54
C.f.	58.1	49.9	51.3	35.2	24.6	37.2	78.4	35.6	70.4	24.2
Rock	MB	MB	MB	MB	MB	MB	MB	MB	MB	MB

Component	Depth (m) 834.0	835.3	835.7	837.1	**846.4**	848.8	850.6	851.3	**852.0**
SiO_2	26.20	43.98	39.08	36.46	39.00	46.65	42.01	11.66	38.50
TiO_2	0.85	0.87	0.35	0.22	0.71	0.91	0.60	0.22	1.35
Al_2O_3	8.13	13.80	11.15	10.63	9.60	19.29	17.82	2.43	11.84
Fe_2O_3	5.36	4.88	0.30	7.64	1.29	2.01	3.72	3.05	1.13

(continued)

Table A.23 (continued)

Component	Depth (m) 834.0	835.3	835.7	837.1	**846.4**	848.8	850.6	851.3	**852.0**
FeO	8.69	8.62	10.48	4.71	12.95	7.89	7.25	3.23	4.99
MnO	0.20	0.13	0.18	0.26	0.25	0.16	0.20	0.07	0.64
MgO	33.02	8.59	26.63	27.25	26.62	7.65	9.18	56.36	27.72
CaO	0.70	13.70	6.17	4.07	6.76	11.82	10.48	0.81	10.37
Na_2O	0.12	2.24	0.42	0.34	0.36	2.10	2.03	0.05	0.29
K_2O	0.10	0.17	0.00	1.14	0.13	0.48	0.30	0.05	0.19
Li_2O	0.01	–	–	–	0.025	–	–	–	0.025
P_2O_5	0.17	0.11	0.03	0.03	0.26	0.07	0.07	0.02	0.23
LOI	15.36	1.92	4.94	6.79	1.78	0.16	2.72	22.29	2.52
Total	98.91	99.01	99.73	99.54	99.74	99.19	96.38	100.24	99.80
H_2O^-	0.68	0.76	–	0.82	0.04	0.17	0.35	–	0.06
S	2.42	–	0.63	1.29	0.51	–	–	0.43	0.03
H_2O^+	11.97	–	–	–	1.81	–	–	–	1.84
C.f.	29.9	61.1	28.8	31.2	34.9	56.4	54.4	10.0	18.1
Rock	MB	MB	MB	MB	MB	MB	MB	MB	MB
	Depth (m)								
Component	853.3	**855.0**	**859.4**	860.0	862.0	862.5	870.0	875.3	884.7
SiO_2	45.62	41.80	29.86	46.44	45.08	41.42	46.00	40.27	44.14
TiO_2	0.60	0.92	0.47	0.98	0.92	0.48	1.05	0.60	0.72
Al_2O_3	19.44	12.12	7.64	17.06	15.30	8.85	15.10	13.67	15.54
Fe_2O_3	5.82	2.04	3.54	3.61	3.73	4.21	2.18	7.60	4.02
FeO	4.11	12.58	3.08	7.57	7.53	7.74	10.26	10.80	8.30
MnO	0.02	0.23	0.12	0.22	0.16	0.28	0.18	0.25	0.28
MgO	7.28	17.69	43.68	8.03	8.37	27.81	7.70	7.39	8.38
CaO	13.27	8.93	3.72	12.76	12.40	6.60	12.20	11.04	11.80
Na_2O	2.16	1.05	0.05	2.16	2.17	0.41	2.04	1.49	1.76
K_2O	0.65	0.37	0.05	0.48	0.57	0.30	0.35	0.30	0.78
Li_2O	–	0.025	–	–	–	–	–	–	–
P_2O_5	0.11	0.23	0.03	0.04	0.10	0.07	0.10	0.07	0.08
LOI	1.29	1.14	7.26	0.78	2.74	2.73	1.86	3.64	2.40
Total	100.37	99.13	99.50	100.13	99.07	100.90	99.02	97.12	98.20
H_2O^-	0.23	0.84	–	0.57	–	0.34	–	0.28	0.48
H_2O^+	–	0.21	–	–	–	–	–	–	–
S	–	0.80	0.80	–	0.22	0.25	0.53	4.87	2.25
C.f.	57.7	20.7	13.2	58.2	57.4	30.1	61.8	71.4	59.5
Rock	MB	MB	MB	TP	TP	TP	TP	TP	TP

Table A.24 Trace element contents of rocks, drill hole KZ-585

Component	Depth (m) 681.7	699.0	727.7	749.6	750.5	764.0	774.2	778.7	783.0	791.2	796.0
S	–	–	–	–	–	2.4	–	–	–	–	–
Cu	167	113	139	–	11,227	1,008	300	3,500	12,896	4,161	2,669
Ni	182	176	141	–	4,800	509	166	2,400	5,096	1,605	1,759
Co	39	21	40	–	220	44	40	119	234	74	114
Cr	188	227	202	–	1,249	394	–	2,800	–	48	681
Pb	b.a.	72	25	–	b.a.	25	b.a.	–	66	36	b.a.
Zn	104	20,730	109	–	122	69	60	–	73	74	88
Pt	0.02	0.009	0.0011	0.043	0.545	0.068	2.7	–	0.123	0.195	0.13
Pd	0.027	0.024	0.0094	0.055	1.2	0.089	6.6	–	0.245	0.275	0.212
Rh	0.00425	0.001	0.001	0.009	0.063	0.0051	0.0054	–	0.011	0.014	0.026
Ru	0.001	0.0073	0.004	0.021	0.025	0.019	0.023	–	0.019	0.005	0.023

(continued)

Table A.24 (continued)

	Depth (m)										
Component	681.7	699.0	727.7	749.6	750.5	764.0	774.2	778.7	783.0	791.2	796.0
Au	0.0085	0.004	0.0034	0.0051	0.092	0.0099	0.226	–	0.027	0.02	0.013
Rock	G	G	G	MB	G	MB	MB	MB	MB	MB	MB

	Depth (m)								
Component	797.6	798.7	805.9	808.2	812.6	816.5	821.0	829.0	**831.0**
S	–	–	–	0.43	2.40	–	–	–	0.54
Cu	8,812	1,551	2,665	2,640	5,278	7,501	2,343	5,601	1,920
Ni	2,466	1,535	1,401	790	1,150	3,388	851	2,919	1,180
Co	143	113	94	–	111	158	108	154	–
Cr	–	–	–	130	253	735	128	1,415	160
Pb	29	b.a.	b.a.	–	b.a.	25	25	b.a.	–
Zn	151	89	96	–	141	120	84	105	–
Pt	0.26	0.125	0.14	0.11	0.2	0.27	0.05	0.2	0.062
Pd	0.54	0.212	0.19	0.29	0.32	0.425	0.0925	0.44	0.24
Rh	0.019	0.0125	0.019	–	0.0056	0.0025	0.0125	0.021	0.0056
Ru	0.01	0.0175	0.007	–	0.001	0.014	0.001	0.001	0.0045
Au	0.084	0.0062	0.016	0.063	0.076	0.036	0.02	0.029	0.034
Ag	–	–	–	0.35	–	–	–	–	–
Rock	MB	MB	MB	MB	MB	MB	MB	MB	MB

	Depth (m)								
Component	838.7	**846.4**	848.0	**852.1**	**859.4**	860.0	868.7	880.0	905.0
S	–	1.81	–	1.84	0.80	–	–	–	–
Cu	–	960	3,007	70	–	172	19,688	101,700	177,410
Ni	2,009	170	1,489	–	160	126	9,020	14,891	47,125
Co	177	–	78	–	–	33	379	238	1,571
Cr	109	820	80	50	70	–	54	136	–
Pb	b.a.	–	b.a.	–	–	b.a.	29	295	61
Zn	55	–	70	–	–	115	144	215	522
Pt	0.22	–	0.1	0.0015	0.045	0.033	0.724	4.5	3.9
Pd	0.72	0.0051	0.37	–	0.015	0.016	2.2	9.26	13.4
Rh	0.012	0.0008	0.028	–	0.0065	0.001	0.035	0.0041	0.74
Ru	0.027	–	0.023	0.0009	0.0145	0.023	0.026	0.026	0.049
Au	0.05	0.007	0.024	0.0084	0.155	0.005	0.107	0.8	0.33
Ag	–	–	–	–	0.025	–	–	–	–
Rock	MB	MB	MB	MB	MB	TP	TP	TP	SV

Table A.25 Chemical composition of olivine, drill hole KZ-585

	Depth (m)									
Component	778.7		796.0				801.7			806.5
SiO_2	41.57	41.43	40.13	39.41	39.75	39.64	39.63	38.83	39.06	39.42
FeO	16.28	16.50	15.03	16.65	16.60	16.71	17.34	18.57	19.51	19.45
CaO	0.04	0.06	0.46	0.28	0.16	0.07	0.09	0.31	0.23	0.11
MgO	42.66	43.22	43.53	43.53	42.03	43.10	42.24	42.43	40.73	40.06
NiO	0.13	0.14	0.09	0.07	0.10	0.07	0.10	0.10	0.10	0.10
Total	100.68	101.35	99.24	99.94	98.64	99.59	99.40	100.24	99.63	99.14
Fo	82	82	84	82	82	82	81	80	79	79
Fa	18	18	16	18	18	18	19	20	21	21
Rock	MB	MB	MB	MB	MB	MB	MB	MB	MB	MB

	Depth (m)									
Component	806.5		809.5			811.5				
SiO_2	39.88	39.50	39.90	39.04	39.23	40.15	39.69	40.13	40.04	39.70
FeO	17.80	19.00	19.93	20.70	19.72	16.36	17.05	16.83	16.26	17.11

(continued)

Table A.25 (continued)

	Depth (m)									
Component	806.5		809.5			811.5				
CaO	0.13	0.10	0.16	0.12	0.13	0.22	0.19	0.13	0.15	0.13
MgO	41.01	40.48	39.17	39.42	41.46	42.07	42.39	41.59	42.59	42.35
NiO	0.08	0.08	0.09	0.89	0.08	0.08	0.09	0.09	0.06	0.08
Total	98.90	99.16	99.25	100.17	100.62	98.88	99.41	98.77	99.10	99.37
Fo	80	79	78	77	79	82	82	82	82	82
Fa	20	21	22	23	21	18	18	18	18	18
Rock	MB	MB	MB	MB	MB	MB	MB	MB	MB	MB

	Depth (m)									
Component	811.5	811.7			819.8		821.0		822.0	
SiO_2	40.18	39.86	39.75	39.43	39.94	39.98	39.19	39.18	39.33	39.53
FeO	19.40	16.98	15.78	17.54	16.34	15.38	19.31	19.73	18.63	18.52
CaO	0.14	0.10	0.09	0.12	0.08	0.07	0.10	0.08	0.09	0.06
MgO	44.06	42.14	43.99	42.49	42.84	44.00	40.85	40.30	41.25	40.91
NiO	0.02	0.08	0.08	0.09	0.06	0.04	0.08	0.12	0.18	0.20
Total	103.80	99.16	99.69	99.67	99.26	99.47	99.53	99.41	99.48	99.22
Fo	80	82	83	81	82	84	79	78	80	80
Fa	20	18	17	19	18	16	21	22	20	20
Rock	MB	MB	MB	MB	MB	MB	MB	MB	MB	MB

	Depth (m)									
Component	822.0				822.1	825.8			827.0	
SiO_2	39.53	39.58	40.02	40.23	42.99	39.47	39.38	38.63	38.51	38.58
FeO	18.21	18.48	18.04	18.18	0.94	17.67	18.98	17.65	22.71	22.59
CaO	0.08	0.08	0.13	0.07	0.23	0.14	0.20	0.16	0.08	0.11
MgO	41.31	40.84	40.38	39.73	54.81	42.27	40.71	41.85	38.18	38.08
NiO	0.18	0.16	0.05	0.06	–	0.02	0.05	0.03	0.12	0.14
Total	99.31	99.14	98.62	98.27	98.97	99.57	99.32	98.32	99.60	99.50
Fo	80	80	80	80	99	81	79	81	75	75
Fa	20	20	20	20	1	19	21	19	25	25
Rock	MB	MB	MB	MB	MB	MB	MB	MB	MB	MB

	Depth (m)									
Component	827.0			831.0			831.8			
SiO_2	39.02	39.50	38.77	40.31	40.19	40.27	38.58	38.50	39.64	39.41
FeO	22.81	20.69	22.07	13.29	13.45	13.73	22.73	22.32	15.52	21.10
CaO	0.13	0.08	0.10	0.08	0.07	0.09	0.12	0.11	0.05	0.10
MgO	36.68	38.27	38.29	45.84	45.93	45.37	38.48	39.06	44.65	37.94
NiO	0.13	0.11	0.13	0.05	0.05	0.03	0.11	0.11	0.07	0.12
Total	98.77	98.65	99.36	99.57	99.69	99.49	100.02	100.10	99.93	98.67
Fo	74	77	76	86	86	85	75	76	84	76
Fa	26	23	24	14	14	15	25	24	16	24
Rock	MB	MB	MB	MB	MB	MB	MB	MB	MB	MB

	Depth (m)									
Component	831.8	834.0	840.6				846.4			851.3
SiO_2	39.92	38.63	37.86	38.04	38.03	38.55	39.10	39.17	37.85	41.84
FeO	16.97	22.31	25.52	24.84	24.14	22.85	18.72	18.36	19.38	1.47
CaO	0.11	0.07	0.14	0.12	0.09	0.10	0.12	0.15	0.10	0.58
MgO	42.00	38.37	36.07	36.50	37.51	37.83	41.91	42.14	39.58	55.35
NiO	0.08	0.14	0.17	0.18	0.15	0.15	0.01	0.02	–	–
Total	99.08	99.52	99.76	99.68	99.92	99.48	99.86	99.84	96.91	99.24
Fo	82	75	72	72	73	75	80	80	78	99

(continued)

Table A.25 (continued)

	Depth (m)									
Component	831.8	834.0	840.6				846.4			851.3
Fa	18	25	28	28	27	25	20	20	22	1
Rock	MB	MB	MB	MB	MB	MB	MB	MB	MB	MB
	Depth (m)									
Component	851.3	852.1		855.0				859.4		
SiO_2	41.63	40.75	40.42	37.93	37.87	37.42	36.87	42.35	41.92	42.37
FeO	1.34	9.08	8.70	24.10	24.76	25.75	26.98	1.68	1.58	1.42
CaO	0.80	0.23	0.18	0.09	0.11	0.09	0.15	0.43	0.67	0.51
MgO	56.09	48.59	49.29	36.82	36.31	37.11	36.02	55.57	54.81	56.23
NiO	–	–	0.02	0.06	0.09	–	–	–	–	0.01
Total	99.86	98.65	98.61	99.00	99.14	100.37	100.02	100.03	98.98	100.54
Fo	99	91	91	73	72	72	70	98	98	99
Fa	1	9	9	27	28	28	30	2	2	1
Rock	MB	MB	MB	MB	MB	MB	MB	MB	MB	MB

Table A.26 Chemical composition of pyroxene, drill hole KZ-585 (partial analysis)

	Depth (m)									
Component	796.0	796.7		806.5	822.0		827.0	831.8		834.0
Mineral	Cpx	Cpx	Opx	Cpx	Cpx	Cpx	Cpx	Cpx	Opx	Cpx
CaO	23.25	23.39	0.41	22.00	21.00	21.10	20.74	20.88	0.07	16.27
MgO	13.38	13.50	32.40	15.53	15.21	15.79	14.95	16.69	23.66	18.01
FeO	4.91	4.87	7.74	5.14	6.07	5.04	7.44	5.31	7.62	10.28
NiO	0.027	0.015	0.007	0.012	0.011	0.013	0.021	0.017	0.053	0.045
f	17.0	16.9	11.7	15.6	20.8	35.1	21.7	15.2	13.7	24.3
Wo	51	51	1	46	44	38	44	43	0	33
En	41	41	87	45	44	40	44	48	86	51
Fs	8	8	12	9	12	22	12	9	14	16
Rock	MB	MB	MB	MB	MB	MB	MB	MB	MB	MB

Eastern Noril'sk Intrusive Branch of the Intrusion Noril'sk-I

Table A.27 Chemical composition of rocks, drill hole PE-35

	Depth (m)										
Component	448.5	449.3	450.6	451.5	453.5	461.2	480.0	495.0	505.0	514.0	515.0
SiO_2	46.40	44.10	41.60	46.30	46.40	49.40	49.76	52.32	49.39	50.84	49.63
TiO_2	1.13	0.70	0.79	0.86	1.83	1.29	2.70	2.08	2.02	1.96	1.99
Al_2O_3	14.30	17.00	17.80	18.00	13.30	14.00	11.12	11.83	12.64	11.64	12.18
Fe_2O_3	5.40	4.47	6.66	4.00	6.51	4.68	8.09	7.24	6.87	5.69	7.07
FeO	5.67	5.07	4.54	5.22	7.19	5.96	9.68	9.05	9.86	10.90	9.50
MnO	0.12	0.12	0.10	0.11	0.19	0.19	0.26	0.21	0.24	0.28	0.25
MgO	7.36	8.64	9.12	7.92	7.53	7.16	3.15	2.76	2.96	3.54	3.15
CaO	11.04	11.22	9.11	10.98	9.56	11.90	6.29	6.29	6.83	6.55	9.31
Na_2O	1.83	1.50	1.42	2.00	2.70	2.57	3.75	3.10	3.48	3.52	2.56
K_2O	0.40	0.28	0.35	0.35	0.68	0.86	1.12	0.82	1.36	0.96	0.72
P_2O_5	0.10	0.06	0.08	0.07	0.18	0.13	0.35	0.46	0.44	0.35	0.36
LOI	4.67	5.01	4.72	3.36	2.74	2.23	2.97	3.14	3.06	2.79	2.49
H_2O^-	0.90	1.09	1.18	0.06	0.05	0.02	–	–	–	–	–
Total	99.32	99.26	97.47	99.23	98.86	100.39	99.24	99.30	99.15	99.02	99.21
F	60.1	52.5	55.1	53.8	64.5	59.8	84.9	85.5	85.0	82.4	84.0

(continued)

Table A.27 (continued)

	Depth (m)										
Component	448.5	449.3	450.6	451.5	453.5	461.2	480.0	495.0	505.0	514.0	515.0
S	–	–	–	–	–	0.06	0.16	0.16	0.20	0.168	0.188
CuO	0.025	0.165	0.025	0.033	0.038	0.013	0.004	0.005	0.004	0.004	0.011
NiO	0.000	0.108	0.000	0.000	0.000	0.007	0.004	0.004	0.004	0.006	0.006
Cr_2O_3	0.97	0.28	2.40	0.52	1.06	0.00	0.004	0.004	0.004	0.004	0.007
Rock	T	T	MD	MD	MD	MD	MD	MD	GP	GP	GP

	Depth (m)										
Component	520.0	533.0	537.0	539.6	539.7	539.9	541.5	563.0	581.0	689.0	702.5
SiO_2	47.50	46.58	47.84	44.24	44.39	45.50	45.73	42.97	46.70	48.00	44.44
TiO_2	2.11	1.93	1.68	2.06	1.93	2.72	1.68	1.28	0.90	1.40	1.37
Al_2O_3	12.32	12.79	13.84	12.61	12.42	11.30	15.41	14.56	13.30	14.75	15.28
Fe_2O_3	5.90	6.54	2.34	7.10	5.07	8.29	2.26	4.47	3.58	1.34	2.08
FeO	11.33	10.89	11.44	12.58	13.49	10.54	11.09	8.96	7.76	11.98	11.62
MnO	0.28	0.22	0.20	0.26	0.26	0.24	0.18	0.18	0.16	0.21	0.20
MgO	3.73	4.40	6.06	5.50	5.91	5.47	6.28	5.13	3.93	6.53	6.66
CaO	8.72	9.26	10.60	8.99	9.82	8.94	11.44	11.50	13.38	10.70	10.86
Na_2O	2.56	2.50	2.38	2.28	2.27	2.33	2.33	1.96	1.95	2.26	2.44
K_2O	1.12	0.19	0.53	0.43	0.43	0.50	0.58	0.39	0.24	0.53	0.58
P_2O_5	0.33	0.21	0.23	0.18	0.16	0.20	0.21	0.18	0.14	0.20	0.25
LOI	3.33	3.48	2.39	3.52	3.17	3.17	2.21	6.22	5.40	1.60	1.56
H_2O^-	–	–	–	–	–	0.24	–	–	–	–	–
Total	99.23	98.99	99.53	99.75	99.32	99.44	99.40	97.80	97.44	99.50	97.34
F	82.2	79.8	69.5	78.2	75.9	77.5	68.0	72.4	74.3	67.1	67.3
S	0.148	0.208	0.180	0.208	0.188	1.290	0.040	0.688	1.28	0.06	0.90
CuO	0.008	0.014	0.026	0.017	0.012	0.02	0.019	0.027	0.011	0.019	0.018
NiO	0.007	0.013	0.022	0.015	0.015	0.00	0.025	0.022	0.01	0.022	0.023
Cr_2O_3	0.020	0.006	0.024	0.007	0.008	0.02	0.019	0.018	0.005	0.027	0.023
Rock	GP	G	G	G	G	G	G	G	G	G	Go

	Depth (m)										
Component	704.0	710.5	720.0	730.0	737.0	750.0	762.0	767.0	769.4	774.0	775.5
SiO_2	45.73	48.40	46.90	46.55	46.80	46.50	47.00	45.90	44.90	39.60	46.46
TiO_2	1.42	0.87	0.77	0.75	0.75	0.76	0.85	0.91	0.90	0.72	0.73
Al_2O_3	15.70	17.10	17.30	17.60	17.45	17.80	16.05	15.50	11.45	8.20	17.70
Fe_2O_3	2.95	2.98	2.42	3.12	2.48	3.43	3.42	3.38	3.41	3.91	3.19
FeO	9.30	6.14	7.54	6.91	7.22	6.18	7.54	6.86	9.77	12.14	6.97
MnO	0.21	0.15	0.15	0.15	0.15	0.15	0.17	0.16	0.17	0.20	0.16
MgO	6.33	7.00	9.34	8.84	9.00	8.84	10.00	10.51	15.68	11.88	9.00
CaO	11.12	12.29	10.67	10.67	11.14	11.37	10.44	10.21	8.35	5.19	11.37
Na_2O	2.47	1.96	1.96	1.80	1.80	1.68	1.80	1.68	1.20	0.75	1.80
K_2O	0.58	0.50	0.42	0.50	0.50	0.33	0.50	0.50	0.42	0.33	0.50
P_2O_5	0.20	0.09	0.08	0.07	0.08	0.07	0.09	0.09	0.08	0.07	0.07
LOI	3.38	1.81	1.76	2.62	2.30	2.43	2.19	3.21	3.17	4.32	2.27
H_2O^-	–	0.25	0.03	0.23	0.16	0.13	0.16	0.28	0.15	0.44	0.17
Total	99.39	99.54	99.34	99.81	99.83	99.67	100.21	99.19	99.65	87.75	100.39
F	65.9	56.6	51.6	53.2	51.9	52.1	52.3	49.4	45.7	57.5	53.0
S	0.08	0.05	0.05	0.05	0.08	0.008	0.020	0.040	0.005	1.18	0.04
CuO	0.021	0.008	0.007	0.01	0.013	0.01	0.01	0.01	0.008	0.40	0.01
NiO	0.022	0.01	0.00	0.03	0.04	0.03	0.04	0.04	0.09	0.46	0.03
Cr_2O_3	0.021	0.12	0.00	0.09	0.06	0.09	0.00	0.14	0.35	0.70	0.05
Rock	Go	Go	O	O	O	O	O	O	P	P	T

	Depth (m)									
Component	776.1	778.0	780.5	782.0	782.8	783.5	789.0	791.0	797.0	801.5
SiO_2	41.60	44.06	42.60	42.70	41.10	43.10	40.50	43.00	43.50	49.40
TiO_2	0.67	0.73	0.77	0.63	0.48	0.72	0.58	0.51	0.76	2.06
Al_2O_3	8.60	9.17	10.40	11.90	14.10	10.15	10.30	11.90	12.90	14.35

(continued)

Table A.27 (continued)

	Depth (m)									
Component	776.1	778.0	780.5	782.0	782.8	783.5	789.0	791.0	797.0	801.5
Fe_2O_3	3.09	3.37	3.81	3.54	4.81	4.81	5.22	3.99	4.49	4.86
FeO	11.35	10.02	10.52	9.96	8.90	10.34	10.24	9.55	9.37	6.97
MnO	0.21	0.19	0.20	0.17	0.14	0.23	0.19	0.18	0.16	0.15
MgO	21.68	21.50	18.52	16.81	13.38	18.01	17.90	16.34	15.09	6.17
CaO	6.26	6.26	6.79	7.58	8.66	7.19	7.14	8.12	8.66	8.81
Na_2O	0.92	1.00	1.17	1.25	1.08	1.00	1.00	1.13	1.38	2.76
K_2O	0.33	0.33	0.33	0.33	0.33	0.33	0.25	0.50	0.42	0.89
P_2O_5	0.11	0.08	0.08	0.08	0.07	0.05	0.07	0.05	0.09	0.26
LOI	3.46	2.65	3.25	3.78	5.44	4.51	4.84	3.16	2.90	2.26
H_2O^-	0.26	0.26	0.41	0.21	0.46	0.15	0.54	0.15	0.15	0.28
Total	98.54	99.62	98.85	98.94	98.95	100.59	98.77	98.58	99.87	99.22
F	40.0	38.4	43.6	44.5	50.6	45.7	46.3	45.3	47.9	65.7
S	0.46	0.15	0.46	0.81	1.79	0.34	1.55	0.87	1.14	0.54
CuO	0.18	0.05	0.186	0.25	0.53	0.15	0.58	0.437	0.15	0.60
NiO	0.22	0.17	0.23	0.27	0.55	0.23	0.55	0.35	0.48	0.28
Cr_2O_3	0.53	0.43	0.69	0.51	0.09	0.63	0.69	0.24	0.07	0.00
Rock	T	T	T	T	T	T	T	T	T	T

Table A.28 Chemical composition of olivine, drill hole PE-35

	Depth (m)									
Component	539.9	718.0		720.0	730.0	740.0	750.0	760.0	762.0	765.0
SiO_2	39.42	35.21	36.44	36.15	36.73	35.34	38.05	36.71	33.48	35.27
FeO	19.22	33.32	28.88	29.36	28.27	32.67	22.84	27.43	37.34	31.39
CaO	0.38	0.23	0.27	0.29	0.30	0.27	0.31	0.29	0.22	0.21
MgO	41.42	32.43	35.34	35.41	35.33	33.24	38.85	36.42	32.10	35.10
NiO	0.24	0.029	0.06	0.066	0.076	0.054	0.156	0.014	0.062	0.088
Total	100.68	101.22	100.99	101.28	100.71	101.57	100.21	100.86	103.20	102.06
Fo	79	63	69	68	69	64	75	70	61	67
Fa	21	37	31	32	31	36	25	30	39	33
Rock	G	O	O	O	O	O	O	O	O	O

	Depth (m)									
Component	767.0	769.4	772.0	775.5	780.5	785.0	785.1			
SiO_2	36.36	38.09	37.93	35.29	37.20	37.26	38.59	38.66	38.83	38.50
FeO	27.95	22.62	23.29	31.87	25.31	25.38	22.02	21.64	21.05	21.73
CaO	0.27	0.22	0.30	0.18	0.15	0.21	0.18	0.21	0.17	0.20
MgO	36.64	39.14	38.67	34.42	37.86	37.69	38.65	38.93	39.29	39.24
NiO	0.14	0.152	0.1	0.122	0.194	0.164	0.186	0.183	0.193	0.176
Total	101.36	100.22	100.29	101.88	100.71	100.70	99.63	99.62	99.53	99.85
Fo	70	76	75	66	73	73	76	76	77	76
Fa	30	24	25	34	27	27	24	24	23	24
Rock	O	P	P	P	T	T	T	T	T	T

	Depth (m)										
Component	785.1		785.3	791.0	797.0						
SiO_2	38.55	38.17	37.47	35.81	35.68	36.67	37.33	37.56	39.62	37.25	38.35
FeO	21.97	21.91	25.17	29.52	29.78	25.61	26.90	28.18	21.54	27.61	23.01
CaO	0.18	0.16	0.21	0.17	0.12	0.15	0.17	0.16	0.21	0.16	0.10
MgO	38.84	39.90	37.42	36.00	36.15	36.44	35.66	33.41	36.66	34.93	37.63
NiO	0.172	0.169	0.16	0.236	0.149	0.138	0.118	0.136	0.147	0.131	0.222
Total	99.71	100.31	100.43	101.74	101.88	99.01	100.18	99.45	98.18	100.08	99.31
Fo	76	76	73	68	68	72	70	68	75	69	74
Fa	24	24	27	32	32	28	30	32	25	31	26
Rock	T	T	T	T	T	T	T	T	T	T	T

Table A.29 Chemical composition of clinopyroxene, drill hole PE-35

	Depth (m)										
	461.2		469.5	469.2	704.1		710.5		718.0		720.0
Component	Core	Rim	Core	Core	Core	Rim	Core	Rim	Core	Rim	Core
SiO_2	50.75	51.96	50.74	51.40	51.78	51.83	52.11	51.28	52.00	51.55	52.28
TiO_2	0.53	0.49	0.69	0.49	0.40	0.57	0.29	0.77	0.28	0.87	0.33
Al_2O_3	2.47	1.97	3.02	2.16	2.22	2.16	2.03	1.64	2.35	1.88	1.92
FeO	9.52	9.24	9.65	9.69	7.42	11.75	6.64	14.99	6.92	13.77	6.81
MgO	15.26	15.91	15.06	15.81	16.55	14.87	17.16	14.28	17.45	15.76	16.98
CaO	19.42	19.69	19.64	19.29	20.42	18.58	19.87	17.06	19.14	15.88	20.22
Na_2O	0.36	0.26	0.33	0.40	0.22	0.30	0.26	0.27	0.30	0.31	0.26
Cr_2O_3	0.03	0.04	0.04	0.03	0.20	0.02	0.52	0.00	0.78	0.02	0.60
Total	98.34	99.56	99.17	99.27	99.21	100.08	98.88	100.29	99.22	100.04	99.40
Fe_2O_3[a]	2.35	1.62	2.11	2.82	2.03	1.07	1.47	1.72	1.68	1.36	1.53
FeO[a]	7.41	7.78	7.75	7.15	5.59	10.79	5.32	13.45	5.41	12.55	5.43
f	25.9	24.6	26.4	25.6	20.1	30.7	17.8	37.1	18.2	32.9	18.4
Wo	40	40	41	40	41	38	41	35	39	33	41
En	44	45	43	45	47	43	49	41	50	45	48
Fs	16	15	16	15	12	19	10	24	11	22	11
Rock	MD	MD	MD	MD	Go	Go	Go	Go	O	O	O

	Depth (m)										
	720.0	730.0		740.5		750.0		760.0		762.0	
Component	Rim	Core	Rim	Core	Rim	Core	Rim	Core	Rim	Core	Rim
SiO_2	50.49	51.04	51.44	52.21	51.43	52.26	51.16	52.09	51.82	51.02	52.14
TiO_2	0.94	0.50	0.71	0.37	0.72	0.31	0.72	0.32	0.57	0.40	0.66
Al_2O_3	2.07	2.67	2.13	1.84	2.26	1.92	2.20	2.09	2.22	3.10	1.60
FeO	14.32	7.14	10.47	6.67	11.02	6.65	10.96	6.41	9.04	6.83	10.76
MgO	15.78	16.12	15.95	16.53	16.92	16.92	16.71	16.68	16.35	16.98	16.18
CaO	15.32	20.29	18.30	20.05	16.88	20.07	16.92	20.37	19.09	19.52	17.98
Na_2O	0.30	0.34	0.28	0.20	0.31	0.30	0.29	0.26	0.24	0.27	0.32
Cr_2O_3	0.01	0.65	0.09	0.57	0.07	0.61	0.08	0.66	0.06	0.99	0.00
Total	99.23	98.75	99.37	98.44	99.61	99.04	99.04	98.88	99.39	99.11	99.64
Fe_2O_3[a]	2.50	2.19	1.66	0.00	2.33	1.27	2.16	0.97	1.42	2.21	1.31
FeO[a]	12.07	5.17	8.97	6.67	8.92	5.50	9.01	5.54	7.76	4.84	9.58
f	33.7	19.9	26.9	18.5	26.8	18.1	26.9	17.7	23.7	18.4	27.2
Wo	32	42	37	41	34	41	35	42	39	40	37
En	45	46	46	48	48	48	48	48	47	49	46
Fs	23	12	17	11	18	11	17	10	14	11	17
Rock	O	O	O	O	O	O	O	O	O	O	O

	Depth (m)										
	765.0		767.0		769.4		762.0	774.0		775.5	
Component	Core	Rim	Core	Rim	Core	Rim	Core	Core	Rim	Core	Rim
SiO_2	51.15	51.61	51.78	50.82	52.45	51.92	51.18	51.09	51.41	52.32	50.38
TiO_2	0.40	0.68	0.34	0.92	0.30	0.81	0.44	0.45	1.17	0.40	0.92
Al_2O_3	2.97	2.29	2.17	2.28	2.19	1.68	2.91	3.25	2.30	1.97	2.33
FeO	6.57	9.81	6.36	10.70	6.13	9.52	6.18	6.02	7.78	7.16	12.36
MgO	16.15	16.73	17.01	15.75	16.73	16.81	16.55	16.31	16.65	16.71	14.72
CaO	20.10	17.92	20.29	18.30	20.39	18.15	20.47	20.49	19.08	19.96	18.13
Na_2O	0.28	0.26	0.29	0.27	0.24	0.32	0.26	0.30	0.41	0.28	0.24
Cr_2O_3	0.98	0.06	0.79	0.05	0.80	0.20	1.05	1.09	0.57	0.48	0.02
Total	98.60	99.36	99.03	99.09	99.23	99.41	99.04	99.00	99.37	99.28	99.10
Fe_2O_3[a]	0.99	1.64	2.00	2.09	0.22	1.57	1.69	1.38	1.42	1.00	2.08
FeO[a]	5.68	8.34	4.56	8.82	5.93	8.11	4.66	4.78	6.50	6.26	10.49

(continued)

Table A.29 (continued)

	Depth (m)										
	765.0		767.0		769.4		762.0	774.0		775.5	
Component	Core	Rim	Core	Rim	Core	Rim	Core	Core	Rim	Core	Rim
f	18.6	24.8	17.3	27.6	17.1	24.1	17.3	17.2	20.8	19.4	32.0
Wo	42	37	42	38	42	37	42	43	39	41	38
En	47	47	48	45	48	48	48	47	48	48	42
Fs	11	16	10	17	10	15	10	10	13	11	20
Rock	O	O	O	O	P	P	P	P	P	T	T

	Depth (m)										
	778.0		780.5	782.8	783.5	785.3	787.2	790.0	791.0	797.0
Component	Core	Rim	Core	Core	Core	Core	Core	Core	Core	Core
SiO_2	50.87	52.50	51.76	51.72	51.72	51.35	51.86	51.22	51.41	51.79
TiO_2	0.42	0.90	0.29	0.50	0.38	0.42	0.34	0.47	0.40	0.39
Al_2O_3	3.20	1.80	2.09	2.13	2.27	3.11	1.99	2.95	3.01	2.05
FeO	5.97	7.57	6.11	7.13	6.60	6.44	6.17	7.00	6.56	6.69
MgO	16.47	16.74	16.76	16.54	16.88	16.72	16.96	16.49	16.33	16.75
CaO	20.42	19.28	20.14	19.38	19.90	19.63	20.33	19.58	19.68	20.04
Na_2O	0.24	0.44	0.25	0.30	0.28	0.26	0.26	0.28	0.24	0.26
Cr_2O_3	1.16	0.41	0.72	0.70	0.81	1.07	0.71	0.84	1.12	0.70
Total	98.75	99.64	98.12	98.40	98.84	99.00	98.62	98.83	98.75	98.67
Fe_2O_3[a]	1.55	0.63	0.88	0.60	1.44	0.91	1.49	1.23	0.18	1.28
FeO[a]	4.57	7.00	5.31	6.59	5.30	5.62	4.83	5.89	6.40	5.54
f	16.9	20.2	17.0	19.5	18.0	17.8	17.0	19.2	18.4	18.3
Wo	42	40	42	40	41	41	42	41	41	41
En	48	48	48	48	48	49	48	48	48	48
Fs	10	12	10	12	11	10	10	11	11	11
Rock	T	T	T	T	T	T	T	T	T	T

[a]Hereinafter calculated data

Table A.30 Chemical composition of spinel group minerals, drill hole PE-35

	Depth (m)			
Component	451.5	453.5	539.9	
TiO_2	6.17	6.04	1.22	2.15
Al_2O_3	7.14	6.90	10.98	15.41
FeO	52.19	50.52	42.04	35.37
MnO	0.65	0.66	0.67	0.59
MgO	5.42	5.18	6.08	8.49
Cr_2O_3	28.76	31.35	36.62	36.20
Total	100.33	100.65	97.61	98.21
Fe_2O_3[a]	24.29	22.13	19.50	14.27
FeO[a]	30.33	30.60	24.50	22.53
Rock	MD	MD	G	G

[a]Hereinafter calculated data

Lower Talnakhsky-Type Intrusions

Lower Talnakh Intrusion

Table A.31 Chemical composition of rocks, drill hole KZ-665

	Depth (m)										
Component	998.8	1,003.0	1,014.0	1,025.0	1,028.0	1,031.2	1,034.8	1,042.0	**1,055.0**	1,087.0	1,087.0
SiO_2	46.02	64.40	48.90	41.76	39.94	41.37	40.09	41.55	40.34	41.27	42.72
TiO_2	0.87	0.77	0.62	0.52	0.39	0.39	0.46	0.32	0.78	0.78	0.34
Al_2O_3	13.98	14.12	14.95	13.59	11.84	12.80	12.71	14.59	13.73	16.24	16.10
Fe_2O_3	3.22	1.93	1.97	3.91	3.20	2.60	4.77	1.33	4.69	4.80	4.00
FeO	13.78	2.50	5.28	7.56	8.55	8.52	6.53	8.51	8.64	7.23	6.00
MnO	0.38	0.11	0.19	0.19	0.20	0.16	0.23	0.15	0.18	0.19	0.14
MgO	7.38	1.47	9.30	16.45	18.97	19.31	19.54	17.95	18.58	15.07	15.62
CaO	4.76	4.84	11.70	8.34	9.00	8.94	8.04	8.94	8.74	10.05	10.34
Na_2O	3.35	7.25	2.17	0.55	1.10	1.50	0.43	0.86	0.86	1.30	1.70
K_2O	0.92	0.11	1.36	0.78	0.91	0.70	0.48	0.48	0.35	0.48	0.60
P_2O_5	0.17	0.16	0.08	0.06	0.10	0.09	0.05	0.05	0.04	0.04	0.09
H_2O^-	–	–	–	–	0.21	0.18	–	–	–	–	0.13
LOI	4.59	1.73	3.12	5.70	5.32	3.50	5.65	3.99	3.03	2.36	2.35
Total	99.42	99.39	99.64	99.41	99.73	100.06	98.98	98.72	99.96	99.81	100.13
C.f.	69.7	75.1	43.8	41.1	38.3	36.5	36.6	35.4	41.8	44.4	39.0
S	0.068	0.176	0.140	0.236	0.120	0.330	0.044	0.088	0.144	0.076	0.230
CuO	0.007	0.006	0.004	0.011	0.025	0.050	0.011	0.017	0.019	0.012	0.025
NiO	0.006	0.003	0.010	0.032	0.025	0.051	0.045	0.048	0.062	0.036	0.064
Cr_2O_3	0.004	0.010	0.029	0.032	0.020	0.002	0.013	0.015	0.015	0.016	0.002
Rock	MD	MD	MD	TR	TR	TR	TR	TR	TR	TR	TR
	Depth (m)										
Component	1,104.0	1104	1,120.0	1,131.4	1,138.9	1,143.0	1,150.0	1,159.0	1,172.0	1,172.3	1,172.3
SiO_2	43.06	42.87	41.14	48.19	41.68	44.94	49.79	48.11	40.56	39.92	40.86
TiO_2	0.47	0.40	0.44	1.54	0.91	0.57	1.75	1.54	0.39	0.81	0.50
Al_2O_3	13.57	16.04	15.12	14.51	11.24	13.60	12.10	14.63	8.54	10.86	10.00
Fe_2O_3	3.04	2.36	3.11	2.95	4.44	1.82	4.69	4.01	3.55	5.38	2.44
FeO	7.55	8.05	9.63	9.29	9.72	8.01	8.86	9.41	9.48	9.08	9.84
MnO	0.13	0.15	0.18	0.21	0.23	0.23	0.23	0.22	0.18	0.21	0.17
MgO	16.95	15.51	15.73	6.35	18.76	16.78	6.48	6.99	23.22	20.29	23.64
CaO	9.66	10.19	9.08	10.50	8.33	8.68	8.80	10.69	7.38	7.58	7.16
Na_2O	0.75	1.34	0.93	2.22	0.70	0.82	3.79	2.17	1.23	0.65	0.75
K_2O	0.36	0.50	0.40	0.43	0.53	0.78	0.31	0.48	0.53	0.34	0.35
P_2O_5	0.12	0.09	0.05	0.18	0.06	0.08	0.18	0.17	0.09	0.05	0.09
H_2O^-	–	0.16	–	–	–	–	–	–	0.12	–	0.15
LOI	3.35	2.70	4.26	2.13	3.59	3.77	2.94	1.45	4.32	4.16	4.30
Total	99.01	100.36	100.07	98.50	100.19	100.08	99.92	99.87	99.59	99.33	100.25
C.f.	38.5	40.2	44.8	65.8	43.0	36.9	67.7	65.8	35.9	41.6	34.2
S	0.300	0.320	1.140	0.080	0.100	0.068	0.040	0.064	0.200	0.276	0.400
CuO	–	0.038	0.081	0.029	0.030	0.008	0.018	0.016	0.013	0.021	0.050
NiO	–	0.050	0.095	0.024	0.067	0.041	0.019	0.024	0.038	0.076	0.050
Cr_2O_3	–	0.001	0.013	0.027	0.021	0.015	0.027	0.040	0.010	0.016	0.002
Rock	TR	TR	TR	A	TR	TR	GP	A	P	TR	P
	Depth (m)										
Component	1,173.9	1,175.2	1,180.0	1,180.5	1,186.6	1,191.5	1,192.8	1,194.5	1,196.3	1,202.3	1,207.1	1,210.1
SiO_2	41.38	40.48	40.22	40.42	43.58	38.77	39.91	45.89	41.46	37.87	38.96	48.05
TiO_2	0.47	0.55	0.42	0.53	0.70	0.48	0.46	2.07	0.51	0.34	0.35	1.03
Al_2O_3	9.79	8.78	8.49	9.94	11.31	10.56	13.66	11.84	13.65	11.01	9.09	14.74
Fe_2O_3	3.75	6.86	4.00	5.13	4.76	0.90	4.93	4.29	5.02	4.94	3.84	2.50

(continued)

Table A.31 (continued)

	Depth (m)											
Component	1,173.9	1,175.2	1,180.0	1,180.5	1,186.6	1,191.5	1,192.8	1,194.5	1,196.3	1,202.3	1,207.1	1,210.1
FeO	9.01	7.89	7.95	8.81	2.84	15.70	7.64	10.43	8.80	13.11	10.34	6.59
MnO	0.19	0.24	0.15	0.20	0.20	0.19	0.20	0.34	0.21	0.24	0.68	0.37
MgO	22.64	22.86	21.89	22.01	12.44	15.98	19.62	5.95	15.70	19.28	21.75	7.85
CaO	7.24	5.78	8.94	6.70	22.72	9.80	7.29	12.13	8.52	6.96	6.83	9.58
Na_2O	0.75	0.49	1.40	0.49	0.16	1.10	1.20	2.49	0.82	0.43	0.32	2.39
K_2O	0.75	0.39	0.92	0.78	0.05	0.90	0.21	0.16	1.21	0.58	0.43	1.99
P_2O_5	0.13	0.06	0.11	0.05	0.03	0.10	0.06	0.32	0.06	0.04	0.10	0.11
H_2O^-	0.16	–	0.18	–	–	0.17	–	–	–	–	–	–
LOI	4.20	5.43	4.95	4.39	0.96	2.38	4.66	4.35	4.52	5.18	7.81	4.60
Total	100.46	99.81	99.62	99.45	99.75	97.03	99.84	100.26	100.48	99.98	100.50	99.80
C.f.	36.1	39.2	35.3	38.8	37.9	51.0	39.1	71.2	46.8	48.4	39.5	53.7
S	0.250	0.076	0.260	0.252	0.104	3.830	0.064	1.664	1.556	2.716	1.056	0.084
CuO	0.025	0.011	0.013	0.012	0.009	0.225	0.013	0.033	0.099	0.231	–	0.056
NiO	0.051	0.057	0.380	0.069	0.011	0.178	0.105	0.034	0.108	0.140	–	0.022
Cr_2O_3	0.002	0.017	0.020	0.018	0.015	0.001	0.092	0.018	0.017	0.024	–	0.013
Rock	P	P	P	P	MB	MB	T	T	T	M	M	M

Table A.32 Chemical composition of olivine, drill hole KZ-665

	Depth (m)								
Component	1,055.0	1,120.0	1,120.0	1,172.0	1,184.0	1,192.8	1,193.5	1,193.5	1,194.0
SiO_2	38.99	38.45	38.54	38.98	38.58	38.78	37.90	38.17	37.97
FeO	20.58	23.15	22.46	20.20	21.89	21.73	26.11	24.34	25.98
CaO	0.17	0.17	0.18	0.22	0.10	0.22	0.17	0.09	0.15
MgO	39.67	37.72	38.40	40.11	39.15	38.68	35.37	37.02	35.36
NiO	0.047	0.053	0.035	0.050	0.020	0.050	0.017	0.033	0.033
Total	99.46	99.54	99.62	99.56	99.74	99.46	99.57	99.65	99.49
Fo	77	74	75	78	76	76	71	73	71
Fa	23	26	25	22	24	24	29	27	29
Rock	TR	TR	TR	TR	TR	TR	P	P	P

Combined Section of the Upper Talnakh and Lower Talnakh Intrusions

Table A.33 Chemical composition of rocks, drill hole SV-16

	Depth (m)									
Component	1,645.1	1,675.0	1,680.1	1,685.0	1,690.0	1,695.0	1,699.9	1,705.0	1,710.0	1,715.0
SiO_2	48.40	43.10	43.20	43.00	49.20	43.60	43.10	42.70	43.70	43.40
TiO_2	0.75	0.39	0.40	0.38	1.35	0.43	0.45	0.46	0.44	0.43
Al_2O_3	15.10	14.80	16.60	16.90	19.80	16.70	16.50	16.20	16.30	16.50
Fe_2O_3	2.50	2.94	3.38	2.86	2.88	2.50	2.93	3.23	2.26	2.30
FeO	5.31	6.18	5.96	5.89	5.60	5.67	6.10	6.10	6.61	7.11
MnO	0.17	0.17	0.18	0.17	0.13	0.15	0.16	0.18	0.17	0.19
MgO	8.24	15.40	13.22	13.07	5.76	13.23	13.23	13.38	13.22	13.07
CaO	10.60	9.30	9.95	10.39	9.09	11.47	10.17	9.95	11.25	11.04
Na_2O	3.13	0.83	0.94	0.78	3.34	0.97	0.97	0.81	0.86	0.84
K_2O	0.73	1.22	1.11	0.95	0.82	0.97	1.50	1.28	0.77	0.34
P_2O_5	0.07	0.03	0.03	0.05	0.33	0.00	0.03	0.04	0.03	0.05
H_2O^-	0.36	0.80	0.50	0.51	0.11	0.36	0.32	0.86	0.43	0.32
LOI	4.26	5.32	4.25	4.91	1.46	4.39	4.94	4.95	4.63	4.45
Total	99.62	100.48	99.72	99.86	99.87	100.44	100.40	100.14	100.67	100.04
C.f.	48.7	37.2	41.4	40.1	59.6	38.2	40.6	41.1	40.2	41.9
CuO	0.003	0.017	0.015	0.017	0.005	0.008	0.022	0.022	0.009	0.009

(continued)

Table A.33 (continued)

Component	Depth (m) 1,645.1	1,675.0	1,680.1	1,685.0	1,690.0	1,695.0	1,699.9	1,705.0	1,710.0	1,715.0
NiO	0.009	0.041	0.028	0.029	0.011	0.035	0.054	0.051	0.029	0.038
CoO	0.003	0.006	0.005	0.005	0.003	0.006	0.007	0.007	0.005	0.006
ZnO	0.007	0.014	0.018	0.012	0.012	0.011	0.012	0.011	0.008	0.012
Cr_2O_3	0.052	0.034	0.038	0.036	0.039	0.037	0.037	0.038	0.036	0.033
V_2O_5	0.040	0.015	0.023	0.019	0.022	0.020	0.019	0.019	0.014	0.017
Rock	D	D	D	D	D	D	D	TR	TR	TR

Component	Depth (m) 1,720.0	1,725.0	1,730.0	1,735.0	1,740.0	1,743.0	1,750.0	1,752.1	1,752.5	1,760.0
SiO_2	42.80	43.20	42.50	43.20	40.90	39.70	38.20	36.70	37.30	40.50
TiO_2	0.44	2.35	0.46	0.49	0.47	0.44	0.35	0.34	0.33	0.48
Al_2O_3	15.60	12.10	14.80	14.30	12.30	11.90	8.40	5.20	5.30	9.60
Fe_2O_3	3.35	3.60	3.40	3.05	3.51	2.27	4.74	7.40	7.82	4.62
FeO	7.25	6.03	6.75	7.25	4.67	10.92	8.69	7.83	7.18	7.18
MnO	0.19	0.22	0.22	0.21	0.18	0.21	0.19	0.21	0.24	0.23
MgO	12.92	9.33	13.69	14.16	14.78	17.27	21.63	26.30	26.30	21.16
CaO	10.82	14.72	10.82	9.96	13.85	8.22	6.28	4.11	4.11	7.57
Na_2O	0.93	1.52	0.70	0.77	0.39	0.70	0.57	0.35	0.29	0.89
K_2O	0.45	0.30	0.59	0.43	0.19	0.55	0.34	0.38	0.39	0.64
P_2O_5	0.04	0.10	0.07	0.00	0.04	0.02	0.00	0.03	0.03	0.04
H_2O^-	0.40	0.19	0.33	0.42	0.94	0.52	0.47	0.47	0.42	0.43
LOI	5.11	6.58	6.01	6.15	7.41	7.49	9.22	10.40	10.49	6.76
Total	100.30	100.24	100.34	100.39	99.63	100.21	99.08	99.72	100.20	100.10
C.f.	45.1	50.8	42.6	42.1	35.6	43.3	38.3	36.7	36.3	35.8
CuO	0.027	0.087	0.015	0.113	0.025	0.087	0.056	0.080	0.051	0.011
NiO	0.044	0.047	0.050	0.088	0.049	0.079	0.089	0.093	0.080	0.043
CoO	0.009	0.010	0.007	0.008	0.007	0.011	0.011	0.012	0.012	0.008
ZnO	0.012	0.010	0.014	0.017	0.009	0.012	0.013	0.013	0.014	0.010
Cr_2O_3	0.037	0.043	0.038	0.072	0.038	0.048	0.049	0.048	0.052	0.048
V_2O_5	0.015	0.040	0.022	0.026	0.017	0.014	0.010	0.015	0.017	0.023
Rock	TR	O	TR	TR	TR	TR	P	P	P	P

Component	Depth (m) 1,762.0	1,768.0	1,775.0	1,780.2	1,785.5	1,795.0	1,805.0	1,810.0	1,815.0	1,825.3
SiO_2	39.10	46.20	44.20	47.20	47.80	48.20	47.70	47.40	47.50	47.00
TiO_2	0.45	1.34	0.68	2.47	1.70	1.05	0.89	0.84	0.56	0.85
Al_2O_3	9.10	14.60	18.20	11.60	12.30	15.60	15.40	17.50	17.60	17.90
Fe_2O_3	5.62	4.70	3.16	7.13	5.65	2.88	3.26	2.30	3.43	2.35
FeO	8.62	7.47	5.53	9.69	9.05	7.40	6.97	6.75	6.10	7.25
MnO	0.24	0.23	0.15	0.24	0.26	0.18	0.18	0.18	0.16	0.16
MgO	21.32	7.62	7.94	5.44	7.31	7.16	8.56	7.31	7.78	8.25
CaO	7.14	11.90	11.47	9.74	7.79	11.47	11.90	12.12	11.90	11.47
Na_2O	0.83	2.32	2.16	3.38	3.38	2.32	2.00	2.26	2.05	1.74
K_2O	0.48	0.65	0.78	0.67	1.00	1.25	0.53	1.00	0.68	0.63
P_2O_5	0.06	0.10	0.08	0.20	0.19	0.11	0.10	0.09	0.13	0.09
H_2O^-	0.46	0.39	0.29	0.27	0.45	0.34	0.20	0.25	0.25	0.24
LOI	6.94	2.96	5.76	2.25	3.63	2.33	2.59	2.47	2.08	2.14
Total	100.36	100.48	100.40	100.28	100.51	100.29	100.28	100.47	100.22	100.07
C.f.	40.0	61.5	52.3	75.6	66.8	58.9	54.4	55.3	55.1	53.8
CuO	0.038	0.020	0.091	0.014	0.011	0.012	0.011	0.010	0.009	0.011
NiO	0.066	0.015	0.050	0.004	0.008	0.010	0.016	0.016	0.019	0.024
CoO	0.012	0.004	0.005	0.004	0.004	0.003	0.004	0.003	0.004	0.005
PbO	–	–	–	–	0.004	–	–	–	–	–
ZnO	0.013	0.021	0.016	0.030	0.024	0.016	0.014	0.016	0.018	0.014
Cr_2O_3	0.051	0.053	0.041	0.042	0.046	0.069	0.103	0.093	0.076	0.069
V_2O_5	0.017	0.050	0.023	0.096	0.060	0.045	0.040	0.036	0.033	0.037

(continued)

Table A.33 (continued)

Component	Depth (m) 1,762.0	1,768.0	1,775.0	1,780.2	1,785.5	1,795.0	1,805.0	1,810.0	1,815.0	1,825.3
Rock	P	M	M	GP	G	G	Go	Go	Go	Go

Component	Depth (m) 1,830.0	1,840.0	1,845.0	1,849.0	1,850.0	1,852.0	1,854.2	1,860.0	1,864.0	1,869.0
SiO_2	45.40	46.60	46.30	37.50	49.80	35.00	36.40	40.10	40.20	46.80
TiO_2	0.87	0.90	0.94	0.59	1.31	0.57	0.72	0.64	0.56	1.12
Al_2O_3	16.80	16.00	15.80	7.90	19.90	7.60	10.20	11.40	8.40	15.00
Fe_2O_3	3.83	3.08	3.73	4.69	2.76	8.89	8.11	5.22	4.68	4.75
FeO	6.82	6.68	6.82	9.19	5.89	10.63	10.52	10.42	11.99	7.25
MnO	0.17	0.18	0.15	0.19	0.11	0.19	0.18	0.19	0.25	0.19
MgO	9.02	9.02	9.18	23.03	5.75	18.67	16.81	16.10	21.01	6.85
CaO	10.39	11.47	10.39	4.11	8.87	4.33	5.84	7.57	6.49	10.82
Na_2O	2.16	2.11	2.11	0.47	3.44	0.70	1.00	1.09	0.85	2.32
K_2O	0.65	0.52	0.72	0.27	0.78	0.34	0.78	0.65	0.78	1.83
P_2O_5	0.08	0.12	0.13	0.07	0.34	0.08	0.09	0.07	0.08	0.18
H_2O^-	0.54	0.28	0.37	1.56	0.06	0.75	0.34	0.29	0.40	0.36
LOI	3.83	3.42	3.35	9.88	1.42	9.53	6.33	4.62	4.16	2.84
Total	100.56	100.38	99.99	99.45	100.43	97.28	97.32	98.36	99.85	100.31
C.f.	54.1	52.0	53.5	37.6	60.1	51.1	52.6	49.3	44.2	63.7
CuO	0.014	0.012	0.009	0.201	0.005	1.770	1.770	0.990	0.420	0.103
NiO	0.031	0.028	0.028	0.250	0.009	0.940	0.770	0.615	0.170	0.035
CoO	0.004	0.005	0.004	0.011	0.004	0.025	0.019	0.016	0.014	0.005
PbO	–	–	–	–	0.003	0.003	0.003	0.006	0.005	0.003
ZnO	0.016	0.010	0.012	0.025	0.012	0.013	0.016	0.016	0.019	0.015
Cr_2O_3	0.070	0.095	0.118	0.430	0.039	0.550	0.480	0.055	0.061	0.056
V_2O_5	0.024	0.037	0.038	0.026	0.026	0.028	0.031	0.026	0.024	0.043
Rock	Go	Go	O	P	LG	P	TR	T	T	Go

Table A.34 Chemical composition of olivine, drill hole SV-16

Component	Depth (m) 1,675.0			1,680.0		1,685.0				
SiO_2	38.44	38.61	38.49	38.58	38.52	38.45	38.76	38.64	38.45	38.46
FeO	21.76	20.48	20.44	21.14	21.19	21.88	20.38	20.91	20.42	20.50
MgO	39.56	40.74	41.04	39.49	39.98	39.38	40.55	40.03	41.38	39.25
CaO	0.16	0.19	0.21	0.17	0.27	0.17	0.17	0.26	0.09	0.15
NiO	0.049	0.058	0.067	0.048	0.038	0.034	0.027	0.061	0.040	0.360
Total	99.97	100.08	100.25	99.43	100.00	99.91	99.89	99.90	100.38	98.72
Fo	76	78	78	77	77	76	78	77	78	77
Fa	24	22	22	23	23	24	22	23	22	23
Rock	TR	TR	TR	TR	TR	TR	TR	TR	TR	TR

Component	Depth (m) 1,695.0	1,699.0	1,699.9		1,705.0			1,710.0	1,715.0	
SiO_2	38.25	38.26	39.65	39.21	38.60	38.86	38.13	38.46	37.80	38.36
FeO	22.43	22.47	17.53	17.57	20.63	18.80	22.76	21.26	24.22	22.03
MgO	39.25	39.17	41.99	43.00	40.62	42.27	39.11	40.25	38.06	39.41
CaO	0.11	0.13	0.11	0.17	0.17	0.19	0.12	0.08	0.13	0.18
NiO	0.038	0.039	0.058	0.049	0.047	0.051	0.041	0.048	0.046	0.041
Total	100.08	100.07	99.34	100.00	100.07	100.17	100.16	100.10	100.26	100.02
Fo	76	76	81	81	78	80	75	77	74	76
Fa	24	24	19	19	22	20	25	23	26	24
Rock	TR	TR	TR	TR	TR	TR	TR	TR	TR	TR

Component	Depth (m) 1,715.0		1,720.0		1,730.0	1,743.0				
SiO_2	38.60	38.45	38.44	38.37	38.43	38.45	37.80	38.47	38.69	39.07

(continued)

Table A.34 (continued)

	Depth (m)									
Component	1,715.0		1,720.0		1,730.0	1,743.0				
FeO	19.70	20.10	21.65	22.02	20.85	21.46	24.25	21.78	21.23	19.89
MgO	41.95	41.75	39.84	39.54	40.92	40.08	37.80	36.61	39.76	40.49
CaO	0.06	0.09	0.07	0.05	0.05	0.06	0.11	0.06	0.07	0.11
NiO	0.038	0.055	0.029	0.034	0.029	0.020	0.236	0.028	0.029	0.013
Total	100.35	100.45	100.03	100.01	100.28	100.07	100.20	96.95	99.78	99.57
Fo	79	79	77	76	78	77	74	75	77	78
Fa	21	21	23	24	22	23	26	25	23	22
Rock	TR	TR	TR	TR	TR	TR	TR	TR	TR	TR

	Depth (m)									
Component	1,743.0								1,750.0	
SiO_2	39.15	38.95	38.70	38.60	38.92	38.63	39.12	38.50	38.83	38.47
TiO_2	–	0.058	0.000	0.058	0.051	0.000	0.000	0.062	–	–
Al_2O_3	–	0.000	0.000	0.000	0.000	0.000	0.000	0.000	–	–
FeO	20.46	20.86	20.97	21.40	20.99	20.03	19.29	20.37	20.48	20.59
MnO	–	0.342	0.483	0.423	0.401	0.400	0.263	0.290	–	–
MgO	39.58	39.55	38.91	38.81	38.45	38.59	40.38	39.16	40.13	41.10
CaO	0.08	0.096	0.004	0.107	0.074	0.046	0.068	0.000	0.11	0.10
Na_2O	–	0.000	0.163	0.000	0.000	0.000	0.000	0.000	–	–
K_2O	–	0.011	0.000	0.000	0.024	0.000	0.000	0.000	–	–
NiO	0.016	0.065	0.000	0.136	0.210	0.220	0.129	0.047	0.033	0.022
Cr2O3	–	0.000	0.000	0.092	0.000	0.036	0.000	0.052	–	–
Total	99.29	99.93	99.23	99.63	99.12	97.95	99.25	98.48	99.58	100.28
Fo	78	77	77	76	77	77	79	77	78	78
Fa	22	23	23	24	23	23	21	23	22	22
	TR	TR	TR	TR	TR	TR	TR	TR	TR	TR

	Depth (m)									
Component	1,752.1		1,752.5	1,760.0			1,762.0			1,795.0
SiO_2	38.48	38.43	38.90	39.30	39.31	38.70	39.54	39.67	39.54	36.46
FeO	19.20	19.28	19.11	20.06	20.10	20.52	21.07	19.76	19.67	33.05
MgO	42.88	42.86	41.89	39.64	39.53	40.59	40.20	38.97	39.45	30.03
CaO	0.07	0.09	0.11	0.11	0.11	0.10	0.17	0.22	0.20	0.20
NiO	0.040	0.044	0.045	0.033	0.042	0.042	0.020	0.029	0.024	0.028
Total	100.67	100.70	100.06	99.14	99.09	99.95	101.00	98.65	98.88	99.77
Fo	80	80	80	78	78	78	77	78	78	62
Fa	20	20	20	22	22	22	23	22	22	38
Rock	P	P	P	P	P	P	P	P	P	Go

	Depth (m)									
Component	1,795.0	1,805.0				1,815.0				1,825.0
SiO_2	36.62	36.81	37.17	36.48	36.70	37.05	36.94	37.77	37.65	37.00
FeO	32.15	32.80	31.14	32.43	31.25	30.26	30.71	28.36	29.10	29.73
MgO	30.71	29.51	30.68	30.74	31.67	32.12	31.83	32.64	32.03	32.90
CaO	0.26	0.15	0.20	0.23	0.23	0.18	0.17	0.24	0.23	0.18
NiO	0.028	0.032	0.034	0.029	0.050	0.044	0.046	0.067	0.058	0.068
Total	99.77	99.30	99.22	99.91	99.90	99.65	99.70	99.08	99.07	99.88
Fo	63	62	64	63	64	65	65	67	66	66
Fa	37	38	36	37	36	35	35	33	34	34
Rock	Go	Go	Go	Go	Go	Go	Go	Go	Go	Go

	Depth (m)									
Component	1,825.0	1,825.3		1,830.0				1,852.0		
SiO_2	37.30	37.61	37.60	38.29	37.74	37.71	36.97	38.76	38.62	39.34
FeO	28.75	28.22	28.24	27.19	28.39	28.64	29.27	25.83	25.93	21.91
MgO	33.43	33.32	33.35	32.94	32.78	32.54	33.60	33.54	33.73	36.88
CaO	0.16	0.17	0.13	0.13	0.14	0.14	0.14	0.07	0.07	0.16
NiO	0.066	0.073	0.067	0.080	0.084	0.089	0.088	0.105	0.126	0.220

(continued)

Table A.34 (continued)

Component	Depth (m)									
	1,825.0	1,825.3		1,830.0				1,852.0		
Total	99.71	99.39	99.39	98.63	99.13	99.12	100.07	98.31	98.48	98.51
Fo	67	68	68	68	67	67	67	70	70	75
Fa	33	32	32	32	33	33	33	30	30	25
Rock	Go	Go	Go	Go	Go	Go	Go	P	P	P

Component	Depth (m)									
	1,852.0									1,854.0
SiO_2	38.97	39.36	39.05	39.25	37.97	38.48	38.30	38.33	38.21	37.85
FeO	22.35	19.33	21.13	22.07	22.58	22.62	21.81	20.83	22.53	24.85
MgO	37.30	40.20	38.71	36.91	39.48	38.18	39.66	40.86	38.93	36.89
CaO	0.09	0.05	0.10	0.14	0.15	0.10	0.07	0.05	0.14	0.11
NiO	0.241	0.279	0.187	0.230	0.224	0.250	0.289	0.282	0.238	0.244
Total	98.95	99.22	99.18	98.60	100.40	99.63	100.13	100.35	100.05	99.94
Fo	75	79	77	75	76	75	76	78	75	73
Fa	25	21	23	25	24	25	24	22	25	27
Rock	P	P	P	P	P	P	P	P	P	P

Component	Depth (m)							
	1,854.0	1,854.2				1,864.0		
SiO_2	37.66	37.75	37.59	37.33	37.85	38.23	37.64	37.83
FeO	25.91	24.99	25.88	27.00	24.64	26.80	27.06	26.09
MgO	35.98	36.94	36.14	35.38	37.15	33.57	34.68	35.49
CaO	0.009	0.16	0.16	0.11	0.17	0.12	0.12	0.13
NiO	0.273	0.208	0.263	0.294	0.199	0.122	0.161	0.121
Total	99.83	100.05	100.03	100.11	100.01	98.84	99.66	99.66
Fo	71	72	71	70	73	69	70	71
Fa	29	28	29	30	27	31	30	29
Rock	P	P	P	P	P	T	T	T

Kruglogorsky Type Intrusions

Olor River Intrusion

Table 35 Chemical composition of rocks, drill hole VKH-4

Component	Depth (m)						
	2,276.0	2,277.1	2,280.9	2,281.4	2,282.2	2,284.4	2,285.0
SiO_2	46.96	45.63	46.77	45.61	44.74	47.15	47.93
TiO_2	1.16	0.84	1.10	1.17	1.12	1.07	1.08
Al_2O_3	16.13	19.19	15.90	14.53	15.01	14.34	14.36
Fe_2O_3	2.80	2.46	3.20	16.57	5.66	14.49	2.95
FeO	8.97	6.56	8.84	–	9.44	–	9.71
MnO	0.18	0.15	0.17	0.21	0.18	0.21	0.20
MgO	8.40	7.40	9.73	6.99	9.40	7.93	10.94
CaO	8.36	9.40	9.66	8.93	8.10	9.99	8.62
Na_2O	3.89	2.80	2.15	2.54	2.09	2.32	1.69
K_2O	0.92	1.26	0.88	0.94	1.00	0.76	0.46
P_2O_5	0.13	0.08	0.11	0.14	0.13	0.12	0.13
Cu	0.024	0.02	0.13	–	0.82	0.055	0.035
Ni	0.02	0.015	0.058	–	0.29	0.028	0.038
Co	0.006	0.005	0.009	–	0.018	0.005	0.013
Cr_2O_3	0.014	0.012	0.016	–	0.012	–	0.015
Ba	–	–	–	0.00	–	0.00	–

(continued)

Table 35 (continued)

LOI	3.52	3.93	1.40	2.71	2.56	1.54	0.51
Total	101.48	99.75	100.12	100.34	100.57	100.01	98.68
C.f.	58.4	54.9	55.3	70.3	61.6	64.6	53.6
S	0.14	0.23	0.36	–	0.13	–	0.13
CO_2	0.35	0.42	0.35	–	0.16	–	0.10
Rock	CG	LG	O	Gqo	O	Go	O

According to Noril'sk geologists with additions from authors

Zubovsky-Type Intrusions

Zub-Marksheidersky Intrusion

Table A.36 Chemical composition of rocks, drill hole AS-28

	Depth (m)								
Component	290.0	315-1	315-2	318.0	340.0	347.0	352.0	355.5	364.1
SiO_2	42.64	44.54	42.52	41.74	40.74	41.78	44.06	40.36	44.56
TiO_2	0.45	0.49	0.45	0.40	0.40	0.60	0.45	0.40	0.96
Al_2O_3	13.45	19.40	14.15	12.06	9.61	9.43	15.46	5.94	14.15
Fe_2O_3	2.75	1.64	1.60	2.47	3.07	2.02	2.19	3.25	2.11
FeO	9.15	7.27	9.44	8.93	10.44	12.03	8.35	11.96	9.08
MnO	0.20	0.13	0.18	0.19	0.21	0.21	0.16	0.21	0.17
MgO	19.47	12.25	17.78	20.55	23.76	23.42	15.91	27.72	17.04
CaO	6.32	9.60	7.69	7.40	6.18	5.77	8.52	4.38	7.55
Na_2O	1.15	1.71	1.25	1.12	1.06	1.12	1.56	0.82	1.50
K_2O	0.27	0.37	0.27	0.26	0.25	0.23	0.29	0.25	0.35
P_2O_5	0.082	0.082	0.082	0.082	0.076	0.091	0.064	0.082	0.10
LOI	4.12	2.01	4.08	4.42	3.99	3.15	2.34	3.80	2.13
Cr_2O_3	–	0.02	0.006	0.03	–	–	–	–	–
Total	100.05	99.51	99.50	99.65	99.79	99.85	99.35	99.17	99.70
H_2O	3.64	1.80	3.60	4.02	3.80	2.68	1.80	3.40	1.74
S	0.24	0.077	0.077	0.14	0.18	0.46	0.21	0.29	0.14
Cu	0.15	0.04	0.06	–	0.049	0.16	0.03	0.06	0.025
Ni	0.016	0.088	0.017	0.10	0.15	0.21	0.091	0.17	0.07
F	<0.1	<0.1	<0.1	<0.1	<0.1	<0.1	<0.1	<0.1	<0.1
C.f.	37.9	42.1	38.3	35.7	36.3	37.5	39.9	35.4	39.6
Rock	TR	TR	TR	P	P	P	P	P	P
	Depth (m)								
Component	376.0	378.0	381.0	384.3	384.6	390.0	395.0	425.5	429.6
SiO_2	42.44	45.10	43.98	43.12	45.58	45.52	45.20	46.12	46.82
TiO_2	0.50	0.80	0.69	0.60	0.72	0.82	0.72	1.03	1.12
Al_2O_3	10.83	15.20	12.41	10.83	15.64	14.85	14.77	14.68	14.59
Fe_2O_3	2.62	2.94	2.34	2.17	2.05	2.24	2.16	2.16	2.07
FeO	12.43	9.84	10.30	11.10	9.51	10.02	10.52	10.66	11.06
MnO	0.21	0.20	0.20	0.22	0.20	0.21	0.21	0.21	0.22
MgO	20.13	14.31	18.09	20.59	13.59	13.80	14.31	11.83	10.27
CaO	6.68	8.11	7.11	6.04	8.39	8.67	7.96	8.39	9.31
Na_2O	1.24	1.66	1.62	1.34	1.90	1.70	1.76	1.92	1.89
K_2O	0.26	0.36	0.41	0.28	0.42	0.42	0.45	0.54	0.54
P_2O_5	0.064	0.10	0.08	0.08	0.10	0.10	0.096	0.13	0.14
LOI	2.31	1.45	2.28	2.88	1.36	1.51	1.35	1.98	1.82
Total	99.71	100.07	99.51	99.25	99.46	99.86	99.51	99.65	99.85
H_2O	2.00	1.28	1.40	2.68	0.90	1.28	1.02	1.24	1.26

(continued)

Table A.36 (continued)

Component	Depth (m) 376.0	378.0	381.0	384.3	384.6	390.0	395.0	425.5	429.6
S	0.06	0.13	0.089	0.077	0.14	0.10	0.15	0.37	0.37
Cu	0.14	0.003	0.01	0.02	0.028	0.03	0.04	0.05	0.04
Ni	0.18	0.03	0.047	0.079	0.033	0.03	0.034	0.027	0.018
F	<0.1	<0.1	<0.1	<0.1	<0.1	<0.1	<0.1	<0.1	<0.1
C.f.	42.8	47.2	41.1	39.2	46.0	47.0	47.0	52.0	56.1
Rock	P	P	P	P	O	O	O	O	O

Table A.37 Chemical composition of olivine, drill hole AS-28

Component	Depth (m) 280.0	355.5	357.3	357.3	376.0	395.0
SiO_2	38.88	38.68	38.95	38.86	38.39	37.57
FeO	20.72	21.27	20.46	20.85	23.49	27.30
CaO	0.08	0.13	0.12	0.12	0.13	0.22
MgO	39.77	39.49	39.80	39.55	37.32	34.46
NiO	0.13	0.17	0.21	0.18	0.18	0.14
Total	99.58	99.74	99.54	99.56	99.51	99.69
Fo	77	77	78	77	74	69
Fa	23	23	22	23	26	31
Rock	TR	P	P	P	P	O

South Pyasinsky Intrusive Branch of Pyasinsko–Vologochansky Intrusion

Table A.38 Chemical composition of rocks, drill hole AS-6

Component	Depth (m) 133.0	158.0	161.0	211.0	230.0	239.2	254.0	271.0
SiO_2	47.58	46.86	46.78	47.66	44.04	43.20	44.02	42.98
TiO_2	1.66	2.12	1.80	1.40	0.94	0.82	0.62	0.65
Al_2O_3	14.95	13.88	15.01	15.28	14.87	12.58	15.36	14.05
Fe_2O_3	4.98	6.38	3.03	3.63	2.17	1.51	1.09	1.77
FeO	9.02	9.02	11.46	9.67	12.41	13.50	12.12	12.41
MnO	0.20	0.24	0.25	0.23	0.22	0.23	0.20	0.20
MgO	6.54	6.10	6.79	7.13	14.51	18.65	16.19	17.61
CaO	9.10	9.99	9.99	11.83	8.42	6.91	7.73	7.39
Na_2O	2.71	2.83	0.41	2.27	1.73	1.48	1.53	1.50
K_2O	0.82	0.72	0.53	0.48	0.25	0.23	0.18	0.19
P_2O_5	0.20	0.21	0.18	0.14	0.096	0.082	0.064	0.064
LOI	2.21	1.56	1.34	0.99	0.42	1.03	0.88	1.02
Total	99.97	99.91	97.57	100.71	100.08	100.22	99.98	99.83
C.f.	68.2	71.6	68.1	65.1	50.1	44.6	44.9	44.6
Rock	Go	Go	Go	Go	TR	P	TR	TR

Component	Depth (m) 300.0	323.0	350.0	370.0	377.0	402.0	412.0
SiO_2	41.68	42.26	43.10	42.20	43.22	46.26	45.14
TiO_2	0.50	0.58	0.76	0.72	0.80	1.09	1.16
Al_2O_3	14.05	14.13	13.97	11.19	11.92	11.92	10.29
Fe_2O_3	2.80	2.38	1.90	2.82	2.13	3.66	3.47
FeO	10.97	11.41	13.21	14.26	14.22	6.28	5.45
MnO	0.20	0.20	0.21	0.23	0.22	0.23	0.21

(continued)

Table A.38 (continued)

Component	Depth (m) 300.0	323.0	350.0	370.0	377.0	402.0	412.0
MgO	17.71	17.71	18.30	16.97	18.45	10.63	10.33
CaO	6.77	7.38	7.53	6.77	6.91	16.83	18.07
Na_2O	1.98	1.50	1.63	1.80	1.68	1.50	1.68
K_2O	0.14	0.17	0.24	0.26	0.50	0.48	0.58
P_2O_5	0.058	0.060	0.086	0.099	0.10	0.12	0.12
LOI	3.02	1.61	1.01	1.38	1.10	2.27	3.21
Total	99.88	99.39	101.95	98.70	101.25	101.27	99.71
C.f.	43.7	43.8	45.2	50.2	47.0	48.3	46.3
Rock	TR	TR	TR	TR	TR	O	O

Table A.39 Chemical composition of olivine, drill hole AS-6

Component	Depth (m) 211.0		230.0	239.0		239.2			300.0	412.0	
SiO_2	35.99	34.45	37.45	36.73	36.65	37.29	38.32	38.21	38.24	38.72	36.15
FeO	34.83	44.53	24.31	27.73	27.29	27.77	25.80	24.42	24.30	25.94	37.03
CaO	0.11	–	–	–	–	–	–	–	–	–	–
MgO	28.99	21.42	36.70	33.95	34.79	34.67	35.78	35.78	37.55	36.51	27.83
NiO	0.056	–	–	–	–	–	–	–	–	–	–
Total	99.98	100.40	98.46	98.41	98.73	99.73	99.90	98.41	100.09	101.17	101.01
Fo	60	46	73	69	69	69	71	72	73	72	57
Fa	40	54	27	31	31	31	29	28	27	28	43
Rock	Go	Go	TR	TR	TR	P	P	P	TR	O	O

Vologochansky Intrusive Branch of Pyasinsko–Vologochansky Intrusion

Table A.40 Chemical composition of rocks, drill hole AS-8

Component	Depth (m) 526.0	537.0	541.0	544.0	547.0	555.0	567.0	701.0	745.0	755.0	782.5	784.5
SiO_2	45.30	44.70	42.80	44.00	43.50	39.80	48.40	45.70	46.50	40.60	33.98	32.70
TiO_2	0.87	0.92	0.89	0.88	0.89	0.70	1.45	2.10	1.08	0.71	0.62	1.68
Al_2O_3	17.80	12.50	9.00	11.00	10.50	12.20	16.00	11.60	17.10	11.56	9.90	10.70
Fe_2O_3	0.68	1.30	2.86	2.86	1.16	6.00	1.95	7.34	1.43	0.99	1.75	7.60
FeO	8.64	10.80	11.10	10.00	10.20	11.80	6.30	7.34	8.90	15.08	5.89	6.00
MnO	0.17	0.23	0.24	0.20	0.15	0.16	0.12	0.26	0.17	0.18	0.16	0.19
MgO	7.70	15.23	19.20	17.00	14.60	12.00	8.64	5.80	8.40	14.99	26.71	16.30
CaO	10.65	8.05	6.50	8.20	8.28	8.00	6.74	9.60	10.65	7.00	7.57	14.30
Na_2O	1.66	1.11	0.90	1.22	1.00	1.44	2.14	2.14	2.00	1.30	0.14	0.55
K_2O	0.70	0.50	0.42	0.38	0.70	0.57	2.29	3.00	0.44	0.67	0.63	1.14
P_2O_5	0.12	0.10	0.10	0.12	0.10	0.10	0.11	0.18	0.17	0.12	0.03	0.10
CuO	0.053	0.024	0.056	0.030	0.080	0.705	0.054	0.022	0.017	–	–	0.007
NiO	0.029	0.073	0.140	0.094	0.110	0.388	0.02	0.010	0.033	0.720	0.029	0.007
CoO	0.007	0.011	0.016	0.014	0.014	0.270	0.008	0.010	0.011	–	–	0.011
Cr_2O_3	–	0.158	0.339	0.304	0.294	0.050	0.038	0.003	0.029	0.015	0.015	0.003
SO_3	0.82	1.45	0.85	0.20	1.60	–	1.40	2.33	–	–	–	3.72
LOI	4.80	2.76	4.80	3.50	7.40	4.91	4.50	3.01	1.84	4.85	12.51	4.50
Total	100.00	99.92	100.21	100.00	100.58	99.09	100.16	100.45	98.77	98.79	99.93	99.51
S	–	–	–	–	–	3.60	–	–	–	3.24	0.54	–
C.f.	54.8	44.3	42.1	43.1	43.8	59.7	48.9	71.7	55.2	51.7	22.2	45.5
Rock	Go	O	P	P	TR	TR	T	MD	Go	T	MB	T

Table A.41 Chemical composition of olivine, drill hole AS-8

Component	Depth (m) 541.0	555.0			695.4	782.5	
SiO_2	33.70	37.87	37.69	37.87	40.89	40.48	41.29
FeO	48.82	26.22	26.99	26.22	7.72	11.13	6.99
CaO	0.15	0.13	0.15	0.13	0.31	0.70	0.63
MgO	16.79	35.22	34.63	35.22	50.48	47.49	50.88
NiO	0.16	0.104	0.142	0.104	0.018	0.01	–
Total	99.62	99.54	99.60	99.54	99.42	99.81	99.79
Fo	38	71	70	71	92	88	93
Fa	62	29	30	29	8	12	7
Rock	P	TR	TR	TR	MD	MB	MB

Morongovsky Intrusive Complex

Mt. Putanaya Intrusion

Table A.42 Chemical composition of rocks, drill hole E-36

Component	Depth (m) 5.0	15.0	17.0	22.0	31.0	35.0	39.0	40.0	48.0	62.8	73.0
SiO_2	44.86	47.08	47.10	45.28	43.58	46.94	45.44	43.26	44.26	41.40	47.86
TiO_2	0.81	1.12	1.20	0.96	0.85	2.24	5.12	0.72	2.60	0.44	1.56
Al_2O_3	16.20	19.04	19.20	13.64	15.04	13.67	10.73	11.60	14.21	11.02	15.66
Fe_2O_3	2.40	2.38	2.36	1.98	2.75	4.28	5.21	3.11	4.16	1.24	4.78
FeO	9.76	9.07	9.00	12.35	12.90	8.86	10.76	13.36	10.11	15.57	6.73
MnO	0.15	0.15	0.15	0.19	0.15	0.15	0.18	0.20	0.15	0.17	0.12
MgO	10.86	5.85	5.86	13.53	13.14	6.43	5.77	18.23	6.44	18.51	6.43
CaO	11.90	9.88	10.68	8.28	9.70	11.08	10.01	6.54	13.44	8.30	9.21
Na_2O	0.10	2.64	2.64	1.83	0.67	2.92	2.78	1.35	1.27	0.18	3.25
K_2O	0.55	0.44	0.44	0.37	0.56	0.56	0.93	0.25	1.67	0.45	1.00
P_2O_5	0.17	0.14	0.13	0.11	0.19	0.20	0.30	0.10	0.30	0.13	0.17
LOI	–	1.84	1.40	1.20	–	2.13	2.44	1.25	–	–	2.93
Total	97.76	99.63	100.16	99.72	99.53	99.46	99.67	99.97	98.61	97.41	99.70
C.f.	52.8	66.2	66.0	51.4	54.4	67.1	73.5	47.5	68.9	47.6	64.2
H_2O^+	–	0.80	1.02	0.96	–	0.98	2.10	1.12	–	–	1.74
S	0.05	0.035	0.032	–	0.10	–	0.03	0.038	–	0.04	0.034
Cl	–	0.18	0.35	0.10	–	0.17	0.10	0.10	–	–	0.19
F	–	0.05	–	0.05	–	0.05	0.05	0.05	–	–	0.05
Cu	0.02	–	–	–	–	–	–	–	0.01	0.02	–
V_2O_5	0.05	–	–	–	–	–	–	–	0.01	0.06	–
Rock	TR	G	G	TR	TR	G	G	TR	PG	TR	PG
Component	Depth (m) 75.0	92.0	110.0	121.0	124.0	152.0	174.0	174.0	183.0	185.0	207.0
SiO_2	48.72	43.06	43.08	41.22	42.80	39.44	40.96	40.00	41.84	41.02	42.16
TiO_2	1.40	0.52	0.18	0.40	0.42	0.48	0.24	0.24	0.24	0.28	0.28
Al_2O_3	11.60	14.19	14.19	14.01	13.93	7.44	13.58	13.52	14.10	12.20	13.93
Fe_2O_3	2.78	1.40	1.47	1.33	1.77	5.04	3.43	1.71	2.14	2.98	1.37
FeO	8.53	12.60	12.89	12.34	11.74	11.68	9.97	11.05	10.80	11.81	12.02
MnO	0.16	0.17	0.19	0.13	0.17	0.20	0.17	0.13	0.16	0.18	0.17
MgO	9.31	13.37	16.79	18.19	18.52	24.37	20.15	20.72	19.76	21.40	20.25
CaO	12.95	7.34	7.48	9.80	6.67	4.54	5.74	8.80	6.81	6.01	6.41

(continued)

Table A.42 (continued)

Component	Depth (m) 75.0	92.0	110.0	121.0	124.0	152.0	174.0	174.0	183.0	185.0	207.0
Na_2O	1.94	1.36	1.36	0.24	1.27	1.18	1.18	0.30	1.09	0.95	1.13
K_2O	0.63	0.21	0.11	0.50	0.25	0.10	0.10	0.38	0.14	0.14	0.17
P_2O_5	0.180	0.057	0.035	0.120	0.057	0.055	0.055	0.100	0.037	0.042	0.046
LOI	1.88	1.52	1.83	–	2.13	5.36	4.00	–	2.86	3.15	1.73
Total	100.08	95.80	99.61	98.28	99.73	99.89	99.58	96.95	99.98	100.16	99.67
C.f.	54.9	51.2	46.1	42.9	42.2	40.7	39.9	38.1	39.6	40.9	39.8
H_2O^+	1.30	1.24	0.58	–	1.88	4.84	3.58	–	2.50	2.78	1.50
S	0.039	0.038	0.022	0.05	0.025	0.029	0.022	–	0.021	0.025	0.023
Cl	0.11	0.10	0.10	–	0.10	0.16	0.14	–	0.14	0.10	0.12
F	0.05	0.05	0.05	–	0.05	0.05	0.05	–	0.05	0.05	0.05
Cu	–	–	–	0.02	–	–	–	0.03	–	–	–
V_2O_5	–	–	–	–	–	–	–	0.06	–	–	–
Rock	TR	TR	TR	TR	TR	P	TR	TR	TR	TR	TR

Component	Depth (m) 226.0	227.0	247.0	271.0	288.0	293.0	300.0	311.0	360.0	373.0	379.0
SiO_2	39.86	42.24	42.16	41.42	40.14	45.12	42.42	42.38	43.90	43.22	47.84
TiO_2	0.30	0.20	0.24	0.24	0.31	0.24	0.31	0.35	0.59	0.77	0.48
Al_2O_3	12.95	14.97	14.36	14.19	11.42	25.12	14.88	15.22	15.74	14.69	13.31
Fe_2O_3	1.53	1.65	2.30	2.51	3.33	1.18	1.60	2.22	3.88	3.53	2.18
FeO	12.05	11.23	10.55	10.63	11.48	3.90	10.92	10.42	9.28	10.40	5.81
MnO	0.14	0.16	0.16	0.16	0.18	0.09	0.15	0.14	0.15	0.14	0.12
MgO	21.61	19.48	19.28	19.66	21.68	6.14	18.42	16.88	13.24	13.00	8.24
CaO	8.64	6.65	7.07	7.07	6.14	13.08	6.94	7.61	8.27	11.48	17.76
Na_2O	0.55	1.09	1.09	1.00	1.14	1.91	1.29	1.29	1.58	0.76	0.53
K_2O	0.60	0.14	0.14	0.23	0.10	0.33	0.25	0.33	0.50	0.85	1.03
P_2O_5	0.100	0.031	0.038	0.020	0.031	0.014	0.018	0.039	0.050	0.170	0.120
LOI	–	1.86	2.60	3.14	4.27	2.93	2.43	3.14	2.71	–	–
Total	98.33	99.70	99.99	100.27	100.22	100.05	99.63	100.02	99.89	99.01	97.42
C.f.	38.6	39.8	40.0	40.0	40.6	45.3	40.5	42.8	49.9	51.7	49.2
H_2O^+	–	1.66	2.26	2.92	3.72	2.36	2.28	2.68	2.44	–	–
S	0.12	0.024	0.023	0.016	0.018	0.023	0.02	0.024	0.025	–	0.10
Cl	–	0.10	0.11	0.10	0.10	0.10	0.10	0.10	0.10	–	–
F	–	0.05	0.05	0.05	0.05	–	0.05	0.05	0.05	–	–
Cu	0.01	–	–	–	–	–	–	–	–	0.02	–
V_2O_5	0.06	–	–	–	–	–	–	–	–	0.05	0.06
Rock	TR	TR	TR	TR	TR	LG	TR	TR	TR	TR	TR

Table A.43 Chemical composition of olivine, drill hole E-36

Component	Depth (m) 17.0	40.0	92.0	110.0	124.0		136.0	
SiO_2	38.56	37.23	37.21	36.36	37.53	37.12	37.54	37.44
FeO	23.48	25.45	25.57	27.43	24.04	24.34	24.44	24.08
CaO	0.13	–	–	–	–	–	–	–
MgO	36.93	35.51	35.19	33.56	36.95	35.89	36.34	36.85
NiO	0.159	0.14	0.14	0.14	0.15	0.13	0.165	0.13
Total	99.26	98.33	98.11	97.49	98.67	97.48	98.49	98.50
Fo	74	71	71	69	73	72	73	73
Fa	26	29	29	31	27	28	27	27
Rock	G	TR	TR	TR	TR	TR	TR	TR

Component	Depth (m) 136.0		152.0		174.0			
SiO_2	37.04	37.30	38.02	38.03	38.06	37.16	36.40	37.20
FeO	24.72	24.02	25.2	24.99	22.55	22.12	22.41	22.45

(continued)

Table A.43 (continued)

	Depth (m)							
Component	136.0		152.0		174.0			
CaO	–	–	0.07	0.08	–	–	–	–
MgO	36.16	34.07	36.17	36.40	38.39	37.94	35.51	37.99
NiO	0.15	0.15	0.153	0.162	0.14	0.143	0.14	0.15
Total	98.07	95.54	99.61	99.66	99.14	97.36	94.46	97.79
Fo	72	72	72	72	75	75	74	75
Fa	28	28	28	28	25	25	26	25
Rock	TR	TR	P	P	TR	TR	TR	TR

	Depth (m)								
Component	185.0	207.0	227.0	247.0	288.0	311.0			
SiO_2	37.20	37.71	37.72	37.93	38.17	37.98	37.94	37.94	37.21
FeO	22.95	22.87	22.23	22.42	22.99	25.6	25.78	25.72	24.5
CaO	–	–	–	–	–	0.09	0.08	0.1	–
MgO	37.83	37.95	38.17	37.23	37.71	35.75	35.63	35.7	36.52
NiO	0.152	0.15	0.14	0.14	0.15	0.14	0.15	0.135	0.14
Total	98.13	98.68	98.26	97.72	99.02	99.56	99.58	99.60	98.37
Fo	75	75	75	75	75	71	71	71	73
Fa	25	25	25	25	25	29	29	29	27
Rock	TR	TR	TR	TR	TR	TR	TR	TR	TR

Mt. Pegmatitovaya Intrusion

Table A.44 Chemical composition of rocks, drill hole E-28

	Depth (m)								
Component	154.0	155.0	160.0	165.0	175.0	182.0	185.0	198.0	208.0
SiO_2	47.20	47.90	47.60	45.60	47.30	49.60	52.10	46.90	48.20
TiO_2	1.23	0.94	1.80	3.11	3.02	3.06	1.34	3.00	1.80
Al_2O_3	14.00	14.90	16.40	15.20	15.20	12.10	15.40	15.80	18.20
Fe_2O_3	1.59	3.22	2.23	4.20	3.70	2.96	2.81	4.20	0.81
FeO	9.91	7.54	9.69	11.85	11.70	14.07	10.34	11.34	11.42
MnO	0.15	0.19	0.21	0.23	0.23	0.24	0.18	0.23	0.19
MgO	10.64	8.72	5.64	4.40	3.49	3.38	2.26	3.95	3.26
CaO	8.07	11.11	11.08	10.06	9.23	8.06	8.30	9.47	10.67
Na_2O	1.92	2.09	2.09	2.64	3.00	3.50	4.25	2.82	3.42
K_2O	0.66	0.50	0.62	0.66	0.46	0.55	0.62	0.81	0.62
P_2O_5	0.11	0.09	0.15	0.15	0.26	0.30	0.66	0.19	0.42
LOI	4.44	2.79	1.51	1.64	2.71	2.25	2.42	1.50	1.46
H_2O^-	0.36	0.08	0.14	0.22	0.14	0.20	0.24	0.12	–
Total	100.28	100.07	99.16	99.96	100.44	100.27	100.92	100.33	100.47
C.f.	51.9	55.2	67.9	78.5	81.5	83.4	85.3	79.7	79.0
S	0.012	0.064	0.048	0.056	0.128	0.516	0.048	0.04	0.00
Rock	CG	CG	T	PG	PG	PG	PG	G	G

	Depth (m)								
Component	228.0	258.0	268.0	276.0	278.0	281.0	287.0	298.0	304.0
SiO_2	48.70	48.40	48.10	47.50	48.90	48.90	47.70	49.00	47.70
TiO_2	1.42	1.08	1.14	1.24	1.00	1.03	1.04	0.98	1.18
Al_2O_3	15.30	15.30	16.10	15.50	17.80	17.10	16.90	16.80	16.00
Fe_2O_3	3.33	2.66	2.59	1.94	1.34	2.78	1.25	1.33	0.65
FeO	9.33	9.48	8.47	9.41	7.97	7.04	8.69	9.33	10.49
MnO	0.20	0.20	0.19	0.18	0.17	0.18	0.18	0.18	0.17
MgO	6.30	7.50	8.42	8.59	6.88	7.13	7.75	7.13	8.33

(continued)

Table A.44 (continued)

Component	Depth (m) 228.0	258.0	268.0	276.0	278.0	281.0	287.0	298.0	304.0
CaO	11.12	11.34	12.00	11.45	11.22	11.22	11.46	11.23	11.93
Na_2O	2.09	2.45	2.27	2.55	2.82	2.36	2.27	2.64	2.12
K_2O	0.48	0.46	0.50	0.37	0.55	0.50	0.50	0.53	0.52
P_2O_5	0.15	0.13	0.13	0.13	0.13	0.11	0.11	0.13	0.26
LOI	1.04	1.52	1.85	1.77	1.19	1.16	1.49	1.60	1.35
H_2O^-	0.02	0.24	0.22	0.18	0.14	0.15	0.36	0.20	–
Total	99.48	100.76	101.98	100.81	100.11	99.66	99.70	101.08	100.70
C.f.	66.8	61.8	56.8	56.9	57.5	57.9	56.2	59.9	57.2
S	0.04	0.016	0.000	0.000	0.008	0.02	0.00	0.00	–
Rock	Go	Go	Go	Go	Go	Go	Go	Go	Go

Component	Depth (m) 310.0	320.0	330.0	360.0	390.0	400.0	410.0	413.0
SiO_2	46.90	46.50	44.90	44.50	44.20	43.80	45.40	46.20
TiO_2	1.16	1.20	0.81	0.88	0.92	0.87	1.53	0.81
Al_2O_3	16.80	12.40	14.00	14.70	12.60	11.90	13.50	15.00
Fe_2O_3	0.13	1.19	1.95	2.25	1.02		2.50	2.09
FeO	10.77	12.07	11.02	10.84	13.57	15.65	12.06	10.27
MnO	0.18	0.22	0.21	0.20	0.22	0.23	0.22	0.21
MgO	9.12	12.49	14.20	14.05	15.50	15.80	12.20	10.10
CaO	12.00	11.41	9.59	8.77	7.95	7.83	9.03	12.75
Na_2O	1.90	1.55	1.58	1.92	1.58	1.42	2.27	1.25
K_2O	0.41	0.41	0.35	0.56	0.37	0.35	0.46	0.42
P_2O_5	0.20	0.13	0.08	0.11	0.13	0.09	0.12	0.09
LOI	1.19	1.22	1.24	1.52	1.79	1.82	1.27	1.52
H_2O^-	–	–	0.04	0.10	0.16	0.10	0.26	0.18
Total	100.76	100.79	99.97	100.40	100.01	99.86	100.82	100.89
C.f.	54.5	51.5	47.7	48.2	48.5	49.8	54.4	55.0
S	–	–	0.036	0.004	0.02	0.176	0.016	0.028
Rock	O	O	TR	TR	TR	TR	CG	CG

Table A.45 Mercury concentration in the rocks, drill hole E-28

Component	Depth (m) 184.0	185.0	187.0	218.0	238.0	273.0	279.0	287.0	293.0	297.0
Hg(ppb)	365	315	365	325	235	220	210	95	250	300

Component	Depth (m) 301.0	304.0	310.0	320.0	340.0	350.0	360.0	370.0	380.0
Hg(ppb)	250	315	215	280	210	315	135	180	185

Table A.46 Chemical composition of olivine, drill hole E-28

Component	Depth (m) 165.0			198.0	208.0	218.0	228.0	248.0	258.0A
SiO_2	32.67	32.68	33.54	32.07	33.60	33.43	32.86	33.14	33.57
FeO	51.07	51.02	47.42	54.89	46.59	46.88	49.67	47.75	45.84
MgO	16.39	16.43	16.70	13.14	19.89	19.87	17.71	19.53	20.93
CaO	0.28	0.27	0.26	0.15	0.19	0.25	0.21	0.21	0.23
NiO	–	–	–	–	0.009	0.006	0.004	0.017	0.020
Total	100.41	100.40	97.92	100.25	100.28	100.44	100.45	100.65	100.59

(continued)

Table A.46 (continued)

	Depth (m)								
Component	165.0			198.0	208.0	218.0	228.0	248.0	258.0A
Fo	36	36	39	30	43	43	39	42	45
Fa	64	64	61	70	57	57	61	58	55
Rock	PG	PG	PG	G	Go	Go	Go	Go	Go

	Depth (m)								
Component	258.0	273.0		278.0		279.0			
SiO_2	32.72	34.63	32.80	34.80	33.69	34.19	34.38	33.86	34.46
TiO_2	–	–	–	–	–	–	0.056	0.278	0.000
Al_2O_3	–	–	–	–	–	–	0.088	0.000	0.000
FeO	47.67	45.11	48.35	42.09	43.68	41.60	38.64	38.90	39.37
MnO	–	–	–	–	–	–	0.516	0.584	0.643
MgO	20.72	21.00	21.34	23.21	23.47	24.88	24.06	23.80	24.18
CaO	0.19	0.12	0.10	0.17	0.13	0.11	0.192	0.148	0.091
Na_2O	–	–	–	–	–	–	0.000	0.000	0.208
K_2O	–	–	–	–	–	–	0.028	0.003	0.000
NiO	0.083	–	–	–	0.009	0.026	0.005	0.061	0.000
CuO	–	–	–	–	–	–	0.047	0.000	0.000
ZnO	–	–	–	–	–	–	0.057	0.000	0.000
Cr_2O_3	–	–	–	–	–	–	0.090	0.209	0.000
Total	101.38	100.86	102.59	100.27	100.98	100.81	98.06	97.84	98.95
Fo	44	45	44	50	49	52	53	52	52
Fa	56	55	56	50	51	48	47	48	48
Rock	Go	Go	Go	Go	Go	Go	Go	Go	Go

	Depth (m)								
Component	287.0	297.0	310.0	320.0	330.0	340.0	350.0	360.0	
SiO_2	34.03	34.91	36.25	36.42	36.72	36.50	36.18	36.02	36.36
TiO_2	–	–	–	–	–	–	–	–	0.000
Al_2O_3	–	–	–	–	–	–	–	–	0.000
FeO	41.60	38.60	33.56	33.21	31.35	31.10	32.45	32.52	31.31
MnO	–	–	–	–	–	–	–	–	0.494
MgO	25.31	26.92	26.97	30.00	31.65	32.38	31.64	31.88	31.82
CaO	0.11	0.12	0.12	0.09	0.08	0.09	0.07	0.10	0.084
Na_2O	–	–	–	–	–	–	–	–	0.000
K_2O	–	–	–	–	–	–	–	–	0.000
NiO	0.014	0.018	0.056	0.076	0.073	0.074	0.056	0.075	0.111
CuO	–	–	–	–	–	–	–	–	0.135
ZnO	–	–	–	–	–	–	–	–	0.168
Cr_2O_3	–	–	–	–	–	–	–	–	0.141
Total	101.06	100.57	96.96	99.80	99.87	100.14	100.40	100.60	100.63
Fo	52	55	59	62	64	65	63	64	64
Fa	48	45	41	38	36	35	37	36	36
Rock	Go	Go	O	O	TR	TR	TR	TR	TR

	Depth (m)									
Component	360.0				370.0	380.0	400.0			410.0
SiO_2	35.58	35.50	36.43	36.20	35.99	35.49	35.85	36.92	36.81	31.15
TiO_2	0.007	0.000	0.000	0.077	–	–	–	0.082	0.000	–
Al_2O_3	0.000	0.000	0.000	0.000	–	–	–	0.001	0.000	–
FeO	31.09	30.87	31.69	30.80	32.66	34.17	32.16	29.17	29.12	45.25
MnO	0.577	0.656	0.364	0.578	–	–	–	0.402	0.323	–
MgO	31.28	30.56	31.85	31.21	31.74	31.16	32.67	32.28	32.67	27.90
CaO	0.076	0.054	0.079	0.119	0.11	0.06	0.100	0.040	0.071	0.07
Na_2O	0.071	0.000	0.018	0.209	–	–	–	0.088	0.186	–
K_2O	0.000	0.000	0.000	0.027	–	–	–	0.008	0.000	–

(continued)

Table A.46 (continued)

Component	Depth (m) 360.0				370.0	380.0	400.0			410.0
NiO	0.157	0.195	0.230	0.237	0.088	0.069	0.097	0.000	0.120	0.055
CuO	0.107	0.117	0.000	0.093	–	–	–	0.079	0.000	–
ZnO	0.026	0.058	0.371	0.006	–	–	–	0.000	0.000	–
Cr_2O_3	0.000	0.000	0.000	0.005	–	–	–	0.000	0.074	–
Total	98.97	98.01	101.03	99.56	100.59	100.95	100.88	99.07	99.37	104.43
Fo	64	64	64	64	63	62	64	66	67	52
Fa	36	36	36	36	37	38	36	34	33	48
Rock	TR	TR	TR	TR	TR	TR	TR	TR	TR	T

Table A.47 Chemical composition of clinopyroxene, drill hole E-28

Component	Depth (m) 165.0		182.0		185.0	198.0	208.0	218.0
SiO_2	50.86	50.84	50.15	49.10	50.98	51.06	50.47	50.65
TiO_2	0.82	0.90	0.06	0.74	0.07	0.85	0.94	0.86
Al_2O_3	1.68	1.96	2.05	3.73	1.08	1.98	2.13	1.90
FeO	15.20	13.54	30.56	23.00	29.19	12.72	13.05	13.85
MgO	13.16	13.94	8.73	10.36	7.51	13.75	14.08	13.63
CaO	17.98	18.92	6.73	9.86	9.72	18.80	18.45	18.20
Na_2O	0.30	0.36	0.55	1.37	0.28	0.30	0.31	0.30
Cr_2O_3	0.02	0.01	0.06	–	0.01	0.02	0.02	0.06
Total	100.02	100.47	98.89	98.16	98.84	99.48	99.45	99.45
f	39.32	35.27	66.26	55.46	68.56	34.17	34.21	36.31
Fe_2O_3[a]	1.70	3.00	0.00	1.19	0.00	0.99	2.25	1.69
FeO[a]	13.67	10.84	30.56	21.92	29.19	11.83	11.02	12.33
Wo	37	39	16	23	23	39	38	38
En	38	40	28	34	24	40	41	40
Fs	25	21	56	43	53	21	21	22
Rock	PG	PG	PG	PG	PG	G	Go	Go

Component	Depth (m) 228.0	248.0	258.0	273.0	278.0	279.0	287.0	297.0
SiO_2	51.26	51.29	51.06	50.98	51.53	51.54	51.29	51.11
TiO_2	0.81	0.80	0.86	0.69	0.64	0.64	0.71	0.70
Al_2O_3	2.29	2.19	2.28	2.37	2.17	2.21	2.27	2.45
FeO	11.82	12.34	11.65	10.30	10.03	10.11	10.54	9.65
MgO	14.56	14.17	14.54	14.88	15.35	14.99	14.87	15.22
CaO	19.20	19.09	19.16	19.81	19.64	19.54	19.33	19.68
Na_2O	0.32	0.29	0.31	0.28	0.29	0.35	0.28	0.31
Cr_2O_3	0.04	0.06	0.06	0.16	0.13	0.10	0.08	0.21
Total	100.30	100.23	99.92	99.47	99.78	99.48	99.37	99.33
f	31.29	32.82	31.01	27.97	26.82	27.45	28.45	26.24
Fe_2O_3[a]	2.04	1.50	1.92	2.14	1.90	1.43	1.27	1.87
FeO[a]	9.99	10.99	9.93	8.38	8.32	8.83	9.40	7.96
Wo	39	39	39	41	40	41	40	41
En	42	41	42	43	44	43	43	44
Fs	19	20	19	16	16	16	17	15
Rock	Go	Go	Go	Go	Go	Go	Go	Go

Component	Depth (m) 310.0	320.0	330.0	340.0	350.0	360.0	370.0	380.0	410.0
SiO_2	51.36	50.51	51.29	52.69	51.23	51.09	50.69	51.06	51.07
TiO_2	0.60	0.57	0.69	0.52	0.94	0.51	1.17	1.15	1.06
Al_2O_3	2.53	2.52	2.57	1.56	2.48	2.31	2.63	2.18	2.48

(continued)

Table A.47 (continued)

Component	Depth (m) 310.0	320.0	330.0	340.0	350.0	360.0	370.0	380.0	410.0
FeO	8.97	8.98	8.94	9.45	10.96	8.57	9.79	11.56	11.14
MgO	15.68	15.42	15.32	15.18	15.90	16.20	14.68	15.73	14.69
CaO	19.77	20.01	19.87	20.65	19.32	19.61	20.04	17.83	18.61
Na_2O	0.34	0.34	0.35	0.32	0.29	0.25	0.32	0.34	0.32
Cr_2O_3	0.30	0.38	0.38	0.01	0.10	0.61	0.04	0.02	0.03
Total	99.55	98.73	99.41	100.38	101.22	99.15	99.36	99.87	99.40
f	24.30	24.62	24.66	25.88	27.89	22.89	27.23	29.19	29.84
Fe_2O_3[a]	2.08	3.08	1.53	1.09	3.66	2.42	1.61	2.15	0.70
FeO[a]	7.10	6.21	7.56	8.47	7.66	6.39	8.34	9.63	10.51
Wo	41	41	41	42	39	40	42	37	39
En	45	44	44	43	44	46	42	45	43
Fs	14	15	15	15	17	14	16	18	18
Rock	O	O	TR	TR	TR	TR	TR	TR	CG

[a]Hereinafter calculated data

Daldykansky Intrusive Complex

Daldykansky Intrusion

Table A.48 Chemical composition of rocks, drill hole NP-49

Component	Depth (m) 1,365.4	1,367.5	1,372.5	1,377.0	1,379.0	1,386.0	1,388.7	1,392.3	1,395.7	1,396.8
SiO_2	46.99	47.94	47.80	46.53	48.16	45.43	46.29	48.14	48.27	48.49
TiO_2	1.893	1.998	1.932	1.857	1.955	2.772	2.671	1.871	2.406	1.738
Al_2O_3	14.01	14.20	14.33	13.40	14.36	12.24	13.28	14.53	13.34	14.88
Fe_2O_3	14.47	15.59	15.31	13.87	15.19	18.48	18.62	14.78	15.63	14.28
MnO	0.235	0.253	0.257	0.228	0.25	0.286	0.297	0.248	0.271	0.236
MgO	6.71	6.50	6.58	6.94	6.77	5.46	4.93	6.48	6.25	6.71
CaO	10.00	10.30	10.22	10.89	10.34	8.94	8.69	10.3	10.33	10.46
Na_2O	2.84	2.63	3.18	2.83	2.31	3.75	3.99	2.85	2.53	2.30
K_2O	0.52	0.64	0.63	0.74	0.60	0.21	0.93	0.59	0.69	0.53
P_2O_5	0.206	0.222	0.202	0.17	0.198	0.251	0.379	0.198	0.239	0.193
Ba	0.014	0.03	0.024	0.014	0.021	0.02	0.041	0.021	0.027	0.017
LOI	1.76	0.48	0.28	2.46	0.02	3.04	0.98	0.42	0.3	0.08
Total	99.65	100.79	100.73	99.93	100.17	100.88	101.1	100.43	100.28	99.91
C.f.	68.3	70.6	69.9	66.7	69.2	77.2	79.1	69.5	71.4	68.1
Cu	231	220	215	800	203	154	421	198	248	209
Ni	99	94	102	110	82	58	44	88	69	99
Co	48	43	50	56	48	43	46	37	43	50
Pb	26	20	26	26	24	17	22	19	31	18
Zn	115	114	111	234	103	89	125	106	156	106
Cr	195	151	226	155	182	59	37	155	90	196
V	252	373	297	480	419	426	300	300	512	400
Ba	142	182	210	115	154	39	237	186	290	226
Sr	243	216	202	387	177	117	256	250	238	238
Pt	0.084	0.054	0.092	0.052	0.06	0.057	0.03	0.041	0.048	0.068
Pd	0.014	0.011	0.01	0.012	0.012	0.014	0.017	0.009	0.013	0.012
Rh	<0.005	<0.005	<0.005	<0.005	<0.005	<0.005	<0.005	<0.005	<0.005	<0.005

(continued)

Table A.48 (continued)

	Depth (m)									
Component	1,365.4	1,367.5	1,372.5	1,377.0	1,379.0	1,386.0	1,388.7	1,392.3	1,395.7	1,396.8
Au	0.016	0.008	0.013	0.01	0.011	0.007	0.005	0.01	0.006	0.01
Ag	0.6	0.31	0.018	0.21	0.18	0.53	0.48	0.015	0.219	0.004
Rock	Go	Go	Go	Go	Go	PG	PG	PG	O	O

	Depth (m)									
Component	1,400.3	1,405.0	1,407.0	1,411.0	1,412.0	1,419.0	1,424.5	1,429.0	1,433.5	1,437.0
SiO_2	47.57	42.44	49.48	48.54	48.17	48.04	48.13	48.12	48.08	47.89
TiO_2	1.845	5.563	2.502	1.381	1.508	1.694	1.499	1.644	1.547	1.508
Al_2O_3	14.96	10.35	12.9	14.54	14	14.4	14.17	13.74	15.09	15.75
Fe_2O_3	13.49	25.87	16.53	13.96	14.48	14.42	14.61	14.59	13.75	13.52
MnO	0.232	0.397	0.264	0.238	0.246	0.246	0.259	0.254	0.239	0.228
MgO	5.38	5.03	4.53	7.61	7.49	7.24	8.06	8.07	7.49	7.40
CaO	9.25	8.39	8.88	11.34	11.15	10.72	11.2	11.24	10.53	10.9
Na_2O	4.64	2.49	3.65	2.11	2.99	3.00	2.36	2.36	2.98	2.73
K_2O	0.59	0.56	0.96	0.40	0.41	0.47	0.43	0.44	0.50	0.42
P_2O_5	0.218	0.352	0.38	0.135	0.158	0.184	0.175	0.207	0.185	0.162
Ba	0.027	0.014	0.003	0.003	0.003	0.003	0.003	0.003	0.003	0.003
LOI	2.54	−1.08	0.56	−0.04	−0.06	−0.18	−0.22	−0.10	−0.06	−0.24
Total	100.74	100.38	100.64	100.22	100.55	100.24	100.68	100.57	100.33	100.27
C.f.	71.5	83.7	78.5	64.7	65.9	66.6	64.5	64.4	64.7	64.6
Cu	204	456	324	138	158	186	175	169	180	158
Ni	60	36	38	108	104	112	121	124	123	121
Co	35	54	39	54	50	50	54	46	48	46
Pb	23	26	23	14	14	12	19	19	14	18
Zn	97	231	122	86	92	97	89	97	108	89
Cr	120	22	34	205	211	215	300	344	256	290
V	375	704	511	242	282	341	319	264	308	253
Ba	233	206	290	108	158	188	117	119	136	119
Sr	309	190	202	238	214	250	202	179	256	250
Pt	0.036	0.072	0.05	0.038	0.082	0.052	0.054	0.04	0.05	0.11
Pd	0.014	0.066	0.011	0.008	0.01	0.011	0.0089	0.01	0.009	0.012
Rh	<0.005	<0.005	<0.005	<0.005	<0.005	<0.005	<0.005	<0.005	<0.005	<0.005
Au	0.007	0.011	0.005	0.004	0.008	0.016	0.006	0.01	0.011	0.09
Ag	0.34	0.15	0.02	0.018	0.017	0.022	1.07	0.01	0.19	0.013
Rock	O	PG	PG	O	O	O	O	O	O	O

	Depth (m)									
Component	1,441.6	1,443.3	1,446.0	1,452.5	1,456.0	1,461.0	1,465.7	1,468.0	1,473.0	1,478.0
SiO_2	48.33	48.49	48.75	48.38	48.04	48.28	48.27	47.84	47.97	48.17
TiO_2	1.452	1.402	1.504	1.544	1.598	1.752	1.782	1.895	1.903	1.768
Al_2O_3	15.72	15.72	15.66	15.51	15.67	14.67	14.82	14.38	14.65	14.64
Fe_2O_3	13.09	13.49	13.06	13.88	14.00	14.85	14.76	15.42	15.15	14.39
MnO	0.22	0.231	0.23	0.227	0.227	0.244	0.241	0.249	0.245	0.237
MgO	7.44	7.50	7.09	7.11	7.25	6.89	6.74	6.95	6.97	6.92
CaO	10.92	11.14	10.98	10.99	11.12	10.79	10.89	10.79	10.81	10.57
Na_2O	2.60	2.10	2.35	2.01	2.01	2.69	2.50	2.31	2.21	2.78
K_2O	0.43	0.43	0.49	0.45	0.42	0.50	0.49	0.47	0.47	0.51
P_2O_5	0.163	0.154	0.177	0.164	0.145	0.199	0.197	0.165	0.189	0.212
Ba	0.003	0.003	0.003	0.008	0.009	0.014	0.019	0.013	0.015	0.012
LOI	−0.08	−0.24	0.10	−0.26	−0.24	−0.24	−0.22	−0.32	−0.32	−0.18
Total	100.29	100.42	100.39	100.01	100.25	100.64	100.49	100.16	100.26	100.03
C.f.	63.8	64.3	64.8	66.1	65.9	68.3	68.7	68.9	68.5	67.5
Cu	158	147	158	170	164	192	192	196	199	210
Ni	124	115	108	121	115	104	97	113	115	105

(continued)

Table A.48 (continued)

Component	Depth (m) 1,441.6	1,443.3	1,446.0	1,452.5	1,456.0	1,461.0	1,465.7	1,468.0	1,473.0	1,478.0
Co	35	35	35	52	48	50	48	46	50	45
Pb	10	18	14	14	26	22	20	20	20	22
Zn	83	89	200	97	86	100	100	107	109	104
Cr	275	253	234	218	215	186	183	194	166	200
V	264	286	264	207	273	273	263	390	382	310
Ba	136	128	116	111	107	136	107	161	169	161
Sr	226	244	207	262	207	201	201	195	200	192
Pt	0.07	0.07	0.038	0.14	0.055	0.06	0.09	<0.01	<0.01	<0.01
Pd	0.009	0.047	0.009	0.012	0.01	0.013	0.019	0.01	0.013	0.014
Rh	<0.005	<0.005	<0.005	<0.005	<0.005	<0.005	<0.005	<0.005	<0.005	<0.005
Au	0.018	0.013	0.009	0.01	0.012	–	–	0.011	–	0.003
Ag	0.009	0.34	0.26	0.48	0.4	–	–	0.36	–	0.46
Rock	O	O	O	O	O	O	O	O	O	O

Component	Depth (m) 1,482.0	1,487.0	1,490.0	1,495.8	1,498.0	1,502.0	1,508.0	1,512.7	1,515.5	1,518.0	1,521.5
SiO_2	48.65	48.38	47.81	48.71	48.3	47.84	47.99	47.87	47.22	47.41	47.52
TiO_2	1.84	1.795	1.901	1.841	1.921	1.97	1.932	1.92	1.907	1.865	1.878
Al_2O_3	14.92	14.44	14.26	15.11	14.68	14.37	14.42	14.9	14.88	14.67	14.6
Fe_2O_3	14.67	14.51	15.26	14.52	14.63	15.41	15.28	14.84	14.62	14.77	15.06
MnO	0.242	0.234	0.247	0.239	0.238	0.26	0.261	0.266	0.265	0.254	0.261
MgO	6.61	6.96	6.88	6.28	6.8	6.72	6.86	5.51	6.75	6.61	6.86
CaO	10.53	10.51	10.55	10.38	10.57	10.47	10.38	10.26	10.37	10.35	10.42
Na_2O	2.31	2.21	2.51	2.21	2.02	2.45	2.26	1.97	2.15	2.72	1.97
K_2O	0.55	0.49	0.51	0.62	0.56	0.55	0.54	0.62	0.66	0.56	0.60
P_2O_5	0.207	0.20	0.227	0.22	0.208	0.191	0.189	0.216	0.194	0.211	0.208
Ba	0.016	0.012	0.012	0.02	0.017	0.018	0.017	0.024	0.021	0.018	0.022
LOI	−0.28	−0.22	−0.30	0.02	−0.22	−0.28	−0.08	0.60	0.86	0.50	0.72
Total	100.27	99.52	99.87	100.17	99.72	99.97	100.05	99.00	99.90	99.94	100.12
C.f.	68.9	67.6	68.9	69.8	68.3	69.6	69.0	72.9	68.4	69.1	68.7
Cu	210	225	221	217	213	217	229	233	189	233	233
Ni	110	102	102	92	92	94	97	97	110	126	118
Co	50	50	45	39	39	45	55	44	44	46	44
Pb	20	17	8	12	17	27	8	7	6	8	5
Zn	106	106	109	109	103	107	110	116	100	115	132
Cr	162	177	181	140	220	165	181	166	156	281	200
V	310	302	342	326	342	358	358	318	342	358	342
Ba	192	164	178	186	193	186	178	207	200	171	193
Sr	212	185	182	222	187	200	197	248	197	187	225
Pt	0.025	<0.01	<0.01	<0.01	<0.01	<0.01	<0.01	<0.01	<0.01	0.11	<0.01
Pd	0.029	0.01	0.015	0.01	0.013	0.012	0.017	0.012	0.011	0.016	0.011
Rh	<0.005	<0.005	<0.005	<0.005	<0.005	<0.005	<0.005	<0.005	<0.005	<0.005	<0.005
Au	0.005	0.0045	0.004	0.008	0.009	0.009	0.022	0.0065	0.006	0.0055	0.0033
Ag	0.16	0.2	0	0.18	0.18	0.5	0.12	0.13	0.79	0.27	0.41
Rock	O	O	O	O	O	O	Go	Go	Go	Go	Go

Character of Composition of Mafic Horizons in Layered Intrusions of the Talnakhsky Ore Junction

Table A.49 Comparison of a chemical composition of rocks from mafic and ultramafic parts of the layered magmatic bodies

	UTI				LTI			
	Picrite		Mafic		Picrite		Mafic	
Component	*X*(15)	*δ*	*X*(31)	*δ*	*X*(49)	*δ*	*X*(10)	*δ*
SiO_2	38.47	2.94	46.60	1.67	41.37	1.97	47.65	2.19
TiO_2	0.56	0.11	1.11	0.65	0.48	0.13	1.27	0.51
Al_2O_3	8.64	2.52	15.35	1.68	12.29	3.14	15.42	3.38
Fe_2O_3	4.58	2.70	3.42	1.46	3.78	1.42	3.04	1.06
FeO	11.76	3.42	7.29	2.16	8.11	2.05	8.31	2.61
MnO	0.21	0.03	0.18	0.06	0.20	0.08	0.21	0.10
MgO	20.21	6.02	8.47	2.54	18.11	4.06	7.28	1.98
CaO	5.37	1.22	10.80	1.96	8.83	2.82	8.27	4.03
Na_2O	0.84	0.36	2.19	0.72	0.82	0.33	2.24	0.88
K_2O	0.41	0.19	1.15	0.77	0.61	0.31	1.66	2.28
P_2O_5	0.08	0.03	0.19	0.28	0.06	0.03	0.15	0.08
LOI	6.44	1.69	2.75	0.96	5.04	1.96	4.28	1.90
Total	97.57		99.5		99.70		99.78	

Note: *X*—an average (in brackets quantity of analyses), *δ*—a standard deviation

Contents of Volume 2

V.V. Ryabov et al., *Trap Magmatism and Ore Formation in the Siberian Noril'sk Region: Volume 1. Trap Petrology; Volume 2. Atlas of Magmatic Rocks*, Modern Approaches in Solid Earth Sciences 3, DOI 10.1007/978-94-007-5022-7,

Index

V.V. Ryabov et al., *Trap Magmatism and Ore Formation in the Siberian Noril'sk Region: Volume 1. Trap Petrology; Volume 2. Atlas of Magmatic Rocks*, Modern Approaches in Solid Earth Sciences 3, DOI 10.1007/978-94-007-5022-7,

MIX
Papier aus verantwortungsvollen Quellen
Paper from responsible sources
FSC® C105338

If you have any concerns about our products,
you can contact us on
ProductSafety@springernature.com

In case Publisher is established outside the EU,
the EU authorized representative is:
Springer Nature Customer Service Center GmbH
Europaplatz 3, 69115 Heidelberg, Germany

Printed by Libri Plureos GmbH
in Hamburg, Germany